# Autodesk Inventor® 8 Essentials Plus

DANIEL T. BANACH
TRAVIS JONES
ALAN J. KALAMEJA

autodesk® Press

Australia • Canada • Mexico • Singapore • Spain • United Kingdom • United States

autodesk® Press

# Autodesk Inventor® 8 Essentials Plus

Daniel T. Banach, Travis Jones, and Alan J. Kalameja

**AutoDesk Press Staff**

**Vice President, Technology and Trades SBU**
Alar Elken

**Editorial Director**
Sandy Clark

**Senior Acquisitions Editor**
James DeVoe

**Senior Development Editor**
John Fisher

**Channel Manager**
Fair Huntoon

**Marketing Coordinator**
Casey Bruno

**Production Director**
Mary Ellen Black

**Production Manager**
Andrew Crouth

**Production Editor**
Stacy Masucci

**Art/Design Specialist**
Mary Beth Vought

**Editorial Assistant**
Katherine Bevington

**Production Services**
***TIPS*** Technical Publishing, Inc.

Cover image courtesy of NASA



Printed in Canada
2 3 4 5 XXX 08 07 06 05 04

For more information contact
Thomson Delmar Learning
Executive Woods
5 Maxwell Drive, PO Box 8007
Clifton Park, NY 12065-8007.
Or find us on the World Wide Web at
http://delmar.com



**Library of Congress Cataloging-in-Publication Data**

ISBN 1-4018-6496-1

# CONTENTS

## CHAPTER 3 Creating and Editing Sketched Features



## CHAPTER 4 Creating Placed Features

## CHAPTER 5 Creating and Editing Drawing Views

## CHAPTER 6 Creating and Documenting Assemblies

## CHAPTER 7 Complex Sketching and Constraining Techniques

## CHAPTER 8 Complex Part Modeling Techniques

## CHAPTER 9 Complex Drawing View Creation and Editing

## CHAPTER 10 Complex Assembly Modeling

## CHAPTER 12 Design Automation Techniques

# ACKNOWLEDGMENTS

The authors would like to thank Jason Ruge for his careful and thoughtful technical edit. His attention to detail has added a great deal to the book.

The authors would also like to thank the following reviewers:

Ralph Hafer, College of the Redwoods, Eureka, CA
Tom Singer, Sinclair Community College, Dayton, OH

We wish to acknowledge the effort demonstrated by the *Electronic Student Workbook* authors, including Neil Munro, Phil Dollan, and Brent Dunn.

We also wish to acknowledge the following individuals who contributed to this book:

| | |
|---|---|
| Paul Lebovitz | Kristen Kent |
| Todd Nicol | George Hudetz |
| Tod Cordill | Christian Maylath |
| Lakshmanan Lakshmanan | Tom McNeil |
| Rodney Kerbyson | Marjorie Elliott |
| Kenneth Hill | Rene Plautz |
| Rockie Lyons Beaman | Steve Hader |
| Matt Pettengill | Jeffrey Rowe |

# INTRODUCTION

Welcome to the *Autodesk Inventor 8 Essentials Plus* manual. This manual provides a thorough coverage of the features and functionalities offered in Autodesk Inventor 8. It is designed to help you become productive quickly, and to encourage you to use the self-paced learning activities in the Autodesk Inventor Design Support System (DSS).

Also included with this manual is the *Autodesk Inventor 8 Electronic Student Workbook* that contains real-world, illustrated, step-by-step exercises. This workbook and all the Autodesk Inventor data files for each exercise are contained in a self-installing file on the *Autodesk Inventor 8 Essentials Plus Exercises CD-ROM* attached to the back cover of your book. Instructions for executing the installation are included in a later section of this introduction.

Each chapter in this manual is organized with the following elements:

**Objectives** Describes the content and learning objectives.

**Topic Coverage** Presents a concise, thorough review of the topic.

**Exercises** Presents the workflow for a specific tool or process through illustrated, step-by-step instructions. All the exercises are provided in the *Electronic Student Workbook* that accompanies this manual.

**Chapter Summary** Summarizes, in table format, the tools and processes discussed in the chapter.

**Applying Your Skills** Checks your skills and understanding of the material covered in the chapter using challenge exercises. These exercises describe a design challenge, but do not provide step-by-step instructions.

**Checking Your Skills** Tests your understanding of the material using True/False or multiple-choice questions.

## NOTE TO THE LEARNER

Autodesk Inventor is designed for easy learning. The integrated Design Support System (DSS) provides you with ongoing support, as well as access to online documentation.

As just described, each chapter in this manual has the same instructional design, making it is easy to follow and understand. Each exercise is task oriented and is based on real-world mechanical engineering examples.

## WHO SHOULD USE THIS MANUAL?

The manual is designed to be used in instructor-led courses, although you may also find it helpful as a self-paced learning tool.

### RECOMMENDED COURSE DURATION

Four days (32 hours) to six days (48 hours) are recommended, although the manual may be used for specific Autodesk Inventor topics that may have a duration of only a few hours.

### USER PREREQUISITES

It is recommended that you have a working knowledge of Microsoft® Windows NT® 4.0, Windows 2000®, or Windows XP®, and a working knowledge of parametric solid modeling concepts.

### MANUAL OBJECTIVES

The primary objective of this manual is to provide instruction on how to create part and assembly models, document those designs with drawing views, and automate the design process.

Upon completion of all chapters in this manual, you will be proficient in the following:

- basic and complex part modeling techniques
- complex assembly modeling techniques
- sheet metal design
- weldments
- complex drawing view creation techniques
- design automation techniques

While working through these materials, we encourage you to make use of the Autodesk Inventor Design Support System (DSS), where you may find solutions to additional design problems that are not specifically addressed in this manual.

### MANUAL DESCRIPTION

This manual provides the foundation for a hands-on course that covers basic and advanced Autodesk Inventor features used to create, edit, document, and print parts and assemblies. You learn about the part and assembly modeling tools through online and print documentation, and through the real-world exercises in this manual.

## ESSENTIALS ELECTRONIC STUDENT WORKBOOK

The *Autodesk Inventor 8 Essentials Plus Electronic Student Workbook* is a computer-based tool that contains the real-world, illustrated, step-by-step exercises presented in the *Essentials Plus* manual.

This *Electronic Student Workbook*, and the associated exercise data files, are supplied in a self-installing file called *Setup.exe* on the *Autodesk Inventor 8 Essentials Plus Exercises* CD-ROM attached to the back cover of your book.

### INSTALLING THE ELECTRONIC STUDENT WORKBOOK

To install the *Electronic Student Workbook* and associated exercise data files:

1. Browse to the *Autodesk Inventor 8 Essentials Plus Exercises* CD-ROM and double-click *Setup.exe*.
2. Follow the onscreen instructions.
3. The installation program tests your system to ensure that Autodesk Inventor 8 is installed. By default, an Essentials Exercises folder will be created in the Inventor 8 folder where you installed Autodesk Inventor 8, unless you use the Browse button to specify a different folder.
4. All required files will be copied to the Essentials Plus Exercise folder and an Autodesk Inventor 8 Essentials Exercises shortcut will be placed on your Desktop for starting the *Electronic Student Workbook*.
5. To start the *Electronic Student Workbook*, double-click the Autodesk Inventor 8 Essentials Plus Exercises shortcut.

## PROJECTS

Most engineers work on several projects at a time, with each project consisting of a number of files. To accommodate this, Autodesk Inventor uses projects to help organize related files and maintain links between files.

Each project has a *project file* that stores the paths to all files related to the project. When you attempt to open a file, Autodesk Inventor uses the paths in the current project file to locate other necessary files.

For convenience, a project file is provided with the *Electronic Student Workbook*.

## USING THE PROJECT FILE

Before starting the exercises in the *Electronic Student Workbook*, you must complete the following steps:

1. Start Autodesk Inventor.
2. In the Getting Started dialog box, click Projects in the What to Do section.
3. In the Projects window, right-click the background and choose Browse. Navigate to the folder where you installed the Essentials Exercises and then double-click the IV8_*Essentials_Exercises.ipj* file.
4. Double-click the IV8_*Essentials_Exercises* project in the Projects window to make it the active project.
5. You can now start using the exercises in the *Electronic Student Workbook*.

**NOTE** Projects are reviewed in more detail in Chapter 1, "Getting Started."

## TERMS AND PHRASES

To help you to better understand Autodesk Inventor, a few of the terms and phrases that will be used in this book are explained below.

**Parametric modeling:** Parametric modeling is the ability to drive the size of the geometry by dimensions, and it is sometimes referred to as "dimension-driven design." For example, if you increase the length of a plate from 5= to 6=, change the 5= dimension to 6= and the geometry will be updated. Think of it as the geometry being along for a ride, driven by the dimensions. This is the opposite of associative dimensions: As lines, arcs, and circles are drawn, they are created to the exact length or size; when they are dimensioned, the dimension reflects the exact value of the geometry. If you need to change the size of the geometry, you stretch the geometry, and the dimension automatically gets updated. Think of this as the dimension being along for a ride, driven by the geometry.

**Feature-based parametric modeling:** Feature-based means that as you create your model, each hole, fillet, chamfer, extrusion, and so on is an independent feature whose dimensional values you can change without having to re-create the feature.

**Adaptivity:** Adaptivity allows parts to have a physical relationship between each other, such as when one plate is created with a dimensioned cutout, and another plate is created with the same cutout but NO dimensions are created. The second cutout can be constrained to match the size of the dimensioned cutout. When the dimensions change on the first cutout, the second cutout will be updated or adapt to show the change.

**Bi-directional associativity:** The model and the drawing views are linked. If the model changes, the drawing views will automatically be updated.

Additionally, if the dimensions in a drawing view change, the model is updated and the drawing views are updated based on the updated part.

**Bottom-up assembly:** This refers to an assembly whose parts were created in individual part files outside of the assembly and then referenced into the assembly.

**Top-down assembly:** This refers to an assembly whose parts were created while working in the assembly file.

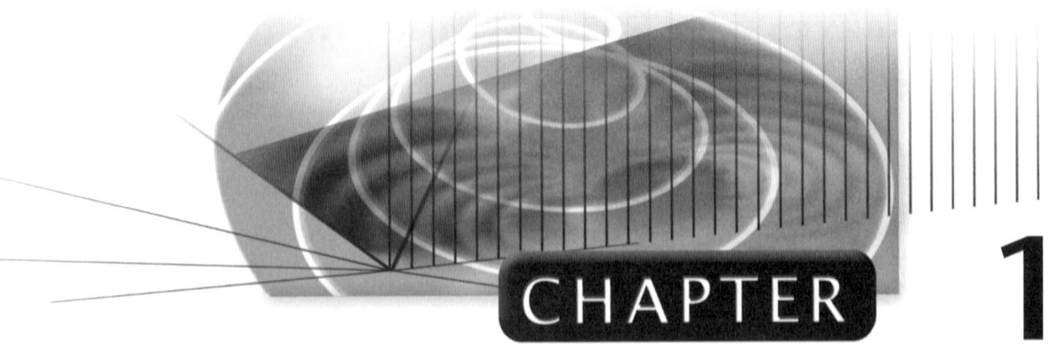

# CHAPTER 1

# Getting Started

This chapter provides a look at the user interface, projects, application options, starting tools (commands), and how to view parts in Autodesk Inventor.

## CHAPTER OBJECTIVES

**In this chapter, you will gain an understanding of**

- The Getting Started screen
- What a project file is used for
- How to create a project file for a single user
- The different file types used in Autodesk Inventor
- Application options
- Help system
- The user interface
- How to issue commands
- The different viewing tools

## GETTING STARTED WITH AUTODESK INVENTOR

The Autodesk Inventor startup screen looks similar to the following image. If the Getting Started screen is not present when you start up the program, click Getting Started from the File menu. There are four startup options on the left side of the screen:

- Getting Started
- New
- Open
- Projects

The information available on the Getting Started page helps reinforce what you learn in this book.

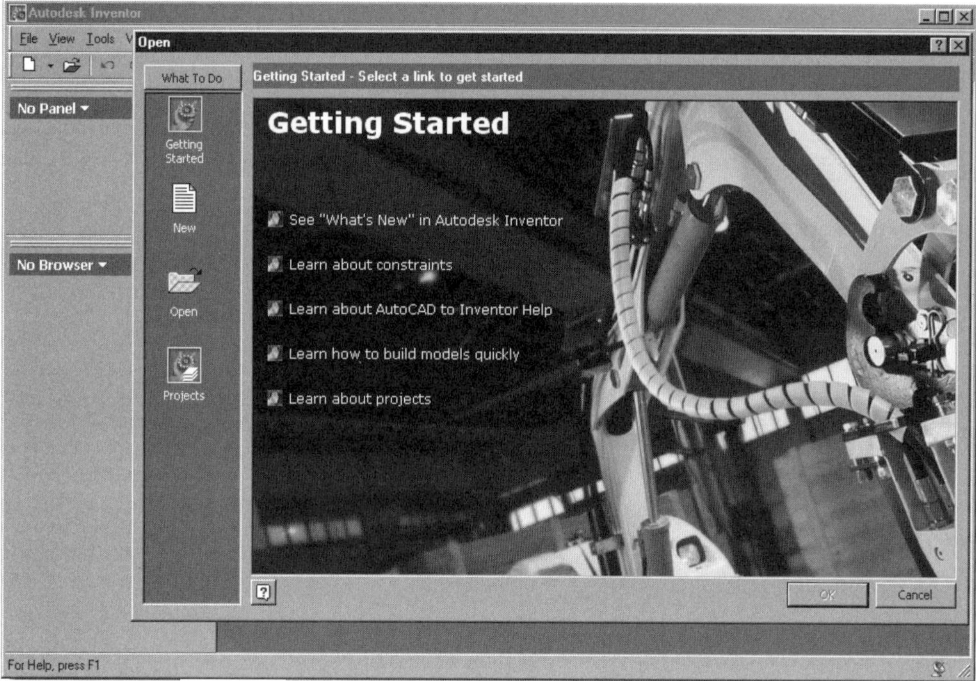


Figure 1-1

### GETTING STARTED

Use Getting Started to find information about:

- What is new in the current release
- Applying geometric constraints
- Transitioning from AutoCAD
- Creating parts
- Creating projects

## NEW

Click New when you want to start a new file. The Open dialog box will display. First, select one of the drafting standards named on the tabs, then select a template for a new part, assembly, presentation file, sheet metal part, or drawing, as shown in the following image. Create a subdirectory in the *Autodesk\Inventor(version number)\templates* directory and add a file to it, and a new templates tab with the same name as the subdirectory is automatically created.

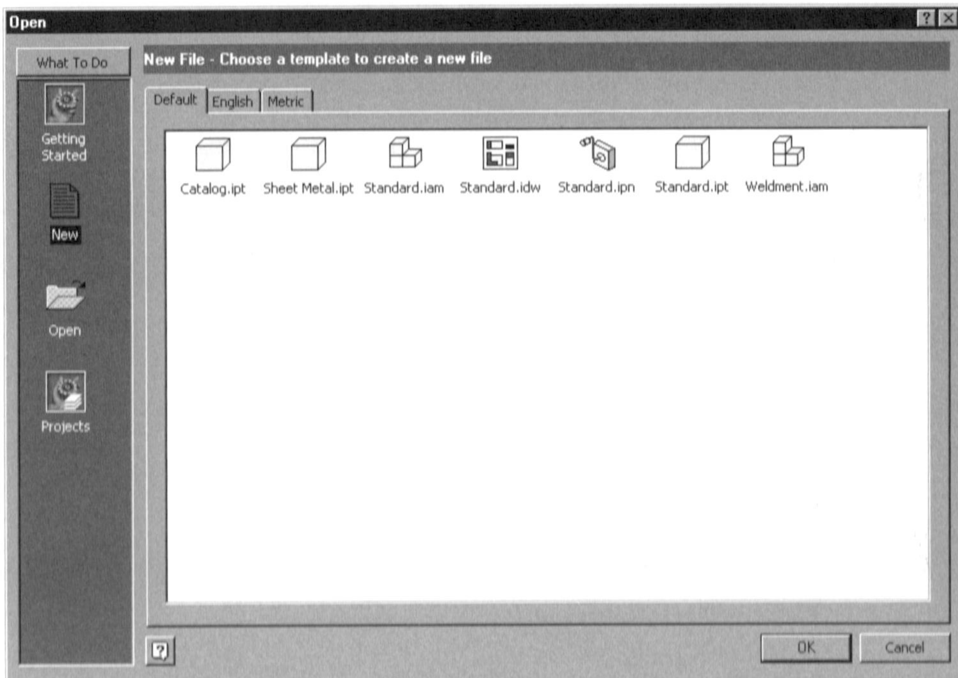


**Figure 1-2**

## OPEN

Click Open to open an existing Autodesk Inventor file. The directory that opens by default is set in the current project file. You can open files from other directories, but it is not recommended since part, drawing, and assembly relationships may not be resolved when you reopen an assembly.

## PROJECTS

Click Projects to create a new project or make an existing project current. Projects are text files that contain search paths to find files that are needed for a given project. With these search paths, all the needed files can be located when an assembly, drawing, or presentation file is opened. There is no limit to the number of projects that can be set up, but only one project can be active at any given time. Projects are covered in the next section.

## PROJECTS IN AUTODESK INVENTOR

Almost every design you create in Autodesk Inventor involves more than a single file. Each part, assembly, presentation, and drawing created from any of these files is stored in a separate file. Each of these files also has its own file extension. There are many times that a design will reference other associated files. An assembly file, for example, will reference a number of individual part files and/or additional subassemblies. When you open the parent or top-level assembly, it must contain information that allows Autodesk Inventor to locate each of the referenced files. Autodesk Inventor uses a project file to organize and manage these file relationships.

You can structure the file locations for a design project in many ways. A single-person design shop has different needs than a large manufacturing company or a design team with multiple designers working on the same project. In addition to project files, Autodesk Inventor includes a basic check-out and check-in file reservation mechanism that controls file access for multi-user design teams. For more information on multi-user environments consult the online help system.

Autodesk Inventor always has a project named Default. The Default project is all that many users need. Specifically, if all the files defining a design are located in a single folder, or in a folder tree where each referenced part is located with its parent or in a subfolder underneath the parent, the Default project is all that is required.

**NOTE** It is recommended that files in different folders never have the same name to avoid the possibility of Autodesk Inventor unintentionally resolving a reference to an incorrect file.

### PROJECT SETUP

Always plan your project folder structure before you start a design to reduce the possibility of file resolution problems later in the design process. A typical project might consist of parts and assemblies unique to the project, standard components that are unique to your company, and off-the-shelf components such as fasteners, fittings, or electrical components.

### PROJECT FILE SEARCH OPTIONS

Before you create a project, you need to understand how Autodesk Inventor stores cross-file reference information and how it resolves that information to find the referenced file. Autodesk Inventor stores the file name, a subfolder path to the file, and a library name (optionally) as the three fundamental pieces of information about the referenced file.

When you use the default project file, the subfolder path is located relative to the folder containing the referencing file. It may be empty, or go deeper in the subfolder hierarchy, but it can never be located at a level above the parent folder.

If you create your own project file and specify project search locations, the subfolder path is initially applied to your specified project locations. If Autodesk Inventor does not find the file there, it will continue to look relative to the folder containing the source file.

A project file can have zero or more file search path locations defined. These paths are searched in a specific order when a document, such as an assembly, needs to find referenced components, such as the part files in the assembly.

Projects are created in the Projects dialog box that is shown in the following image. The Projects dialog box is divided into two areas. The top portion lists the project files that are available.

Double-click on a project's name to make it the active project. Only one project file can be active in Autodesk Inventor at a time. The bottom portion reflects information about the active project, an explanation of which follows.

You can access project file options from the Options area at the bottom of the dialog box.

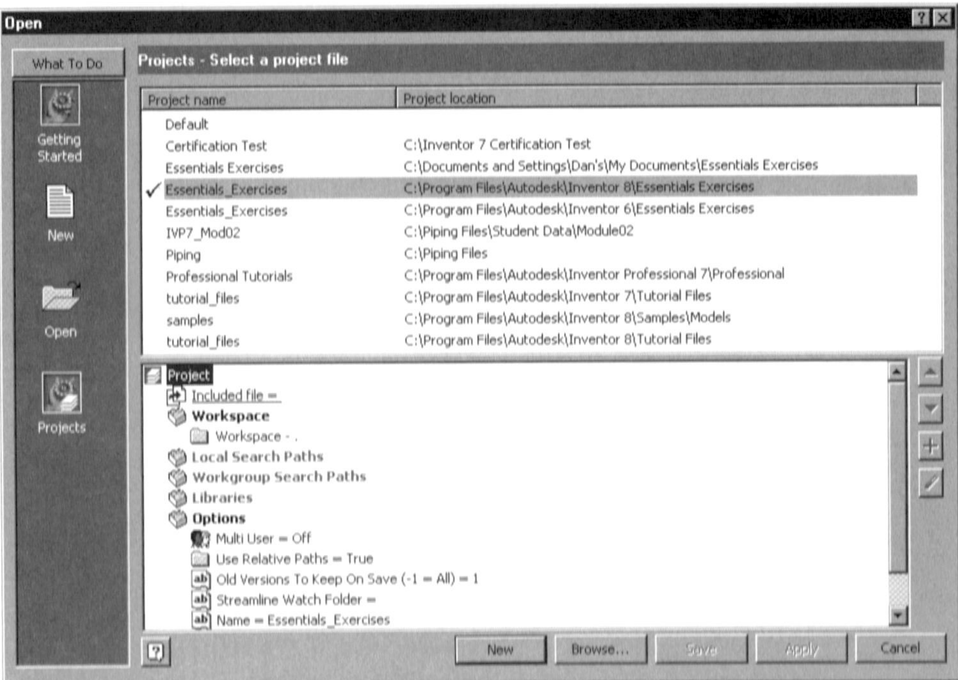


Figure 1-3

## INCLUDED FILE, WORKSPACE, AND LOCAL SEARCH PATHS

The Included file, Workspace, and Local Search Paths are used when implementing a multi-user design environment. These values are typically not defined outside of multi-user design environments. For more information on multi-user environments, consult the online help system.

### WORKGROUP SEARCH PATHS

Some design environments require more file management sophistication than the Default project provides. The files that compose the design, for example, may not be collocated in a single folder tree and may even be located on different disks or servers. You may also want to keep your drawings in one folder, presentations in another, main assemblies in another, etc. A project addresses this need by allowing you to define a search path for each location where files are to be stored.

When defining a path to a folder on another machine, it is better to define a full path starting with the server name (*server*\...) and not to use shared network drives.

**NOTE** You do not need a separate workgroup location defined for each folder; rather, it is recommended that you only define one workgroup location for the root folder of a folder tree containing the Autodesk Inventor files for one project or design. In most cases where the Default project file is insufficient, a project file with only one workgroup location will suffice. It gives you more flexibility in organizing the subfolders of your project than the Default.

### LIBRARY SEARCH PATHS

Library search paths are treated differently than other search paths. They are used to store standard company and third-party models that are not normally edited. In a networked environment, you store these standard components on the network and define the location as a library search path. Files loaded from a library search path cannot be modified; they are recognized by Autodesk Inventor as read-only files. If you create a reference to a file that exists under a library search path, Autodesk Inventor will store the library name with the reference information. It uses the existence of that name to scope the reference specifically to the associated library path.

**NOTE** The folders need not exist prior to adding them to the project file search paths. Named paths added in the Project File Editor are automatically created if they do not exist. Consult the online help system for more information on projects for collaborative environments and search paths.

### PROJECT FILE SEARCH ORDER

Project file search paths are named paths to local or network folders that contain the documents associated with the project. When opening a document that references other files, Autodesk Inventor searches the paths in the following order and loads the first file that matches the reference information stored in the parent document.

#### Library Components

- Library Path. Library components are treated differently than other referenced files. The library name is stored along with the file name in the referencing file. When the referencing file is opened, only the path associated with this library name is searched for the referenced component. If the file is not found there, no other project locations are searched.

### All Other Referenced Components

- Workspace
- Local Search Paths in the order listed in the project file
- Workgroup Search Paths in the order listed in the project file

In all cases, if the file is not found by the above search, Autodesk Inventor searches using the folder containing the parent document as the base instead. Also, with each location searched, Autodesk Inventor appends the subfolder information stored with the reference to the search folder location and looks for the file there. If not found there, Autodesk Inventor then looks for the file in the search folder location.

## FILE RESOLUTION

When a component (part or subassembly) is placed in a higher-level assembly, the information required to locate the referenced file is saved in the assembly document. The assembly uses this information to locate the referenced components the next time the assembly is opened. The same is true when placing a view of a component or presentation in a drawing, creating a presentation of an assembly, or deriving a part from another component. If the file cannot be located, the Resolve Link dialog box appears as shown in the following image. From here, you can manually locate the file or skip the file.

**NOTE** The search order for referenced files follows a defined set of rules. If there are multiple referenced files of the same name in the searched paths, Autodesk Inventor will load the first instance found.

**NOTE** Before creating a new project, close all Autodesk Inventor files and use descriptive names for search paths and directories.

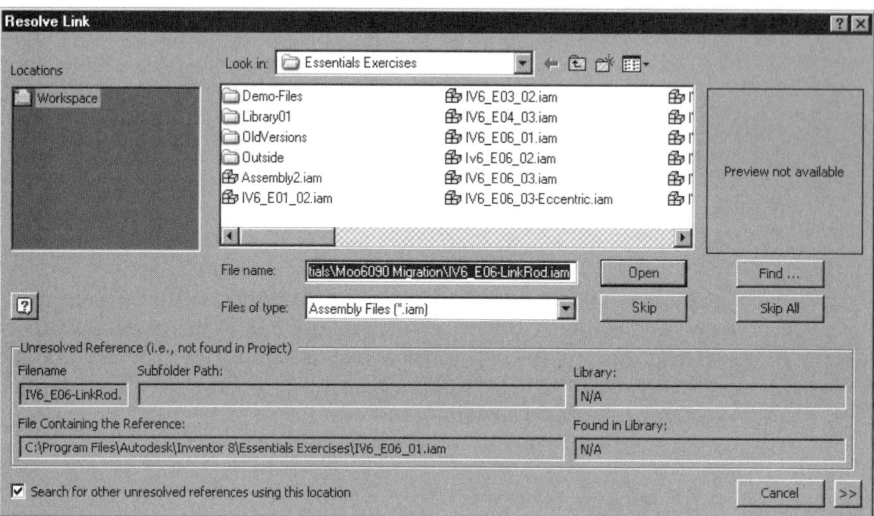


Figure 1-4

## CREATING PROJECTS

To create a new project, or edit an existing project, use the Autodesk Inventor Project File Editor. The Project File Editor displays a list of active projects. A project file has an *.ipj* file extension and is stored in the home folder for the design project, while a shortcut is stored in the Projects Folder. The Projects Folder is specified on the Files tab of the Options dialog box, as shown in the following image. All projects with a shortcut in the Projects Folder are listed in the top pane of the Project File Editor.

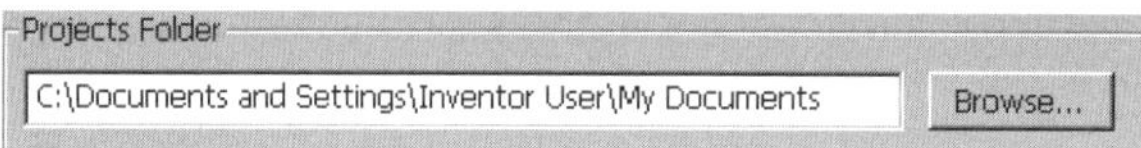


Figure 1-5

To create a project, follow these steps:

1 Click the Project icon in the What To Do area of the Getting Started menu, or click Projects on the File menu.

**TIP** You can also create a new project when the Autodesk Inventor application is not running. Click the Microsoft Windows Start menu, and then click Programs > Inventor (version number) > Tools > Project Editor.

2 In the Projects dialog box, click the New button at the bottom of the dialog box.

3 In the Inventor project wizard, follow the prompts to the following questions:

### What type of project are you creating?

Here you define the type of project that you want to create, as shown in the following image.

### New Vault Project

This project type is used with Autodesk Vault and is grayed out until Autodesk Vault is installed. It creates a project with one workspace and any needed library location(s), and sets the multi-user mode to Vault.

### New Semi-Isolated Master Project

This project type is used where files will be shared among multiple users. It creates a workgroup and any needed library location(s), and sets the multi-user mode to Semi-isolated.

### New Semi-Isolated Workspace

This project type is used for users that are working on a master project; the master project is referenced in this project. The workgroup and library locations are defined from the master project. This project type creates a workspace and sets the multi-user mode to Semi-isolated.

### New Shared Project

This project type is used where multiple users use the same files. It creates one workgroup (where the shared files are stored) and any needed library location(s), and sets the multi-user mode to Shared. No workspace is defined.

### New Single User Project

This is the default project type, which is used when only one user will reference Autodesk Inventor files. It creates one workspace (where Autodesk Inventor files are stored) and any needed library location(s), and sets the multi-user mode to Off. No workgroup is defined.

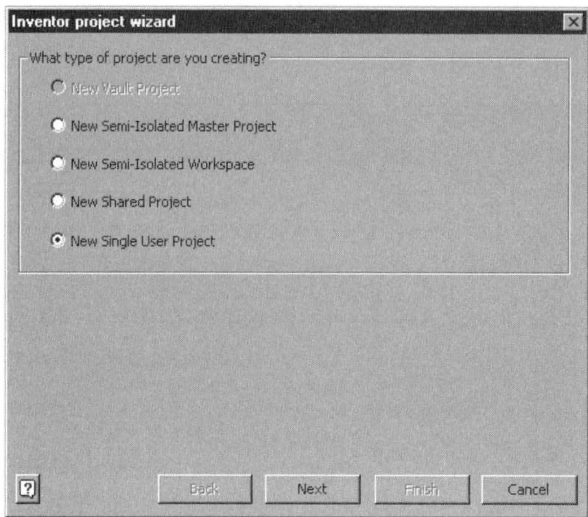


**Figure 1-6**

The next section covers the steps for creating a new single-user project. For more information on projects, consult the online help system or the book on projects that ships in the box with Autodesk Inventor.

## Creating a New Single-User Project

After clicking New Single User Project, click the Next button and specify the project name and location on the second page of the Autodesk Inventor project wizard.

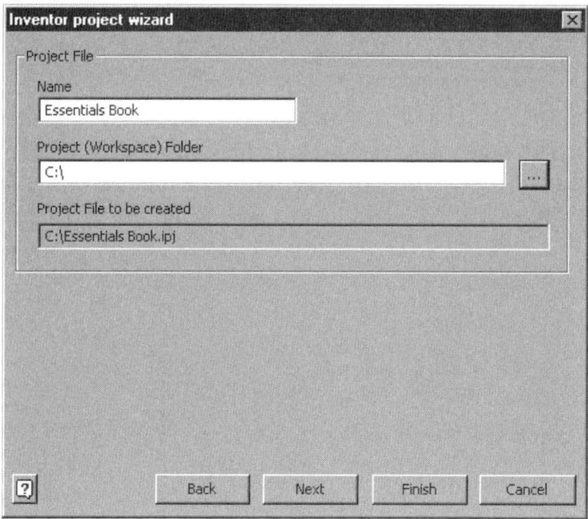


**Figure 1-7**

**Name** Enter a descriptive project file name in the Name edit box. The project file will use this name with an *.ipj* file extension, and will be placed in the folder indicated in the Project (Workspace) Folder edit box. A shortcut to the project will be placed in the Projects folder specified on the Files tab of the Application Options dialog box.

**Project (Workspace) Folder** Specifies the path to the home or top-level folder for the project. You can accept the suggested path, enter a path, or click the Browse button to manually locate the path. The default home folder is a subfolder under My Documents (or to wherever you last browsed) named to match the project file name.

**Project File to be created** The full path name of the project file is displayed below the Location edit box.

**NOTE** You can specify the home or top-level folder as the location of your workspace, but it is preferable that the project file (*.ipj*) be the only Autodesk Inventor file stored in the home folder because it makes it easier to create other project locations (such as library and workspace folders) as subfolders without creating nesting situations that can lead to confusion.

Click the next button at the bottom of the Inventor project wizard and specify the project name and location on the third page of the Inventor project wizard.

You can add library search paths from existing project files to this new project file. The library search paths from every project with a shortcut in your Projects Folder are listed on the left in the Inventor project wizard dialog box, as shown in the following image. You can add and remove libraries from the New Project area by clicking on their names in either the All Projects or New Project area and then clicking the arrows in the middle of the dialog box. The libraries listed by default in the New Project list will match those in the project file you selected prior to starting the New Project process.

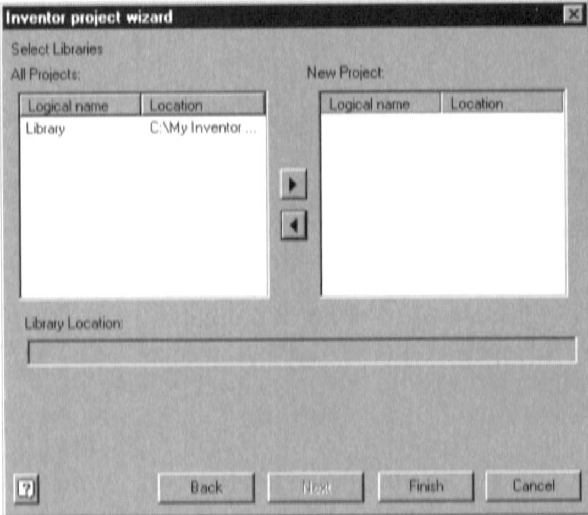


Figure 1-8

Click the Finish button to create the project. The new project will appear in the Open dialog box. Double-click on a project's name in the Open dialog box to make it the active project. A check mark will appear to the left of the active project, as shown in the following image.

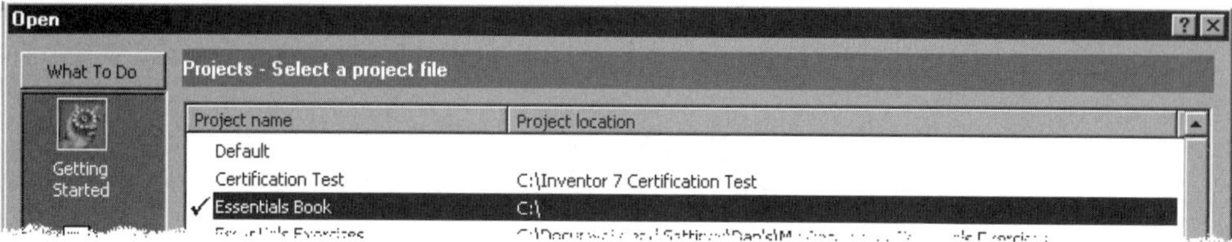


Figure 1-9

## EDITING PROJECTS

To edit a project, follow these steps.

1. Click the Project icon in the What To Do area of the Getting Started screen. If you are already working in Autodesk Inventor, click Projects on the File menu (any open Autodesk Inventor files need to be closed). If Autodesk Inventor is not running, you can edit a project from the standalone Inventor Project Editor. To start the editor, click the Microsoft Windows Start menu and then click Programs > Inventor (version number) > Tools > Project Editor, or in Microsoft Windows Explorer, right-click an *.ipj* file and click Edit.
2. In the top portion of the Project Editor dialog box, double-click the name of the project to make it the current project.
3. In the lower portion of the Project Editor dialog box, right-click the section to edit and click an option on the menu. You can also select the area to edit, and then click the Add or Edit icon on the right side of the dialog box.
4. To edit the order in which directories are searched, select the directory path to move, and click the up or down arrow on the right side of the dialog box.

# EXERCISE 1-1 Projects

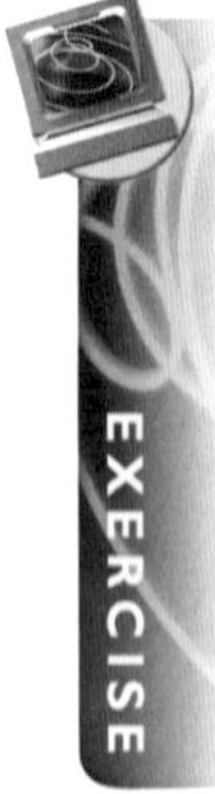


In this exercise, you create a project file for a single-user project. You then examine how the project file provides access to referenced files in an Autodesk Inventor assembly. To navigate to the exercise in the *Electronic Student Workbook*,

1 Click Chapter 1 from the Main TOC.
2 Click Projects from the Chapter 1 TOC.

The following image illustrates the completed exercise.

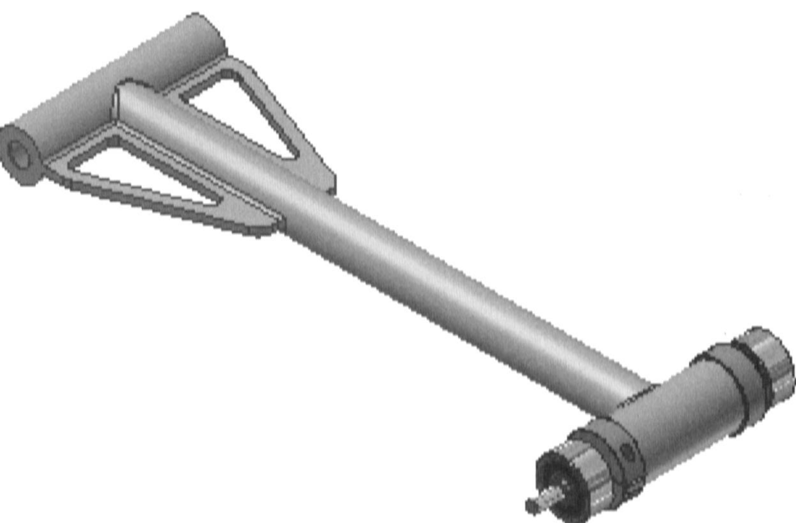

Figure 1-10

## FILE INFORMATION

While creating parts, assemblies, presentation files, and drawing views, that data is stored in separate files with different file extensions. This section describes the different file types and the options for creating them.

### FILE TYPES

The following are the main file types that you can create in Autodesk Inventor, the file extension for each file type, and a description of what they are used for.

#### Part (.ipt)

Part files contain only one part, which can be either 2D or 3D.

#### Assembly (.iam)

Assembly files can consist of a single part, multiple parts, or subassemblies. The parts themselves are saved to their own part file and are referenced (linked) in the assembly file. See Chapters 6 and 10 for more information about assemblies.

#### Presentation (.ipn)

Presentation files show parts of an assembly, exploded in different states. A presentation file is associated with an assembly, and any changes made to the assembly will be updated in the presentation file. A presentation file can be animated, showing how parts are assembled or disassembled. The presentation file extension is IPN, but you save animations as AVI files. See Chapter 6 for more information about presentation files.

#### Sheet Metal (.ipt)

Sheet metal files are part files that have the sheet metal environment loaded. In the sheet metal environment, you can create sheet metal parts and flat patterns. You can create a sheet metal part while in a regular part. This requires that you load the sheet metal environment manually. See Chapter 11 for more information about creating sheet metal parts.

#### Drawing (.idw)

Drawing files can contain 2D projected drawing views of parts, assemblies, and/or presentation files. You can add dimensions and annotations to drawing views. The parts and assemblies in drawing files are linked, like the parts and assemblies in assembly and presentation files. See Chapters 5 and 9 for more information about drawing views.

#### Project (.ipj)

Project files are ASCII-based text files that contain search paths to locations of all the files in the project. The search paths are used to find the files in a project.

### iFeature (.ide)

iFeature files can contain complete parts, 3D features, or 2D sketches that can be inserted into a part file. You can place size limits and ranges on iFeatures to enhance their functionality. See Chapter 12 for more information about creating iFeatures.

### Design Views (.idv)

Design Views are configuration files in which the following information about an assembly file is saved: component visibility, component selection status, color settings, zoom magnification, and viewing angle. See Chapter 10 for more information about creating Design Views.

## MULTIPLE DOCUMENT ENVIRONMENT

You can open multiple Autodesk Inventor files at the same time in a single Autodesk Inventor session. To switch between the open documents, click the file on the Windows menu, or the files can be arranged to fit the screen or cascaded. If the files are arranged or cascaded, click a file to activate it. Only one file can be active at a time.

## SAVE OPTIONS

On the File menu there are three options for saving your files—Save, Save Copy As, and Save All—as shown in the following image.

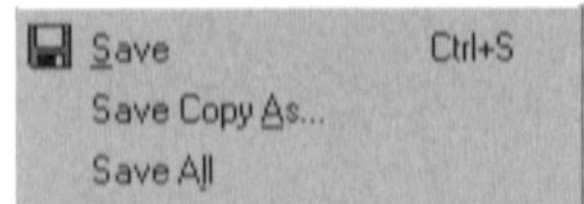


**Figure 1-11**

### Save

The Save command saves the current document with the same name and to the location in which it was created. If this is the first time that a new file is saved, you are prompted for a file location.

**NOTE** To run the Save command, click the Save icon on the Standard toolbar, use the shortcut keys CTRL-S, or click Save on the File menu.

### Save Copy As

Use the Save Copy As command to save the active document with a new name and location, if required. A new file is created, but is not made active.

### Save All

Use the Save All command to save the active document and all of its dependents. The files are saved with the same name, to the location where they were created. The first time that a new file is saved, you will be prompted for a file location.

## APPLICATION OPTIONS

Autodesk Inventor can be customized to your preferences. On the Tools menu, click Application Options to open the Options dialog box, as shown in the following image. You set options on each of the tabs to control specific actions in the Autodesk Inventor software. Each tab is covered in the pertinent sections throughout this book. For more information about application options, see the online help system.

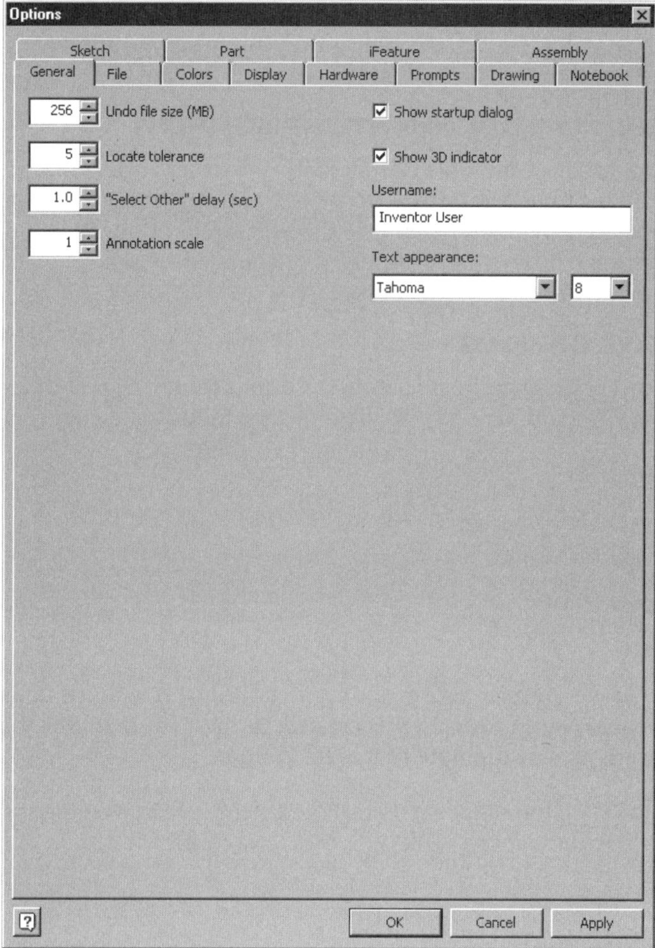


Figure 1-12

**General** Set general options for how Autodesk Inventor operates.

**File** Set where files are located.

**Colors** Change the color of the background on your screen using options on the Colors tab.

**Display** Adjust how the parts look. Your video card and your requirements affect the appearance of parts on your screen. Experiment with different settings to achieve maximum video performance.

**Hardware** Adjust the interaction between your video card and the Autodesk Inventor software. The software is dependent on your video card. Take time to make sure you are running a supported video card and the recommended video drivers. If you experience video-related issues, experiment with the options on the Hardware tab. For more information about video drivers, click Graphics Drivers on the Help menu.

**Prompts** Modify the response given to messages that are displayed.

**Drawing** Specify the way drawings are created.

**Notebook** Specify how the Engineer's Notebook is displayed.

**Sketch** Modify how sketch data is created and displayed.

**Part** Change how parts are created.

**iFeature** Adjust where iFeatures data is stored.

**Assembly** Specify how assemblies are controlled.

## DESIGN SUPPORT SYSTEM

The online help system in Autodesk Inventor goes beyond basic command definition by offering assistance while you design. The following help mechanisms make up the Design Support System:

- Help Topics
- Visual Syllabus
- What's New
- Tutorials
- Design Doctor
- Sketch Doctor
- Autodesk Online (Skill Builders)

To access the online help system, use one of these methods.

- Press the F1 key and the help system gives you help with the operation that is active.
- Click an option on the Help menu.
- Click a Help option on the right side of the Standard toolbar.
- In any dialog box, click the "?" icon.

The visual syllabus and tutorials guide you through the process to complete chosen tasks. To Start the visual syllabus, click Visual Syllabus on the Standard toolbar. To start a tutorial, pick Tutorials on the Help menu. The following image shows the first dialog box that is displayed after you click Visual Syllabus.

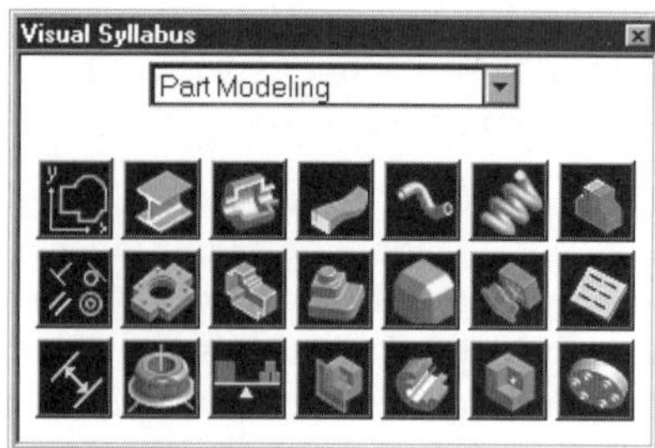


Figure 1-13

The Design Doctor and Sketch Doctor are displayed on your screen when a problem exists with the current operation. The following image shows the Sketch Doctor examining a problem sketch.

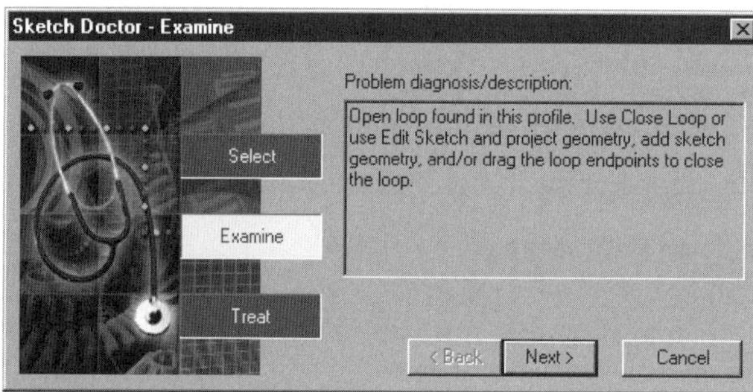


Figure 1-14

Another way to get into the Design and Sketch Doctor is to click on the cross (+) on the right side of the Standard toolbar when it appears red, or when the red cross appears in a dialog box. This alerts you to a problem. The Design or Sketch Doctor opens and provides options for resolving the problem.

## USER INTERFACE

The default part environment in the Autodesk Inventor application window looks similar to the following image.

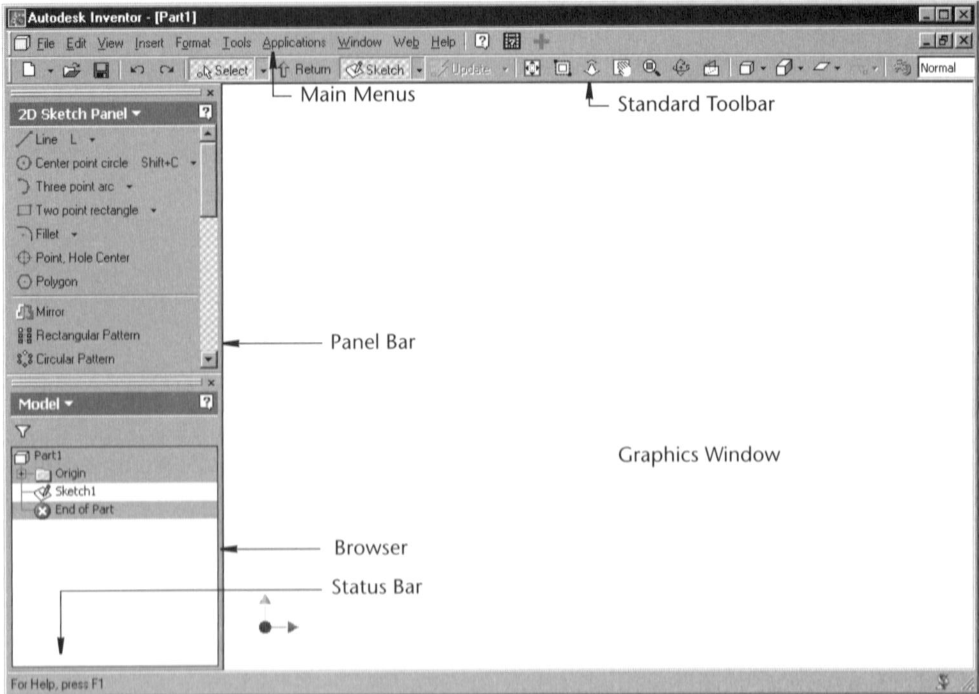


**Figure 1-15**

The screen is divided into the following areas:

**Main Menus** Access tools via text menus.

**Standard Toolbar** Access basic Windows and Autodesk Inventor tools.

**Command Bar** View the current settings. Perform quick edit of objects.

**Panel Bar** Activate most of the Autodesk Inventor tools. The set of tools in the Panel Bar changes to reflect the environment in which you are working.

**Browser Toolbar** Change how the Browser looks. The default settings are used for this book.

**Browser** Shows the history of how the contents in the file were created. The Browser can also be used to edit objects.

**Graphics Window** Displays the graphics of the current file.

**Status Bar** View text messages about where you are.

## TOOLBARS AND PANEL BAR TOOLS

Toolbars and the Panel Bar can be moved to another docked or undocked location. To move a toolbar, click the two horizontal lines on the top of the toolbar, keep the mouse button depressed, and drag to a new location. You get a preview of what the toolbar looks like in the new location. To turn toolbar visibility on and off, click Toolbar on the View menu.

The tools in the Panel Bar adjust to the environment you are working in—such as part, assembly, or sheet metal. You can manually change the status of the Panel Bar. Click the title line of the Panel Bar that has a down arrow, or right-click in the Panel Bar and click another environment on the menu.

The tool icons in the Panel Bar are displayed in non-expert mode by default, with a line of text describing the icon. In Expert mode, the text is removed. To turn Expert mode on (text is removed), either click the title area of the panel, or right-click in the Panel Bar, and click Expert, as shown in the following image.

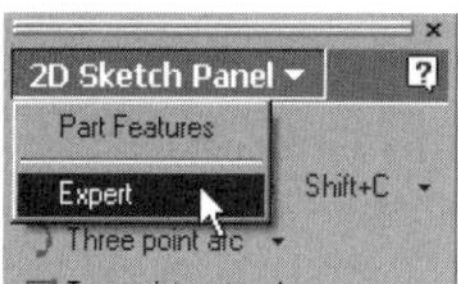


**Figure 1-16**

### COMMAND ENTRY

There are several methods within Autodesk Inventor to issue commands. In the following sections you will learn how to start a command. There is no right or wrong method for starting a command; with experience, you will develop your own preference.

### TOOLBARS AND PANEL BAR

In the last section, you learned how to control toolbars and the Panel Bar. To use a tool from a toolbar, move the cursor over the desired tool and a tool tip appears with the name of the tool. The following image shows the Line tool from the Sketch toolbar, and the tooltip. Some of the icons in the Panel Bar have a small down arrow in the lower right corner, Select the arrow to see additional tools. To activate a tool, move the cursor over a tool icon and release the mouse button.

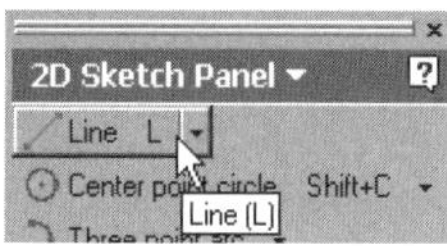


**Figure 1-17**

## SHORTCUT MENUS

Autodesk Inventor also uses shortcut menus. Shortcut menus are text menus that pop up when you press the right mouse button. The shortcut menus are context sensitive; you get a menu with tools or options that are relevant to the current task.

## WINDOWS SHORTCUTS

Another method for activating tools is to use Windows shortcut keys. Windows shortcut keys use a two-key combination. Select the two keys at the same time to activate the operation.

CTRL-C Copy

CTRL-N Create a new document

CTRL-O Open a document

CTRL-P Print the active document

CTRL-S Save the current document

CTRL-V Paste

CTRL-Y Redo

CTRL-Z Undo

## AUTODESK INVENTOR SHORTCUT KEYS

Autodesk Inventor has keystrokes called shortcut keys that are preprogrammed. To start a command via a shortcut key, press the desired preprogrammed key(s). The following chart shows a few of the predefined shortcut keys. In the next section you will learn how to modify shortcut keys.

**Shortcut Keys**

| Key | Description |
|---|---|
| F1 | Help. |
| F2 | Pan the screen. |
| F3 | Zoom in or out on the objects on the screen. |
| F4 | Rotate the objects on the screen. |
| F5 | Return to the previous view. |
| SHIFT+F5 | Next view. |
| B | Add a balloon to a drawing. |

| Shortcut Keys (*continued*) | |
|---|---|
| **Key** | **Description** |
| C | In the assembly environment, add an assembly constraint using the Assembly Constraint dialog box. |
| D | Add a general dimension to a sketch or drawing. |
| E | Extrude a sketch. |
| F | Add a feature control frame to a drawing. |
| H | Create holes using the Hole dialog box. |
| L | Draw a line. |
| O | Add an ordinate dimension. |
| P | Place a component into the current assembly. |
| R | Revolve a sketch. |
| S | Make a plane the active sketch. |
| T | Add a tweak to a part in the current presentation file. |
| ESC | Abort command. |
| DELETE | Delete the selected objects. |
| BACKSPACE | Clears the last sketch selection as long as the command is active. |
| SHIFT+right-click | Selection tool menu. |
| SHIFT+Rotate tools | Auto rotate, puts the contents of the screen into hands free rotation. Left-click to stop the rotation. |
| SPACE BAR | While in the rotate command, toggles the common view (glass box) on and off. |
| TAB | Alternate between input fields. |
| CTRL+SHIFT | Add and remove objects from a selection set. |
| CTRL+ENTER | Disable inferencing when committing precise input sketch points. |
| CTRL | Override the creation of constraints while sketching. |

## CUSTOMIZING MENUS AND TOOLBARS

You can customize the Autodesk Inventor user interface to suit your working style by modifying or adding toolbars and menus. There are specific environments that become active when working with Autodesk Inventor. These environments are based on the type of file that is being edited and can be customized along with toolbars and commands. When you design or edit a part file, for example, the part environment is active. If an assembly file is open or is the current editing target, the assembly environment is active.

Customization of Autodesk Inventor is performed by selecting Customize from the Tools menu, as shown in the following image.

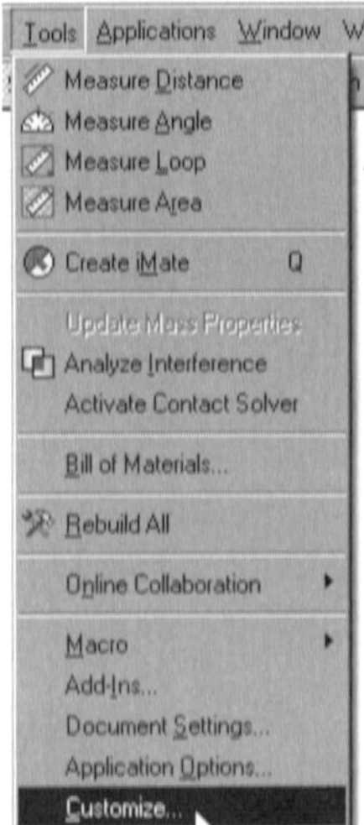


**Figure 1-18**

The Customize dialog box, illustrated in the following image, is displayed.

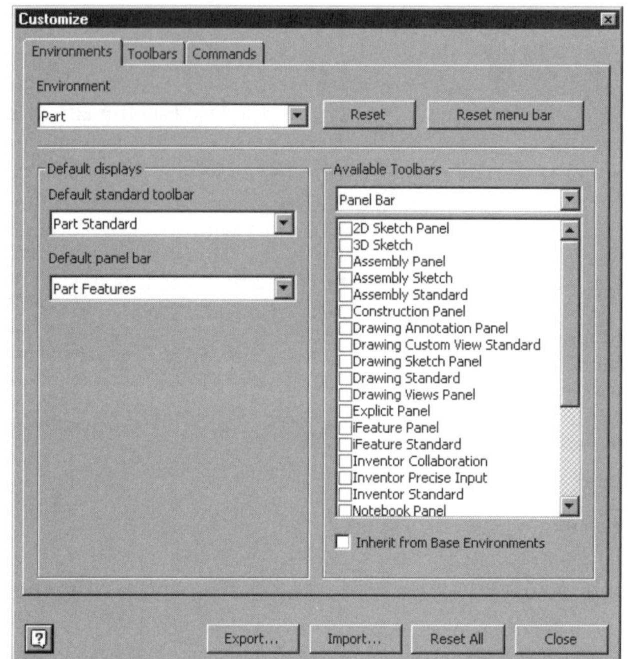


**Figure 1-19**

The Environment drop-down list shows all registered environments on the system, including Autodesk Inventor and any third-party environments. The default display area is where you specify the default behavior when an environment becomes active. When an environment is activated, the basic user interface components take on the specified settings. The Default Standard sets the Standard toolbar that is available. Default Panel Bar sets the toolbar that is displayed in the Panel Bar for the selected environment.

The Available Toolbars section is where you can further customize the interface by changing a menu or the Panel Bar. When Panel Bar is selected, as shown in the previous image, you can select the toolbars that you want available in the Panel Bar drop-down list. The following image, for example, shows three items selected in the available toolbars area, and the Panel Bar is the active item to be customized. The Part Features toolbar is set to be displayed in the Panel Bar by default. The other two selected options, Part Standard and Part Test, are made available by selecting the arrow next to the name of the Panel Bar.

Base environments have unique toolbars, Panel Bars, and browsers associated with them. Some base environments have secondary environments with additional toolbars. After selecting a toolbar in the Available Toolbars list, click the checkbox next to Inherit from Base Environments if you want to use the base environments settings.

If you are making changes to the current environment, you need to exit the environment and re-enter it to see the changes. For example, while in the Sketch environment, if you customize a toolbar, you will not see the changes until you exit the sketch and then go back into it.

### TOOLBAR CREATION

The Toolbars tab, shown in the following image, allows you to customize the content of toolbars, add and remove user-defined toolbars, and define which toolbars are displayed outside of the Panel Bar.

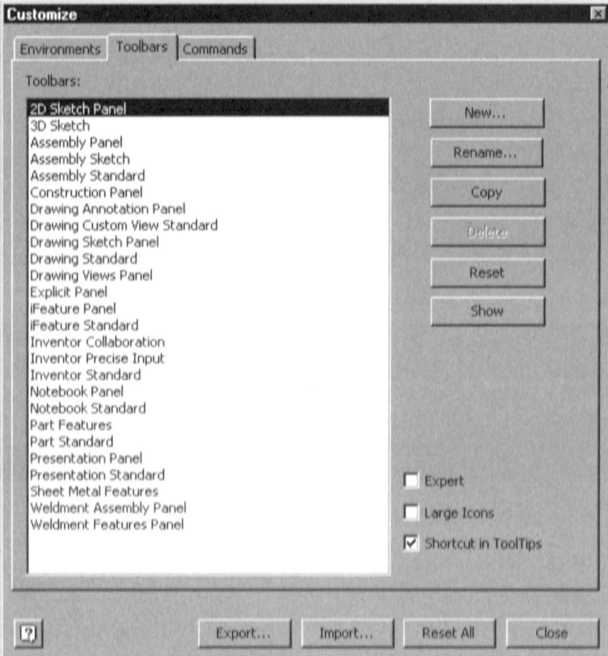


**Figure 1-20**

Custom toolbars are added using the New button and removed using the Delete button. When the New button is selected, a dialog box appears requesting the name for the new toolbar. When you select the Delete button, you will be asked for confirmation before the toolbar is removed.

You can also rename or copy toolbars using the appropriate buttons. The Show button displays the selected toolbar, and the Reset button returns the selected toolbar to its default state.

The three checkboxes located at the bottom-right corner of the dialog box affect the behavior of the Panel Bar in the following ways:

### Shortcut in Tooltips

This checkbox displays shortcuts when the cursor is placed on top of a tool in the Panel Bar, as shown in the following image.

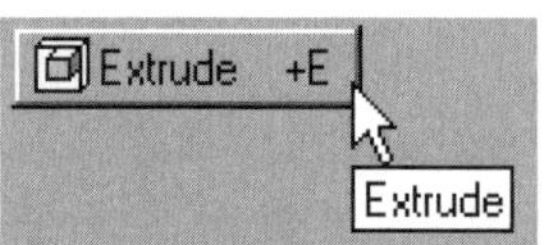


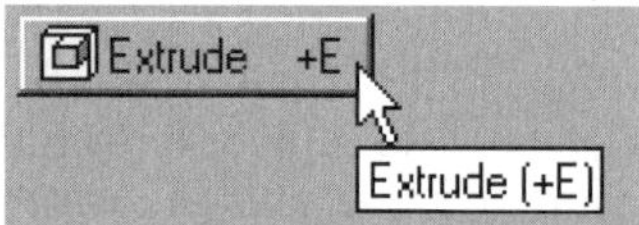


**Figure 1-21**

### Expert

This checkbox toggles the Panel Bar between Learning and Expert modes, as shown in the following image.

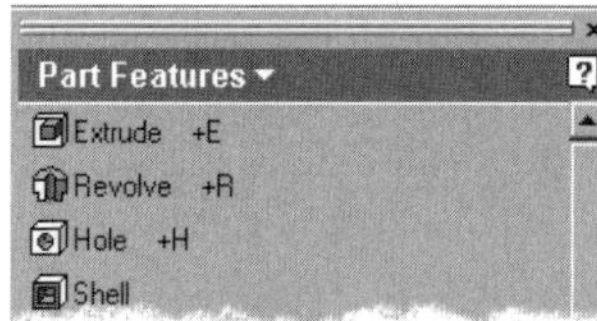


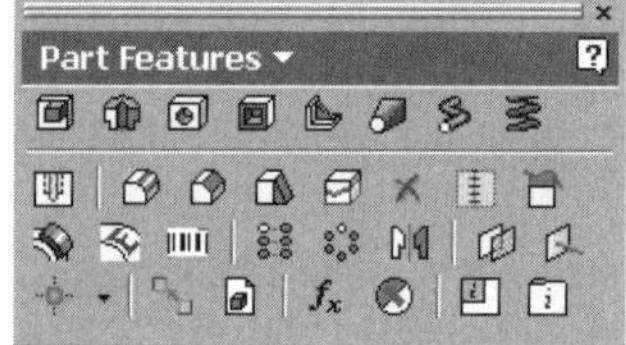


**Figure 1-22**

### Large Icons

This checkbox toggles the icon display between small and large icons.

## COMMANDS TAB

The Commands tab gives you access to all of the tools that are available in Autodesk Inventor, and allows you to modify the shortcut that is used to start the tool. The following image shows the Commands tab of the Customize dialog box.

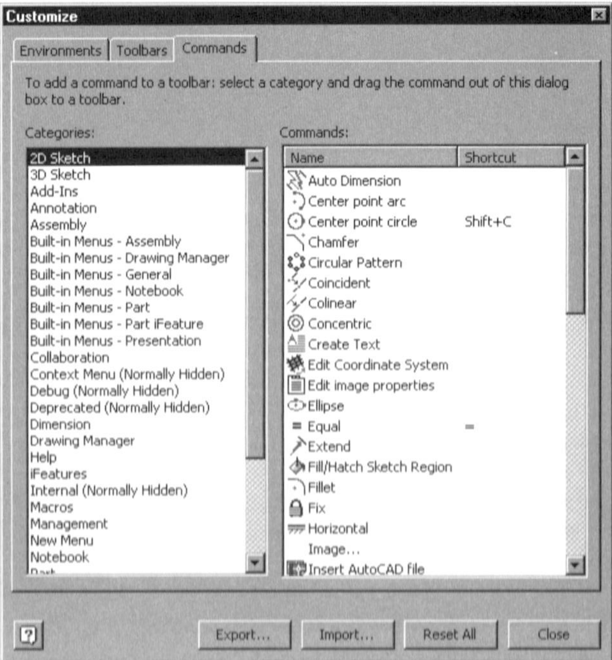


**Figure 1-23**

When you select an item from the Categories list, the tools that are available are displayed on the right side of the dialog box. A black triangle next to the tool indicates that it has additional tools available in a fly-out menu. The Constraints tool, for example, contains all of the constraint types. You can add tools to a toolbar by dragging them from the Commands list and dropping them onto the toolbar. Dragging the tool off of a toolbar while the Customize dialog box is active will remove the tool from the toolbar.

To add a tool to a toolbar, follow these steps:

1. Select Customize from the Tools menu.
2. Select the Toolbars tab.
3. Choose the toolbar that you want to customize, or create a new toolbar to modify.
4. Click the Show button from the Toolbars tab to display the toolbar.
5. Click the Commands tab.
6. Choose a category from the Categories list.
7. Drag the desired tool onto the toolbar. When a vertical bar is displayed, as shown in the following image, release the mouse button.

Figure 1-24

To add a new shortcut or change an existing shortcut, follow these steps:

1 Click in the Shortcut area of the tool that you want to change. The following image shows what the dialog box looks like when you click in the Shortcut area of the Chamfer tool.
2 On the keyboard, press the key(s) that will be the shortcut. A single letter or number can be used. You can also use the SHIFT, CTRL, and ALT keys and a letter or number. The ALT key can be combined with the SHIFT and/or CTRL key(s) and a letter or number. The ALT key CANNOT be used with only a letter.
3 Press ENTER to create the shortcut.

To delete a shortcut, click on the shortcut and then press DELETE or BACKSPACE on your keyboard to remove the shortcut.

**NOTE** To switch an existing shortcut to a different, new tool, you must first delete the shortcut from the existing tool.

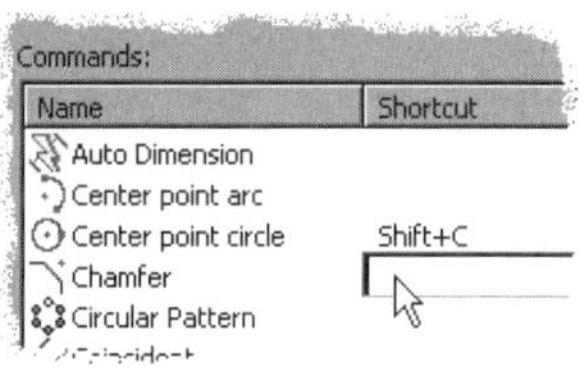


Figure 1-25

## General Options

The Export, Import, Reset All, and Close buttons reside at the bottom of the dialog box and are available regardless of which tab is active. These buttons function as follows:

**Export** Exports an XML file containing customized settings.

**Import** Imports an XML file containing customized settings. Importing can take place only after a fresh installation of Autodesk Inventor.

**Reset All** Resets all customized settings to their original installation setting.

**Close** Closes the dialog box.

### ROLLUP COMMAND DIALOGS

You can control whether dialog boxes appear normal or rolled up to show only the name of the dialog box when the mouse is not located in the dialog box itself. To roll up a dialog box, click the middle icon (push pin) in the upper right corner of the dialog box. The dialog box shows only the horizontal title bar, as shown in the following image. To maximize the dialog box, move the mouse over it.

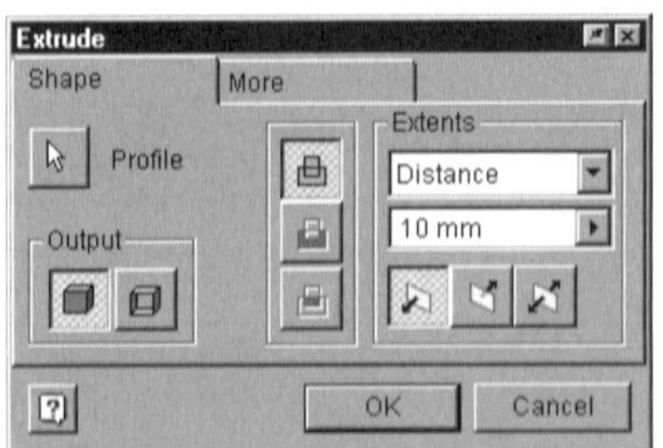


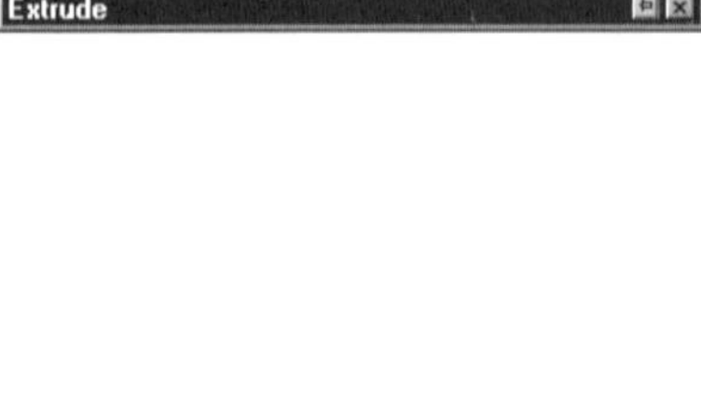


Figure 1-26

### UNDO AND REDO

You may want to undo an action that you just performed, or undo an undo. The Undo tool backs up Autodesk Inventor one function at a time. If you Undo too far, you can use the Redo tool to move forward one step at a time. The Zoom, Rotate, and Pan tools do not affect the Undo and Redo tool. To start the tools, either use the shortcut keys—CTRL-Y for Redo or CTRL-Z for Undo—or select the tool from the Standard toolbar, as shown in the following image. The Undo tool is to the left, and the Redo tool is to the right.

**NOTE** To set the Undo file size allocation, click Tools ➤ Application Options. On the General tab of the Options dialog box, change the "Maximum size of Undo file (Mb)."

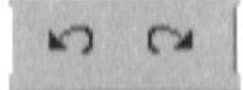

Figure 1-27

## VIEWING MODELS FROM DIFFERENT VIEWPOINTS

When you work on a 2D sketch, the default view is looking straight down at the XY plane (plan view). When you work in 3D, it is helpful to view objects from a different viewpoint, and to zoom in and out or pan the objects on the screen. The next section guides you through the most common methods for viewing objects from different perspectives and viewpoints. For each of the tools, the physical objects are not moving. Your perspective or viewpoint of the objects is what creates the movement of the part. If you are in an operation while a viewing command is issued, the operation resumes after the transition to the new view.

### ISOMETRIC VIEW

Change to an isometric viewpoint by either pressing the **F6** key or right-clicking in the graphics window, and then selecting Isometric View from the menu, as shown in the following image. The view on the screen transitions to a predetermined isometric view. The isometric view can be redefined within the Common View (Glass Box) option. Common View is covered later in this section.

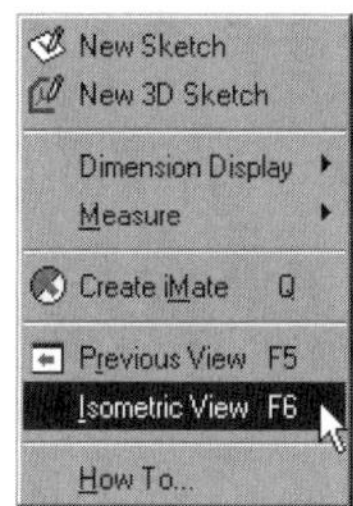


**Figure 1-28**

### CAMERA VIEWS

You can set the camera viewpoint to be either orthographic (lines are projected perpendicular to the plane of projection) or perspective (geometry on the screen converges to point similar to the human eye sees). By default, the orthographic camera is set. To change the camera, select either Orthographic Camera from the Standard toolbar, as shown in the following image, or the Perspective Camera that is the option just below Orthographic Camera. For clarity, this book shows all the images in the Orthographic Camera. For more information about options for setting the Perspective Camera, see the Help system.

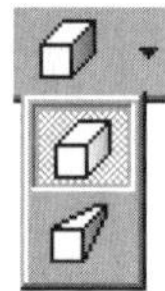

**Figure 1-29**

### SHADOW

To give your model a realistic look, you can choose to have shadows displayed on the ground or not. From the Standard toolbar, click either Ground Shadow (as shown in the following image), No Ground Shadow or X-Ray Ground Shadow. The default is No Ground Shadow. With shadows on, only model features cast shadows. Work geometry, sketches, origin indicators, engineer's notes, and trails in presentation documents do not cast shadows.

X-Ray Ground Shadows are the same as Ground Shadows but they also show detail information about hidden features.

**Figure 1-30**

## VIEW TOOLS

To help zoom, pan, and rotate the geometry on the screen, you use the View tools on the Standard toolbar. Descriptions of the View tools follow.

**Viewing Tools**

| Button | Tool | Function |
|---|---|---|
| | Zoom All | Maximizes the screen with all parts that are in the current file. The screen transitions to the new view. Issue the Zoom All tool or press the Home key. |
| | Zoom Window | Zoom in on an area that is designated by two points, issue the Zoom Window tool or press **Shift+F3**, and select the first point. With the mouse button depressed, move the cursor to the second point. A rectangle appears representing the window. When the correct window is displayed on the screen, release the mouse button and the view transitions to it. |
| | Zoom In-Out | To zoom in or out from the parts, issue the Zoom In-Out tool, or press the **F3** key. Then, in the graphics window, press and hold the left mouse key. Move the mouse toward you to make the parts appear larger, and away from you to make the parts appear smaller. If you have a mouse with a wheel, roll the wheel toward you and the parts appear larger; roll the wheel away from you and the parts appear smaller. |
| | Pan View | Moves the view to a new location. Click the Pan View tool or select the **F2** key. Press and hold the left mouse button and the screen moves in the same direction that the cursor moves. If you have a mouse with a wheel, hold down the wheel, and the screen moves in the same direction that the cursor moves. |
| | Zoom Selected | Fills the screen with the maximum size of a selected face or faces. Either select the face or faces and then issue the Zoom Selected tool or select the **End** key, or launch the Zoom Selected tool or select the **End** key and then select the face or faces to which to zoom. |

**Viewing Tools (*continued*)**

| Button | Tool | Function |
|---|---|---|
|  | Dynamic Rotate | Rotates objects dynamically. Click the Dynamic Rotate tool on the Standard toolbar; a circular image with lines at the quadrants and center appears. To rotate the parts freely, click a point inside the circle and keep the mouse button pressed as you move the cursor. The model rotates in the direction of the cursor movement. When you release the mouse button, the model stops rotating. To accept the view orientation, either press the **ESC** key or right-click and click Done from the menu. Click the outside of the circle to rotate the model about the center of the circle.<br>To rotate the parts about the vertical axis, click one of the horizontal lines on the circle and, with the mouse button pressed, move the cursor sideways. To rotate the parts about the horizontal axis, click one of the vertical lines on the circle and, with the mouse button pressed, move the cursor upward or downward. |
|  | Look At | Changes your viewpoint so you are looking parallel to a plane, or rotate the screen viewpoint to be horizontal to an edge. Click the Look At tool on the Standard toolbar or press the **Page Up** key, then select a plane or edge. The Look At tool can also be issued by selecting a plane or edge, and then right-clicking while the cursor is in the graphics window and selecting Look At from the Standard toolbar. |
| 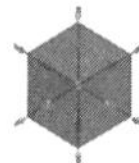 | Common View (Glass Box) | Changes the viewpoint to a predetermined viewpoint. Click the Dynamic Rotate tool and then either press the **SPACE** bar or right-click and select Common View from the menu. A cube appears with arrows pointing at each corner and face. Click one of the arrows on the cube and the viewpoint rotates to that perspective. If you are looking at the cube from a plan view, you can click an edge of the cube to rotate the cube 90°. To accept the view orientation, either press **ESC** or right-click and select Done from the menu. To change the default viewpoint that is used when Isometric View is selected from the menu, click the arrow on a corner that you want to be the default isometric view. Then right-click and select Redefine Isometric from the menu. |
| F4 | F4 Shortcut to Dynamic Rotation | Rotates objects dynamically. While working, press and hold down the **F4** key. The circular image appears with lines at the quadrants and center. With the **F4** key depressed, rotate the model. When you finish rotating the model, release the **F4** key. If you are performing an operation while the **F4** key is pressed, that operation will resume after you release the **F4** key. |
| F5 | Previous View | Returns to the view that was previously on the screen. Press the **F5** key or right-click in the graphics window and select Previous View from the menu. Continue this process until the view to which you want to return is on the screen. |

**Viewing Tools (*continued*)**

| Button | Tool | Function |
|---|---|---|
| | Display Options | Accesses the three options from the Standard toolbar to display 3D parts: Shaded Display, Hidden Edge Display, and Wireframe Display. You can choose which mode works best for you and switch between the modes as you see fit. Each display mode is described next. |
| | Shaded Display | Shades objects in the color or material that was assigned to them. Parts or faces that are behind other parts or faces are not displayed. |
| | Hidden Edge Display | Shades objects and display the edges that are behind other parts or faces. In complex parts and assemblies, this display can be confusing. |
| | Wireframe Display | Displays only the outline of objects. In wireframe display, objects are not shaded. |

EXERCISE

# EXERCISE 1-2 Viewing a Model

In this exercise, you use View Manipulation tools that make it easier to work on your designs. To navigate to the exercise in the *Electronic Student Workbook*, do the following:

1 Click Chapter 1 from the Main TOC.

2 Click Viewing a Model from the Chapter 1 TOC.

The completed exercise is shown in the following image.

**Figure 1-31 Completed exercise**

## CHAPTER SUMMARY

| To | Do This | Tool |
|---|---|---|
| Create a project | Select Projects from either the Autodesk Inventor File menu or from the What To Do area of the Getting Started page. |  Projects |
| Save the active document | Click Save on the File menu or on the Standard toolbar. | |
| Save a copy of the active document with a new name | Click Save Copy As on the File menu. | |
| Save the active document and all of its dependents | Click Save All on the File menu. | |
| Set Autodesk Inventor preferences | Click Application Options on the Tools menu. | |
| Get help | Press the **F1** key, click one of the options on the Help menu, or click the Help Topics or Visual Syllabus icon on the Main menu bar. |  |
| Undo or redo a command | Click the Undo or Redo icon on the Standard toolbar. |  |
| Change the current viewpoint to an isometric viewpoint | Right-click in the graphics window, and select Isometric View from the menu. | |
| Zoom or pan | On the Standard toolbar, click one of the Zoom or Pan tools, or use the wheel on a wheel mouse. |  |
| Dynamically rotate a part | Click the Dynamic Rotate tool on the Standard toolbar, or use the **F4** key. | |
| Change the display | Click the Shaded Display, Hidden Edge Display, or Wireframe Display icon on the Standard toolbar. |  |

## CHECKING YOUR SKILLS

Use these questions to test your knowledge of the material covered in this chapter.

1 Explain what a project file is used for.

______________________________________________

______________________________________________

2 **True___ False___** Only one project can be active at any time.

3 List the sequence of the locations searched when a file is opened in Autodesk Inventor.

______________________________________________

______________________________________________

______________________________________________

4 **True___ False___** Autodesk Inventor stores the part, assembly information, and related drawing views in the same file.

5 **True___ False___** Press and hold down the **F4** key to dynamically rotate a part.

6 **True___ False___** The Save Copy As command saves the active document with a new name and then makes it current.

7 List four ways to access the help system.

______________________________________________

______________________________________________

______________________________________________

______________________________________________

**8** Explain how to create a shortcut.

____________________________________________

____________________________________________

____________________________________________

____________________________________________

**9** **True___** **False___** The Look At tool changes the viewpoint to an isometric view.

**10** **True___** **False___** You can only edit a part while it is in shaded display.

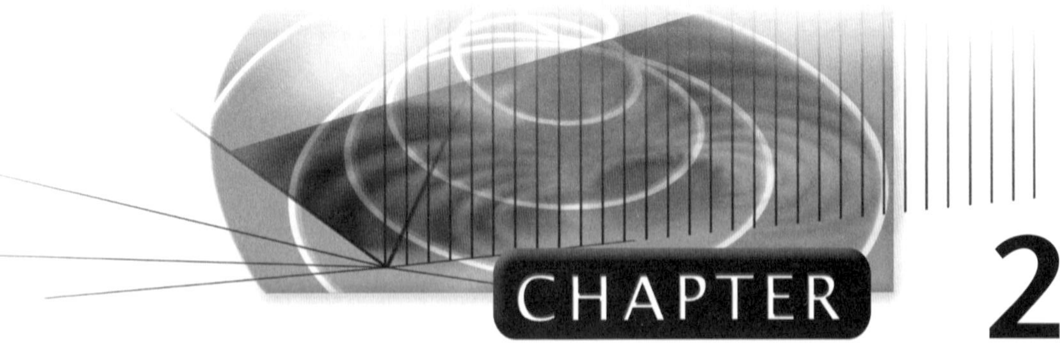


# 2

# Sketching, Constraining, and Dimensioning

Most 3D parts in Autodesk Inventor start from a 2D sketch. This chapter first provides a look at the application options for sketching and part creation. It then covers the three steps in creating a 2D parametric sketch—sketching a rough 2D outline of a part, applying geometric constraints, and then adding parametric dimensions.

## CHAPTER OBJECTIVES

**In this chapter, you gain an understanding of**

- Sketch and part options
- Sketching an outline of a part
- Creating geometric constraints
- Dimensioning a sketch
- Changing a dimension's value in a sketch

## SKETCHING AND PART APPLICATION OPTIONS

Before you create a sketch, examine the sketch and part options in Autodesk Inventor that will affect sketching and part modeling. While learning Autodesk Inventor, refer back to these option settings to determine which ones work best for you—there are no right or wrong settings.

### SKETCH OPTIONS

Autodesk Inventor sketching options can be customized to your preferences. Click Tools > Application Options and then click the Sketch tab, as displayed in the following image. A description of the Sketch options follows. These settings are global and affect all open and new Autodesk Inventor documents.

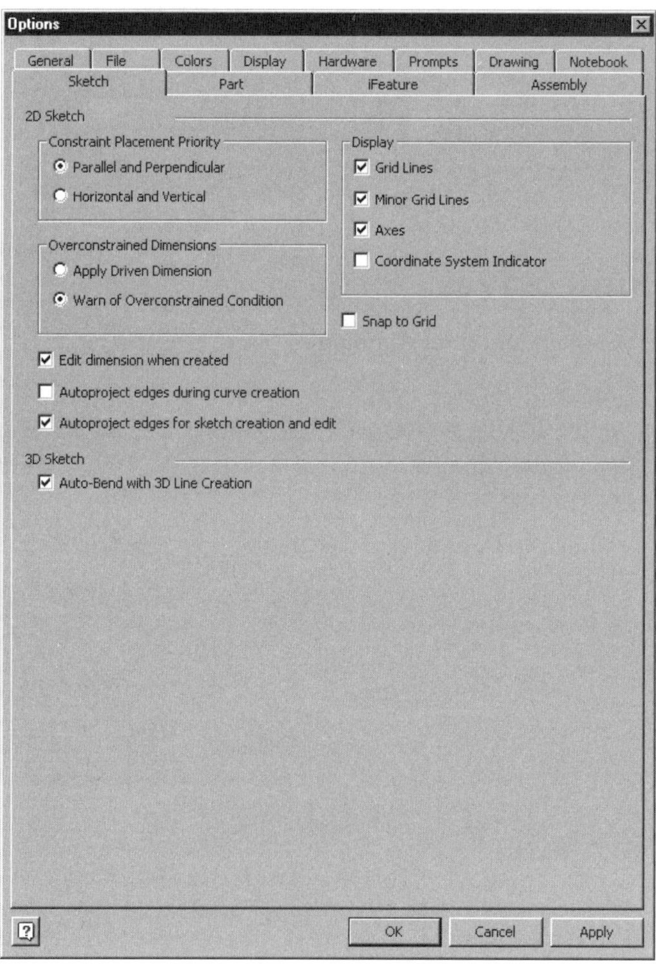


Figure 2-1

### Constraint Placement Priority

- Parallel and Perpendicular

  When checked, and a parallel or perpendicular condition exists while sketching, a parallel or perpendicular constraint will be applied before any other possible constraints that affect the geometry being created.

- Horizontal and Vertical

  When checked, and a horizontal or vertical condition exists while sketching, a horizontal or vertical constraint will be applied before any other possible constraints that affect the geometry being created.

### Display

- Grid Lines

  Toggles both minor and major grid lines on the screen on and off. To set the grid distance, click Tools > Document Settings and on the Sketch tab of the Document Settings dialog box, change the Snap Spacing and Grid Display.

- Minor Grid Lines

  Toggles the minor grid lines on the screen on and off.

- Axes

  Toggles the lines that represent the *X*- and *Y*-axis of the current sketch on and off.

- Coordinate System Indicator

  Toggles the icon on and off that represents the *X*-, *Y*-, and *Z*-axis at the 0,0,0 coordinates of the current sketch.

### Overconstrained Dimensions

- Apply Driven Dimensions

  When checked, and dimensions are added that would over-constrain the sketch, the dimension is added as a driven (reference) dimension.

- Warn of Overconstrained Condition

  When checked, and dimensions are added that would over-constrain the sketch, the Dimension dialog box appears warning of the condition, and then the dimension is not added.

### Snap to Grid

When checked, endpoints of sketched objects will snap to the intersections of the grid as the cursor moves over them.

### Edit dimensions when created

When checked, the values of dimensions will be edited in the Edit Dimension dialog box immediately after the dimension is positioned.

### Autoproject edges during curve creation

When checked, and while sketching objects that are in a different plane, *scrubbing* an object will automatically project it onto the current sketch. AutoProject can also be toggled on and off while sketching by right-clicking and selecting AutoProject from the menu.

### Autoproject edges for sketch creation and edit

When checked, all of the edges that define that plane will be automatically projected onto the sketch plane as reference geometry when you create a new sketch.

### 3D Sketch

- Auto-Bend with 3D Line Creation

  When checked, a tangent arc will automatically be placed between two 3D lines as they are sketched. To set the radius of the arc, click Tools > Document Settings, and on the Sketch tab of the Document Settings dialog box change the Auto-Bend Radius.

## PART OPTIONS

Autodesk Inventor part options can be customized to your preferences. Click Tools > Application Options, and click the Part tab, as displayed in the following image. Following is a description of the Part options. These setting are global—they will affect all open and new Autodesk Inventor documents.

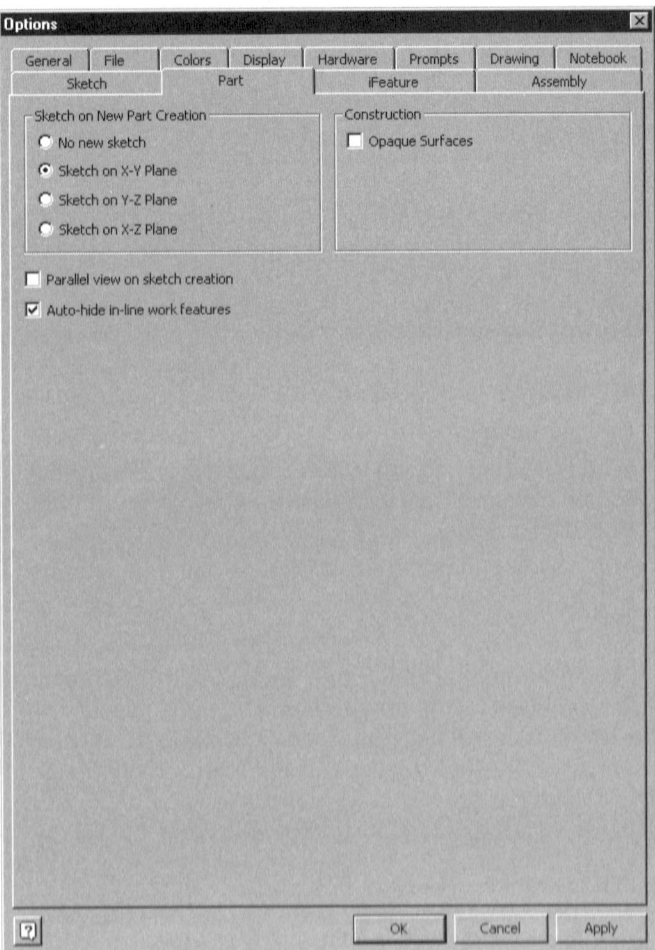


**Figure 2-2**

## Sketch on New Part Creation

- No New Sketch

  When checked, no sketch plane is set when a new part is created.

- Sketch on X-Y Plane

  When checked, the X-Y plane is set as the current sketch plane when a new part is created.

- Sketch on Y-Z Plane

  When checked, the Y-Z plane is set as the current sketch plane when a new part is created.

- Sketch on X-Z Plane

  When checked, the X-Z plane is set as the current sketch plane when a new part is created.

### Parallel View on Sketch Creation

When checked, the view orientation will automatically be changed to look directly at the sketch plane.

### Auto-Hide In-Line Work Features

When checked, work features that are consumed by another work feature will automatically be hidden.

### Construction

- Opaque Surfaces

  When checked, and as a surface is created, it will be opaque—otherwise it will be translucent. After a surface is created, its translucency can be controlled from a menu.

## UNITS

Autodesk Inventor uses a default unit of measurement for every part, assembly, and drawing file. The default unit is set from the template file from which it was created. When specifying numbers in dialog boxes with no unit, the default unit will be used. The default unit can be changed in the active part or assembly document by clicking Tools > Document Settings and on the Units tab, as displayed in the following image. The unit system values are changed for all of the existing values in that file.

**NOTE** In a drawing file, the dimension style and active drafting standard establish how the unit will appear. Drawing settings will be covered in Chapter 5.

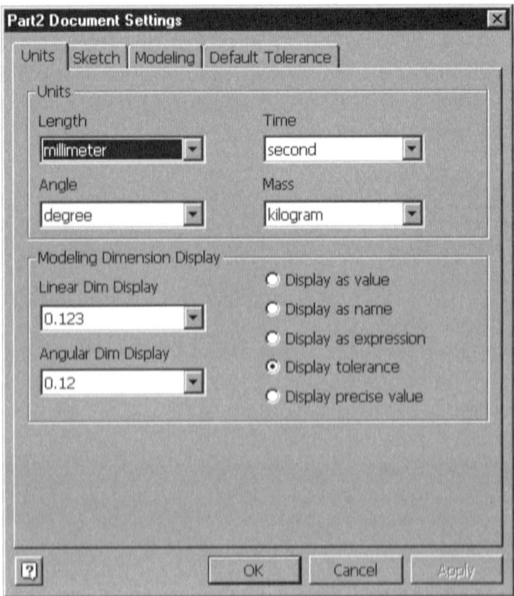


**Figure 2-3**

The default unit for any value can be overridden by entering in the desired unit. If you were working in a mm file, for example, and placed a horizontal dimension whose default value was 50 mm, you could enter 2 in. Dimensions are displayed on the screen in the default units.

For the previous example, 50.8 mm would be displayed on the screen. When a dimension is edited, the overridden unit will be displayed in the Edit Dimension dialog box as shown in the following image.

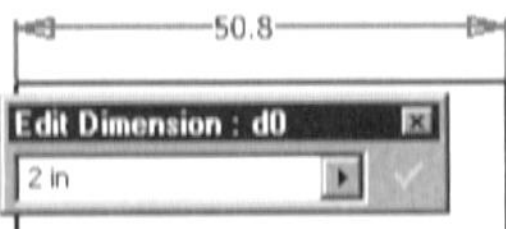


**Figure 2-4**

## TEMPLATES

As you create new files, they are each created from a template. You can modify or add your own templates. As you work, make note of changes that are made to each file. You then create a new file or modify an existing file that contains all of the changes, and save that file to your template directory—which by default is located at *\Autodesk\Inventor (version number)\templates* directory—or create a new subdirectory under the templates folder and place the file there. After you create the new template subdirectory, a new tab will appear with the directory name. You can place any file in this new directory and it will be available as a template.

If you want to share template files among many users, you can modify the template location by clicking Tools > Application Options, clicking the File tab, and modifying the templates location as shown in the following image. The template location will need to be modified for each user who needs access to these common templates.

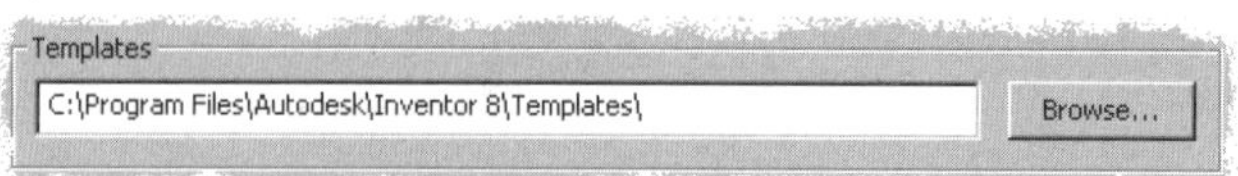


**Figure 2-5**

**NOTE** Template files have file extensions like other files but are located in the template directory. Template files should not be used as production files.

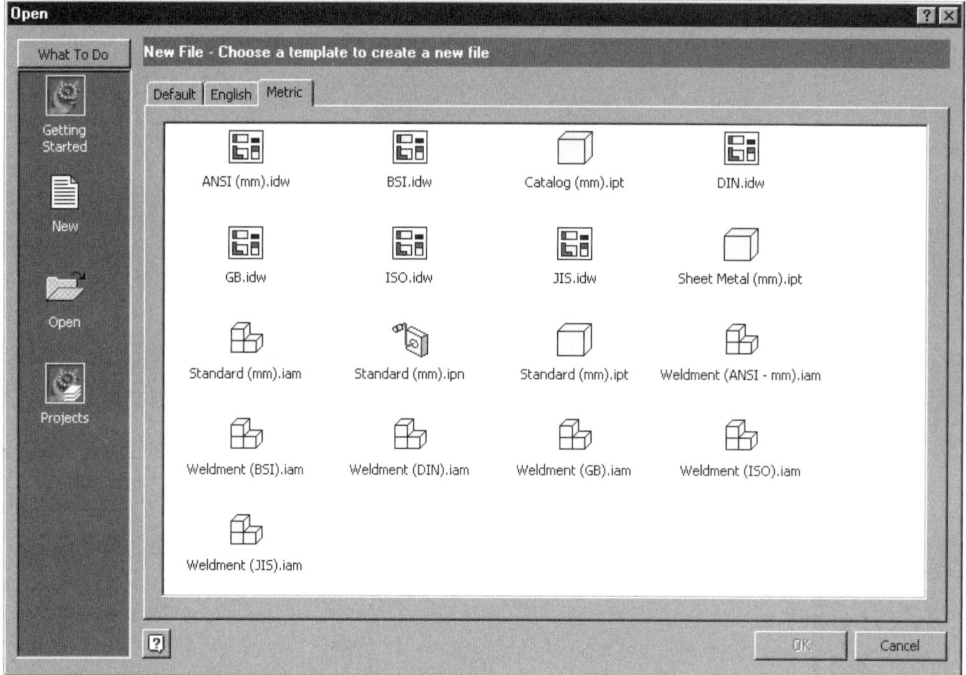


**Figure 2-6**

## CREATING A PART

The first step in creating a part is to start a new part file or create a new part file in an assembly. The first four chapters in this book deal with creating parts in a new part file, and Chapter 6 covers creating and documenting assemblies. The following methods can be used to creating a new part file:

- Click the New icon from What To Do and then click the *Standard.ipt* icon from the Default tab, as shown in the following image, or click *Standard (unit).ipt* from one of the other tabs.
- Click New from the File menu, and then click the *Standard.ipt* icon from the Default tab, as shown in the following image. You can also click *Standard (unit).ipt* from one of the other tabs.
- Use the shortcut key, CTRL-N, and then click the *Standard.ipt* icon from the Default tab, as shown in the following image. You can also click *Standard (unit).ipt* from one of the other tabs.
- Click the down arrow of the New icon and select Part from the left side of the Standard toolbar.

After starting a new part file using one of the previous methods, Autodesk Inventor's screen will change to reflect the part environment.

**NOTE** The units for the files located in the Default tab are based upon the unit that was selected when Autodesk Inventor was installed.

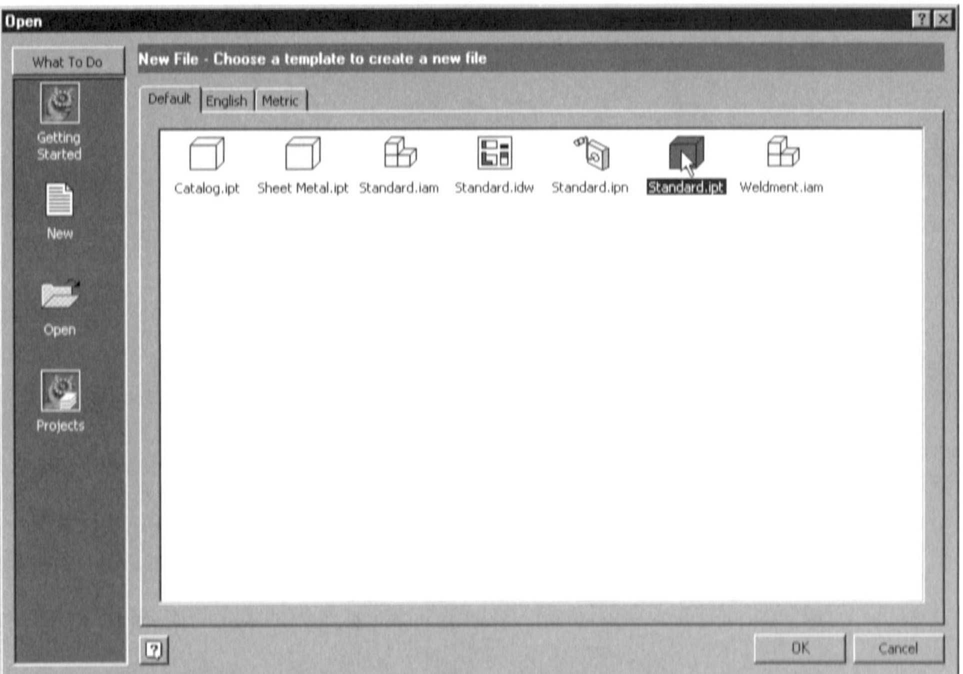


Figure 2-7

## SKETCHES AND DEFAULT PLANES

Before you start sketching, you must have an active sketch on which to draw. A sketch is a plane on which 2D objects are sketched. Any planar face or work plane on a part can be made the active sketch. A sketch is automatically set by default when you create a new part file. The default plane on which the sketch is created can be changed by selecting Tools > Application Options, and clicking the sketch plane on the Part tab to which new parts should default.

Each time a new Autodesk Inventor part is created, there are three planes (XY, YZ, and XZ), three axes (*X*, *Y*, and *Z*), and the center (origin) point at the intersection of the three planes. These default planes can become the active sketch. By default, visibility is turned off to these planes, axes, and center point. To see the planes, axes, or center point, expand the Origin entry in the Browser by clicking on the + to the left side of the text. You can then move the cursor over the names and they will appear in the graphics area. The following image shows the default planes, axes, and center point in the graphics window shown in an isometric view. The Browser shows the Origin menu expanded. To leave the visibility of the planes or axes on, right-click in the Browser while the cursor is over the name and select Visibility from the menu.

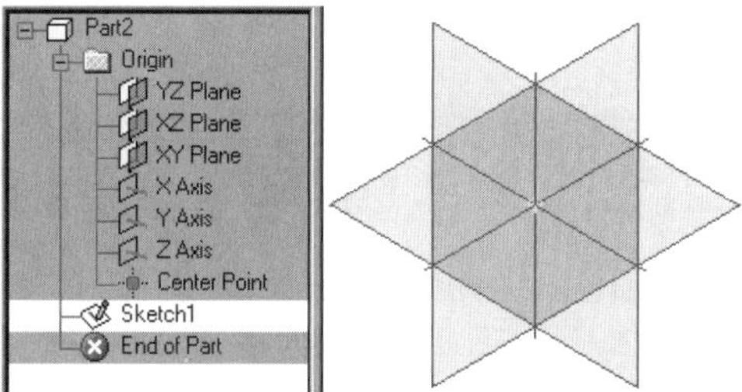


**Figure 2-8**

## NEW SKETCH

Issue the Sketch tool to change a planar face, a work plane, or a non-active sketch in the active part to the active sketch. To make a sketch active, follow one of these methods:

- Click the Sketch tool from the Standard toolbar as shown in the following image. Then click a face, a work plane, or an existing sketch from the Browser.
- Click a face, a work plane, or an existing sketch from the Browser. Then click the Sketch tool from the Standard toolbar as shown in the following image.
- Press the hot key **S** and click a face of a part, a work plane, or an existing sketch from the Browser.
- Click a face of a part, a work plane, or an existing sketch from the Browser and then press the hot key **S**.

- While not in the middle of an operation, right-click in the graphics window and click New Sketch from the menu. Then click a face, a work plane, or an existing sketch from the Browser.
- While not in the middle of an operation, click a face of a part, a work plane, or an existing sketch from the Browser. Then right-click in the graphics window and select New Sketch.

After the sketch has been activated, the *X*- and *Y*-axes will automatically align to this plane and then you can begin to sketch.

**NOTE** This book assumes that when you installed Autodesk Inventor you selected mm as the default unit. If inch was selected as the default unit, then select the *Standard (mm) .ipt* template file from the Metric tab. This book will use the XY plane as the default sketch plane.

## STEP 1—SKETCH THE OUTLINE OF THE PART

As stated at the beginning of the chapter, 3D parts usually start with a 2D sketch of the outline shape of the part. A sketch can be created with lines, arcs, circles, splines, or with any combination of these elements. The next section will cover sketching strategies, tools, and techniques.

### SKETCHING OVERVIEW

When deciding what outline to start with, analyze what the finished shape will look like. Look for the shape that best describes the part. When looking for this outline, try to look for a flat face. It is usually easier to work on a flat face than on a curved edge. As you gain modeling experience, you can reflect on how the model was created and think about other ways that the model could have been built. There is usually more than one way to generate a given part.

When sketching, draw the geometry so that it is close to the desired shape and size—you do not need to be concerned about exact dimensional values. Autodesk Inventor allows islands in the sketch (closed objects that lie within another closed object). An example would be a circle that is drawn inside of a rectangle. When the sketch is extruded, the island will become a void in the solid. A sketch can consist of multiple closed objects that are coincident.

The following guidelines will help you generate good sketches:

- Select an outline that best represents the part. It is usually easier to work from a flat face.
- Draw the geometry close to the finished size. If you want a 20 mm square, for example, do not draw a 200 mm square.
- Create the sketch proportionately in size to the finished shape. When drawing the first line segment, use the Precise Input dialog to sketch the first line to size. This will give you a guide on how to sketch the rest.

- While sketching, notice that the distance and angle of the object being sketched will appear at the bottom of the screen. Use this information as a guide.
- Draw the sketch so that it does not overlap. The geometry should start and end at the same point.
- Do not allow the sketch to have a gap—all of the connecting endpoints should be coincident.
- Keep the sketches simple. Leave out fillets and chamfers. They can easily be placed as features after the sketch is turned into a solid. The simpler the sketch, the fewer the number of constraints and dimensions that will be required to constrain the model.
- If you want to create a solid, the sketch must form a closed shape. If it is open, it can only be turned into a surface.

## SKETCHING TOOLS

Before you start sketching the part, examine the 2D sketching tools that are available. By default, the 2D sketch tools are displayed in the Panel Bar with *expert mode* turned off (text descriptions shown). As you become more proficient with Autodesk Inventor, you can turn off the text, as shown in the following image, by either selecting on the title area of the menu or right-clicking in the Panel Bar and clicking on Expert. You can also use the 2D Sketch toolbar to access the 2D sketch tools.

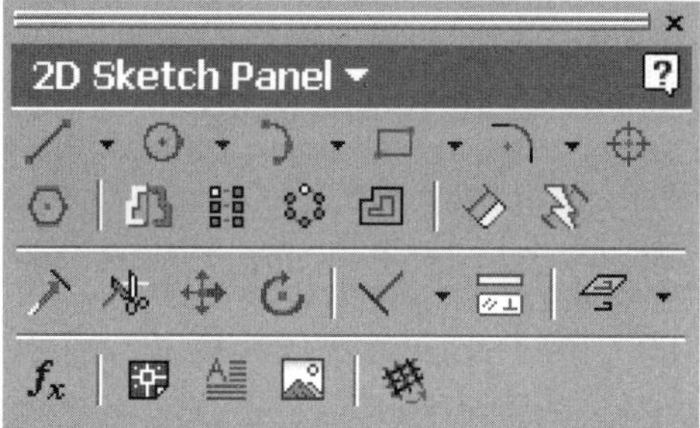


Figure 2-9

**Sketch Tools**

| Button | Tool | Function |
|---|---|---|
| | Line | Creates line segments. An arc segment can be drawn from the endpoint of a line. The next section will cover the sketching techniques. |
| | Spline | Draws a spline by clicking points that lie on the spline. |
| | Center Point Circle | Creates a circle by clicking a center point for the circle and then a point on the circumference of the circle. |

**Sketch Tools *(continued)***

| Button | Tool | Function |
|---|---|---|
| | Tangent Circle | Creates a circle that will be tangent to three lines or edges by clicking the lines or edges. |
| | Ellipse | Creates an ellipse by clicking a center point for the ellipse, a point on the first axis, and another point that will lie on the ellipse. |
| | Three Point Arc | Creates an arc by clicking a start and end point and then a point that will lie on the arc. |
| | Tangent Arc | Creates an arc that is tangent to an existing line or arc by clicking the endpoint of a line or arc and then clicking a point for the other endpoint of the arc. |
| | Center Point Arc | Creates an arc by clicking a center point for the arc and then clicking a start and end point. |
| | Two Point Rectangle | Creates a rectangle by clicking a point and then clicking another point to define the opposite side of the rectangle. The edges of the rectangle will be horizontal and vertical. |
| | Three Point Rectangle | Creates a rectangle by clicking two points that will define an edge and then clicking a point to define the third corner. |
| | Fillet | Creates a fillet between two nonparallel lines, two arcs, or a line and arc at a specified radius. If two parallel lines are selected, a fillet is created between them without specifying a radius. |
| | Chamfer | Creates a chamfer between lines. There are three options to create a chamfer: both sides equal distances, two defined distances, or a distance and angle. |
| | Point, Hole Center | Creates a hole center that will be used to place a hole or points that can be used as vertices for other objects, like splines. |
| | Polygon | Creates an inscribed or a circumscribed polygon with the number of faces you specify. |
| | Mirror | Mirrors the selected objects about a centerline. A symmetry constraint will be applied to the mirrored objects. |
| | Rectangular Pattern | Creates a rectangular array of a sketch with a number of rows and columns you specify. |

| Sketch Tools *(continued)* | | |
|---|---|---|
| **Button** | **Tool** | **Function** |
|  | Circular Pattern | Creates a circular array of a sketch with a number of copies and spacing you specify. |
|  | Offset | Creates a duplicate of the selected objects that are a given distance away. By default, an equal distance constraint is applied to the offset objects. |
|  | General Dimension | Creates a parametric or driven dimension. |
|  | Auto Dimension | Automatically places dimensions and constraints on a selected profile. |
| 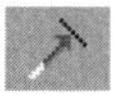 | Extend | Extends the selected object to the next object it finds. Click near the end of the object that you want extended. While in the Extend tool, hold down the **SHIFT** key to trim objects. |
|  | Trim | Trims the selected object to the next object it finds. Click near the end of the object that you want trimmed. While in the Trim tool, hold down the **SHIFT** key to extend objects. |
| 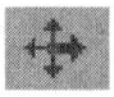 | Move | Moves the selected profile from one point to another point. A copy of the sketch can be made by selecting Copy in the Move dialog box. If the moved sketch has objects that are constrained to it, they will also be moved. |
|  | Rotate | Rotates the selected sketch about a specified point. A copy of the sketch can be made by selecting Copy in the Rotate dialog box. If the rotated sketch has objects that are constrained to it, they will also be rotated. |
| 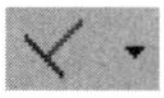 | Constraints | Constraints tools (accessed from the context menu) place geometric constraints on the geometry. Constraints will be covered later in this chapter. |
|  | Show Constraints | Click the object(s) whose constraints you want to see. When the constraints are visible, they can be deleted by right-clicking on a constraint and then clicking Delete on the menu. Showing constraints will be covered later in this chapter. |

**Sketch Tools (*continued*)**

| Button | Tool | Function |
|---|---|---|
| | Project Geometry | Projects selected edges, vertices, and work features onto the current sketch. The projected edges, vertices, and work features can be used as part of the sketch or dimensions can be placed on them. The projected edges, vertices, and work features are constrained to the original edge, vertex, or work feature. If the parent object changes or moves, so will the projected object. |
| | Project Cut Edges | Projects edges that lie on the section plane onto the sketch of the new part. Geometry is only projected if the uncut part would intersect the sketch. |
| | Project Flat Pattern | Selected areas of a flat pattern of a sheet metal part can be projected onto the part. Sketches can then be dimensioned to both the part and projected flat. |
| | Parameters | Creates or modifies a parameter. |
| | Insert AutoCAD File | Inserts an AutoCAD 2D file onto the active sketch. |
| | Create Text | Adds text to the active sketch. |
| | Insert Image | Inserts a file to the active sketch. |
| | Edit Coordinate System | Realigns the sketch coordinate system to existing objects. Arrows on the screen will indicate the directions of the *X*-axis and the *Y*-axis. |

## USING THE SKETCH TOOLS

After starting a new part, use the sketch tools to draw the shape of the part. As you are sketching, Autodesk Inventor gives you visual feedback about what is happening on the screen and there are text messages in the bottom area of the Status Bar. The following image shows the message when the Line tool is activated. To start sketching, issue the sketch tool that you need, and then in the graphics window click a point and follow the prompt in the Status Bar. The following sections will introduce a technique that can be used to create a sketch.

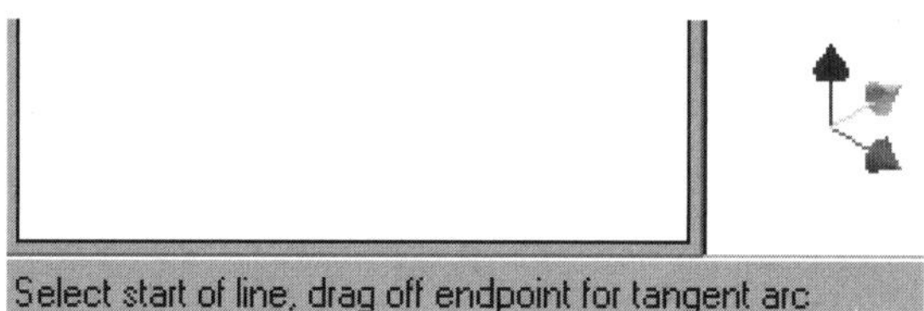


Figure 2-10

## Line Tool

The Line tool is one of the most powerful tools that you will use to sketch. Not only can you draw lines with the Line tool, but you can also draw an arc from the endpoint of a line segment. After issuing the Line tool, you will be prompted to click a first point, select a point in the graphics window, and then click a second point. You can continue drawing line segments or, from the endpoint of a line segment or arc, move the cursor over the endpoint and a small circle will appear at that endpoint. The following image shows how the endpoint of a line segment looks when the cursor is moved over it.

Figure 2-11

Click on the small circle, and with the left mouse button pressed down, move the cursor in the direction that you want the arc to go. Depending upon how you move the mouse, up to eight different arcs can be drawn. The arc will be tangent to the horizontal or vertical edges that are displayed from the selected endpoint. The following image shows an arc being drawn that is normal to the sketched line.

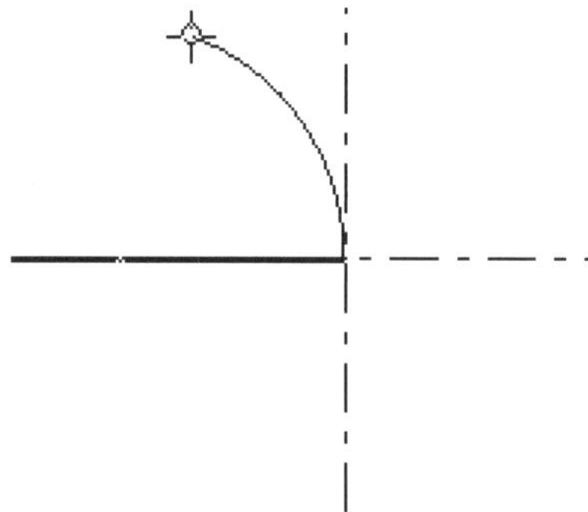

Figure 2-12

**TIP** When sketching, look at the bottom right corner of the Status Bar (bottom of the screen) to see the coordinates, length, and angle of the objects that you are drawing. The following image shows the Status Bar when a line is being drawn.

| -12.917 mm, 33.313 mm | Length=17.834 mm | Angle=43.88 deg |
|---|---|---|

**Figure 2-13**

## Inferred Points

While sketching, dashed lines will appear on the screen. These dashed lines represent the endpoints of lines and arcs that represent their horizontal, vertical, or perpendicular positions. As the cursor gets close to these inferred points, it will snap to that location. If that is the point that you want, click that point; otherwise, continue to move the cursor until it reaches the desired location. When inferred points are selected, no constraints (geometric rules such as horizontal, vertical, collinear, etc.) are applied from them. Using inferred points helps create more accurate sketches. The following image shows the inferred points from two endpoints that represent their horizontal and vertical position.

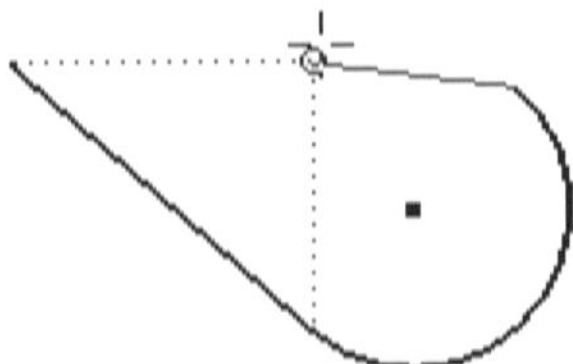

**Figure 2-14**

## Automatic Constraints

While sketching, small constraint symbols appear that represent geometric constraint(s) that will be applied to the object. If you do not want a constraint to be applied, hold down the CTRL key when the point is selected. The following image shows a line being drawn from the arc, tangent to the arc and parallel to the angled line. The symbol will be shown near the object from which the constraint is coming. Constraints will be covered in the next section.

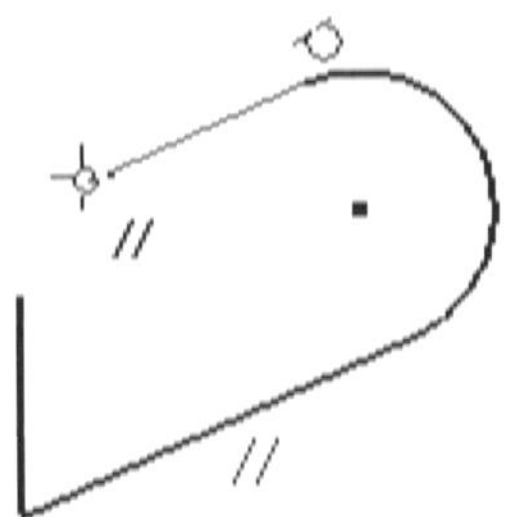

**Figure 2-15**

### Scrubbing

While sketching, you may prefer to have a different constraint applied than the one that automatically appears on the screen. You may want a line to be perpendicular to a given line, for example, instead of being parallel to a different line. The technique to change the constraint is called *scrubbing*. To place a different constraint while sketching, move the cursor so it touches (scrubs) the other object to which the constraint should be related. Move the cursor back to its original location and the constraint symbol changes to reflect the new constraint. The same constraint symbol will also appear near the scrubbed object, representing that it is the object to which the constraint is matched. Continue sketching as normal. The following image shows the top horizontal line being drawn with a perpendicular constraint that was scrubbed from the left vertical line. Without scrubbing the left vertical line, the applied constraint would have been parallel to the bottom line.

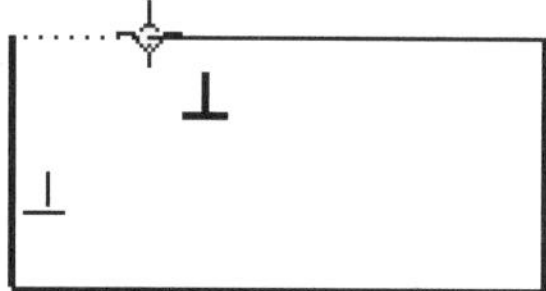

Figure 2-16

## PRECISE INPUT

While sketching, it may be easier to draw objects at a specified length or angle. These values can be entered into the Precise Input dialog box. To open the Precise Input dialog box, click View > Toolbars and click Precise Input, or right-click on either the Standard toolbar or Menu Bar and click Precise Input from the menu. Before entering values into the dialog box, start a sketch tool—like Line. Click the method of input and then enter the values into the Precise Input dialog box followed by **ENTER**. The following image shows the Precise Input dialog box with Expert display mode on and with the input options displayed.

Figure 2-17

## SELECTING OBJECTS

After sketching objects, you may need to move, rotate, or delete some or all of the objects. To edit an object, it must be part of a selection set. There are two methods to place objects into a selection set.

1 Objects can be selected individually by clicking on them. To select multiple individual objects, hold down the **CTRL** or **SHIFT** key while clicking the objects. Selected objects can be removed from a selection set by holding

down the **CTRL** or **SHIFT** key and reselecting them. As objects are selected, their color will change to represent that they have been selected.

2 Multiple objects can be selected by defining a selection window. To define the window, click a starting point. With the left mouse button depressed, move the cursor to define the box. If the window is drawn from left to right, only the objects that are fully enclosed in the window will be selected. If the window is drawn from right to left, as shown in the following image, all of the objects that are fully enclosed in the window *and* the objects that are touched by the window will be selected.

3 You may use a combination of Steps 1 and 2 to create a selection set.

When an object is selected, its color will change according to the color style that you are using. To remove all of the objects from the selection set, click in a blank area of the graphics window.

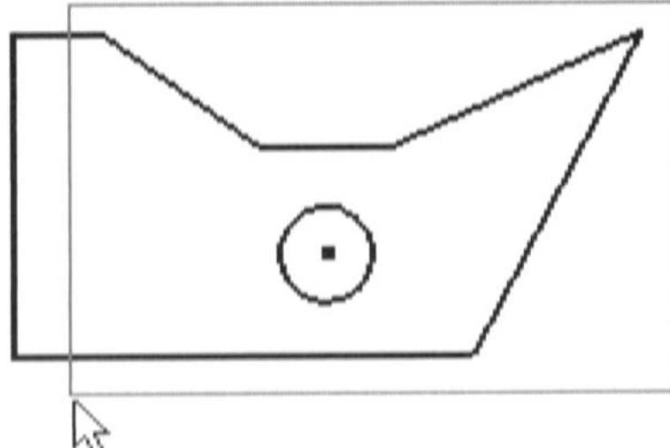

**Figure 2-18**

## DELETING OBJECTS

To delete objects, first select them and then either press the **DELETE** key or right-click and select Delete from the menu. The following image shows the Delete menu.

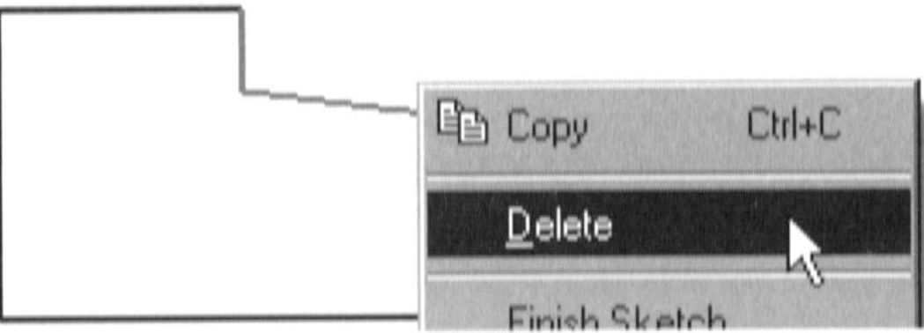


**Figure 2-19**

EXERCISE

# EXERCISE 2-1 Creating A Sketch With Lines

In this exercise, you create a new part file, and then you create sketch geometry using basic construction techniques. To navigate to the exercise in the *Electronic Student Workbook* do the following:

1 From the Main TOC page, click Chapter 2.
2 From the TOC for Chapter 2, click Creating a Sketch With Lines.

The following image illustrates the completed exercise.

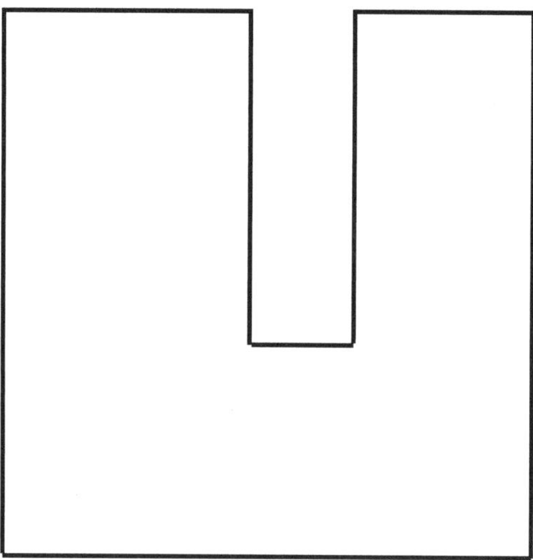

**Figure 2-20 Completed exercise**

# EXERCISE 2-2
# Creating a Sketch With Tangencies

In this exercise, you create a new part file, and then you create a simple profile consisting of lines and tangential arcs.

This exercise also illustrates how you can use the Autodesk Inventor Design Support System to assist in the design process.

To navigate to the exercise in the *Electronic Student Workbook* do the following:

1 From the Main TOC page, click Chapter 2.
2 From the TOC for Chapter 2, click Creating a Sketch With Tangencies.

The completed exercise is shown in the following image.

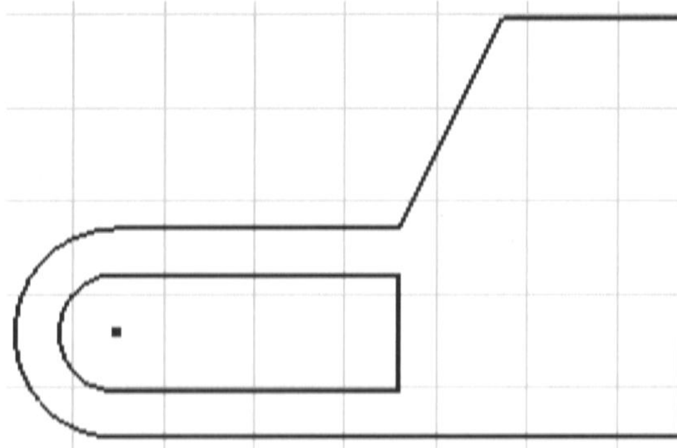

**Figure 2-21 Completed exercise**

A part created from the sketch is shown in the following image.

**Figure 2-22 Part created from sketch**

## STEP 2—CONSTRAINING THE SKETCH

After the sketch is drawn, you may want to add geometric constraints to the sketch. Geometric constraints apply behavior to a specific object or create a relationship between two objects. An example of a constraint is applying a vertical constraint to a line so that it will always be vertical. You could apply a parallel constraint between two lines to make them parallel to one another; then, as one of the line's angles would change so would the other. A line and an arc, or two arcs, can have a tangent constraint applied to them.

When a constraint is added, the number of constraints or dimensions that are required to fully constrain the sketch will be decreased. A fully constrained sketch is a sketch whose objects cannot move.

To help you see which objects are constrained, Autodesk Inventor will change the color of constrained objects if a fix constraint has been applied to the sketch. If a fix constraint has not been applied, the color of the constrained objects will not change. The fix constraint also prevents a sketch from moving in the sketch plane. A point, points, edge, or edges can be fixed using the fix constraint. If no fix constraint has been applied to the sketch, objects are free to move in their sketch plane.

**NOTE** Autodesk Inventor does not force you to fully constrain a sketch. It is recommended to fully constrain a sketch, however, as this will allow you to better predict how a part will react when dimensions values are changed.

### CONSTRAINT TYPES

Autodesk Inventor has eleven geometric constraints that you can apply to a sketch. The following image shows the constraint types and the symbols that represent them. The following are descriptions of each constraint:

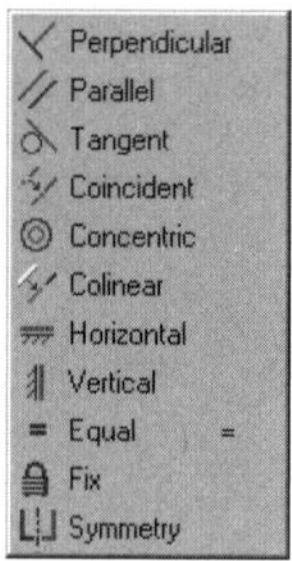


**Figure 2-23 Constraints menu**

**Constraint Tools**

| Button | Tool | Function |
|---|---|---|
| | Perpendicular | Lines will be repositioned at 90° angles to one another—the first line sketched will stay in its position and the second will rotate until the angle between them is 90°. |
| | Parallel | Lines will be repositioned so they are parallel to one another—the first line sketched will stay in its position and the second will move to become parallel to the first. |
| | Tangent | An arc or circle and a line will become tangent to another arc or circle. |
| | Coincident | A gap between two endpoints of arcs and/or lines will be closed. |
| | Concentric | Arcs and or circles will share the same center point. |
| | Collinear | Two selected lines will line up along a single line—if the first line moves, so will the second. The two lines do not have to be touching. |
| | Horizontal | Lines are positioned parallel to the *X*-axis, or a horizontal constraint can be applied between the center points of arcs or circles. The center points will share the same horizontal axis. |
| | Vertical | Lines are positioned parallel to the *Y*-axis, or a vertical constraint can be applied between the center points of arcs or circles. The center points will then share the same vertical axis. |
| | Equal | If two arcs or circles are selected, they will then have the same radius or diameter. If two lines are selected, they will become the same length. If one of the objects changes, so will the other object to which the Equal constraint has been applied. If the Equal constraint is applied after one of the arcs, circles, or lines have been dimensioned; the second arc, circle, or line will take on the size of the first one. |

**Constraint Tools (*continued*)**

| Button | Tool | Function |
|---|---|---|
| | Fix | Applying a fixed point or points will prevent the endpoints or edges of objects from moving. The fixed point overrides any other constraint. Any endpoint or segment of a line, arc, circle, spline segment, or ellipse can be fixed. Multiple points in a sketch can be fixed. If you select near the endpoint of an object, the endpoint will be locked from moving. If you select near the midpoint of a segment the entire segment will be locked from moving. If applying constraints, and the profile is moving in directions that are undesirable, you can apply Fix constraints to hold the endpoints of the objects in place. You can remove a fix constraint as needed. Deleting constraints will be covered later in this chapter. |
| | Symmetry | Selected geometry will be symmetric about another line, centerline, or edge. |

**NOTE** To fully constrain a sketch, fix a point in the sketch.

## ADDING CONSTRAINTS

As stated previously in this chapter, constraints can be applied while the objects are being sketched. You may also apply additional constraints after the sketch is drawn. While adding constraints, Autodesk Inventor will not allow you to over-constrain the sketch or add duplicate constraints. If you add a constraint that would conflict with another, you will be warned with the message, "Adding this constraint will over-constrain the sketch." If you try to add a vertical constraint to a line that already has a horizontal constraint, for example, you will be alerted in a dialog box with "Adding this constraint will over-constrain the sketch." To apply a constraint, follow these steps:

1. Click a constraint from the constraint pop-up menu in the Sketch Panel Bar or Sketch toolbar, or right-click in the graphics window, click Create Constraint, and choose the specific constraint from the menu as shown in the previous image.
2. Click the object or objects to apply the constraints.

## SNAPS

Another method to place geometry with a coincident constraint is to use the snaps—midpoint, center, and intersection. After using a snap, a coincident constraint will be applied. The coincident constraint will maintain the relationship that was selected. If a midpoint snap is used, for example, the sketched point will always be in the middle of the selected object, even if the selected object's length changes.

To use snaps, follow these steps:

1 Select a sketching tool, and right-click in the graphics window when one of the snaps is needed.
2 From the menu, click the specific snap.
3 Click on the object to which the sketched object will be constrained. For the intersection snap select two objects.

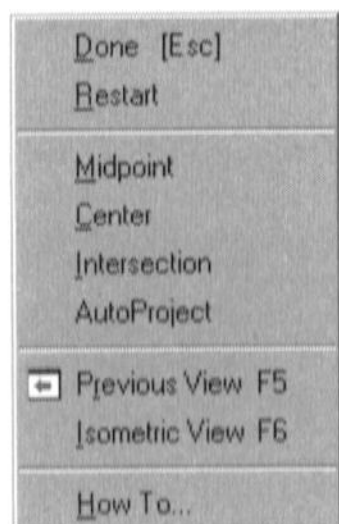


Figure 2-24

### DRAGGING A SKETCH

To help determine if an object is constrained, you can drag it to a new location. While not in a command, click a point or an edge on the sketch. With the left mouse button depressed, drag it to a new location. If the geometry stretches, it is underconstrained. If dimensions are on the object, they too will prevent the object from stretching. For example, if you draw a rectangle that has two horizontal and vertical constraints applied to it and then drag a point on one of the corners the size of the rectangle will change but the lines will maintain their horizontal and vertical behavior.

### SHOWING AND DELETING CONSTRAINTS

To see the constraints that are applied to an object, use the Show Constraints tool from the Sketch Panel Bar, as shown on the left side of the following image. After issuing the Show Constraints tool, select an object and a row of constraint icons will appear similar to what is shown on the right side of the following image.

Figure 2-25

As you move the cursor over a constraint icon, the objects that are linked to that constraint will change color.

To delete the constraint, either click on it and then right-click, or right-click while the cursor is over the constraint and select Delete on the menu. The following image shows the parallel constraint being deleted.

To close constraint icons, click the X on the right side of the constraint symbols.

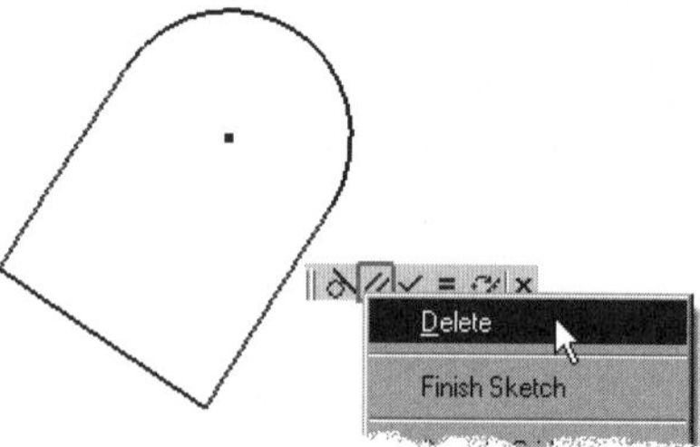


Figure 2-26

To show all the constraints for all of the objects in the sketch, do the following:

- While not in an operation, right-click in the graphics window and from the menu click Show All Constraints, as shown in the following image. To hide all the constraints, right-click in the graphics window and select Hide All Constraints from the menu.

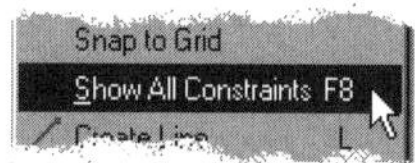


Figure 2-27

# EXERCISE 2-3
# Adding and Displaying Constraints

In this exercise, you add geometric constraints to sketch geometry to control the shape of the sketch. To navigate to the exercise in the *Electronic Student Workbook* do the following:

1 From the Main TOC page, click Chapter 2.
2 From the TOC for Chapter 2, click Adding and Displaying Constraints.

The following image illustrates the opened exercise.

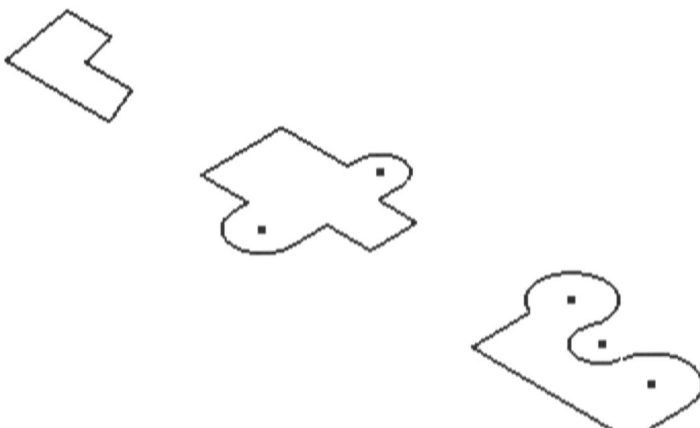

**Figure 2-28 Opened exercise**

The following image illustrates the completed exercise.

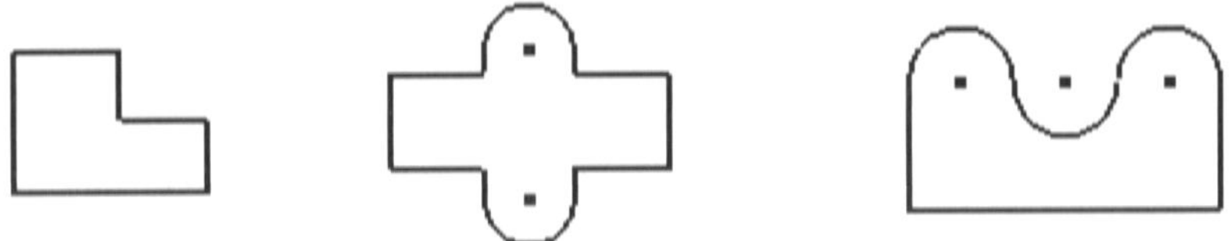

**Figure 2-29 Completed exercise**

## STEP 3—ADDING DIMENSIONS

The last step to constrain a sketch is to add dimensions. The dimensions you place will control the size of the sketch and will also appear in the drawing views when they are generated. When placing dimensions, try to avoid having extension lines go through the sketch as this will require more clean-up when drawing views are generated. When placing dimensions, click near the side from which you anticipate the dimensions will originate in the drawing views.

All dimensions that are created are parametric, which means that the dimension will change the size of the geometry. All parametric dimensions are created with either the General Dimension or Auto Dimension tools. The General Dimension tool will be covered in the next section.

### GENERAL DIMENSIONING

The General Dimension tool can create linear, angle, radial, or diameter dimensions one at a time. To start the General Dimension tool, follow one of these techniques:

- Click the Create Dimension tool from the Sketch toolbar as shown in the following image.
- Right-click in the graphics window and click Create Dimension from the menu.
- Press D.

**Figure 2-30**

When placing a dimension, the extension line of the dimension will automatically snap to the nearest endpoint of a selected line—when an arc or circle is selected, it will snap to its center point. To dimension to a quadrant of an arc or circle, see "Dimensioning to a Quadrant" later in this chapter.

After the General Dimension tool has been selected, follow these steps to place a dimension:

1. Click a point or points to locate where the dimension is to start and end.
2. After selecting the point(s) to dimension, a preview image will appear on the screen showing what type it is and where it will be placed. If the dimension type is not what you want, right-click and then click the correct style from the menu. The previous image shows the menu that appears when a vertical dimension is previewed. After changing the dimension type, the dimension preview will change to reflect the new style.
3. Click a point on the screen to place the dimension.

The next sections cover how to dimension specific objects and how to create specific types of dimensioning with the General Dimension tool.

### Dimensioning Lines

There are two techniques for dimensioning a line. First, start the General Dimension tool then do one of the following:

- Click near two endpoints, move the cursor until the dimension is in the correct location, and click.
- To dimension the length of a line, click anywhere on the line (the two endpoints will be selected automatically), move the cursor until the dimension is in the correct location, and click.

### Dimensioning an Angle

To create an angular dimension, start the General Dimension tool, click near the midpoint of two lines that you want the angle dimension to be between, move the cursor until the dimension is in the correct location, and click.

### Dimensioning Arcs and Circles

To dimension an arc or a circle, start the General Dimension tool, click on its circumference, move the cursor until the dimension is in the correct location, and then click. By default, when an arc is dimensioned, the result is a radius dimension. When a circle is dimensioned, the default is a diameter dimension. To change the radial dimension to a diameter or a diameter to radial, right-click before the dimension is placed and select the other style from the menu.

#### Diametric Dimensions

Diametric dimensions are used to create diameter dimensions for sketches that represent a quarter outline of a revolved part. To create a diametric dimension, follow these steps:

1. Draw a sketch that represents a quarter section of the finished part.
2. Draw a line, if needed, around which the sketch will be revolved. This line can be on the closed profile of the sketch.
3. Start the General Dimension tool.
4. Click the line (not an endpoint) that will be the axis of rotation.
5. Click the other point to be dimensioned.
6. Right-click and click Linear Diameter from the menu.
7. Move the cursor until the diameter dimension is in the correct location and click.

The image on the left of the following image shows a sketch with the menu for changing the dimension to Linear Diameter. The image on the right shows the placed diametric dimensions—the left vertical line will be the axis of rotation. More options for creating diametric dimensions will be covered in the "Revolve" section of Chapter 3.

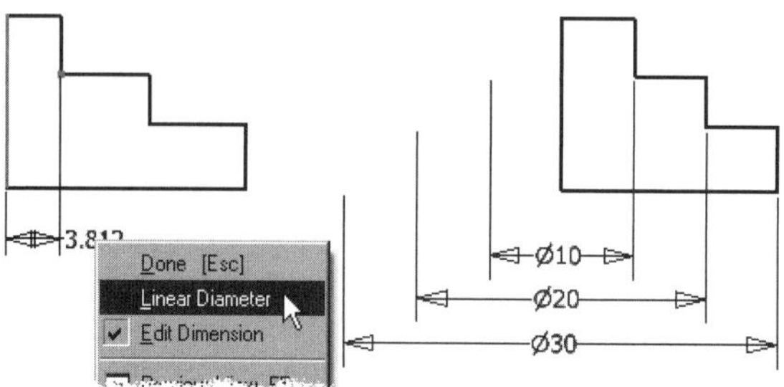


Figure 2-31

## Dimensioning to a Quadrant

To dimension to a quadrant of an arc or circle, follow these steps:

1. Start the General Dimension tool.
2. Click a line that is parallel to the quadrant.
3. Move the cursor over the quadrant that should be dimensioned.
4. Move the cursor over the quadrant until the constraint symbol changes to reflect a quadrant, as shown on the left of the following image.
5. Click and then move the cursor until the dimension is in the correct location, and click as shown on the right side of the following image.

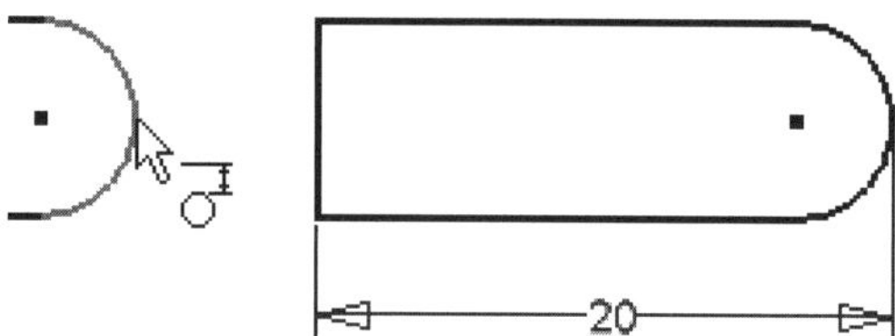


Figure 2-32

To dimension to two quadrants, follow these steps:

1. Start the General Dimension tool.
2. Click an arc or circle that includes one of the quadrants to which it will be dimensioned.
3. Move the cursor over the quadrant of the second arc or circle to which it will be dimensioned.
4. Move the cursor over the quadrant until the constraint symbol changes to *quadrant.*
5. Click and then move the cursor until the dimension is in the correct location, and click as shown in the following image.

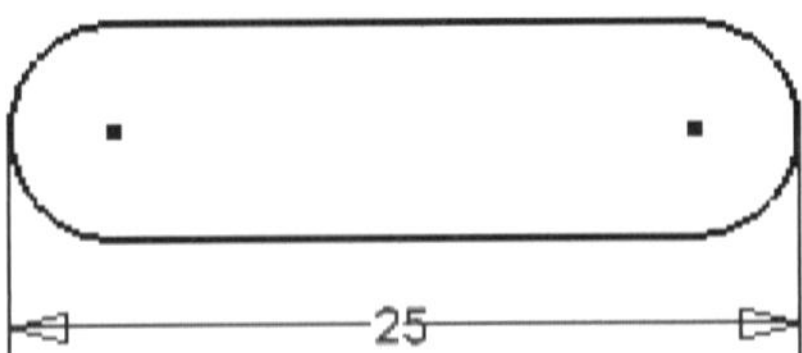


Figure 2-33

### ENTERING AND EDITING A DIMENSION VALUE

After placing the dimension, you can change the value of the dimension. Depending upon your setting for editing dimensions when they are created, the Edit Dimension dialog box may or may not automatically appear after the dimension is placed. To set the Edit Dimension option, do the following:

1 Click Tools > Application Options.
2 On the Sketch tab of the Options dialog box, select or deselect Edit dimension when created as shown on the left of the following image.

If the Edit dimension when created option is selected, the Edit Dimension dialog box will appear automatically after the dimension is placed; otherwise, the dimension will be placed with the default value and you will not be prompted for a different value. This option can also be set by right-clicking in the graphics window while placing a dimension and selecting or deselecting Edit Dimension from the menu, as shown to the right in the following image.

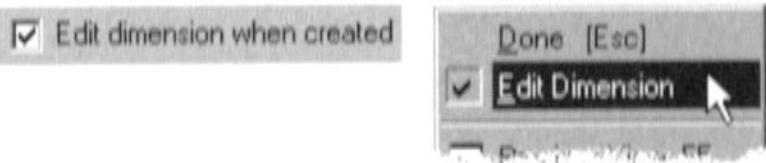


Figure 2-34

To edit a dimension that has already been created, double-click on the value of the dimension and the Edit Dimension dialog box will appear, as shown in the following image. Enter the new value and unit for the dimension; then either press ENTER or click the check mark in the Edit Dimension dialog box. If no unit is entered, the units that the file was created with will be used. When inputting values, enter the exact value—do not round up or down. The accuracy shown in the dimension is from the current dimension style. Autodesk Inventor parts are accurate to six decimal places; for example, 1.0625 is more accurate than 1.06. When placing dimensions, it is recommended that you place the smallest dimensions first—this will help prevent the geometry from flipping in the opposite direction.

Figure 2-35

### Repositioning a Dimension

Once a dimension is placed, it can be repositioned but the origin points cannot be moved. Follow these steps to reposition a dimension:

1 Exit the current operation by either pressing **ESC** twice, or right-clicking and then clicking Done from the menu.

2 Move the cursor over the dimension until the move symbol appears as shown in the following image.

3 With the left mouse button depressed, move the dimension or value to a new location and release the button.

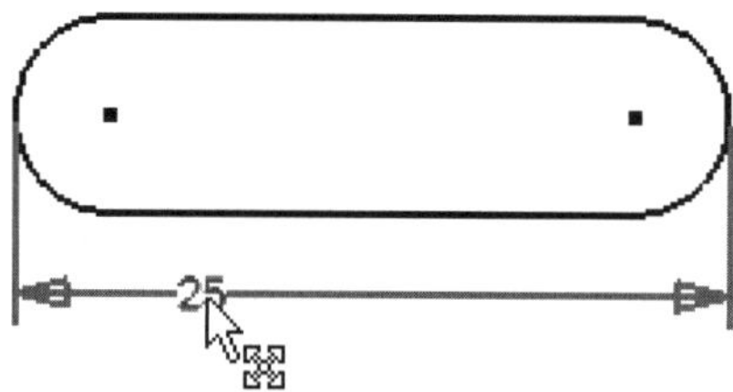


**Figure 2-36**

### Over-Constrained Sketches

As stated in the "Adding Constraints" section, Autodesk Inventor will not allow you to over-constrain a sketch or add duplicate constraints. The same is true when adding dimensions. If you add a dimension that will conflict with another constraint or dimension, you will be warned that this dimension will over-constrain the sketch or that it already exists. You will then have an option to either not place the dimension or place it as a driven dimension.

A driven dimension is a reference dimension. It is not a parametric dimension—it just reflects the size of the points to which it is dimensioned. A driven dimension will appear with parentheses around the dimensions value, like (30). When you place a dimension that will over-constrain a sketch, a dialog box will appear similar to the one in the following image. You can either Cancel the operation (and no dimension will be placed) or Accept the warning (and a driven dimension will be created).

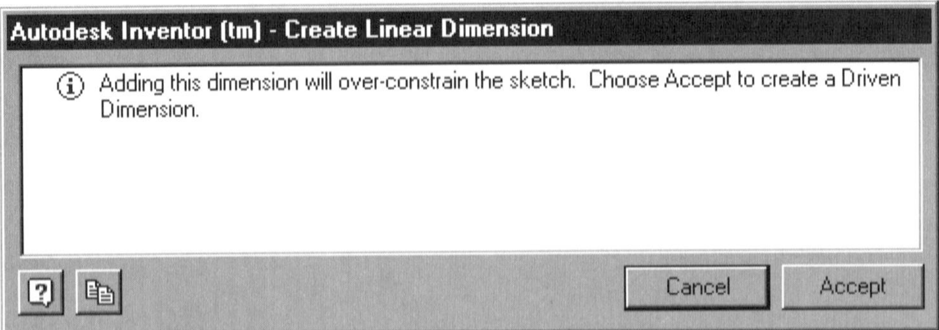


Figure 2-37

Autodesk Inventor gives you an option for handling over-constrained dimensions. To set the over-constrained dimensions option, click Tools > Application Options and on the Sketch tab of the Options dialog box change the Over-constrained Dimensions option as shown in the following image.

- Apply Driven Dimension

  When checked, a driven dimension will automatically be created without warning you of the condition.

- Warn of Overconstrained Condition

  A dialog box will appear stating that this dimension will over-constrain the sketch. Click Cancel to NOT place a driven dimension and click Accept to place a driven dimension.

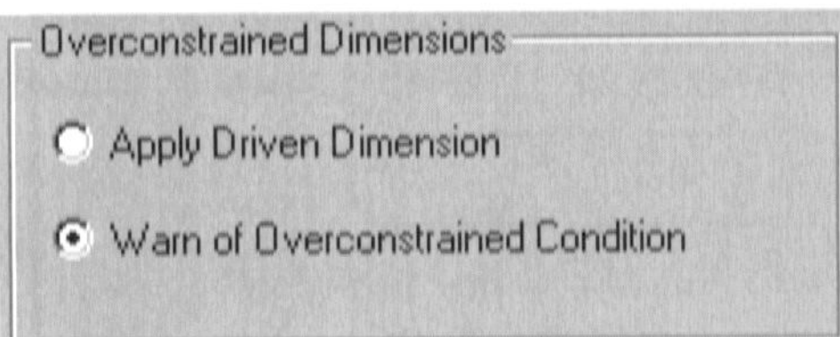


Figure 2-38

# EXERCISE 2-4 Dimensioning A Sketch

In this exercise, you add dimensional constraints to a sketch. To navigate to the exercise in the *Electronic Student Workbook* do the following:

1 From the Main TOC page, click Chapter 2.
2 From the TOC for Chapter 2, click Dimensioning a Sketch.

The following image illustrates the opened exercise.

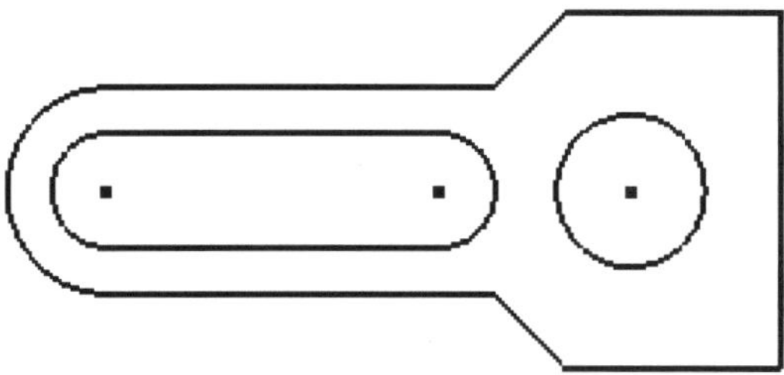

**Figure 2-39 Opened exercise**

The following image illustrates the completed exercise.

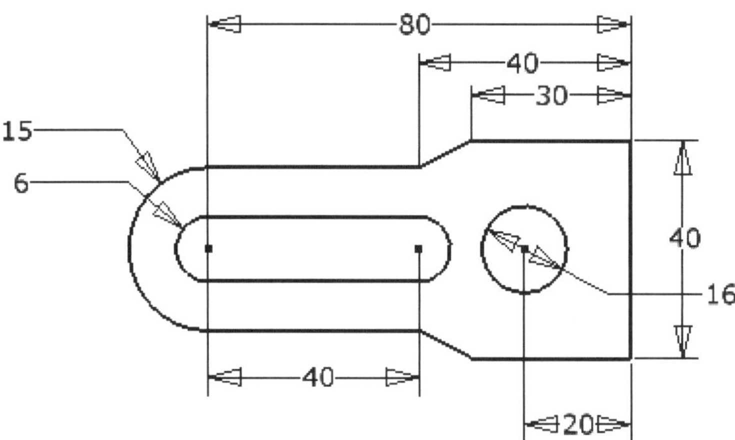


**Figure 2-40 Completed exercise**

## CHAPTER SUMMARY

| To | Do This | Tool |
|---|---|---|
| Modify the Sketch options of Autodesk Inventor | Click Tools > Application Options and click the Sketch tab. | |
| Modify the Part options of Autodesk Inventor | Click Tools > Application Options and click the Part tab. | |
| Create a new part file | Click the New icon in What To Do and then click the *Standard.ipt* icon from the Default tab, or click *Standard (unit).ipt* from one of the other tabs. | Standard.ipt |
| Make a planar face, a work plane, or a non-active sketch in the active part the active sketch | Click the Sketch tool from the Standard toolbar then click a face, a work plane, or an existing sketch from the Browser. Or, click a face, a work plane, or an existing sketch from the Browser, and then click the Sketch tool from the Standard toolbar or press **S**. | |
| Sketch the outline of the part | Use the 2D Sketch tools from either the Panel Bar or the Sketch toolbar. | Sketch |
| Add geometric constraints to a sketch | Click a constraint from the constraint pop-up menu in the Sketch Panel Bar or Sketch toolbar, or right-click in the graphics window and click Create Constraint and choose the specific constraint from the menu. | |
| Add parametric dimensions to a sketch | Click the General Dimension tool from the Sketch toolbar, or right-click in the graphics window and from the menu click Create Dimension or press **D**. | |

EXERCISE

# Applying Your Skills

## SKILL EXERCISE 2-1

In this exercise, you create a sketch and then add geometric and dimensional constraints to control the size and shape of the sketch. To navigate to the exercise in the *Electronic Student Workbook* do the following:

1. From the Main TOC page, click Chapter 2.
2. From the TOC for Chapter 2, click Exercise 1 under Applying Your Skills.

The following image illustrates the completed exercise.

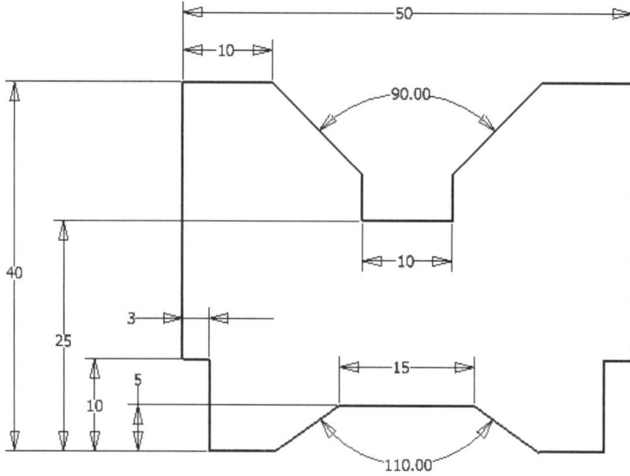


**Figure 2-41 Completed exercise**

### SKILL EXERCISE 2-2

In this exercise, you create a sketch with linear and arc shapes, and then add geometric and dimensional constraints to fully constrain the sketch. To navigate to the exercise in the *Electronic Student Workbook* do the following:

1. From the Main TOC page, click Chapter 2.
2. From the TOC for Chapter 2, click Exercise 2 under Applying Your Skills.

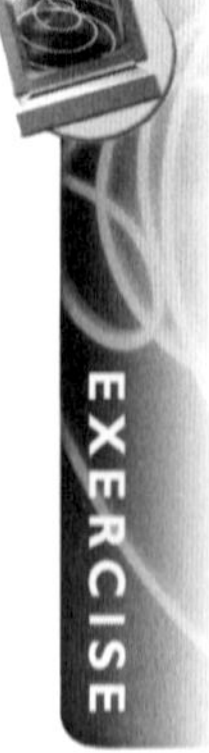


The following image illustrates the completed exercise.

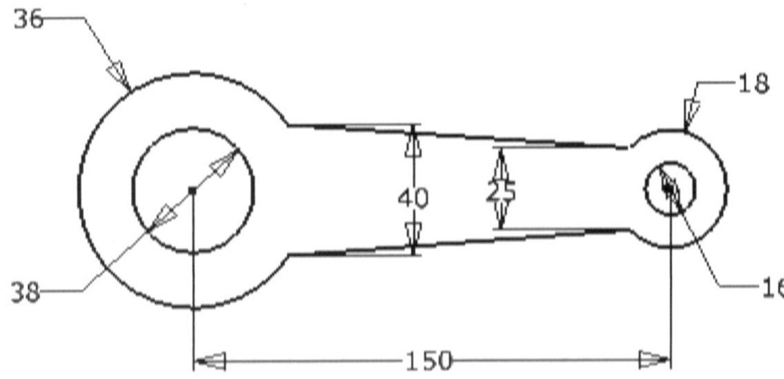


**Figure 2-42 Completed exercise**

## CHECKING YOUR SKILLS

Use these questions to test your knowledge of the material in this chapter.

1 **True___ False___** When sketching, constraints are not applied to the sketch by default.

2 **True___ False___** When sketching and a point is inferred, a constraint is applied to represent that relationship.

3 **True___ False___** A sketch does not need to be fully constrained.

4 **True___ False___** When working on an mm part, you cannot use English units.

5 **True___ False___** After a sketch is fully constrained, a dimension's value cannot be changed.

6 **True___ False___** A driven dimension is another name for a parametric dimension.

7 Explain how to draw an arc while still in the Line command.

______________________________________________________________

______________________________________________________________

8 Explain how to remove a geometric constraint from a sketch.

______________________________________________________________

______________________________________________________________

9 Explain how to change a vertical dimension to an aligned dimension while it is being created.

______________________________________________________________

______________________________________________________________

10 Explain how to create a dimension between two quadrants of two arcs.

______________________________________________________________

______________________________________________________________

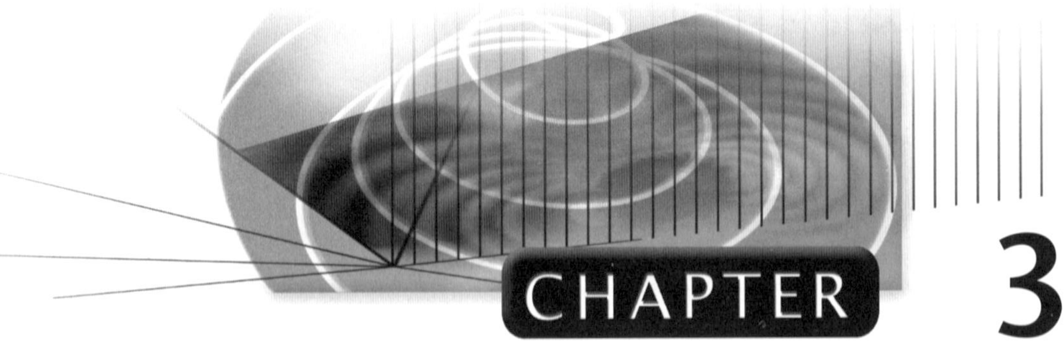


# Creating and Editing Sketched Features

After you have drawn, constrained, and dimensioned a sketch, your next step is to turn the sketch into a 3D part. This chapter takes you through the process to create and edit sketched features.

## CHAPTER OBJECTIVES

**After completing this chapter, you will be able to**

- Understand what a feature is
- Use the Autodesk Inventor Browser to edit parts
- Extrude a sketch into a part
- Revolve a sketch into a part
- Edit features of a part
- Edit the sketch of a feature
- Make an active sketch on a plane
- Create sketched features using one of the three operations: cut, join, or intersect

## UNDERSTANDING FEATURES

After a sketch has been created, constrained, and dimensioned, the next step in creating a model is to turn the sketch into a 3D feature. The first sketch of a part that is used to create a 3D feature is referred to as the base feature. In addition to the base feature, you can create sketched features, where you draw a sketch on a planar face or work plane and either add or subtract material to or from existing features in a part. Use the Extrude, Revolve, Sweep, or Loft tools to create sketched features in a part. You can also create placed features such as fillets, chamfers, and holes by applying them to features that have been created. Placed features will be covered in Chapter 4. Features are the building blocks that help create a part.

A plate with a hole in it, for example, would have a base feature representing the plate and a hole feature representing the hole. As features are added to the part, they appear in the Browser, showing the history of the part or assembly (the order in which the features are created or the parts are assembled). Features can be edited, deleted, or reordered from the part as required.

### CONSUMED AND UNCONSUMED SKETCHES

Any sketch can be used as a profile in feature creation. A sketch that has not yet been used in a feature is called an unconsumed sketch. When you turn a 2D sketched profile into a 3D feature, the feature consumes the profile. The following image shows an unconsumed sketch in the Browser on the left and a consumed sketch in the Browser on the right.

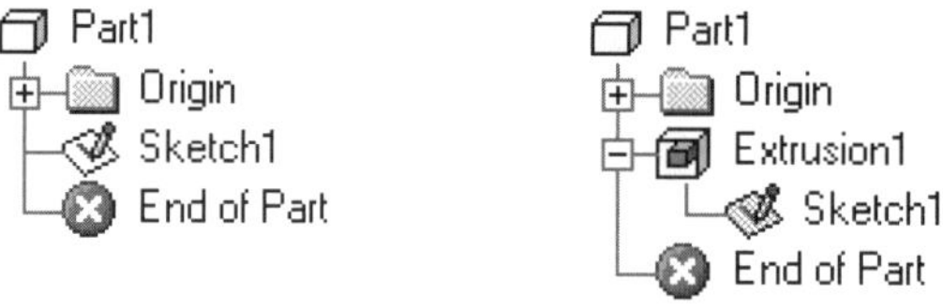


**Figure 3-1**

Although a consumed sketch is not visible as you view the 3D feature, you may need to access sketches and change their geometric or dimensional constraints in order to modify their associated features. A consumed sketch can be accessed from the Browser by either double-clicking on its name or icon or right-clicking and selecting Edit Sketch from the menu.

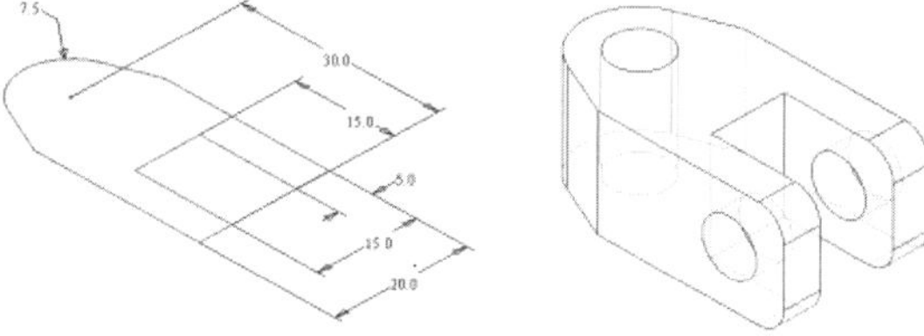

**Figure 3-2**

## USING THE BROWSER FOR CREATING AND EDITING

The Autodesk Inventor Browser, by default, is docked along the left side of the screen and displays the history of the file. In the Browser you can create, edit, rename, copy, delete, and reorder features or parts. The Browser can be expanded or collapsed to display the history of the part(s) (the order in which the features were created) by clicking the + and – on the left side of the part name in the Browser. An alternate method of expanding the Browser is to place your cursor so that it is on top of an icon (do not click) of a feature to be expanded. After a couple of seconds the item in the Browser automatically expands. You can also expand or collapse all children by right-clicking the appropriate option from the menu while in the Browser.

The following image shows the Browser with all of the features expanded.

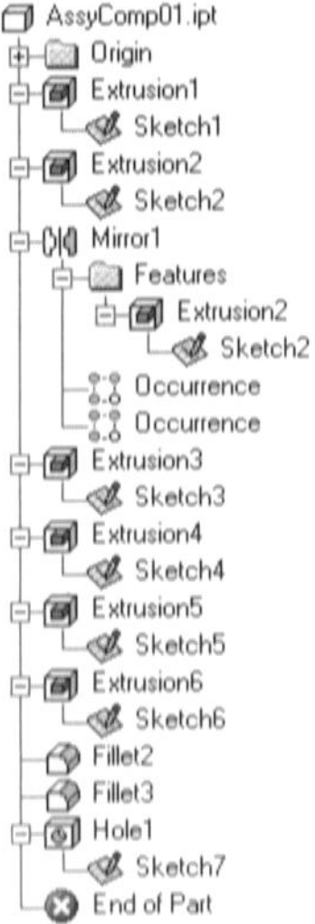


**Figure 3-3**

As parts grow in complexity, so will the information found in the Browser. Dependent features are indented to show that they are related to the item listed above it. This is referred to as a *parent-child relationship*. The child cannot exist without the parent and is dependent upon the parent. If a hole is created in an extruded rectangle, for example, and the extrusion is then deleted, the hole will also be deleted. To help filter out some of the object types that appear in the Browser, you can click the icon that looks like a funnel from the top of the Browser. After clicking the funnel, you can select objects to hide from the drop-down list, as shown in the following image.

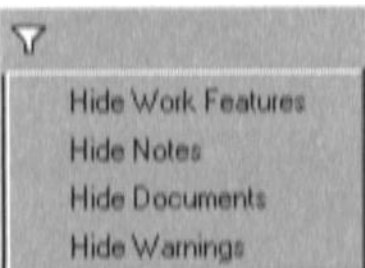


**Figure 3-4**

Each feature in the Browser is given a default name. The first extrusion, for example, will be named Extrusion1 and the number will sequence as you add similar features. The Browser can also help you locate parts and features in the graphics area. To highlight a feature or part in the graphics window, simply move the mouse over the feature or part name in the Browser.

To zoom in on a selected feature, right-click on the feature's name in the Browser and click Find in Window on the menu. The Browser itself functions similarly to a toolbar, except that it can be resized while docked. To close the Browser, click the X in the upper-right corner of the Browser. If the Browser is not visible on the screen, you can display it by clicking Browser Bar from the Toolbar pop-up menu on the View menu or right-clicking while over a toolbar and clicking Browser Bar from the list.

Specific functionality of the Browser will be covered throughout the book in the sections where it pertains. A basic rule, however, is to either right-click or double-click on the feature's name to edit or perform a function on the feature. A sample Browser is shown in the following image, illustrating some of the topics described above.

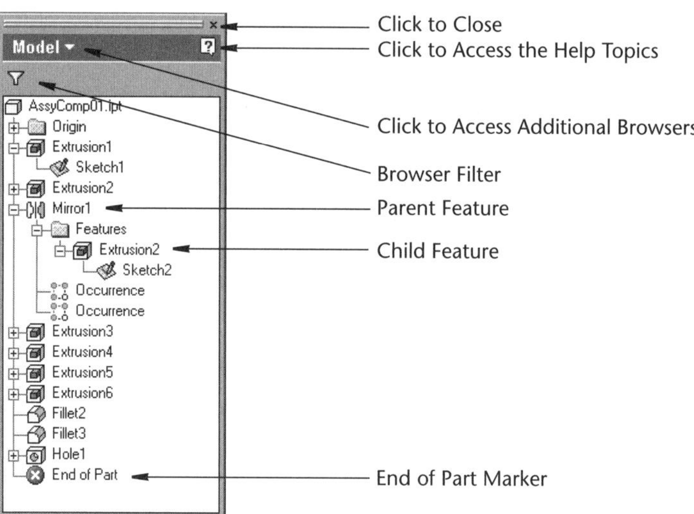


**Figure 3-5**

## SWITCHING ENVIRONMENTS

Up to this point, you have been working in the sketch environment where the work is done in 2D. The next step is to turn the sketch into a feature. To turn a sketch into a feature you need to exit the *sketch environment* and enter the *part environment*. There are a number of methods that can be used to accomplish this.

- Click the Return tool on the left side of the Standard toolbar shown in the following image.

**Figure 3-6**

- Click the arrow on the Panel Bar near 2D Sketch Panel and click Part Features from the drop-down list as shown in the following image.

**Figure 3-7**

**NOTE** If you exit the sketch environment with the Return tool, this is done automatically.

- Right-click in the graphics area and click Create Feature from the menu (as shown in the following image), and then click the tool you need to create the feature. Only the tools that are applicable to the current situation will be available.

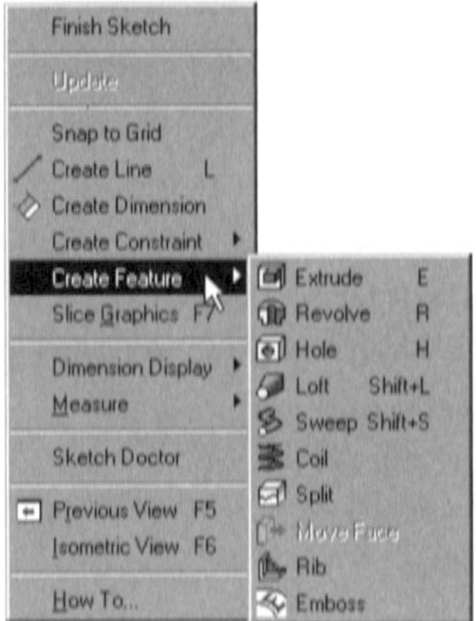


**Figure 3-8**

- Enter a shortcut key to initiate one of the feature tools.

## FEATURE TOOLS

Once you are in the part environment, the Panel Bar icons will change to the part feature tools. The following image shows the part features Panel Bar with the Expert mode turned on.

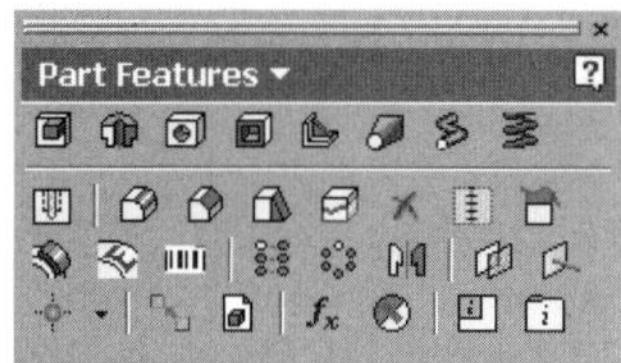


Figure 3-9

## Feature Tools

| Button | Tool | Function |
|---|---|---|
| | Extrude | Extrudes a sketch in the positive or negative *Z*-axis. This extrusion can form a base feature or add or remove material from a part. Extruding is covered later in this chapter. |
| | Revolve | Rotates a sketch around a straight edge or axis at a specified angle. This revolution can form a base feature or add or remove material from a part. Revolving is covered later in this chapter. |
| | Hole | Creates a drilled, tapped, countersunk, or counterbore hole feature in a part. Creation of holes is covered in Chapter 4. |
| | Shell | Removes material from the part, leaving a specified wall thickness. Shelling is covered in Chapter 4. |
| | Rib | Creates a thin-walled extrusion from a 2D sketch. |
| | Loft | Creates a lofted base feature or loft feature by blending between sketches that lie on different planes. |
| | Sweep | Creates a swept feature by sweeping a sketch about a defined path. |
| | Coil | Creates a 3D helical part or feature by revolving a sketch around a centerline. The specifics of the 3D helical are determined by data entered into a dialog box. |
| | Thread | Creates a thread feature in a hole or on a cylinder. The thread is displayed in the graphic area as well as in drawing views. Creating a thread feature is covered in Chapter 4. |
| | Fillet | Creates a fillet feature on an edge or edges—the feature can be a fillet or a round. Creating fillet features is covered in Chapter 4. |
| | Chamfer | Creates a chamfer feature on a selected edge or edges. Creating chamfer features is covered in Chapter 4. |
| | Face Draft | Creates a face draft feature that angles a selected planar face in or out. Creating face draft features is covered in Chapter 4. |

**Feature Tools** ***(continued)***

| Button | Tool | Function |
|---|---|---|
| | Split | Allows a face to be split into two faces or splits a part into two. |
| | Delete Face | Creates a delete face feature in the Browser. You can delete faces and have the option to include *Healing*, and delete lumps or voids from a part. When a face is deleted the parametric model is converted to a surface model. |
| | Knit Surface | Knits two or more surfaces together to form a single surface. When this tool is activated in the construction environment it can be used to analyze edge conditions of surfaces. Surfaces can be knitted together to form a quilt or you can knit a combination of faces and surfaces to change the model from a surface model to a solid model. |
| | Replace Face | Replaces existing faces of features in a part by merging them with a surface boundary that is selected. |
| | Thicken/ Offset | Creates a thicken feature by adding material to a selected set of faces or surfaces. You can also choose to offset a face or surface. Depending on whether you choose to thicken or offset, the part can be changed to a surface model or a surface model can be changed to a solid part. |
| | Emboss | Creates a profile that physically alters a face by raising or lowering regions of the face. |
| | Decal | Creates a decal, or silkscreen, type feature. Decals are usually separate parts, such as stickers. |
| | Rectangular Pattern | Selected feature(s) will be arrayed in a rectangular pattern. Creating rectangular patterns is covered in Chapter 4. |
| | Circular Pattern | Selected feature(s) will be arrayed in a circular pattern around a centerline. Creating circular patterns is covered in Chapter 4. |
| | Mirror Feature | Selected feature(s) will be mirrored about a plane. |
| | Work Plane | Creates a work plane feature that is based on user input. A work plane can be used as a sketch plane or as a plane for the mirror feature tool. Creating work planes is covered in Chapter 4. |
| | Work Axis | Creates a work axis feature that can be used as an axis for a circular pattern or used to create work planes. Creating a work axis is covered in Chapter 4. |

**Feature Tools *(continued)***

| Button | Tool | Function |
|---|---|---|
| | Work Points | Creates a work point feature that can be used to help create work planes and other tools that require points to be selected. Creating work points is covered in Chapter 4. |
| | Grounded Work Points | Creates a work point feature that is initially locked into position. Its position can be modified using a 3D Manipulator. Creating work points is covered in Chapter 4. |
| | Promote | Stitches an imported surface model into a single surface and then promotes it so the surface can be used for creating features. A part can also be promoted as surfaces that can be used in another part file. This is typically done when working in the assembly environment. |
| | Derived Component | Creates a part based on another part or assembly. Any changes to the source will also appear in the derived part. |
| | Parameters | Displays and defines parameters used in the model. You can create user-defined parameters, rename, add equations, and link parameters to a Microsoft Excel file. |
| | Create iMate | Creates a predefined constraint or group of constraints (Composite iMates) on a component to specify how parts will connect when they are inserted into an assembly. |
| | Insert iFeature | Places an iFeature on the current part. |
| | View Catalog | Opens Windows Explorer to let you browse any iFeatures that have been created. |

## EXTRUDING A SKETCH

The most common method for creating a feature is to extrude a sketch and give it depth along the *Z*-axis. Before extruding, it is helpful to view the part in an isometric view. Autodesk Inventor previews the extrusion depth and direction in the graphics window.

To extrude a sketch, click the Extrude tool from the Part Features Panel Bar, press the hot key E, or right-click in the graphics screen and click Create Feature > Extrude on the menu. After you issue the tool, the Extrude dialog box is displayed as shown in the following image.

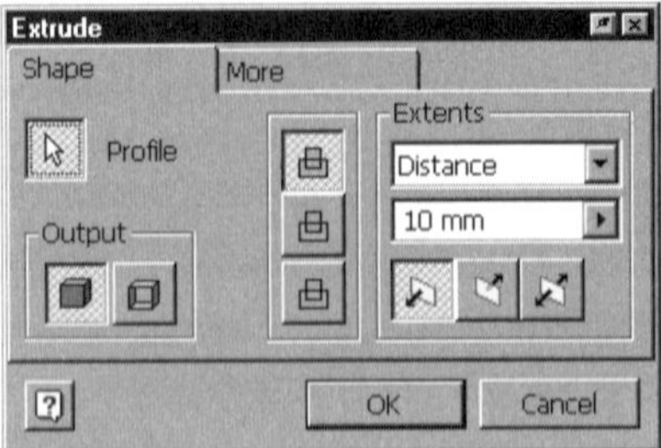


**Figure 3-10**

The Extrude dialog box has two tabs: Shape and More. When changes are made in the dialog box, the shape of the sketch will change in the graphics area to represent these values and options. When you have entered the values and the options you need, click the OK button to create the extruded feature.

## SHAPE

The Shape tab is where you specify the profile to use, the operation (middle column), extents, and output type. It contains several options.

Profile — Click this button to choose the sketch to extrude. If there are multiple closed profiles, you will need to select which sketch area you want to extrude. If there is only one possible profile, Autodesk Inventor will select it for you and you can skip this step. If the wrong profile or sketch area is selected, reselect the Profile button and choose the new profile or sketch area.

### Operation

This is the middle column of buttons that is not labeled. If this is the first sketch that you are working with, it is referred to as a base feature and only the top button is available. The operation defaults to Join (top button). Once the base feature has been established, you can then extrude a sketch adding or removing material from the part using the join or cut options; or keep what is common between the existing part and the completed extrude operation using the intersect option.

Join — Adds material to the part.

Cut — Removes material from the part.

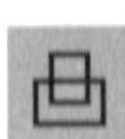

Intersect — Keeps what is common to the part and the new feature.

### Extents

Extents determines the type and distance that the extrusion will go. This section has three areas: termination, distance, and direction.

#### Termination

The termination determines how the sketch will be extruded. There are five options from which to choose. As in the operation section, this section has options that are not available until a base feature exists.

**Distance** The sketch will be extruded a specified distance.

**To Next** The sketch will be extruded until it reaches a plane or face. The sketch must be fully enclosed in the area that it is projecting to, otherwise use the *To* termination with the Extend to Surface option. Click the Direction button to determine the extrusion direction.

**To** The sketch will be extruded until it reaches a selected face or plane. To select a plane or face to end the extrusion, click the Select Surface button, as shown in the following image, and then click a face or plane at which the extrusion should terminate.

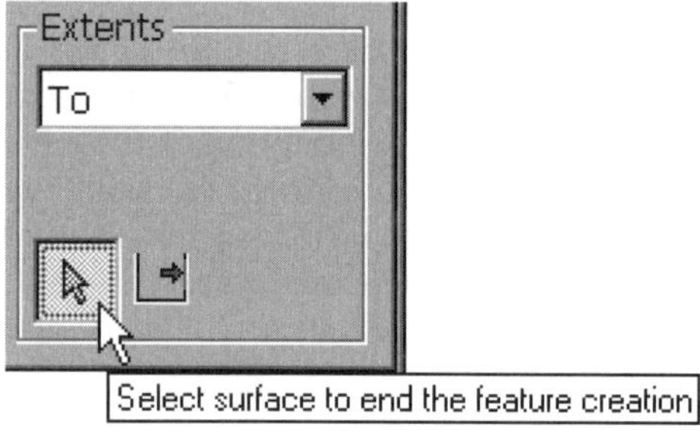


**Figure 3-11**

**From To** The extrusion will start at a selected plane or face and stop at another plane or face. To select a plane or face to start and end the extrusion, click the Select the surface to start feature creation button, and then click the face or plane where the extrusion will start. You should then click the Select the surface to end the feature creation button, and then click the face or plane where the extrusion will terminate.

**All** The sketch will be extruded all the way through the part in one direction.

#### Distance

If distance is selected as the termination, enter a value at which the sketch will be extruded, click the arrow to the right and measure two points to determine a value, or select from the list of the most recent values that were used. After a value is entered, a preview image is displayed in the graphics area representing how the extrusion will look. Another method is to click the edge of the extrusion shown in the graphics area and drag it.

A preview image is displayed in the graphics area as well as the corresponding value, which is displayed in the distance area.

If values and units appear red as they are being entered, it means that something is incorrect and needs to be fixed.

If two decimal places were entered (i.e., **2.12.5**) or MMM were entered for the unit, as shown in the following image, for example, the values are displayed in red and need to be corrected before the extrusion can be completed.

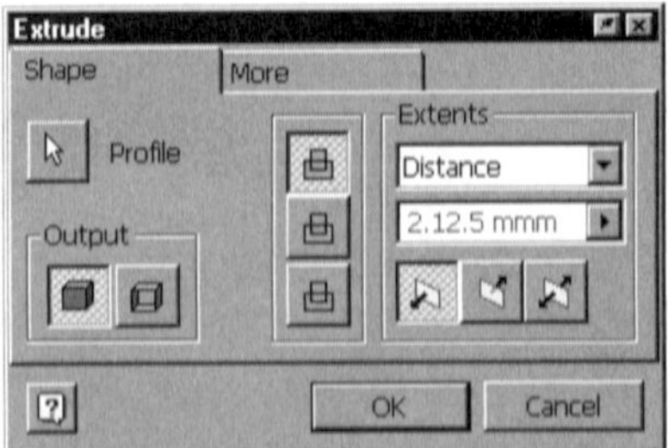


**Figure 3-12**

**NOTE** When extruding a sketch a given distance, drag the sketch to get a preview of how it will look at different distances.

### Direction

There are three buttons to choose from for determining the direction. Choose the first two to flip the extrusion direction or click the last button to have the extrusion go equal distances in the negative and positive directions. If the extrusion distance is 2", for example, the extrusion will go 1" in both the negative and positive Z directions when using the mid-plane option.

## Output

There are two options available to select the type of output that will be generated by the Feature tool.

| | |
|---|---|
| Solid | Extrudes the sketch and the result is a solid body. |

| | |
|---|---|
| Surface | Extrudes the sketch and the result is a surface. |

When you convert a sketch into a part or feature, the dimensions on the sketch disappear. When the feature is edited, the dimensions reappear. The dimensions are also displayed when drawing views are made. For more information on editing parts or features, see "Editing 3D Parts" later in this chapter.

## MORE

The More tab, as shown in the following image, contains additional options to refine the feature that is being created:

**Taper** Taper extrudes the sketch and applies a taper angle to the feature.

**Alternate Solution** Alternate Solution terminates the feature on the most distant solution for the selected surface. An example is shown in the following image.

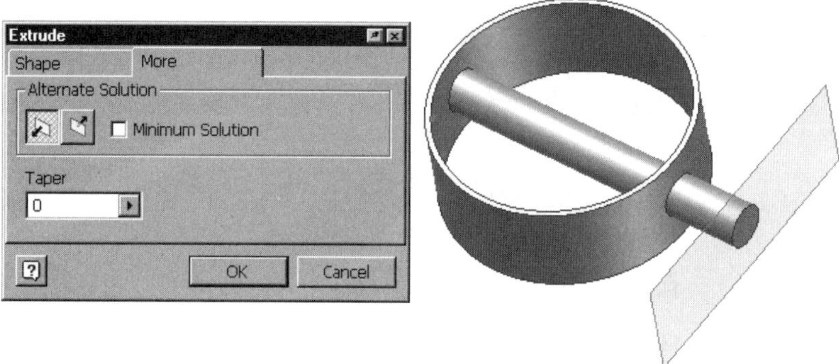


Figure 3-13

**Minimum Solution** Minimum Solution terminates the feature on the first possible solution for the selected surface. An example is shown in the following image.

**TIP** To extend the taper angle out from the part, give the taper angle a positive number, also known as reverse draft.

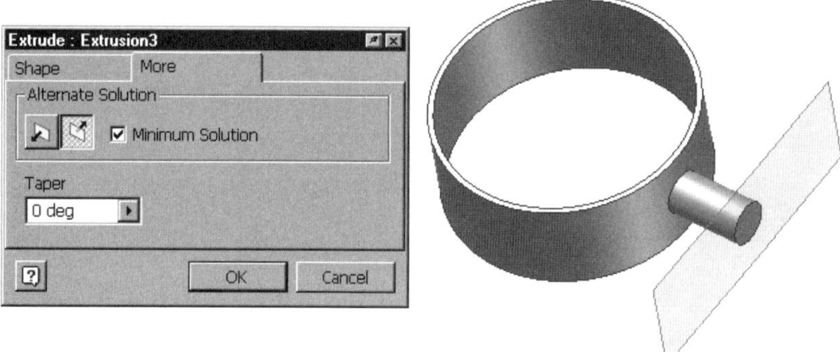


Figure 3-14

# EXERCISE 3-1 Extruding a Sketch

In this exercise, you create a sketch and extrude the profile. Two more extrusions are created to complete the part. To navigate to the exercise in the *Electronic Student Workbook,* do the following:

1 From the Main TOC page, click Chapter 3.
2 From the TOC for Chapter 3, click Extruding a Sketch.

The following image illustrates the completed exercise.

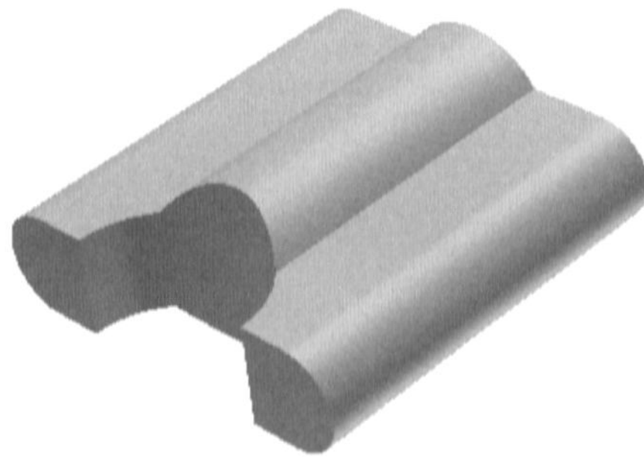

**Figure 3-15 Completed exercise**

## REVOLVING A SKETCH

Another method for creating a part is to *revolve* a sketch around a straight edge or axis (centerline). Revolve can be used to create cylindrical parts or features. To revolve a sketch, you will follow the same steps you did to extrude a sketch (create the sketch and add constraints and dimensions), then click the Revolve tool from the Part Features Panel Bar, then press **R** or right-click in the graphics window and click Create Feature > Revolve on the menu. The Revolve dialog box is displayed, as shown in the following image.

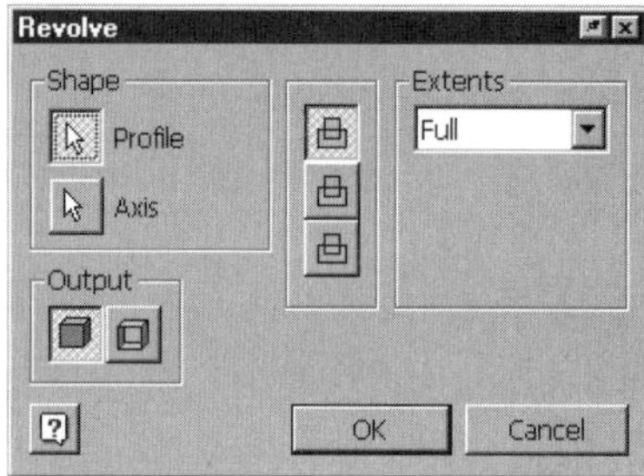


**Figure 3-16**

The Revolve dialog box has four sections: Shape, Operation (middle column), Extents, and Output. When changes are made in the dialog box, the shape of the sketch changes in the graphics area to represent the values and options. When you have entered the values and the options you need, click the OK button to create the revolve feature.

### SHAPE

This section has two options: Profile and Axis.

Profile — Click this button to choose the profile to revolve. If the Profile button is shown depressed, this is telling you that a profile or sketch needs to be selected. If there are multiple closed profiles you will need to select the profile you want to revolve. If there is only one possible profile, Autodesk Inventor will select it for you and you can skip this step. If the wrong profile or sketch area is selected, click the Profile button and choose the new profile or sketch area.

Axis — Click a straight edge or centerline about which the sketch should be revolved. The edge does not need to be part of the sketch. If a centerline is used, it needs to be part of the sketch. See the section below on how to create a centerline and create diametric dimensions.

## OPERATION

This is the middle column of buttons that is not labeled. If this is the first sketch that you are working with, it is referred to as a base feature and only the top button is available. The operation defaults to Join (top button). Once the base feature has been established, you can then revolve a sketch adding or removing material from the part using the join or cut options; or keep what is common between the existing part and the completed revolve operation using the intersect option.

Join — Adds material to the part.

Cut — Removes material from the part.

Intersect — Keeps what is common to the part and the new feature.

## EXTENTS

The Extents area determines if the sketch will be revolved 360° or another specified angle.

**Full** Full is the default option and will revolve the sketch 360° about a specified edge or axis.

**Angle** Click this option from the drop-down list and the Revolve dialog box displays additional options, as shown in the following image. Enter an angle for the sketch to be revolved. There are three buttons below the degree area that will determine the direction of the revolve. Choose the first two to flip the revolve direction or click the last button to have the revolve go equal distances in the negative and positive directions. If the angle was set to 90°, for example, the revolve will go 45° in both the negative and positive directions.

Figure 3-17

### OUTPUT

There are two options available to select the type of output that will be generated by the Revolve tool.

Solid Revolves the sketch, and the result is a solid body.

Surface Revolves the sketch, and the result is a surface.

## CENTERLINES AND DIAMETRIC DIMENSIONS

When revolving a sketch, you will want to specify diametric dimensions, instead of radial dimensions, the majority of the time. Sketches that are revolved are usually a quarter section of the completed part. The following image shows a sketch that represents a quarter section of the completed part with a centerline and the diametric dimensions.

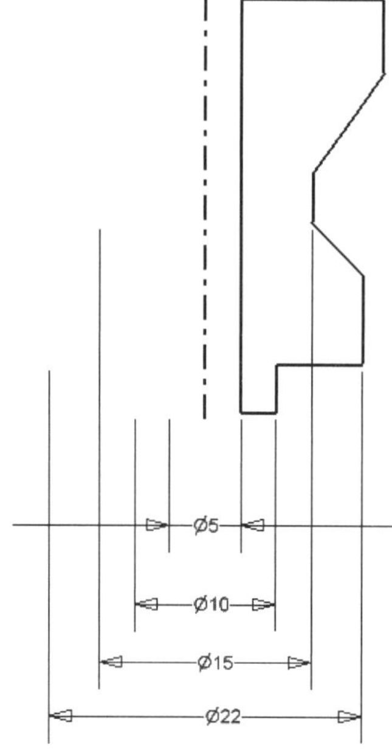


**Figure 3-18**

The dimensions that are placed on a sketch are used for drawing views. If you want to place a diametric dimension on the sketch, a centerline must be selected. To create a centerline, you must first draw a line that will become the centerline. You then exit the Line tool, click the line (to select it), then click Centerline from the Style drop-down menu on the Command Bar, as shown in the following image.

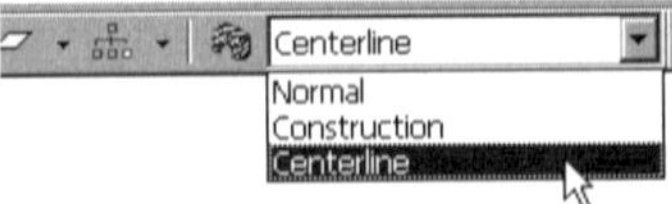


**Figure 3-19**

The line is changed to a centerline. To create a diametric dimension, use the General Dimension tool and select either the centerline and the other point or line to be dimensioned, or click a point or edge and *then* the centerline to place the diametric dimension. When selecting the centerline, make certain to select the entire centerline—not just an endpoint of the centerline.

# EXERCISE 3-2 Revolving a Sketch

In this exercise, you open an existing clutch assembly then create a revolved feature to complete the assembly. To navigate to the exercise in the *Electronic Student Workbook* do the following:

1 From the Main TOC page, click Chapter 3.

2 From the TOC for Chapter 3, click Revolving a Sketch.

The following image illustrates the opened exercise.

**Figure 3-20 Opened exercise**

The following image illustrates the completed exercise.

**Figure 3-21 Completed exercise**

## EDITING A FEATURE

After a feature is created, the feature consumes all of the dimensions that were visible in the sketch. If you need to change the dimensions' values, taper, operation, termination, or output type that were entered in the dialog box, you will need to edit the sketch or feature. To edit the information that was entered in the dialog box while the feature was created, follow these steps:

1 Right-click on the feature's name in the Browser.

2 Select Edit Feature from the menu.

The following image shows the menu that is displayed after right-clicking on Extrusion1.

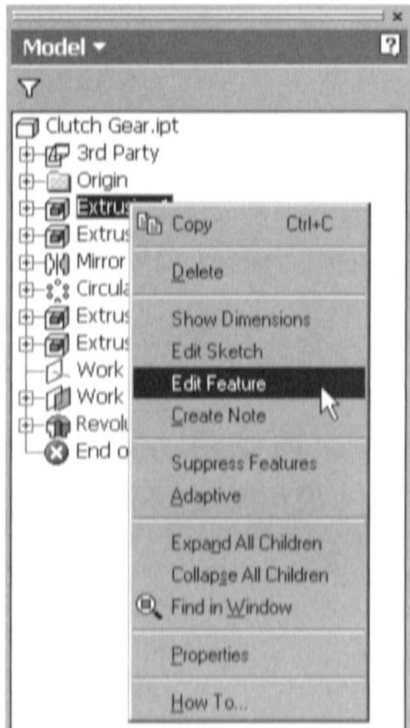


**Figure 3-22**

3 In the dialog box, enter new values or change the settings.

4 Click the OK button to complete the edit.

Everything can be changed in the dialog box except the join operation on a base feature.

### EDITING FEATURE SIZE

To change the dimensional values of a feature, you need to edit the feature so the dimensions are visible. There are multiple methods that can be used to edit the dimensions. There is no preferred method—use the method that works best for your workflow.

- In the Browser, double-click on the feature's name or icon.
- In the Browser, right-click on the feature's name and choose Edit Sketch from the menu that is displayed.
- In the Browser expand the children of the feature, right-click on the name of the sketch, and choose Edit Sketch from the menu as shown in the following image.

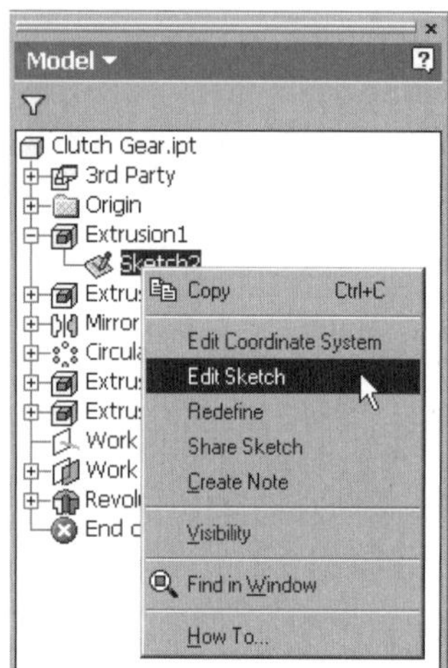


**Figure 3-23**

- From the Command Bar, click the down arrow on the Select button and choose Feature Priority from the drop-down menu as shown in the following image. Then double-click on the feature that you want to edit and the dimensions will appear on the part.

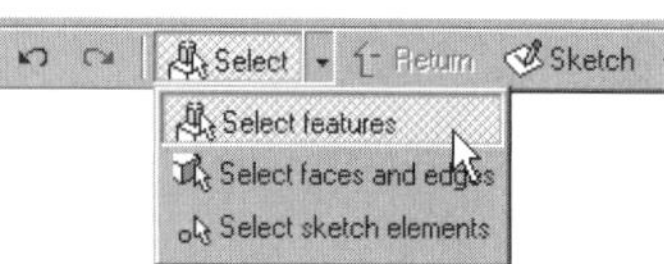


**Figure 3-24**

After the dimensions are visible on the screen, double-click on the dimension text that you want to edit. The Edit Dimension dialog box is displayed. Enter a new value and then click the check mark in the dialog box, or press the ENTER key. Continue to edit the dimensions and, when complete, click the Return, Sketch, or the Update buttons on the Command Bar as shown in the following image. The dimensions will disappear and the new values will be used to regenerate the part.

Figure 3-25

## RENAMING FEATURES AND SKETCHES

By default, each feature is given a name. These feature names may not help you when trying to locate a specific feature of a complex part. The feature names will not be descriptive to your design intent. The first extrusion, for example, is given the name Extrusion1 by default, whereas the design intent may be that the extrusion is the thickness of a plate. All feature names can be edited. To rename a feature, slowly double-click the feature name. Enter a new name (spaces are allowed). For the example above, the first extrusion could be renamed Plate Thickness.

## FEATURE COLOR

When parts become complex, you may want to change colors of specific features. This is also useful when you want to differentiate between a cast and a machined surface. To change a feature color, right-click on the feature name in the Browser and click Properties on the menu. In the Feature Properties dialog box, select a new feature color from the drop-down menu as shown in the following image, and then click OK to complete the operation. In the top area of the Feature Properties dialog box, the feature's name can also be changed.

Figure 3-26

## DELETING A FEATURE

You may choose to delete a feature after it has been placed. To delete a feature, right-click on the feature name in the Browser and click Delete on the menu, as shown in the following image. The Delete Features dialog box will then appear, and you should choose what you want to delete from the list. Multiple features can be deleted by holding down the CTRL key, clicking their names in the Browser, and then clicking Delete on the right-click menu.

Figure 3-27

### FAILED FEATURES

If, after updating the part, the feature in the Browser turns red, it is an alert that the new values or settings were not successfully regenerated. You MUST click the Undo button on the Standard toolbar until the part reappears correctly. You can then edit and enter new values or select different settings.

## EDITING A FEATURE SKETCH

In the last section, you learned how to edit the dimensions and the settings in which the feature was created. In this section, you will learn how to add and delete constraints, dimensions, or geometry in the original 2D sketch. To edit the 2D sketch of a feature, do the following:

1 In the Browser, right-click on the feature's name or the name of the sketch that you want to edit and choose Edit Sketch from the menu, as shown in the following image.

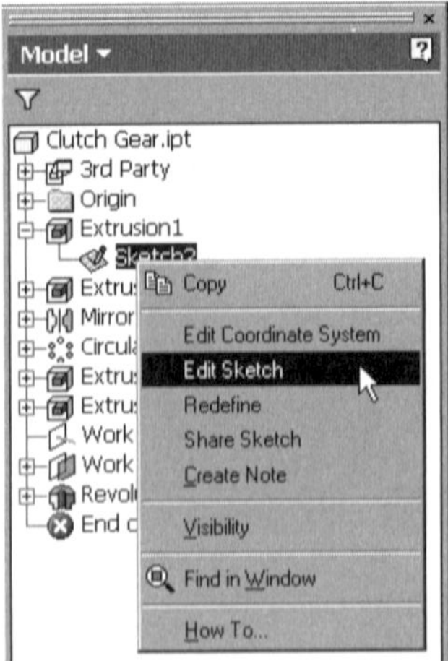


**Figure 3-28**

2 Then add or remove geometry from the sketch and add or delete constraints and dimensions, as you learned in Chapter 2.

While editing the sketch, you can both add and remove objects in the sketch. When adding geometry to the sketch; lines, arcs, circles, and splines can be added. Objects can also be deleted. To delete an object, select it and either right-click and select Delete on the menu, or press the DELETE key. If you erase an object from the sketch that has dimensions associated with it, the dimensions are no longer valid for the sketch and will be deleted. The entire sketch can also be deleted and replaced with an entirely new sketch. When replacing entire sketches, other features that would be consumed by the new objects should be deleted first and recreated. Once the sketch has been modified, update the part by clicking the Return, Sketch, or the Update button on the Command Bar.

**NOTE** If you get an error after updating the part, make sure that the sketch forms a closed profile. If the appended sketch forms multiple closed profiles, you will need to reselect the profile area.

# EXERCISE 3-3
# Editing Features and Sketches



In this exercise, you edit a consumed sketch in an extrusion and update the part. You also edit a feature termination, and recover from a modeling failure introduced during the edit of a feature. To navigate to the exercise in the *Electronic Student Workbook* do the following:

1 From the Main TOC page, click Chapter 3.

2 From the TOC for Chapter 3, click Editing Features and Sketches.

The following image illustrates the opened exercise.

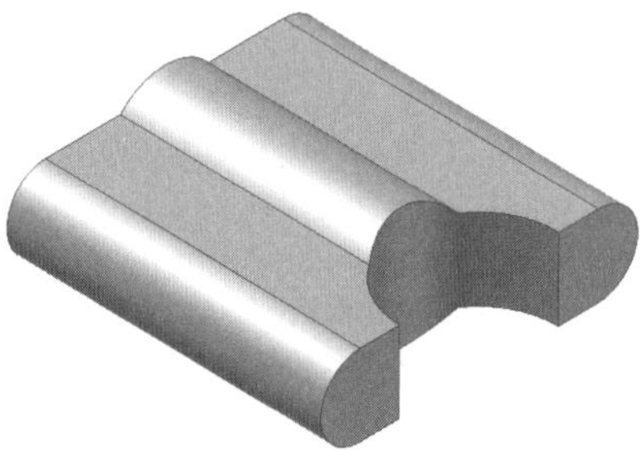

**Figure 3-29 Opened exercise**

The following image illustrates the completed exercise.

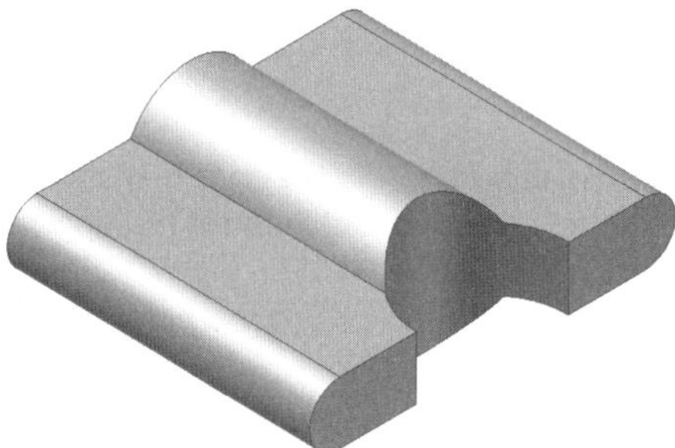

**Figure 3-30 Completed exercise**

## SKETCHED FEATURES

A sketched feature is a feature that you draw on a plane and add or remove material from the part. The basic steps to create a sketched feature are the following:

1. Create or make an existing sketch active.
2. Draw the geometry that defines the sketch.
3. Add constraints and dimensions.
4. Perform a Boolean operation that will either add or remove material from the part or keep whatever is common between the part and the completed feature.

There are no limits to the number of sketched features that can be added to a part. Each sketched feature, however, needs to be on its own plane. Even if multiple sketched features will exist on the same plane of the part, a new sketch will need to be created for each sketched feature. In this section, you will learn how to assign a plane to the active sketch and then how to work with sketched features.

## DEFINING THE ACTIVE SKETCH PLANE

As was stated in the last section, each sketch must exist on its own plane. The active sketch has a plane on which the sketch is drawn. To assign a plane to the active sketch, there are three requirements.

1. The part in which the plane will be placed must be an Autodesk Inventor part.
2. The part must be active.
3. A plane on which the sketch will be created must be a planar face or a work plane. The planar face does not need to have a straight edge. A cylinder has two faces, for example, one on the top and the other on the bottom of the part. Neither has a straight edge, but a sketch can be placed on either face.

To make a sketch active, use one of the following methods:

- Start the Sketch tool from the Command Bar as shown in the following image, and then click the plane where you want to place the sketch.

**Figure 3-31**

- Press the hot key S, and then click the plane where you want to place the sketch.
- Click a plane that will contain the active sketch and issue the Sketch tool from the Command Bar.
- Click a plane that will contain the active sketch, and press the hot key S.

- Click a plane that will contain the active sketch, right-click in the graphics area, and click New Sketch from the menu.
- Start the Sketch tool from the Command Bar, expand the Origin folder in the Browser, and click one of the default work planes.
- Expand the Origin folder in the Browser, right-click on one of the default work planes, and click New Sketch on the menu.
- To make a sketch that has already been created the active sketch, start the Sketch tool and then click the sketch name on the Browser.

Once a sketch is created, it is displayed in the Browser with the name Sketch#, and sketch tools are displayed in the 2D Sketch Panel Bar. The number will sequence for each new sketch that is created. In the Browser, slowly double-click the existing name and enter a new one to rename a sketch. After a new sketch has been created on a plane, it is sometimes easier to work in a plan view (looking straight at the current plane). This can be done by using the Look At tool and then clicking the plane. Now that the sketch has been created, you can place sketch curves, apply constraints, and apply dimensions, exactly as you did with the first sketch. In addition to constraining and dimensioning the new sketch, you can also constrain the new sketch to the existing part. You can place dimensions to geometry that do not lie on the current plane—the dimensions, however, will be placed on the current plane. When you look at a part from different viewpoints, you will see arcs and circular edges appearing as lines. Remember that they are still circular edges. When you constrain or dimension them, the constraints and dimensions will be placed on their center points. After constraining and dimensioning the sketch, the sketch can be extruded, revolved, or swept. To delete a sketch, exit the sketch environment by clicking the sketch name in the Browser and then either right-clicking with the mouse in the graphics area and clicking Finish Sketch, or clicking the Return tool in the Command Bar. There can only be one active sketch at a time.

**TIP** Rename the sketches to better explain the purpose for which they are being used.

## FACE CYCLING

Autodesk Inventor has dynamic face highlighting to help you select the correct face to make the active sketch. As you move the mouse over a given face, it highlights. If you continue to move the mouse, different faces are highlighted as the mouse passes over them.

To cycle to a face that is behind another one, move the mouse over the face that is in front of the one that you want to select. Hold the mouse still and the Select Other tool is displayed (as shown in the following image). Select the left or right arrow to cycle through the faces until the correct face is highlighted, and then press the left mouse button or click the green rectangle in the middle of the Select Other tool. You can also access the Select

Other tool by right-clicking the Graphics Window and clicking Select Other from the menu.

**Figure 3-32**

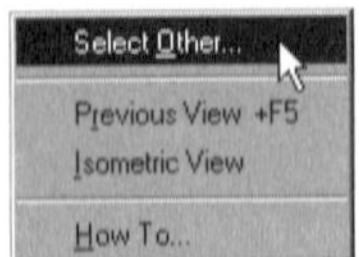


**Figure 3-33**

You can specify the amount of time before the Select Other tool will automatically display on the General tab after selecting Application Options from the Tools menu. The time delay can be specified in tenths of a second. If you do not want the Select Other tool to open automatically, enter the word **OFF** in the edit box. The default value is 1.0 seconds.

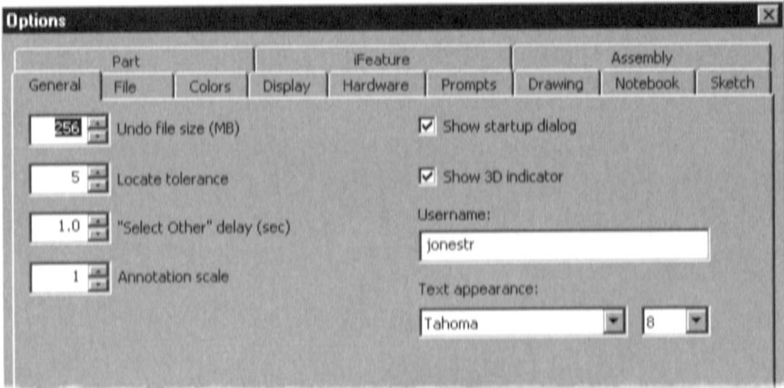


**Figure 3-34**

# EXERCISE 3-4 Sketch Planes

In this exercise, you create a sketch plane on the angled face of a part then create a slot using the new sketch plane. To navigate to the exercise in the *Electronic Student Workbook* do the following:

1 From the Main TOC page, click Chapter 3.

2 From the TOC for Chapter 3, click Sketch Planes.

The following image illustrates the opened exercise.

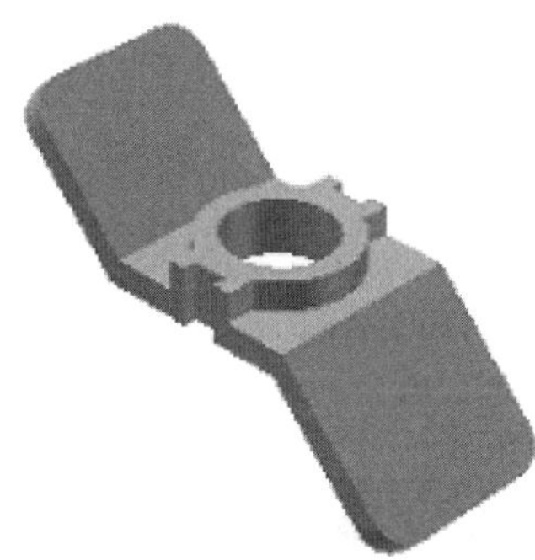

**Figure 3-35 Opened exercise**

The following image illustrates the completed exercise.

**Figure 3-36 Completed exercise**

## PROJECTING PART EDGES

Building parts partially based on existing geometry is done often, and you will frequently need to reference faces, edges, or loops of features or parts that have been created.

While in a part file, you can project an edge, face, or loop onto a sketch. Projected geometry maintains an associative link to the original geometry that is projected. If you project the face of a feature onto another sketch, for example, and the parent sketch is modified, the projected geometry will update to reflect the changes.

### DIRECT MODEL EDGE REFERENCING

While you sketch, you can use direct model edge referencing to

- Automatically project edges of the part to the sketch plane as you sketch a curve
- Create dimensions and constraints to edges of the part that do not lie on the sketch plane
- Control the automatic projection of part edges to the sketch plane

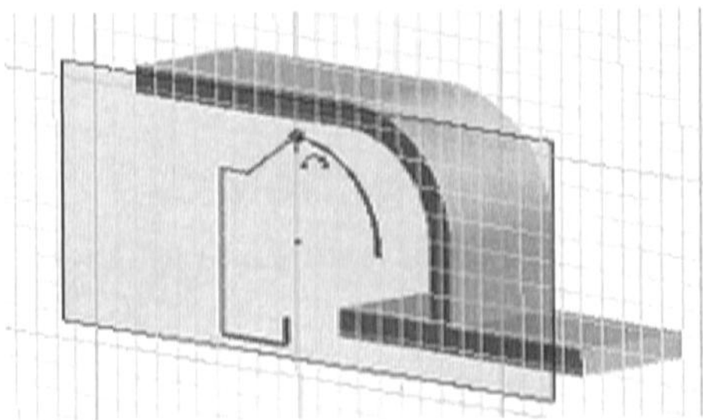

**Figure 3-37**

#### Creating Reference Geometry

There are two ways to automatically project part edges to the sketch plane.

1 Rub the cursor on an edge of the part while sketching a curve.

2 Click an edge of the part while creating a dimension or constraint.

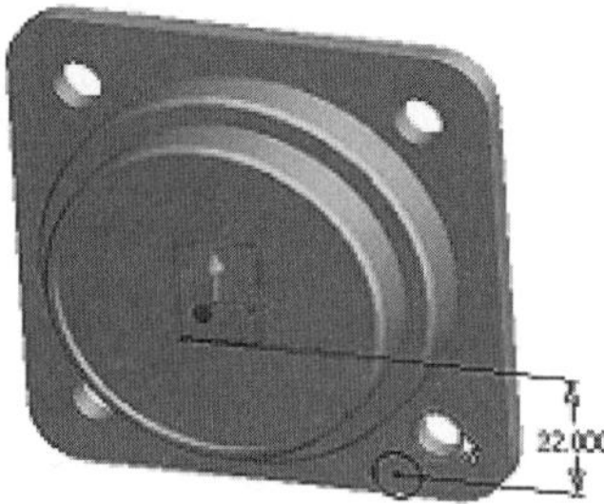


Figure 3-38

On the Sketch tab in the Options dialog box you can use

- Autoproject edges during curve creation—Controls the ability to rub and project edges while sketching a curve
- Automatic edges for sketch creation and edit—Controls the automatic projection of part edges coplanar with the sketch plane

**NOTE** Neither of these options disables the ability to reference part edges when creating dimensions and constraints.

## PROJECT LOOP AND CHAIN OPTION

To project edges from another part or feature and use them in the current sketch, you use the Project Geometry tool discussed in Chapter 7. After the tool has been activated, you can click a single edge to project it onto the current sketch. You can use the Select Other tool to cycle between a single edge and the entire outer loop of the face being projected.

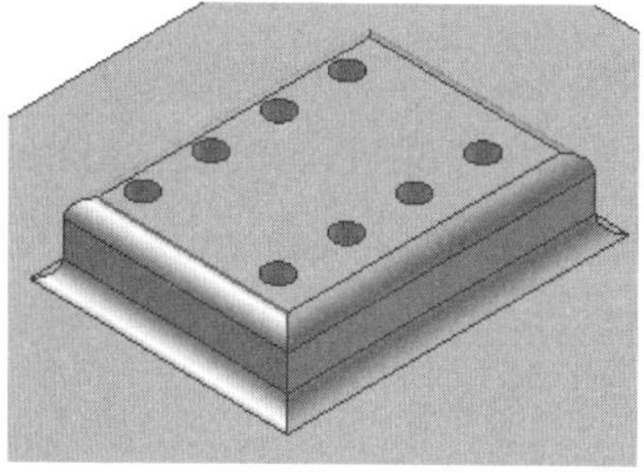
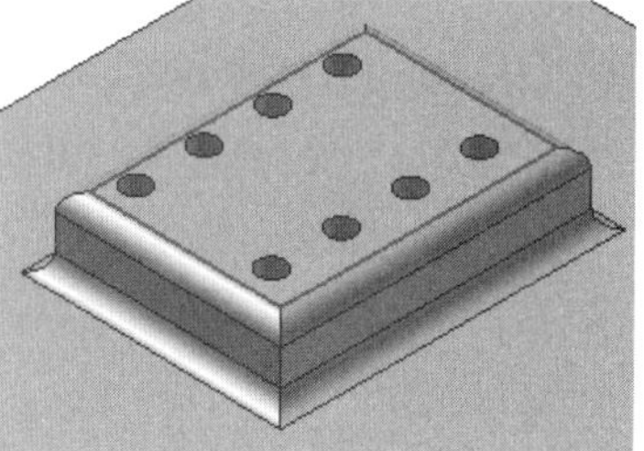

Figure 3-39

If a loop is clicked for the projection, the sketch is updated to reflect the modification when any part of the profile changes. A face, and any internal loops contained on the face, can also be projected. If a face is projected, the internal islands that are defined on the face are also projected and will update accordingly. For example, if the face of the above image is projected, the outer loop of the face and all of the circles that define the hole pattern on the face will be projected and updated if modified.

## CHAPTER SUMMARY

| To | Do This | Tool |
|---|---|---|
| Place a sketch plane | Click the Sketch tool on the command bar and click an existing face or work plane. | Sketch |
| Edit sketch features | Right-click the sketch icon in the Browser, and then click Edit Sketch. | |
| Define relationships to existing features | Click the Join, Cut, or Intersect operation. | |
| Create an extruded feature | Click the Extrude tool in the Panel Bar or from the Features toolbar. | |
| Create a revolved feature | Click the Revolve tool in the Panel Bar or from the Features toolbar. | |
| Display sketch geometry | Right-click the Sketch icon in the Browser, and then click Visibility. | |
| Project faces, loops, or edges from existing features | Click the Project Geometry tool in the Panel Bar or from the Sketch toolbar. | |

EXERCISE

# Applying Your Skills

### SKILL EXERCISE 3-1

In this exercise, you create a bracket from a number of extruded features. To navigate to the exercise in the *Electronic Student Workbook* do the following:

1. From the Main TOC page, click Chapter 3.
2. From the TOC for Chapter 3, click Exercise 1 under Applying Your Skills.

The following image illustrates the completed exercise.

**Figure 3-40 Completed exercise**

## SKILL EXERCISE 3-2

In this exercise, you create a connecting rod part, and add draft to extrusions during feature creation. To navigate to the exercise in the *Electronic Student Workbook* do the following:

1. From the Main TOC page, click Chapter 3.
2. From the TOC for Chapter 3, click Exercise 2 under Applying Your Skills.

The following image illustrates the completed exercise.

**Figure 3-41 Completed exercise**

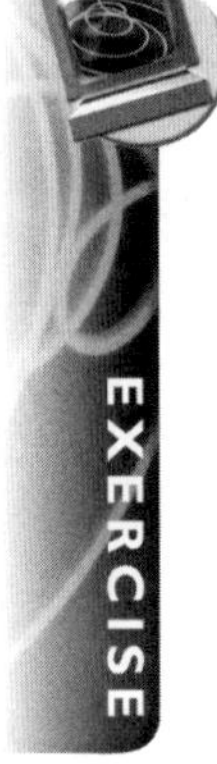


### SKILL EXERCISE 3-3

In this exercise, you create a pulley using a revolved feature. To navigate to the exercise in the *Electronic Student Workbook* do the following:

1. From the Main TOC page, click Chapter 3.
2. From the TOC for Chapter 3, click Exercise 3 under Applying Your Skills.

The following image illustrates the completed exercise.

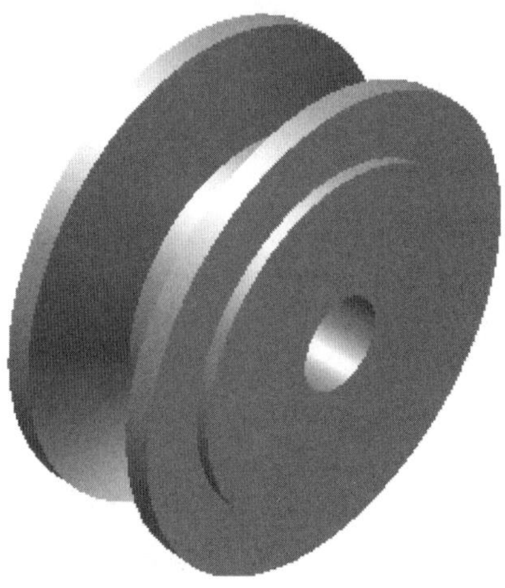

**Figure 3-42 Completed exercise**

## CHECKING YOUR SKILLS

Use these questions to test your knowledge of the material covered in this chapter.

1 What is a base feature?

______________________________________________

______________________________________________

2 **True___ False___** When creating a feature with the Extrude or Revolve tool, you can drag the sketch to define the distance or angle.

3 Which objects can be used as an axis of revolution?

______________________________________________

______________________________________________

4 Explain how to create a diametric dimension on a sketch.

______________________________________________

______________________________________________

5 Name two ways to edit an existing feature.

______________________________________________

______________________________________________

6 **True___ False___** Once a sketch becomes a base feature, you cannot delete or add constraints, dimensions, or objects to the sketch.

7 Name three operation types used to create sketched features.

______________________________________________

______________________________________________

______________________________________________

8 **True___ False___** A cut operation cannot be performed before a base feature is created.

9 **True___ False___** Once a sketched feature exists, its termination cannot be changed.

10 **True___ False___** Geometry that is projected from one feature to a sketch that defines another feature will automatically update based on changes to the original projected geometry.

# CHAPTER 4

# Creating Placed Features

In Chapter 3, you learned how to create and edit base and sketched features. In this chapter you will learn how to create *placed* features. Placed features are features that are predefined except for specific values and only need to be located. Placed features can be edited from the Browser like sketched features. When a placed feature is edited, either the dialog box that was used to create it will open, or feature values are displayed on the part.

When creating a part, it is usually better to use placed features instead of sketched features wherever possible. To make a through hole as a sketched feature, for example, you can draw a circle profile and dimension it, then extrude it with the cut operation, using the All extension. You can also create a hole as a placed feature where you can select the type of hole, size it, and then place it using a dialog box. When drawing views are then generated, the type and size of the hole can be easily annotated and automatically updates if the hole type or values change.

## CHAPTER OBJECTIVES

**After completing this chapter, you will be able to**

- Create fillets
- Create chamfers
- Create holes
- Create internal and external threads
- Shell a part
- Add face draft to a part
- Create work axes
- Create work points
- Create work planes
- Pattern features

## FILLETS

Fillet features consist of fillets and rounds. Fillets add material to interior edges to create a smooth transition from one face to another. Rounds remove material from exterior edges.

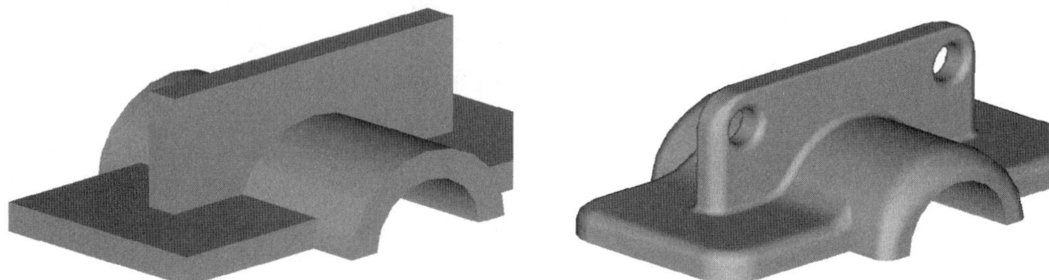

Figure 4-1

When creating fillets in 3D, you select the edge that needs to be filleted and the fillet is created between the two faces that share the edge. This is different from placing a fillet in 2D. In the 2D environment, you click two objects and a fillet is created between them. When creating a part, it is good practice to create fillets and chamfers as one of last features in the part. Fillets add complexity to the part, which in turn adds to the size of the file and removes edges that may be needed to place other features.

To create a fillet feature, click the Fillet tool from the Part Features Panel Bar as shown in the following image.

Figure 4-2

After you click the tool, the Fillet dialog box is displayed as shown in the following image.

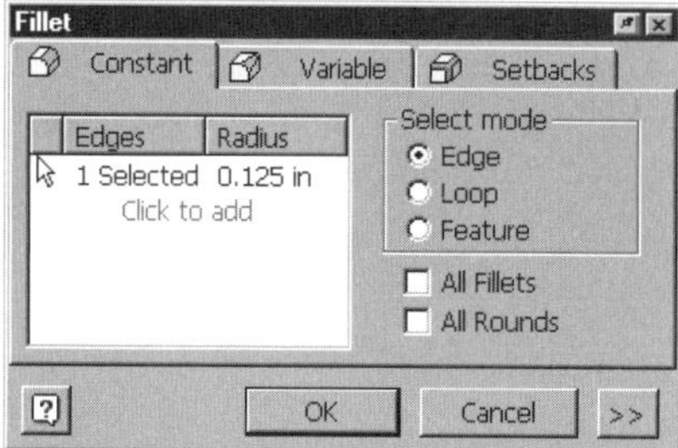


Figure 4-3

The Fillet dialog box has three tabs: Constant, Variable, and Setbacks. Each tab creates a different type (or style) of fillet. The options for each of the tabs are described in the following sections.

Before we look at the tabs, let's look at the methodology that will be used to create fillets. You can either click an edge or edges to fillet or select the type of fillet to create. If an edge is clicked before the Fillet feature tool is issued, it is placed in the first selection set. Selection sets contain the edges that will be filleted when the OK button is clicked to create the feature.

Each fillet feature can contain multiple selection sets, each having its own unique fillet value. There is no limit to the number of selection sets that can exist in a single instance of the feature. An edge, however, can only exist in one selection set. All of the selection sets are shown as a single fillet feature in the Browser. To add to the first selection set, select the edges that you want to have the same radius. To create another selection set, select Click to add and then click the edges that will be part of the next selection set. To remove an edge that has been selected, select the selection set that the edge is part of and the edges will highlight. Hold down the CTRL key and click the edge to be removed from the selection set. After clicking the edge(s) to fillet, click the type of fillet that you want to create in the Fillet dialog box, and then enter the values for the fillet. As changes are made in the dialog box, a representation of the fillet is previewed in the graphics window. When the fillet type and value are correct, click the OK button to create the fillet.

To edit a fillet's type and radius, follow these steps:

1 Start the Edit Feature tool by right-clicking on the fillet's name in the Browser and clicking Edit Feature from the menu. The Fillet dialog box is displayed with all of the settings that were used to create it.

2 Change the fillet settings as needed.

3 To edit only the dimensional value of a fillet, you can double-click on the fillet's name or icon in the Browser or change the select priority to Feature Priority and double-click on the fillet that is on the part.

4 The dimension(s) are displayed on the part. Double-click the dimension to change it and enter new values in the Edit Dimension dialog box. Then click the check mark in the dialog box or click Enter.

5 Click the Update tool to update the part.

### CONSTANT TAB

With the options on the Constant tab, shown in the previous image, you create fillets that have the same radius from beginning to end. There is no limit to the number of edges that can be filleted with a constant fillet. The edges can be selected as a single set or selected in multiple sets—each set can have its own radius value. The order in which the edges are selected is not important. The edges that are to be filleted need to be selected individually, and the use of the window or crossing selection method is not allowed. If you change the value of a group, all of the fillets in that group will change. To remove an edge from a group, choose the group in the select area and then hold down the CTRL key and click the edge. Below is a description of the options that are available on the Constant tab.

### Select Edge and Radius

By default, after issuing the Fillet tool, you can click edges and they are displayed in the first selection set. To create another selection set, select Click to add and then click the edges that will be part of the next selection set. After clicking an edge, a preview image of the fillet is displayed on the edge that reflects the current values, as shown in the following image.

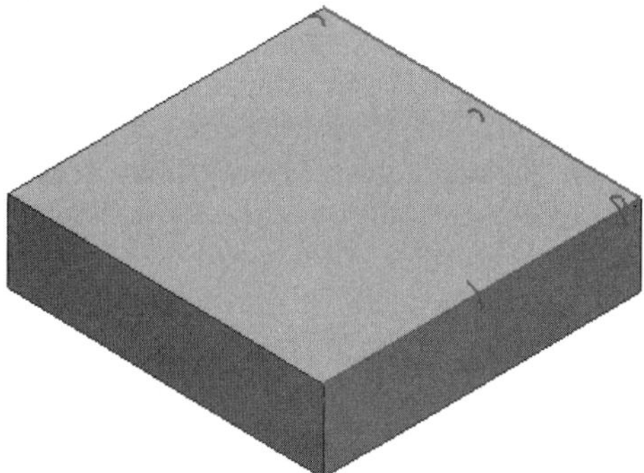

Figure 4-4

### Select Mode

**Edge** Click the Edge mode to select individual edges to fillet. By default, any edge that is tangent to the clicked edges is also selected. If you do not want to have tangent edges automatically selected, uncheck the Automatic Edge Chain option in the More (>>) section.

**Loop** Click the Loop option to have all of the edges filleted that form a closed loop with the selected edge.

**Feature** The Feature option will select all of the edges of a selected feature.

**All Fillets** The All Fillets option will select all concave edges of a part that have not already been filleted. The All Fillets option adds material to the part and requires a separate edge selection set.

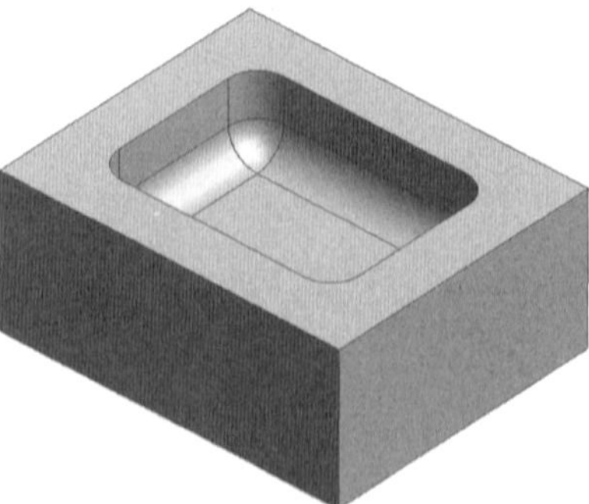

Figure 4-5

**All Rounds** The All Rounds option will select all convex edges of a part that have not already been filleted. The All Rounds option removes material from the part and requires a separate edge selection set.

Figure 4-6

## More Options

**Roll along sharp edges** When clicked, a constant radius is maintained and adjacent faces are extended as needed.

**Rolling ball where possible** When clicked, a fillet is created around a corner that looks like a ball has been rolled along the edges that define the corners, as shown on the left in the following image. When the rolling ball solution is possible but has not been selected, a blended solution is used, as shown on the right in the following image.

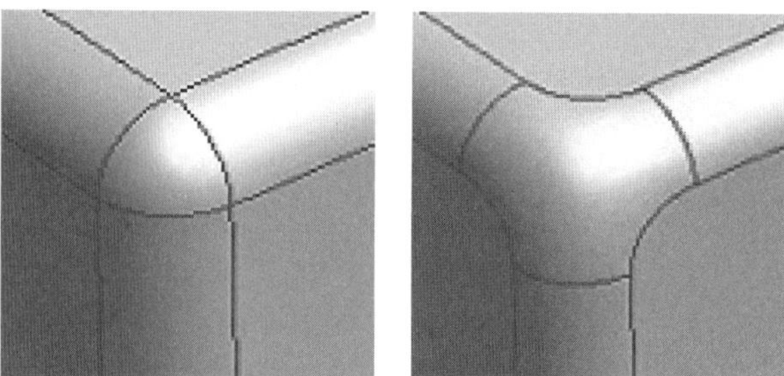

Figure 4-7

**Automatic Edge Chain** When clicked, tangent edges are automatically selected after clicking an edge.

**Preserve All Features** When clicked, all features that intersect with the fillet are checked and their intersections are calculated during the fillet operation. If the check box is cleared, only the edges that are part of the fillet operation are calculated during the operation.

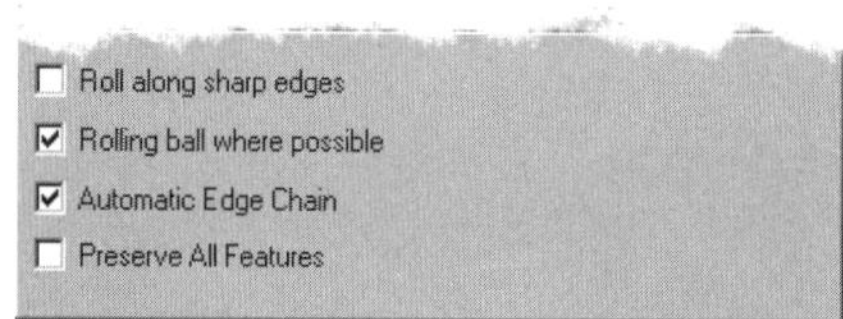


Figure 4-8

### VARIABLE TAB

With the options under the Variable tab, shown in the following image, you will create a fillet that has a different start and end radius.

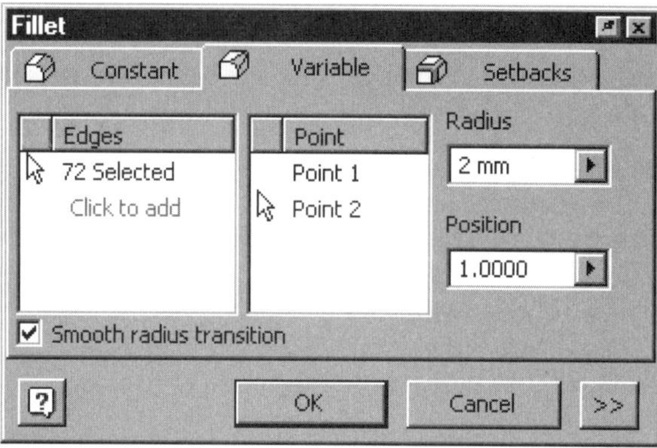


Figure 4-9

**Edges** By default, after clicking the fillet tool, you can click an edge and it is displayed in the first selection set. For a variable fillet, only one edge can exist per selection set. After clicking an edge, a preview image of the fillet is displayed on the edge that reflects the current values. To add another point, click on the edge and its name is displayed in the Point column.

**Point** To determine where the start or end point is on the selected edge, select the Start or End text and a filled cyan circle is displayed at the point on the edge to represent it.

**Radius** After clicking a point's name, enter a radius.

**Position** If another point is added, enter a value between 0.000001 and 1. This number represents a percentage between the start and end point relative to the start point.

**Smooth Radius Transition** The fillet blends from the start to the end radius as a smooth transition, similar to a cubic spline. If unchecked, the fillet blends from the start to the end radius in a straight line.

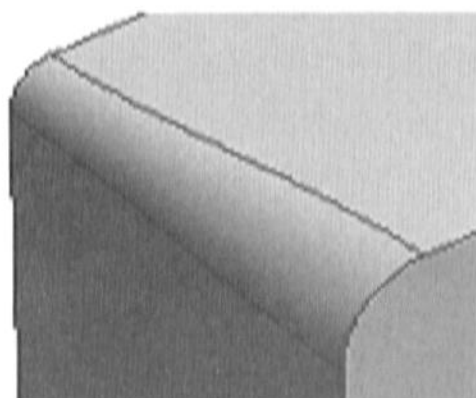

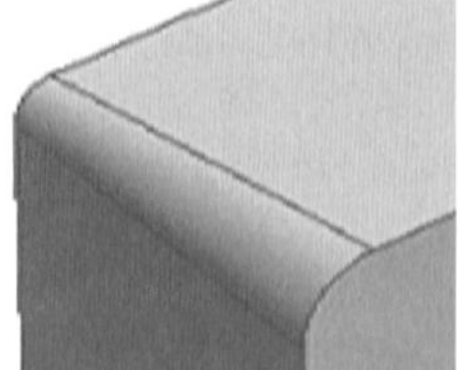

**Figure 4-10**

### SETBACKS TAB

You can specify the distance that a fillet starts its transition from a vertex with the options on the Setbacks tab, as shown in the following image.

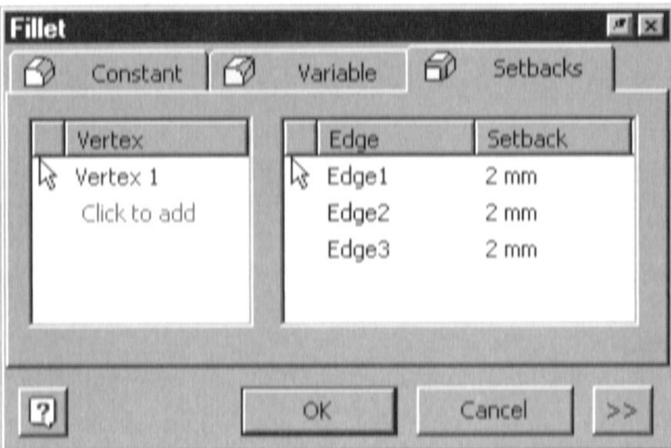


**Figure 4-11**

Using these options, you can model special fillet applications where more than three edges converge, as shown in the following image. You can choose a different radius for each converging edge, if needed. Setbacks can only be used where three filleted edges form a vertex.

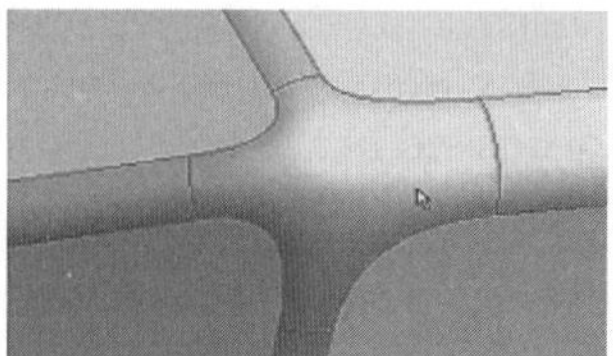

Figure 4-12

**TIP** If you get an error when creating or editing a fillet, try to create it with a smaller radius. If you still get an error after trying to use a smaller radius, you can try to create the fillets in a different sequence or create multiple fillets in the same operation.

## CHAMFERS

Chamfers are similar to fillets except that the edge is beveled rather than rounded. When you create a chamfer on an interior edge, material is added to your model. When you create a chamfer on an exterior edge, material is cut away from your model.

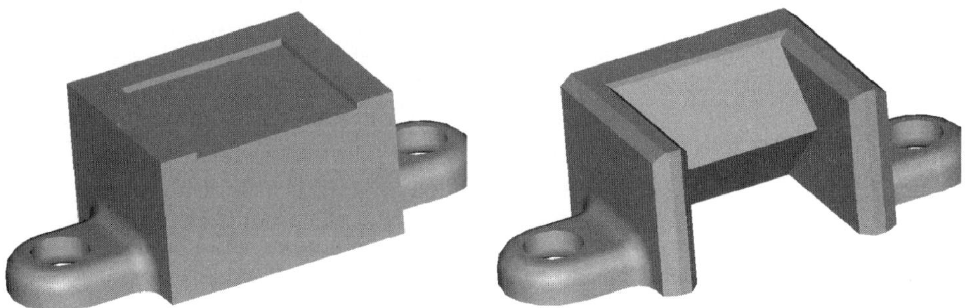

Figure 4-13

To create a chamfer feature, follow the same steps as you did to create the fillet features. Click the common edge and the chamfer is created between the two faces sharing the edge. To create a chamfer feature, click the Chamfer tool from the Part Features Panel Bar, as shown in the following image.

Figure 4-14

After you click the tool, the Chamfer dialog box is displayed, as shown in the following image. As with fillet features, you can select multiple edges to be included in a single chamfer feature. From the dialog box, click a method, enter a distance or angle, click the edge or edges to chamfer, and then click OK to create the chamfer.

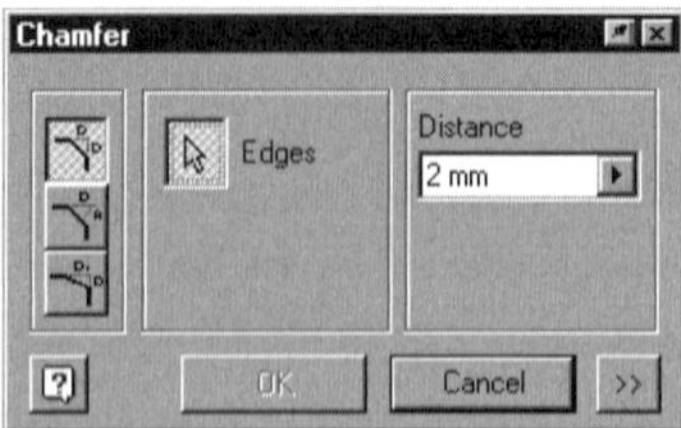


**Figure 4-15**

To edit the type of chamfer feature or distances, use one of the following methods:

- Double-click the feature's name or icon in the Browser.
- Right-click the chamfer's name in the Browser and click Edit Feature from the menu.
- Change the Select Priority to Feature Priority and then double-click the chamfer on the part. The Chamfer dialog box is displayed with all the settings that were used to create the feature. Change the settings as needed.

## METHOD

**Distance** The Distance option creates a 45° chamfer on the selected edge. The size of the chamfer is determined by typing a distance in the dialog box. The value is offset from the two common faces. A single edge, multiple edges, or a chain of edges can be selected. A preview image of the chamfer is displayed on the part. If the wrong edge is selected, hold down the CTRL key and select the edge to remove.

**Distance and Angle** The Distance and Angle option creates a chamfer offset from a selected edge on a specified face, at an angle from the number of degrees specified. In the dialog box, enter an angle and distance for the chamfer, then click the face that the angle is based on and specify an edge to be chamfered. One or multiple edges can be selected. The edges must lie on the selected face. A preview image of the chamfer is displayed on the part. If the wrong face or edge was selected, click on the Edge or Face button and choose a new face or edge.

**Two Distances** The Two Distances option will create a chamfer offset from two faces, each being the amount that you specify. Click an edge first and enter a value for Distance1 and Distance2. A preview image of the chamfer is displayed. To reverse the direction of the distances, select the Flip button. When the correct information about the chamfer is in the dialog box, click the OK button to create the chamfer. Only a single edge or chained edges can be used with the Two Distance option.

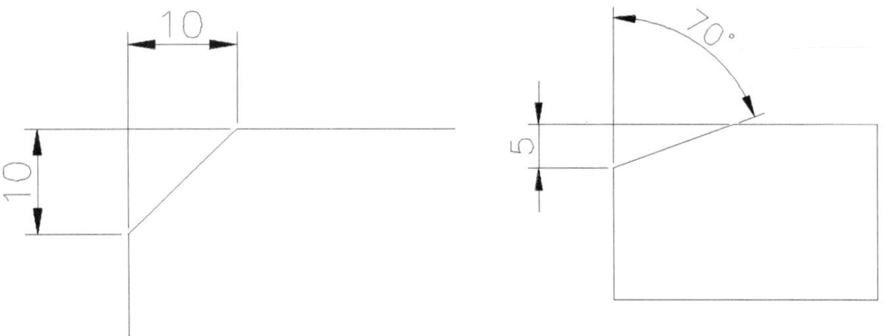


Figure 4-16 Distance for the image on the left. Distance and angle for the image on the right.

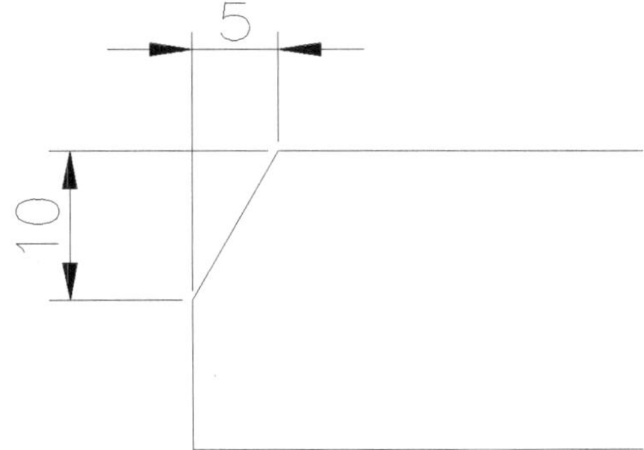


Figure 4-17 Two distances

## EDGE AND FACE

**Edge** Click an edge or edges to be chamfered.

**Face** Click a face from which the chamfer will be based.

**Flip** Click the button to reverse the direction of the distances for the Two Distance chamfer.

## DISTANCE AND ANGLE

**Distance** Enter a distance that will be used for the offset.

**Angle** Enter a value that will be used for the angle if creating the Distance-Angle chamfer type.

## EDGE CHAIN AND SETBACK

**Edge Chain** Click if tangent edges will be selected automatically after clicking an edge.

**Setback** When the Distance method is used and three chamfers meet at a vertex, you can choose to have the intersection of the three chamfers form a flat edge (left button), or the intersection can meet at a point as though the edges were milled (right button).

**Preserve All Features** When clicked, all features that intersect with the chamfer are checked and their intersections are calculated during the fillet operation. If the check box is cleared, only the edges that are part of the fillet operation are calculated during the operation.

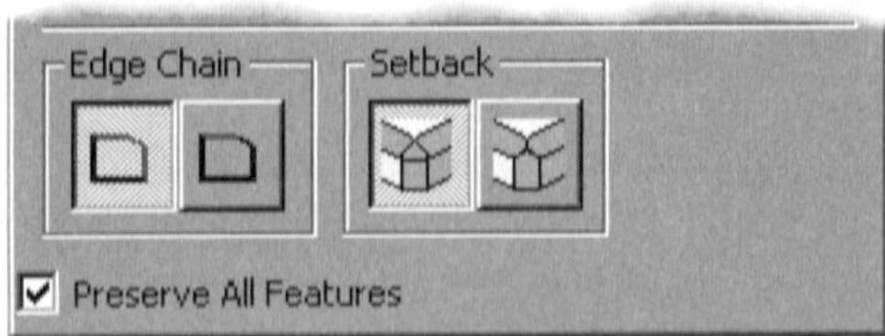


**Figure 4-18**

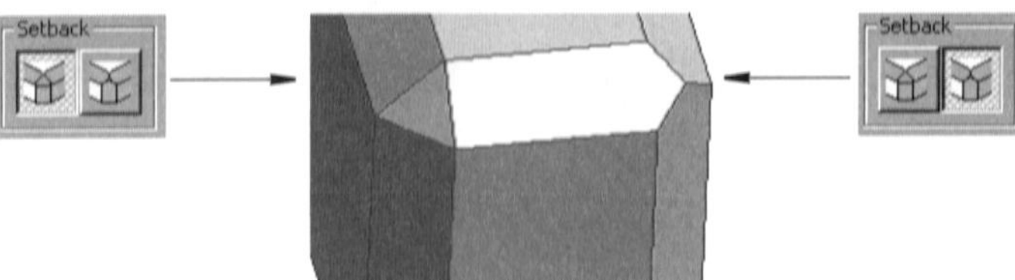


**Figure 4-19**

# EXERCISE 4-1
# Creating Fillets and Chamfers

In this exercise, you create constant radius fillets, variable radius fillets, and chamfers. To navigate to the exercise in the *Electronic Student Workbook* do the following:

1 From the Main TOC page, click Chapter 4.
2 From the TOC for Chapter 4, click Creating Fillets and Chamfers.

The following image illustrates the opened exercise.

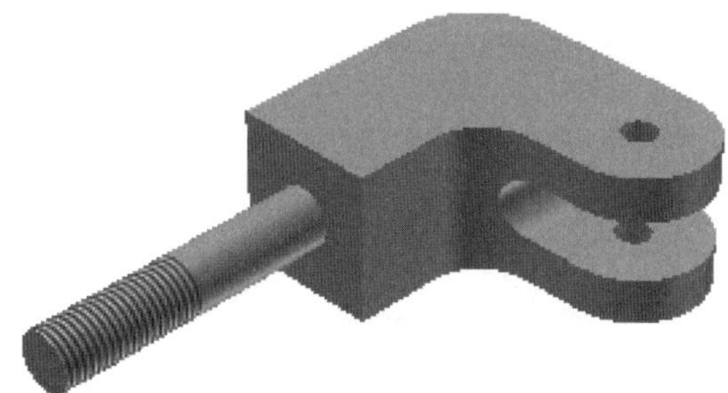

**Figure 4-20 Opened exercise**

The following image illustrates the completed exercise.

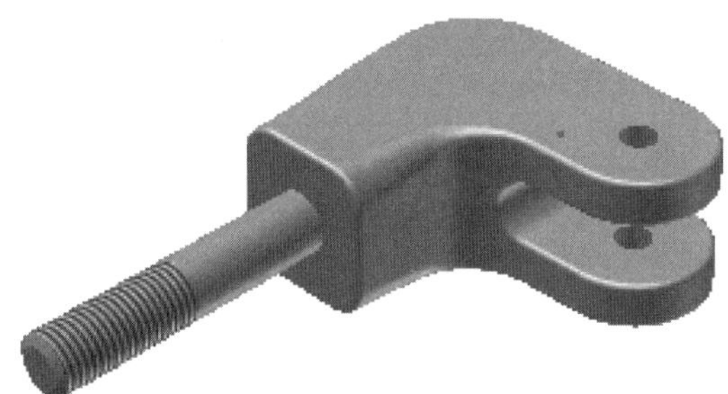

**Figure 4-21 Completed exercise**

## HOLES

There are four basic hole types that Autodesk Inventor can create: drilled, counterbore, countersink, and tapped holes.

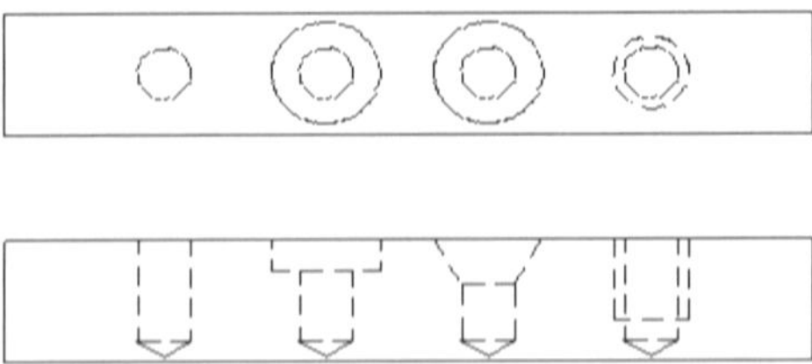

**Figure 4-22**

To create a hole feature, follow these steps:

1. Before creating a hole, create a new sketch on which the hole will exist.
2. After activating the sketch, create hole centers to represent the center of the hole(s). If a hole is going to be located between two edges, you will need to place a hole center on the active sketch. If a hole that you are going to create will be concentric to a circular edge that lies on the active sketch, a hole center is automatically created at the center points or the circular edges. Hole centers are covered in the next section.
3. After the hole center(s) have been placed, click the Hole tool from the Part Features Panel Bar (shown in the following image), or press the hotkey **H**.

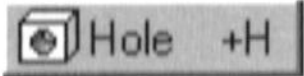


**Figure 4-23**

4. The Holes dialog box is displayed as shown in the following image. It has four tabs containing options for creating holes. Each of the four tabs is explained below. The Centers button is depressed by default—click any hole centers where you want to create a hole. If multiple hole centers are selected, they are grouped together as a single feature in the Browser. To deselect a hole center, hold down the CTRL key and click it. After the hole centers have been selected, select the options that you need from the Holes dialog box. As the options are changed, a preview image of the hole(s) is displayed on the part. When you are done making changes, click the OK button to create the hole(s).

**NOTE** Any endpoint of a line or sketch curve that is on the active sketch can also be used as a hole center.

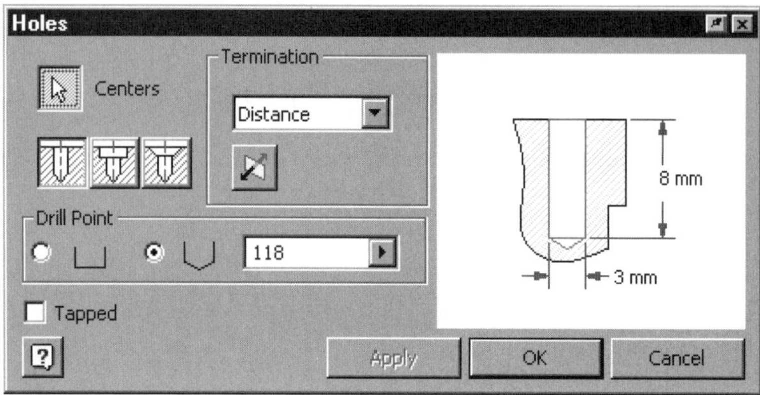


Figure 4-24

To edit the type of hole feature or distances, use one of the following methods:

- Double-click the feature's name or icon in the Browser.
- Right-click the hole's name in the Browser and click Edit Feature from the menu.
- Change the select priority to Feature Priority and then double-click the hole feature on the part in the graphics window.

After issuing the Hole tool, the Holes dialog box is displayed with all of the settings that were used to create it. Change the settings as needed.

## HOLES DIALOG BOX

In the Holes dialog box, shown in the above image, you establish the type of hole, its termination, and additional options such as type of drill point, angle, and tapped properties.

**Centers** Select the hole center point where you want to create a hole.

**Hole Type** Click the type of hole that you want to create: drilled, counterbore, or countersink.

**Termination** Select how the hole will terminate.

*Distance*—Specify a distance for the depth of the hole.

*Through All*—The hole will extend through the entire part in one direction.

*To*—Select a plane at which the hole will stop.

*Flip*—Reverse the direction that the hole will travel.

**Drill Point** Select either a flat or angle drill point. If angle is selected, you can specify the angle of the drill point.

**Dimensions** To change the diameter or depth of the hole, click the dimension in the dialog box and enter a new value.

### TAPPED

Click this box if the hole is tapped. When selected, thread information is displayed in the dialog box (shown in the image below) after you determine if a hole is tapped and specify its values.

**Thread Type** From the drop list, select the standard that the hole is based on: ANSI Unified Screw Threads, ANSI Metric M Profile, ISO Metric Profile, ISO Metric Trapezoidal Threads, ISO Pipe Threads, JIS Pipe Threads, or DIN Pipe Threads.

**Full Depth** Click this box if the threads extend the full depth of the hole.

**Nominal Size** Select the nominal size for the thread from the drop list.

**Pitch** Select the pitch size for the thread from the drop list.

**Class** From the drop list, select the class of thread.

**Diameter** From the drop list, select the nominal diameter type for the hole. The available types are: Minor, Pitch, Major, or Tap Drill.

**Right Hand** Click this option to create threads that wind clockwise and recede.

**Left Hand** Click this option to create threads that wind counterclockwise and recede.

**Dimensions** To change the depth of the thread, click the dimension in the dialog box and enter a new value.

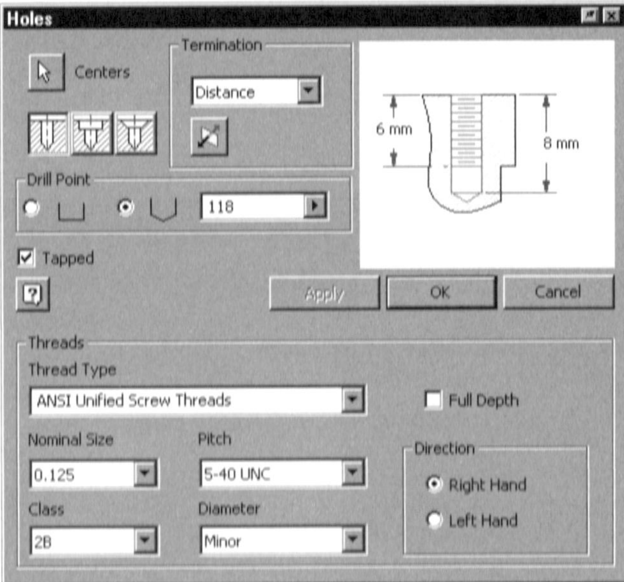


Figure 4-25

### HOLE CENTERS

Hole centers are sketched entities that are used to create holes that lie on a plane. Hole centers are also used as points to which to sketch. To create a hole center, follow these steps:

1. Make a sketch active.
2. Click the Point, Hole Center tool from the 2D Sketch Panel Bar, as shown in the following image.
3. Click a point where the hole will be placed, move the cursor over objects until a coincident constraint symbol is displayed, move the cursor over a line and it will snap to the endpoint or midpoint, or right-click and select Midpoint, Center, or Intersection from the menu to snap to the desired point.

**Figure 4-26**

**Figure 4-27**

# EXERCISE 4-2 Creating Holes

In this exercise, you add drilled, tapped, and counterbored holes to a cylinder head. To navigate to the exercise in the *Electronic Student Workbook* do the following:

1 From the Main TOC page, click Chapter 4.

2 From the TOC for Chapter 4, click Creating Holes.

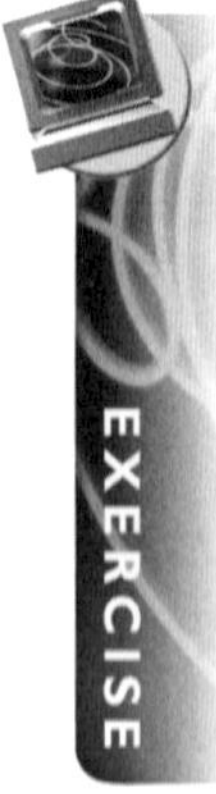


The following image illustrates the opened exercise.

**Figure 4-28 Opened exercise**

The following image illustrates the completed exercise.

**Figure 4-29 Completed exercise**

## THREADS

Thread features are used to create both internal and external threads. The threads are displayed on the parts with a graphical representation, and when a drawing view is created, the thread can be called out per drafting standards. Since the threads are graphical representations, they do not physically exist on the part—if a model was to be cast directly from the part, no threads would exist on the finished model. A thread feature can be added to any hole, or internal or external cylinder. If you are adding threads to a hole, it can be done using either the Hole or Thread tool—if it is added using the Thread tool, it is a separate feature in the Browser. For this reason, it is recommended that the Hole tool be used with the tapped option when creating a tapped hole.

To create external threads, follow these steps:

1. Create a cylinder and dimension it to the size that will represent the major diameter of the thread (this value must lie between the maximum and minimum major diameter).
2. To create the thread feature, issue the Thread tool from the Part Features Panel Bar, as shown in the following image.
3. Enter the data into the Thread Feature dialog box as needed.

**Figure 4-30**

There are two tabs in the Thread Feature dialog box—each is described below. The thread data comes from an Excel spreadsheet named *Thread.xls*. By default, the spreadsheet is located in the Program Files\Autodesk\Inventor<version>\Design Data directory. This spreadsheet can be modified to match your company's standards. The thread data is used to display the thread on the part and when the thread is annotated in a drawing view. The thread data is not associative to the threads on existing parts. When changes are made to the spreadsheet, the new values are only used when new threads are created. If a dimensional change is made to the diameter of the cylinders where a thread has been placed, a warning dialog box is displayed when the part is updated. The dialog box notifies you that an inappropriate thread size is used. Accept the warning message and then modify the thread feature to a size that fits the corresponding diameter.

### LOCATION TAB

**Face** Click this button, and then select a cylindrical or conical face on or in which to place a thread.

**Display in Model** When checked, a visual representation of the thread is displayed on the part. If not checked, the thread is only displayed when a drawing view is created.

### Thread Length

**Full Length** When checked, the thread continues the entire length of the selected face.

**Flip** Click this button to reverse the direction of the thread.

**Length** When Full Length is not checked, enter a value for the length of the thread.

**Offset** When Full Length is not checked, enter a value that the thread will be offset. The offset distance is from the closest plane in relation to where you selected the face.

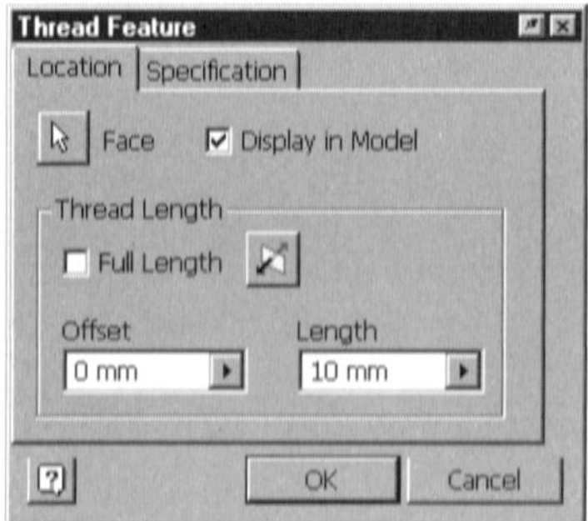


Figure 4-31

#### SPECIFICATION TAB

**Thread Type** From the drop list, select the type—the types are defined by the tabs in the *Thread.xls* file.

**Nominal Size** From the drop list, select the nominal size for the thread.

**Pitch** From the drop list, select the pitch for the thread.

**Class** From the drop list, select the class of thread.

**Right hand or Left hand** Select the direction for the thread, thread size, and its representation. The representation does not change by clicking the other direction.

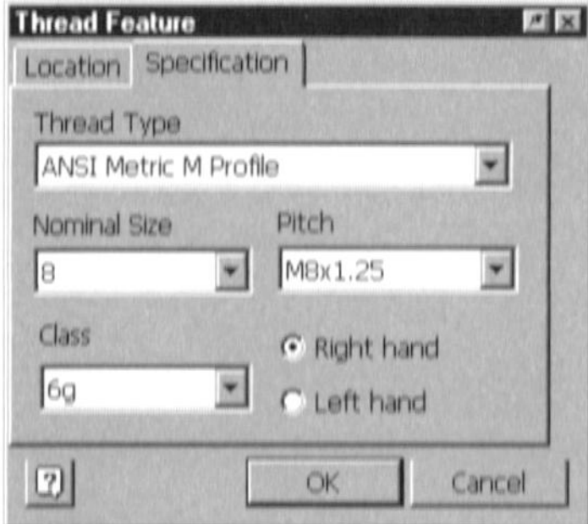


Figure 4-32

# EXERCISE 4-3 Creating Threads

In this exercise, you use the Thread tool to create custom threads on mating faces of a plastic bottle and cap. To navigate to the exercise in the *Electronic Student Workbook*, do the following:

1 From the Main TOC page, click Chapter 4.
2 From the TOC for Chapter 4, click Creating Threads.

The following image illustrates the opened exercise.

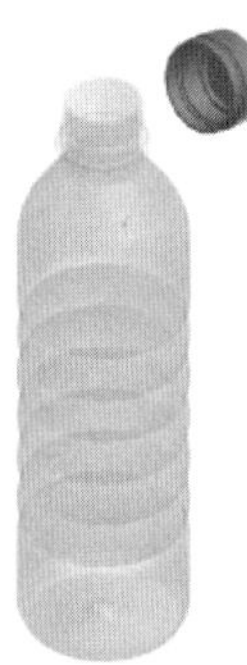

**Figure 4-33 Opened exercise**

The completed exercise is shown in the following image.

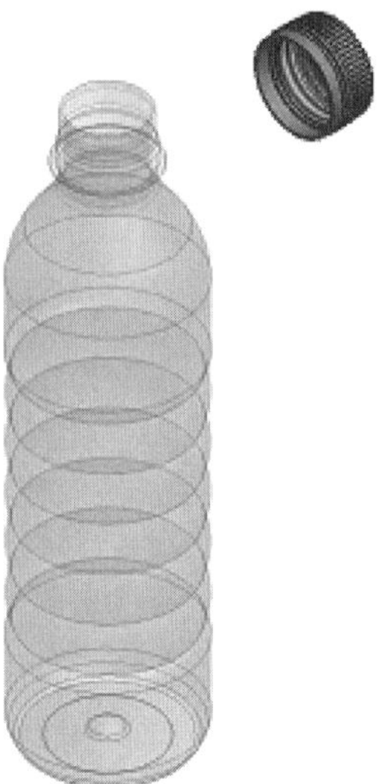

**Figure 4-34 Completed exercise**

## SHELLING

While designing parts, you may need to create a model that is made up of thin walls. The easiest way to create a thin-walled part is to create the main shape and then use the Shell tool to remove material.

The term *shell* refers to giving a thickness (wall thickness) to the outside shape of a part and removing the remaining material—like scooping out the inside of a part, leaving the walls a specified thickness. The wall thickness can be offset in, out, or evenly in both directions. If the part that is shelled contains a void, such as a hole, the feature will have the thickness built around it.

A part may contain more than one shell feature and individual faces of the part can have different thicknesses. If a wall has a different thickness than the shell thickness, it is referred to as a unique face thickness. If a face that you select for a unique face thickness has faces that are tangent to it, those faces will also have the same thickness. Faces can be removed from being shelled and these faces are left open. If no face is removed, the part is hollow on the inside.

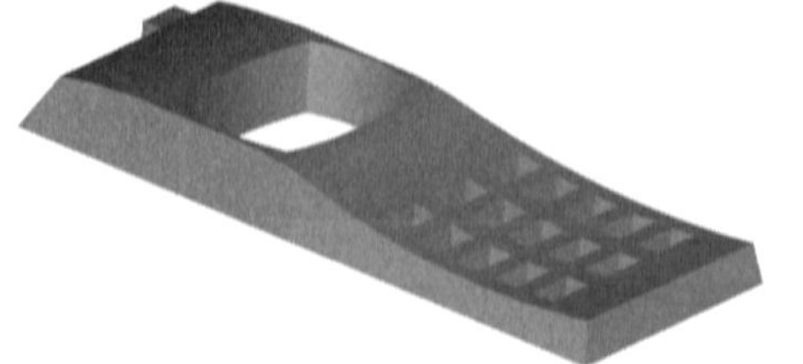

**Figure 4-35**

To create a shell feature, follow these steps:

1. Create a part that will be shelled.
2. Click the Shell tool from the Part Features Panel Bar, as shown in the following image.

**Figure 4-36**

3. The Shell dialog box is displayed, as shown below. Enter the data as needed.
4. After filling in the information in the dialog box, click OK and the part will be shelled.

To edit a shell feature, use one of the following methods:

- Double-click the feature's name or icon in the Browser.
- Right-click the name of the shell feature in the Browser and click Edit Feature from the menu.
- Change the Select Priority to Feature Priority and double-click the shell feature on the part.

The Shell dialog box is displayed with all of the settings that were used to create the feature. Change the settings as needed.

Figure 4-37

The following section explains the options that are available for the Shell tool.

### REMOVE FACES

Click the Remove Faces button and then click the face or faces that will be left open. To deselect a face, click the Remove Faces button and hold down the CTRL key while you click the face.

### THICKNESS

Enter a value or select a previously used value from the drop list to be used for the shell thickness.

### DIRECTION

**Inside** Offsets the wall thickness by the given value into the part.

**Outside** Offsets the wall thickness by the given value out of the part.

**Both** Offsets the wall thickness evenly into and out of the part by the given value.

### UNIQUE FACE THICKNESS

Unique face thickness is available by clicking the More (>>) button that is located on the lower right corner of the dialog box, as shown in the following image.

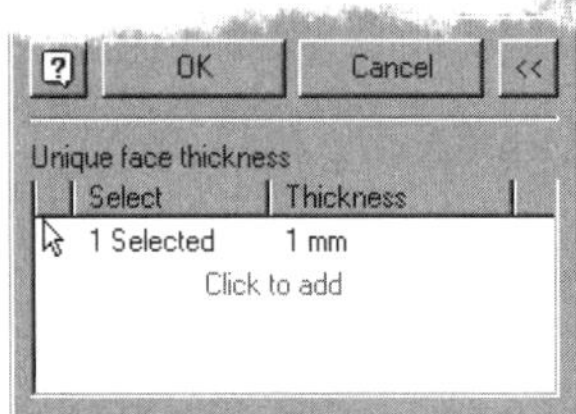


Figure 4-38

To give a specific face a thickness, select Click to add and then click the face and enter a value. A part may contain multiple faces that have a unique thickness.

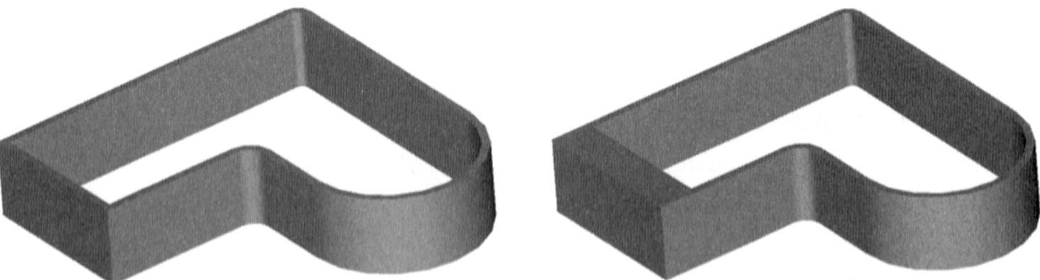

**Figure 4-39**

EXERCISE

# EXERCISE 4-4 Shelling a Part

In this exercise, you remove material from a split part interior using the Shell tool. To navigate to the exercise in the *Electronic Student Workbook* do the following:

1 From the Main TOC page, click Chapter 4.
2 From the TOC for Chapter 4, click Shelling a Part.

The following image illustrates the opened exercise.

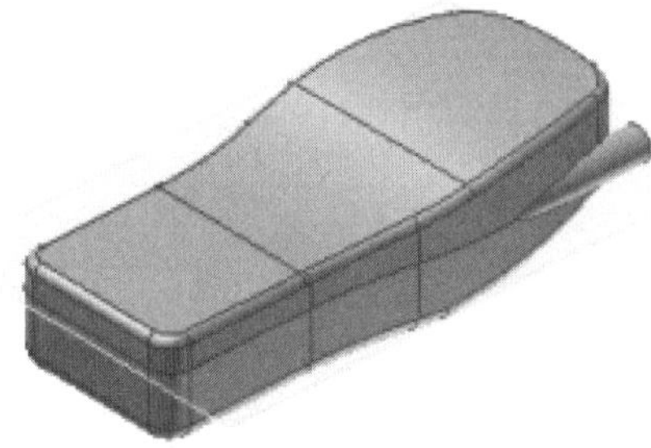

**Figure 4-40 Opened exercise**

The following image illustrates the completed exercise.

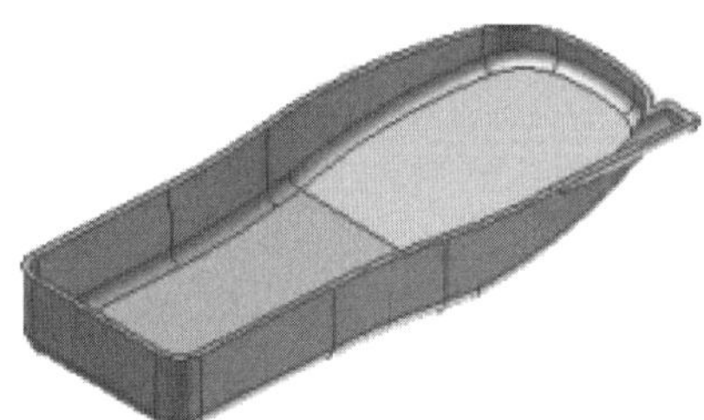

**Figure 4-41 Completed exercise**

## FACE DRAFT

Face draft is a feature that applies an angle to a face. Face draft can be applied to any specified internal or external face, including shelled parts. When face draft is applied, any tangent face also has the face draft applied to it.

To create a face draft feature, follow these steps:

1 Issue the Face Draft tool from the Part Features Panel Bar, as shown in the following image.

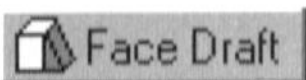


Figure 4-42

2 The Face Draft dialog box is displayed, as shown in the following image. The options in the Face Draft dialog box are explained in the next section.

3 Click the Pull Direction that shows how the mold will be pulled from the part.

4 Then click a face or faces to which the face draft will be applied. If an incorrect face is selected, it can be deselected by holding down the CTRL key and clicking the face.

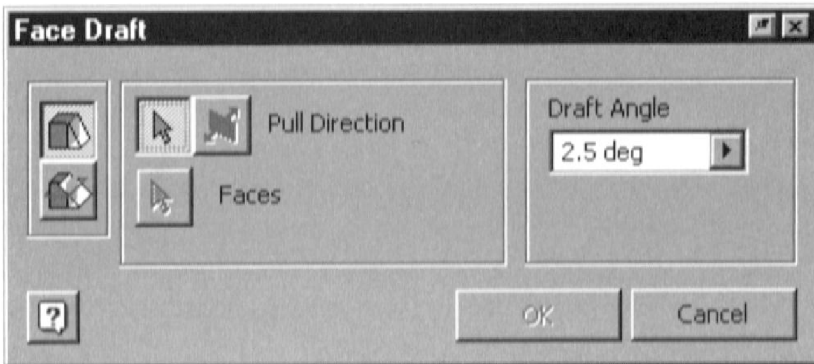


Figure 4-43

### DRAFT TYPE

Click the type of draft that you want to create. There are two types of draft available.

### FIXED EDGE

Allows draft to be created from an edge or series of contiguous edges.

### FIXED PLANE

Allows draft to be created from a selected plane. The plane that is selected is used to specify both the pull direction and the fixed plane.

### PULL DIRECTION AND FIXED PLANE

Click the direction that the mold will be pulled from the part. The angle will expand in this direction. When the Fixed Plane draft type is selected, click a planar face or work plane from which the faces will be drafted.

**Direction** Click the arrow and move the mouse around the part. A dashed line is displayed—this line points 90° from the highlighted face and shows the direction that the mold will be pulled. When the correct direction is displayed on the screen, left-click.

**Flip** Click the flip button to reverse the pull direction 180°.

### FACES

Click the face or faces to which the face draft will be applied. As you move the mouse over the face, a symbol with an arrow is displayed that shows how the draft will be applied from the nearest edge. As the mouse is moved to different edges, the arrow direction shows how the draft will be applied to that specific edge.

### DRAFT ANGLE

Enter a value for the draft angle, or select a previously used draft angle from the drop list.

# EXERCISE 4-5 Creating Face Drafts

In this exercise, you split the faces on a model of a wireless phone handset, and then apply draft angles to the faces. To navigate to the exercise in the *Electronic Student Workbook* do the following:

1 From the Main TOC page, click Chapter 4.
2 From the TOC for Chapter 4, click Creating Face Drafts.

The following image illustrates the opened exercise.

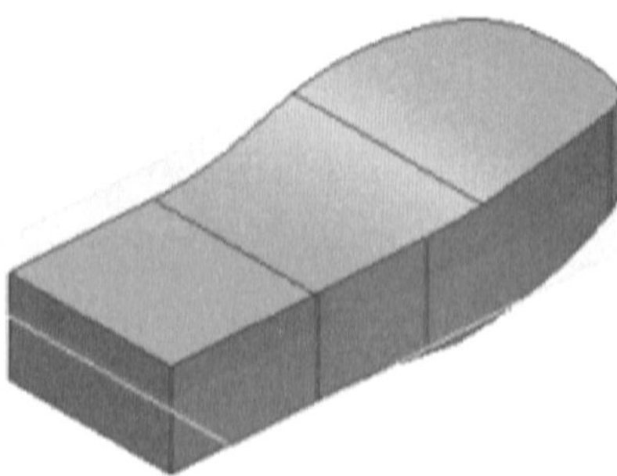

**Figure 4-44 Opened exercise**

The following image illustrates the completed exercise.

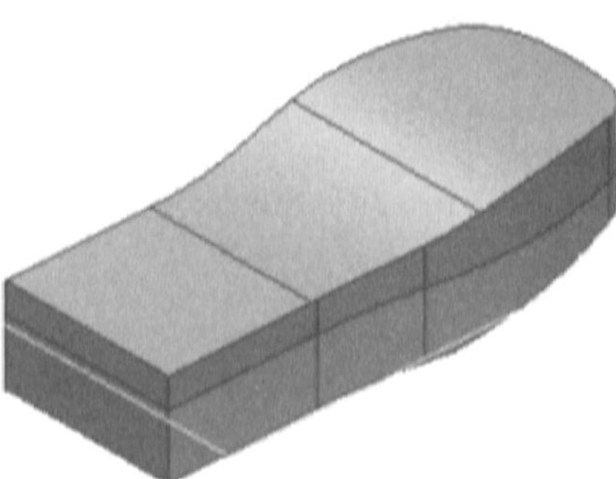

**Figure 4-45 Completed exercise**

## WORK FEATURES

When you create a parametric part, you define how the features of the part relate to one another so that a change in one feature results in appropriate changes in all related features. Work features are special construction features that are parametrically attached to parts. You typically use work features to help you position and define new features in your model. There are three types of work features: work planes, work axes, and work points.

Use work features in the following situations:

- To position a sketch for new features when a part face is not available.
- To establish an intermediate position that is required to define other work features. You can create a work plane at an angle to an existing face, for example, and then create another work plane at an offset value from that plane.
- To establish a plane or edge from which parametric dimensions and constraints can be placed.
- To provide an axis or point of rotation for revolved features and patterns.
- To provide an external feature termination plane off the part (i.e., a beveled extrusion edge) or an internal feature termination plane in cases where there are no existing surfaces.

### CREATING A WORK AXIS

A work axis is a feature that acts like a construction line. It is infinite in length and can be used to help create work planes. Work axes can also be used as an axis of rotation for polar arrays or can be constrained using assembly constraints. Their length always extends beyond the part—as the part changes size, the work axis also changes size. A work axis is parametrically tied to the part. As changes occur to the part, the work axis will maintain its relationship to the points or edge from which it was created. To create a work axis, use the Work Axis tool from the Part Features Panel Bar, as shown in the following image.

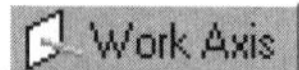


**Figure 4-46**

Use one of the following methods to create a work axis:

- Click a cylindrical face, and a work axis is created along the axis of revolution.
- Click two points on a part, and a work axis is created through both points.
- Click an edge on a part, and the work axis is created on top of the edge.
- Click a work point or sketch point and a plane or face, and a work axis is created that is normal to the selected plane or face and will pass through the point.
- Click two nonparallel planes, and a work axis is created at their intersection.

## CREATING WORK POINTS

A work point is a feature that can be created on the active part or in 3D space. A work point can be created any time a point is required. To create a work point, use the Work Point tool from the Part Features Panel Bar, as shown in the following image. You can also use the Work Plane or Work Axis tool—right-click and select Create Point when one of these work feature tools are active. If the tool is issued while creating a work feature, the tool will be exited after the work point is created.

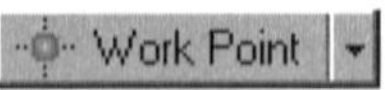


**Figure 4-47**

Use one of the following methods to create a work point:

- Click an endpoint or the midpoint of an edge, click an edge and axis, and a work point is created at the intersection (or theoretical intersection) of the two.
- Click an edge and plane, and a work point is created at the intersection (or theoretical intersection) of the two.
- Click three nonparallel faces or planes, and a work point is created at their intersection (or theoretical intersection).

### Grounded Work Point

You can create grounded work points that are positioned in 3D space. They are not associated with the part or any other work features, including the original locating geometry. When surrounding geometry is modified, the grounded work point remains in the specified location. To create a grounded work point, use the Grounded Work Point tool from the Part Features Panel Bar, located by clicking the arrow next to the Work Point tool, as shown in the following image.

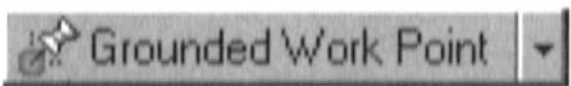


**Figure 4-48**

After clicking the Grounded Work Point tool, you select a vertex, midpoint, sketch point, or a work point on the model. When the vertex or point has been selected and a portion of the triad is clicked, the 3D Move/Rotate dialog box and a triad are displayed, as shown in the following image. The initial orientation of the triad matches the principle axes of the part.

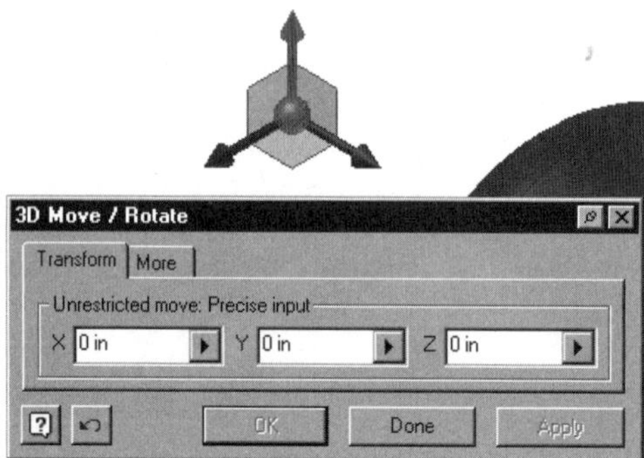


Figure 4-49

Enter values in the 3D Move/Rotate dialog box to precisely position the grounded work point, or select areas of the triad to move the triad and locate the grounded work point in the desired direction, as described below.

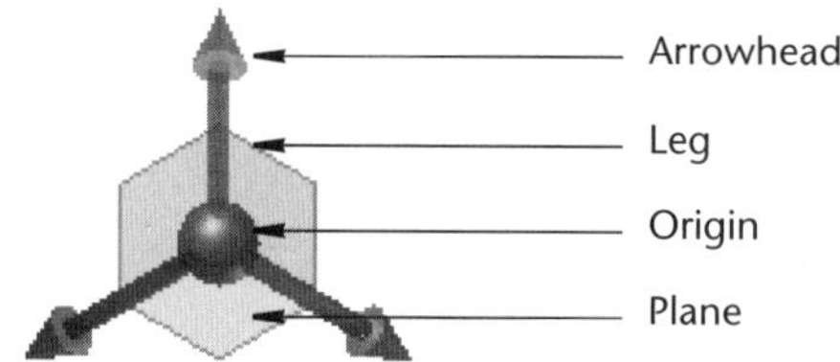


Figure 4-50

**Arrowheads** Select an arrowhead to specify a position along a particular axis.

**Legs** Select a leg to rotate about that axis.

**Origin** Click to move the triad freely in 3D space.

**Planes** Select a plane to move the triad in both axes of that plane.

Once the triad is properly positioned, click Apply or OK in the 3D Move/Rotate dialog box to create the grounded work point. A grounded work point can be identified in the Browser by the thumbtack icon that is placed on the work point, as shown in the following image.

Figure 4-51

# EXERCISE 4-6 Creating Work Axes

In this exercise, you create a work axis to position a circular pattern. To navigate to the exercise in the *Electronic Student Workbook* do the following:

1. From the Main TOC page, click Chapter 4.
2. From the TOC for Chapter 4, click Creating Work Axes.

The following image illustrates the opened exercise.

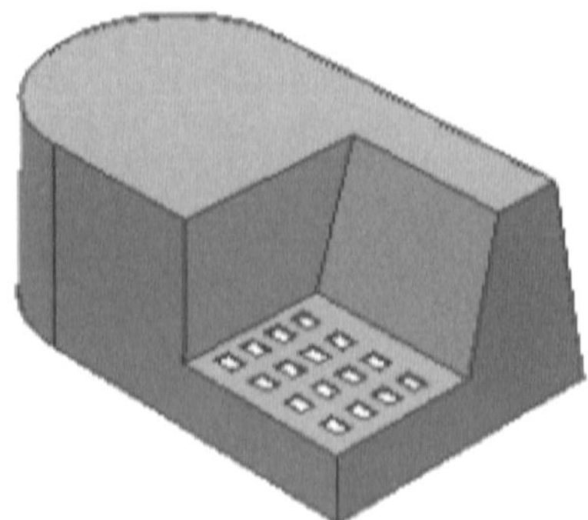

**Figure 4-52 Opened exercise**

The following image illustrates the completed exercise.

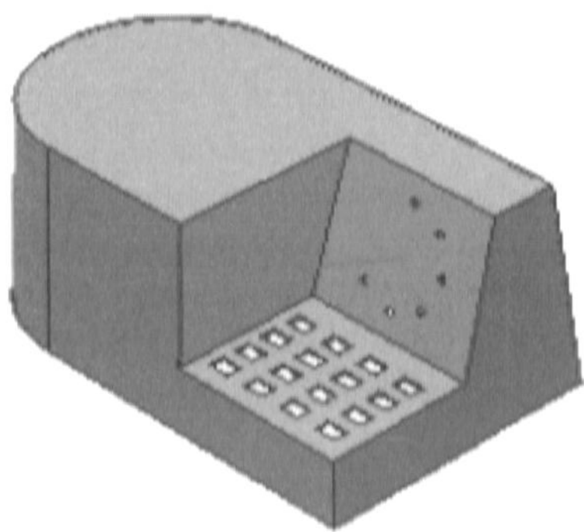

**Figure 4-53 Completed exercise**

## CREATING WORK PLANES

Before introducing work planes, it is important that you understand when you need to create a work plane. You can use a work plane when you need to sketch a feature and there is no planar face at the location you need to sketch. If you want a feature to terminate at a plane, and there is no existing plane, you will need to create a work plane. If you want to apply an assembly constraint to a plane on a part and there is no existing plane, you will need to create a work plane. If a plane exists in any of these scenarios, you should use it and not create a work plane. A new sketch can be created on a work plane.

A work plane is a feature that looks like a rectangular plane. It is parametrically tied to the part and will always be larger than the part. If the part moves, the work plane will also move. If a work plane is tangent to the outside face of a 1" diameter cylinder and the cylinder diameter changes to 2", for example, the work plane moves with the outside face of the cylinder. You can create as many work planes on a part as needed, and any work plane can be used to create a new sketch. A work plane is a feature and must be edited and deleted like any other feature.

Before creating a work plane, ask yourself where this work plane needs to exist and what you know about its location. You might want a plane to be tangent to a given face and parallel to another plane, for example, or to go through the center of two arcs. Once you know what you want, select the appropriate options and create a work plane. There are times when you may need to create an intermediate (construction) work plane before creating the final work plane. You may need to create a work plane, for example, that is at 30° and tangent to a cylindrical face. You should first create a work plane that is at a 30° angle and located at the center of the cylinder, and then create a work plane parallel to the angled work plane that is also tangent to the cylinder.

To create a work plane, click the Work Plane tool from the Part Features Panel Bar as shown in the following image. Depending upon what, where, and how you select, a work plane will then be created. Work planes can also be based on the default reference planes that exist in every part. The default planes have their visibility initially turned off, but can be turned on by expanding the Origin folder in the Browser and then right-clicking on a plane or planes and clicking Visibility from the menu. These default planes can be used to create a new sketch or to create other work planes.

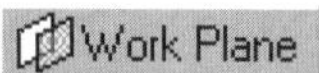


**Figure 4-54**

Use one of the following methods to create a work plane:

- Click three points.
- Click a plane and a point.

- Click a plane and an edge or axis.
- Click two edges, two axes, or an edge and axis.
- Create a work plane that is tangent to a face. Click a plane and face, and the resulting work plane is parallel to the selected plane and tangent to the selected face.
- To create an angled plane, click a plane and an edge—a dialog box is displayed for you to enter the angle.
- To create an offset work plane from a surface, click a plane and then drag the new work plane to a selected location. While dragging the work plane, an Offset dialog box is displayed that displays the offset distance. Click a point, then enter a value for the offset distance, and click the check mark in the dialog box or press the ENTER key on the keyboard.
- Click parallel planes to create a work plane at the midpoint of the two selected planes.

When creating a work plane, and more than one solution is possible, the Select Other tool is displayed. Click the forward or reverse arrows from the Select Other tool until the correct solution is displayed. Click the check mark in the selection box. If a midpoint on an edge is clicked, the resulting work plane will be linked to the midpoint. If the selected edge's length changes where the midpoint was selected, the location of the work plane will adjust to the new midpoint.

**NOTE** The order in which points or planes are selected is irrelevant.

Use the Show Me animations located in the Visual Syllabus (shown in the following image) to view animations that display how to create certain types of work planes.

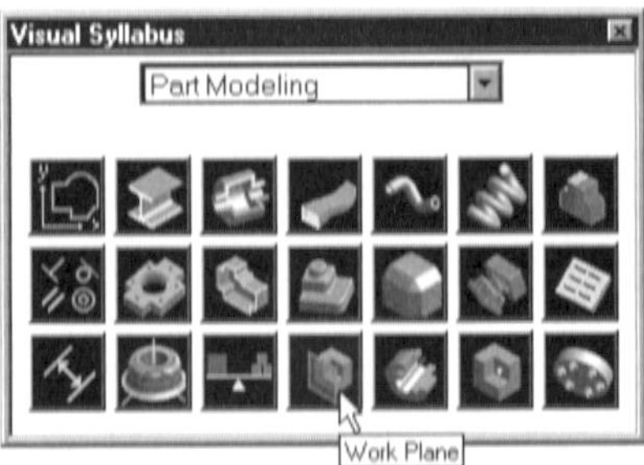


**Figure 4-55**

## Types of Work Planes

The following is a list of work planes that are used in the modeling process:

- Angled work plane
- Edge and face normal work plane

- Edge and tangent work plane
- Offset work plane
- Point and face normal work plane
- Point and face parallel work plane
- Sketch geometry work plane
- Tangent and face parallel work plane
- Tangent work plane
- 3-point work plane
- 2-edge or 2-axis work plane
- Through line end point, perpendicular to line
- Normal to arc at a point on the arc
- Normal to a spline or work curve at a point on the spline or curve
- Midway between two parallel planes

## FEATURE VISIBILITY

The visibility of the origin planes, origin axes, origin point or user work planes, user work axes, user work points, and sketches can be controlled by either right-clicking on them in the graphics window or on their name in the Browser, and clicking Visibility from the menu. The visibility can also be controlled for ALL origin planes, origin axes, origin point or user work planes, user work axes, user work points, and sketches from the Object Visibility option in the View menu. Visibility can be checked to turn visibility on or cleared to turn it off, as shown in the following image.

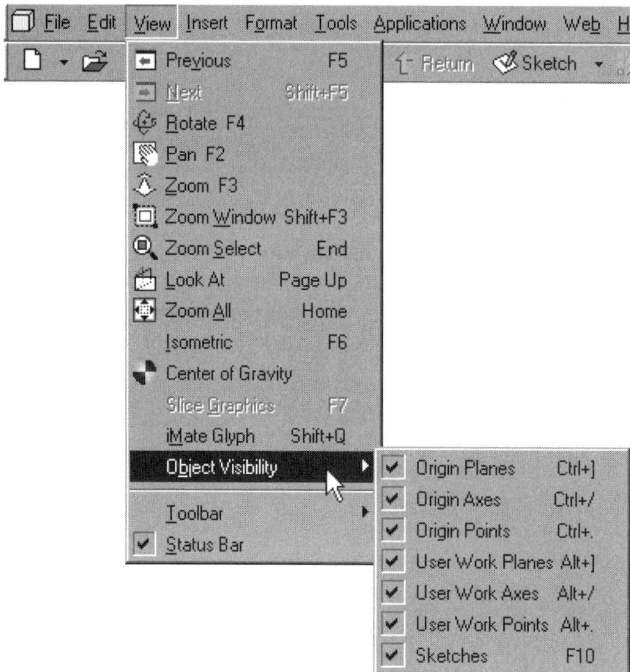


Figure 4-56

# EXERCISE 4-7 Creating Work Planes

In this exercise, you create work planes in order to create a boss on a cylinder head and a slot in a shaft. To navigate to the exercise in the *Electronic Student Workbook* do the following:

1 From the Main TOC page, click Chapter 4.
2 From the TOC for Chapter 4, click Creating Work Planes.

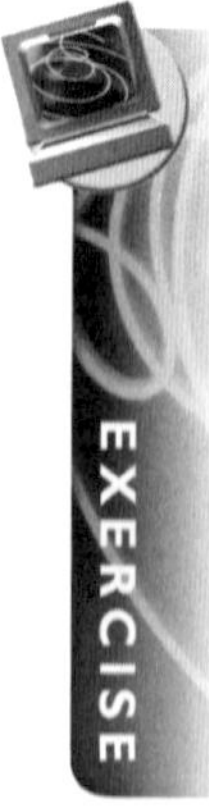


The following image illustrates the opened exercise.

**Figure 4-57 Opened exercise**

The following image illustrates the completed exercise.

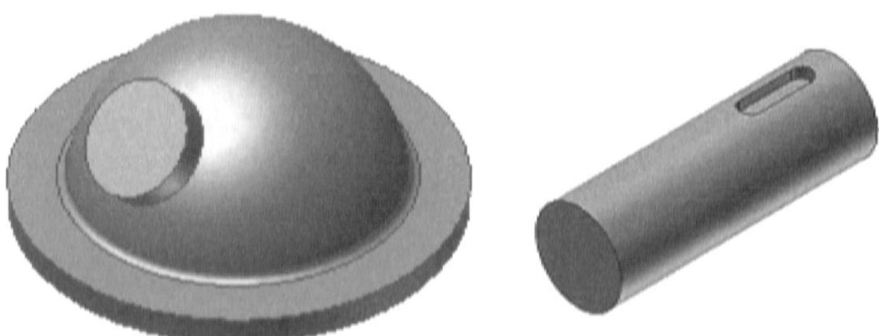

**Figure 4-58 Completed exercise**

## PATTERNS

Once a feature is created, you will find many instances in which you need to duplicate the feature multiple times with a set distance or angle between them. You can use the pattern tools to expedite this process.

Figure 4-59

There are two types of available patterns: rectangular and circular. The pattern is held together as a single feature in the Browser, but the individual occurrences are listed under the pattern feature. The entire pattern or individual occurrences can be suppressed. Both rectangular and circular patterns have a child relationship to the parent feature(s) that was patterned. If the size of the parent feature changes, all of the child features will also change. If a hole is patterned, and the parent hole type changes, the child holes also change. Because a pattern is a feature, it can be edited like any other feature. The base part or feature, as well as patterns, can be patterned. A rectangular pattern repeats the selected feature(s) along the direction set by two edges on the part (these edges do not need to be horizontal or vertical as shown in the following image) and a circular pattern repeats the feature(s) around an axis or edge.

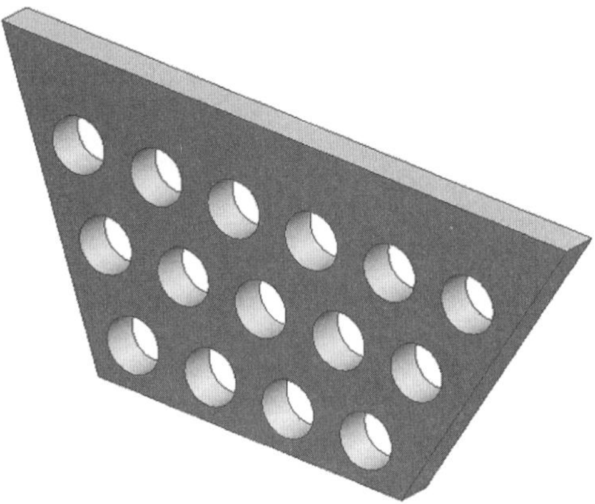

Figure 4-60

When creating a rectangular pattern, you define two directions by clicking an edge that defines each alignment. Before creating a circular pattern, you must have a work axis, a part edge, or a circular face about which the feature will be rotated. After clicking the Rectangular Pattern or Circular Pattern tool from the Part Features Panel Bar, as shown in the following images, a Pattern dialog box is displayed. Click a feature to pattern and then enter the values, and click the edges and axis as needed. If you move the dialog box away from the part, a preview image of the pattern is displayed.

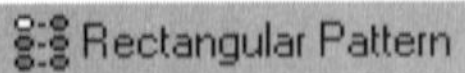


**Figure 4-61**

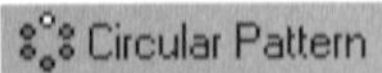


**Figure 4-62**

Below is a description of the options available for rectangular and circular patterns.

## RECTANGULAR PATTERNS

The options in the Rectangular Pattern dialog box, shown in the following image, are explained in this section.

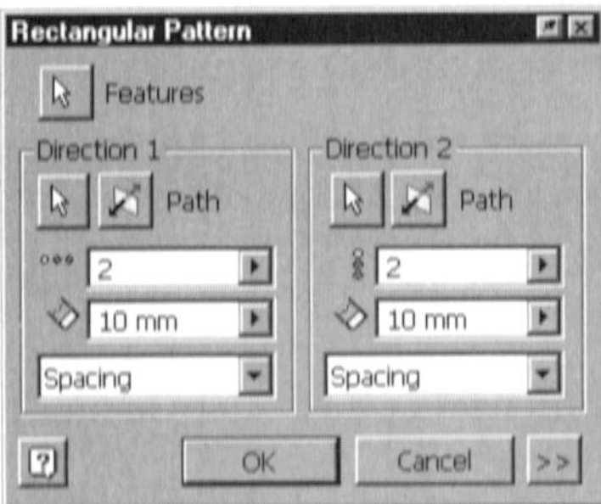


**Figure 4-63**

**Features** Click this button and then click a feature or features to be patterned from either the graphics window or the Browser. Features can be removed from the selection set by holding down the CTRL key and clicking them.

### Direction 1

In Direction 1, you define the first direction for the alignment of the pattern—it can be an edge, an axis, or a path.

**Path** Click this button and then click an edge that defines the alignment along which the feature will be patterned.

**Flip** If the preview image shows the pattern going in the wrong direction, click this button to reverse its direction.

**Column Count** Enter a value or click the arrow to choose a previously used value that represents the number of times the feature(s) will be duplicated.

**Column Spacing** Enter a value or click the arrow to choose a previously used value that represents the distance between the patterned features.

**Distance** Creates the occurrences of the pattern at an equal distance within the specified range.

**Curve Length** Create the occurrences of the pattern at an equal distance along the length of the selected curve.

## Direction 2

In Direction 2, you will define the second direction for the alignment of the pattern—it can be an edge, an axis, or path, but cannot be parallel to Direction 1.

**Path** Click this button and then click an edge that defines the alignment along which the feature will be patterned.

**Flip** If the preview image shows the pattern going in the wrong direction, click this button to reverse its direction.

**Row Count** Enter a value or click the arrow to choose a previously used value that represents the number of times the feature(s) will be duplicated.

**Row Spacing** Enter a value or click the arrow to choose a previously used value that represents the distance between the patterned features.

**Distance** Creates the occurrences of the pattern at an equal distance within the specified range.

**Curve Length** Create the occurrences of the pattern at an equal distance along the length of the selected curve.

Options for the Start point of the direction, Termination Method and Orientation Method, as shown in the following image, are available by clicking the More (>>) button located in the bottom right corner of the Rectangular Pattern dialog box.

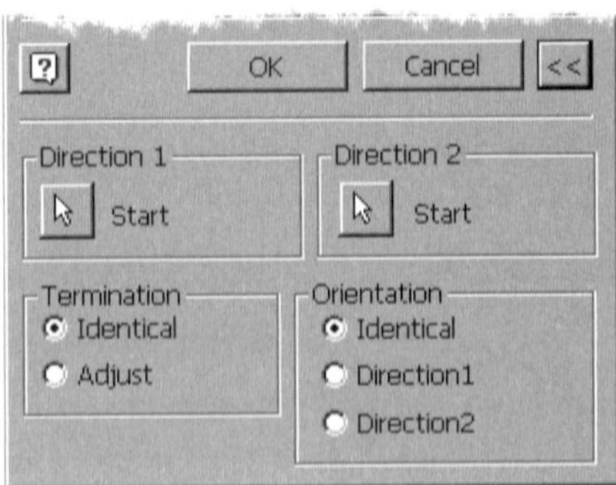


Figure 4-64

### Start

Specifies where the start point for the first occurrence of the pattern will be placed. The pattern can begin at any selectable point on the part. You can select the start point for both Direction 1 and Direction 2.

### Termination

**Identical** When Identical is selected, all of the occurrences in the pattern will use the same termination as the parent feature(s). This is the default option.

**Adjust** When Adjust is selected, the termination of each occurrence is calculated individually. Since each occurrence is calculated separately, the processing time can be increased. This option must be used if a parent feature terminates to a face or plane.

### Orientation

**Identical** When Identical is selected, all of the occurrences in the pattern are oriented the same as the parent feature(s). This is the default option.

**Direction1** When selected, the position of the patterned features is controlled by the selected direction. Each occurrence of the pattern is rotated to maintain proper orientation with the 2D tangent vector of the path.

**Direction2** When selected, the position of the patterned features is controlled by the selected direction. Each occurrence of the pattern is rotated to maintain proper orientation with the 2D tangent vector of the path.

## CIRCULAR PATTERNS

The options in the Circular Pattern dialog box, shown in the following image, are explained in this section.

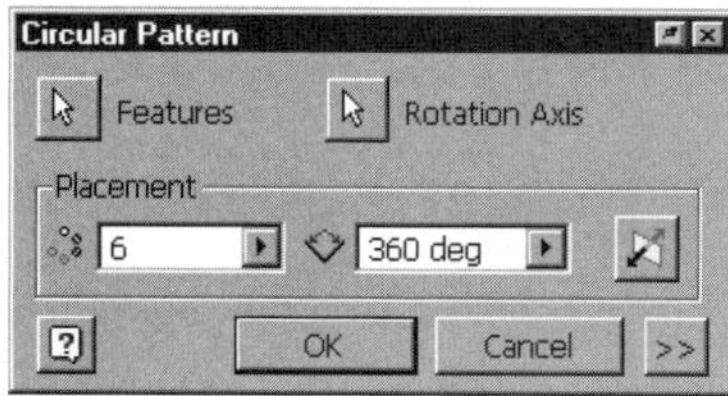


Figure 4-65

**Features** Click this button and then click a feature or features to be patterned. Features can be removed from the selection set by holding down the CTRL key and clicking them.

**Rotation Axis** Click the button and then click an edge, axis, or circular face (center) that defines the axis about which the feature(s) will be rotated.

## Placement

**Occurrence Count** Enter a value or click the arrow to choose a previously used value that represents the number of times the feature(s) will be duplicated.

**Occurrence Angle** Enter a value or click the arrow to choose a previously used value that represents the angle that will be used to calculate the spacing of the patterned features.

**Flip** If the preview image shows the pattern going in the wrong direction, click this button to reverse its direction.

By clicking the More (>>) button located in the bottom right corner of the Circular Pattern dialog box, you can access options for the Creation Method and Positioning Method of the feature, as shown in the following image.

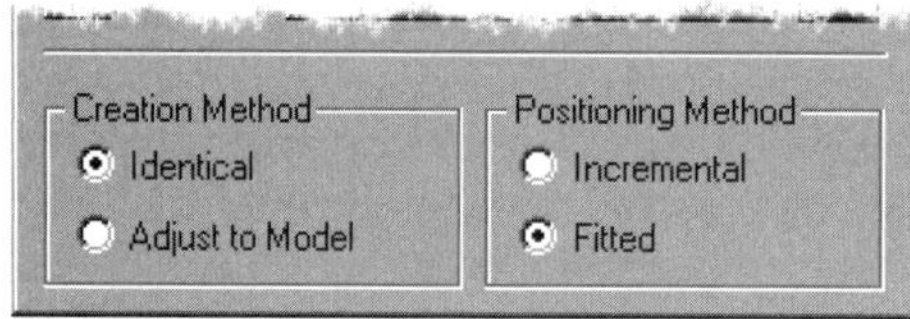


Figure 4-66

## Creation Method

**Identical** When Identical is selected, all of the occurrences in the pattern use the same termination as the parent feature(s). This is the default option.

**Adjust to Model** When Adjust to Model is selected, each occurrence termination is calculated individually. Becasue each occurrence is calculated separately, the processing time can be increased. This option must be used if a parent feature terminates to a face or plane.

### Positioning Method

**Incremental** When Incremental is selected, each occurrence is separated by the number of degrees specified in Angle in the dialog box.

**Fitted** When Fitted is selected, each occurrence is evenly spaced within the angle specified in Angle in the dialog box.

**TIP** A work axis, an edge, or a cylindrical face, about which the feature will be rotated, must exist before creating a circular pattern.

## LINEAR PATTERNS—PATTERN ALONG A PATH

There are many modeling cases in which you need to create a pattern that follows a path. A path can be defined by either a complete or partial ellipse, an open or closed spline, or a series of curves (including lines, arcs, splines, etc.).

To pattern along a path, use the Path button and the options described above for rectangular patterns. The path that is used can be either 2D or 3D.

To create a pattern along 3D paths, follow these steps:

1. Create a 3D sketch.
2. Create a path (include model edges and work curves).
3. Create the pattern.

# EXERCISE 4-8
# Creating Rectangular Patterns

In this exercise, you add a rectangular pattern of holes to a plastic cover plate. To navigate to the exercise in the *Electronic Student Workbook* do the following:

1 From the Main TOC page, click Chapter 4.
2 From the TOC for Chapter 4, click Creating Rectangular Patterns.

The following image illustrates the opened exercise.

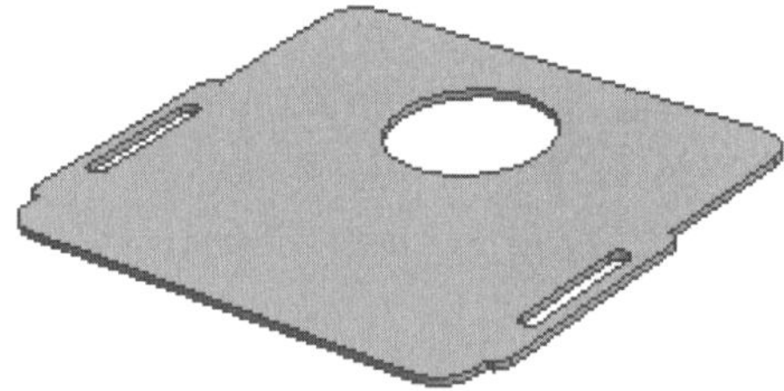

**Figure 4-67 Opened exercise**

The following image illustrates the completed exercise.

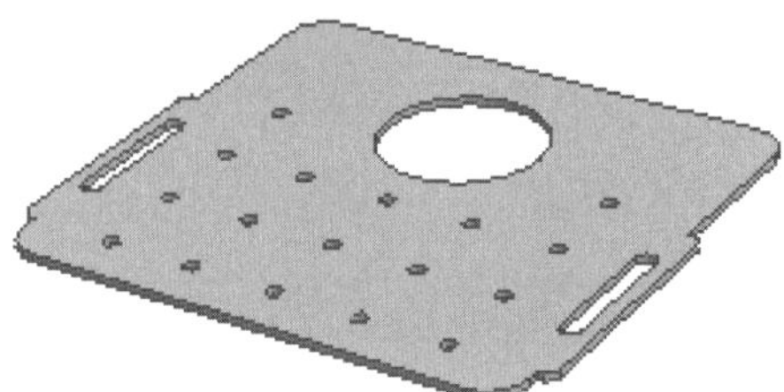

**Figure 4-68 Completed exercise**

# EXERCISE 4-9 Creating Circular Patterns

In this exercise, you create a circular pattern of eight counterbored holes on a flange. To navigate to the exercise in the *Electronic Student Workbook* do the following:

1 From the Main TOC page, click Chapter 4.
2 From the TOC for Chapter 4, click Creating Circular Patterns.

The following image illustrates the opened exercise.

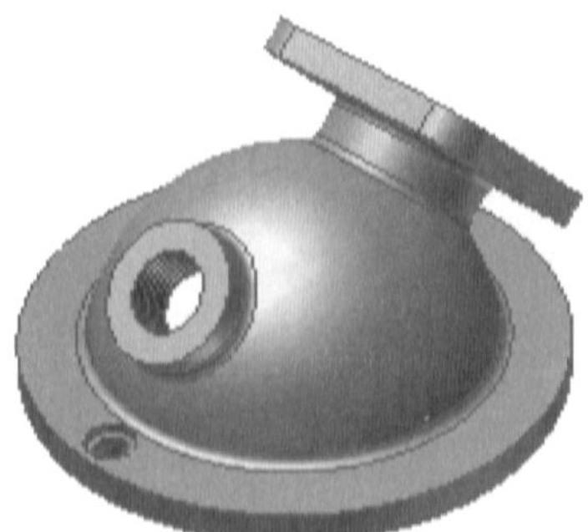

**Figure 4-69 Opened exercise**

The following image illustrates the completed exercise.

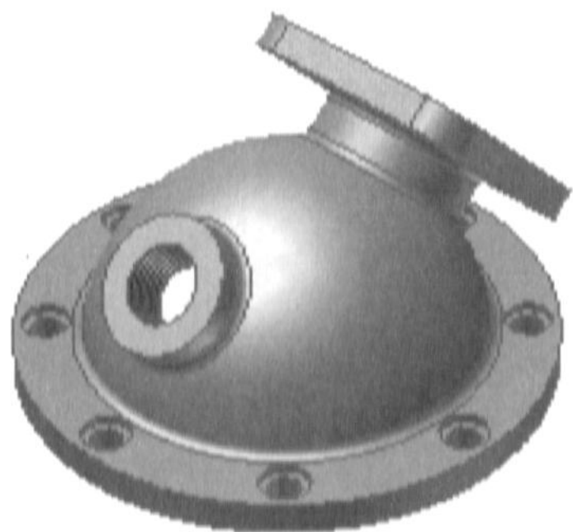

**Figure 4-70 Completed exercise**

# EXERCISE 4-10 Creating Path Patterns

In this exercise, you pattern a boss and hole along a non-linear path. To navigate to the exercise in the *Electronic Student Workbook* do the following:

1 From the Main TOC page, click Chapter 4.

2 From the TOC for Chapter 4, click Creating Patterns Along a Path.

The following image illustrates the opened exercise.

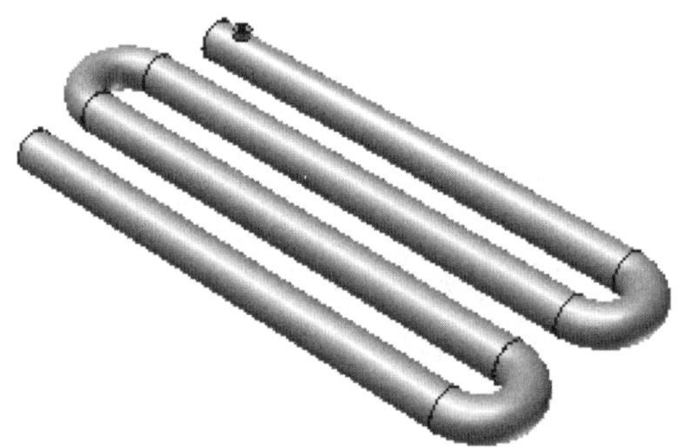

**Figure 4-71 Opened exercise**

The following image illustrates the completed exercise.

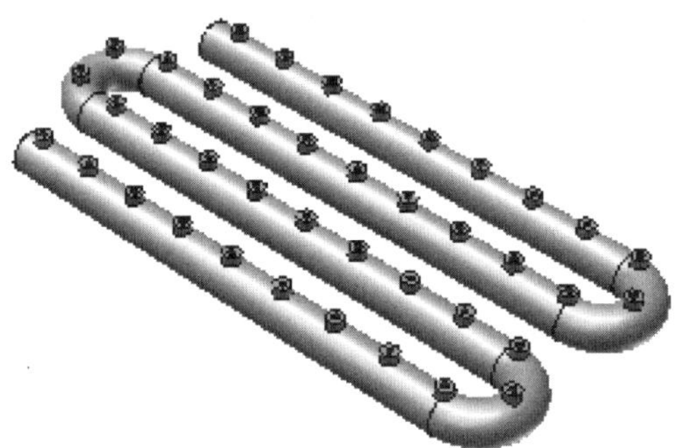

**Figure 4-72 Completed exercise**

## CHAPTER SUMMARY

| To | Do This | Tool |
|---|---|---|
| Create a hole feature | Click the Point, Hole Center tool on the 2D Sketch toolbar to sketch a hole center.<br>Click the Hole tool on the Part Features toolbar or from the Panel Bar. |   |
| Create a fillet feature | Click the Fillet tool on the Part Features toolbar or from the Panel Bar. |  |
| Create a chamfer feature | Click the Chamfer tool on the Part Features toolbar or from the Panel Bar. |  |
| Create a shell feature | Click the Shell tool on the Part Features toolbar or from the Panel Bar. |  |
| Create a rectangular pattern | Click the Rectangular Pattern tool on the Part Features toolbar or from the Panel Bar. |  |
| Create a circular pattern | Click the Circular Pattern tool on the Part Features toolbar or from the Panel Bar. |  |
| Create a thread feature | Click the Thread tool on the Part Features toolbar or from the Panel Bar. |  |
| Create face draft | Click the Face Draft tool on the Part Features toolbar or from the Panel Bar. |  |
| Create a work plane | Click the Work Plane tool on the Part Features toolbar or Panel Bar. |  |
| Create a work axis | Click the Work Axis tool on the Part Features toolbar or Panel Bar. |  |
| Create a work point | Click the Work Point tool on the Part Features toolbar or Panel Bar. |  |
| Create a grounded work point | Click the Grounded Work Point tool on the Part Features toolbar or Panel Bar. |  |

# Applying Your Skills

### SKILL EXERCISE 4-1

In this exercise, you create a drain plate cover. To navigate to the exercise in the *Electronic Student Workbook* do the following:

1. From the Main TOC page, click Chapter 4.
2. From the TOC for Chapter 4, click Exercise 1 under Applying Your Skills.

The following image illustrates the completed exercise.

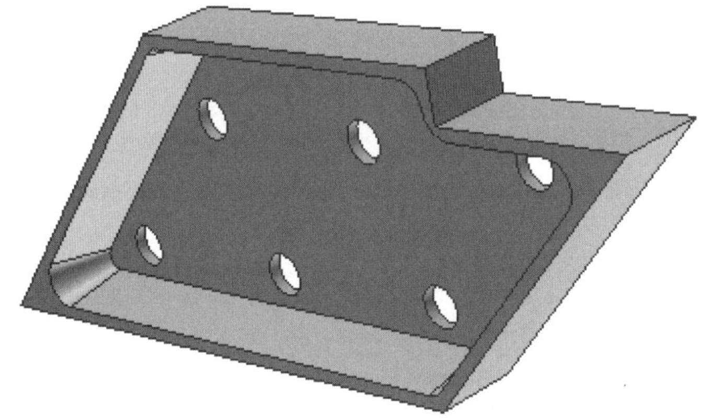

**Figure 4-73 Completed exercise**

### SKILL EXERCISE 4-2

In this exercise, you create a connector part. To navigate to the exercise in the *Electronic Student Workbook* do the following:

1. From the Main TOC page, click Chapter 4.
2. From the TOC for Chapter 4, click Exercise 2 under Applying Your Skills.

The following image illustrates the completed exercise.

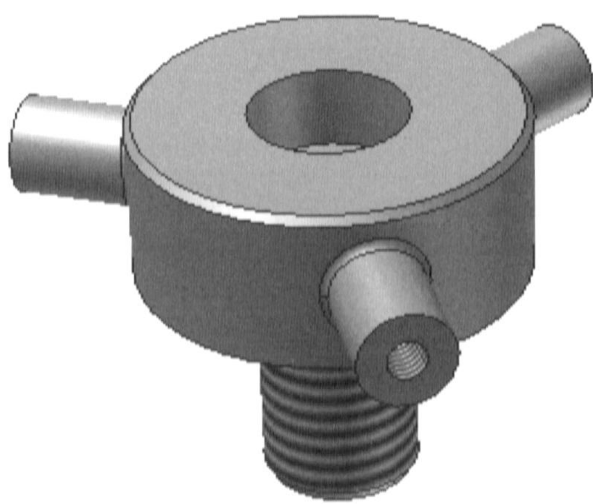

**Figure 4-74 Completed exercise**

## CHECKING YOUR SKILLS

Use these questions to test your knowledge of the material covered in this chapter.

1 **True___ False___** When creating a fillet feature that has more than one selection set, each selection set is displayed as an individual feature in the Browser.

2 In regards to creating a fillet feature, what is a smooth radius transition?

________________________________________

________________________________________

3 **True___ False___** When creating a fillet feature with the All Fillets option, material is removed from all concave edges.

4 **True___ False___** When creating a chamfer feature with the Distance and Angle option, only one edge can be chamfered at a time.

5 **True___ False___** When creating a hole feature, you do not need to have an active sketch.

6 What is a Hole Center used for?

________________________________________

________________________________________

7 **True___ False___** Thread features are graphically represented on the part and will be annotated correctly when drawing views are generated.

8 **True___ False___** A part may contain only one shell feature.

9 When creating a face draft feature, what is the definition of pull direction?

________________________________________

________________________________________

10 **True___ False___** The only method to create a work axis is by clicking a cylindrical face.

11 **True___ False___** Every new sketch needs to be derived from a work plane feature.

**12** Explain the steps to create an offset work plane.

________________________________________________

________________________________________________

**13 True___ False___** Work planes cannot be based on the default work planes.

**14 True___ False___** When creating a rectangular pattern, the directions that the features are duplicated along must be horizontal or vertical.

**15 True___ False___** When creating a circular pattern, only a work axis can be used as the axis of rotation.

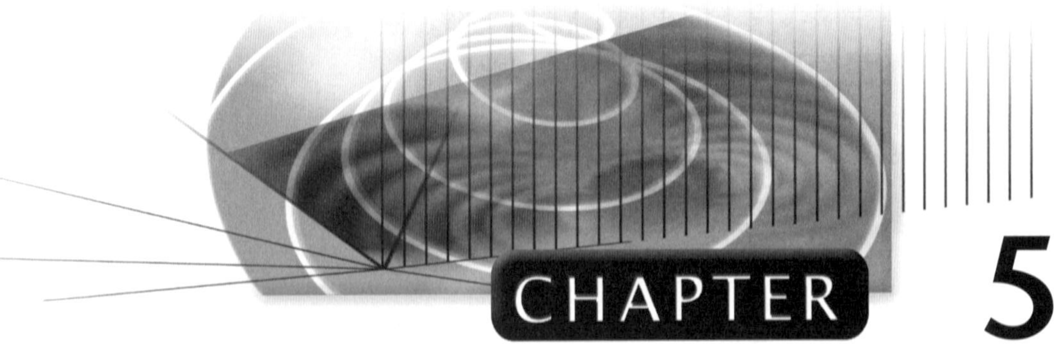


# Creating and Editing Drawing Views

After creating a part or assembly, the next step is to create two-dimensional (2D) drawing views that represent that part or assembly. To create drawing views, you start a new drawing file, select a three-dimensional (3D) part or assembly on which to base the drawing views, insert or create a drawing sheet with a border and title block, project orthographic views from the part or assembly, and then add annotations to the views. Drawing views can be created at any point after a part exists. The part does not need to be complete because the part and drawing views are associative in both directions (bidirectional). This means that if the part changes, the drawing views will automatically be updated when you return to the drawing views. If a parametric dimension changes in a drawing view, the part will get updated before the drawing views get updated. This chapter will guide you through the steps for setting up drawing standards and creating borders and title blocks, drawing views of a single part, cleaning up the dimensions, and adding annotations.

## CHAPTER OBJECTIVES

**After completing this chapter, you will be able to**

- Understand drawing options
- Create and edit drawing borders and title blocks
- Set up drawing standards
- Create base and projected drawing views from a part
- Create auxiliary, section, and broken views
- Edit the properties and location of drawing views
- Retrieve model dimensions to use in drawing views
- Edit, move, and hide dimensions
- Select drawing objects using a window or a crossing window
- Measure distances and angles of drawing objects
- Add automated centerlines
- Add reference dimensions
- Add annotations such as GD&T, surface finish symbols, weld symbols, and datum identifiers

## DRAWING OPTIONS, CREATING A DRAWING, AND DRAWING TOOLS

In this first section, you will learn about the drawing options that are available, learn how to create a new drawing, and learn about the tools that are used to create drawing views.

### DRAWING OPTIONS

Before drawing views are created, the drawing options should be set to your preferences. To set the drawing options, click Application Options from the Tools menu. The Options dialog box will appear. Click the Drawing tab and then your screen should resemble the following image. Make any changes to the options before creating the drawing views or the changes may not affect drawing views that have already been created. The following sections are descriptions of the drawing options.

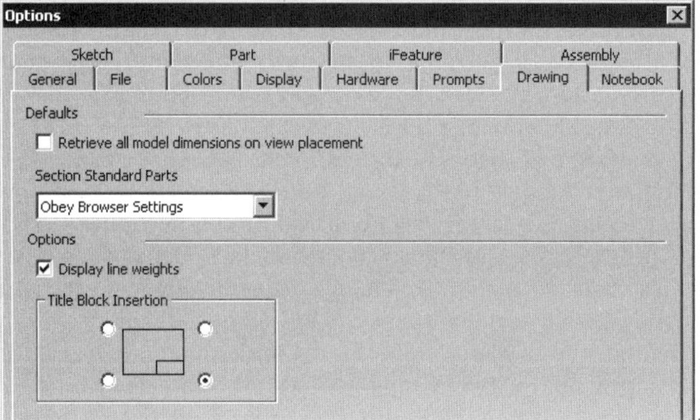


**Figure 5-1**

#### Retrieve All Model Dimensions on View Placement

When this option is checked, applicable model dimensions are added to drawing views when they are placed. If the box is not checked, no model dimensions will be placed automatically. You can override this by manually selecting All Model Dimensions in the Drawing View dialog box when creating views.

#### Section Standard Parts

This area of the Drawing tab allows you to control whether standard parts such as nuts, bolts, and washers are sectioned. Three options are available in this area: Always, Never, and Obey Browser Settings.

#### Display Line Weights

When checked, visible lines will be displayed in drawings with the line weights defined in the active drafting standard. If the box is not checked, all visible lines are displayed with the same weight.

**NOTE** This setting does not affect line weights when the views are printed.

### Title Block Insertion

Selects the location where the title block will be placed.

## CREATING A DRAWING

The first step in creating a drawing from an existing part or assembly is to create a new drawing IDW file by either clicking the New icon from the What To Do section of the Getting Started page and then clicking the *Standard.idw* icon from the Default template tab (as shown in the following image), or clicking New from the File menu, or clicking the down arrow on the New icon from the left side of the Standard toolbar and clicking Drawing.

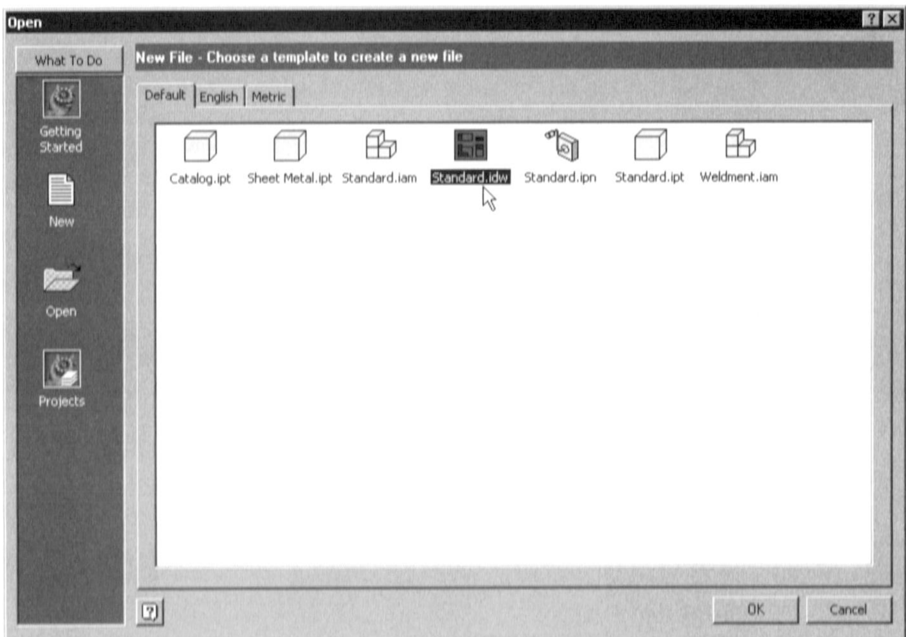


Figure 5-2

## DRAWING TOOLS

After starting a new drawing, Autodesk Inventor's screen will change to reflect the new drawing environment. There are three toolbars available in the Panel Bar that you will use to create drawings: Drawing Views, Sketch, and Drawing Annotation. By default, when a new drawing is created, the Panel Bar will show the Drawing Views toolbar. These tools allow you to create drawing views and sheets. The Sketch tools enable you to add geometry and dimensions to draft views. The Drawing Annotation tools enable you to add annotations to existing drawing views. The three sets of tools will be explained throughout this chapter.

## DRAWING SHEET PREPARATION

When you start a drawing file, the program displays a default drawing sheet with a default title block and border. The template IDW file that is selected determines the default drawing sheet, title block, and border. The drawing sheet represents a blank piece of paper on which the border, title block, and drawing views are placed. There is no limit to the number of sheets that can exist in the same drawing, but you must have at least one drawing sheet. To create a new sheet, click the New Sheet tool from the Drawing Views Panel Bar as shown in the following image or right-click in the Browser and select New Sheet from the menu.

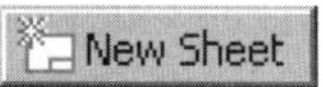


**Figure 5-3**

A new blank sheet will appear in the Browser and the new sheet will be shown in the graphics window without a border or title block. The next section discusses how to add a border and title block. The sheet can be renamed by slowly double-clicking on its name in the Browser and then entering a new name, or right-clicking on the sheet in the Browser and clicking Edit Sheet from the menu. A new name can then be entered in the Edit Sheet dialog box. To create a sheet of a different size, you can click Sheet Formats from Drawing Resources in the Browser, as shown in the following image.

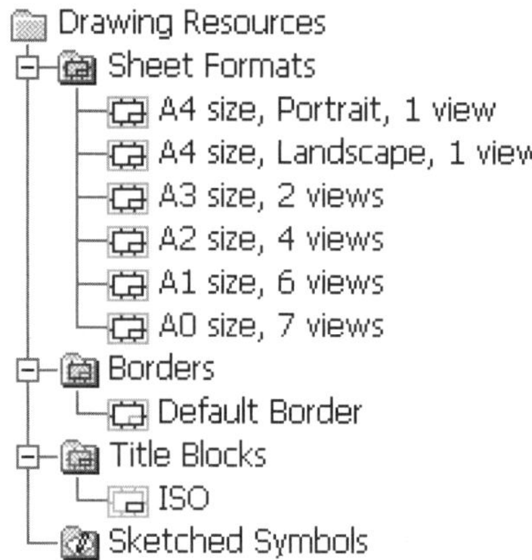


**Figure 5-4**

If a sheet is selected from the Sheet Formats list, predetermined drawing views will be created. If the necessary sheet size is not on the list, right-click on the sheet name in the Browser and click Edit Sheet from the menu. Then click in the Size in the Edit Sheet dialog box, and select a size from the list as shown in the following image. To use your own values, select Custom Size from the list and then enter values for height and width.

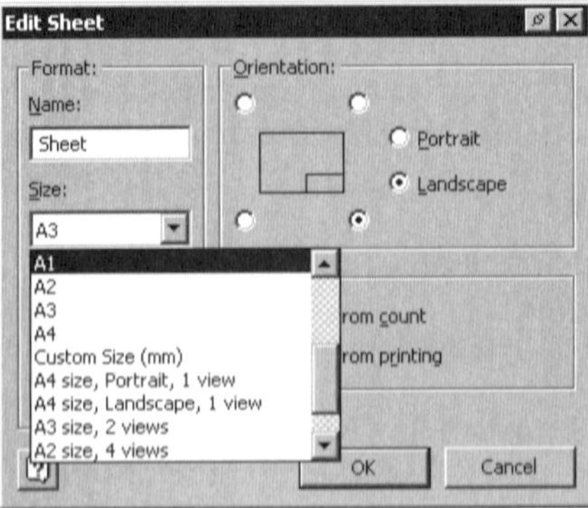


Figure 5-5

**NOTE** The sheet size is inserted full scale (1=1) and should be plotted at 1=1. The drawing views will be scaled to fit the sheet size.

## BORDER CREATION

The four lines and zone labels that surround the edges of the sheet make up the border. To insert the default border in a blank drawing sheet, expand Drawing Resources in the Browser and double-click on Default Border. To insert a customized border with specified margins, expand Drawing Resources in the Browser, right-click on Default Border, and click Insert Drawing Border. The Default Drawing Border Parameters dialog box will appear, similar to the following image. Select the options and enter the values that you want and then click the OK button. The border will then appear on the sheet. The sheet margins will be offset from the edges of the sheet. To delete an existing border, expand the sheet name in the Browser, right-click on the border's name, and click Delete from the menu.

Figure 5-6

## TITLE BLOCK CREATION

To add a title block to the drawing sheet, you can either insert a default title block or you can construct a customized title block and insert it into the drawing sheet.

### DEFAULT TITLE BLOCK

To insert a default title block, follow these guidelines:

- Make the sheet active, then place the title block by double-clicking on its name in the Browser.
- Insert the title block by expanding Drawing Resources>Title Blocks in the Browser and then either double-clicking on the title block's name, as shown in the following image, or right-clicking on the title block's name and selecting Insert from the menu.

**Figure 5-7**

The title block will be inserted on the current sheet. By default, the information in the title block will automatically be filled in from the properties of the drawing file. The file properties for the drawing can be filled in by either right-clicking on the top-level drawing name in the Browser and clicking Properties, or by clicking iProperties from the File menu. The Drawing Properties dialog box will appear and you can fill in the information as needed. The following image shows the Drawing Properties dialog box and the result in the title block.

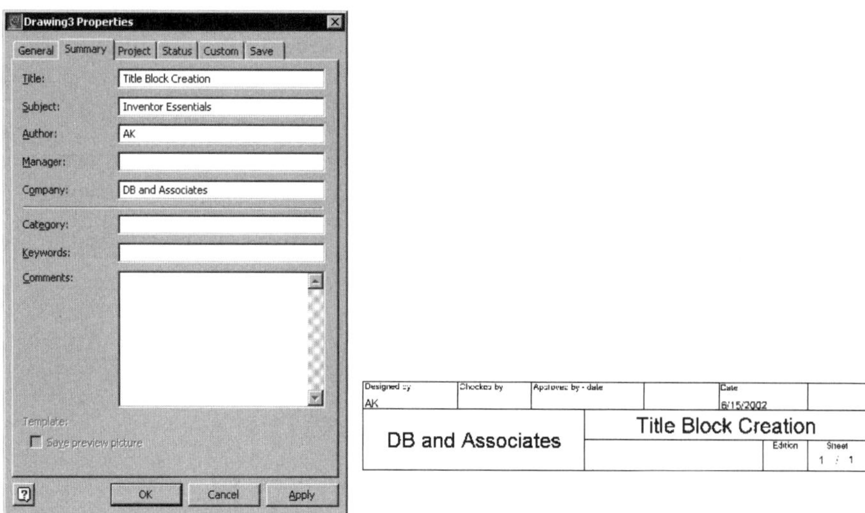


**Figure 5-8**

## CREATING A NEW TITLE BLOCK

In this section, you will be introduced to the methods of creating a customized title block. The book, however, will utilize the default title blocks. If you want to create a customized title block, you can either edit an existing title block or create one from scratch. You also have the ability to bring in AutoCAD data.

To edit an existing title block, right-click on the title block name in the Browser under Drawing Resources>Title Block and click Edit from the menu. The title block will appear with all of its attributes shown. Edit as needed and when complete, right-click and click Save Title Block and you will have the option to save it as another name. If the title block is saved as another name, the new title block's name will appear in the Browser under Drawing Resources>Title Block. The new title block will not be available to new drawings unless it is saved in a template file.

If you want to create a new title block from scratch, right-click on Title Blocks in the Browser under Drawing Resources, and click Define New Title Block. Next, using the tools from the Sketch toolbar, create the title block. Create the title block using the sketch tools in the same fashion as you would when creating a sketch. To create text that is populated from the file properties, use the Property Field tool. When you are done creating the title block, right-click and click Save Title Block from the menu. The title block will appear in the Browser under Drawing Resources>Title Block. The new title block will not be available to new drawings unless it is saved in a template file. See Help to learn how to customize a title block's properties.

# EXERCISE 5-1
## Sheets, Borders, and Title Blocks

In this exercise, you set up an A2 drawing sheet with a border, title block, and a new drawing standard. To navigate to the exercise in the *Electronic Student Workbook* do the following:

1. From the Main TOC page, click Chapter 5.
2. From the TOC for Chapter 5, click Sheets, Borders and Title Blocks.

The following image illustrates the completed exercise.

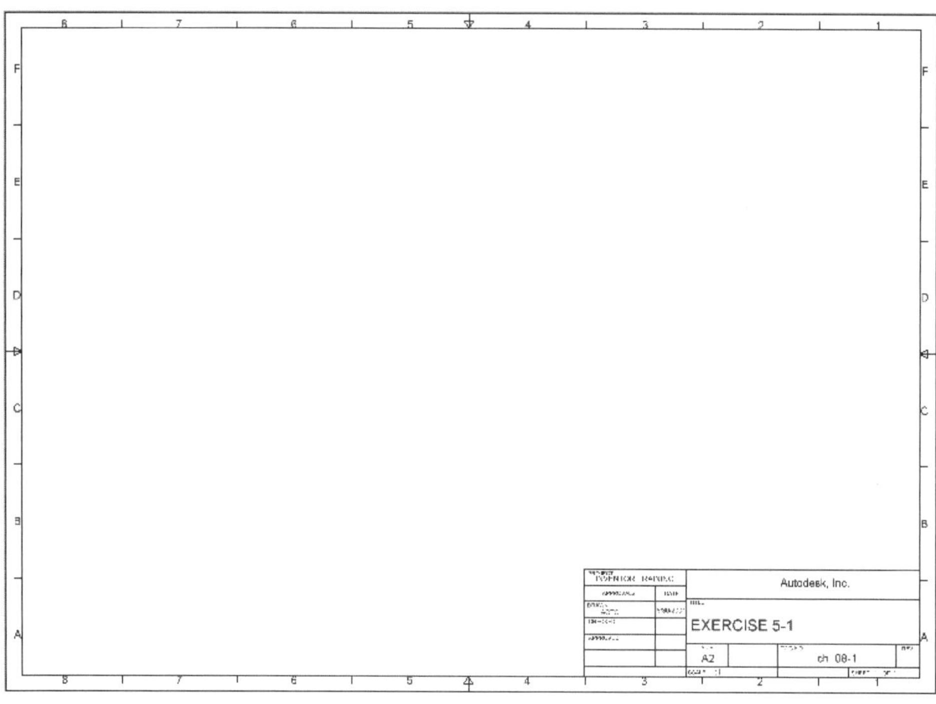


**Figure 5-9 Completed exercise**

# SETTING UP DRAFTING AND DIMENSION STANDARDS

Before you create drawing views, you can set up drafting and dimension standards.

## DRAFTING STANDARDS

There are six drafting standards that are already configured and included with Autodesk Inventor: ANSI (American National Standards Institute), BSI (British Standards Institute), DIN (Deutsche Industrie Norm—The German Institute for Standardization), GB (Guojia Biaozhun—The Chinese National Standard), ISO (International Organization for Standardization), and JIS (Japan Industrial Standard). You can choose to use one of these standards or create your own standard based on one of them. To make a standard current, click Standards from the Format menu. The Drafting Standards dialog box will appear as shown in the following image. From the Select Standard list, select the standard that you want to use.

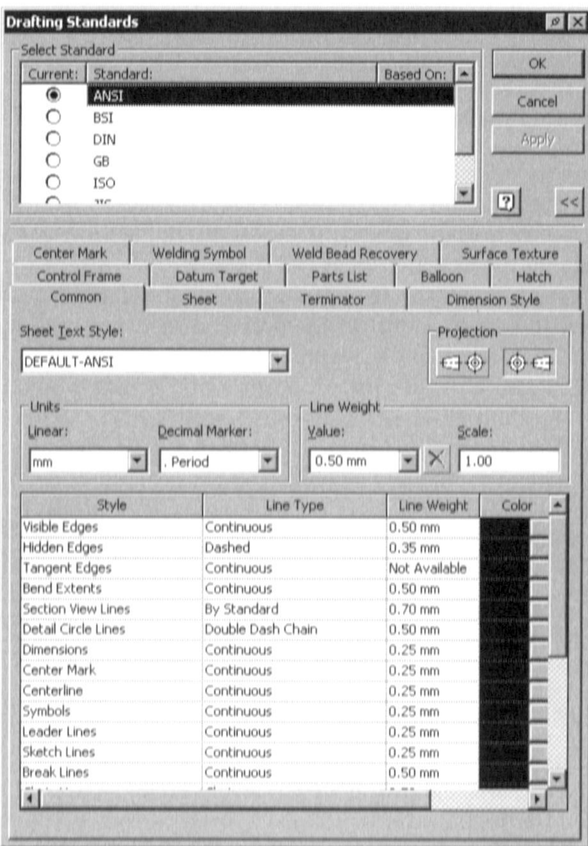


Figure 5-10

## CREATING A NEW DRAFTING STANDARD

The default drafting standards cannot be modified. If you need to modify a drafting standard, you will need to create a new drafting standard based on one of the included standards. If a new drafting standard is created, the changes will apply only to that drawing. To make the new drafting standard available for other new drawing files, make the changes in the template that you use to create new drawings and then save the file. To automatically save the new drafting standard as the default, set it as the *active* standard.

To create a new drafting standard, follow these steps:

1 Click Standards from the Format menu.

2 In Select Standard in the Drafting Standards dialog box, select Click to add new standard at the bottom of the Standards list.

3 In the New Standard dialog box, enter a name for the new standard and select the standard on which the new style will be based.

4 In the Drafting Standards dialog box, click the >> button to expand the dialog, if necessary, and thirteen tabs will appear. Make the changes as needed, and when done, click the Apply button.

The following is a description of the tabs for the Drafting Standards dialog box:

- Common—Controls the text style, projection direction, units of measurement, and line style
- Sheet—Controls the sheet and view labels in the Browser bar and the color scheme of the drawing sheet
- Terminator—Controls the style of arrows and datum references
- Dimension Style—Controls the default dimension style and defines which characters will be available to use in dimension text
- Centermark—Controls the proportion of the dashes used in a centermark
- Welding Symbol—Controls the weld symbols
- Weld Bead Recovery—Controls the display characteristics for caterpillar and end treatment weld symbols.
- Surface Texture—Controls the surface texture symbols
- Control Frame—Controls the control frame of geometric tolerances
- Datum Target—Controls the datum target of geometric tolerances
- Parts list—Controls the styles of the parts list
- Balloon—Controls the styles of the balloons
- Hatch—Controls the styles of the hatch patterns

## DIMENSION STYLES

You set the properties of a dimension style to control how dimensions will be displayed when a drawing view is created. There are six dimension styles (based on the six drafting standards) that are already configured and included with Autodesk Inventor: ANSI, BSI, DIN, GB, ISO, and JIS. You can choose to use one of these dimension standards or create your own standard based on one of them. To make a dimension style current, click Dimension

Style from the Format menu. The Dimension Styles dialog box will appear, as shown in the following image. From the Standards list, select the standard that you want to use. Depending upon the standard that you choose, there may be a style to choose from under Style Name. Examples would be DEFAULT-ISO-Method 1b or X.XXX radial.

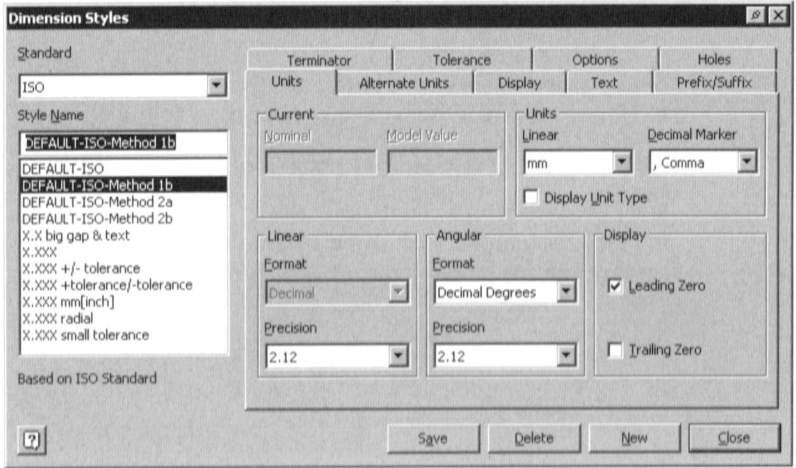


**Figure 5-11**

## CREATING A NEW DIMENSION STYLE

The default dimension styles cannot be modified. If you need to modify a dimension style you will need to create a new dimension style based on one of the included standards. After creating a new dimension style that will become the default style, set it as the *active* dimension style in the Drafting Standards dialog box under the Dimension Style tab. To create a new dimension style, follow these steps:

1 Click Dimension Style from the Format menu.

2 In the Dimension Styles dialog box, select a standard that the new style will be based on.

3 Click the New button and enter a new name for the style.

4 Modify the style by changing the items under the tabs in the Dimension Style dialog box.

5 When done modifying the new dimension style, click the Save button.

The following is a description of the tabs for the Dimension Style dialog box:

- Units—Enables the units of measure that will be used for the dimensions in the drawing.
- Alternate Units—Controls if and how alternate units will be displayed.
- Display—Controls the display characteristics for dimension lines in a drawing.

- Text—Controls the font, size, and format style for the default text.
- Prefix/Suffix—Controls the symbol, text, and location for a prefix/suffix before and after, or above and below, dimension text.
- Terminator—Controls the default arrowhead style for the drafting standard.
- Tolerance—Controls how a tolerance will be displayed.
- Options—Controls dimension style options. The options you select define the appearance of dimension lines and leaders in your drawing. The list of options available depends on the method and type you selected. Select as many options as desired from the list.
- Holes—Controls the type of hole or thread note displayed in a drawing.

### APPLYING DIMENSION STYLES TO SELECTED DIMENSIONS

You can modify individual dimensions for special annotation purposes. You might, for example, want some annotations to appear as reference dimensions or have special tolerances as seen in the following image.

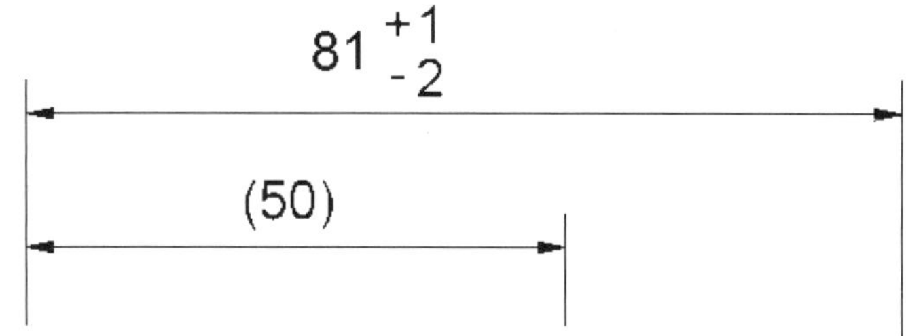


Figure 5-12

To override dimension style settings, right-click on a dimension and choose Options or Tolerance. When an override setting is common to one found in the Drafting Standards or the Dimension Style dialog boxes, the rules of thumb are the following:

- Override settings supersede the Dimension Style settings.
- Dimension Style settings supersede the Drafting Standard settings.

**NOTE** If you apply a dimension style to a dimension, any overrides on that dimension will be lost.

## TEMPLATES

After you have created a drawing sheet, border, and title block, and have set the drafting standard and dimension style, you can save the file as a template and start new drawing files based on these settings. To save a file as a template, click Save Copy As from the File menu, enter a new template name, and save the file in the Autodesk Inventor template folder. Use that file when creating new drawings.

## CREATING DRAWING VIEWS

After the drawing sheet and standards have been set, drawing views can be created from an existing part, assembly, or presentation file. The file from which the views will be created does not need to be opened when a drawing view is created. It is suggested, however, that both the file and the associated drawing file be stored in the same directory and that the directory be contained in the project file. When creating drawing views, you will find that there are many different types of views that can be created.

**Base View** The first drawing view of an existing part, assembly, or presentation file.

**Projected View** An orthographic or isometric view that is generated from an existing drawing view.

- Orthographic (Ortho View)—A drawing view that is projected horizontally or vertically from another view.
- Iso View—A drawing view that is projected at a 30° angle from a given view. An isometric view can be projected to any of the four quadrants.

**Auxiliary View** A drawing view that is perpendicular to a selected edge of another view.

**Section View** A drawing view that represents the area defined by slicing a plane through a part or assembly.

**Detail View** A drawing view in which a selected area of an existing view will be generated at a specified scale.

**Broken View** A drawing view that shows a section of the part removed, but the ends remain. Any dimension that spans over the break will reflect the correct length.

**Breakout View** A drawing view that has a defined area of material removed in order to expose internal parts or features.

## USING THE DRAWING VIEW DIALOG BOX

The Drawing View dialog box is used to create the various drawing views just described and is activated by clicking the Base View tool from the Drawing Views Panel Bar (as shown in the following image) or by right-clicking in the graphics window and clicking Base View from the menu.

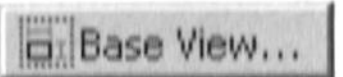


**Figure 5-13**

The Drawing View dialog box shown in the following image is divided into two tabbed areas—Component and Options. Each will be described in this section.

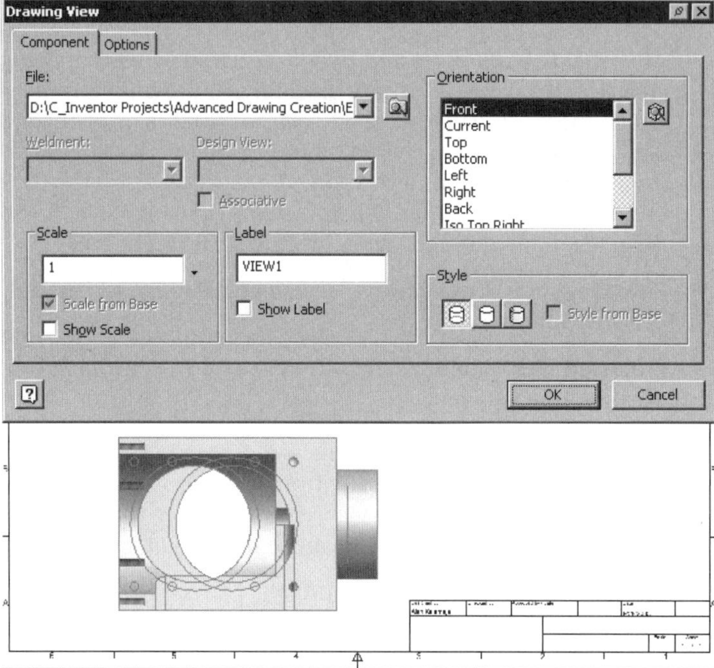


Figure 5-14

## The Component Tab

The following categories are available under the Drawing View Component tab:

**File** Any open part, assembly, or presentation files will appear in the drop-down list. You can also click the Explore directories icon and navigate to and select a part, assembly, or presentation file.

**Design View / Presentation View** After selecting a file that has Design Views, the names of the design views will appear in the drop list. If a file has Presentation Views, the names of the presentation views will appear in the drop list.

**Orientation** After selecting the file, you can choose the orientation in which the view will be created. After selecting an orientation, a preview image will appear in the graphics window. If the preview image does not show the view orientation that you want, select a different orientation view by selecting another option. The following list shows the different view orientations that are available:

- Front
- Current (current orientation of part, assembly, or presentation in its file)
- Top

- Bottom
- Left
- Right
- Back
- Iso Top Right
- Iso Top Left
- Iso Bottom Right
- Iso Bottom left

**Scale** Enter a number for the scale in which the view will be created. Note that the drawing will be plotted at full scale (1=1) and the drawing views are scaled as needed. The scale of the views can be edited after the views have been generated.

**Scale from Base** This area sets the scale of a dependent view to be the same as that of its base view. When selected, the dependent view maintains the same scale as its base view. To change the scale of a dependent view, clear the check box.

**Weldment** This area is enabled only when selecting a document that is a weldment. Weldments have four states—Assembly Only; Weld Preparations; Preparations & Welds; and Preparations, Welds, & Machining Features.

**Label** This area allows you to change the label for the selected view. When you create a view, a default label is determined by the active drafting standard. To change the label, select the label in the box and enter the new label.

**Show Label** This area displays or hides the view label. Clicking the check box displays the label and leaving the box unchecked hides the label.

**Style** In the Style area, choose how the view will be displayed. There are three choices—Hidden Line, Hidden Line Removed, and Shaded. The preview image will not update to reflect the style choice. When the view is created, the chosen style will appear. The style can be edited after the view has been generated.

**Style from Base** This features sets the display style of a dependent view to be the same as that of its base view. When the check box is selected, the dependent view uses the same display style as its base view. To change the display style, of a dependent view, clear the check box.

## The Options Tab

The Options tab of the Drawing View dialog box is displayed in the following image.

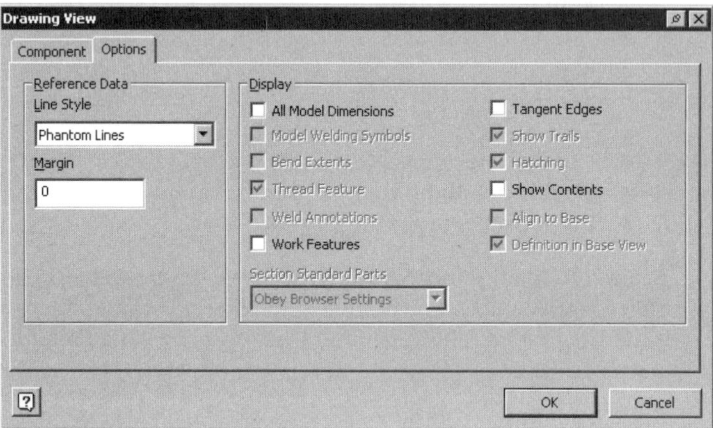


**Figure 5-15**

Areas to be found in the Options tab include the following:

## Reference Data

**Line Style** Three options are contained in this drop list. When Visible Edges is selected, reference data will have no special display characteristics. When Phantom Lines is selected (for wireframe view types), the reference data and product data will not hide each other. The reference data will be displayed on top of product data, and the reference data will be displayed with tangent edges turned off and all other edges as the phantom line type. When Phantom Lines is selected (for shaded view types), the reference data will be displayed as wireframe with tangent edges turned off and all other edges as phantom lines. The reference data will be displayed on top of the product data that is shaded. When Hidden Edges is selected, reference data will not be displayed.

**Margin** To see more reference data, set the value to expand the boundaries by a specified value on all sides.

## Display

**All Model Dimensions** Check the box to see model dimensions in the view when it is created. If the box is not checked, model dimensions will not automatically be placed upon view creation. When checked, only the dimensions that are parallel to this view, and have not been retrieved in existing views on the sheet, will display.

**Model Welding Symbols** This box is active only if creating a weldment. If checked, welding symbols placed in the model will be retrieved and used in the drawing.

**Bend Extents** This box is active only if creating a view of a sheet metal part. It controls the visibility of bend extent lines/edges.

**Thread Feature** This box is active only if creating a view of an assembly model.

**Weld Annotations** This box is active only if creating a view of a weldment. It controls the display of weld annotations.

**Work Features** This feature is enabled for models capable of having work features.

**Tangent Edges** Sets the visibility of tangent edges in a selected view. Checking the box displays tangent edges, while not checking the box hides them.

**Show Trails** Controls the display of trails in views based on drawings created from presentation files.

**Hatching** This area sets the visibility of the hatch lines in the selected section view. Checking the box displays hatch lines, while not checking the box hides them.

**Align to Base** Use this feature to remove the alignment constraint of a selected view to its base view. When the box is checked, alignment of views exists. Clearing the check box breaks the alignment.

**Definition in Base View** This area displays or hides the projection line for the view. Checking the box displays the line, while not checking the box hides the line.

### BASE VIEWS

A base view is the first view that is created from the selected part, assembly, or presentation file. When you create a base view, the scale is set in the dialog box. From the base view, other drawing views can be projected. Projected drawing views will be explained in the next section. There is no limit to the number of base views that can be created in a drawing (based on different parts, assemblies, or presentation files). When creating a base view, the orientation of that view can be selected from the Orientation list found under the Component tab of the Drawing View dialog box.

To create a base view, follow these steps:

1 Click the Base View tool from the Drawing Views Panel Bar, or right-click in the graphics window and click Base View from the menu. The Drawing View dialog box will appear.
2 Under the Component tab, click the Explore directories icon to navigate to and select the part, assembly, or presentation file from which the drawing views will be created or projected. After making the selection, a preview image will appear in the graphics area. DO NOT place the view until all the options have been set.
3 Select the type of view to generate from the Orientation list.
4 Select the scale for the view.

5 Select the style for the view.

6 Locate the view by selecting a point in the graphics area.

## PROJECTED VIEWS

A projected view can be an orthographic or isometric view that is projected from a base view or any other existing view. When creating a projected view, a preview image will appear, showing the orientation of the view that will be created as the mouse is moved to a location on the drawing. There is no limit to the number of projected views that can be created. To create a projected drawing view, follow these steps:

1 Click the Projected View tool from the Drawing Views Panel Bar, as shown in the following image, or right-click inside the bounding area of an existing view box (shown as dashed lines when the mouse is moved into the view) and click Create View>Projected from the menu.

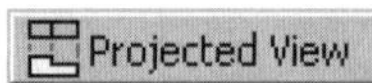


**Figure 5-16**

2 If the Projected View tool is selected from the Drawing Views Panel Bar, click inside the desired view to start the projection.

3 Move the mouse horizontally, vertically, or at an angle to get a preview image of the view that will be generated. Keep moving the mouse until the preview matches the view that you want to create, then press the left mouse button. Continue placing projected views.

4 When finished, right-click and click Create from the menu.

# EXERCISE 5-2
# Creating a Multiview Drawing

In this exercise, you will create an independent view to serve as the base view and then you will add projected views to create a multiview orthographic drawing. Finally, you will add an isometric view to the drawing. To navigate to the exercise in the *Electronic Student Workbook* do the following:

1 From the Main TOC page, click Chapter 5.

2 From the TOC for Chapter 5, click Creating a Multiview Drawing.

The completed exercise is shown in the following image.

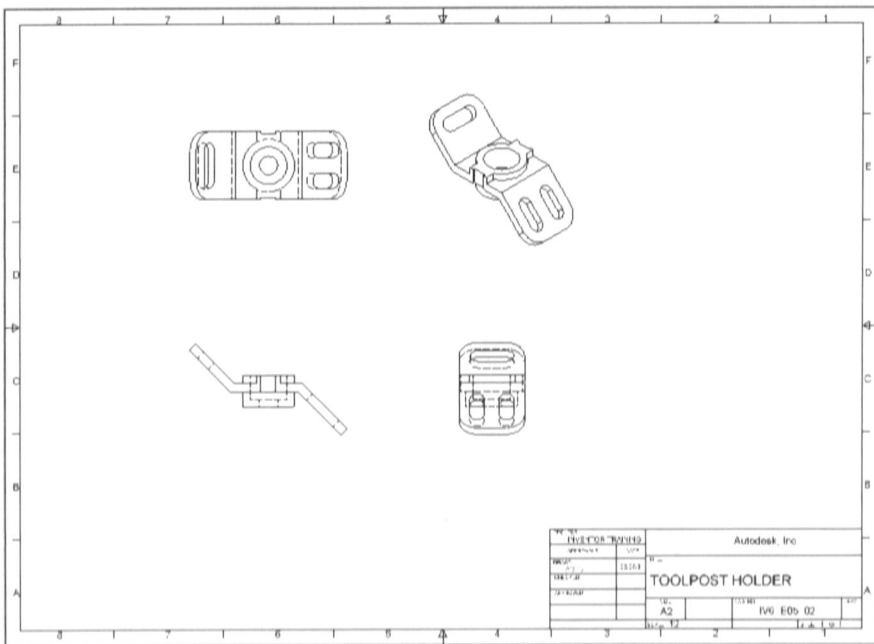


**Figure 5-17 Completed exercise**

## AUXILIARY VIEWS

An auxiliary view is a view that is projected perpendicularly to a selected edge or line. It is primarily designed to view the true size and shape of a surface that appears foreshortened in other views.

To create an auxiliary drawing view, follow these steps:

1 Choose the Auxiliary View tool from the Drawing Views Panel Bar, as shown in the following image. You can also right-click inside the bounding area of an existing view box (shown as a dotted box when the mouse is moved into the view) and then select Create View>Auxiliary from the menu.

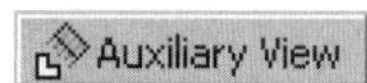


Figure 5-18

2 Click inside the view from which the auxiliary view will be projected. The Auxiliary View dialog box will appear, as shown in the following image. Type in a name for the Label and a value for the Scale, and then select a Style (hidden, hidden line removed, or shaded).

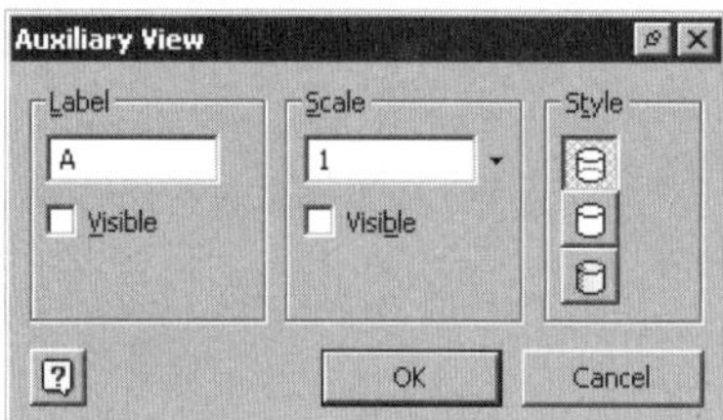


Figure 5-19

3 In the selected drawing view, select an edge or line from which the auxiliary view will be perpendicularly projected, as shown in the following image.

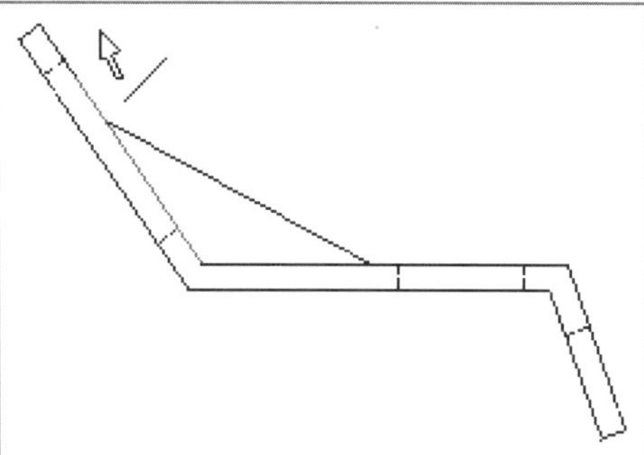

Figure 5-20

4 Move the mouse to position the auxiliary view, as shown in the following image. Notice that the view takes on a shaded appearance as it is being positioned.

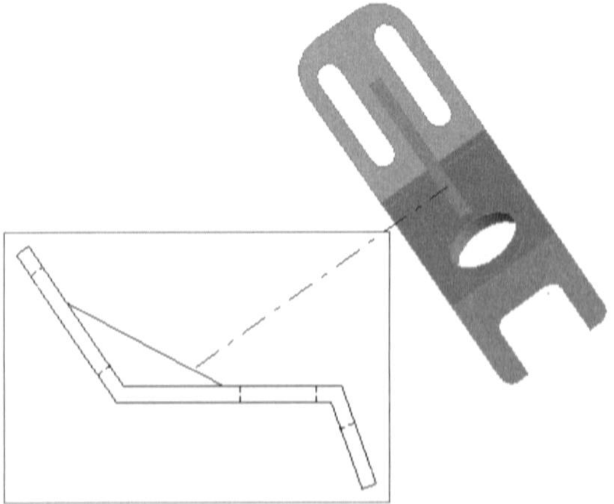

**Figure 5-21**

5 Click a point on the drawing sheet to create the auxiliary view. The following image illustrates the completed auxiliary view layout.

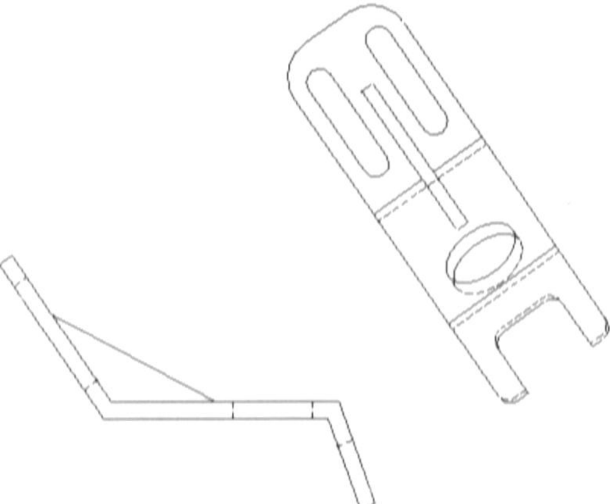

**Figure 5-22**

## SECTION VIEWS

A section view is a view that is created by sketching a line or multiple lines that will define the plane(s) that will be cut through a part or assembly. The view will represent the surface area of the cut. When defining a section, you will sketch line segments that can be horizontal, vertical, or at an angle. Arcs, splines, and circles cannot be used to define section lines. When you sketch the section line(s), geometric constraints will automatically be applied between the line being sketched and the geometry in the drawing view. You can also infer points by moving the mouse over certain geometry locations (such as centers of arcs, endpoints of lines, and so on) and then moving the mouse away to display a dotted line showing that you are inferring or tracking that point. To place a geometric constraint between the drawing view geometry and the section line, click in the drawing when a green circle appears; the glyph for the constraint will appear. If you do not want the section lines to automatically have constraint(s) applied to them when they are created, hold down the CTRL key when sketching the line(s). Because the area in the section view that is solid material is displayed with a hatch pattern, you should set the hatching style before the section view is created.

To create a section drawing view, follow these steps:

1 Choose the Section View tool from the Drawing Views Panel Bar, as shown in the following image. You can also right-click inside the bounding area of an existing view box (shown as a dotted box when the mouse is moved into the view) and select Create View>Section from the menu.

Figure 5-23

2 If you selected the Section View tool from the Drawing Views Panel Bar, click inside the view from which the section view will be created.

3 Sketch the line or lines that define where and how you want the view to be cut. In the following image, a vertical line is sketched through the center of the object, from the top to the bottom.

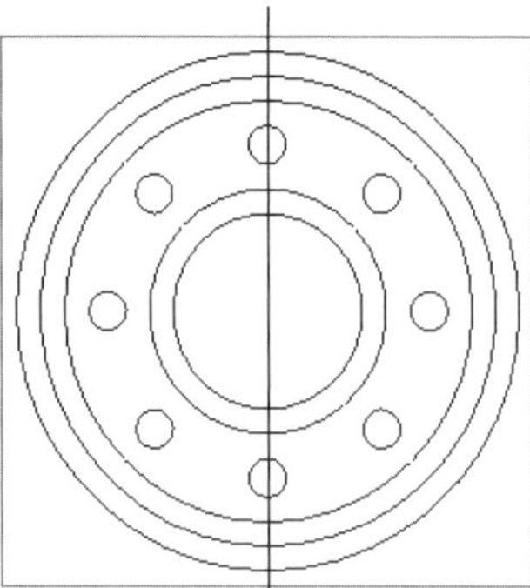

Figure 5-24

4 When you finish sketching the section line, right-click and select Continue from the menu.

5 The Section View dialog box will appear as in the following image. Fill in the information for how you want the Label, Scale, and Style to appear in the drawing view.

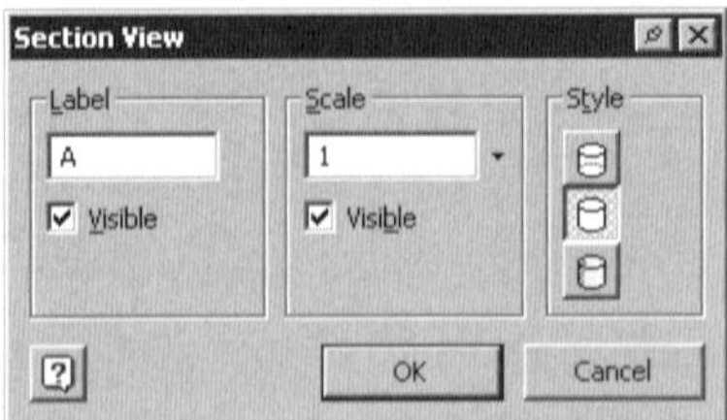


**Figure 5-25**

6 Move the mouse to position the section view, as shown in the following image.

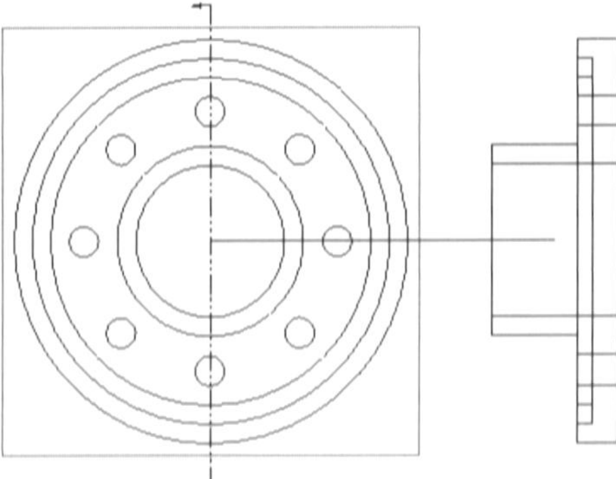

**Figure 5-26**

7 After the view has been created, constraints and/or dimensions can be added or deleted to the section line(s) to better define the section cut. To add or delete a constraint or dimension, select the section line, right-click, and select Edit from the menu, as shown in the following image.

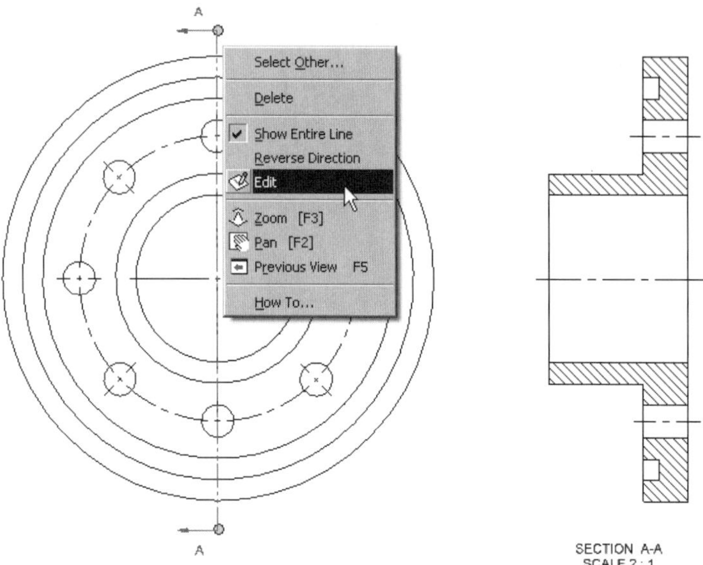


Figure 5-27

8 From the Sketch toolbar, use the General Dimension and Constraint tools to add or delete constraints and dimensions.

9 When you are done editing the section line, right-click and select Finish Sketch or Return on the Command toolbar.

**NOTE** If the section lines were not constrained, you can drag them to show the section in a new location. After you drag the section lines, the dependent section view will be updated to reflect the change.

10 The completed section view is illustrated in the following image. Since the cutting plane cuts through the entire view, this is called a *full* section. Also in the image, centerlines have been added to the section view and to the bolt circles in the front view.

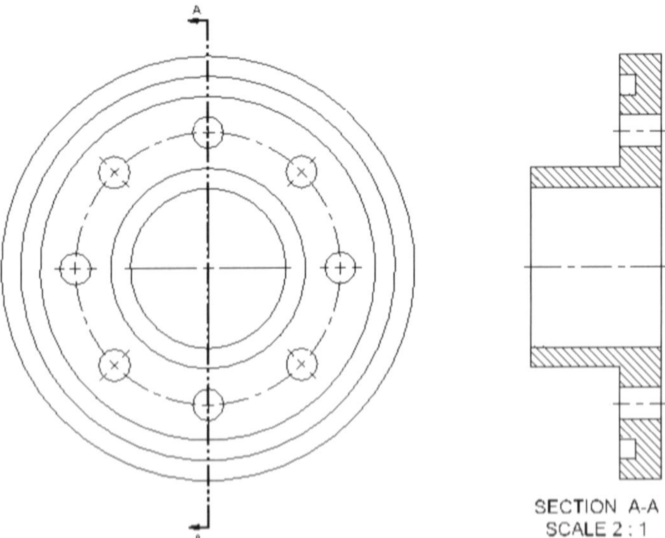


**Figure 5-28**

## Half Sections

The cutting plane line can be constructed in various configurations, depending on the desired results. When the cutting plane line cuts halfway through the object (as in the following image), this is referred to as a *half* section. The section view displays crosshatching through half of the surfaces.

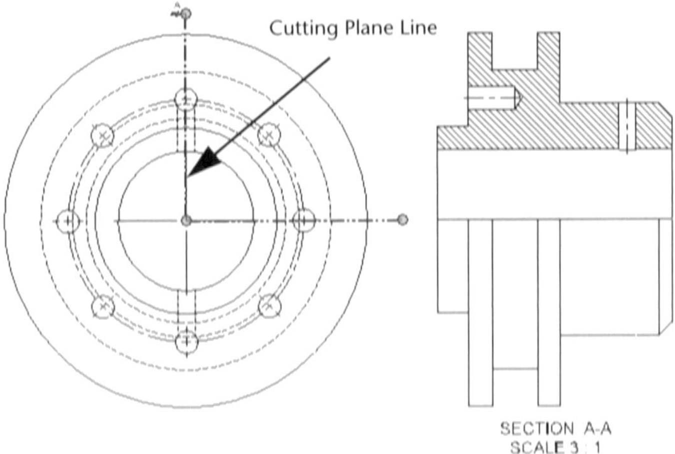


**Figure 5-29**

## Aligned Sections

Aligned sections take into consideration the angular position of features in a drawing. Instead of the cutting plane line being drawn vertically through the object, it is angled or aligned with the same angle as the feature (in this case, the hole), as shown in the following image. This allows aligned sections to produce a clearer drawing. Also notice in the following image the creation of a solid line that runs through the center of the section view. This line is generated by the change in direction of the cutting plane line. If you want to add a centerline to this area, right-click on the solid horizontal line and turn off its visibility.

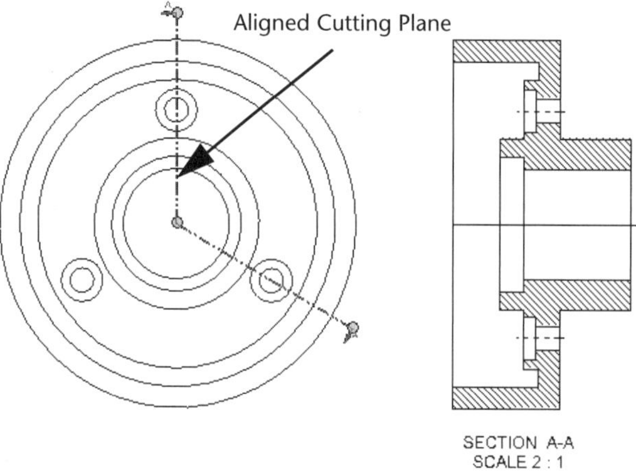


**Figure 5-30**

## Offset Sections

Offset sections take their name from the offsetting of the cutting plane line that passes through certain details in a given view. If the cutting plane passes straight through an object, some features are exposed while others remain hidden. By offsetting the cutting plane line as shown in the following image, you can control the cutting plane line direction and the specific features of a part it passes through.

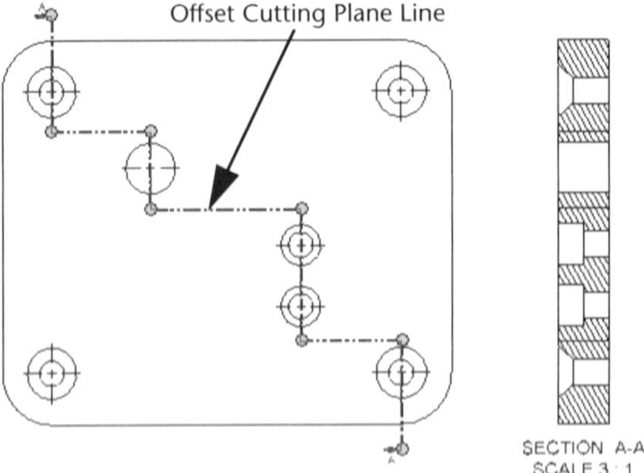


Figure 5-31

## Modifying a Hatch Pattern

You can edit the hatch pattern by right-clicking on the hatch pattern in the section view and selecting Modify Hatch from the menu. This will launch the Modify Hatch Pattern dialog box shown in the following image. Make the desired changes in the Pattern, Scale, Angle, Shift, Double, and Color mode areas. Click the OK button to have the changes reflected in the drawing.

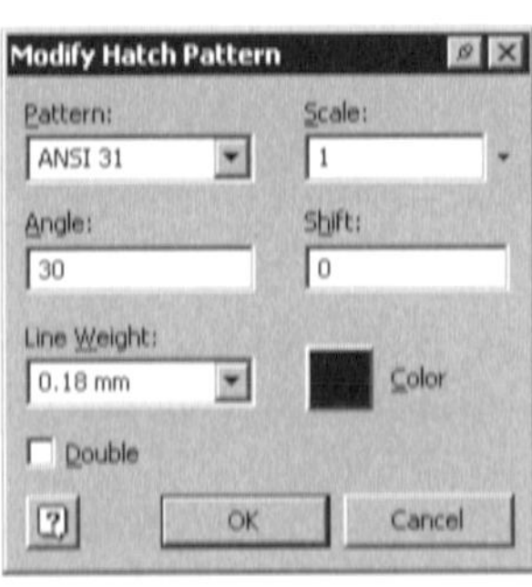


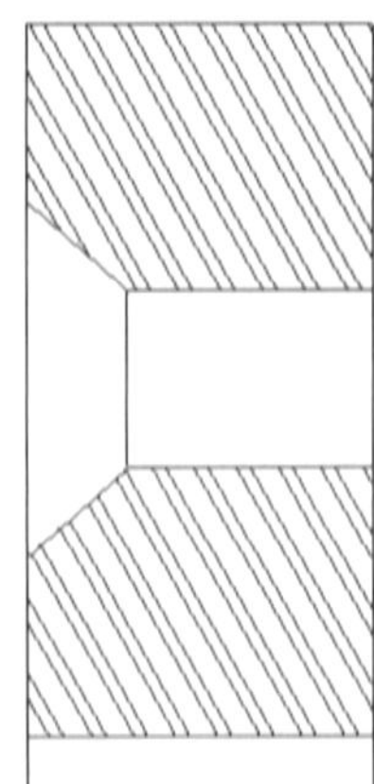

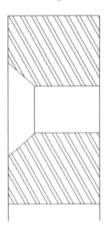

Figure 5-32

**NOTE** The Shift mode of the Modify Hatch Pattern dialog box shifts the hatch pattern to offset it slightly from the hatch pattern on a different part. This option would be ideal for creating sections that involve assembly models. Enter the distance for the shift.

## DETAIL VIEWS

A detail view is a drawing view that enlarges an area of an existing drawing view by a specified scale. The enlarged area is defined by a circle or rectangle and can be placed anywhere on the sheet. To create a detail drawing view, follow these steps:

1 Select the Detail View tool from the Drawing Views Panel Bar, as shown in the following image. You can also right-click inside the bounding area of an existing view box (shown as dashed lines when the mouse is moved into the view) and select Create View>Detail from the menu.

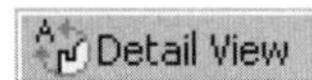


**Figure 5-33**

2 If you selected the Detail View tool from the Drawing Views toolbar, click inside the view from which the detail view will be created.

3 The Detail View dialog box will appear as shown in the following image. Fill in information for how you want the Label, Scale, and Style to appear in the drawing view. Do not click the OK button at this time, as doing so will end the operation.

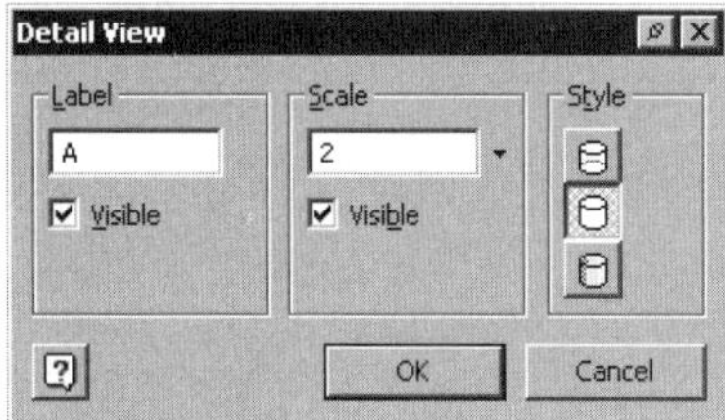


**Figure 5-34**

4 In the selected view, select a point that will be used as the center of the circle that will describe the detail area (as shown in the following image).

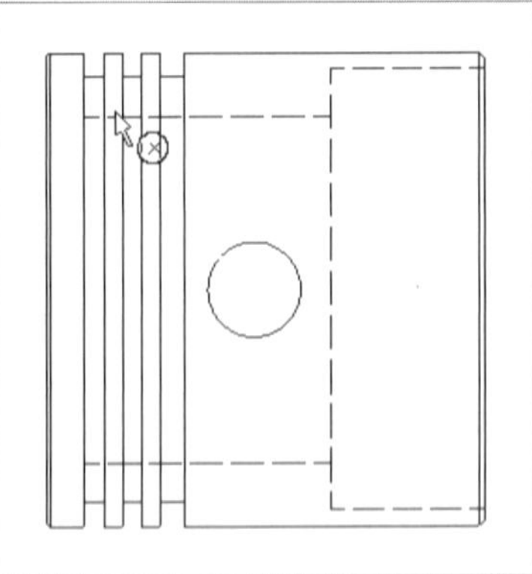

**Figure 5-35**

**NOTE** You can also right-click at this point and choose Rectangular Fence to change the bounding area of the detail area definition.

5 Select another point that will define the radius of the detail circle. As you move the mouse, a circle of the specified size will appear, as shown in the following image.

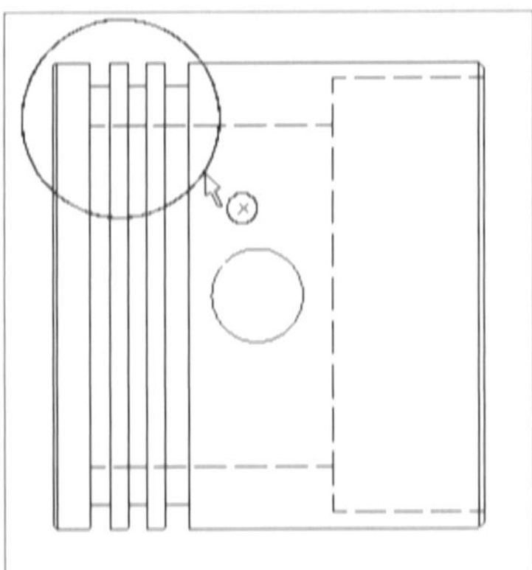

**Figure 5-36**

6 Select a point on the sheet where you want the view to be placed. There are no restrictions on where the view can be placed. The completed detail view is illustrated in the following image.

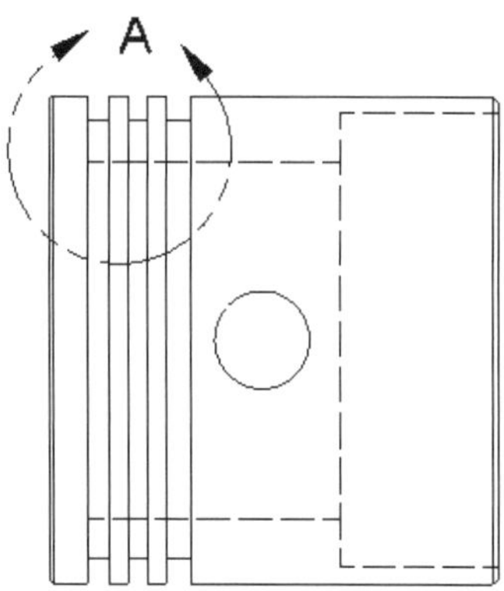


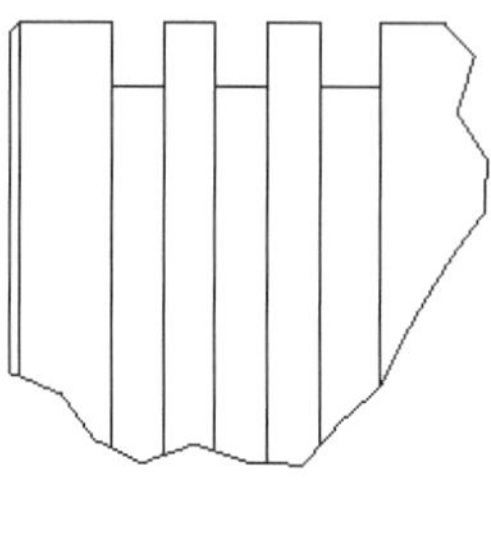

DETAIL A
SCALE 2 : 1

Figure 5-37

## EDITING DRAWING VIEWS

After creating the drawing views, you may need to move, edit properties of, or delete a drawing view. Each will be discussed in this section.

### MOVING DRAWING VIEWS

To move a drawing view, first pick inside the view until a bounding box consisting of red dotted lines appears. Then move your cursor to the red dotted boundary, press and hold down the left mouse button, and move the view to its new location. Release the mouse button when you are finished. As the view is moved, a rectangle will appear that represents the bounding box of the drawing view. If you move a base view, any projected (children or dependent) views will also move with it. If an orthographic or auxiliary view is moved, it will only be able to be moved along the axis in which it was projected. Detail and isometric views can be moved anywhere in the drawing sheet.

### EDITING DRAWING VIEW PROPERTIES

After creating a drawing view, you may need to change the label, scale, style, or hatching visibility; remove the alignment constraint to its base view; or control the visibility of the view projection lines for a section or auxiliary view. To edit a drawing view, follow one of these guidelines:

- Double-click in the bounding area of the view.
- Double-click on the icon of the view you want to edit in the Browser.
- Right-click in the drawing view's bounding area or on its name, and click Edit View from the menu.
- Right-click on the drawing view's name in the Browser and click Edit View from the menu.

When performing the operations listed above, the Drawing View dialog box will appear as shown in the following image. This is the same dialog box used to create drawing views. Depending on the view that is selected, certain options may be grayed-out from the dialog box. Make the necessary changes and then click the OK button to complete the edit. When the scale is changed in the base view, all of the dependent views will be scaled as well.

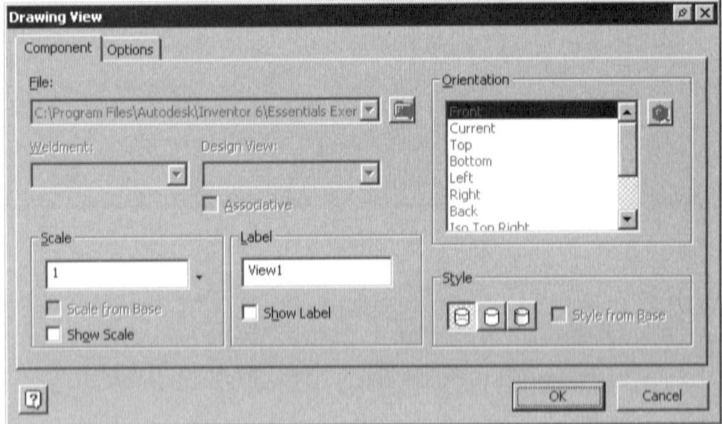


**Figure 5-38**

See Using the Drawing View Dialog Box on page 175 for detailed descriptions of Component and Options tab elements.

## DELETING DRAWING VIEWS

To delete a drawing view, either right-click in the bounding area of the drawing view or on its name in the Browser and click Delete from the menu. You can also click in the bounding area of the drawing view and press the DELETE key on the keyboard. A dialog box will appear asking to confirm the deletion of the view. If the selected view has a view that is dependent on it, you will be asked if the dependent views should also be deleted. By default, the dependant view will be deleted. To exclude a dependant view from the delete operation, expand the dialog box and click on the Delete cell to change the selection to No.

# EXERCISE 5-3 Editing Drawing Views

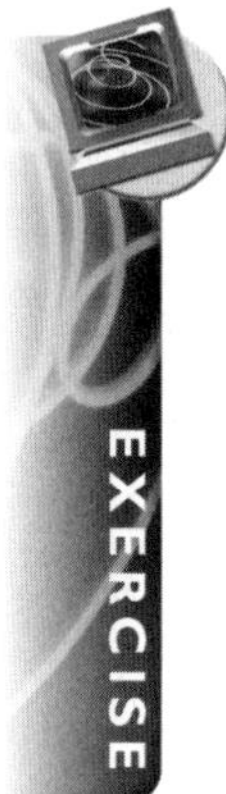


In this exercise, you edit drawing views in an existing drawing. To navigate to the exercise in the *Electronic Student Workbook* do the following:

1. From the Main TOC page, click Chapter 5.
2. From the TOC for Chapter 5, click Editing Drawing Views.

The following image illustrates the opened exercise.

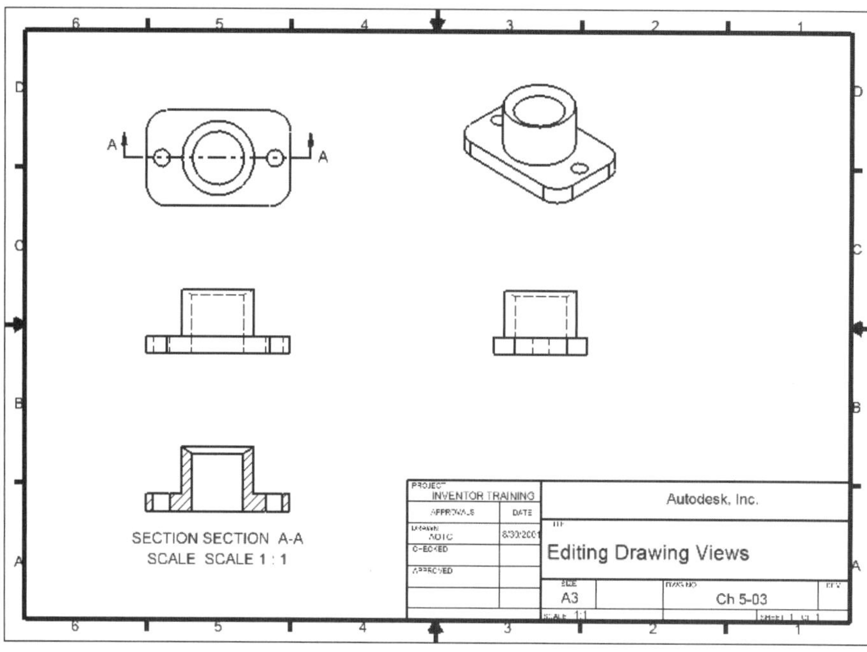


**Figure 5-39 Opened exercise**

The following image illustrates the completed exercise.

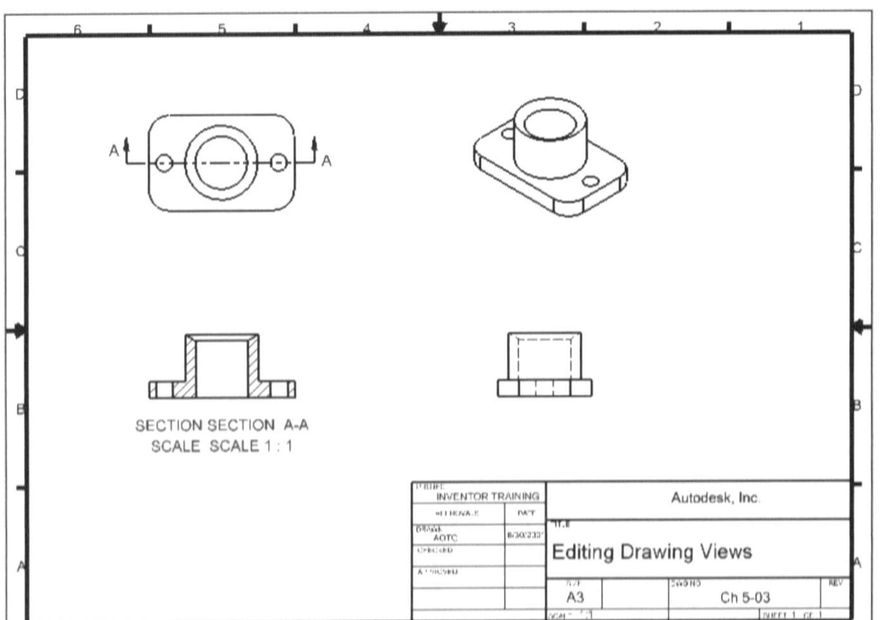


**Figure 5-40 Completed exercise**

## DIMENSIONS

Once the drawing view(s) has been created, you may need to change the value of a model dimension. If the dimensions were not displayed when the view was created, you may want to get the model dimensions so they appear in a view, hide certain dimensions, add drawing (reference) dimensions, or move dimensions to a new location. All of these operations will be covered in the next section.

### RETRIEVE DIMENSIONS

Model dimensions will not automatically appear once a drawing view has been created. You can use the Retrieve Dimensions tool to select valid model dimensions for display in a drawing view. Only those dimensions that were placed in the model on a plane parallel to the view will be displayed.

To activate this command, use one of the following three methods:

- Right-click in the bounding area of a drawing view and select Retrieve Dimensions.
- Right-click on the name of the view in the Browser.
- Click on the Retrieve Dimension button located at the bottom of the Drawing Annotations toolbar.

All three methods are shown in the following image.

The appropriate dimensions for the view that were used to create the part will appear.

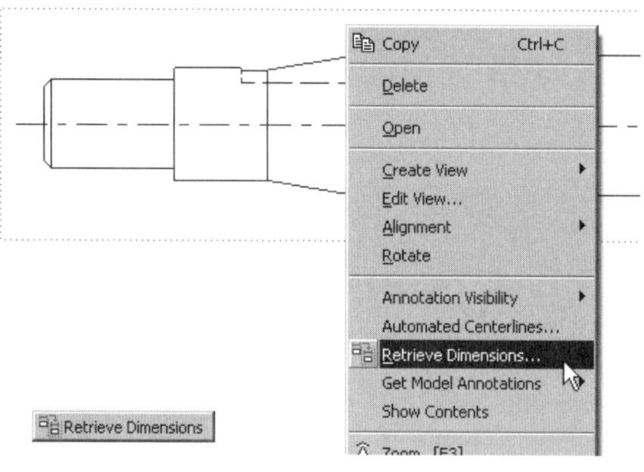


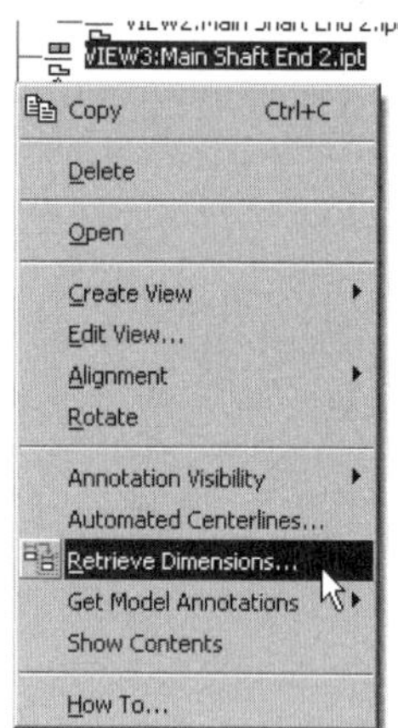


**Figure 5-41**

Selecting Retrieve Dimensions will display the Retrieve Dimensions dialog box shown in the following image. Through this dialog box you first select a drawing view in which to retrieve the model dimensions. The Select Source area of the dialog box allows you to selectively retrieve model dimensions based on a part's feature. The Select Parts option allows you to retrieve all model dimensions based on a part. When either of these methods are used, only valid model dimensions are displayed in preview mode. Once you have previewed the model dimensions, use the Select Dimensions button to pick the dimensions you want to retrieve.

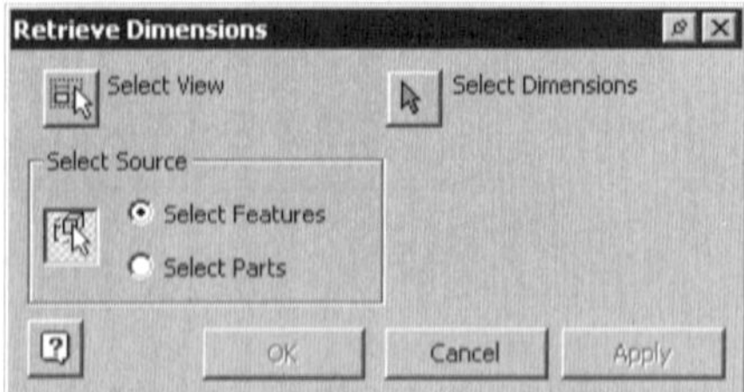


**Figure 5-42**

To retrieve dimensions, follow these steps:

1 Select Retrieve Dimensions from the Drawing Annotations Panel bar.
2 Select the drawing view in which to retrieve model dimensions.
3 To retrieve model dimensions based on one or more features, click the Select Features radio button in the Retrieve Dimensions dialog box. Then select the desired features.
4 If you want to retrieve model dimensions based on the entire part in a drawing view, click the Select Parts radio button. Then select the desired part or parts. You can select multiple objects by holding down the SHIFT or CTRL keys. You can also select objects with a selection window or crossing selection box.
5 When the model dimensions appear in the drawing view in preview mode, click the Select Dimensions button and select the desired dimensions. The selected dimensions will be highlighted.
6 Click the Apply button to retrieve the model dimensions.
7 Click the OK button to dismiss the Retrieve Dimensions dialog box.

The following illustration shows the effects of retrieving model dimensions using the Select Features mode. In this example, the large rectangle of the engine block was selected. This resulted in the horizontal and vertical model dimensions being retrieved.

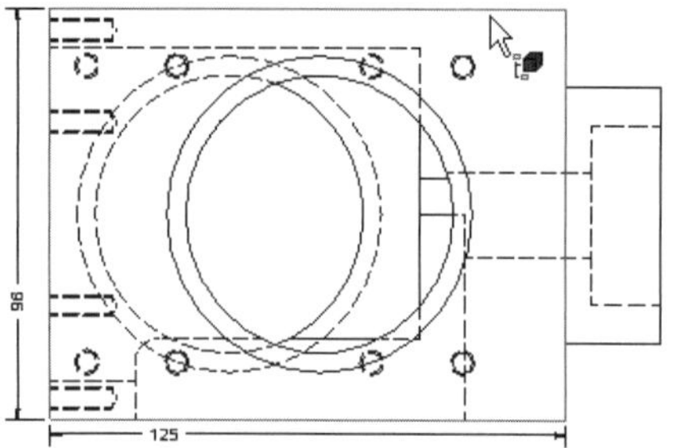


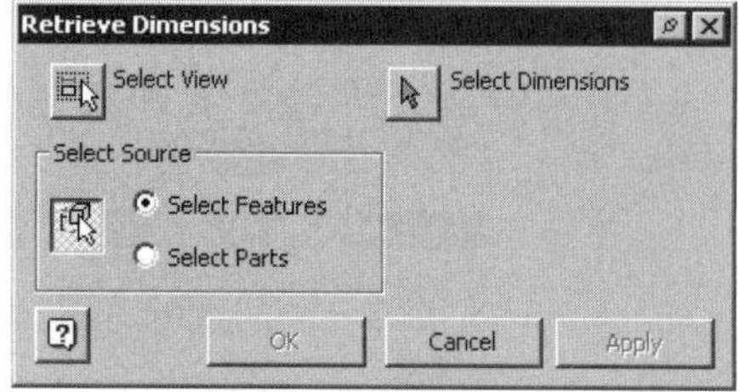


**Figure 5-43**

The following illustration shows the effects of retrieving model dimensions using the Select Parts mode. In this example, the entire engine block was selected. This resulted in all model dimensions parallel to this drawing view being retrieved.

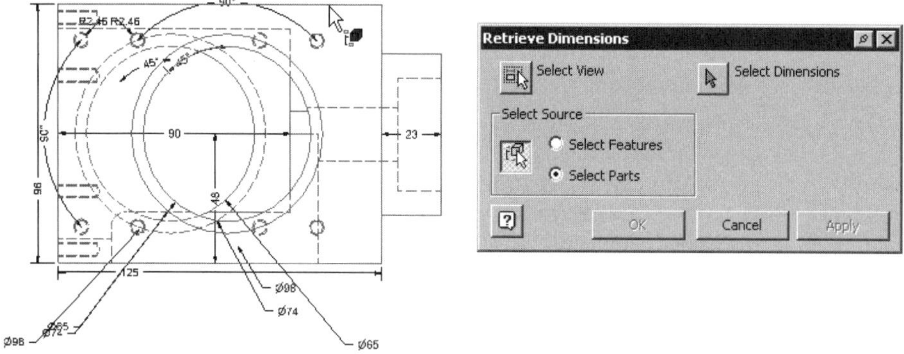


**Figure 5-44**

Model dimensions can also be automatically retrieved during the drawing view creation process. To perform this task, follow these steps:

1. Launch the Options dialog box by selecting Tools > Application Options.
2. Click on the Drawing tab and locate the option Retrieve all model dimensions on view placement.
3. Place a check in the box to automatically retrieve model dimensions in all placed views.
4. Clear the check box if you want drawing views created without model dimensions being automatically retrieved.
5. When you want to retrieve model dimensions in a base drawing view, place a check in the All Model Dimensions option, which is found under the Options tab of the Drawing View dialog box.

## DIMENSION VISIBILITY

To hide all of the model or drawing dimensions for a drawing view, right-click in the bounding area of a drawing view and click Annotation Visibility from the menu, as shown in the following image. The appearance of a check means the specific feature is turned on. Clicking the check adjacent to Model Dimensions will remove the check and turn off the model dimensions in a drawing. Use the same operation to turn off drawing dimensions.

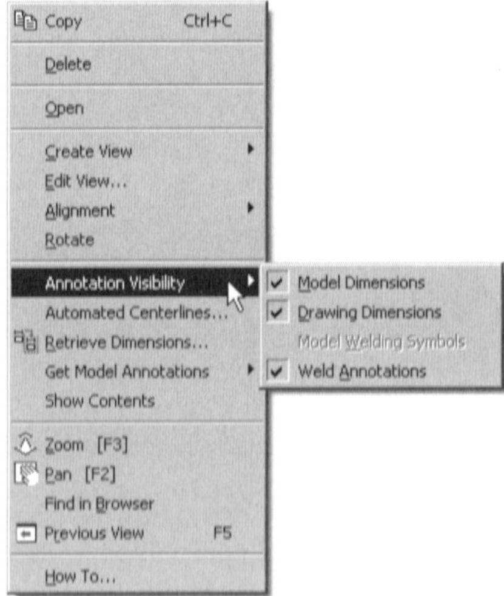


**Figure 5-45**

## CHANGING MODEL DIMENSION VALUES

While working in a drawing view, you may find it necessary to change the model (parametric) dimensions of a part. You can open the part file, change the dimension's value and save the part, and the change will then be reflected in the drawing views. You can also change a dimension's value in the drawing view by right-clicking on the dimension and clicking Edit Model Dimension from the menu, as shown in the following image.

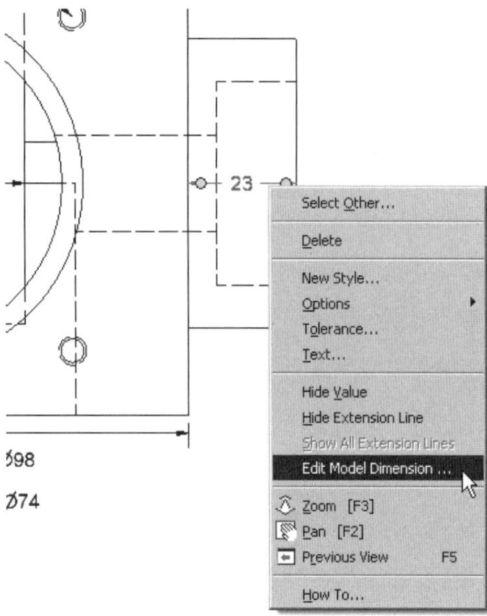


**Figure 5-46**

The Edit Dimension text box will appear. Enter a new value and click the check mark or press ENTER. The associated part will be updated and saved and the associated drawing view(s) will be updated automatically to reflect the new value.

**Figure 5-47**

**NOTE** It is possible to edit model dimensions found in drawing views. This feature is made possible when you install Autodesk Inventor and enable the option to edit model dimensions from the drawing. It is considered, however, poor practice to make major dimension changes from the drawing. It is highly recommended that all dimension changes be made through the part model.

## DRAWING (REFERENCE) DIMENSIONS

After laying out the drawing views, you may find that another dimension is required to better define the part. You can go back and add a parametric dimension to the sketch if the part was underconstrained, add a driven dimension if the sketch was fully constrained, or add a drawing (reference) dimension to the drawing view. A drawing dimension is not a parametric dimension; it is associative to the geometry to which it is referenced. The drawing dimension reflects the length of the geometry being dimensioned. After a drawing dimension is created, and the value of the geometry that was dimensioned changes, the drawing dimension will get updated to reflect the change. A drawing dimension is added with the General Dimension tool from the Drawing Annotation Panel Bar, as shown in the following image. Create a drawing dimension in the same way you would create a parametric dimension.

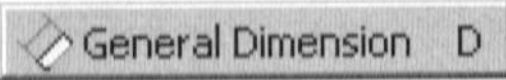


**Figure 5-48**

## SELECTING DRAWING OBJECTS

Managing drawing objects can be a tedious task, especially when the drawings are large and cumbersome. At times, you may have a need to move a number of views to a new location; or you may need to change a number of dimensions to a different dimension style. Rather than selecting the objects or views individually, you may use a window selection mode or crossing window selection mode to assist you in the selection of multiple objects.

In the example shown in the following image, you want to select all drawing objects that make up the shaft without selecting any dimensions. Clicking and dragging the cursor from the left at A to the right at B will form a solid box representing a window. This window will select all drawing objects that are completely enclosed within it. The selected geometry will highlight to distinguish it from unselected geometry.

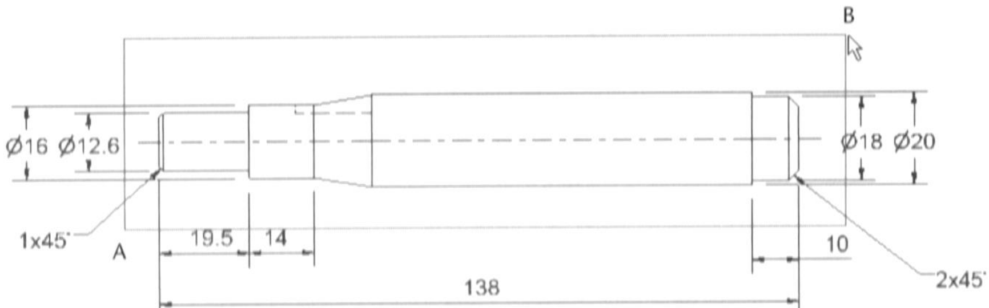


**Figure 5-49**

The results of selecting drawing objects by a window are illustrated in the following image. The selected geometry will be highlighted to distinguish it from unselected geometry.

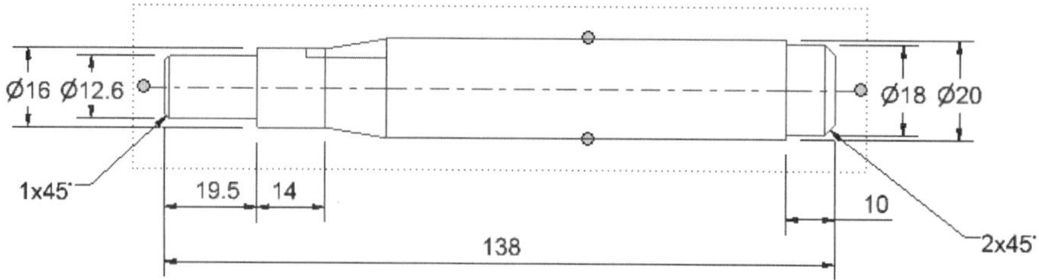


**Figure 5-50**

The example illustrated in the following image shows how objects and dimensions can be selected by a crossing window. If you click and drag the cursor from the right at A to the left at B, a dashed box will be formed. This dashed window will select all drawing geometry that is completely surrounded by and intersected by it.

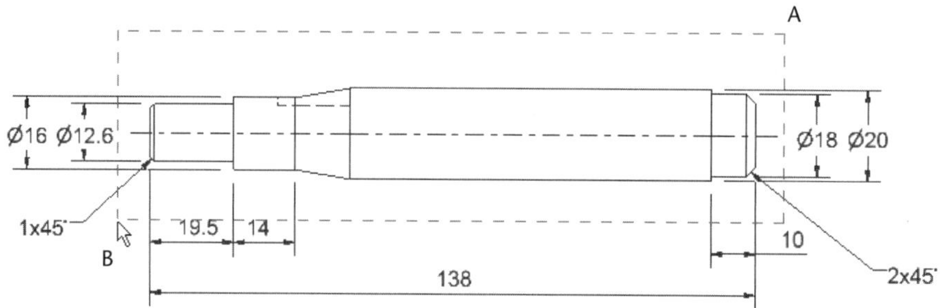


**Figure 5-51**

The results of using a crossing window are illustrated below, with all drawing view lines and dimensions selected.

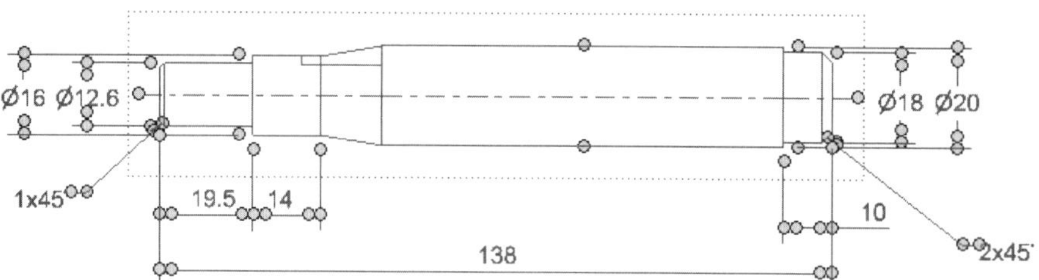


**Figure 5-52**

**NOTE** If you accidentally select an object, you can deselect it by holding down the **SHIFT** or **CTRL** keys as you select the object. You can also create multiple window and crossing window selections by pressing and holding down the **SHIFT** or **CTRL** keys as you construct the selection box.

## MEASURING CAPABILITIES IN DRAWING MODE

As a means of checking the accuracy of drawing information, two measuring tools—Measure Distance and Measure Angle—are available in Drawing Mode, as shown in the following image.

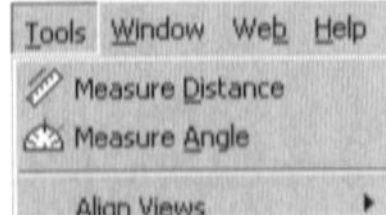


Figure 5-53

One method of measuring a distance is to pick on the desired object at A, as shown in the following image. Doing so will display the distance of the object as shown in the Measure Distance dialog box.

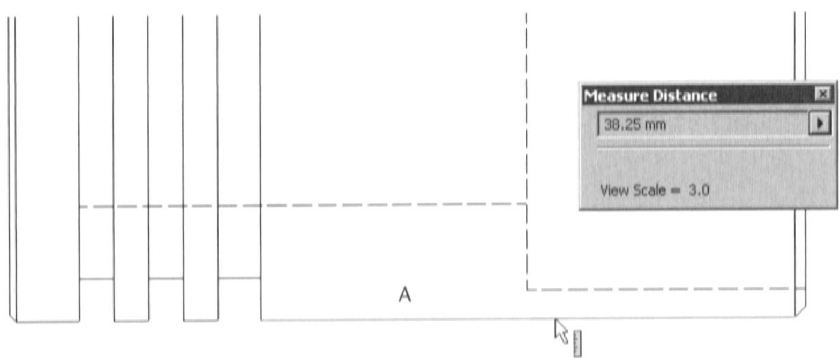


Figure 5-54

Another way to measure distance is by picking two points (A and B) as shown in the following image. Additional information, such as the Sheet Position and Delta X and Y values is also displayed in the Minimum Distance dialog box.

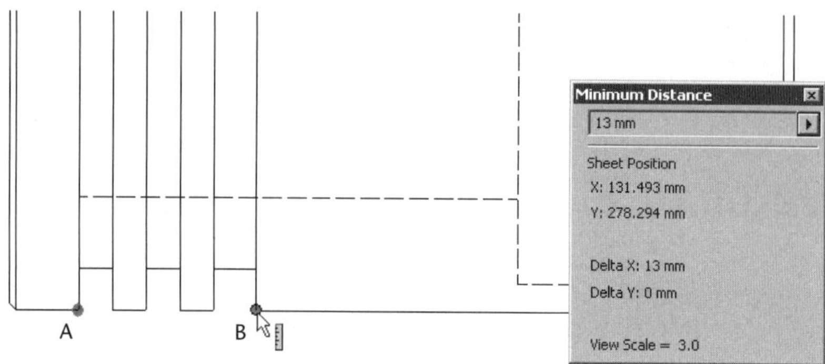


**Figure 5-55**

When measuring angles, you can pick two lines (at A and B) to calculate the angle. An arc is drawn, as shown in the following image, to signify the angle being calculated. The calculated value in degrees is displayed in the Angle dialog box.

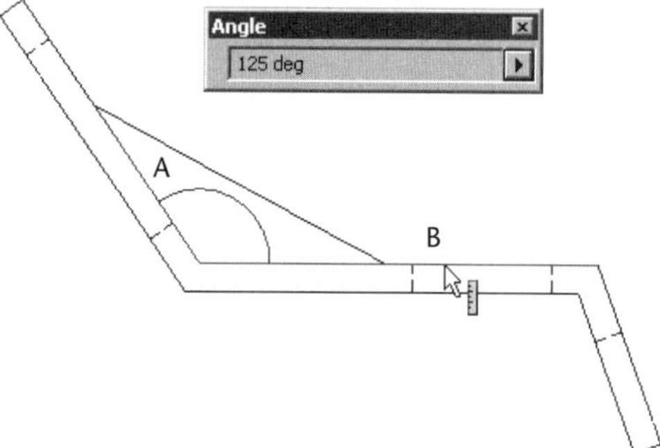


**Figure 5-56**

You could also measure an angle by using the following sequence, which corresponds to the following image:

1 Pick on the endpoint of one leg of the angle at A.
2 Pick on the vertex of the angle at B.
3 Pick on the endpoint of the other leg of the angle at C.

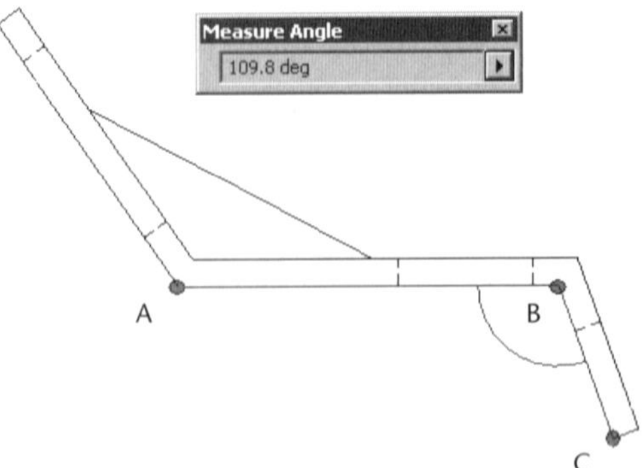


**Figure 5-57**

## ANNOTATIONS

To complete an engineering drawing, you must add annotations such as centerlines, surface texture symbols, welding symbols, geometric tolerance symbols, text, bill of materials, and balloons. Before adding annotations to a drawing, make the drawing standard active and make modifications to it as needed.

### CENTERLINES

When you need to annotate the centers of holes, circular edges, or the middle (center axis) of two lines, there are four ways to construct the needed centerlines. Use the Center Mark, Center Line Bisector, Center Line, and Centered Pattern tools from the Drawing Annotation Panel Bar as shown in the following image. The centerlines are associated to the geometry that is selected when it is created. If the geometry changes or moves, the centerlines will automatically update to reflect the change. This section outlines the guidelines for creating the different types of centerlines.

To edit a center mark, centerline bisector, centerline, or center pattern, position the mouse over it. When the green circles appear correctly, right-click and select from the corresponding Edit options from the menu.

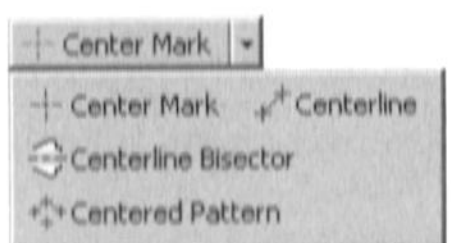


**Figure 5-58**

To add a center mark, follow these steps:

1 Click the Center Mark tool from the Drawing Annotation Panel Bar.

2 In the graphics window, select the geometry in which you want to place a center mark.

3 Continue placing center marks by selecting geometry.
4 To complete the operation, right-click and select Done from the menu.

**NOTE** If the features form a circular pattern, the center mark for the pattern is automatically placed when you have selected all of the members.

To add a centerline bisector, follow these steps:

1 Click the Centerline Bisector tool from the Drawing Annotation Panel Bar.
2 In the graphics window, select two lines between which you want to place the centerline bisector.
3 Right-click and select Create from the menu to place the centerline bisector.
4 Continue placing centerline bisectors by selecting geometry.
5 To complete the operation, right-click and select Done from the menu.

To add a centerline, follow these steps:

1 Click the Centerline tool from the Drawing Annotation Panel Bar.
2 In the graphics window, select a piece of geometry for the start of the centerline.
3 Click a second piece of geometry for the ending location.
4 Right-click and select Create from the menu to create the centerline. The centerline will be attached to the midpoints of the selected geometry.
5 Continue placing centerlines by selecting geometry.
6 To complete the operation, right-click and select Done from the menu.

To add a centered pattern, follow these steps:

1 Click the Centered Pattern tool from the Drawing Annotation Panel Bar.
2 In the graphics window, select the defining feature for the pattern to place its center mark.
3 Click the first feature of the pattern.
4 Continue selecting features in a clockwise direction until all of the features are added to the selection set.
5 Right-click and select Create from the menu to create the centered pattern.
6 To complete the operation, right-click and select Done from the menu.

### AUTOMATED CENTERLINES FROM MODELS

The ability to automatically create centerlines can automatically can eliminate a considerable amount of work. You can control what features get centerlines and marks and in what views these occur.

Automated centerlines and marks can be set as a drawing template default by picking the Automated Centerline Settings button in the Document Settings dialog box (as shown in the following image). Automated centerlines can also be set for an individual drawing view by right-clicking on the view background to display a menu. To display the Centerline Settings dialog box, click Automated Centerlines.

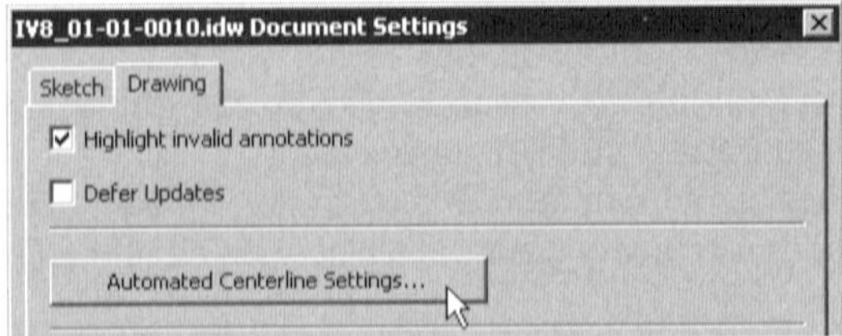


**Figure 5-59**

## Centerline Settings

The Centerline Settings dialog box (see the following image) allows you to set the type of feature to which automated centerlines will be applied (such as holes, fillets, cylinders, etc.), and the projection type (plan or profile). Each of these areas will be discussed in greater detail.

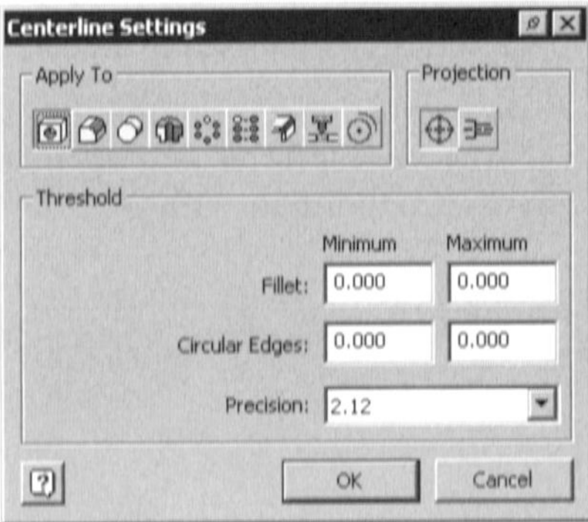


**Figure 5-60**

## Apply To

This area controls the feature type to which you want automated centerlines applied. Feature types include hole, fillet, cylinder, revolved circular patterns, rectangular patterns, sheet metal bends, punches, and circular geometry. Click on the appropriate button, and automated centerlines will be applied to all features of that type in the drawing. You can click on multiple buttons to apply automated centerlines to multiple features. To disable centerlines in a feature, click on the feature button a second time.

## Projection

Click on the projection buttons to have automated centerlines applied to plan (axis normal) and/or profile (axis parallel) views.

## Threshold

Thresholds are minimum and maximum value settings and are provided for fillet features, arcs, and circles. Any object residing within a range should get the appropriate center mark. The values are based upon the model values, not the drawing values. This allows you to know what will or will not receive a centerline regardless of the view scale in the document. For example, if you set a minimum value of 0.50 for the fillet feature, a fillet that has a radius of 0.495 will not receive a center mark. A zero value on both threshold settings (min/max) denotes no restriction. This means that center marks will be placed on all fillets regardless of size.

**NOTE** Centerline property defaults are stored in documents or templates. To store centerline property defaults in the active document, click Tools > Document Settings > Drawing. When you create automated centerlines, the defaults are displayed. You can make adjustments to the defaults before you create the centerlines. These adjusted settings apply only to the active view, and are not stored as defaults.

## TEXT POSITIONING

At times, you may need a way to align general note text, field text, and sketch text objects with each other to concatenate multiple text box objects.

To begin the process of positioning text, first select multiple text objects by pressing the **SHIFT** or **CTRL** keys as you select the text. Next, right-click to display the menu containing the Align option (as shown in the following image).

**Figure 5-61**

Selecting Align will display the Align Text dialog box (as shown in the following image).

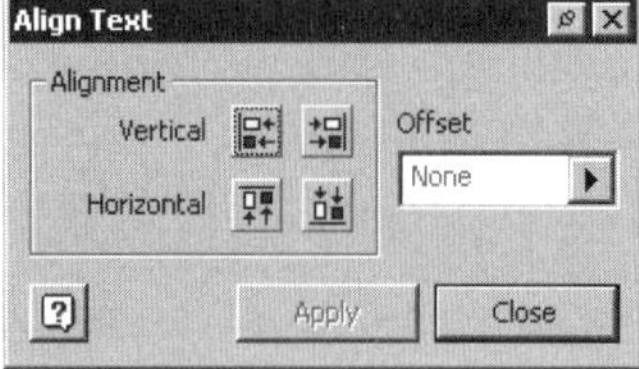


**Figure 5-62**

### Alignment

Two buttons for vertical and horizontal alignment are available to assist in the positioning of text. The Vertical alignment buttons consist of left- and right-justified controls. The Horizontal alignment buttons consist of top- and bottom-justified controls.

### Offset

This edit box allows you to apply a line-spacing value in the form of an offset to the horizontal and vertical text positions.

The first selected text object becomes the anchor point for the alignment of the other text objects. When positioning text vertically, the top text object in the following image was selected first, followed by the others (after the first text string is selected, the order of selecting the other text strings is not important).

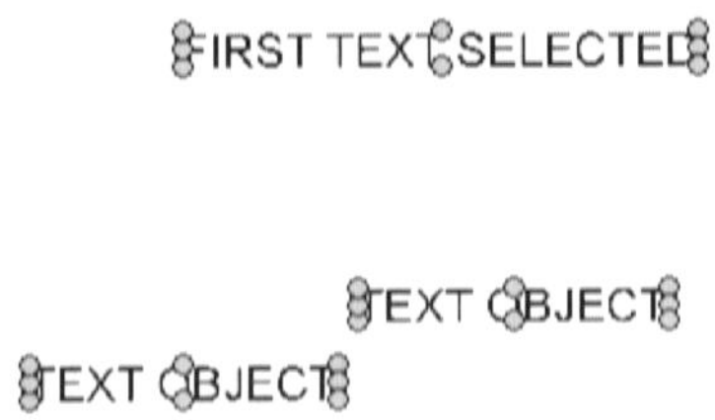

FIRST TEXT SELECTED

TEXT OBJECT

TEXT OBJECT

**Figure 5-63**

Clicking on the Vertical Align Left button displays the results in the following image. In this image, the base point of the first selected text object becomes the new base point for all other selected text objects.

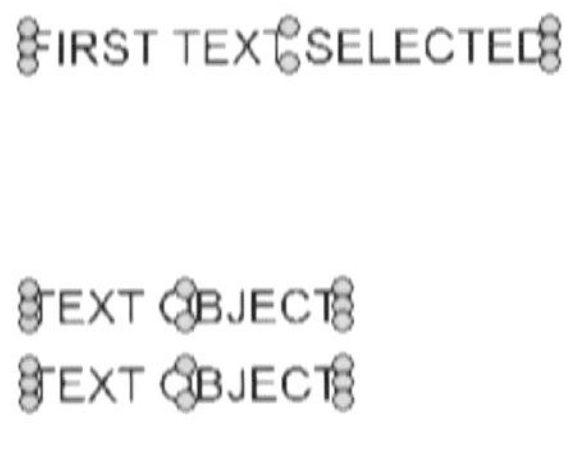

FIRST TEXT SELECTED

TEXT OBJECT

TEXT OBJECT

**Figure 5-64**

If you enter an offset value such as 0.30 and again click on the Vertical Align Left button, all selected text objects share the vertical spacing between each other, as illustrated in the following image.

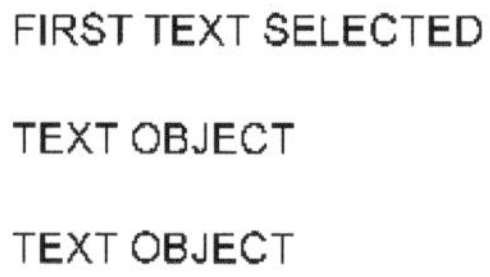

FIRST TEXT SELECTED

TEXT OBJECT

TEXT OBJECT

**Figure 5-65**

Horizontal positioning of text is similar to vertical, in that the first selected text object becomes the base point for all other selected text objects. In the following image, the top text object is selected first, followed by the others.

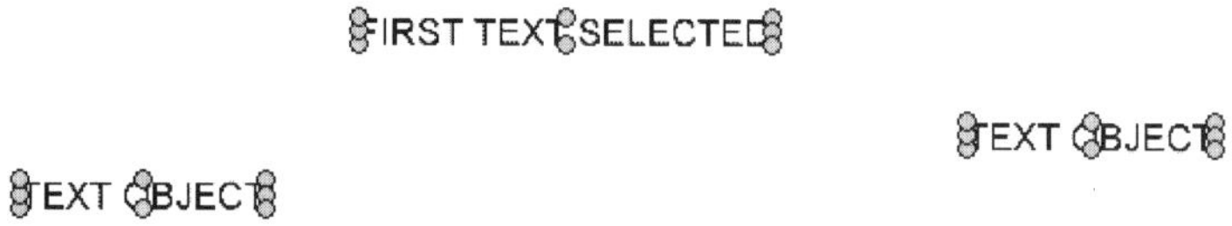

FIRST TEXT SELECTED

TEXT OBJECT

TEXT OBJECT

**Figure 5-66**

Clicking the Horizontal Top Align button displays the results in the following image. A Bottom Align button is also available if dealing with lower case letters.

TEXT OBJECT FIRST TEXT SELECTED TEXT OBJECT

**Figure 5-67**

## SURFACE TEXTURES, WELDING SYMBOLS, AND FEATURE CONTROL FRAMES

To add more detail annotations to your drawing, you can add surface texture symbols, welding symbols, and feature control frames by clicking the corresponding tool from the Drawing Annotation Panel Bar, as shown in the following image.

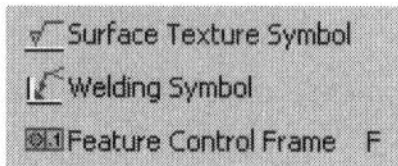


**Figure 5-68**

You should then follow these steps:

1 Click on the tool from the Drawing Annotation Panel Bar.
2 Select a point at which the leader will start.
3 Continue selecting points to position the extension lines.
4 Right-click and select Continue from the menu.

5 Fill in the information as needed in the dialog box.
6 When done, click the OK button in the dialog box.
7 To complete the operation, right-click and select Done from the menu.
8 To edit a symbol, position the mouse over it. When the green circles appear, right-click and click the corresponding Edit option from the menu.

### TEXT AND LEADERS

To add text to the drawing, click either the Text or Leader Text tool from the Drawing Annotation Panel Bar as shown in the following image.

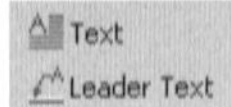


**Figure 5-69**

The Text tool will add text while the Leader Text tool will add a leader and text. When placing text through the Format Text dialog box, as shown in the following image, select the orientation and text style as needed and type in the text; click the OK button to place the text in the drawing. To edit the text or text leader position, move your cursor over it. When the green circles appear, right-click and click the corresponding Edit option from the menu.

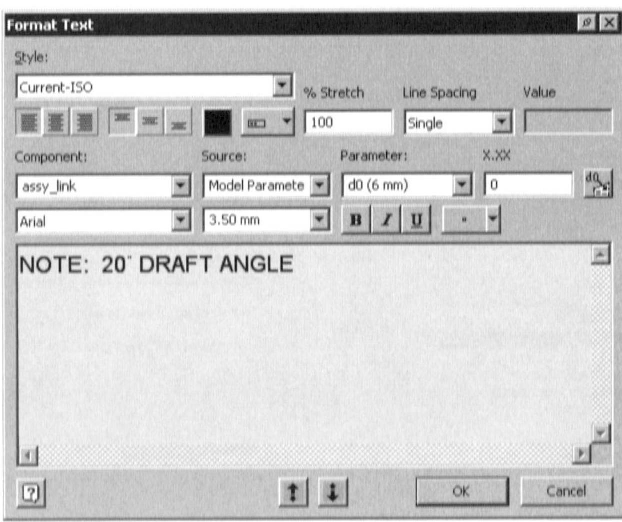


**Figure 5-70**

## HOLE AND THREAD NOTES

Another annotation that can be added is a hole or thread note. Before a hole or thread note can be placed in a drawing, a hole or thread feature must exist (extruded circles cannot be annotated as a hole), as well as a drawing view that shows the hole in a plan view. If the hole or thread feature changes, the note will be updated automatically to reflect the change.

To create a hole or thread note, follow these steps:

1 Click the Hole/Thread Notes tool from the Drawing Annotation Panel Bar as shown in the following image.

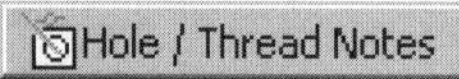


**Figure 5-71**

2 Select the hole or thread feature that will be annotated.

3 Select a second point to locate the leader and the note.

4 To complete the operation, right-click and select Done from the menu.

5 To edit a note, position the mouse over it. When the green circles appear, right-click and click the corresponding Edit option from the menu.

## HOLE NOTE STYLES

Hole note formatting enables you to define a style for global formatting of a hole note, as well as edit existing hole notes and apply specific formatting on an as-needed basis. This will allow the flexibility to conform or create your own standard for the formatting of hole notes.

To create a hole note style, follow these steps:

1 Access the Hole Note Style tab through Format/Dimension Styles/Holes (see the following image).

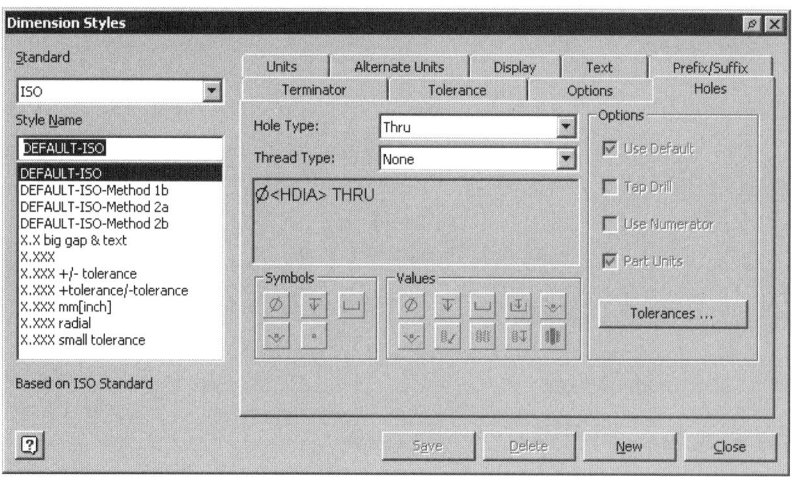


**Figure 5-72**

2 The default Standard style is modifiable by clicking New on the Holes tab.

3 Under the Options area of the Holes tab, remove the check from the box next to Use Default. This will activate the Hole Note editor.

4 Select the hole type to format a note from the type list.

5 Change the Value or Symbol depending on the desired hole format.

6 If desired, apply tolerance values to the Hole Note Edit dialog box through the Hole Tolerances dialog box (as shown in the following image).

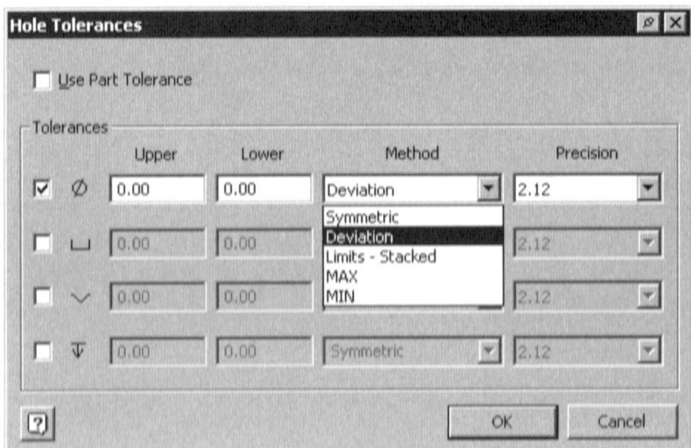


**Figure 5-73**

7 Click the Save button to save the newly created hole note style.

## EDITING HOLE NOTES

1 A hole note edit is invoked by right-clicking on an existing hole note to display the menu shown in the following image.

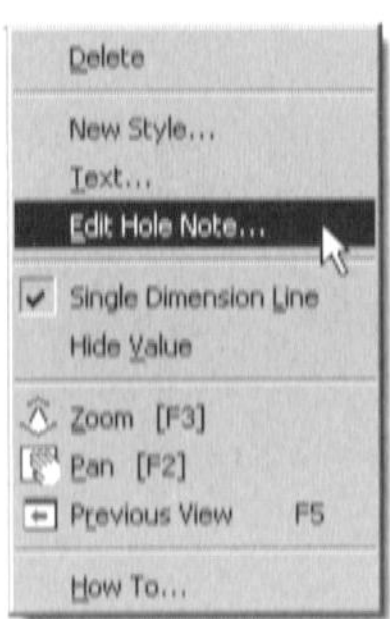


**Figure 5-74**

2 This will display the Hole Note Edit dialog box illustrated in the following image. Edits can be performed on a single hole note.

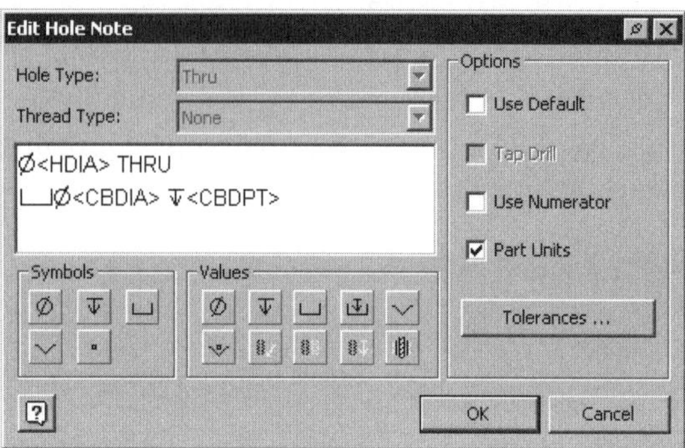


**Figure 5-75**

# EXERCISE 5-4
# Creating Text and Dimension Styles

In this exercise, you create new text and dimension and styles. To navigate to the exercise in the *Electronic Student Workbook* do the following:

1 From the Main TOC page, click Chapter 5.
2 From the TOC for Chapter 5, click Creating Text and Dimension Styles.

The following image illustrates the completed exercise.

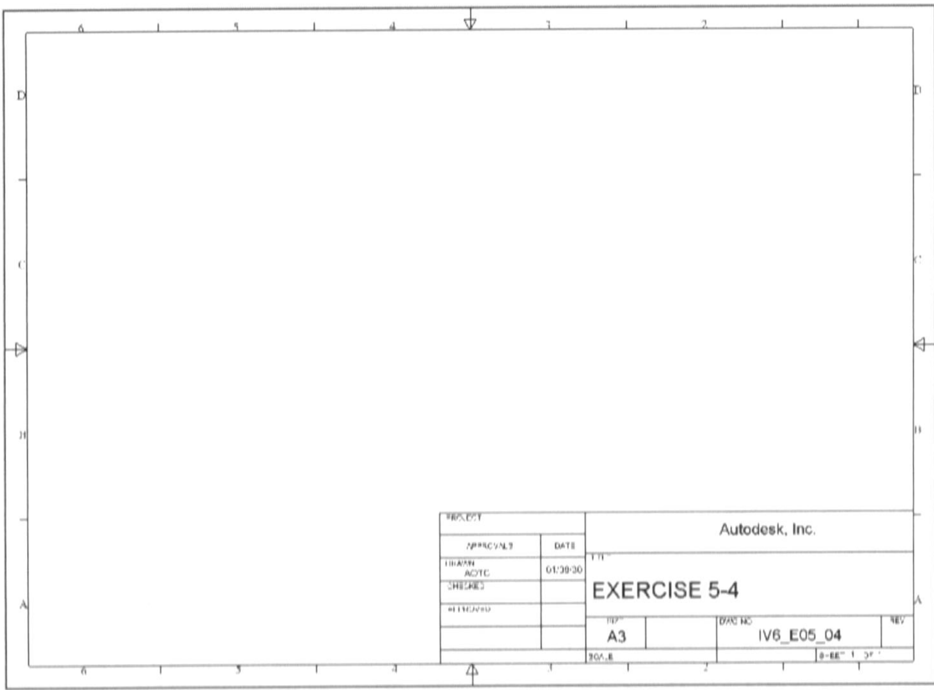

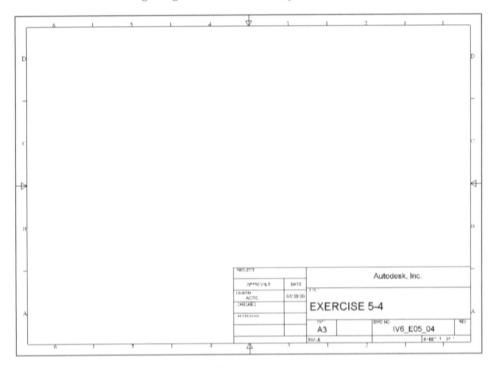

**Figure 5-76 Completed exercise**

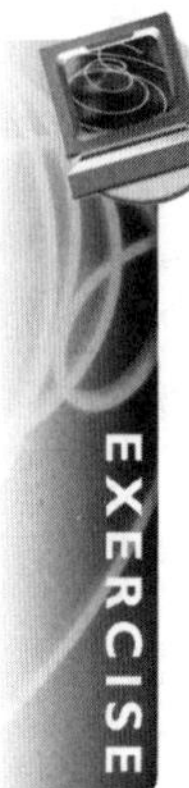


# EXERCISE 5-5
# Adding Dimensions and Annotations

In this exercise, you add dimensions and annotations to a part drawing. To navigate to the exercise in the *Electronic Student Workbook* do the following:

1 From the Main TOC page, click Chapter 5.

2 From the TOC for Chapter 5, click Adding Dimensions and Annotations.

The following image illustrates the opened exercise.

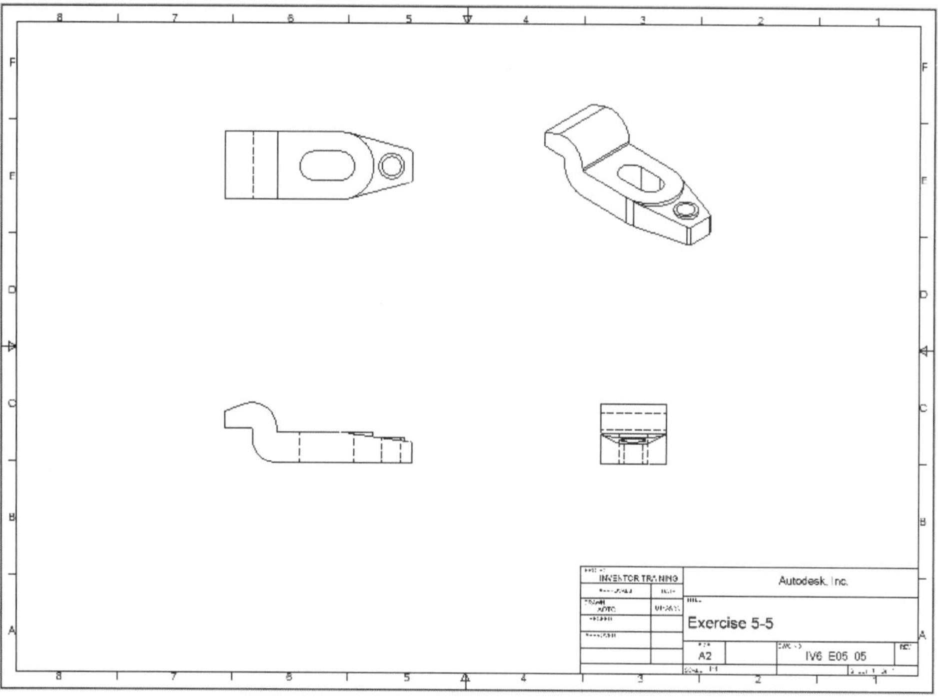


**Figure 5-77 Opened exercise**

The following image illustrates the completed exercise.

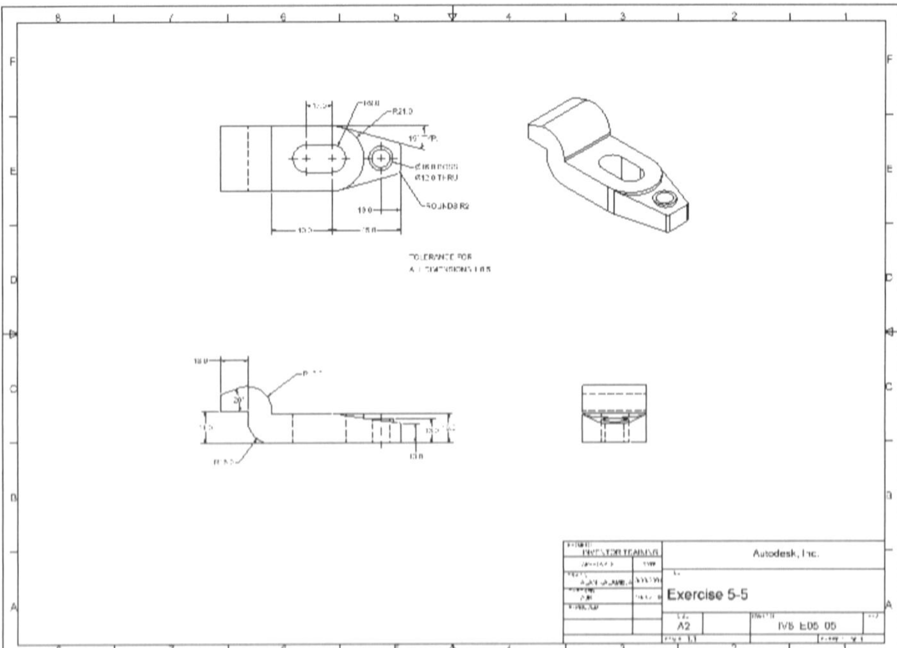

**Figure 5-78 Completed exercise**

## REVIEW SEQUENCE FOR CREATING DRAWING VIEWS

The following is a brief outline reviewing the steps that are needed to create a new sheet with drawing views.

1. Create a new sheet or select an existing drawing sheet from the Browser.
2. Add a border.
3. Add a title block.
4. Create drawing views based on a part, assembly, or presentation file.
5. Get model dimensions if not done in conjunction with Step 4.
6. Edit drawing views, if necessary.
7. Add additional dimensions and annotations.

## CHAPTER SUMMARY

| To | Do This | Tool |
|---|---|---|
| Set drafting standards | Select Format > Standards. | |
| Set sheet size | Right-click the sheet and select Edit Sheet. | |
| Insert a border | Double-click the border in the Browser. | |
| Insert a title block | Double-click the title block in the Browser. | |
| Edit a title block | Right-click the title block in the Browser's Drawing Resources section and select Edit. | |
| Create an independent view | Click the Base View tool in the Drawing Views Panel Bar or from the Drawing Views toolbar. |  |
| Create a projected view | Click the Projected Views tool in the Drawing Views Panel Bar or from the Drawing Views toolbar.<br>Select the base view, right-click and select Create View > Projected. |  |
| Create an auxiliary view | Click the Auxiliary View tool from the Drawing Views Panel Bar. |  |
| Create a section view | Click the Section View tool from the Drawing Views Panel Bar. |  |
| Create a detail view | Click the Detail View tool from the Drawing Views Panel Bar. |  |
| Retrieve a model dimension | Right-click a view and select Retrieve Model Dimensions. |  |
| Place a drawing dimension | Click the General Dimension tool in the Drawing Annotation Panel Bar or from the Drawing Annotation toolbar. |  |
| Place center marks and centerlines | Click one of the four tools in the Drawing Annotation Panel Bar or from the Drawing Annotation toolbar. |  |
| Place a text note | Click the Text tool in the Drawing Annotation Panel Bar or from the Drawing Annotation toolbar. |  |
| Place a leader | Click the Leader Text tool in the Drawing Annotation Panel Bar or from the Drawing Annotation toolbar. | 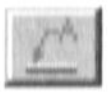 |
| Enter drawing properties | Select File > iProperties. | |
| Print a drawing | Select File > Print. | |

# Applying Your Skills

## SKILL EXERCISE 5-1

In this exercise, you create a working drawing of a part. To navigate to the exercise in the *Electronic Student Workbook* do the following:

1. From the Main TOC page, click Chapter 5.
2. From the TOC for Chapter 5, click Exercise 1 under Applying Your Skills.

The following image illustrates the completed exercise.

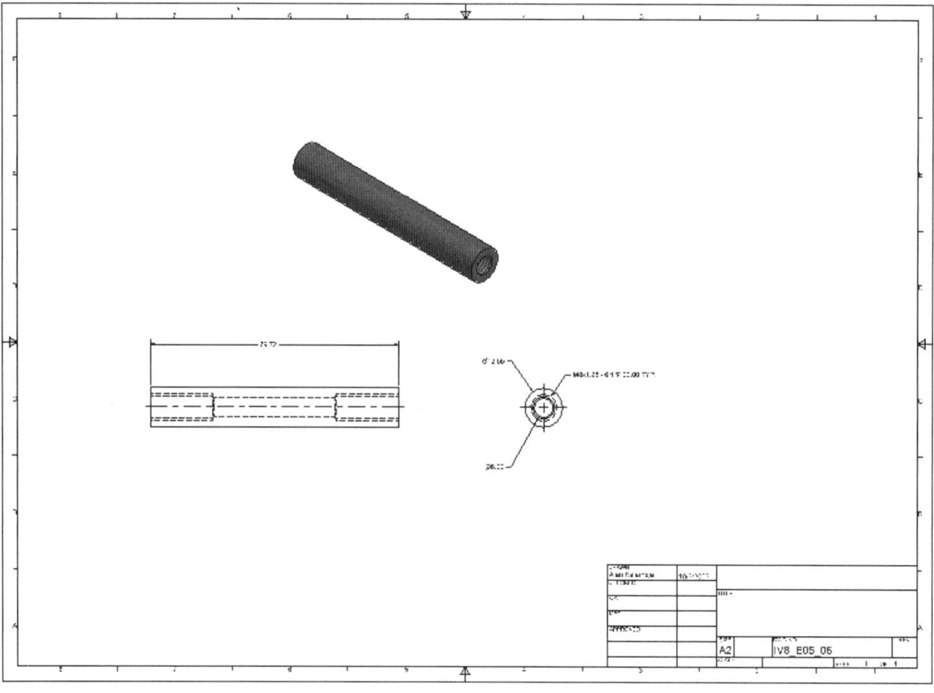


**Figure 5-79 Completed exercise**

## SKILL EXERCISE 5-2

In this exercise, you create a drawing for a drain plate cover. To navigate to the exercise in the *Electronic Student Workbook*:

1 From the Main TOC page, click Chapter 5.

2 From the TOC for Chapter 5, click Exercise 2 under Applying Your Skills.

The following image illustrates the completed exercise.

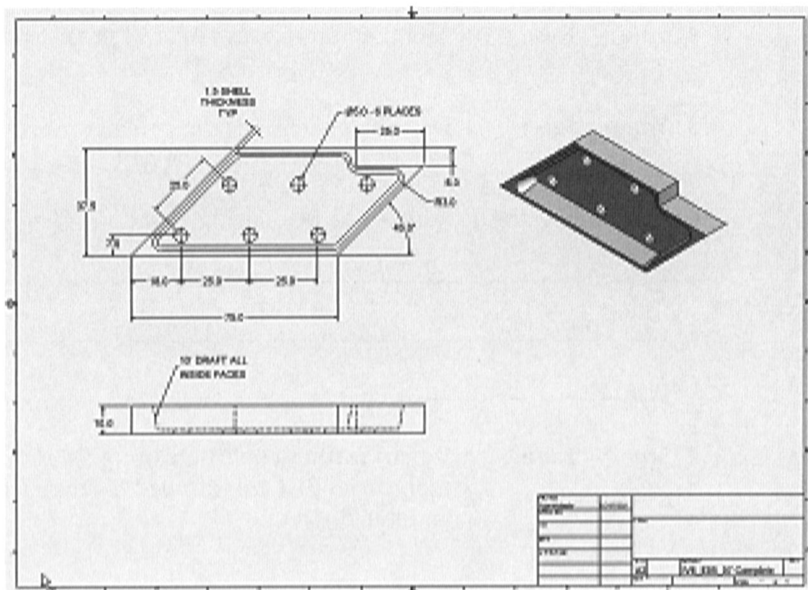

**Figure 5-80 Completed exercise**

## CHECKING YOUR SKILLS

1 **True___ False___** A drawing can have an unlimited number of sheets.

2 **True___ False___** A drawing's sheet size is normally scaled to fit the size of the drawing views.

3 **True___ False___** There can only be one base view per sheet.

4 **True___ False___** An isometric view can only be projected from a base view.

5 **True___ False___** Drawing dimensions can parametrically drive dimensional changes back to the part.

6 Explain how to shade an isometric drawing view.

_______________________________________________

_______________________________________________

_______________________________________________

7 **True___ False___** When creating a hole note using the Hole/Thread Notes tool, circles that are extruded to create a hole can be annotated.

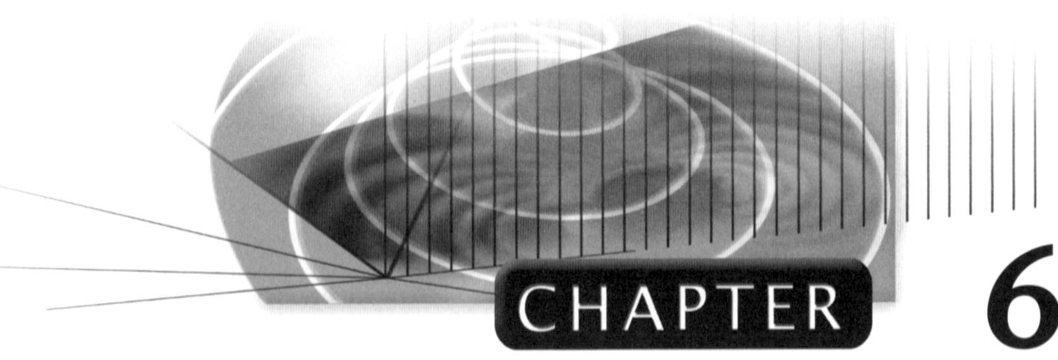

# CHAPTER 6

# Creating and Documenting Assemblies

In the first three chapters, you learned how to create a component in its own file. In this chapter, you will learn how to place individual component files into an assembly file. This process is referred to as bottom-up assembly modeling. You will also learn to create components in the contents of the assembly file, which is referred to as top-down assembly modeling. After creating components, you will then learn to constrain the components to one another using assembly constraints, edit the assembly constraints, check for interference, and learn how to create presentation files that show how the components are assembled or disassembled.

## CHAPTER OBJECTIVES

**After completing this chapter, you will be able to**

- Understand the assembly options
- Create bottom-up assemblies
- Create top-down assemblies
- Create subassemblies
- Constrain components together using assembly constraints
- Edit assembly constraints
- Check for interference
- Create a presentation file
- Create balloons
- Create a parts list of an assembly

## CREATING ASSEMBLIES

As you previously learned, component files have the IPT extension and a component file can only have one component in it. In this chapter, you will learn how to create assembly files (IAM file extension). An assembly file holds the information that is needed to assemble the components together. All of the components in an assembly are *referenced in*, meaning that each component exists in its own component IPT file and its definition is linked into the assembly. You can edit the components while in the assembly, or you can open the component file and edit it. When the changes are made to a component and the component is saved, the changes will be reflected in the assembly after the assembly is opened or updated. There are three methods for creating assemblies: bottom-up, top-down, and a combination of both top-down and bottom-up techniques. *Bottom-up* refers to an assembly in which all of the components were created in individual component files and are referenced into the assembly. A top-down approach refers to an assembly in which the components are created from within the context of the assembly. In other words, the user creates each component from within the top-level assembly. Each component in the assembly is saved to its own component (IPT) file. Both the bottom-up and top-down assembly techniques will be covered in the following sections.

To create a new assembly file, select the New icon from What To Do and then select the *Standard.iam* icon from the Default template tab, as shown in the following image. There are two other methods for creating a new assembly: you can select New from the File menu and then select the *Standard.iam* icon or you can select the down arrow of the New icon from the left side of the Standard toolbar and select Assembly. After issuing the new component operation, Autodesk Inventor's tools will change to reflect the new assembly environment. The assembly tools will be shown in the Panel Bar. They will be covered throughout this chapter.

**NOTE** There is no correct or incorrect way to create an assembly. You will determine which method works best for the assembly that you are creating based on experience. An assembly can be created using both the bottom-up and top-down assembly techniques. Whether you place or create the components in the assembly, all of the components will be saved out to their own individual IPT files and the assembly will be saved as an IAM file.

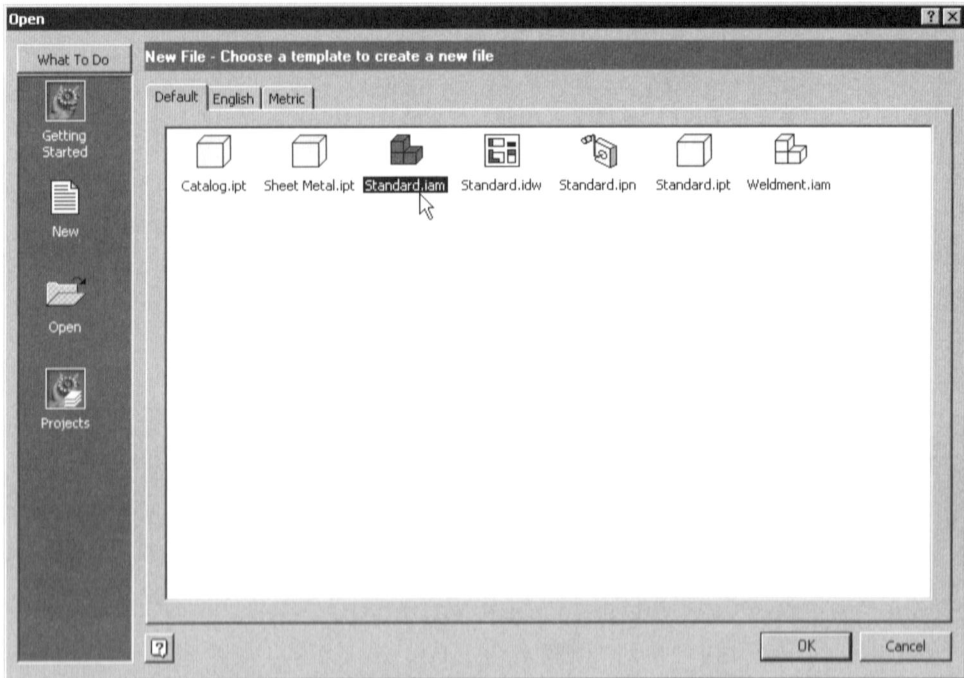


Figure 6-1

## ASSEMBLY OPTIONS

Before creating an assembly, let's examine the assembly option settings. Under the Tools menu, select Application Options and the Options dialog box will appear. Select the Assembly tab and your screen should look similar to the following image. The following items describe the various assembly settings. These settings are global and will affect how new components are created or placed in the assembly.

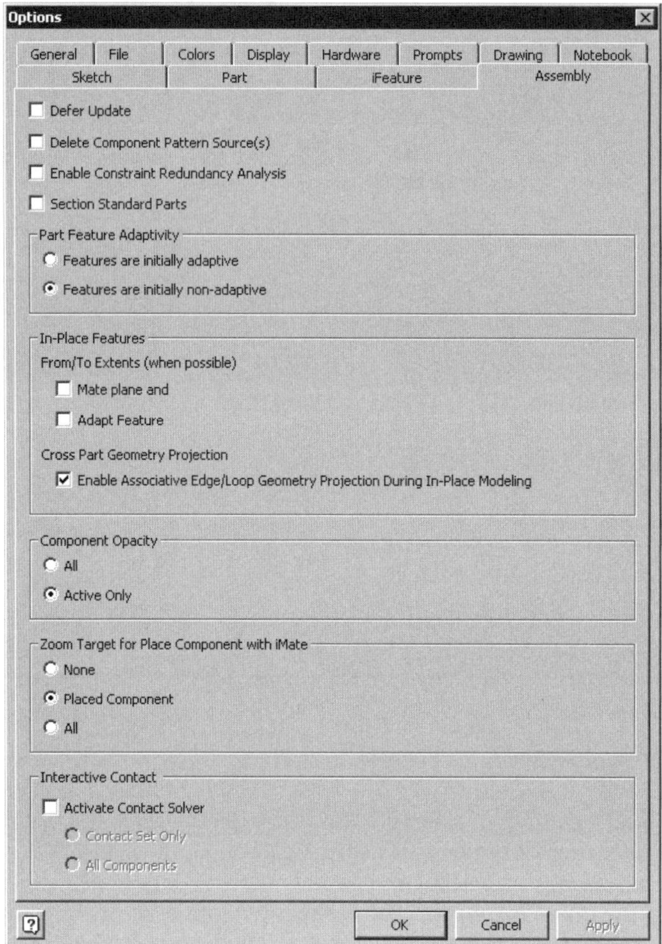


**Figure 6-2**

**Defer Update** When checked and a component is changed that will affect the assembly, you will have to manually click the Update button to get the assembly up-to-date. If unchecked, the assembly will automatically update to reflect the change to a component.

**Delete Component Pattern Source(s)** When checked, the source component will automatically be deleted when pattern elements are deleted. If unchecked, the source component will not be deleted when the pattern elements are deleted.

**Enable Constraint Redundancy Analysis** When checked, a secondary analysis of all assembled components for adaptivity adjustments is performed. You are notified when redundant constraints exist. Also, the Degrees of Freedom are updated but not displayed.

**Section Standard Parts** When checked, standard parts will be sectioned when a section drawing view is created. To prevent the sectioning of standard parts, leave this box unchecked.

**NOTE** Part Feature Adaptivity, In-Place Features, and Cross Part Geometry Projection will be covered later in this chapter under the heading of Adaptivity.

### Component Opacity

In this section, you determine if all or only the active component will be opaque when an assembly cross-section is displayed.

**All** When checked, all components in the assembly will be opaque.

**Active Only** When checked, only the active component in the assembly will be opaque; the others will be dimmed.

### Zoom Target for Place Component with iMate

This section sets the default zoom behavior for the graphics window when placing components with iMates.

**None** When checked, no zooming is performed. This leaves the graphics display as is.

**Placed Component** Zooms in on the placed part so that it fills the graphics window.

**All** Zoom the assembly so that all elements in the model fit the graphics window.

**NOTE** The Interactive Contact section of the Options dialog box will be covered in Chapter 10.

## THE ASSEMBLY BROWSER

The Assembly Browser in the following image displays the hierarchy of all component occurrences in the assembly. Each occurrence of a component is represented by a unique name. From the Browser, you can select a component for editing, move components between assembly levels, control component status, rename components, edit assembly constraints, and manage design views.

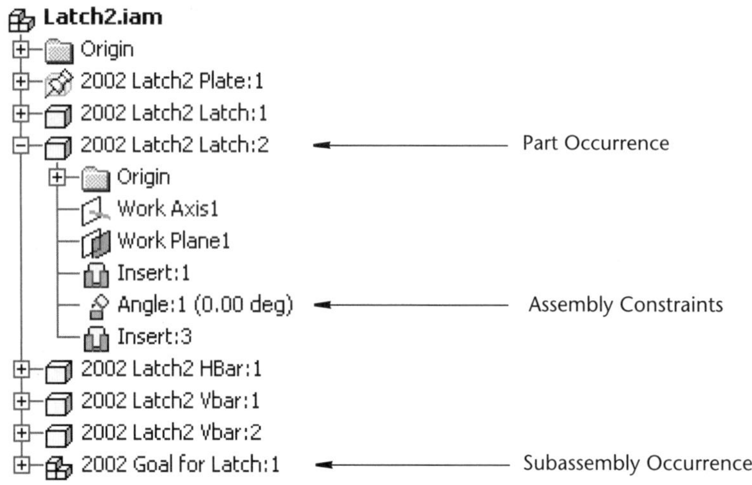


**Figure 6-3**

## BOTTOM-UP APPROACH

The *bottom-up assembly approach* uses files that are referenced to an assembly file. To create a bottom-up assembly, create the components in their individual files. If an assembly file is placed into another assembly, it will be brought in as a *subassembly.* Any drawing views created for these files will not be brought into the assembly file. You will want to make sure that the path(s) for the file location(s) of the placed component(s) is represented in the project file on which the assembly is based; otherwise you will not be able to locate the component when the assembly is reopened.

To insert a component into the current assembly, select the Place Component tool from the Assembly Panel Bar, as shown in the following image, press the hot key **P**, or right-click and select Place Component from the menu.

**Figure 6-4**

The Open dialog box will appear, as shown in the following image. You then pick a point in the assembly where you want the component located. If multiple occurrences of the component are needed in the assembly, continue selecting points. When done, press the **ESC** key or right-click and select Done from the menu.

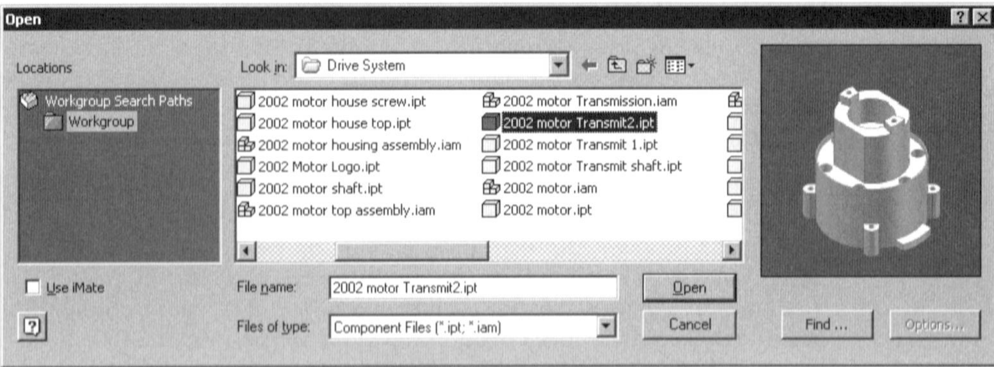


Figure 6-5

## TOP-DOWN APPROACH

You can create new components while in an assembly. This method is referred to as the *top-down assembly approach*. To create a new component in the current assembly, select the Create Component tool from the Assembly Panel Bar as shown in the following image and press the shortcut key **N**, or right-click and select Create Component from the menu.

Figure 6-6

The Create In-Place Component dialog box will appear, as shown in the following image. Enter in a new name, file type, a location where it will be saved, the template file on which to base it, and determine if the component will be constrained to a face on another component. If the Constrain sketch plane to selected face or plane box is checked, a flush constraint will be applied between the selected face and the new plane. If it is unchecked, no constraint will be applied. You will still select a face, however, on which to start sketching the new component.

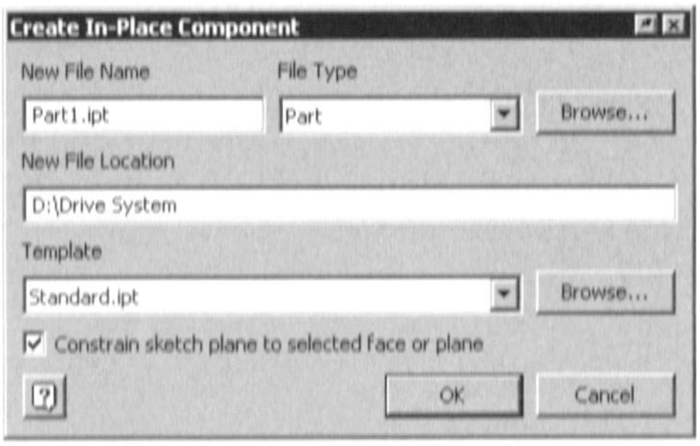

Create In-Place Component
New File Name: Part1.ipt
File Type: Part
Browse...
New File Location: D:\Drive System
Template: Standard.ipt
Browse...
Constrain sketch plane to selected face or plane
OK
Cancel

Figure 6-7

## OCCURRENCES

An *occurrence* is a copy of an existing component and has the same name as the original component, but the number after the colon is sequenced. If the original component is named Bracket:1, for example, the occurrence will be Bracket:2 (see the following image). If the original component or an occurrence of the component changes, all of the components will reflect the change.

To create an occurrence, you can place the component multiple times—if the component already exists, you can select the component's icon in the Browser and drag an occurrence into the assembly. The *copy and paste* method can also be used to place components by right-clicking on the component name in the Browser or on the component itself and then right-clicking and selecting Copy from the menu. You can then right-click and select Paste from the menu, or press the CTRL and V keys at the same time. If you want an occurrence of the original component to have no relationship with its source component, you will need to make the original component active and then issue the Save Copy As tool from the File menu and enter a new name. The new component will have no relationship to the original and can be placed into the assembly.

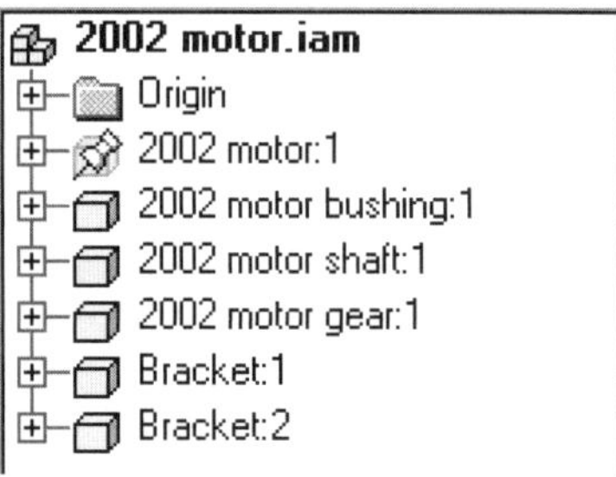


**Figure 6-8**

## MULTIPLE DOCUMENTS AND DRAG & DROP COMPONENTS

In Autodesk Inventor you can open as many files as needed. This is referred to as a *multiple document environment*. The screen can be split as needed to show all of the files that are open by using options under the Window menu. You can switch between the open files to model, edit, or interrogate the files as needed. With multiple documents open, you can drag a component from one file into an assembly. To do so, both the assembly and the component files must be open and the screen split so both files are visible. Then, select the component's name in the Browser with the left mouse button and drag it into the assembly file. Release the mouse button.

**NOTE** You can press the **CTRL** and **TAB** keys to switch between various files that are currently open.

## ACTIVE COMPONENT

To edit a component while in the assembly, you need to use the active component. Only one component in the assembly can be active at a time. To make a component active, you double-click on the component in the graphics area—you can double-click on the file name or an icon in the Browser, or you can right-click on the component name in the Browser and select Edit from the menu. Once the component is active, the other component names will be grayed out, as shown in the following image. If Component Opacity—from Application Options under the Assembly tab—is set to Active Only, the other components in the assembly will be dimmed.

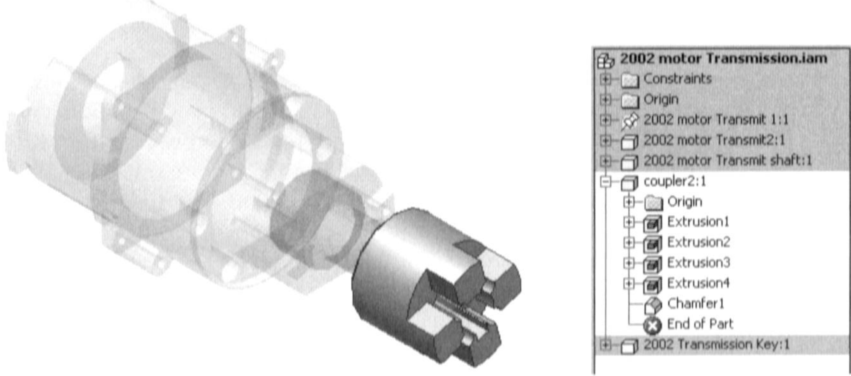


Figure 6-9

You can then edit the component as you learned to in previous chapters and save the changes by using the Save tool. Only the active component will be saved. To make the assembly active, either click the Return button from the Command Bar (as shown in the following image), double-click on the assembly name in the Browser, or right-click in the graphics window and select Finish Edit from the menu.

Figure 6-10

## OPEN AND EDIT

Another method for editing a component in the assembly is to open the component in another window. This can be done by selecting the Open tool from the Standard toolbar, select Open from the File menu, or right-click on the component's name in the Browser and select Open from the menu. The component will appear in a new window—edit the component as needed, save the changes, and then activate the assembly file and the changes will appear in the assembly.

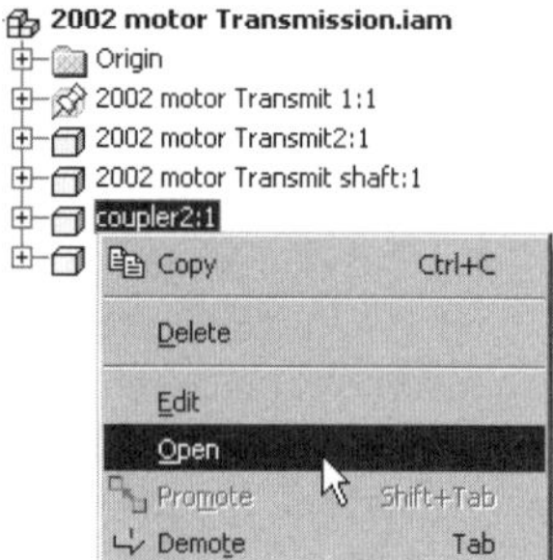


Figure 6-11

## GROUNDED COMPONENTS

When assembling components together, you may want to have a component—or multiple components—that are grounded (stationary), meaning that they will not move. When applying assembly constraints, the other components will be moved to the grounded component(s). By default, the first component in the assembly is grounded. There is no limit to how many components can be grounded. It is recommended that at least one component in the assembly be grounded, otherwise the assembly will be able to move. A grounded component is represented with a pushpin before its name in the Browser. To ground or *unground* a component, right-click on the part's name in the Browser and select (or deselect) Grounded from the menu, as shown in the following image.

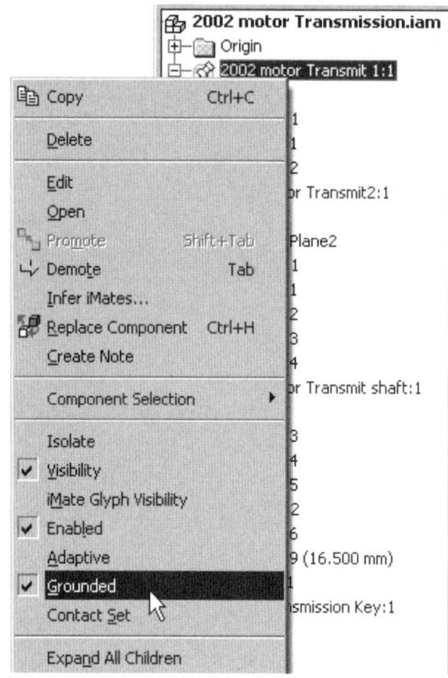


Figure 6-12

## SUBASSEMBLIES

While working, you may want to group components together into a subassembly. Two methods are used to create a subassembly. An existing assembly can be placed into another assembly or a new assembly can be created from within the assembly. To create a subassembly from *within* the assembly, issue the Create Component tool then change the file type to "Assembly" as shown in the following image. Any subsequent component that is placed or created will be a component of the subassembly.

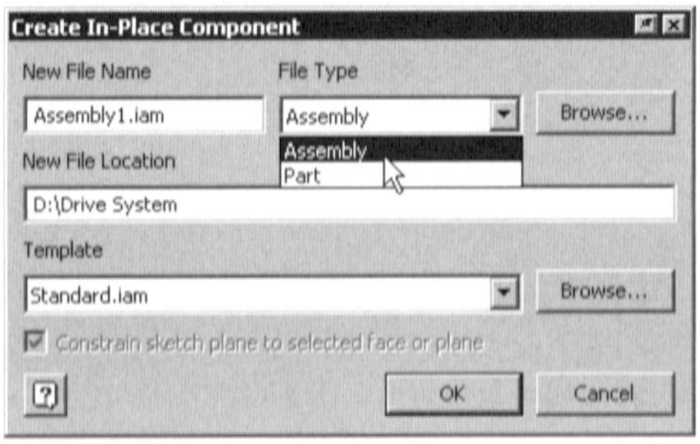

**Figure 6-13**

To make a subassembly active, double-click on the subassembly's name in the Browser. To make the top-level assembly active, select the Return tool from the Command Bar or double-click on the top-level assembly's name in the Browser. When working in the top level of the assembly, any subassembly will act as a single component when selected. Components can also be promoted into a subassembly or demoted from a subassembly by selecting their name in the Browser and selecting either Promote or Demote from the menu shown in the following image.

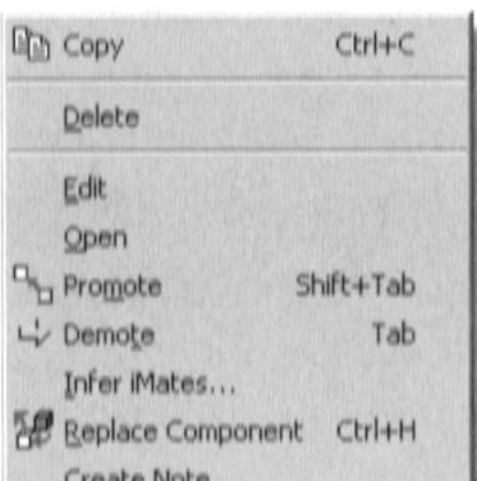


**Figure 6-14**

## RESTRUCTURING COMPONENTS

After components are in an assembly, they can be restructured or moved from the main assembly to another subassembly, from subassembly to subassembly, or from subassembly to the main assembly. To restructure a component, select the component to be restructured in the Browser, press and hold the left mouse button, and move the component up or down in the

Browser. A line appears to show where it will be placed. Keep moving the mouse until the line is positioned in the correct place, then release the left mouse button. The following image shows how the Browser looks when a component is being restructured in a subassembly. When restructuring components into or from a subassembly, a dialog box may appear stating that assembly constraints may be lost. After you complete the restructuring, examine the assembly constraints to verify that they are still valid.

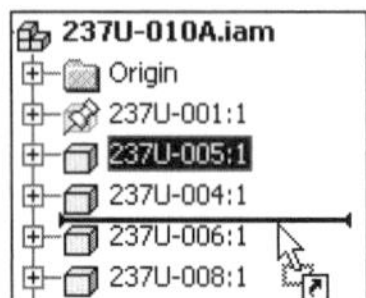


Figure 6-15

## ASSEMBLY CONSTRAINTS

In previous sections, you have learned how to create assembly files, but the components did not have any relationship to each other except for the relationship that was set when a component was created in reference to a face on another component. If a bolt was placed in a hole, for example, and the hole moved, the bolt would not move to the new hole position. Assembly constraints are used to create relationships between components. If the hole moves, for example, the bolt will move to the new hole location. In a previous chapter, you learned about geometric constraints.

When geometric constraints are applied to sketches, they reduce the number of dimensions or constraints required to fully constrain a profile. When assembly constraints are applied, they reduce the degrees of freedom (DOF) that allow the components to move freely in space.

There are six degrees of freedom: three translational and three rotational. Translational means that a component can move along an axis: *X*, *Y*, or *Z*. Rotational means that a component can rotate about an axis: *X*, *Y*, or *Z*. As assembly constraints are applied, the number of DOF will be decreased.

Autodesk Inventor does not require components to be fully constrained. By default, the first component created or added to the assembly will be grounded and will have zero DOF. More than one component can be grounded, as discussed earlier in this chapter. Other components will move in relation to the grounded component(s). To see a graphical display of the DOF remaining on all of the components in an assembly, select Degrees of Freedom from the View menu, as shown in the following image.

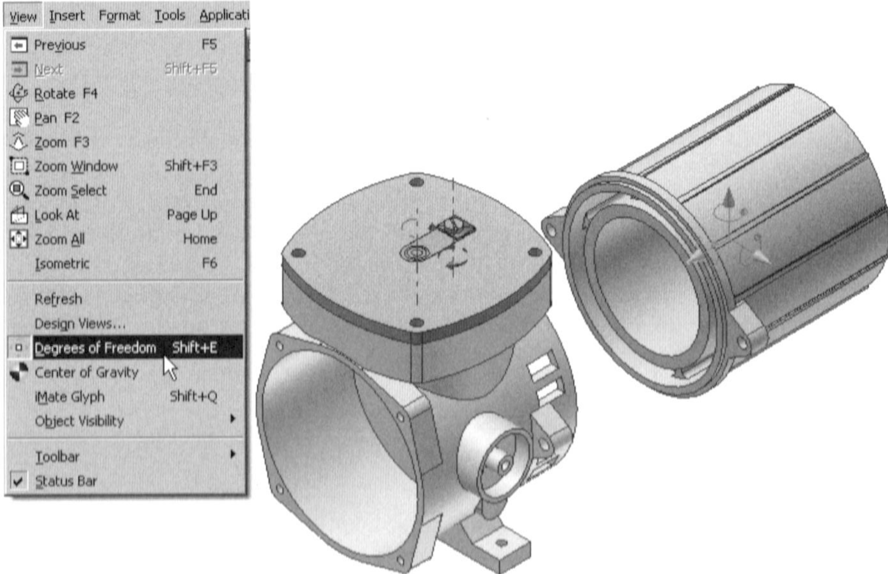


**Figure 6-16**

An icon will appear in the center of the component that shows the DOF remaining on the component. The line and arrows represent translational freedom, and the arc and arrows represent rotational freedom. To turn off the DOF icons, again select Degrees of Freedom from the View menu.

**TIP** You can turn on the DOF symbols for a single (or a few) component(s) by right-clicking on the component's name in the Browser, selecting Properties from the menu, then selecting the Degrees of Freedom checkbox on the Occurrence tab. Doing the same steps will toggle the symbols off once they have been turned on.

When placing or creating components in an assembly, it is recommended to have them in the order in which they will be assembled. The order will be important when placing assembly constraints and creating presentation views. When constraining components to one another, you will need to understand the terminology. The following is a list of terminology used with assembly constraints:

**Line** Can be the centerline of an arc or circular edge, a selected edge, or a work axis.

**Normal** A vector that is perpendicular to the outside of a planar face.

**Plane** Can be defined by the selection of a plane or face: two non-collinear lines, three non-linear points, or one line and a point that does not fall on the line. When edges and points are used to select a plane, this is referred to as a construction plane.

**Point** Can be an endpoint, the midpoint of a line, or the center of an arc or circular edge.

**Offset** The distance between two selected lines, planes, points, or any combination of the three.

### TYPES OF CONSTRAINTS

Autodesk Inventor has four types of assembly constraints (mate, angle, tangent, and insert), two types of motion constraints (rotation and rotation-translation), and a transitional constraint. The constraints can be accessed through the Constraint tool from the Assembly Panel Bar, as shown in the following image, by right-clicking and selecting Constraint from the menu, or by using the hot key C.

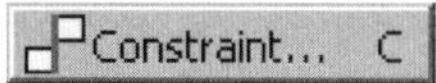


**Figure 6-17**

The Place Constraint dialog box will appear as shown in the following image. The dialog box is divided into four areas (the following sections describe these areas). Depending upon the constraint type, the option titles may change.

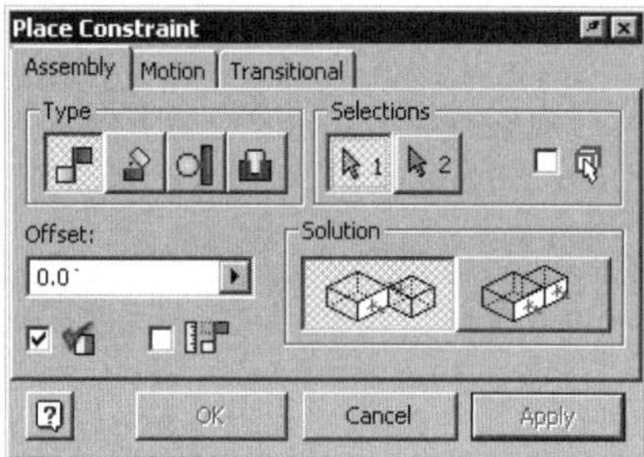


**Figure 6-18**

## The Assembly Tab

**Type** Select the type of assembly constraint to apply: mate, angle, tangent, or insert.

**Selections** In this area, click the button with the number 1 and then select a component's edge, face, point, etc. on which to base the constraint type. Then click the button with the number 2, and select a component's edge, face, point, etc. on which to base the constraint type. By default, the second arrow will become active after you have selected the first input. You can edit an edge, face, point, etc. of an assembly constraint that has already been applied, by clicking the number button that corresponds to the constraint and then selecting a new edge, face, point, etc. While working on complex assemblies, you can check the box on the right side of Selections (called Pick part first). If it is checked, you will need to select the component before selecting a component's edge, face, point, etc.

**Offset/Angle** Enter, or select, a value for the offset or angle from the drop list.

**Solution** In this area, you can select how the constraint will be applied; either the normals will be pointing in the same or opposite directions.

**Show Preview** When this box is checked and constraints are applied to two components, you will see the underconstrained components move to their constrained positions. If this box is left unchecked, you will not see the components assembled until you click the Apply button.

**Predict Offset and Orientation** When this box is checked, the offset distance between two constrained components will be displayed. This allows you to either accept this offset distance or enter a new offset distance in the edit box.

### The Motion Tab

**Type** Select the type of assembly constraint to apply: rotation or rotation-translation.

**Selections** In this area, click the button with the number 1 and then select a component's face or axis on which to base the constraint type. You will see a glyph showing the direction of rotation motion. Then, click the button with the number 2 and select the component's axis or face on which to base the constraint type. A second glyph appears showing the direction of rotation.

**Ratio** Enter, or select, a value for the ratio from the drop list.

**Solution** In this area, you can select how the constraint type will be applied. The components will either rotate in the same or opposite directions.

### The Transitional Tab

A transitional constraint will maintain contact between the two selected faces. A transitional constraint can be used between a cylindrical face and a set of tangent faces on another part.

**Type** Select *transitional* as the type of assembly constraint to apply.

**Selections** In this area, click the First Selection button and select the first face on the part that will be moving. Click the Second Selection button and then select a face around which the first part will be moving. If there are tangent faces, they will automatically become part of the selected face.

## CONSTRAINT TYPES

This section explains each of the assembly constraint types.

### Mate

There are three types of mate constraints: plane, line, and point.

## Mate Plane

The mate plane constraint assembles two components so that the surface normals on the selected planes will be opposite one another. In the following image, the mate condition is being applied (it is selected from Solution in the Place Constraint dialog box).

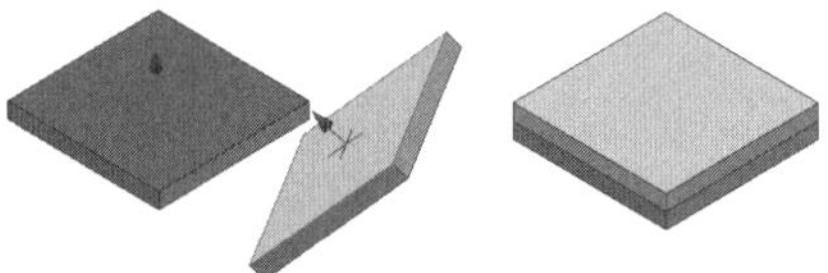

Figure 6-19

## Mate Line

This subset of the mate constraint assembles the edges of lines in the following image.

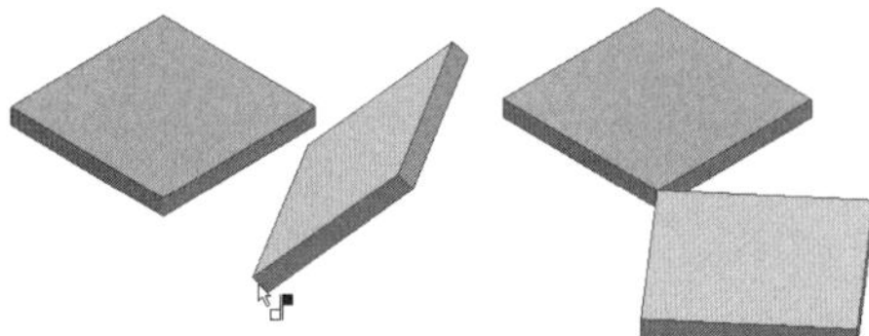

Figure 6-20

The mate line constraint can also be used effectively to assemble the center of a cylinder with a matching hole feature, as in the following image.

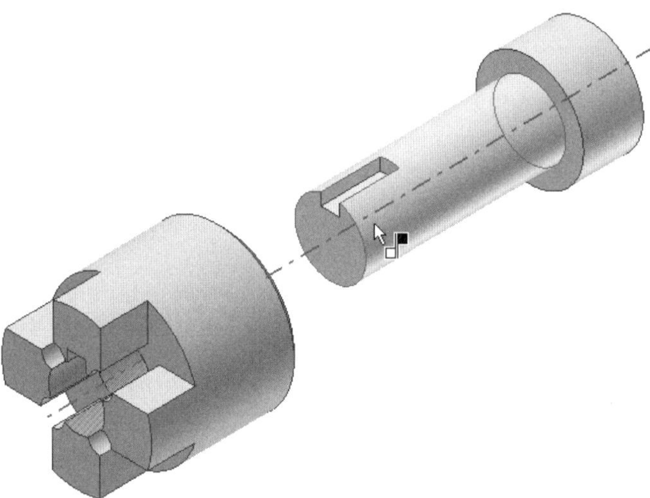

Figure 6-21

### Mate Point

The mate point constraint assembles two points (center of arcs and circular edges, endpoints and midpoints) together, as shown in the following image.

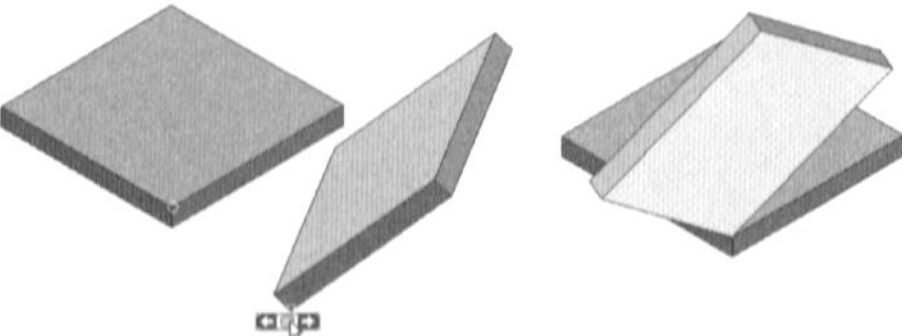

Figure 6-22

### Mate Flush Solution

The mate flush solution constraint is used to align two components so that the selected planes face the same direction, or have their surface normals pointing in the same direction, as shown in the following image. Faces are the only geometry that can be selected for this constraint.

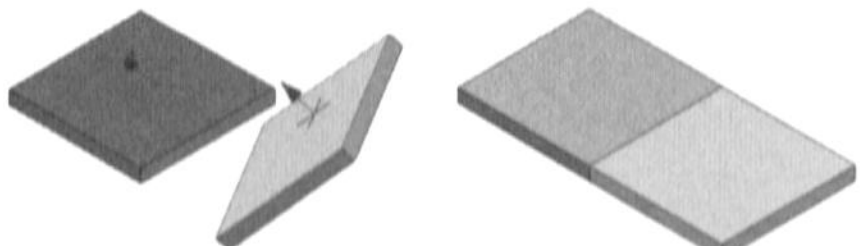

Figure 6-23

### Angle

Using the angle constraint, you will specify the degrees between the selected planes. The following image illustrates the angle constraint with two planes selected and a 30° angle applied.

Two solutions are available when placing an angle constraint; namely Directed Angle and Undirected Angle. You can experiment with both of these solutions especially when driving the angle constraint and observing the behavior of the assembly.

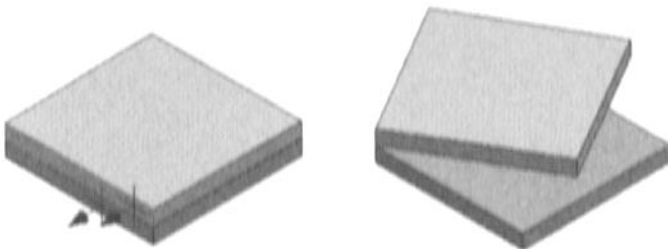

Figure 6-24

### Tangent

Select the tangent constraint if you need to define a tangent relationship between planes, cylinders, spheres, cones, and ruled splines. At least one of the selected faces needs to be a curve, and the tangency may be inside or outside the curve.

The following image illustrates the tangent constraint with two outside curved faces selected and the outside solution applied.

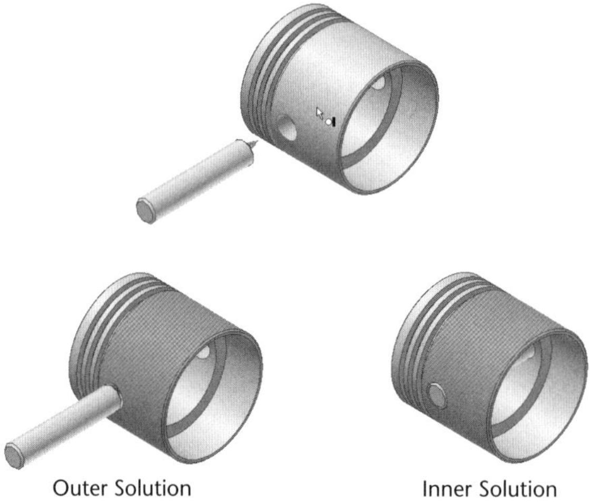


Figure 6-25

## Insert

Select the circular edges of two different components. The centerlines of the components will be aligned and a mate constraint will be applied to the planes defined by the circular edges. The insert constraint takes away five DOF with one constraint, but it only works with components that have circular edges. Circular edges define a centerline/axis and a plane. The following image illustrates the insert constraint with two circular edges selected and the opposed solution applied.

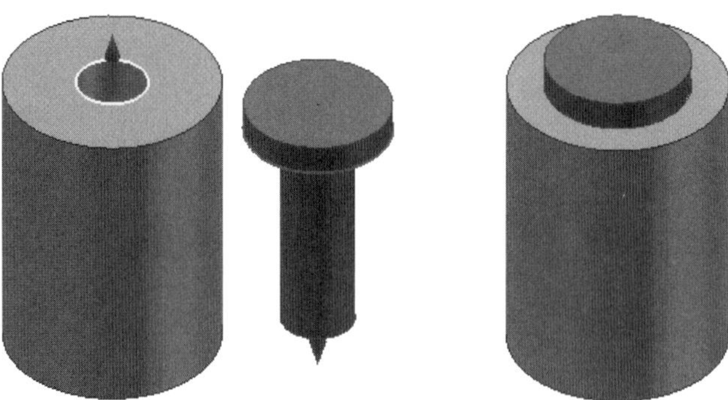

Figure 6-26

## Motion

There are two types of motion constraints: rotation and rotation/translation, as illustrated in the following image. Motion constraints provide the ability to animate the motion of gears, pulleys, rack and pinions, and other devices. By applying motion constraints between two or more components, you can drive one component and cause the others to move accordingly.

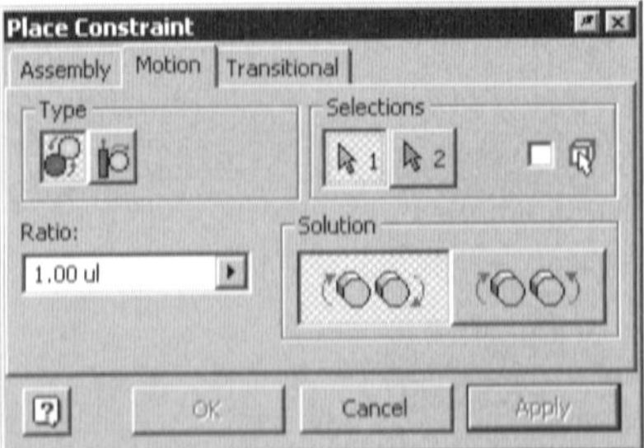


Figure 6-27

Both types of motion constraints are secondary constraints, which means they do not maintain positional relationships between components. It is also advised that you fully constrain your components before you apply motion constraints. You can then suppress constraints that restrict the motion of the components you want to animate.

## Rotation

Use the rotation constraint to define a component that will rotate in relation to another component, by specifying a ratio for the rotation between the two components. Use this constraint for showing the relationship between gears and pulleys. Selecting the tops of the gear faces displays the rotation glyph in the following image. You may also have to change the solution type from Forward to Backward, depending on the desired results.

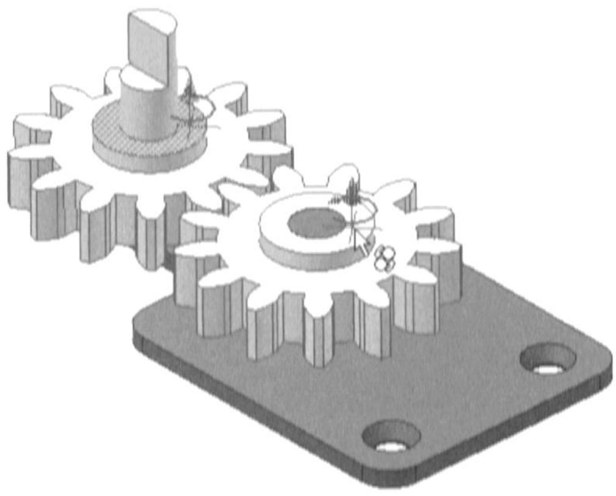

Figure 6-28

## Rotation-Translation

Use the rotation-translation constraint to define the rotation relative to translation of a second component. This type of constraint is well suited for showing the relationship between rack and pinion gear assemblies. In the following image of the rack and pinion assembly, select the top face of the pinion and one of the front faces of the rack. Supply a distance the rack will travel based on the pitch diameter of the pinion gear. You can then drive the constraints and test the travel distance of the mechanism.

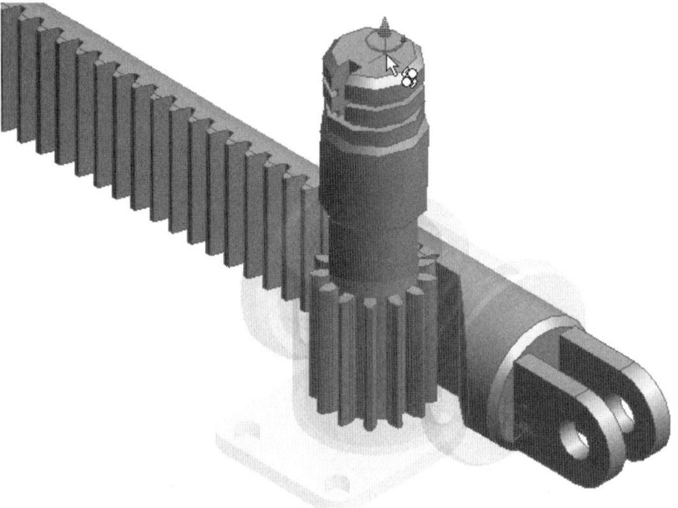

Figure 6-29

## Transitional

A transitional constraint specifies the intended relationship between (typically) a cylindrical part face and a contiguous set of faces on another part, such as a cam in a slot. A transitional constraint maintains contact between the faces as you slide the component along open DOF. This constraint type is accessed through the Transitional tab of the Place Constraints dialog box, as shown in the following image.

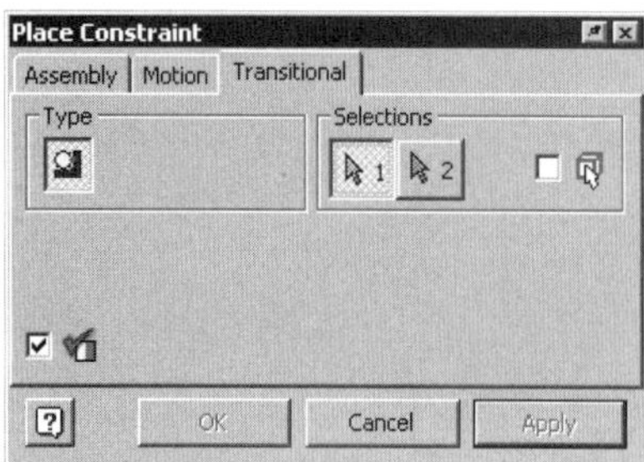


Figure 6-30

Select the moving face first, followed by the transitional face, as in the following image of the cam and follower.

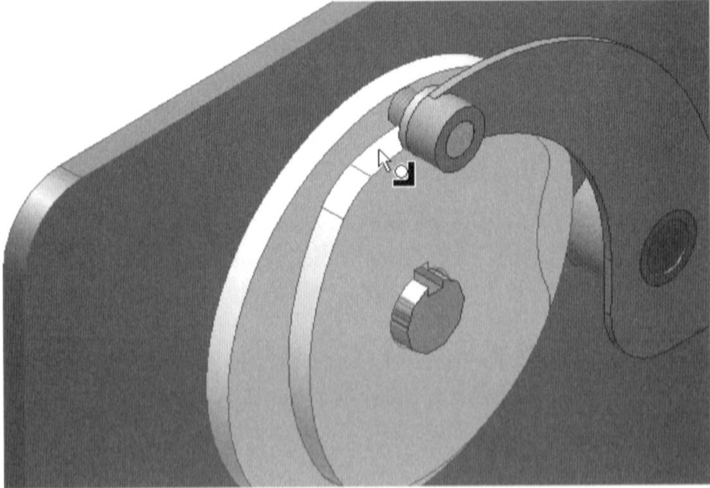

Figure 6-31

## APPLYING ASSEMBLY CONSTRAINTS

After selecting the assembly constraint that you want to apply, the Selections 1 button will become active—or you can click the button if it does not. Position the mouse over the face, edge, point, etc., to apply the first assembly constraint.

You may need to cycle through the selection set using the Select Other tool in the following image until the correct location is highlighted. Cycling is performed by clicking on the left or right arrows in the Select Other tool until you see the desired constraint condition. You then press the left mouse button (or the green rectangle in the Select Other tool box) to place the constraint.

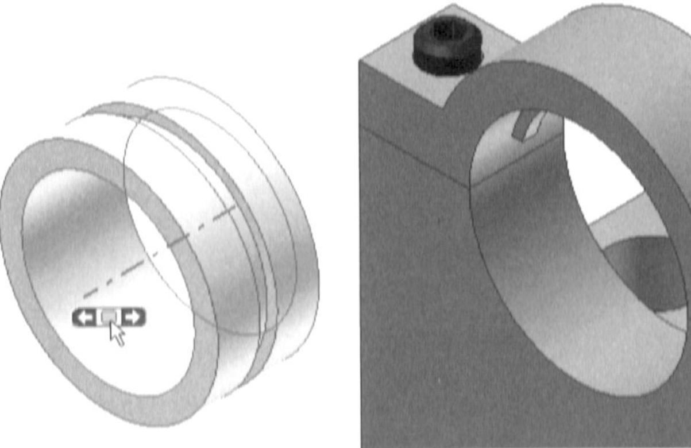

Figure 6-32

The next step is to position the mouse over the face, edge, point, etc. to apply the second assembly constraint. Again, you must cycle through the selection set until the correct location is highlighted. If the Show Preview option is selected in the dialog box, the components will move to show how the assembly constraint will affect the components, and a snapping sound will be heard when the constraint is previewed. To change either selection, click on the number 1 or 2 button and then select the new input. Enter a value as needed for the offset or angle and select the correct Solution option until the desired outcome is shown. Click the Apply button to complete the operation.

**NOTE** If your mouse is equipped with a rotating middle wheel, you can roll the wheel when the Select Other tool displays to cycle through the selection set of faces, edges, or points in a more efficient manner.

## ALT-Drag Constraining

Another method of applying an assembly constraint is to hold down the ALT key while dragging a part to another part—no dialog box will appear. The key to dragging and applying a constraint is to select the correct area on the part. Selecting an edge will create a different type of constraint than selecting a face will create. If a circular edge is selected, for example, an insert constraint will be applied. To apply a constraint while dragging a part, you cannot have another tool active.

To apply an assembly constraint, follow these steps:

- Hold down the ALT key, then select the face, edge, and so on, on the part that will be constrained.
- Select a planar face, linear edge, or axis to place a mate or flush constraint.
- Select a cylindrical face to place a tangent constraint.
- Select a circular edge to place an insert constraint.
- Drag the part into position. As the part is dragged over features on other parts, the constraint type will be previewed. If the face you need to constrain to is behind another face, pause until the Select Other tool appears. Cycle through the possible selection options, and then click the center dot to accept the selection.

To change the constraint type that is being previewed while the part is being dragged, release the ALT key and enter one of the shortcut keys as follows:

**M or 1** Changes to a mate constraint. Press the Spacebar to flip to a flush solution.

**A or 2** Changes to an angle constraint. Press the Spacebar to flip angle direction on the selected component.

**T or 3** Changes to a tangent constraint. Press the Spacebar to flip between an inside and outside tangent solution.

**I or 4** Changes to an insert constraint. Press the Spacebar to flip the insert direction.

**R or 5** Changes to a rotation motion constraint. Press the Spacebar to flip the rotation direction.

**S or 6** Changes to a rotation-translation constraint. Press the Spacebar to flip the translation direction.

**X or 8** Changes to a transitional constraint.

**NOTE** You can create theoretical or construction planes or edges by moving the mouse cursor over an edge and cycling through the geometry, and then moving it to the next edge and cycling through in the same manner until the plane or edge is defined. A work plane can also be used as a plane with assembly constraints—a work axis can be used to define a line.

## MOVING AND ROTATING COMPONENTS

Use the Move Component tool, shown in the following image, to drag individual components in any linear direction in the viewing plane.

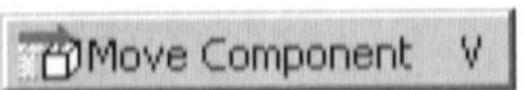


Figure 6-33

To perform a move operation on a component, click and hold the left mouse button on the component to drag it to a new location. Drop the component at its new location by releasing the button.

Moved components will follow these guidelines:

- An unconstrained component remains in the new location when moved until you constrain it to another component.
- A partially constrained component adjusts its location to comply with a constraint that has already been applied.
- You can force a grounded component to move. After the move, the grounded component will remain grounded in its new location. Components constrained to the grounded component will adjust and update to its new location.

Use the Rotate Component tool found on the Assembly Panel Bar, as shown in the following image, to rotate an individual component. This is very useful when constraining faces that are hidden from your view.

Figure 6-34

Follow these steps for rotating a component in an assembly:

1 Select the component to rotate. In the following image, notice the appearance of the 3D rotate symbol on the selected component.

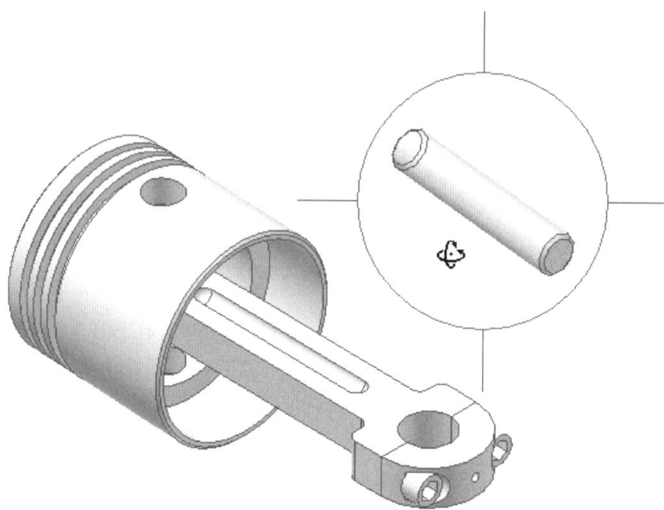

**Figure 6-35**

2 Drag your mouse until you see the desired view of the component.

- For free rotation, click inside the Dynamic Rotate tool and drag in the desired location.
- To rotate about the horizontal axis, click the top or bottom handle of the Dynamic Rotate tool and drag your mouse vertically.
- To rotate about the vertical axis, click the left or right handle of the Dynamic Rotate tool and drag your mouse horizontally.
- To rotate planar to the screen, hover over the rim until the symbol changes to a circle, click the rim, and drag in a circular direction.
- To change the center of rotation, click inside or outside the rim to set the new center.

3 Release the mouse button to drop the component in the new, rotated position.

**NOTE** If you click the Update button when moving or rotating components in an assembly, any components constrained to a grounded component will snap to their constrained positions in the new location.

## EDITING ASSEMBLY CONSTRAINTS

After an assembly constraint has been placed, you may want to edit, suppress, or delete it to reposition the components. There are two methods for editing assembly constraints.

Both methods are executed through the Browser. In the Browser, activate the assembly or subassembly that contains the component you want to edit. Expand the component name and you will see the assembly constraints. Double-click on the constraint name and an Edit Constraint dialog box will appear, allowing you to edit the constraints. The second method is to right-click on the assembly constraint's name in the Browser and select Edit, Suppress, or Delete from the menu, as shown in the following image.

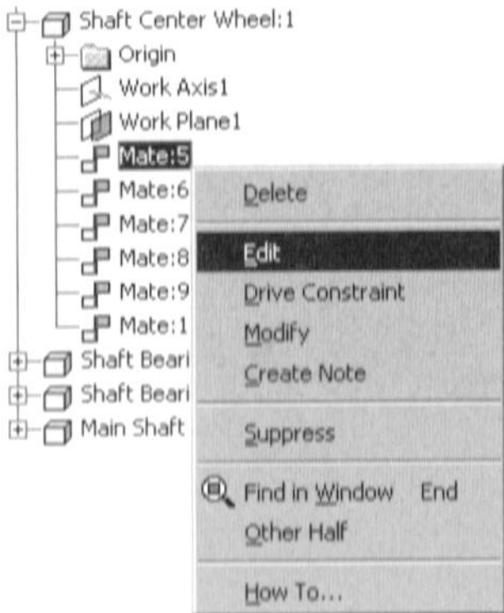


**Figure 6-36**

If you select Edit, the Edit Constraint dialog box will appear. If you select Suppress, the assembly constraint will not be applied. Select Drive Constraint to drive a constraint through a sequence of steps, simulating mechanical motion. Select Delete and the assembly constraint will be deleted from the component. If you try to place or edit an assembly constraint and it cannot be applied, an alert box will appear that explains the problem. You will have to either select new options for the operation, or suppress or delete another assembly constraint that is conflicting with it.

If an assembly constraint is conflicting with another, a small yellow icon with an exclamation point will appear in the Browser, similar to the following image. To edit a conflicting constraint in the Browser, either double-click on its name or right-click on its name and select Recover from the menu, as shown in the following image. The Design Doctor will appear and walk you through the steps to fix the problem.

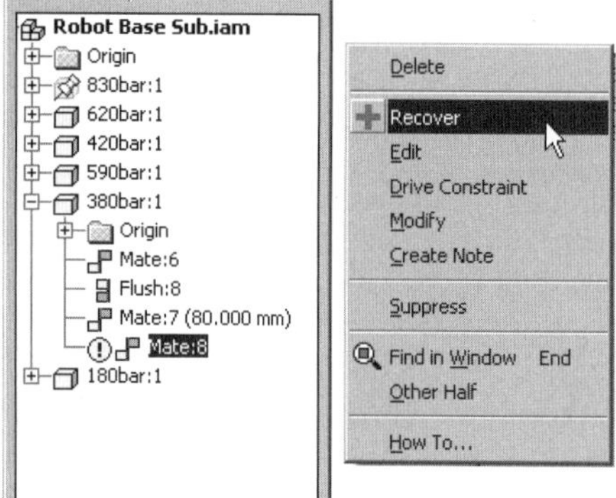


Figure 6-37

Another case involving the editing of constraints is when too many constraints are added and Autodesk Inventor fails to generate error messages. As you view a list of constraints in the Browser, you may notice a few constraints that are preceded by a small circle with the letter "i" inside. The appearance of this icon means you have applied an unnecessary or redundant constraint. While your assembly will not suffer from the presence of these constraints, it is considered good practice to clean up all redundant constraints by deleting them from the Browser.

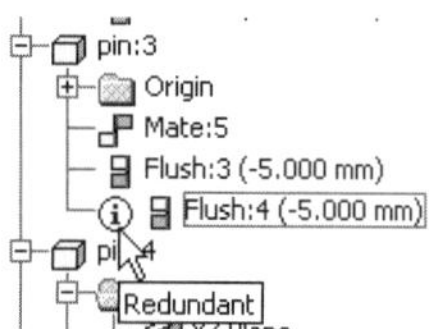


Figure 6-38

# EXERCISE 6-1 Assembling Parts

In this exercise, you assemble a lift mechanism. To navigate to the exercise in the *Electronic Student Workbook*, do the following:

1 From the Main TOC page, click Chapter 6.
2 From the TOC for Chapter 6, click Assembling Parts.

The following image illustrates the opened exercise.

**Figure 6-39 Opened exercise**

The following image illustrates the completed exercise.

**Figure 6-40 Completed exercise**

## DESIGNING PARTS IN-PLACE

Most components created in the assembly environment are created in relation to existing components in the assembly. When creating an in-place component, you can sketch on the face of an existing assembly component or a work plane. You can also click the graphics window background to define the current view orientation as the XY plane. If the YZ or XZ plane is the default sketch plane, you must reorient the view to see the sketch geometry. Select Application Options from the Tools menu, and then click the Part tab to set the default sketch plane.

When you create a new component, you can select an option in the Create In-Place Component dialog box to automatically constrain the sketch plane to the selected face or work plane. After you specify the location for the sketch, the new part immediately becomes active, and the Browser, Panel Bar, and toolbars switch to the part environment (see the following image).

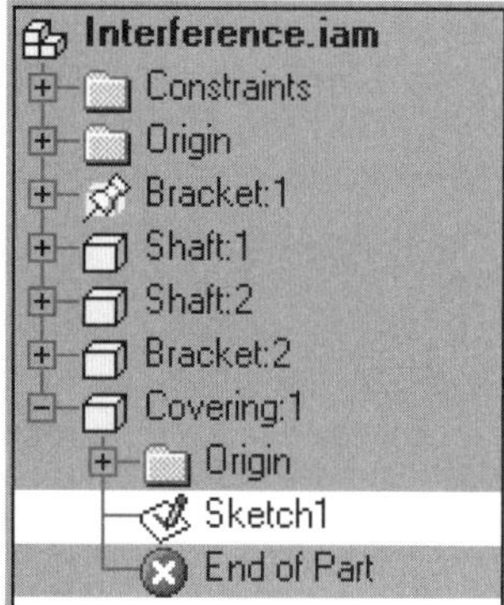


**Figure 6-41**

Notice also that the 2D Sketch panel shown in the following image is available to create the first sketch of your new part.

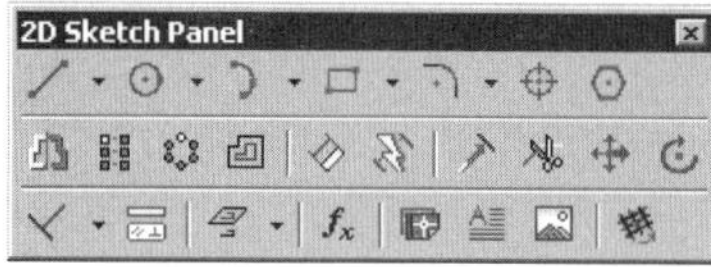


**Figure 6-42**

After you create the base feature of your new part, you can define additional sketches based on the active part or other parts in the assembly. When defining a new sketch, you can click a planar face of the active part or another part to define the sketch plane on that face. You can also click a planar face and drag the sketch away from the face to automatically create the sketch plane on an offset work plane. When you create a sketch plane based on a face of another component, Autodesk Inventor automatically generates an adaptive work plane and places the active sketch plane on it. The adaptive work plane moves as necessary to reflect any changes in the component on which it is based. When the work plane adapts, your sketch moves with it. Features based on the sketch then adapt to match its new position.

After you finish creating a new part, you can return to assembly mode by double-clicking the assembly name in the Browser. In assembly mode, assembly constraints become visible in the Browser. If you selected the Constrain Sketch Plane to Selected Face option when you created your new part, a flush constraint will appear in the Assembly Browser, but it can be deleted at any time. No flush constraint is generated if you create a sketch by clicking in the graphics window.

# EXERCISE 6-2 Designing Parts in the Assembly Context

In this exercise, you create a lid for a container based on the geometry of the container. The lid updates to match changes to the container. To navigate to the exercise in the *Electronic Student Workbook*, do the following:

1. From the Main TOC page, click Chapter 6.
2. From the TOC for Chapter 6, click Designing Parts in the Assembly Context.

The following image illustrates the opened exercise.

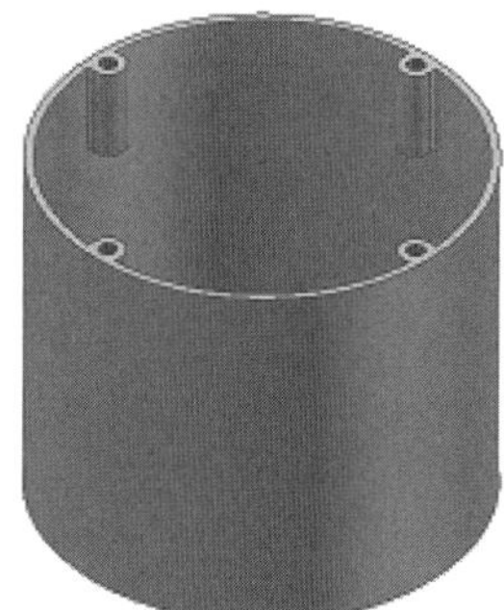

**Figure 6-43 Opened exercise**

The following image illustrates the finished exercise.

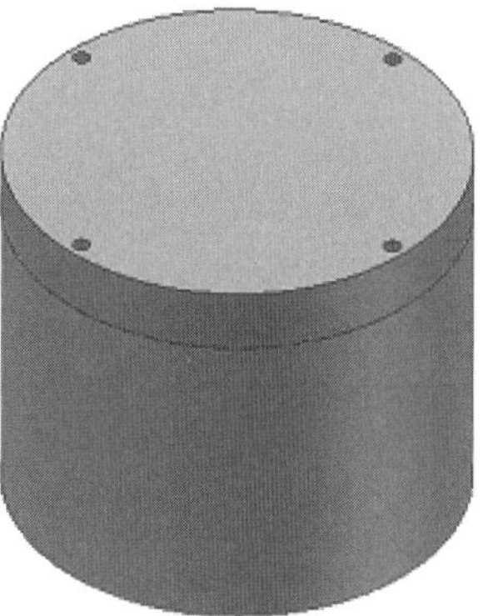

**Figure 6-44 Completed exercise**

## ADAPTIVITY

Adaptivity is the functionality in Autodesk Inventor that allows the size of a part to be determined by setting up a relationship between the part and another part in the assembly. Adaptivity allows underconstrained sketches—features that have undefined angles or extents; hole features; and subassemblies, which contain parts that have adaptive sketches or features—to be adaptive. The adaptivity relationship is acquired by applying assembly constraints between an adaptive sketch or feature and another part. If a sketch is fully constrained, it cannot be made adaptive, but the extruded length or revolved angle can be. A part can only be adaptive in one assembly at a time. In an assembly that has multiple placements of the same part, only one occurrence can be adaptive. The other occurrences will reflect the size of the adaptive part.

An example of adaptivity would be to determine the diameter of a pin from the size of a hole. In the same example, you could determine the diameter of the hole from the size of the pin. Adaptivity can be turned on and off as needed. Once a part's size is determined through adaptivity, you may want to turn its adaptivity *off*. If you want to create adaptive features, there are options within Autodesk Inventor that will speed up the process of creating them. From the Tools menu, select Application Options. On the Assembly tab, there are three areas that relate to adaptivity, as shown in the following image.

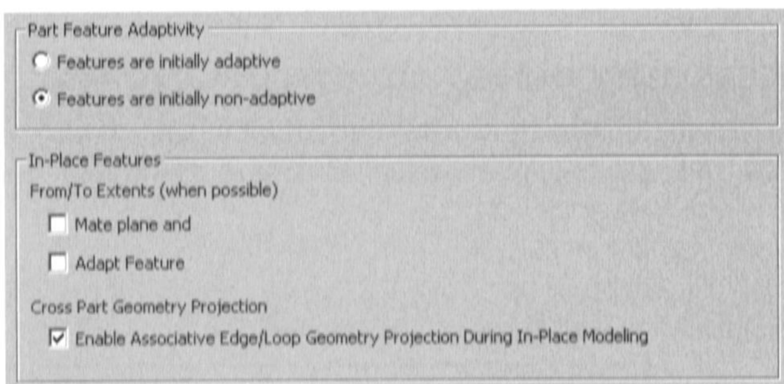


**Figure 6-45**

### ASSEMBLY TAB OPTIONS

The following sections describe the adaptivity options found on the Assembly tab.

#### Part Feature Adaptivity

In this section, you will indicate if new features will be adaptive or non-adaptive when they are created. The adaptivity of a feature can also be changed *after* it has been created. Adaptivity means that a feature will be able to change its size according to the relationship it has with another component.

**Features are initially adaptive** When checked, features will be adaptive when they are created. This is useful when you are creating many adaptive features.

**Features are initially non-adaptive** When checked, features will not be adaptive when they are created.

### In-Place Features From/To Extents (when possible)

Here you will determine if a feature will be adaptive when the To or From/To option is selected for the Extent. If both options are selected, Autodesk Inventor will try to make the feature adaptive. If it cannot, it will terminate at the selected face.

**Mate plane and** Select this option when you create a new component and have a mate constraint applied to the plane on which it was constructed—it will not be adaptive.

**Adapt Feature** Select this option when you want to create a new component and have it adapt to the plane on which it was constructed.

### Cross Part Geometry Projection

**Enable Associative Edge/Loop Geometry Projection During In-Place Modeling** When this box is checked, and geometry is projected from another part onto the active sketch, the projected geometry is associative (sketch associativity) and will update when changes are made to the parent part. Projected geometry can be used to create a sketched feature.

## UNDERCONSTRAINED ADAPTIVE FEATURES

The following images show how parts adapt when assembly constraints are applied. In the following image, a rectangular sketch for a small plate is not dimensioned along its length.

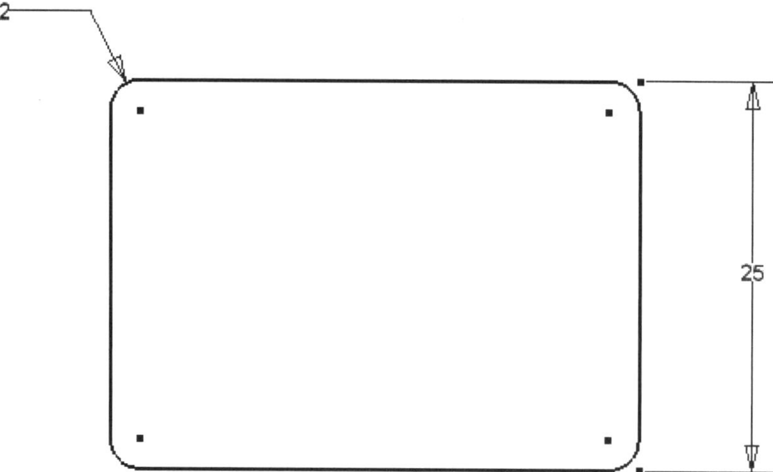


Figure 6-46

The extruded feature is then defined as *adaptive* in the following image by right-clicking on the extrusion in the Browser to display the appropriate menu. Selecting Adaptive from the menu applies the adaptive property to the sketch and the extrusion.

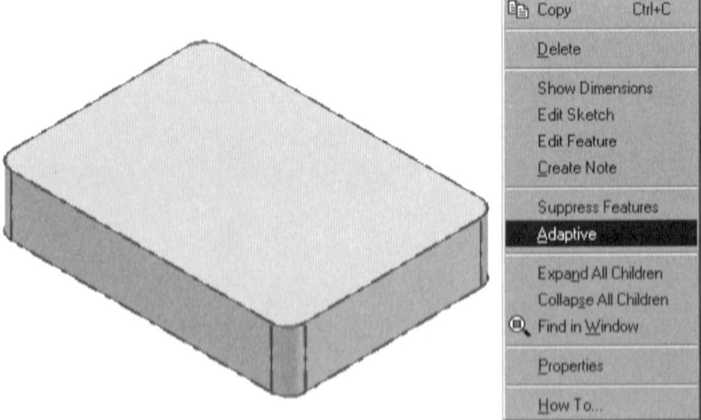


**Figure 6-47**

Once the parts have been placed, assembly constraints are applied between the two plates. Right-clicking on the small box in the Browser activates a menu. Selecting Adaptive from the menu adds a symbol consisting of two arcs with arrows next to the small box part, as shown in the following image. The presence of this symbol in the Browser means the small box part is now adaptive. The flush constraint placed along the edge faces of the plates will force the small plate to adapt its length to meet the length of the large plate.

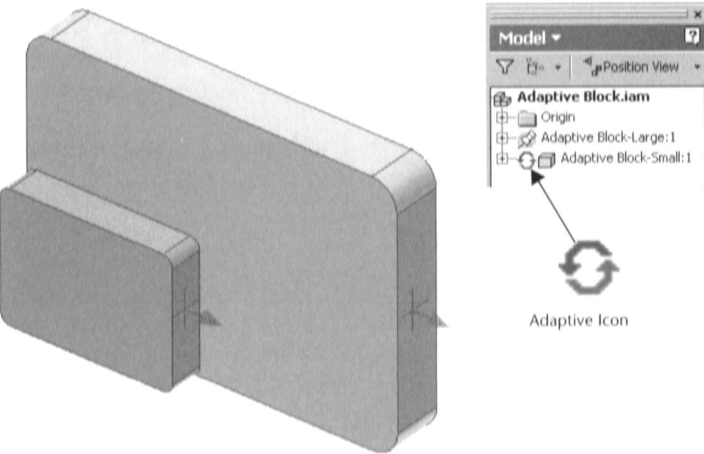


**Figure 6-48**

The results are shown in the following image.

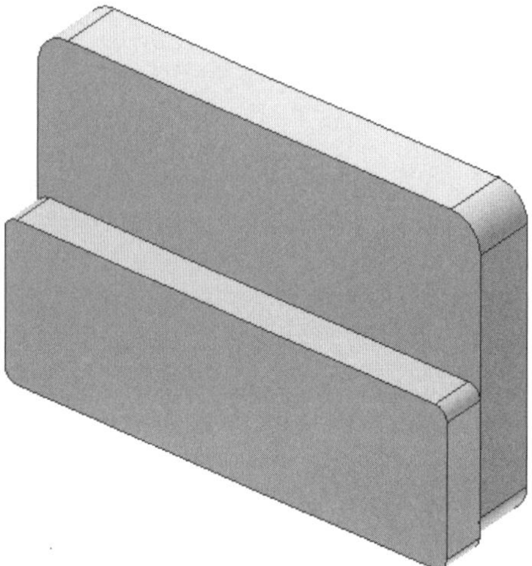

Figure 6-49

When a part containing adaptive features is placed or created in an assembly, it is not initially adaptive. To specify the part as adaptive in the assembly context, right-click the component in the Assembly Browser or graphics window, and select Adaptive from the menu. You can also use the Assembly Browser to specify a feature inside the part as adaptive, and the part will become adaptive automatically.

When you constrain an adaptive part to fixed features on other components, underconstrained features on the adaptive part resize when the assembly is updated.

Only one occurrence of a part can define its adaptive features. If you use multiple placements of the same part in an assembly, all occurrences are defined by the one adaptive occurrence—including placements in other assemblies. If you want to adapt the same part to different assemblies, save the part file with a unique name using Save Copy As, before defining any occurrences as adaptive.

## ENABLED COMPONENTS

When working in an assembly, you may want to hide a component so that other components can be seen. The visibility of a component can be turned on or off by selecting its name in the Browser, or clicking on the part itself and right-clicking and selecting Visibility from the menu. When the visibility of a component is *off* in an assembly, that component can still be used in other assemblies and can be opened and edited. Autodesk Inventor gives you another option, called Enabled, for controlling how components look in an assembly. By default, all components are enabled when they are placed or created—their visibility is on, and they are displayed in the current display mode.

When a component is enabled, it slows down the graphics regeneration speed because it has to be calculated each time a view is dynamically rotated or panned. A component can be disabled. When a component is disabled, it is displayed as a wire frame and cannot be selected in the graphics area. A disabled component can, however, have geometry projected from it. To disable a component, select its name in the Browser, or click on the component itself and right-click and select Enabled from the menu, as shown in the following image.

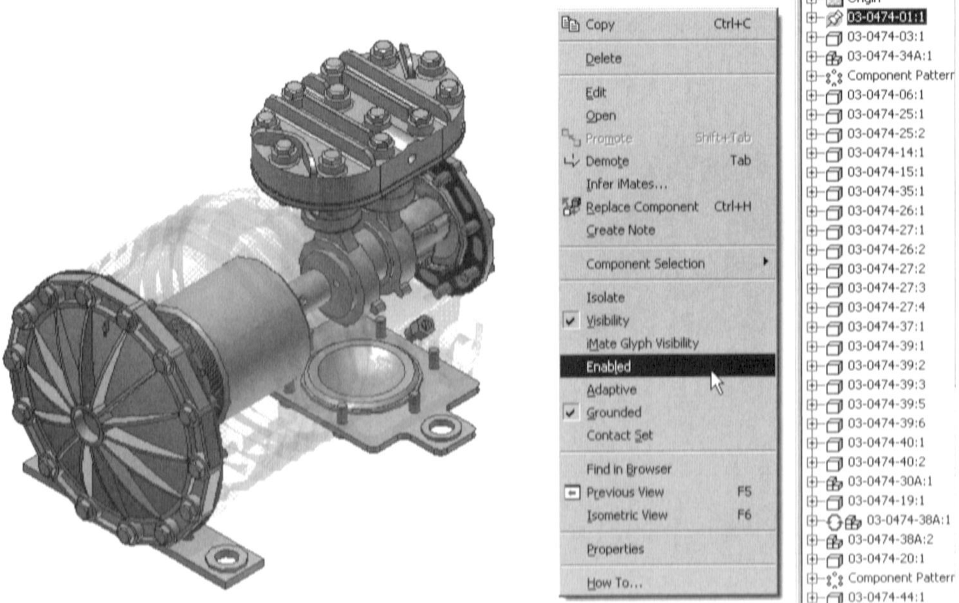


**Figure 6-50**

EXERCISE

# EXERCISE 6-3 Creating Adaptive Parts

In this exercise, you create a link arm that adapts to fit between existing components in an assembly. To navigate to the exercise in the *Electronic Student Workbook*, do the following:

1 From the Main TOC page, click Chapter 6.
2 From the TOC for Chapter 6, click Creating Adaptive Parts.

The completed exercise is shown in the following image.

**Figure 6-51 Completed exercise**

## ANALYSIS TOOLS

Various tools are available that assist in analyzing sketch, part, and assembly models. The Measure tool actually consists of four separate tools. You can measure distances, angles, and loops, and perform area calculations. You can also calculate the center of gravity of parts and assemblies.

### THE MEASURE TOOL

The Measure tool is selected from the Tools menu, as shown in the following image. Each of these tools will be discussed in greater detail.

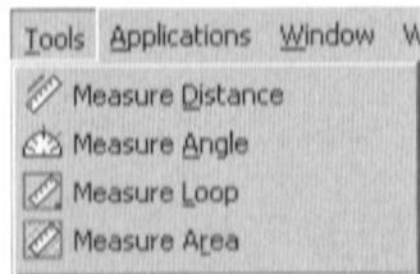


Figure 6-52

**Measure Distance** Measures the length of a line, length of an arc, distance between the points, radius and diameter of a circle, or the position of elements relative to the active coordinate system. A measurement box displays the measurement for the selected length, as shown in the following image. Also, a temporary line designating the measured distance will appear.

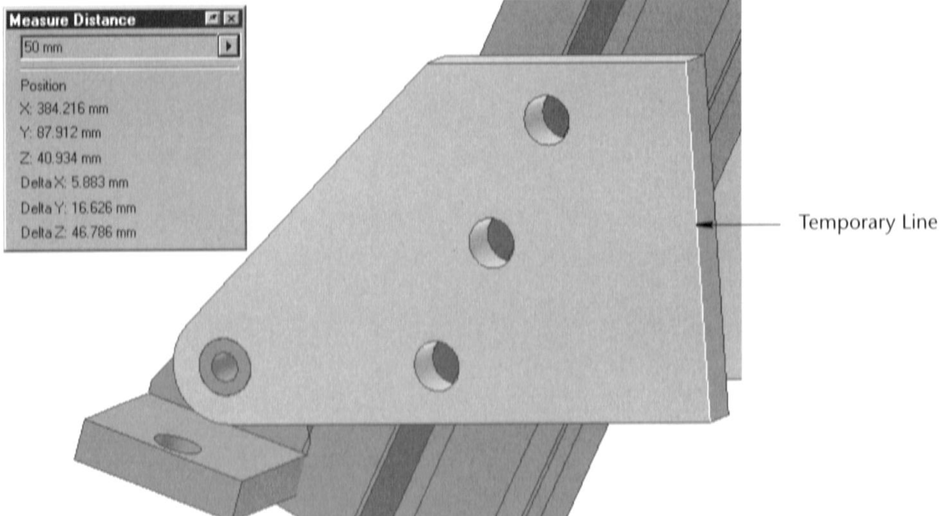


Figure 6-53

**Measure Angle** Measure the angle between two lines, edges, or points. The measurement box (see the following image) displays the angle based on the selection of two edges of the plate.

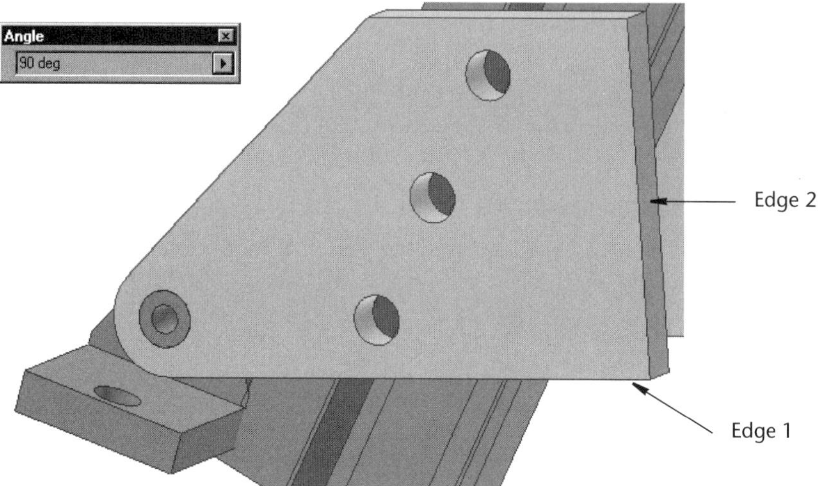


Figure 6-54

**Measure Loop** Measures the length of closed loops defined by face boundaries or other geometry. When moving your cursor over an edge, all edges that form a closed loop will highlight. Clicking on this edge, as shown in the following image, will calculate the closed loop or perimeter of the shape.

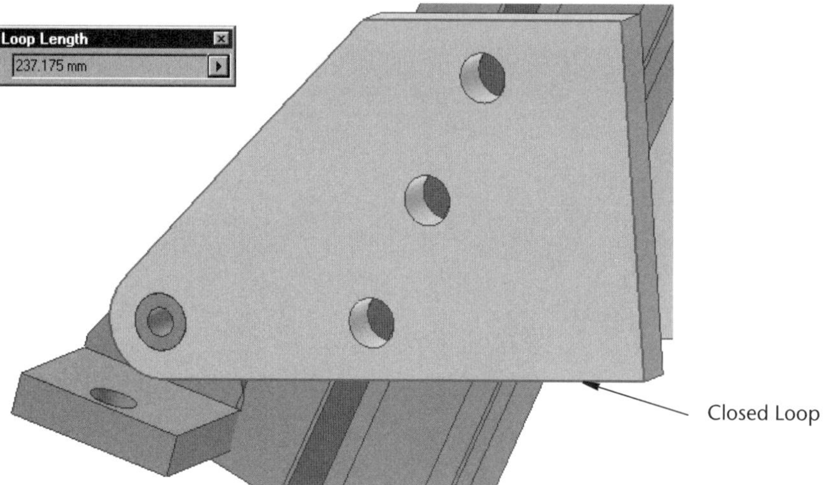


Figure 6-55

**Measure Area** Measures the area of enclosed regions, as shown in the following image. Moving your cursor inside the closed outer shape will cause the outer shape and all holes referred to as islands to also highlight. Clicking inside this shape will calculate the area of the shape.

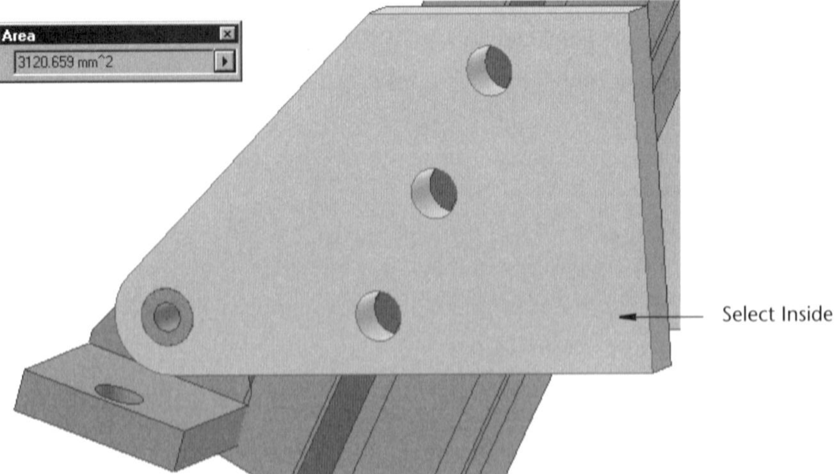


**Figure 6-56**

When you click the arrow beside the measurement box, a menu will appear similar to the following image.

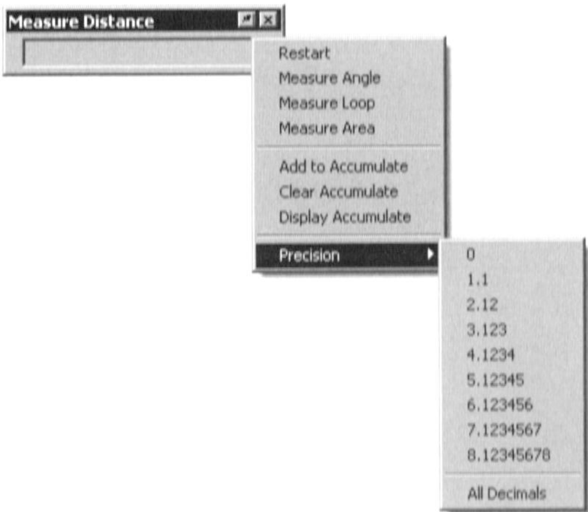


**Figure 6-57**

The following are brief explanations of each option:

**Restart** Clears the measurement from the measurement box so that you can make another measurement.

**Measure Angle** Changes the measurement mode to Measure Angle.

**Measure Loop** Changes the measurement mode to Measure Loop.

**Measure Area** Changes the measurement mode to Measure Area.

**Add to Accumulate** Adds the measurement in the measurement box to accumulate a total measurement.

**Clear Accumulate** Clears all measurements from the accumulated sum, resetting the sum to zero.

**Display Accumulate** Displays the sum of all measurements you have added to the accumulated sum.

**Precision** Changes the decimal display between showing all decimal places and showing the number of decimal places specified in the document settings for the active part or assembly.

## CENTER OF GRAVITY

At times, you may need the ability to get a visual fix on the current coordinates of the center of gravity during the design process. During the design process, the center of gravity must be kept within a particular region for the effectiveness of the overall assembly in which the part will be used. Having the center of gravity available will provide the output in a more real-time fashion for the center of gravity coordinates.

A center of gravity placed in an assembly could be critical to the overall design process of that assembly, whether it is used in the next assembly or in the main assembly.

Select the Center of Gravity tool from the View menu, as shown in the following image.

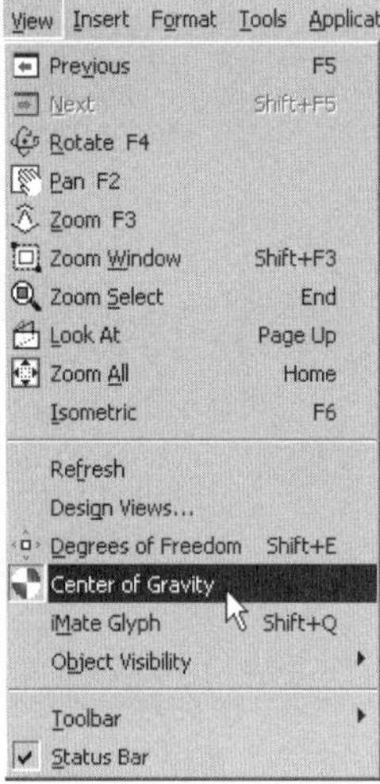


**Figure 6-58**

The following image illustrates the center of gravity icon applied to an assembly model.

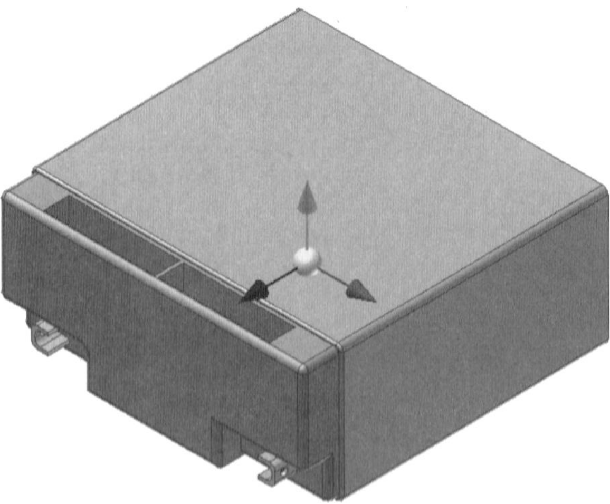

Figure 6-59

## INTERFERENCE CHECKING

You can check for interference in an assembly using one or two sets of objects. To check the interference among sets of stationary components, make active the assembly or subassembly that you want to check for interference. Then select the Analyze Interference tool from the Tools menu, as shown in the following image.

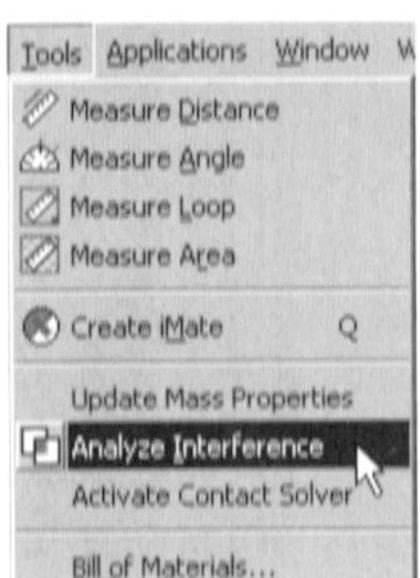


Figure 6-60

The Interference Analysis dialog box will appear, as shown in the following image.

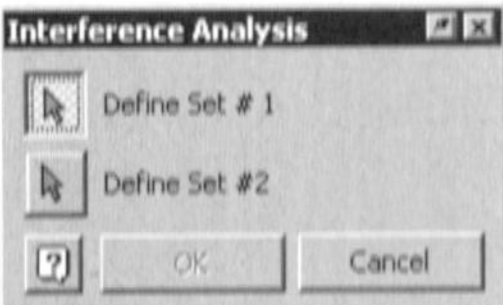


Figure 6-61

Click on Define Set #1 and select the components that will define the first set. Then click on Define Set #2 and choose the components that will define the second set. A component can exist in only one set. To add or delete components from either set, first select the button that defines the set that you want to edit. Then click components to add to the set, or press the CTRL key while selecting components to remove from the set.

**NOTE** Use only Define Set #1 if you want to check for interference in a single group of objects.

Once the sets are defined, click the OK button. The order in which the components are selected has no significance. If interference is found, the Interference Detected dialog box will appear, as shown in the following image.

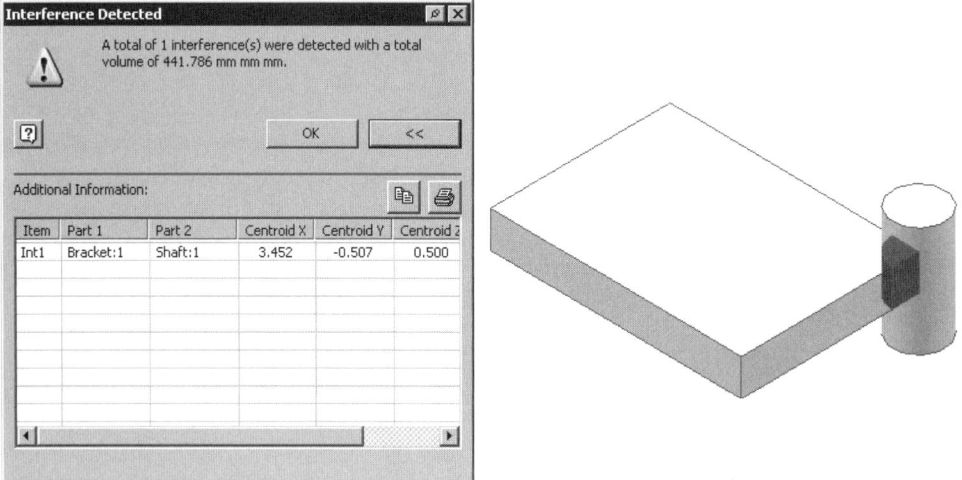


**Figure 6-62**

The information in the dialog box defines the *X, Y, Z* coordinates of the centroid of the interfering volume, and lists the volume of the interference. A temporary solid will also be created in the graphics window that represents the interference. The interference report can be copied to the clipboard or printed from the tools in the Interference Detected dialog box. When the operation is complete, click the OK button and the interfering solid will be removed from the screen. After analyzing and finding interference, edit the components to remove the interference.

**NOTE** Interference can also be detected when driving constraints.

# EXERCISE 6-4 Analyzing an Assembly

In this exercise, you analyze a partially completed assembly for interference between parts and check the physical properties to verify design intent. To navigate to the exercise in the *Electronic Student Workbook*, do the following:

1 From the Main TOC page, click Chapter 6.

2 From the TOC for Chapter 6, click Analyzing an Assembly.

The following image illustrates the completed exercise.

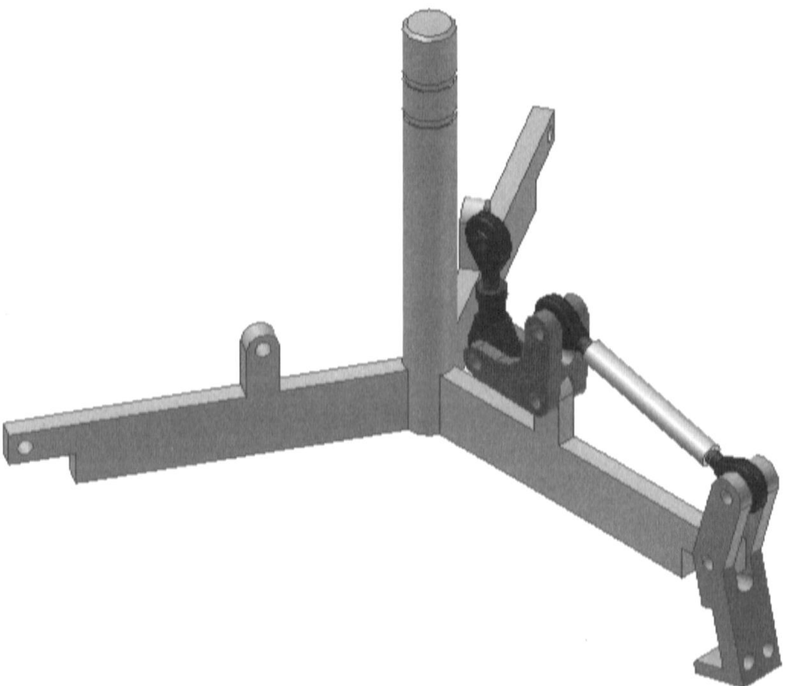

Figure 6-63 Completed exercise

## DOCUMENTING ASSEMBLIES USING PRESENTATION FILES

After creating an assembly, you can create drawing views based on how the parts are assembled. If you need to show the components in different positions—like an exploded view—hide specific components, or create an animation that shows how to assemble or disassemble the components, you need to create a presentation file. A presentation file is a separate file from an assembly and has a file extension of *.ipn*. The presentation file is associated with the assembly file, and changes made to the assembly file will be reflected in the presentation file. No components are created in a presentation file. To create a new presentation file, select the New icon from What To Do and then select the *Standard.ipn* icon from the Default tab, as shown in the following image.

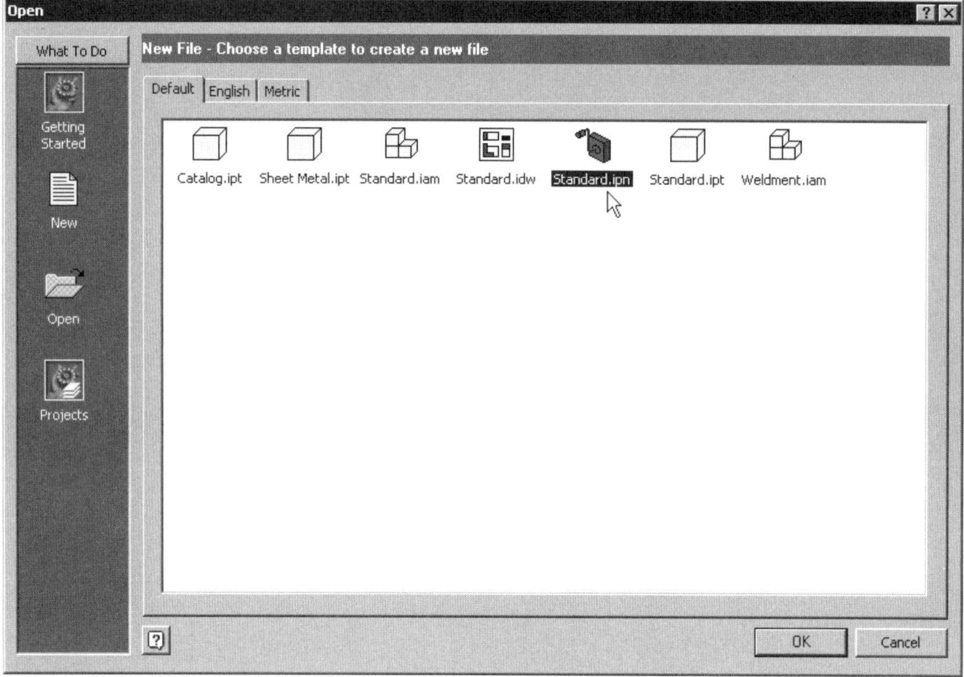


Figure 6-64

By default, the Panel Bar will show the presentation tools available to you (see the following image). The following are explanations of each of these tools.

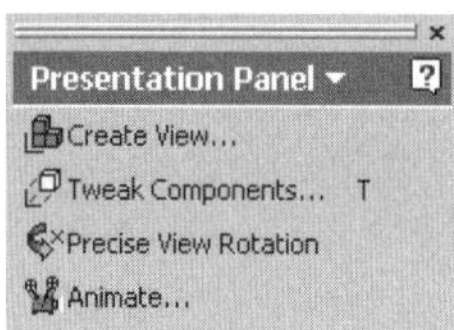


Figure 6-65

**Create View** Creates views of the assembly. There is no limit to the number of views that can exist in a presentation file, but all of the views are based on the same assembly. Once created, a view will be listed as an explosion in the Browser.

**Tweak Components** Moves and/or rotates parts in the view.

**Precise View Rotation** Rotates the view by a specified angle direction using a dialog box.

**Animation** Creates an animation of the assembly—an AVI file can be output.

The main steps for creating a presentation view are to reposition (tweak) the parts in specific directions and then create an animation if needed. Each of the steps will be discussed in the following sections.

## CREATING PRESENTATION VIEWS

The first step in creating a presentation is to create a presentation *view*. Issue the Create View tool from the Presentation Panel Bar or right-click and select Create View from the menu. The Select Assembly dialog box will appear, as shown in the following image. The dialog box is divided into two sections—Assembly and Explosion Method.

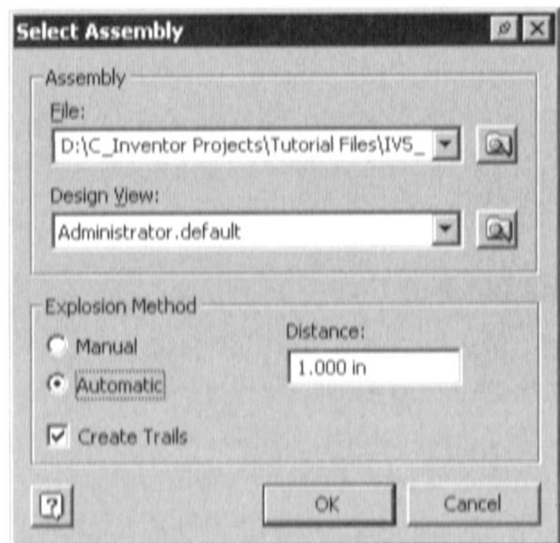


**Figure 6-66**

### Assembly

In this section, you will determine on which assembly and design view the presentation view will be based.

**File** Enter or select an assembly file on which the presentation view will be based.

**Design View** In this area, the default design view will be displayed. If you want to use a different design view, enter its name or select it.

### Explosion Method

In this section, you can choose to manually or automatically explode the parts (separate the parts a given distance).

**Manual** If selected, the components will not be exploded automatically. After the presentation view is created, tweaks can be added that will move or rotate the parts.

**Automatic** Select Automatic if you want the part to be exploded automatically with a given value.

**Create Trails** If Automatic is selected, you can choose to have visible trails (lines that show how the parts are being exploded).

**Distance** If Automatic is selected, enter a value that the parts will be exploded.

After making your selection, click the OK button to create a presentation view. The components will appear in the graphics window and a presentation view will appear in the Browser. If the Automatic option was checked in Explosion Method, the parts will be exploded automatically. To determine how the parts will be exploded, Autodesk Inventor analyzes the assembly constraints. If parts were constrained using the mate plane option and the normals (arrows perpendicular to the plane) for both parts were pointing outward, they will be exploded away from each other. Think of the mating parts as two magnets that want to push themselves in opposite directions. The grounded component in the Browser will be the component that stays stationary, and the other components will move away from it. If trails are created automatically, they will generally come from the center of the part—not necessarily from the center of the holes. After expanding all of the children in the Browser, you can numerically see how the parts exploded. Once the parts are exploded, the distance is referred to as a *tweak*. The number by each tweak reflects the distance that the component is moved from the base component.

## TWEAKING COMPONENTS

After creating the presentation view, you may choose to edit the tweak of a component or move or rotate the component beyond the original explosion factor. To edit an individual tweak, double-click on its value in the Browser, as shown in the following image. Then enter a new value in the text box in the lower-left corner of the Browser and press ENTER to use the new value.

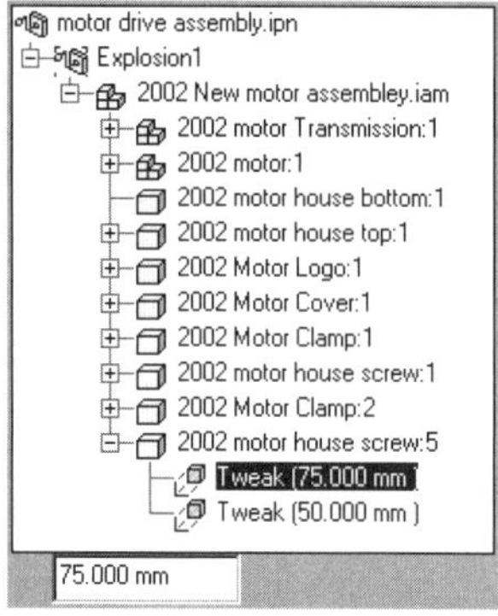


Figure 6-67

To extend all of the tweaks the same distance, right-click on the assembly's name in the Browser and select Auto Explode from the menu, as shown in the following image. Then, enter a value and then click the OK button, and all of the tweaks will be extended the same distance. Negative values cannot be used with the Auto Explode method.

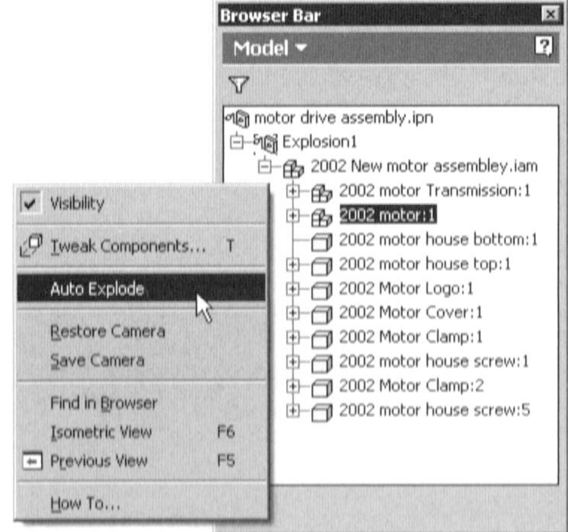


**Figure 6-68**

Another method of manually repositioning the components is to tweak them using the Tweak tool. To manually tweak a component, start the Tweak Component tool from the Presentation Panel Bar, press the hot key **T** or right-click and select Tweak Component from the menu. The Tweak Component dialog box will appear, as shown in the following image. The Tweak Component dialog box is broken into two sections—Create Tweak and Transformations.

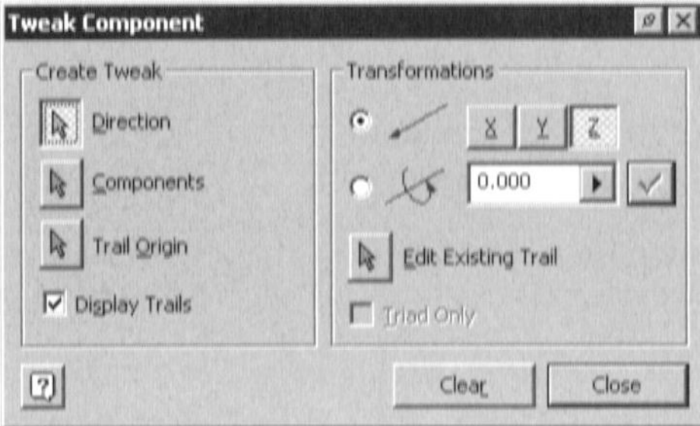


**Figure 6-69**

## Create Tweak

In this section, you will select the components to tweak and set the direction and origin of the tweak.

**Direction** Here you will determine the direction or axis of rotation for the tweak. After clicking the Direction button, select an edge, face, or feature of any component in the graphics window to set the direction triad (X, Y, and Z) for the tweak. The edge, face, or feature that is selected does not need to be on the components that are being tweaked.

**Components** Here you will select the components to tweak. Click the Components button and then select the components in the graphics window or Browser that will be tweaked. If a component was selected when you started the Create Tweak operation, it will automatically be included in the components. To remove a component from the group, press and hold the CTRL key and select the component.

**Trail Origin** Here you will set the origin for the trail. Click the Trail Origin button, then click in the graphics window to set the origin point. If you do not specify the trail origin, it will be placed at the center of mass for the part.

**Display Trails** Check the Display Trails box if you want to see the tweak trails for the selected components. Clear the check box to hide the trails.

## Transformations

In this section, you will set the distance and type of tweak.

**Linear** To move the component in a linear fashion, click the button next to the arrow and line.

**Rotation** To rotate the component, click the button next to the arrow and arc.

**X, Y, Z** Click the X, Y, or Z coordinate button to determine the direction for a linear tweak or the axis for a rotational tweak, or select the arrow on the triad that represents the X, Y, or Z direction.

**Text Box** Enter a positive or negative value for the tweak distance or rotation angle, or pick a point in the graphics area and move the mouse with the mouse button pressed down to set the distance.

**Apply** After making all of the selections, click the Apply button to complete the tweak.

**Edit Existing Trail** To edit an existing tweak, click the Edit Existing Trail button. Select the tweak in the graphics window and change the desired settings.

**Triad Only** To rotate the direction triad without rotating selected components, select the Triad Only option, enter the angle of rotation, and then click the Apply button. After the triad is rotated, you can use it to define tweaks.

**Clear** Click the Clear button in the dialog box to remove all of the settings and set up another tweak.

To tweak a component, follow these steps:

1. Issue the Tweak Component tool.
2. Determine the direction or axis of rotation for the tweak by clicking the Direction button and selecting an edge, face, or feature.
3. Select the components to tweak by clicking the Components button and selecting the components in the graphics window or Browser that will be tweaked.
4. Select a trail origin point.
5. Determine whether or not you want trails visible.
6. Set the type of tweak to linear or rotation.
7. Click the X, Y, or Z coordinate button to determine the direction for a linear tweak, or the axis for a rotational tweak.
8. Enter a value for the tweak in the text box, in the Tweak Component dialog box, or select a point on the screen and drag the part into its new position.
9. Click the Apply button in the Tweak Component dialog box.

### ANIMATION

After the components have been repositioned, you can animate the components to show how they assemble or disassemble. To create an animation, select the Animate tool from the Presentation Panel Bar, or right-click and select Animate from the menu. The Animation dialog box will appear, as shown in the following image. The Animation dialog box is broken into three sections—Parameters, Motion, and Animation Sequence.

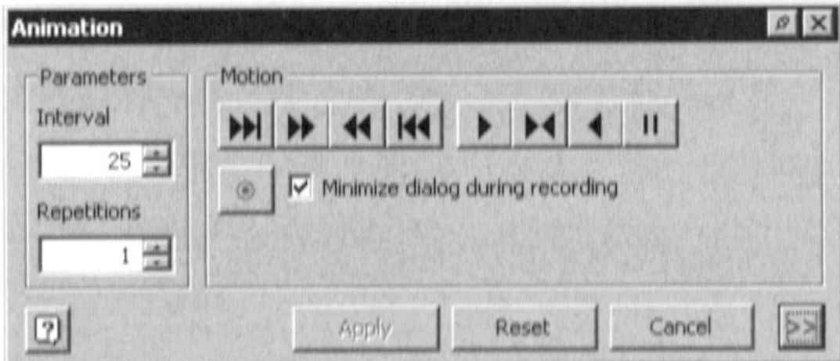


**Figure 6-70**

#### Parameters

In this section, you will specify the playback speed and the number of repetitions for the animation.

**Interval** This value represents the playback speed for the animation in frames. The higher the number, the greater the time delay between frames. A smaller number will speed up the animation.

**Repetitions** Here you will set the number of times to repeat the playback. Enter the desired number of repetitions or use the up or down arrow to select the number.

**NOTE** To change the number of repetitions, click the Reset button on the dialog box and then enter a new value.

## Motion

In this section, you will play the animation for the active view.

**Forward by Tweak** Drives the animation forward one tweak at a time.

**Forward by Interval** Drives the animation forward one interval at a time.

**Reverse by Tweak** Drives the animation in reverse one tweak at a time.

**Reverse by Interval** Drives the animation in reverse one interval at a time.

**Play Forward** Plays the animation forward for the specified number of repetitions. Before each repetition, the view is set back to its starting position.

**Auto Reverse** Plays the animation for the specified number of repetitions. Each repetition plays start to finish, then in reverse.

**Play Reverse** Plays the animation in reverse for the specified number of repetitions. Before each repetition, the view is set back to its ending position.

**Pause** Pauses the animation playback.

**Record** Records the specified animation to a file so that you can play it back later.

To animate components, follow these steps:

1. Issue the Animate tool.
2. Set the number of repetitions.
3. Adjust the tweaks as needed.
4. Click one of the play buttons to view the animation in the graphics window.
5. To record the animation to a file, click the Record button, and then click one of the Play buttons to start recording.

### CHANGING THE ANIMATION SEQUENCE

In this section, you can change the sequence in which the tweaks happen, select the tweak, and then select the needed operation. Expanding the Animation dialog box displays the following image.

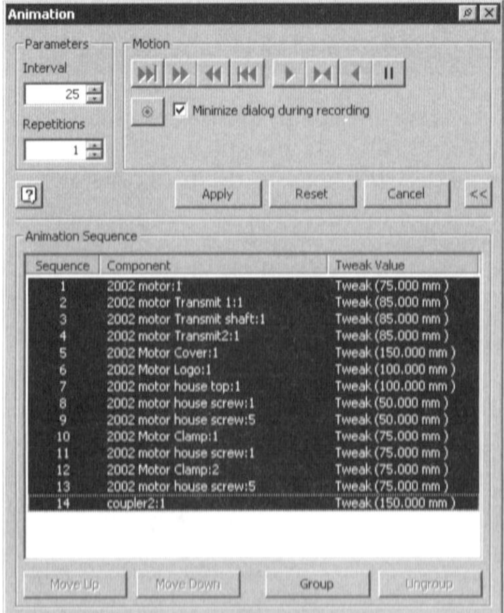


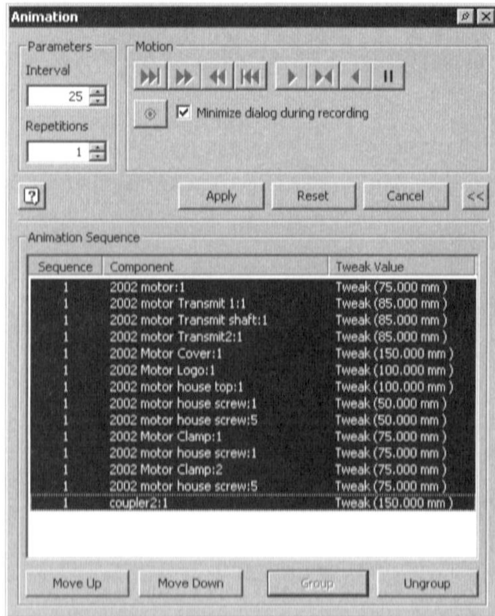


Figure 6-71

**Move Up** Moves the selected tweak up one place in the list.

**Move Down** Moves the selected tweak down one place in the list.

**Group** Select a number of tweaks and then click the Group button. When tweaks are grouped, all of the tweaks in the group will move together as you change the sequence. When tweaks are grouped, the group assumes the sequence order of the lowest tweak number.

**Ungroup** After selecting a tweak that belongs to a group, you can click the Ungroup button, and it can then be moved individually in the list. The first tweak in the group assumes a number that is one higher than the group. The remaining tweaks are numbered sequentially following the first.

# EXERCISE 6-5 Creating Presentation Views

In this exercise, you create a presentation view of an existing assembly. You add tweaks to the assembly components to create an exploded view. To navigate to the exercise in the *Electronic Student Workbook*, do the following:

1. From the Main TOC page, click Chapter 6.
2. From the TOC for Chapter 6, click Creating Presentation Views.

The following image illustrates the completed exercise.

**Figure 6-72 Completed exercise**

## CREATING DRAWING VIEWS FOR ASSEMBLIES AND PRESENTATION FILES

After creating an assembly, you may want to create drawings that utilize the data. Drawing views can be created from assemblies using information from an assembly or presentation file. From them, you can select a specific design view or presentation view. To create drawing views based on assembly data, start a new drawing file or open an existing drawing file. Then use the Base View tool and select the IAM or IPN file from which to create the drawing. If needed, specify the design view or presentation view in the dialog box. The following image shows a presentation view being selected after a presentation file was selected. The drawing can contain as many drawing views, based on different IAM and IPN files, as needed. The same rules apply to creating drawing views from assemblies or presentation files as part (IPT) files.

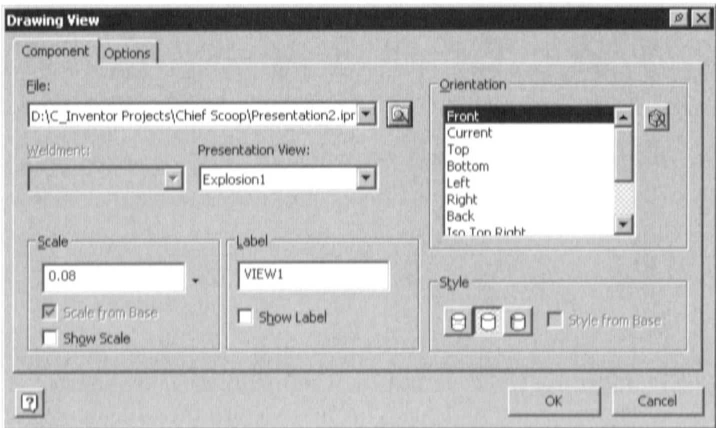


Figure 6-73

## CREATING BALLOONS

After the drawing view(s) is created, you can add balloons to the parts. Before ballooning your drawing, make changes to the balloon style by selecting the Standards tool from the Format menu. Select the desired style from the Balloon tab. These changes will only be used in *this* drawing file. Save the changes to a template file if new drawings will use the same style.

To add balloons to a drawing, follow these steps:

1 Activate the Drawing Annotation Panel Bar by clicking on Drawing Views in the Panel Bar's title and selecting Drawing Annotation from the menu.

2 From the Drawing Annotation Panel Bar, click the arrow next to the Balloon tool, as shown in the following image, and two icons will appear—the first for ballooning components one at a time and the second for ballooning all of the components in a view in a single operation. The hot key **B** can also be used to create balloons for single components.

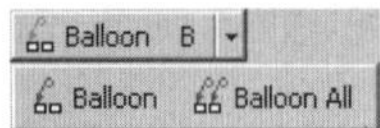


Figure 6-74

3 To balloon a single component, select the Balloon tool from the Drawing Annotation Panel Bar or use the hot key **B**.

4 Select a component to balloon.

5 The Parts List Item Numbering dialog box will appear, as shown in the following image.

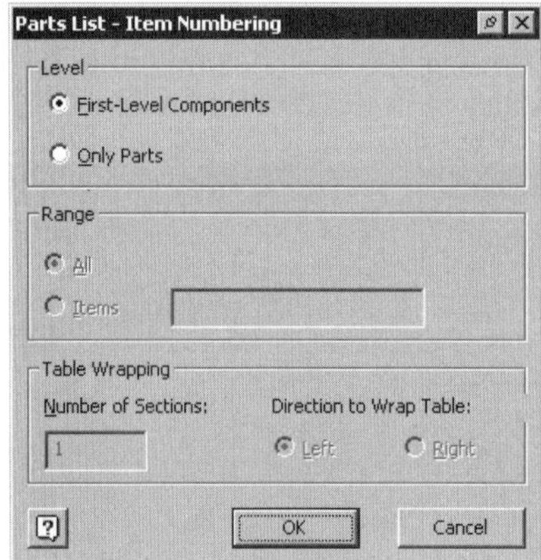


Figure 6-75

6 From within the Parts List Item Numbering dialog box, select which level, range, and format to use. The next section provides descriptions of the options available in this dialog box.

**NOTE** The same dialog box will be used to create a parts list.

A preview image of the balloon will appear.

7 Position the mouse to place the second point for the leader, and then press the left mouse button.

8 Continue to select points to add segments to the balloon's leader.

9 When done adding segments to the leader, right-click and select Continue from the menu to create the balloon.

10 Select the Done option from the menu to cancel the operation without creating the balloon or select the Backup option to undo the last step that was created for the balloon.

11 To balloon all of the components in a view, click the second icon (Balloon All) from the Balloon tool.

12 Click a point in the view in which the balloons will be created. The Parts List Item Numbering dialog box will appear.

13 From within the Parts List Item Numbering dialog box, select which level, range, and format to use.

14 After making your selection, click the OK button and the balloons will be created in the selected view.

### PARTS LIST ITEM NUMBERING DIALOG BOX OPTIONS

**Level** Specifies which level should be included when ballooning components.

**First-Level Components** Select this option if you only want to balloon the top-level components of the assembly in the selected view. Subassemblies and parts that are not part of a subassembly will be ballooned, but parts that are *in* a subassembly will not be.

**Only Parts** Select this option to balloon only the parts of the assembly in the selected view. Subassemblies will not be ballooned, but the parts in the subassemblies will be.

**Range** This section is not used for creating balloons, only for creating parts lists.

**All** Creates a parts list for all of the components in the selected view.

**Items** Creates a parts list for the specified range of parts. Select items and enter the part numbers, separated by commas, or click the parts in the graphics window to include them in the list.

**Format** Specifies the number of columns into which to split the parts list. Splitting parts lists into multiple columns is useful when parts lists are long and do not fit on the drawing sheet as desired.

**Columns** Creates a parts list with the specified number of columns. Enter the number of columns into which you want to split the parts list.

**Left** Attaches the cursor to the top-right corner of the parts list while placing the list on the drawing sheet. This allows you to use the top-right corner of the parts list as an insertion point for the list. You will want to use this option if placing the parts list on the right side of the drawing sheet.

**Right** Attaches the cursor to the top-left corner of the parts list while placing the list on the drawing sheet. This allows you to use the top-left corner of the parts list as an insertion point for the list. You will want to use this option if placing the parts list on the left side of the drawing sheet.

To edit a balloon after it has been created, move the mouse over the balloon and a small green dot will appear at its vertices. Right-click on one of the green dots to access a menu. You can delete, change the arrowhead's appearance, attach another balloon to it, remove an attached balloon, or add a segment or vertex.

**NOTE** A parts list can be created without first defining balloons.

## EDITING BALLOONS

Balloons may be edited by first right-clicking on a balloon. This will activate the menu, as shown in the following image.

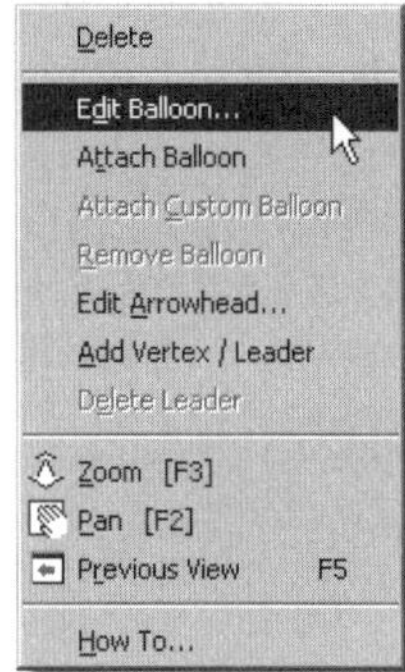


Figure 6-76

Clicking on Edit Balloon in the menu will activate the Edit Balloon dialog box, as shown in the following image.

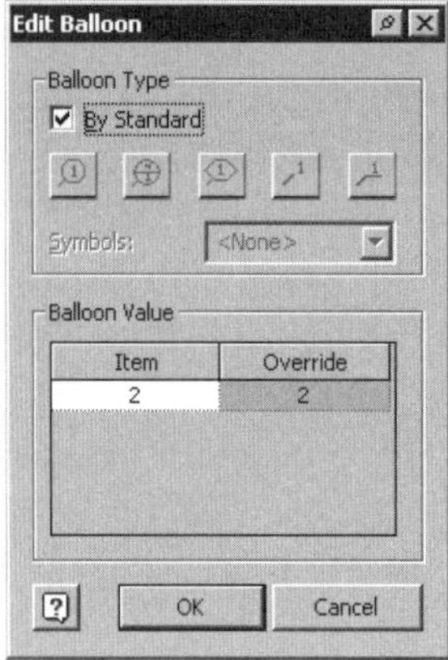


Figure 6-77

Use this dialog box to make changes to a balloon on an individual basis. The following items may be edited: Balloon Type and Balloon Value. The P/L Value (Parts List value) denotes the current value located in the parts list.

### Editing the Balloon Type

To activate the other balloon types, remove the check inside of the box labeled By Standard. You can then select a new balloon type from the group supplied. Changes made to the balloon type affect the individual balloon being edited.

To make changes to all of the balloon types globally, the Balloon tab of the Drafting Standards dialog box, as shown in the following image, can be used.

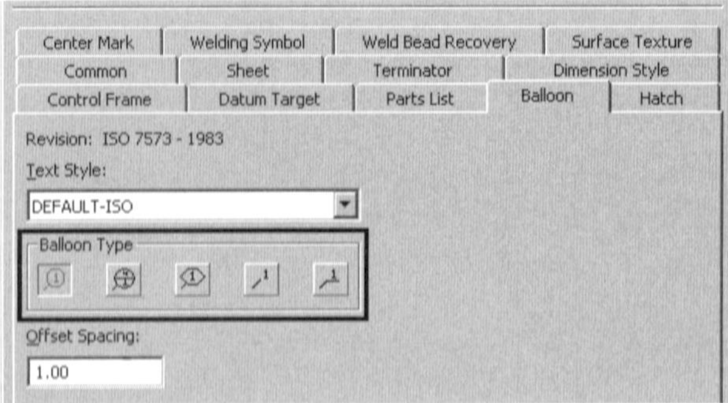


Figure 6-78

## Editing the Balloon Value

Entering a new value in the P/L Value column of the Edit Balloon dialog box will be reflected in the parts list. An edited value in the Item column of the Parts List dialog box will also be reflected in the P/L Value column of the Edit Balloon dialog box. The balloon will grow horizontally, but not vertically, to accommodate extra text.

**NOTE** Any changes performed with the Balloon Edit dialog box will be reflected in all parts lists associated with that view.

## Custom Value Input Override

A user may enter a value in a balloon that is not reflected in the parts list. This is the purpose of a *custom value override*. This option is only available through the Edit Balloon dialog box.

## Aligning Balloons Vertically or Horizontally

Balloons may be aligned horizontally or vertically. Select a number of balloons to align. The first balloon selected will act as an anchor point for the alignment of the other balloons (see the following image).

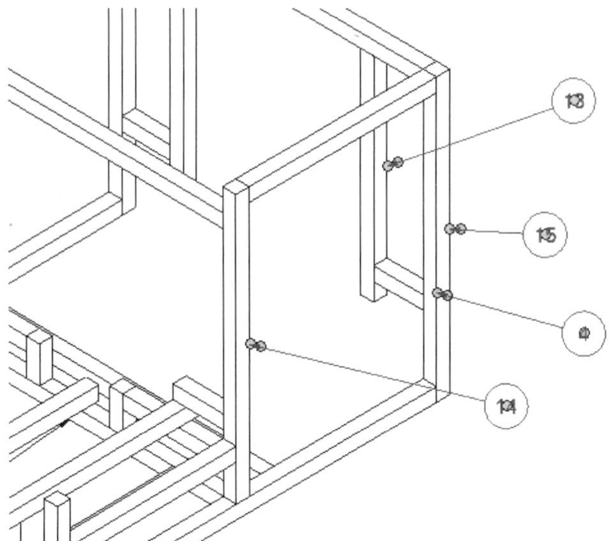

Figure 6-79

Once multiple balloons have been selected, right-click on any of the selected balloons to display the menu and the Align option, as shown in the following image.

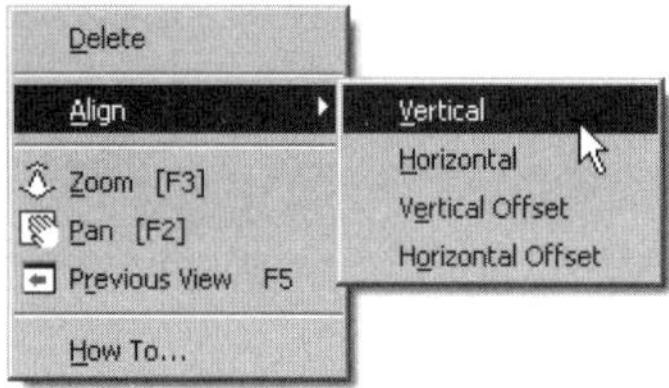


Figure 6-80

Select either the Horizontal or Vertical options to perform the alignment. If Horizontal is selected, the balloons will be aligned along the *X*-axis. If Vertical is selected, the balloons will be aligned along the *Y*-axis as shown in the following image.

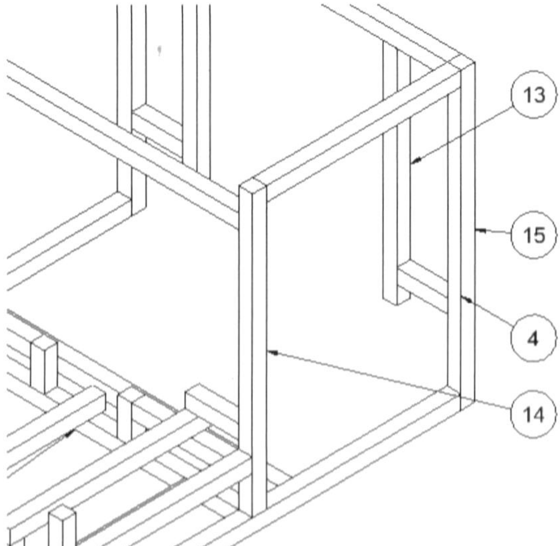


**Figure 6-81**

You can also add an offset spacing value between balloons when performing alignment operations. The same rules apply for balloon offsets as for balloon alignment—namely, the first balloon selected will be the point or anchor for the other balloons. Right-click on any balloon to display the menu, as shown in the following image.

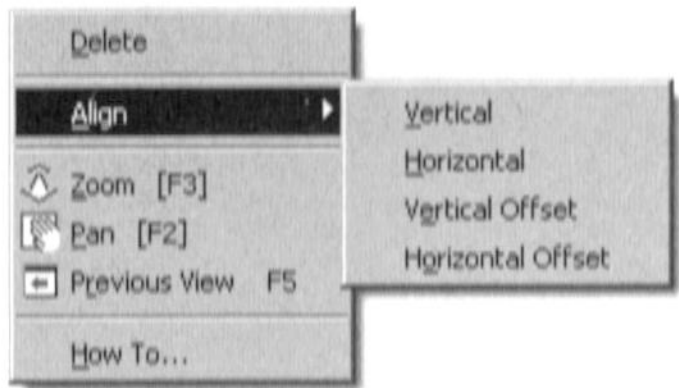


**Figure 6-82**

Select the Horizontal Offset or Vertical Offset options to create a uniform space between balloons. The offset distance is based on the diameter of the balloons placed. The offset distance is a multiplier found in the Drafting Standards dialog box under the Offset Spacing option in the Balloon tab, as shown in the following image.

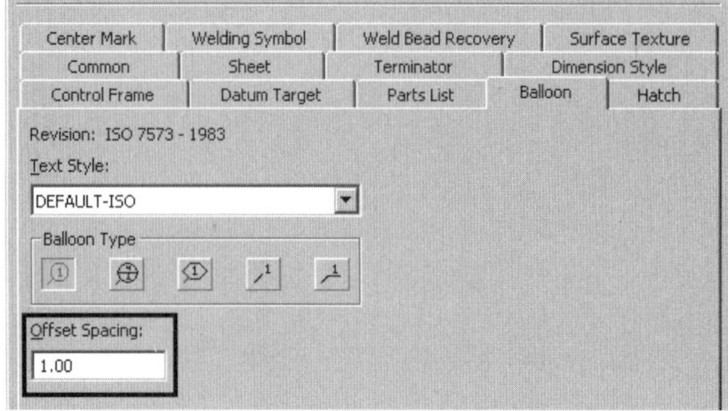


Figure 6-83

A multiplier of 1 is equal to the current balloon diameter. If applied, the gap distance between balloons will be as large as the diameter of one balloon. The default gap multiplier is 1.

## Text Styles

The text style of balloons may be changed in the Balloon tab of the Drafting Standards dialog box, as shown in the following image.

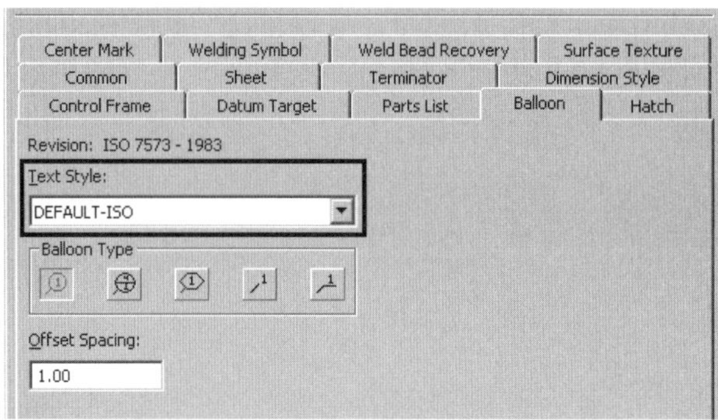


Figure 6-84

## Attaching a Custom Balloon

If you have predefined a custom part, you can select it from a list of custom parts for custom balloon content. To attach a custom balloon, right-click on the balloon to display the menu and select Attach Custom Balloon, as shown in the following image. If there is no custom part available, it will be grayed out.

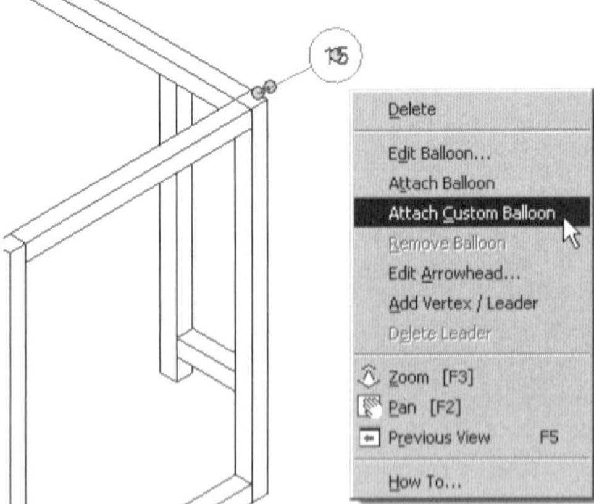


Figure 6-85

The dialog box in the following image will prompt you for your selection of custom balloons. Only one selection is allowed at a time.

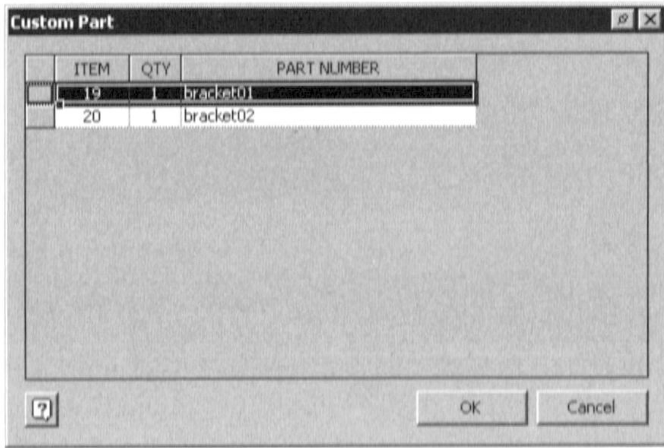


Figure 6-86

After selecting the custom part, click the OK button to return to the drawing. Drag the mouse around for placement of the attached balloon. Click to place the balloon in the desired location, as illustrated in the following image.

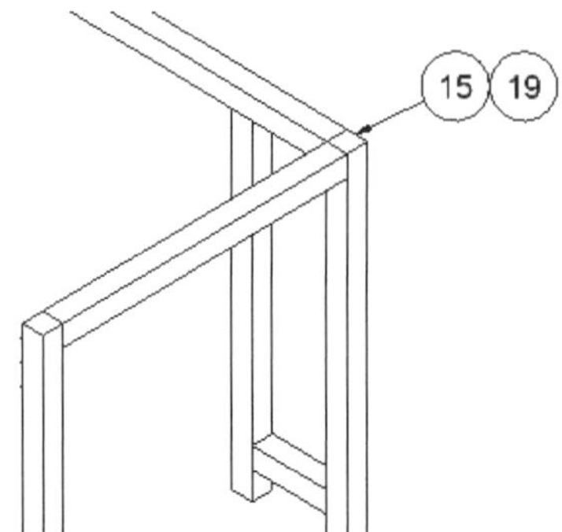


Figure 6-87

## PARTS LISTS

After the drawing views are created and balloons placed, you can also create a parts list, as shown in the following image. As noted earlier in this section, components do not need to be ballooned before creating a parts list.

| Parts List | | | |
|---|---|---|---|
| ITEM | QTY | PART NUMBER | DESCRIPTION |
| 1 | 1 | LNV145 | Body |
| 2 | 1 | LNV146 | Nozzle |
| 3 | 1 | LNV147 | Actuator |
| 4 | 1 | LNV148 | Arm |

Figure 6-88

To create a parts list, click the Parts List tool (see following image) from the Drawing Annotation toolbar. Select a view on which the parts list will be based. If balloons were not created, the Parts List Item Numbering dialog box will appear as was previously discussed in the balloon creation section. After specifying your options, click the OK button and a parts list will appear. If the components were already ballooned, the list will appear without first seeing the Parts List Item Numbering dialog box. Click a point on which to locate the parts list. The information in the list is populated from the properties of each component. A long parts list can also be split into multiple columns.

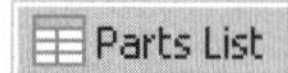


Figure 6-89

Follow these steps to split a parts list that has already been placed on the drawing.

1 Right-click on the parts list and select Edit Parts List from the menu.

2 From the Edit Parts List dialog box, select the row after which the split will occur.

3 Right-click and select Wrap Table at Row from the menu, as shown in the following image.

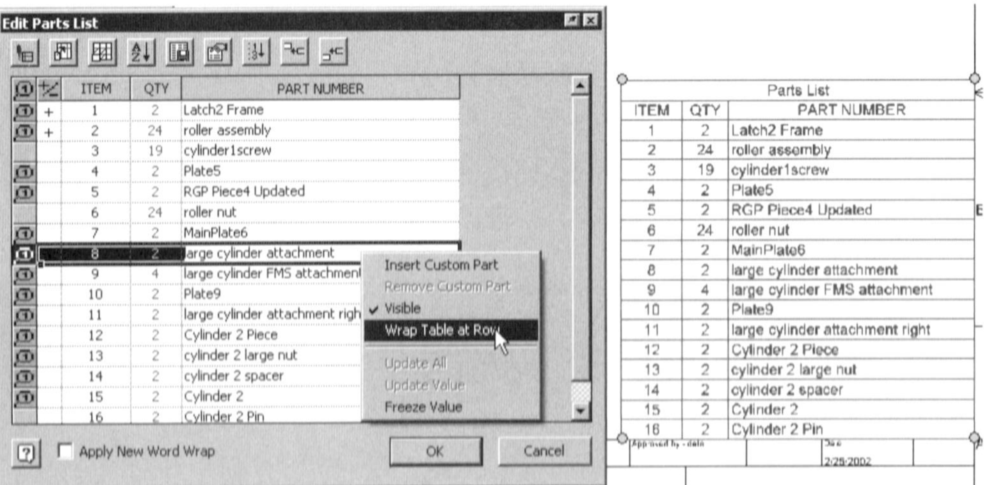


Figure 6-90

4 To complete the operation, click the OK button. The results of performing a column split are illustrated in the following image.

| Parts List | | | Parts List | | |
|---|---|---|---|---|---|
| ITEM | QTY | PART NUMBER | ITEM | QTY | PART NUMBER |
| 9 | 4 | large cylinder FMS attachment | 1 | 2 | Latch2 Frame |
| 10 | 2 | Plate9 | 2 | 24 | roller assembly |
| 11 | 2 | large cylinder attachment right | 3 | 19 | cylinder1screw |
| 12 | 2 | Cylinder 2 Piece | 4 | 2 | Plate5 |
| 13 | 2 | cylinder 2 large nut | 5 | 2 | RGP Piece4 Updated |
| 14 | 2 | cylinder 2 spacer | 6 | 24 | roller nut |
| 15 | 2 | Cylinder 2 | 7 | 2 | MainPlate6 |
| 16 | 2 | Cylinder 2 Pin | 8 | 2 | large cylinder attachment |

Figure 6-91

# EXERCISE 6-6 Creating Assembly Drawings

In this exercise, you create drawing views of an assembly and then add balloons and a parts list. To navigate to the exercise in the *Electronic Student Workbook*, do the following:

1 From the Main TOC page, click Chapter 6.

2 From the TOC for Chapter 6, click Creating Assembly Drawings.

The following image illustrates the completed exercise.

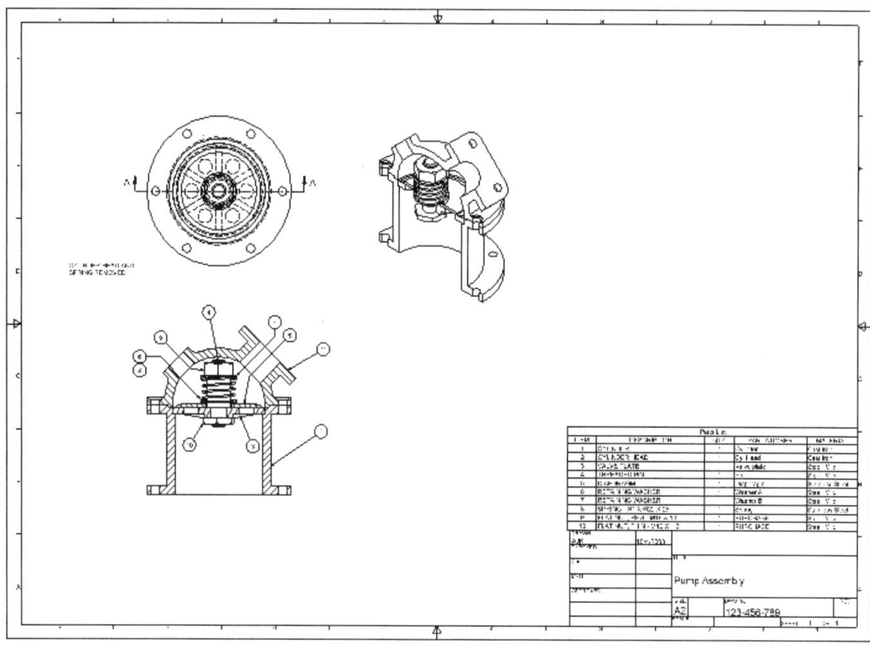


**Figure 6-92 Completed exercise**

## CHAPTER SUMMARY

| To | Do This | Tool |
|---|---|---|
| Place a component in an assembly | Click the Place Component tool in the Panel Bar or from the Assembly toolbar. | |
| Create a component in an assembly | Click the Create Component tool in the Panel Bar or from the Assembly toolbar. | |
| Place an assembly constraint | Click the Constraint tool in the Panel Bar or from the Assembly toolbar. | |
| Edit an assembly constraint | Double-click the constraint in the Browser. | |
| Edit a component in place | Double-click the component in the graphics window or Browser. | |
| Control component visibility | Right-click the component in the graphics window or Browser and click Visibility. | Visibility |
| Check component interference | Select Tools > Analyze Interference from the menu bar. | Analyze Interference |
| Move an assembly through a range of motion | Drag a constrained component. | |
| Create a presentation file | Create a new file with the *.ipn* extension and click the Create View button. | Standard.ipn |
| Create a balloon | Click the Balloon tool in the Drawing Annotation Panel Bar. | |
| Create a parts list | Click the Parts List tool in the Drawing Annotation Panel Bar. | |

EXERCISE

# Applying Your Skills

### SKILL EXERCISE 6-1

In this exercise, you create a new component for a charge pump, then assemble the pump. To navigate to the exercise in the *Electronic Student Workbook*, do the following:

1 From the Main TOC page, click Chapter 6.
2 From the TOC for Chapter 6, click Exercise 1 under Applying Your Skills.

The following image illustrates the completed exercise.

**Figure 6-93 Completed exercise**

## SKILL EXERCISE 6-2

In this exercise, you create an assembly drawing, parts list, and balloons for a charge pump. To navigate to the exercise in the *Electronic Student Workbook*, do the following:

1. From the Main TOC page, click Chapter 6.
2. From the TOC for Chapter 6, click Exercise 2 under Applying Your Skills.

The following image illustrates the completed exercise.

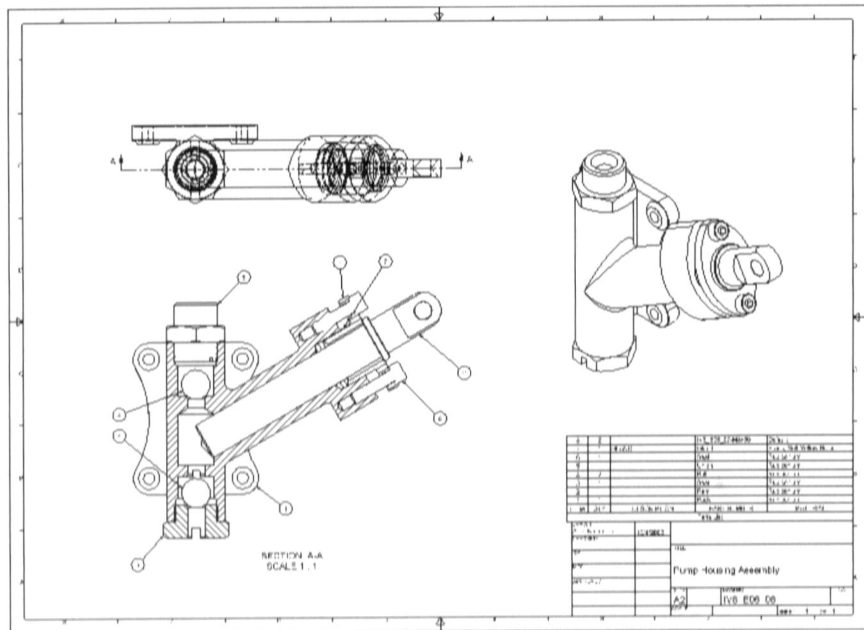

**Figure 6-94 Completed exercise**

## CHECKING YOUR SKILLS

1 **True___ False___** The only way an assembly can be created is by placing existing parts into it.

2 Explain top-down and bottom-up assembly techniques.

________________________________________

________________________________________

3 **True___ False___** An occurrence is a copy of an existing component.

4 **True___ False___** Only one component can be grounded in an assembly.

5 **True___ False___** Autodesk Inventor does not require components in an assembly to be fully constrained.

6 **True___ False___** A sketch must be fully constrained to adapt.

7 What is the purpose of creating a presentation file?

________________________________________

________________________________________

8 **True___ False___** A presentation file is associated to the assembly file on which it is based.

9 **True___ False___** When creating drawing views from an assembly, you can create views from multiple presentation views or design views.

# CHAPTER 7

# Complex Sketching and Constraining Techniques

In the first few chapters, you learned how to create sketches using basic techniques. That experience has given you a good foundation of knowledge on creating basic sketches. In this chapter, you will learn how to utilize existing AutoCAD "DWG" data, create more complex sketches and learn about tools that will reduce the number of steps needed to create a constrained sketch, use construction geometry to better control the sketch, set up relationships between dimensions, and create parts that are driven from a table.

## CHAPTER OBJECTIVES

**After completing this chapter, you will be able to**

- Import AutoCAD DWG data
- Use construction geometry to help constrain sketches
- Create and constrain an ellipse
- Create a 2D spline
- Create a pattern of sketch geometry
- Share a sketch
- Utilize both the symmetry constraint and mirror tool
- Slice the graphics window
- Sketch on another parts face
- Create dimensions using the automatic dimensioning tool
- Change the display of dimensions
- Create relationships between dimensions
- Create parameters
- Create a part that is driven from a Microsoft Excel spreadsheet
- Create parts with dimensional tolerances

## OPENING AUTOCAD FILES

Many Autodesk Inventor users have a large amount of legacy data in AutoCAD DWG format. Instead of redrawing this data, you can import this data into Autodesk Inventor drawings or into a part feature sketch. You can also import AutoCAD files containing 3D solids. When you import an AutoCAD file with 3D solids, each solid is translated to an Autodesk Inventor part file. If multiple solids exist in the AutoCAD file, an Autodesk Inventor assembly file is created and all translated parts are inserted into the assembly. Parts are placed in the same position and orientation as in the AutoCAD file and are given the same name as the AutoCAD file, with a sequenced number at the end of the part name. An AutoCAD file named *Piston.dwg* that contained two solids, for example, would be imported into an Autodesk Inventor assembly named *Piston.iam*, and the parts would be named *Piston1.ipt* and *Piston2.ipt*. When importing a DWG file, there is an import wizard that guides you through the process. In the following sections you will be introduced to the options available when importing AutoCAD data into Autodesk Inventor.

### OPENING AN AUTOCAD FILE OR AUTOCAD MECHANICAL FILE

One method for importing DWG data is to open a drawing file. The DWG File Import wizard guides you through a series of dialog boxes and enables the data to be imported into the following destinations:

**New Drawing** Creates a new Autodesk Inventor drawing file and imports geometry into a draft view. Dimensions and annotations are imported; blocks are translated to sketched symbols in the Drawing Resources folder in the Browser. The geometry can be copied to the clipboard and pasted onto a sketch in a part's *.ipt* file.

**New Part** Creates a new part file and imports geometry to Autodesk Inventor sketch geometry. Dimensions, text, and other annotations are not imported.

**Title Block** Creates a new Autodesk Inventor drawing file and imports geometry, including annotations, into a new title block.

**Border** Creates a new Autodesk Inventor drawing file and imports geometry, including annotations, into a new border.

**Symbol** Creates a new Autodesk Inventor drawing file and imports geometry, including annotations, into a new sketched symbol. The sketched symbol is stored in the drawing, and the symbol is placed on the drawing sheet.

When you open an AutoCAD file, a wizard guides you through the import process. The next time you open the same file type, the previous wizard settings are automatically used to import the file. To open an AutoCAD file into a new Autodesk Inventor file, follow these steps:

1 Select File > Open from the main menu or from the Standard toolbar.

2 In the Open dialog box, from the Files of type list, click DWG Files (*.*dwg*) as shown in the following image.

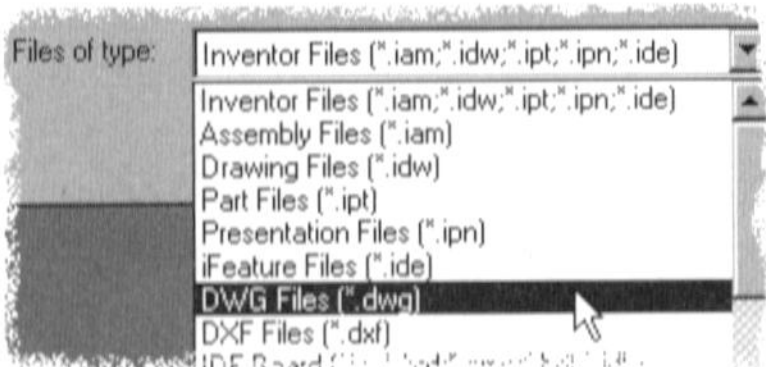


**Figure 7-1**

3 Browse to and pick the desired DWG file.

The Open dialog box will appear, and a preview image of the DWG file will appear in the lower-left corner of the dialog box, as shown in the following image.

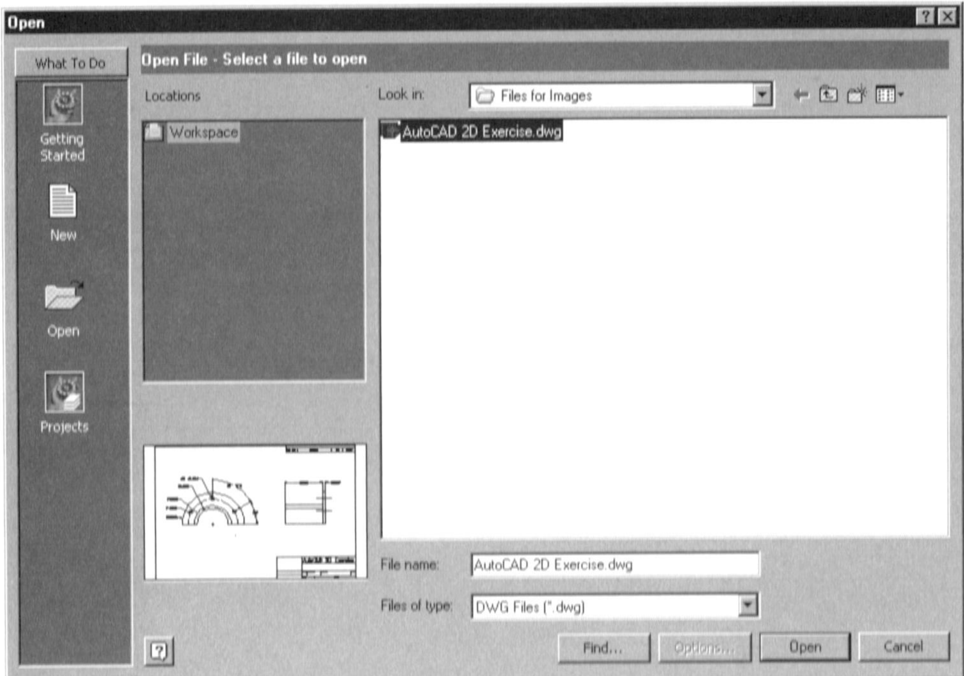


**Figure 7-2**

4 Click Open to start the DWG File Import Options wizard, whose first screen is shown in the following image.

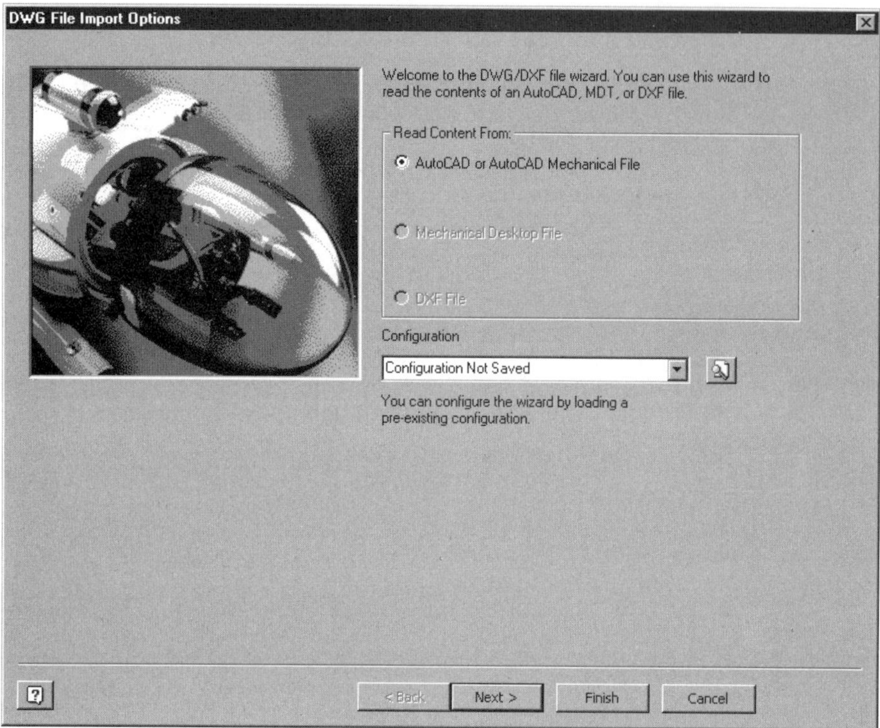


**Figure 7-3**

The type of DWG file—AutoCAD, AutoCAD Mechanical, or Mechanical Desktop—is automatically selected depending upon the data found in the DWG file. You can select a configuration file that will fill in the prompts based on previous responses to the DWG File Import Options wizard.

5 Click Next. The Layers and Objects Import Options dialog box is displayed, as shown in the following image.

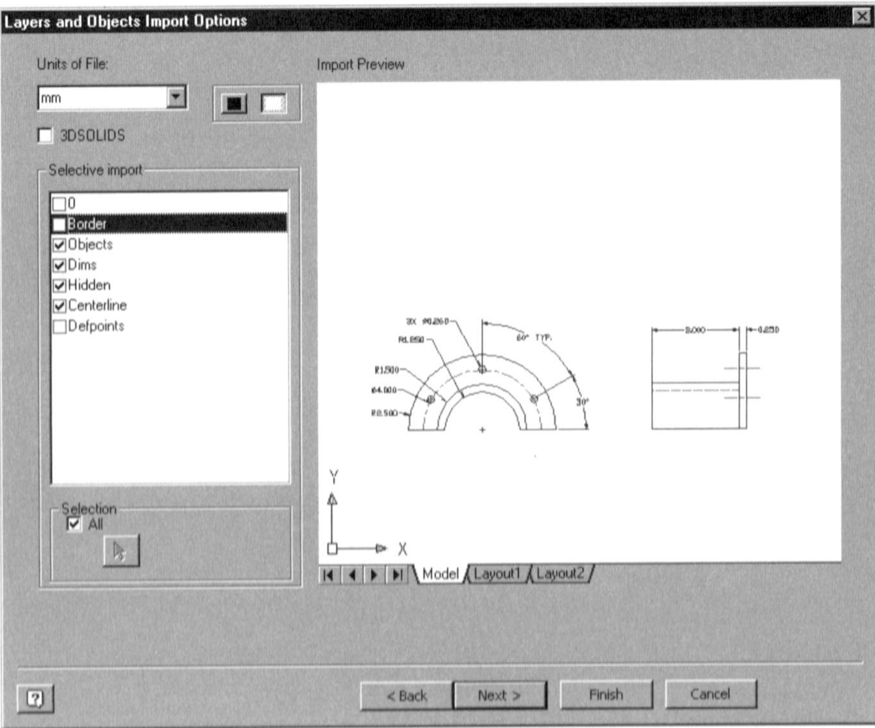


**Figure 7-4**

6 Select the unit in which the file was created by selecting the unit from the drop-down list in the Units of File area.

7 To change the background color of the preview image, click the black or white icon at the top of the Layers and Objects Import Options dialog box.

8 Check the 3D SOLIDS area if you want 3D solids to be imported. Otherwise, no solids will be imported. 3D solids must exist in the file for this option to be valid.

9 Import objects from Model Space or from a Layout within the DWG file by clicking the different tabs at the bottom of the screen. The names of the tabs are identical to the tab names in the DWG file.

10 In the Selective import area, check the layers that will be imported from the preview. As the layers are selected, the preview image will be updated to reflect the change.

11 To select objects in the Import Preview window, uncheck the All area in the Selection area of the dialog box and then select objects in the Import Preview window.

12 Click Next. The Import Destination Options dialog box is displayed, as shown in the following image.

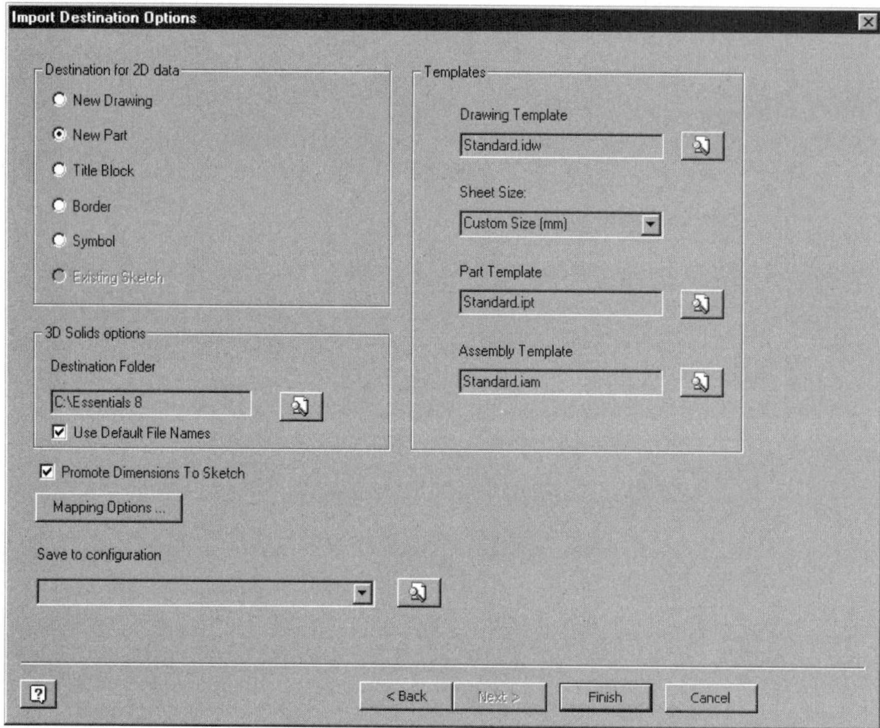


**Figure 7-5**

13 Select the destination for the imported data. You can import 2D data into a draft view in a new drawing or as the contents of a sketch in a new part file. The data can also be imported into a title block, border, symbol, or an existing sketch.

When importing 3D solids, you can change the Destination Folder in which the new files will be saved by clicking the Browse button and then navigating to and selecting a folder. The area is grayed out if 3D Solids are not imported.

Also, when you import 3D solids, you can make the file names be automatically based on the AutoCAD file name by checking the Use Default File Names. If you uncheck this option, you will be prompted for the file names for each 3D solid.

14 Click the Mapping Options button to specify mapping options for how you want the contents of each AutoCAD layer to be imported, or choose how you want AutoCAD fonts to be mapped to Autodesk Inventor Fonts.

15 In the Templates area, specify the template to which the geometry will be imported.

16 To save these configuration settings for future use, click the Save the Configuration button and enter a name.

17 To complete the operation, click the Finish button.

## IMPORTING AUTOCAD DATA INTO A SKETCH

Another method that utilizes existing AutoCAD 2D data inserts the AutoCAD 2D data into the active sketch (in a part or drawing). To insert AutoCAD data into the active sketch, follow these steps:

1. Make a sketch active in either a part or a drawing file.
2. Click the Insert AutoCAD file icon on the 2D Sketch Panel Bar, as shown in the following image. Alternatively, in a drawing, click the Insert AutoCAD file tool on the Drawing Sketch Panel Bar.

Figure 7-6

3. Browse to and pick the desired DWG file.

   The Open dialog box will appear, and a preview image of the DWG file will appear on the right side of the dialog box, as shown in the following image.

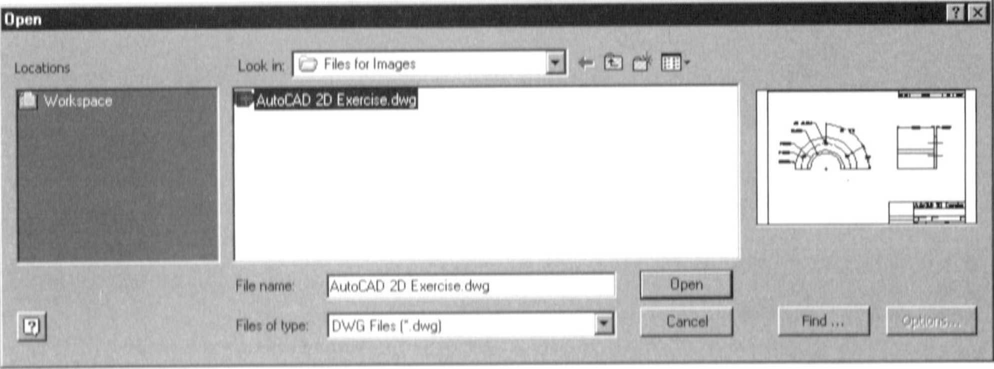


Figure 7-7

4. Click the Open button. Layers and Objects Import Options dialog box will appear, as shown in the following image.

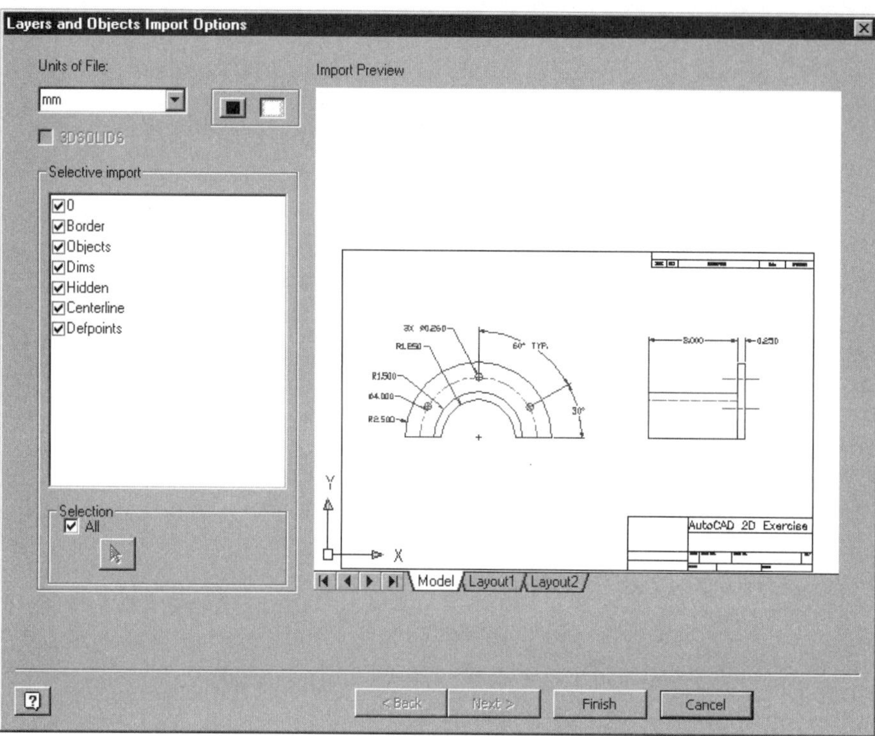


Figure 7-8

5 Select the unit in which the file was created by selecting the unit from the drop-down list in the Units of File area.

6 You can change the background color of the preview image by clicking the black or white icon at the top of the Layers and Objects Import Options dialog box.

7 Check the 3D SOLIDS area if you want 3D Solids to be imported. Otherwise, no solids will be imported. 3D solids must exist in the file for this option to be valid.

8 Import objects from Model Space or from a layout within the DWG file by clicking the different tabs at the bottom of the screen. The names of the tabs are identical to the tab names in the AutoCAD file.

9 In the Selective import area, check the layers that will be imported from the preview. As the layers are selected, the Import Preview window will be updated to reflect the change.

10 To select objects in the preview window, uncheck the All option in the Selection area of the dialog box, and then select objects in the Import Preview window.

11 To complete the operation, click the Finish button.

## IMPORTING OTHER FILE TYPES

Autodesk Inventor can also import parts and assemblies exported from other CAD systems. Autodesk Inventor models created from these formats are base solids or surface models, and no feature histories or assembly constraints are generated when you import a file in any of these formats. You can add features to imported parts, edit the base solids using Autodesk Inventor's solids editing tools, and add assembly constraints to the imported components. To open file types such as SAT, STEP, PRO/E, DXF, and IGES, select File > Open from the main menu or click Open on the Standard toolbar.

In the Open dialog box, from the Files of type list, click the desired file format. The available formats are described as follows.

### SAT

SAT is a neutral file format available from most CAD systems based on the ACIS modeling kernel. CAD systems based on other kernels can often export parts and assemblies in this format. SAT files can be imported either as solid bodies or as surface bodies using the Insert > Import menu option in the part environment.

### STEP

The STEP format is widely used to translate 3D solid models between CAD systems. Autodesk Inventor can import models saved in AP-203 or AP-214 STEP formats.

### PRO/E

Autodesk Inventor can automatically translate and import native Pro/E files up to Version 20. Complex solid models and Pro/E surfaces can cause import problems. Pro/E files exported in STEP format are more reliably imported. Pro/E 2000i and newer files must be exported to another format for import into Autodesk Inventor.

### DXF

The Drawing Interchange File format, DXF, is a neutral file format supported by many CAD systems. Import a DXF file into Autodesk Inventor using the same techniques that you would use to import a DWG file.

### IGES

Early data translation between CAD systems almost always used the IGES file format, and it is still commonly used today. IGES files can contain many different types of data but typically consist of a set of surfaces representing the boundary of the part. An IGES file is imported into a Construction folder as a group of surfaces. You can then select surfaces and stitch them into a surface quilt. If the surfaces can be combined into a surface quilt with no gaps, the quilt can be promoted to an Autodesk Inventor base solid. You can promote individual surfaces and surface quilts to grounded Autodesk Inventor surfaces. Surface healing of IGES data is not done. Gaps between surfaces due to file imprecision or other factors will prevent an imported IGES file from being promoted to a base solid. You can view the source of the IGES file by selecting the Custom tab, which is made available when you access the properties of the file after the data has been imported.

## CONSTRUCTION GEOMETRY

Construction geometry can help you create sketches that would be difficult to create without their aid. Construction geometry can be constrained and dimensioned like normal geometry, but the construction geometry will not be seen in the part when the sketch is turned into a feature. When you sketch, the sketches, by default, have a normal geometry style, meaning that the sketch geometry is visible in the feature. Construction geometry can reduce the number of constraints and dimensions that are required to fully constrain a sketch and can help define the sketch. A construction circle inside a hexagon can drive the size of the hexagon, for example. Without construction geometry, the hexagon would require six constraints and dimensions; it would require only three constraints and dimensions with construction geometry—the circle will have tangent or coincident constraints applied to it and the hexagon. Construction geometry is created by changing the line style before or after geometry is sketched, in one of the following two ways:

- Before sketching, select Construction from the Style drop-down list on the Standard toolbar, as shown in the following image.
- After creating the sketch, select the geometry that you want to be created, and then select Construction from the Style drop-down list on the Standard toolbar, as shown in the following image.

**Figure 7-9**

After turning the sketch into a feature, the construction geometry will disappear. When you edit a feature's sketch that was created with construction geometry, the construction geometry will reappear during editing and disappear when the part is updated. Construction geometry can be added or deleted to a sketch just like any geometry that has a normal style. In the graphics window, construction geometry will be lighter in color and thinner in width than normal geometry. The image on the left of the following image shows a sketch with a construction line for the angled line. The angled line has a coincident constraint applied to it at every point that touches it. The image on the right shows the sketch after it has been extruded. Note that the construction line was not extruded.

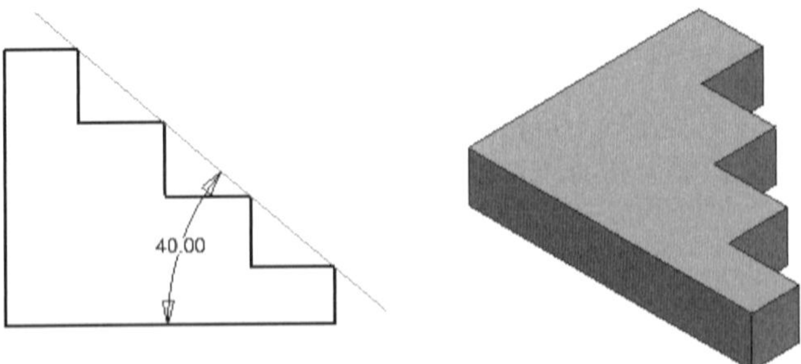


Figure 7-10

## ELLIPSES

Ellipses can be created by selecting the Ellipse tool from under the Circle tool on the Sketch Panel Bar, as shown in the following image.

Figure 7-11

To create an ellipse, follow these steps:

1 Start the Ellipse tool.
2 Click a center point for the ellipse.
3 Click a point that will define the first axis and fall on the outside of the ellipse.
4 Move the cursor until the desired shape of the ellipse is found.
5 Click the point to create the ellipse.

The ellipse can be trimmed, extended, and dimensioned. To dimension an ellipse, follow these steps:

1 Start the General Dimension tool.
2 Click on the ellipse.
3 Move the cursor and click a point to either define the major or minor distances.
4 Repeat the process to dimension the other major or minor distance.

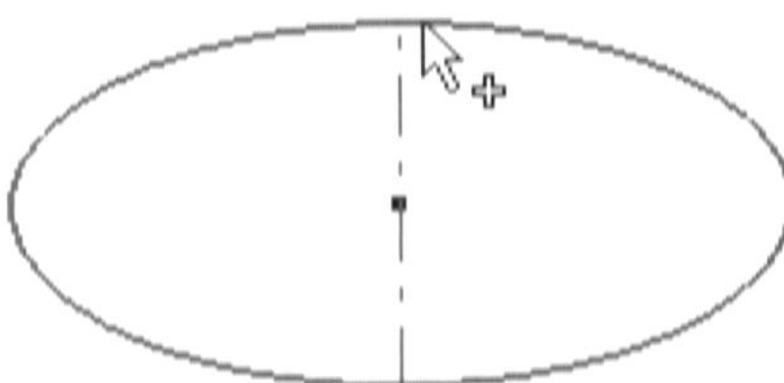

Figure 7-12

An ellipse can also be offset. To offset an ellipse, start the offset tool and then, depending upon where you select, you can create one of the two kinds of ellipses. If you click near the major or minor axis (a axis will appear as shown in the image above) a concentric ellipse will be created. A concentric ellipse will have the same center point (a coincident constraint) as the ellipse from which it was offset. If you select the ellipse away from the major or minor axis, a spline profile will be created that can be freely moved or manipulated.

## 2D SPLINES

A spline is geometry that is sketched as free flowing in shape and can be used to create complex shapes. Points are used to define the shape of the spline; these points can be constrained as needed. The points that define a spline will be visible when a spline is created or selected. To create a spline, follow these steps:

1. Start the Spline tool from the Sketch Panel Bar under the Line tool, as shown in the following image.
2. Either click arbitrary points in the graphics window, or click existing points in the sketch.
3. To exit the command and keep the spline, press **ENTER**, or right-click and select Continue from the menu.

**Figure 7-13**

Once a spline has been created, it can be edited and controlled by right-clicking on a point that lies on the spline or in between the points on the spline. If a point is selected, you will only get menu options that pertain to that point. When you right-click on the spline, the spline options will be displayed, as shown in the following image. Each option will be described in the following sections.

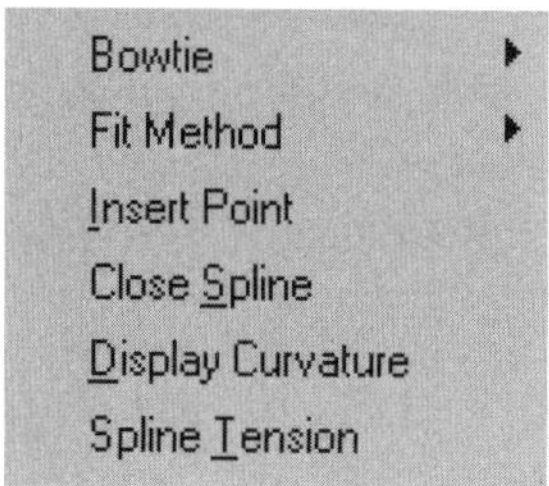


**Figure 7-14**

### BOWTIE

The Bowtie option allows you to precisely modify the spline's shape at the shape point. There are three options under Bowtie—Handle, Curvature, and

Flat. These options control the curvature of the spline and will be described in the next section.

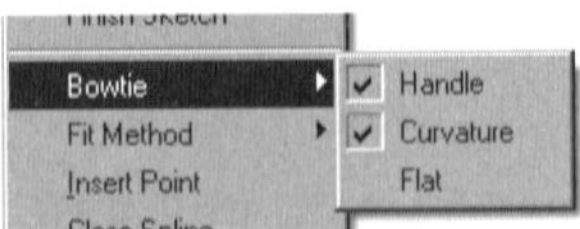


**Figure 7-15**

## Handle

The handle (also referred to as a handlebar) is a line segment that has nodes at each end and is always tangent to the spline. The handle's visibility is controlled by clicking on a point on the spline, right-clicking, and selecting Bowtie > Handle. If Handle is checked, and the spline is selected, the handle will be visible. Once the handlebar is visible, you can click a node and either rotate or change the length of the handle. Rotating the handle will change the tangent angle at that point. The longer the handle length, the longer the spline will be tangent to it. Handlebar lengths can be controlled with dimensions. Handlebar lengths have no units; they can be thought of as a weight. A length (or weight) of 1 is the most typical or "natural." In a loose sense (depending on the actual geometry), a length of 2 causes the spline to hug the curve twice as long, while a length of 1/2 causes it to hug the curve half as long. These are not hard and fast rules. If the spline is a straight line, for example, changing the length of the tangent has little or no effect except that if it is too large the spline will overshoot the endpoint and double back on itself.

**Figure 7-16**

## Curvature

Curvature will control the radius of a spline at the point where the curvature exists. The curvature is represented by an arc. The curvature's visibility is controlled by clicking on a point on the spline, right-clicking, and selecting Bowtie > Curvature—the curvature arc will then appear. If the Curvature box is checked and the spline is selected, the curvature and handle will be visible. To adjust the curvature, click and drag one of the curvature nodes (as shown in the following image) or constrain it to existing geometry. If, after changing the radius of curvature, the spline appears to overshoot the intended direction, the length of the handlebar can be reduced to compensate.

You can create second-order contact between a spline and a circle (or a spline and another spline) by constraining all three elements of the spline (point-coincident, handlebar-tangent, curvature bar-concentric) to the other geometry.

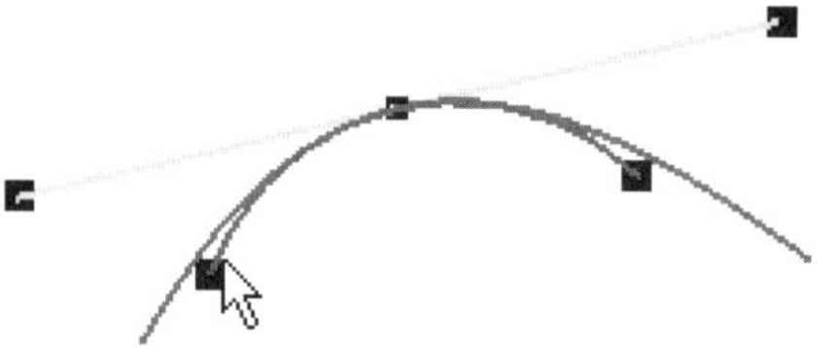

Figure 7-17

### Flat

When the Flat option is selected, handle and curvature nodes will not be seen and a line with two nodes representing the flat will appear. Clicking one of the nodes and dragging it to a new location can adjust the length of the flat, as shown in the following image.

Figure 7-18

## FIT METHOD

The Fit Method is used to control how the spline will be generated between the points. There are three methods to choose from—Smooth, Sweet, and AutoCAD—as shown in the following image.

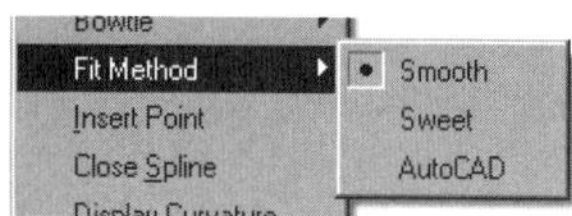


Figure 7-19

**Smooth** A lightweight representation of the spline. It approximates the properties of the sweet splines with fewer control points, leading to smaller file sizes and faster recompute speeds. You cannot control the tension of smooth splines (they will be automatically converted to sweet splines if you try). Compared with sweet splines on the same set of points, smooth splines tend to look more aesthetically pleasing—hence the use of the term "smooth" (an industrial designer's term for nice-looking curves).

**Sweet** Also called "minimum energy splines," sweet splines are formulated to seek a balance between tension and curvature. They are loosely modeled on the bending and stretching properties of materials. A very stiff material such as a steel plate bends slowly, for example, distributing its curvature over its entire length. A flexible piece of rubber, on the other hand, localizes its bending to the positions where the bend is forced, and stays somewhat flat elsewhere. While sweet splines have nice geometric and aesthetic properties, they should be used with caution. Models with many surfaces generated from sweet splines may have larger file sizes, longer load times, and longer recompute times.

**AutoCAD** AutoCAD splines allow compatibility with AutoCAD. These splines are most appropriate for two-way interaction with AutoCAD files, but have two primary drawbacks: first, they have reduced continuity (visible with curvature combs), which may cause surfaces derived from them to have poor reflective qualities (these reflective qualities can be seen using zebra stripes); second, they do not support bowties.

To change the spline method, right-click on the spline and click Smooth, Sweet, or AutoCAD from Fit Method. The next three images show the same spline generated using the three Fit Methods—Smooth, Sweet, and AutoCAD.

**Figure 7-20** Smooth spline

**Figure 7-21** Sweet spline

**Figure 7-22** AutoCAD spline

### INSERT POINT

The Insert Point option will insert a point into the spline. This new point can be controlled and edited like any other point on the spline. To insert a point, right-click on the spline and select Insert Point from the menu. Move the cursor to where the point should be located on the spline, and click. The following image shows a point being added to the right side of the spline.

Figure 7-23

### CLOSE SPLINE

The Close Spline option will close the first and last point of an open spline. To close an open spline, right-click on the spline and click Close Spline from the menu. The following image shows an open spline on the left and the same spline closed on the right.

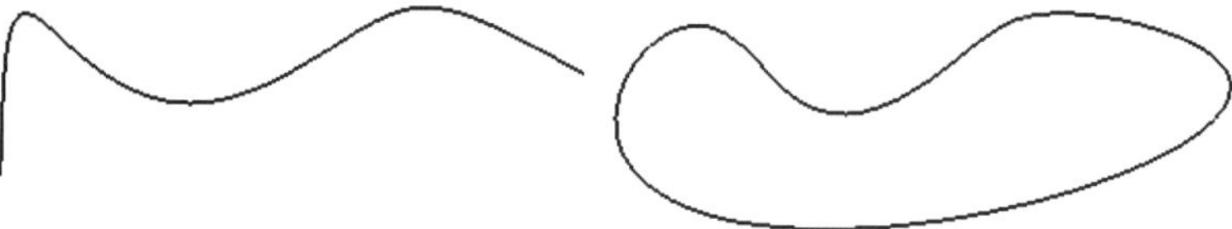

Figure 7-24

### DISPLAY CURVATURE

To better see the curvature of the spline, you can turn on the visibility of curvature lines. These display curvature lines cannot be used to edit the spline. As the spline has more curvature, the lines will be longer. To display the curvature of a spline, right-click on the spline and click Display Curvature from the menu. The main value of displaying the curvature comb is to visualize the continuity of the curve and to get an idea of the quality of surfaces generated from such a curve. Curvature combs magnify discontinuities in a curve. Sudden changes in curvature, for example, show up as sharp changes in the path of the teeth of the curvature comb.

The following image shows a spline with the display curvature lines visible.

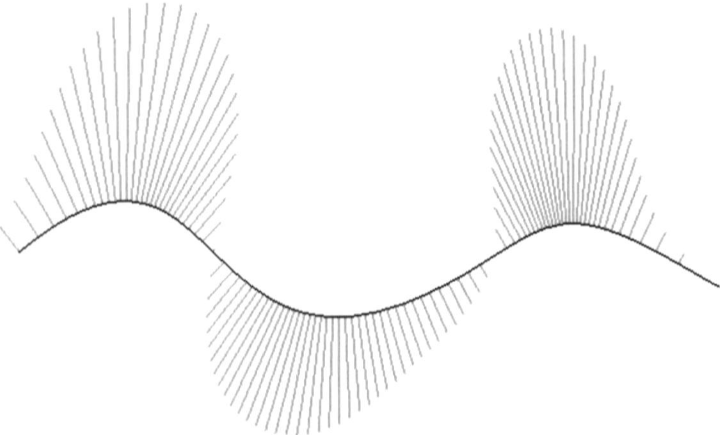

Figure 7-25

## SPLINE TENSION

Another method for controlling a spline is to adjust its tension. The tension of a spline can be adjusted between 0 and 100. The lower the number is, the more curvature it will have and the higher number will be straighter between the points. The default for new splines is set to 0. As the tension is adjusted, a preview image will appear in a lighter color. To adjust the tension of a spline, right-click on the spline, click Spline Tension from the menu and slide the pointer to the desired value. The following image shows a spline with the Spline Tension dialog box.

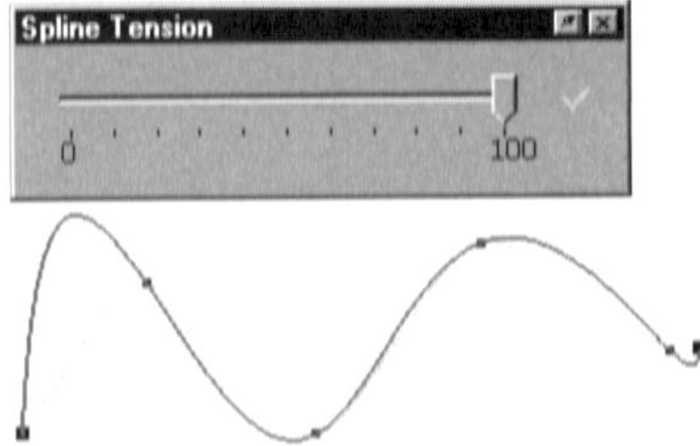


Figure 7-26

## CONSTRAIN AND DIMENSION A SPLINE

To add constraints to a spline, they need to be added to any visible handlebars, curvature arc, or flat of any point. The following constraints can be added: concentric, equal, collinear, horizontal, perpendicular, parallel, tangent, and vertical. Dimensions can be added to any visible handlebars, curvature, flat, or any point. The following image shows a spline that has dimensions to the handlebar, and curvature arc, and the constraints shown on the handlebar.

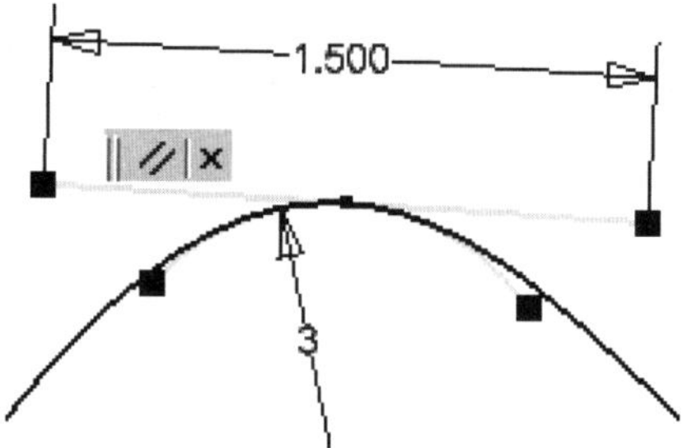


Figure 7-27

# EXERCISE 7-1 Complex Sketching

In this exercise, you work with spline curve controls to create the body of a toaster. To navigate to the exercise in the *Electronic Student Workbook*, do the following:

1. From the Main TOC page, click Chapter 7.
2. From the TOC for Chapter 7, click Complex Sketching.

The following image illustrates the opened exercise.

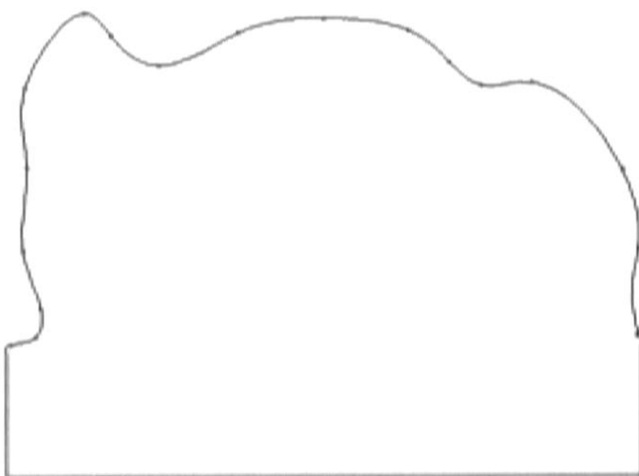

**Figure 7-28 Opened exercise**

The following image illustrates the completed exercise.

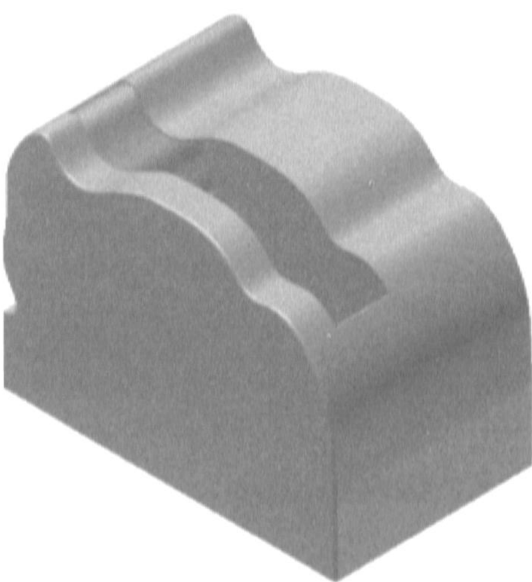

**Figure 7-29 Completed exercise**

## PATTERN SKETCHES

When creating a sketch that will have a rectangular or a polar pattern, you can pattern the geometry in the sketch instead of creating multiple sketches or waiting to pattern the feature. The rectangular pattern does not need to be orientated in a horizontal or a vertical alignment. To pattern a sketch, follow these steps:

1 Click either the Rectangular Pattern or Circular Pattern tool from the Sketch Panel Bar, as shown in the following image.

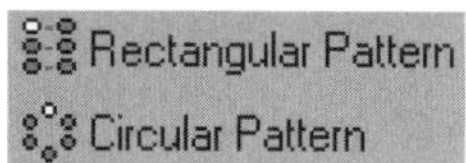


Figure 7-30

The Rectangular or Circular Pattern dialog box will appear. The following image shows the Rectangular Pattern dialog box on the left and the Circular Pattern dialog box on the right.

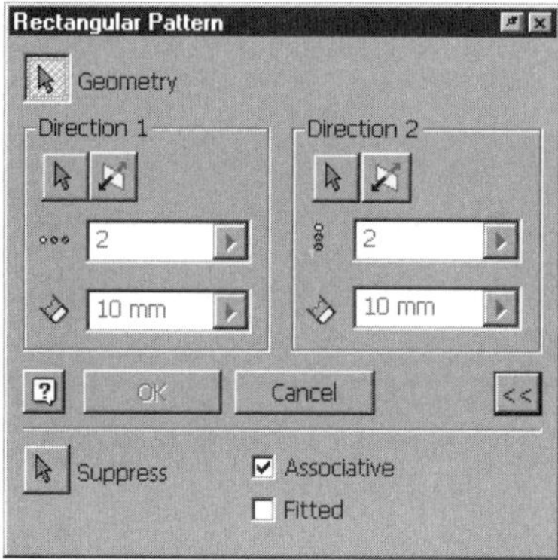


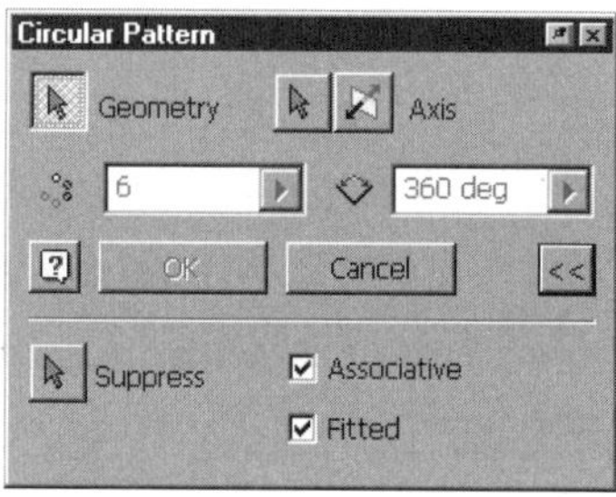


Figure 7-31

2 Click the geometry button and select the geometry that will be patterned. Multiple objects can be selected.

3 Define the direction (1 or 2) or the axis by clicking the arrow and then selecting an object. For a rectangular pattern, the selected object must be a line in the active sketch.

4 If needed, click the Reverse Direction button shown in the following image to change the direction of the pattern.

Figure 7-32

5 Enter the information for the count, spacing, and angle.

6 For a rectangular pattern, you can select a second direction, if needed.
7 Click the OK button to create the sketch pattern.

In the bottom pane of the Pattern dialog boxes are the following options: Suppress, Associative, and Fitted.

**Suppress** Will keep the selected object(s) from appearing in the pattern. When a patterned occurrence is suppressed, it will be represented in a hidden linetype in the graphics window.

**Associative** When checked, the pattern will reflect any changes that are made to the geometry that was patterned. If not checked, any changes made to the geometry that is patterned will not be reflected in the patterned objects.

**Fitted** When checked, the objects that are being patterned will be equally spaced within the specified angle or distance. If not checked, the pattern spacing will specify the angle or distance between the patterned occurrences.

Once the pattern sketch has been created, the pattern will be grouped together. The patterned objects are referred to as occurrences or elements. The individual occurrences can be edited or the entire pattern can be deleted. To edit the patterned sketch, move the cursor over an occurrence and right-click. A menu will appear that gives you the option to suppress or unsuppress elements (depending upon the occurrence the cursor is over when you right-click), and either delete or edit the pattern. The following image shows the edit pattern options while the cursor is over an unsuppressed occurrence of a feature.

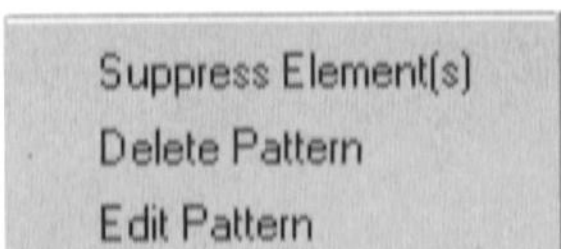


**Figure 7-33**

- Click Suppress Element(s) and then select the object(s) that will be suppressed. To unsuppress objects that are suppressed, reselect them.
- Click Delete Pattern to delete the entire pattern. The original object(s) that were patterned will not be deleted.
- Click Edit Pattern to change the count, spacing, angle, suppress/unsuppress occurrences, and toggle the Associative and Fitted options.

## SHARED SKETCHES

When creating a feature that will use the same sketch geometry, same dimensions, and will lie on the same face of the sketch of an existing feature, you can share the sketch instead of creating a new one. A shared sketch is an associated copy of the original sketch and is placed above the original feature in the Browser. The shared sketch has the same name as the original sketch, and any modifications to an original sketch will be updated to all the features that use the shared sketch. Once a sketch has been shared, the shared sketch can be selected as the profile for additional sketched features. There is no limit to the number of features that can use the shared sketch. The visibility of the shared sketch is on by default and can be turned on and off as needed by right-clicking on the shared sketch in the Browser, and clicking Visibility from the menu. To share a sketch, follow these steps:

1. Locate the feature in the Browser that contains the sketch you want to share.
2. Click the plus sign to the left of that feature to expand the sketch.
3. Right-click on the sketch name and select Share Sketch from the menu, as shown in the following image.
4. Issue the feature tool that you need (extrude, revolve, etc.), and then select the shared sketch as the profile for the feature. The image on the right in the following image shows a shared sketch that has been extruded down to form a rectangle (the shared sketch has also been extruded up with face draft).

**NOTE** A shared sketch that is consumed by a feature cannot be deleted.

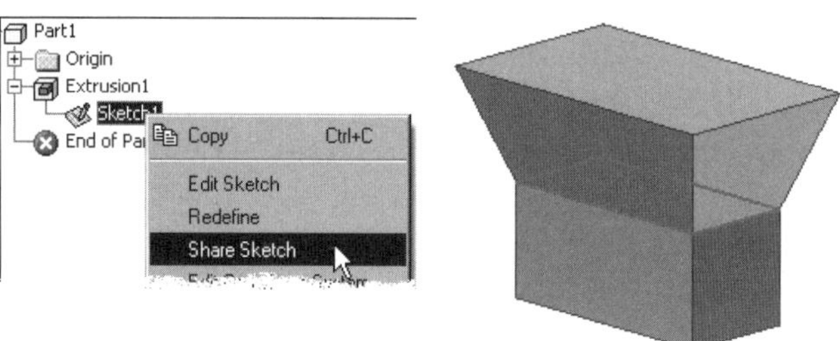


Figure 7-34

## MIRROR SKETCHES AND SYMMETRY CONSTRAINT

When you need to create a symmetrical part, you can use the Mirror sketch tool and the symmetry constraint to reduce the number of constraints and dimensions that are required to fully constrain the sketch. The Mirror tool will automatically apply a symmetry constraint(s) between the selected geometry and the mirrored geometry. As the geometry on one half changes, so will the mirrored geometry on the other side. Follow these guidelines to create a mirrored sketch:

1. Analyze what the finished sketch will look like and determine where the line of symmetry will be.
2. Add constraints and dimension to the sketch either before or after the mirror operation. Draw half of the finished sketch and the line of symmetry. The line of symmetry must be a single line but its style can be normal, construc-

tion, or centerline. The following image shows a sketch that will be mirrored upon the left vertical construction line.

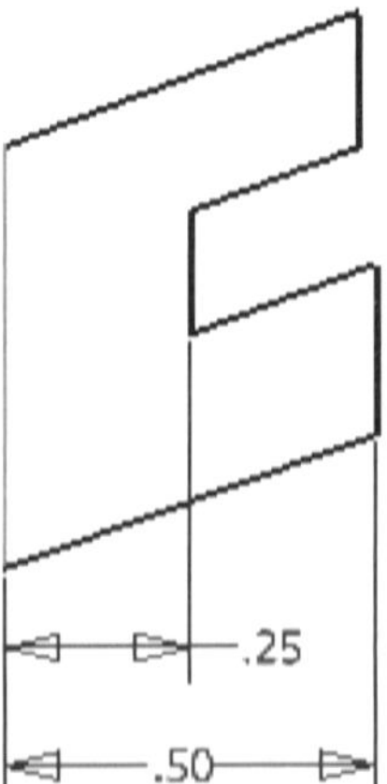


**Figure 7-35**

3 Click the Mirror tool (shown in the following image) from the Sketch Panel Bar.

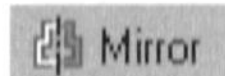


**Figure 7-36**

4 The Mirror dialog box will appear as shown in the following image.

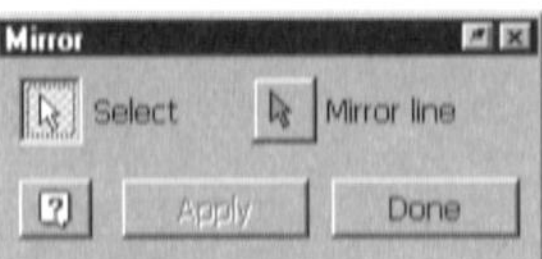


**Figure 7-37**

5 Select the objects in the graphics window that will be mirrored—by default, the Select option is active.

6 Click the Mirror line button in the Mirror dialog box and select the line in the graphics window that will be the line of symmetry.

7 To complete the operation, click the Apply button.

8 A symmetry constraint will automatically be applied between both halves. Any change to a dimension or constraint on the first half will be reflected on the mirrored side. The symmetry constraint can be deleted like any other geometric constraint. The following image shows the geometry on the left mirrored across the vertical construction line with the constraints shown.

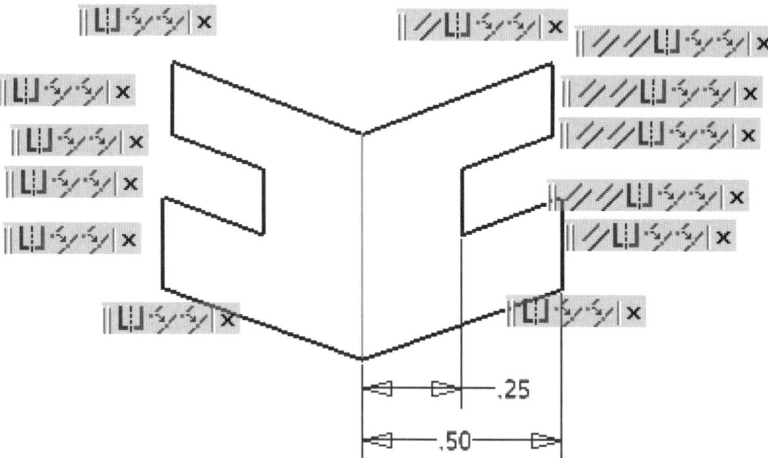


Figure 7-38

A symmetry constraint can also be added manually between two sets of objects. The object(s) must be the same object type (line, arc, circle, etc.) and must lie on opposite sides of a line of symmetry that must exist. After the geometries are drawn, add the Symmetry constraint from the Sketch Panel Bar under the sketch constraints menu shown in the following image. Select the two objects that will be symmetrical and then select the line of symmetry.

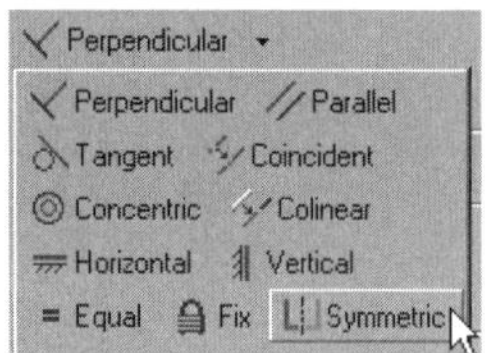


Figure 7-39

## SLICE GRAPHICS

While creating parts, you may need to sketch on a plane that is difficult to see because other features are obscuring the view. The Slice Graphics option will temporarily slice away the portion of the model that obscures the plane on which you want to sketch. The following image shows a part that is shelled out (in the middle)—the image on the left is shown in a normal state, and the image on the right is a menu shown with the Slice Graphics option. Follow these steps to temporarily slice the graphics screen.

1 Make a plane that the graphics will be sliced to the active sketch.
2 Rotate the model so the correct side will be sliced (the side of the model that faces you will be sliced away).

3 While editing the sketch, right-click and select Slice Graphics from the menu (as shown in the following image) and then press the F7 key, or select Slice Graphics from the View menu. The model will be sliced on the active sketch plane.

4 Use sketch tools from the Sketch toolbar to create geometry on the active sketch.

5 To restore the sliced graphics, right-click and select Slice Graphics, select Slice Graphics from the View menu, or click the Sketch or Return button from the Command Bar to leave sketch mode.

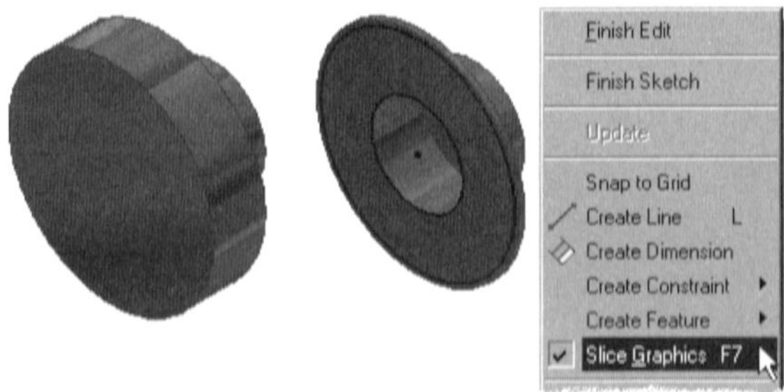


**Figure 7-40**

**NOTE** When working in an assembly, there are different slice graphics tools (Assembly Section Views) available from the Assembly toolbar.

## PROJECT EDGES

As was discussed in Chapter 3, you can automatically project edges of the part to the sketch plane as you sketch a curve. In this section, you will learn about using the Project tool that can project selected edges, vertices, work features, curves, or the silhouette edges of another part in the assembly or of other features in the same part to the active sketch. If the setting Enable Associative Edge/Loop Geometry Projection During In-Place Modeling is checked from the Assembly tab in the Options dialog box (as shown in the following image), the projected edges and vertices will be adaptive (dependent) on the model from which they are projected and maintain the relationship when the originating geometry changes. If this setting is not checked, the projected geometry will not be associative to the geometry from which it was projected, and changes to the original part will not affect the projected geometry.

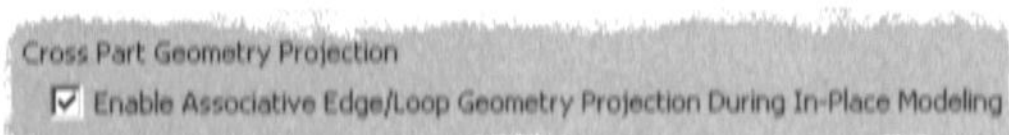


**Figure 7-41**

There are three project tools that are available from the Sketch Panel Bar, as shown in the following image. The three tools are Project Geometry, Project Cut Edges, and Project Flat Pattern.

**Project Geometry** Used to project geometry from a sketch or feature onto the active sketch.

**Project Cut Edges** Used to project edges that lie on the section plane of a part on the active sketch. The geometry is only projected if the uncut part would intersect the sketch plane.

**Project Flat Pattern** Projects a selected face or faces of the flat pattern onto the sheet metal part (Chapter 11 will cover creating sheet metal parts).

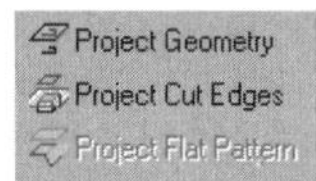


**Figure 7-42**

To project geometry, follow these steps:

1. Make a plane the active sketch.
2. Click the Project Geometry tool from the2D Sketch Panel Bar.
3. Select the geometry that will be projected onto the active sketch. If you want to project all edges that are tangent to an edge, use the Select Other tool to cycle through until they are all highlighted as shown in the following image.
4. To exit the operation, press the ESC key or click another tool.

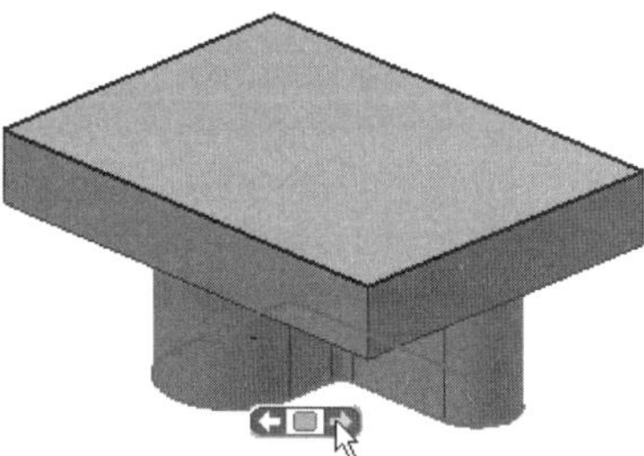

**Figure 7-43**

To project cut edges, follow these steps:

1. Create a section view in an assembly with one or more components, making sure that all the components you need are visible.
2. Make the part active or create a new part that the projected edges will be projected onto.
3. Make a plane that lies on the plane that will be sectioned the active sketch.
4. Click the Project Cut Edges tool from the Sketch toolbar.

5 If you are in a part file, all the edges that touch the active sketch plane will be projected onto it.

6 If you are in an assembly, you need to select all the parts whose edges that touch the active sketch plane will be projected onto the active sketch. In the following image, the model on the left shows an assembly with all the parts being projected onto an active sketch that lies in the middle of the parts. The image on the right shows the projected edges with the parts shown transparently.

7 Press the ESC key or click another tool to exit the operation.

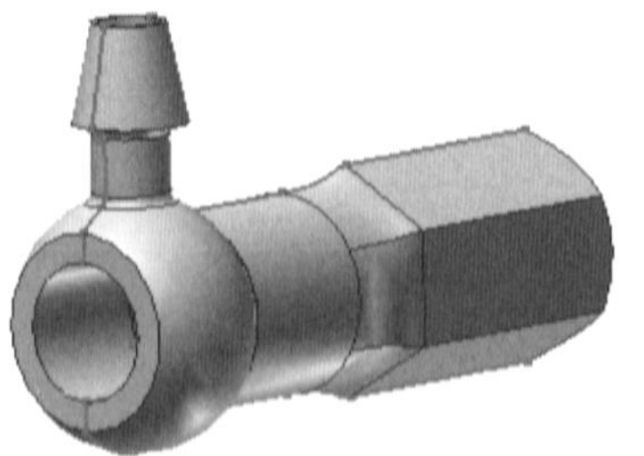
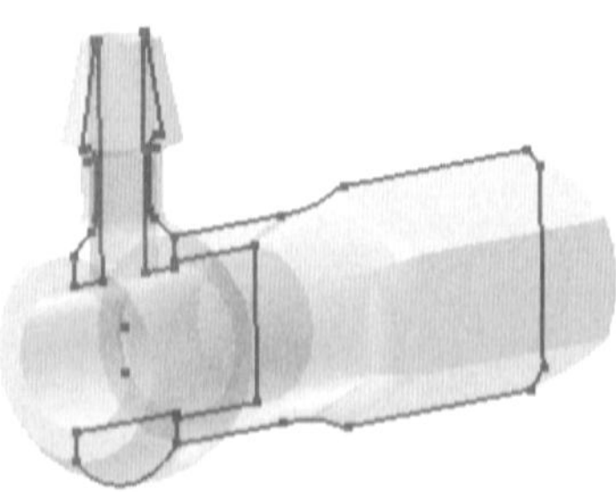

**Figure 7-44**

**NOTE** If the projected geometry is adaptive to another part, the adaptive link between a projected edge and its parent part in the assembly can be broken by expanding the adaptive part in the Browser. Right-click on the sketch's name and uncheck Adaptive. The parts will then be independent of one another.

## SKETCH ON ANOTHER PART'S FACE

In Chapter 6, you learned that when a new part is created in an assembly you have the option to create the first sketch on any planar face or plane on any part, or have an assembly constraint automatically applied between the selected face on the other part and the first sketch. This relationship constrains the sketch to the selected face—if the face moves on the selected part, so will the sketch. The option for creating a sketch on a face or plane of another part is not limited to the first sketch. Anytime a sketch is created, it can be placed on a face or plane of another part. This sketch will also be associated to the selected face or plane. When creating a sketch, click the planar face or plane on the part on which you want the sketch to be created; a workplane will then be created and constrained to the selected face or plane of the other part, and a sketch will be created on the new workplane. The workplane will be "adaptive," meaning that if the other part's face or plane moves, so will the sketch that is tied to the workplane. For more information on adaptivity, refer to Chapters 6 and 10. The following image shows, on the right, a sketch that was created on the angle part—notice that the workplane in the Browser is adaptive.

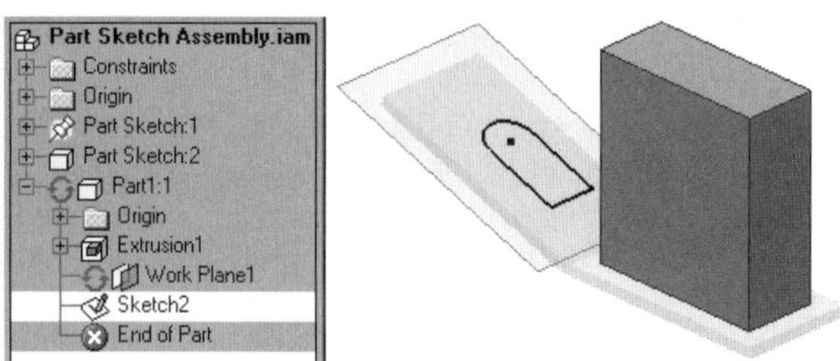
Part Sketch Assembly.iam
Constraints
Origin
Part Sketch:1
Part Sketch:2
Part1:1
Origin
Extrusion1
Work Plane1
Sketch2
End of Part

Figure 7-45

# EXERCISE 7-2 Projecting Edges and Sketching On Another Part's Face

In this exercise, you sketch on another part's face and project edges to create a new part. To navigate to the exercise in the *Electronic Student Workbook*, do the following:

1 From the Main TOC page, click Chapter 7.

2 From the TOC for Chapter 7, click Projecting Edges and Sketching On Another Part's Face.

The completed exercise is shown in the following image.

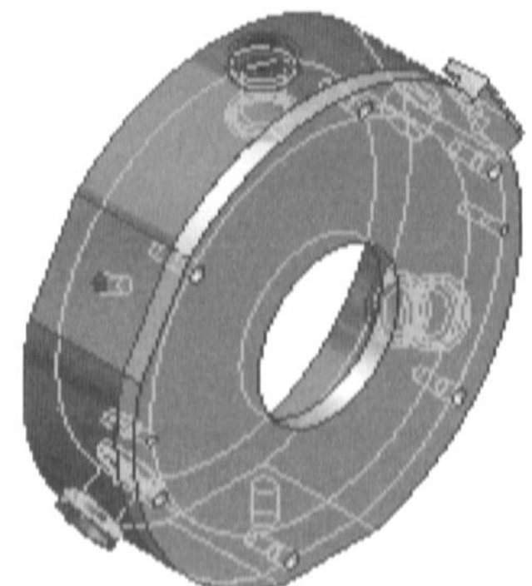

**Figure 7-46 Completed exercise**

## AUTO DIMENSION

Adding constraints and dimensions to or removing dimensions from a sketch can be a time-consuming task. To automate this process, you can use the Auto Dimension tool to automatically create or remove dimensions or add constraints to selected geometry. Before using the Auto Dimension tool, critical constraints and dimensions should be applied. The Auto Dimension tool will not override or replace any existing constraint or dimension. Click the Auto Dimension tool from the Sketch Panel Bar, as shown on the left in the following image and then the Auto Dimension dialog box will appear as shown on the right.

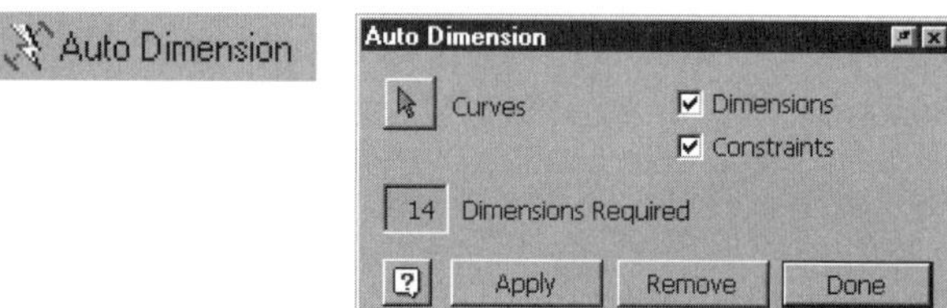


**Figure 7-47**

To use the Auto Dimension tool, proceed with the following steps:

1. Start the Auto Dimension tool from the Sketch Panel Bar.
2. The number of constraints and dimensions required to fully constrain the sketch appear in the lower left corner of the dialog box.
3. Determine if you want to create dimensions/constraints or remove the dimensions/constraints that were previously added using the Auto Dimension tool.
4. Click the Dimensions box and/or Constraints box.
5. Click the Curves option and then select the objects in the graphics window that you want to work with.
6. Click the Apply button to create the dimensions and/or constraints to the selected curves, or
7. Click the Remove button to delete the selected dimensions and/or constraints.
8. After the dimensions are placed, you can change their values by double-clicking on the numbers and entering in a new value.

**NOTE** If you use the Auto Dimension tool on the first sketch in the part, two dimensions or constraints will be required to fully constrain the sketch. Use the Fix constraint to remove these two required dimensions.

## DIMENSION DISPLAY, RELATIONSHIPS, AND EQUATIONS

In Chapter 2, you learned how to create independent dimensions that had no relationship to other dimensions. When creating parts, you may want to set up dimensional relationships between them. The length of a part may need to be twice that of its width, for example, or a hole may always need to be in the middle of the part. In Autodesk Inventor, there are a few different methods that can be used to set up relationships between dimensions. Each method will be covered in the following sections.

### DIMENSION DISPLAY

When each dimension is created, it is automatically tagged with a label that starts with the letter "d" and a number, for example "d0" or "d27." The first dimension created for each part is given the label "d0." Each dimension that is placed for subsequent part sequences goes up one number at a time. If a dimension is erased, the next dimension does not go back and reuse the erased value—instead, it keeps sequencing from the last value on the last dimension created. While creating dimensional relationships, you may want to view the dimensions display style to see the underlying label of the dimension. There are five options for displaying dimension's display style:

**Display as value** The default dimension display style. When selected, the actual value of the dimensions will be displayed on the screen.

**Display as name** When selected, dimensions on the screen will be displayed as the dimension label or actual parameter name (i.e., d12 or Length).

**Display as expression** When selected, dimensions on the screen will be displayed in the format of label = value, showing each actual value (i.e., d7 = 20 mm or Length = 50 mm).

**Display as tolerance** When selected, dimensions on the screen that have a tolerance style will be displayed as the tolerance (i.e., 40 ± .3).

**Display precise value** When selected, dimensions on the screen that have a tolerance style will be displayed to its exact value with the tolerance applied (i.e., 40.3).

To change the dimension display style, click Tools > Document Settings and, from the Units tab, change the display style as displayed in the following image. After you select a dimension display style, all visible dimensions that were not individually changed to a display style will change to that style. As dimensions are created, they will reflect the current dimension display style.

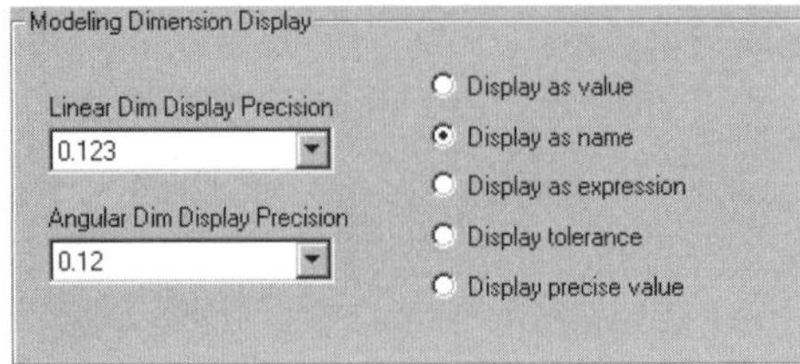


Figure 7-48

To change an individual dimension's display style, right-click on a dimension in the graphics window and then click Dimension Properties on the menu. From the Document Settings tab, you may then select the style from the Modeling Dimension Display drop-down list, as shown in the following image.

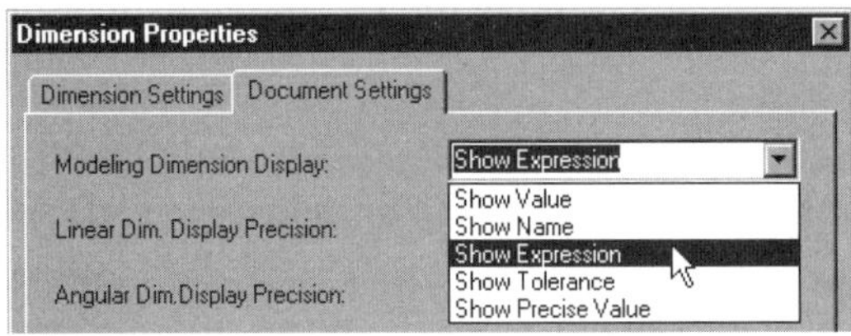


Figure 7-49

## DIMENSION RELATIONSHIPS

Setting a dimensional relationship between two dimensions requires setting a relationship between the dimension you are creating and an existing dimension. When entering text in the Edit Dimension dialog box, enter the dimensions label (d#) of the other dimension, or click the dimension that you want to set the relationship to in the graphics window. The following image shows the Edit Dimension dialog box after selecting the 10 mm dimension to which the new dimension will be related.

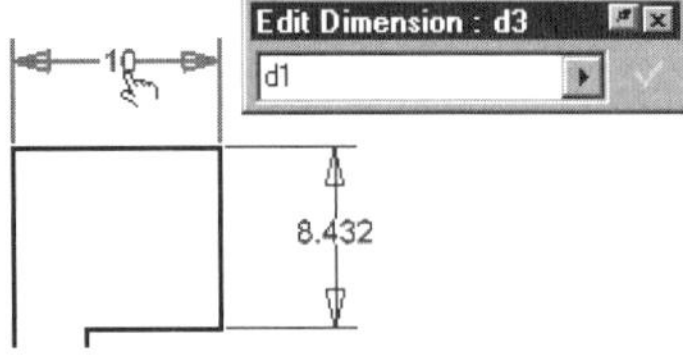


Figure 7-50

## EQUATIONS

Equations can also be used whenever a value is required. A couple of examples would be: (d9/4)*2, or 50 mm + 19 mm. When creating equations, Autodesk Inventor allows prefixes, precedence, operators, functions, syntax, and units. To see a complete listing of valid options, use the Help system and navigate to the topic of equations and the title Edit box reference. Numbers can be entered with or without units; when no unit is entered, the default unit will be assumed. As you enter an equation, Autodesk Inventor calculates it. An invalid expression will be shown in red and a correct expression will be shown in black. For best results while using equations, include units for every term and factor in the equation.

To create an equation in any edit box, follow these steps:

1 Click in the edit field.
2 Enter any valid combination of numbers, parameters, operators, or built-in functions. The following image shows an example of an equation that uses both millimeters and inches for the units.
3 Press ENTER or click the green check mark to accept the expression.

**NOTE** Use ul (unitless) where a number does not have a unit; for example, use a unitless number when dividing, multiplying, or specifying values for a pattern count.

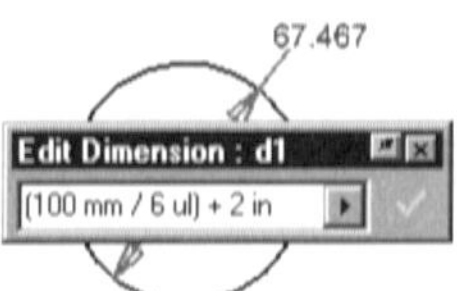


Figure 7-51

## PARAMETERS

Another method of setting up relationships between dimensions is to use parameters. A parameter is a user-defined name that is assigned a numeric value—either explicitly or through equations. Multiple parameters can be used in an equation and parameters can be used to define another parameter, such as depth = length – width. A parameter can be used anywhere a value is required. There are four types of parameters: model parameters, user parameters, reference parameters, and linked parameters.

**Model Parameters** Automatically created and assigned a name when a sketch dimensions; feature parameters such as extrusion distance, draft angle, or coil pitch; and the offset, depth, or angle value of assembly constraint is created. Autodesk Inventor assigns a default name to each model parameter as it is created. The default name format is a "d" followed by an integer incremented for each new parameter. Model parameters can be renamed via the Parameters dialog box.

**User Parameters** Manually created in the Parameters dialog box.

**Reference Parameters** Automatically created when a driven dimension is created. Autodesk Inventor assigns a default name to each reference parameter as it is created. The default name format is a "d" followed by an integer increased by one for each new parameter. You can rename reference parameters via the Parameters dialog box.

**Linked Parameters** Created via a Microsoft Excel spreadsheet and linked into a part or assembly file.

Parameters are created and/or edited by clicking the Parameters tool from the Sketch, Part Features, or Assembly Panel Bars, as shown in the following image.

*fx* Parameters

Figure 7-52

After starting the Parameters tool, the Parameters dialog box will appear. The following image shows an example with a few Model, User, and Reference parameters created. The Parameters dialog box is divided into two sections: Model Parameters and User Parameters. A third section named Reference Parameters will be displayed if driven dimensions exist. The Model Parameters section is automatically filled with the values from dimensions or assembly constraints that are used in the active document. The User parameters section will be manually defined. For both types of parameters, the names and equation can be changed and comments can be added by double-clicking in the cell and entering the new information. The column names for both Model and User Parameters are the same and are defined in the following section.

**Parameter Name** In this cell, the name of the parameter will be displayed. To change the name of an existing parameter, click in the box and enter a new name. When creating a new User Parameter, enter a new name after clicking the Add button. When you update the model, all dependent parameters update to reflect the new name.

**Units** In this cell, enter a new unit of measurement for the parameter. With Autodesk Inventor, you can build equations that include parameters of any unit type. All length parameters are stored internally in centimeters; angular parameters are stored internally in radians. This becomes important when parameters having different units are combined in equations.

**Equation** In this cell, the equation will be displayed, and from the equation the value of the parameter will be determined. If the parameter is a discrete value, the value is displayed in rounded form to match the precision setting for the document. To change the equation, click on the existing equation and enter the new equation.

**Nominal Value** Displays the nominal tolerance result of the equation and can only be modified by editing the equation.

**Tol.** From the drop-down list, select a tolerance condition; nominal, upper or lower. Tolerances will be covered later in this chapter.

**Model Value** In this cell, the calculated value of the equation will be displayed in full precision.

**Export Parameters Column** Check this box if you want the parameter to be exported to the Custom tab of the Properties dialog box. The parameter will also be available in the bill of materials and parts list Column Chooser dialog boxes.

**Comment** If desired, enter a comment for the parameter in this cell. To add a comment, click in the cell and enter the comment.

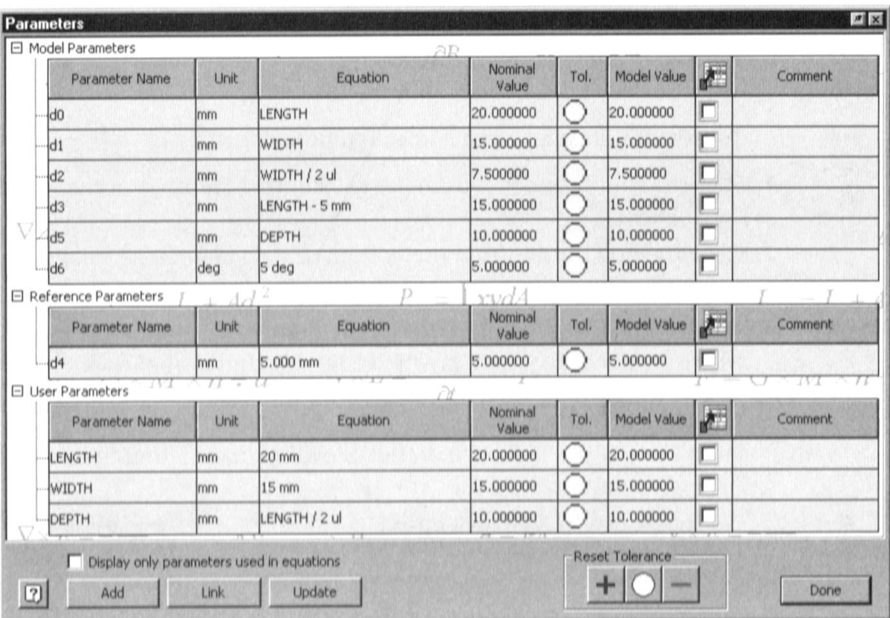


**Figure 7-53**

## USER PARAMETERS

User parameters are parameters that you define in a part or an assembly file. Parameters defined in one environment are not directly accessible in the other environment. If parameters are to be used in both environments, use a linked parameter via a common Microsoft Excel spreadsheet. A parameter can be used any time a numeric value is required. When creating parameters, follow these guidelines:

- Assign meaningful names to parameters; other designers may edit the part file and will need to understand your thought process.
- The parameter name cannot include spaces, mathematical symbols, or special characters, but they can be used to define the equation.
- The parameter name cannot consist of only numbers. It must include at least one alphabetic character, and the alphabetic character must appear

first. W1 or Width1, for example, would be valid. The names 123 and 1W would be invalid parameter names.

- Autodesk Inventor detects capital letters and uses them as unique characters. Length, length, and LENGTH are three different parameter names.
- When entering a parameter name where a value is requested, the upper and lower case of the letters must match the parameter name.
- When defining a parameter equation, you cannot use the parameter name to define itself (i.e., Length = Length/2 would be invalid).
- If the same User parameter name is used in multiple part files, it should be defined in a template file. When new parts are created based on that template, the parameter will already be defined.
- Duplicate parameter names are not allowed. Model, User, and Spreadsheet driven parameters must have unique names.

To create and use a User parameter, follow these steps:

1. Start the Parameters tool from the Sketch, Part Features, or the Assembly Panel Bars.
2. Click the Add button from the bottom of the Parameters dialog box.
3. Enter in the information for each of the cells.
4. After creating the parameter(s), you can enter the parameters name anywhere a value is required; or when editing a dimension, click the arrow on the right and click List Parameters from the menu as shown on the left of the following image. All the available parameters will be displayed in a list similar to that on the right side of the following image. From the list, click the desired parameter.

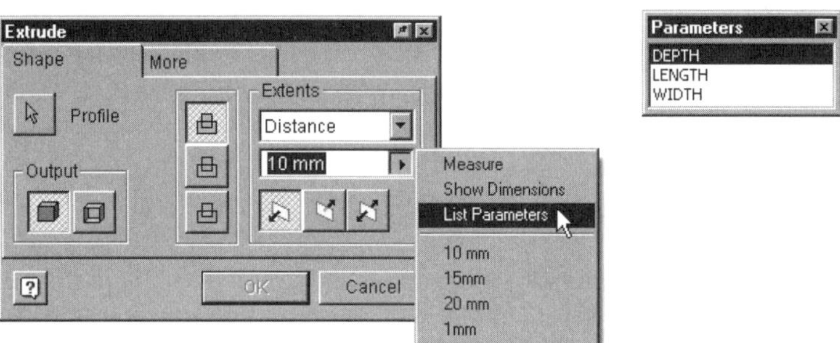


Figure 7-54

## LINKED PARAMETERS

If you want to use the same parameter name and value for multiple parts, you can create a spreadsheet using Microsoft Excel. You can then either embed or link the spreadsheet into a part or assembly file through the Parameters dialog box. When a Microsoft Excel spreadsheet is embedded, there is no link between the spreadsheet and the parameters in the Autodesk Inventor file, and any changes to the spreadsheet will not be reflected in the Autodesk Inventor file. When a Microsoft Excel spreadsheet is *linked* to an Autodesk

Inventor part or assembly file, any changes in the spreadsheet will also update the parameters in the Autodesk Inventor file. More than one spreadsheet can be linked to an Autodesk Inventor file, and each spreadsheet can be linked to multiple part and assembly files. By linking a spreadsheet to an assembly file and to the part files that comprise the assembly, parameters from both environments can be driven from the same spreadsheet. There is no limit to the number of part or assembly files that can reference the same spreadsheet. Each linked spreadsheet appears in the Browser under the 3rd Party folder.

When creating a Microsoft Excel spreadsheet with parameters, follow these guidelines:

- The data in the spreadsheet can start in any cell but must be specified when linking or embedding the spreadsheet.
- The data can be in rows or columns, but they must be in this order: parameter name, value or equations, unit of measurement, and (if needed) a comment.
- The parameter name and value are required, but the other items are optional.
- The parameter name cannot include spaces, mathematical symbols, or special characters. These can be used to define the equation.
- Parameters in the spreadsheet must be in a continuous list. A blank row or column between parameter names eliminates all parameters after the break.
- If you do not specify a unit of measurement for a parameter, the default units for the document will be assigned when the parameter is used. To create a parameter without units, enter ul (unitless) in the units cell.
- Only those parameters defined on the first worksheet of the Microsoft Excel spreadsheet are linked to the Autodesk Inventor file.
- You can include column or row headings or other information in the spreadsheet, but they must be outside the block of cells that contains the parameter definitions.

On the left side of the following image is an example of three parameters that were created in rows with a name, equation, unit, and comment. The image on the right shows the same parameters created in columns.

| | A | B | C | D |
|---|---|---|---|---|
| 1 | Length | 50 | mm | Length of Plate |
| 2 | Width | Length/2 | mm | Width of Plate |
| 3 | Depth | 10 | mm | Depth of Plate |

| | A | B | C |
|---|---|---|---|
| 1 | Length | Width | Depth |
| 2 | 50 | Length/2 | 10 |
| 3 | mm | mm | mm |
| 4 | Length of Plate | Width of Plate | Depth of Plate |

**Figure 7-55**

After the spreadsheet is created and saved, you can create parameters from it by following these steps:

1 Start the Parameters tool from the Sketch, Part Features, or the Assembly Panel Bars.

2 Click Link on the bottom of the Parameters dialog box.

3 The Open dialog box will appear similar to that shown in the following image.

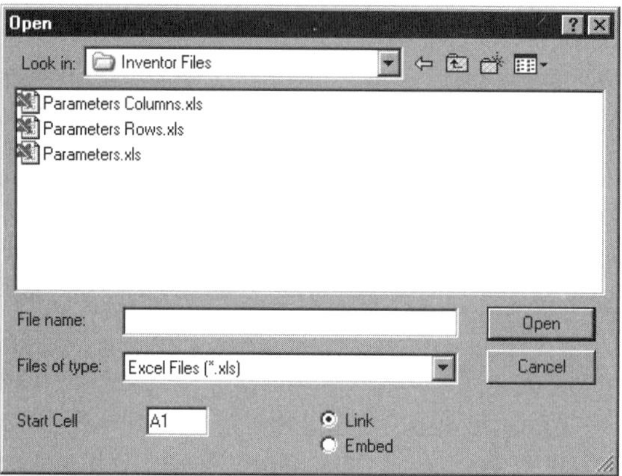


Figure 7-56

4 Navigate to and select the Microsoft Excel file that will be used.

5 In the lower left corner of the Open dialog box, enter the start cell.

6 Select whether the spreadsheet will be linked or embedded.

7 Click the Open button (a section showing the parameters is added to the Parameters dialog box, as shown in the following image). If the spreadsheet was embedded, the new section will be entitled Embedding #.

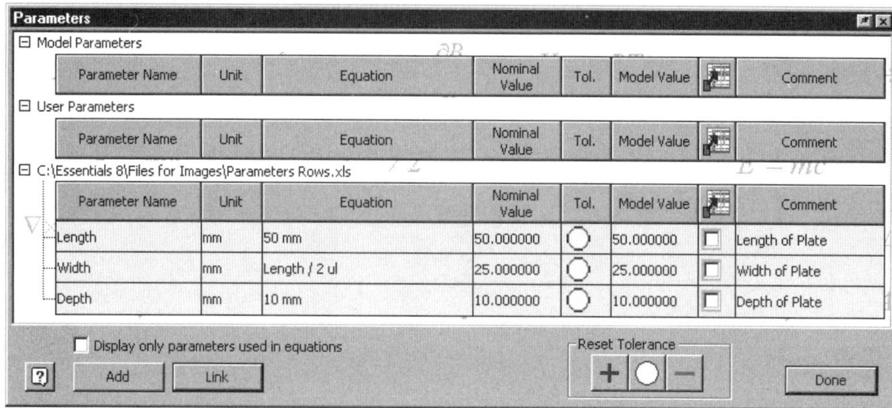


Figure 7-57

8 Click the Done button to complete the operation of linking a spreadsheet.

To edit the parameters that are linked or embedded, follow these steps:

1. Open the Microsoft Excel file.
2. Make the required changes.
3. Save the Microsoft Excel file.
4. Open the Autodesk Inventor part or assembly file that uses the spreadsheet.
5. Click the Update tool from the Standard Toolbar.

or

1. Open the Autodesk Inventor part or assembly file that uses the spreadsheet.
2. Expand the 3rd Party folder from the Browser and either double-click the spreadsheet or right-click and click Edit from the menu, as shown in the following image.

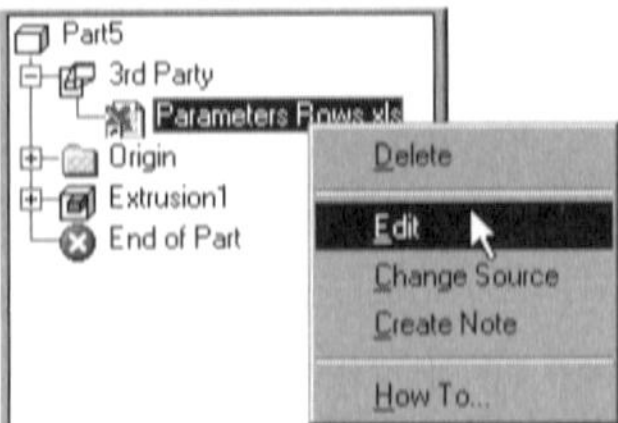


**Figure 7-58**

The Microsoft Excel spreadsheet will open in a new window for editing.

3. Make the required changes.
4. Save the Microsoft Excel file.
5. Make active the Autodesk Inventor part or assembly file that uses the spreadsheet.
6. Click the Update tool from the Standard toolbar.

**NOTE** If the spreadsheet was embedded, the changes will not be saved back to the original file but will only be saved internally to the Autodesk Inventor file.

## PART DIMENSIONAL TOLERANCES

When creating parts, you may want to analyze them for form, fit, and function when the part(s) are set to either a minimum, nominal, or maximum dimensional tolerance condition. Both sketch dimensions and values of features can have a tolerance applied to them. Tolerances can only be set for decimal-placed values—not fractions. The following is a definition of the terms that are used when creating dimensional tolerances.

**Nominal** The dimensional value that is entered as a value or parameter. This is the *perfect* dimension.

**Tolerance** The acceptable range of values for a particular dimension.

**Min** The smallest dimensional value that is allowed for a parameter.

**Max** The largest dimensional value that is allowed for a parameter.

**Maximum Material Condition (MMC)** This is the maximum enclosed volume of the part. In the case of a hole, MMC is the Min diameter and the Min depth. For an extrusion, MMC is the Max value of the dimensions.

**Least Material Condition (LMC)** This is the minimum enclosed volume of a part. In the case of a hole, LMC is the Max diameter and the Max depth. For an extrusion, LMC is the Min value of the tolerances.

**Fits** The relationship between two mating parts.

There are two methods for applying tolerances; Standard Tolerances and Override Tolerances.

**Standard Tolerance** Applies a tolerance to every dimension or feature created after a tolerance style has been created. The values decimal precision must match the existing tolerance style. This method should be used cautiously as numerous values will have a tolerance automatically applied to them while they are being created.

**Override Tolerances** Involves applying a tolerance to individual values, this gives you control over which dimension or feature dimensions should have a tolerance applied to it.

**NOTE** When a part file is saved and all tolerances in the file are not set to Nominal, you will be warned. If a part is manufactured to a minimum or a maximum tolerance, this may lead to inaccurate parts.

Both methods for using tolerances will be outlined in the next sections.

## STANDARD TOLERANCES

There are three basic steps to creating and using Standard Tolerances. Each step will be described in the following section.

### Step 1 – Define the Default Tolerance Style

The standard or default tolerance(s) will be defined in the Document Settings. Each standard tolerance can be defined for both linear and angular values and is set for each decimal precision for which you need a tolerance. A tolerance setting (style) can be set for each display precision (decimal places) for which you want a tolerance used. Every value that is created with the same Tolerance Style decimal place of precision will have a tolerance automatically applied to it. To set the standard tolerance for each decimal place for which you want a tolerance, follow these steps:

1 Click Tools > Document Settings.

2 From the Default Tolerance tab of the Document Settings dialog box, check Use Standard Tolerancing Values.

3 From the Linear section of the dialog box, set the Precision to which the tolerance will be accurate. From the Precision drop-down list, click the number of decimal places to which the tolerance will be accurate. The following image shows the Precision drop-down list. Each number represents a decimal place.

4 From the Linear section of the dialog box, enter a value for the Tolerance (+/-).

5 Repeat Steps 3 and 4 for each decimal place of linear and angular precision that will be used for every value in this document that matches the number of decimal places.

6 If desired, check the Export Standard Tolerance Values box so you can use the tolerance styles to set properties in a drawing (*.idw*) file in which the part will be annotated. An example of this is a user accessing these properties to document the Standard Tolerance values in the title block.

**NOTE** If the same Standard Tolerance values are to be used in multiple files, set the Standard Tolerance style in a template file.

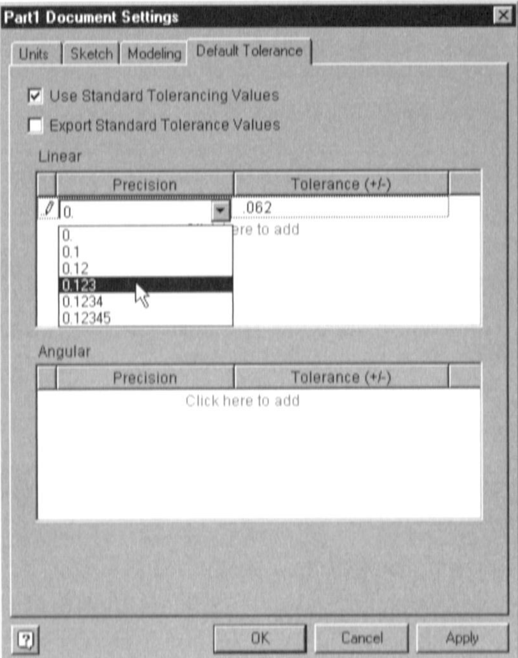


Figure 7-59

## Step 2 – Setting the Default Decimal Precision and Display

1 When working with standard tolerances, you need to set the default display precision (decimal places) for both linear and angular dimensions. If the precision matches the tolerance style precision, the standard tolerance will auto-

matically be applied to it. To set the decimal precision for both linear and angular dimensions as they are placed, click Tools > Document Settings and from the Units tab change the Linear Dim Display Precision and Angular Dim Display Precision by selecting a value from the drop-down list, as shown in the following image. After placing a dimension or feature, this value can also be changed. This technique is covered in the Override Tolerance section.

2 Change how the tolerances will be displayed on the screen by selecting either Display tolerance or Display precise value from the Units tab of the Document Setting dialog box, as shown in the following image.

- Display tolerance will display the value with its maximum and minimum tolerance (i.e., 70.000 ± 0.062).
- Display precise value will display dimensions to their exact value with the maximum, nominal, or minimum tolerance condition (i.e., 70.062).

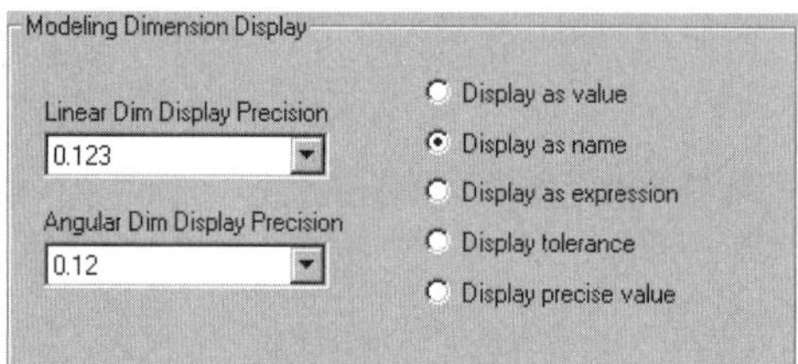


**Figure 7-60**

## Step 3 – Applying a Tolerance Condition

To change the tolerance condition to upper, nominal, or lower, choose one of the following methods:

1 Right-click on a dimension and click Dimension Properties from the menu, as shown on the left in the following image.

2 The Dimension Properties dialog box will appear as shown on the right in the following image.

3 On the right side of the dialog box, click either the upper, nominal, or lower tolerance condition from the Evaluated Size section of the Dimension Settings tab.

4 To finish the operation, click OK.

5 After changing the tolerance condition, click the Update tool on the Standard toolbar to have the model reflect the tolerance change.

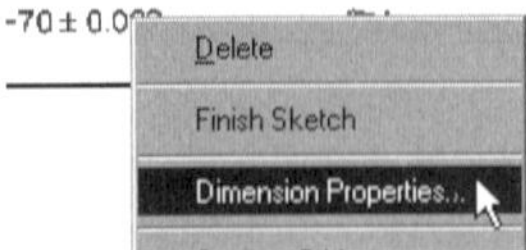


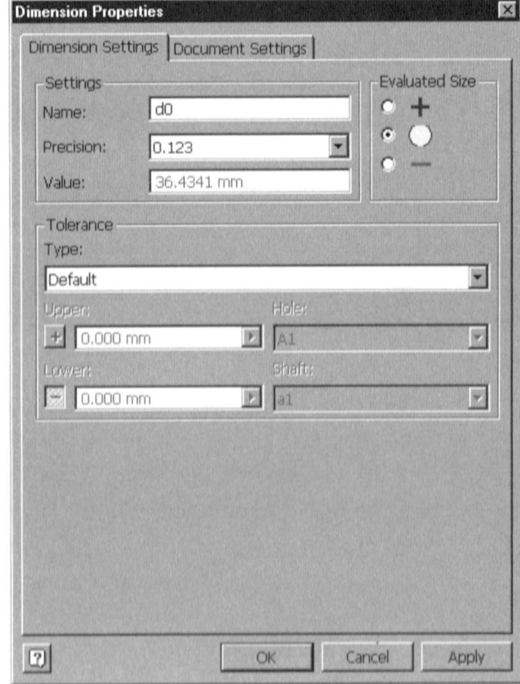


**Figure 7-61**

**NOTE** If the condition is set to a minimum or maximum condition, the value will be underlined as shown in the following image.

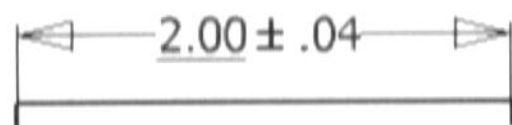


**Figure 7-62**

Another way to set the tolerance condition is from the Parameters dialog box, as described in the following steps:

1 Start the Parameters tool. Select the tolerance condition for each dimension from the Parameters dialog box by selecting the tolerance condition from the drop-down list of the Tol. Column, as shown in the following image.

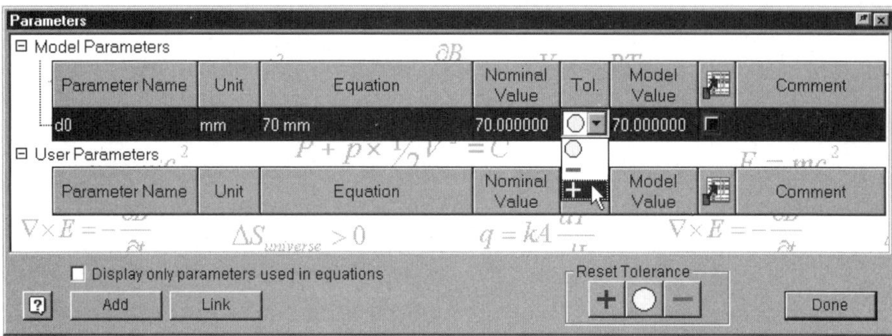


Figure 7-63

2 All the values for a part can also be reset to the same tolerance condition by clicking the upper, nominal, or lower tolerance conditions from the bottom right corner of the Parameters dialog box.

3 To finish the operation, click Done.

4 After changing the tolerance condition, click the Update tool from the Standard toolbar to have the model reflect the tolerance change.

Figure 7-64

### OVERRIDE TOLERANCES

This option allows you to enter the upper and lower tolerances for individual dimensions and features. The techniques for overriding dimensions and features values will be covered in the next two sections.

To override an individual dimensions tolerance value, follow these steps:

## Step 1 – Setting the Tolerance

1 Right-click on a dimension and click Dimension Properties on the menu as shown in the following image.

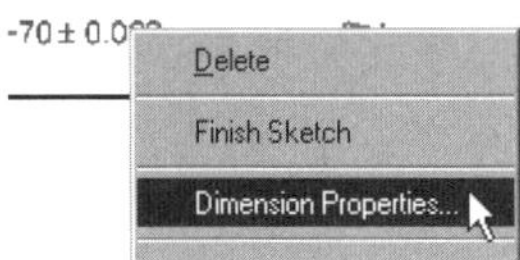


Figure 7-65

2 The Dimension Properties dialog box will appear as shown in the following image. Here, you modify the values—name and precision (decimal place)—as well as set the Tolerance Type. To change the Tolerance Type, select the type from the drop-down list. The following is a description of the Tolerance Types.

**Default** There is no tolerance applied to a dimension.

**Symmetric** The Min and Max tolerances are the same.

**Deviation** The Min and Max tolerances are different.

**Limits-Stacked** The maximum tolerance condition is displayed on top, and the minimum condition is at the bottom.

**Limits-Linear** The minimum, and then the maximum, tolerance condition is displayed on the same line.

**Max** Only the upper limit is shown. There is no nominal or minimum value.

**Min** Only the lower limit is shown. There is no nominal or upper value.

**Limits/Fits-Stacked** Tolerance data from a standard is applied to a nominal dimension and the tolerance is displayed in a fraction with the maximum condition on top and the minimum condition on the bottom.

**Limits/Fits-Linear** Tolerance data from a standard is applied to a nominal dimension and the tolerance is displayed with the minimum and then maximum condition on the same line.

**Limits/Fits-Show Size Limits** Hole/Shaft fit data is provided while also displaying the tolerance range.

**Limits/Fits-Show Size Tolerance** Hole/Shaft fit data is provided while also displaying the actual tolerance values.

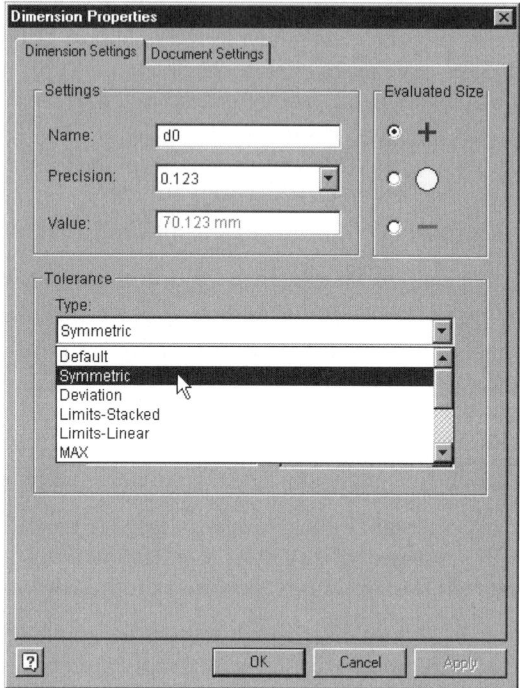


Figure 7-66

3 Depending upon the Tolerance Type that was selected, enter a value for the upper, lower, hole, or shaft tolerance. The following image shows a value set for the symmetric tolerance type.

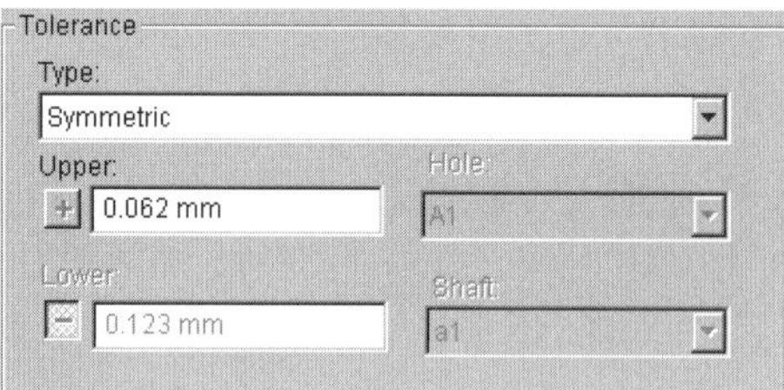


Figure 7-67

## Step 2 - Applying a Tolerance Condition

To change the tolerance condition to upper, nominal, or lower, use one of these methods.

1 Right-click on a dimension and click Dimension Properties from the menu, as shown on the left in the following image.

2 The Dimension Properties dialog box will appear, as shown on the right in the following image. Click the upper, nominal, or lower tolerance condition from the Evaluated Size section of the Dimension Settings tab.

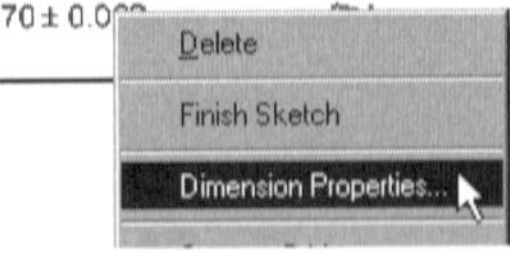


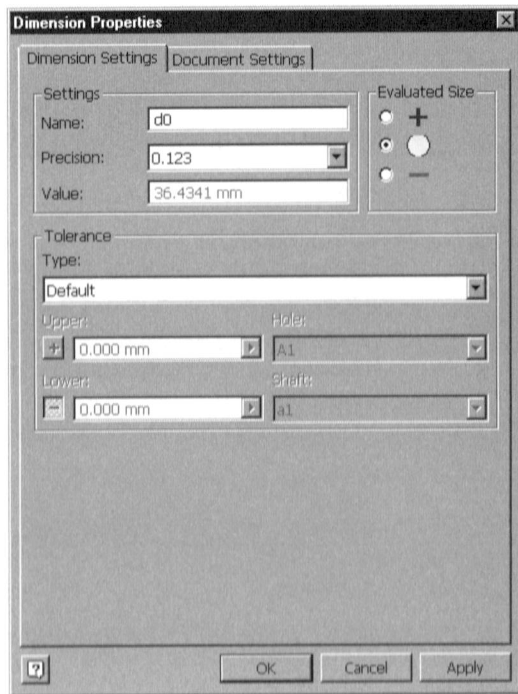


**Figure 7-68**

3 Another way to set the tolerance condition is with the Parameters dialog box. Start the Parameters tool. From the Parameters dialog box, select the tolerance condition for each dimension by selecting the tolerance condition from the drop-down list of the Tol. column, as shown in the following image.

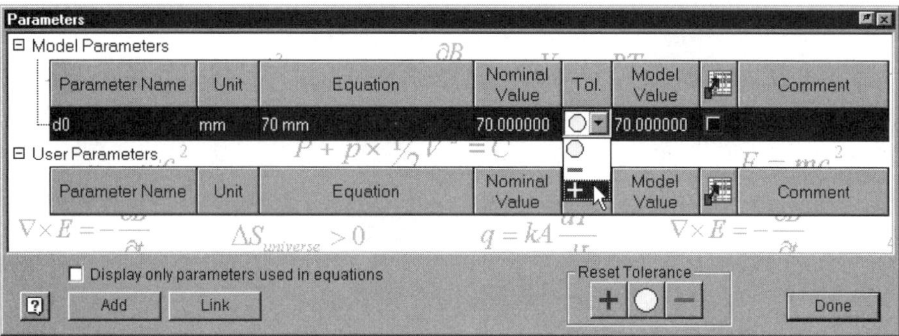


Figure 7-69

4 All the values for a part can also be reset to the same tolerance condition by clicking either maximum, nominal, or minimum from the bottom-right corner of the Parameters dialog box, as shown on the left in the following image.

5 After changing the tolerance condition, click the Update tool from the Standard toolbar, as shown on the right in the following image, to have the model reflect the tolerance change.

Figure 7-70

## APPLYING A TOLERANCE TO A FEATURE'S VALUE

To apply a tolerance to a feature's value, follow these steps:

1 Edit the feature's dimensions to which you wish to apply a tolerance and either right-click, or click the arrow, and click Tolerance from the menu as shown in the following image. A tolerance cannot be added to values in a dialog box. Only dimensions visible in the graphics window can have a tolerance added.

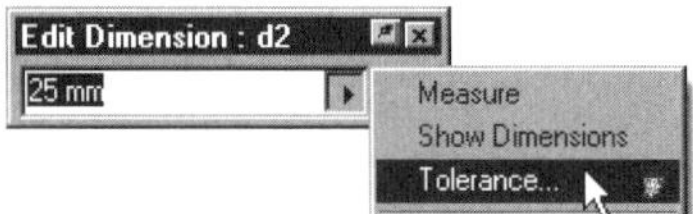


Figure 7-71

2 From the Tolerance dialog box, set the name and precision, as well as the Tolerance Type as needed.

3 Click an option from the Precision drop-down list, as shown in the following image.

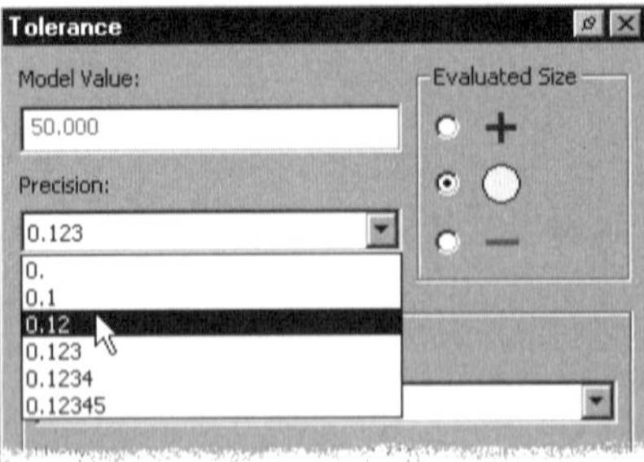


Figure 7-72

4 Depending upon the Tolerance Type that was selected, enter a value for the upper, lower, hole, or shaft tolerance.

5 From the Evaluated Size area in the upper-right corner of the Tolerance dialog box, select maximum, nominal, or minimum, as shown in the previous image. The Evaluated Size can also be set for individual values from the Parameters dialog box.

6 After changing the tolerance condition, click the Update tool from the Standard toolbar, as shown in the following image, to have the model reflect the tolerance change.

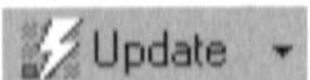


Figure 7-73

7 To change the appearance of the dimensions in the file, click Tools > Document Settings and set the Modeling Dimension Display from the Units tab, as shown in the following image. These options are defined in the Dimension Display section earlier in this chapter.

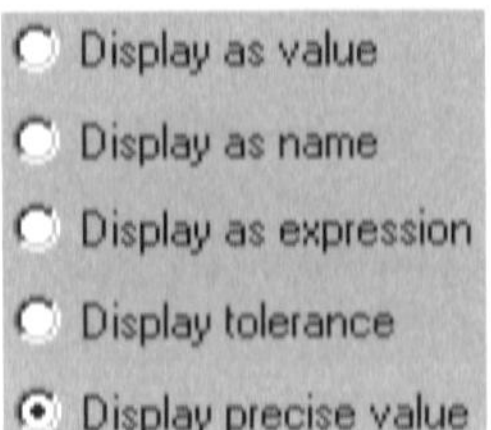


Figure 7-74

## ANNOTATING TOLERANCES IN DRAWINGS

After creating a part with tolerances, you can create drawing views and the dimensions with tolerances will be automatically displayed in the Tolerance Type in which they were last displayed—use the Retrieve Dimensions option in each view in which you want dimensions displayed.

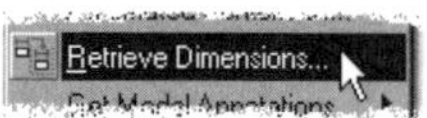


Figure 7-75

Once the dimensions are displayed in a view, its tolerance style can be overridden by double-clicking on the dimension or right-clicking while the cursor is over a dimension and clicking Tolerance from the menu, as shown in the following image. In the Dimension Tolerance dialog box, you can select the tolerance style and decimal place precision to be displayed, as shown in the next image. The overridden tolerance style will not change the tolerance settings in the part file.

Figure 7-76

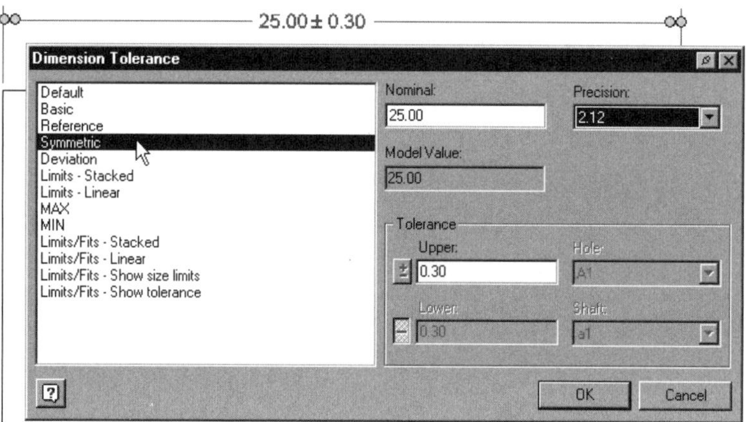


Figure 7-77

# EXERCISE 7-3 Auto Dimension, Relationships, and Parameters

In this exercise, you create dimensional relationships and use parameters to control sketch geometry. You then use the Auto Dimension tool to fully constrain the sketch. To navigate to the exercise in the *Electronic Student Workbook*, do the following:

1 From the Main TOC page, click Chapter 7.

2 From the TOC for Chapter 7, click Auto Dimension, Relationships, and Parameters.

The following image illustrates the completed exercise.

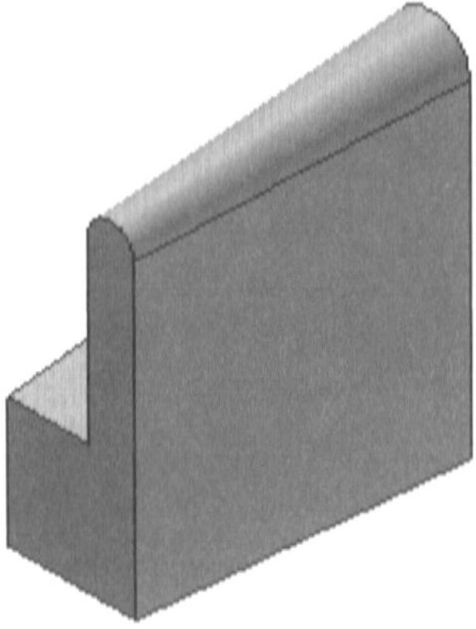

**Figure 7-78 Completed exercise**

## CHAPTER SUMMARY

| To | Do This | Tool |
|---|---|---|
| Import AutoCAD data | Either open the file using the Open tool or click the Insert AutoCAD file tool from the 2D Sketch Panel Bar or Drawing Sketch Panel Bar | |
| Create construction geometry | Change the line style to Construction from the Style area on the Standard toolbar. | Normal<br>Normal<br>Construction |
| Create splines | Click the Spline tool from the 2D Sketch Panel Bar. | |
| To pattern a sketch object | Click either the Rectangular Pattern or Circular Pattern tool from the 2D Sketch Panel Bar. | |
| Share a sketch | Right-click on the sketch name in the Browser and select Share Sketch from the menu. | |
| Mirror a sketch | Click the Mirror tool on the 2D Sketch Panel Bar. | |
| Temporarily slice away a portion of the model that obscures the plane | After making a sketch active, right-click and select Slice Graphics from the menu. | |
| Project selected edges, vertices, work features, curves, or silhouette edges | Click the Project Geometry tool from the 2D Sketch Panel Bar. | |
| Sketch on a plane of another part | Click the 2D Sketch tool from the Standard toolbar and click on any planar face or plane of another part. | Sketch |
| Automatically create constraints and dimensions | Click the Auto Dimension tool from the 2D Sketch Panel Bar. | |
| Create parameters | Click the Parameters tool from the Sketch, Part Features, or the Assembly Panel Bar. | $f_x$ |
| To add tolerances to dimensions or a feature's value | Either set a standard tolerance or override a dimension or feature's value. | |

# Applying Your Skills

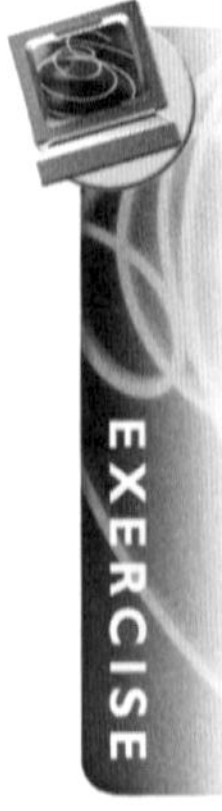


## Skill Exercise 7-1

In this exercise, you use complex sketching and constraining techniques to create a 2D layout for an electronic remote control. To navigate to the exercise in the *Electronic Student Workbook*, do the following:

1 From the Main TOC page, click Chapter 7.

2 From the TOC for Chapter 7, click Exercise 1 under Applying Your Skills.

The following image illustrates the completed exercise.

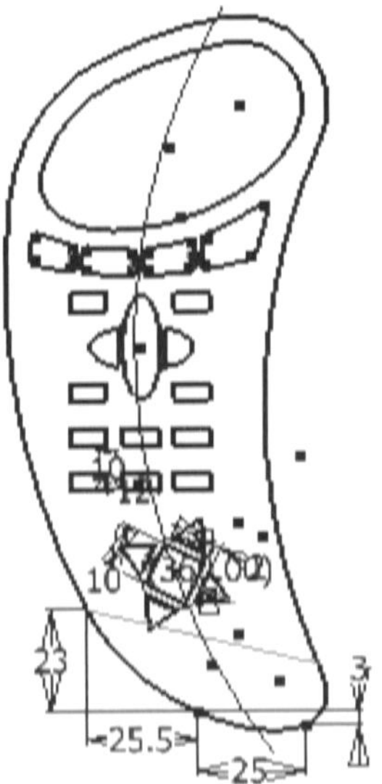


**Figure 7-79 Completed exercise**

## CHECKING YOUR SKILLS

Use these questions to test your knowledge of the material covered in this chapter.

1 **True___ False___** Only 2D AutoCAD data can be imported into Autodesk Inventor.

2 **True___ False___** Geometry that uses the construction style cannot be dimensioned to.

3 **True___ False___** Splines cannot have geometric constraints applied between them and other geometry.

4 **True___ False___** Modifications to a shared sketch will update all the features that use that shared sketch.

5 **True___ False___** Slice Graphics will permanently slice away a portion of the model.

6 **True___ False___** The Project tool can project vertices, work features, curves, or silhouette edges of another part in an assembly to the active sketch.

7 **True___ False___** If the Auto Dimension tool is used on the first sketch in the part, the sketch will be fully constrained.

8 **True___ False___** When creating parameters in a spreadsheet, the data items must be in the following order: parameter name, value or equations, unit of measurement and, if needed, a comment.

9 Explain how to suppress a patterned occurrence.

______________________________________________

______________________________________________

______________________________________________

10 What is the difference between a Model Parameter and a User Parameter?

______________________________________________

______________________________________________

______________________________________________

**11** Explain how to set all the values that have tolerances in a part to the same tolerance condition in a single operation.

________________________________________________________

________________________________________________________

________________________________________________________

________________________________________________________

# CHAPTER 8

# Complex Part Modeling Techniques

In this chapter, you will learn how to use advanced modeling techniques. Using advanced modeling techniques, you can create transitions between parts that would otherwise be difficult to create. Advanced features can be edited like other sketched and placed features, but typically require more than one unconsumed sketch in order to be created. You will also be introduced to techniques in this chapter that will help you be more productive in your modeling.

## CHAPTER OBJECTIVES

**After completing this chapter, you will be able to**

- Extrude an open profile
- Create ribs, webs, and rib networks
- Emboss text
- Create sweep features
- Create coil features
- Create loft features
- Split a part or split faces of a part
- Copy features within a part
- Reorder part features
- Mirror model features
- Suppress features of a part
- Work with file properties
- Change the color of a face
- Create a decal feature
- Use the visualization tools to change the appearance of parts

## USING OPEN PROFILES

In Chapter 3, you learned how to extrude a closed profile. In this section, you will learn how to extrude an open profile. An open profile is a sketch that does not form a closed area. An open profile is extruded bidirectionally—in a positive or negative direction normal to the profile plane and in a fill direction. A fill direction will extend the profile until it touches a termination face (in all directions), creates a closed area, and encloses any existing feature. The following image shows the original part on the left with a line, based on a work plane, in the middle of the sketch. The part in the middle shows the open profile (the line) extruded to the left (the boss in the middle was enclosed). The part on the right shows the open profile extruded to the right.

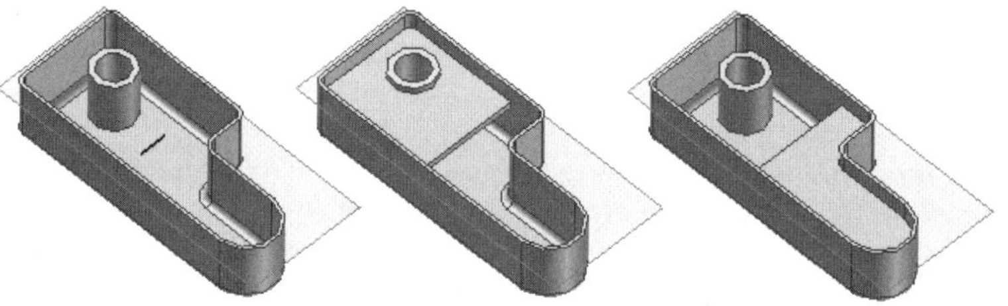

Open Profile    Open Profile Extruded to Left    Open Profile Extruded to Right

**Figure 8-1**

When working with an open profile, only the Extrude tool can be used. The Revolve and Sweep tools cannot use open profiles. The open profile can be consumed inside or extend beyond the part, or a combination of the two. When the open profile is extruded, it will either be extended or trimmed back to be enclosed with the part. An open profile can be constrained and dimensioned like any other profile.

To extrude an open profile, follow these steps:

1. Make a sketch active.
2. Draw an open profile.
3. Add geometric constraints and dimensions as needed.
4. Start the Extrude tool.
5. Click the open profile on the part.
6. Click on the side of the part that will be filled in. The following image shows the left side of the part selected as the fill side. The second image shows the right side selected.
7. Change the Extents as needed: Distance, To Next, To, From To, and All. Then fill in the distance and the extrusion direction.
8. To complete the operation, click OK.

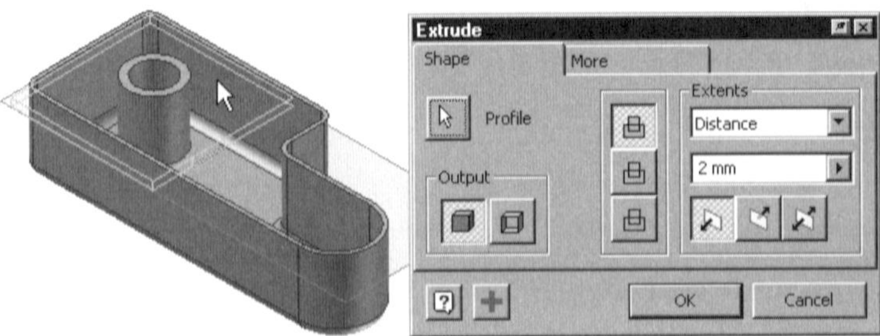


Figure 8-2

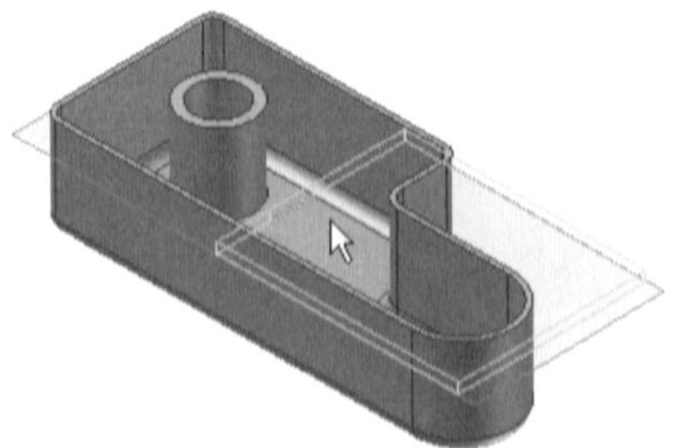

Figure 8-3

## RIB AND WEB FEATURES

Ribs and webs are primarily used to reinforce or strengthen features in mold and cast parts, but can also be used in machined parts and in other cases where additional support and minimal weight are required. The following example shows a part as a sketch and then with a rib and a web.

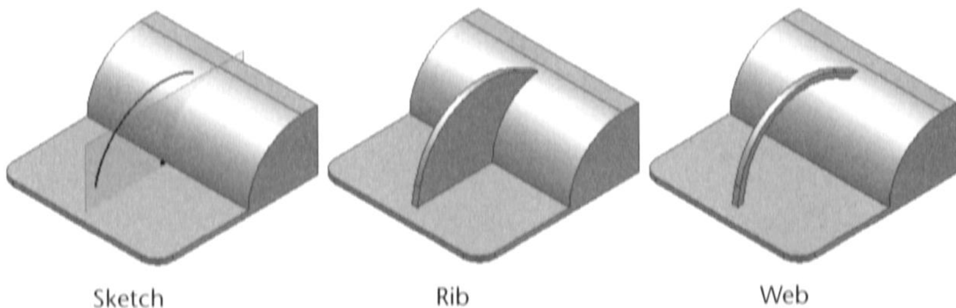


Figure 8-4

Using the Rib tool located on the Part Features Panel Bar, as shown in the following image, you can create ribs, webs, and rib networks.

- A rib is a thin-walled feature that is typically closed.
- A web is similar in width but usually has an open shape.
- A rib network consists of a series of thin-walled support features.

Figure 8-5

A rib or web feature is defined by a single, open, unconsumed profile that is then refined using the options in the Rib dialog box. If there is no unconsumed sketch in the part file, Autodesk Inventor will warn you with the message: "No unconsumed visible sketches." After starting the Rib tool, the Rib dialog box will appear as shown in the following image. Each option is described in the next section.

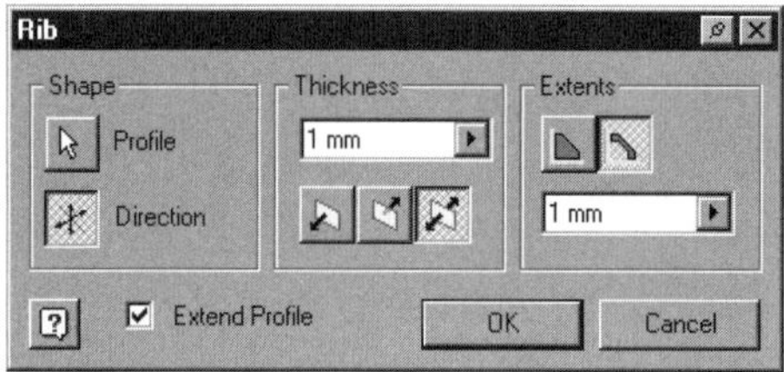


Figure 8-6

## SHAPE

**Profile** — Select this button to choose the sketch to extrude. The sketch can be an open profile or you can select multiple-intersecting or nonintersecting profiles to define a rib or web network.

**Direction** — Select this button, then in the graphics window position the cursor around the open profile to specify whether the rib will extend parallel or perpendicular and in which direction relative to the sketch geometry. Once the correct direction is displayed, click in the graphics window. The following images show a rib in all four directions.

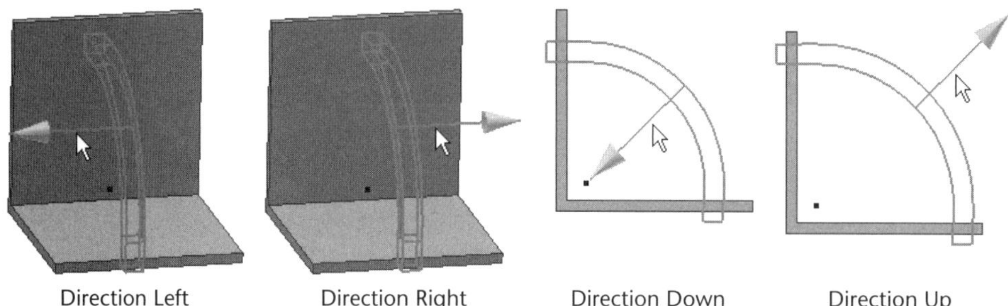


Figure 8-7

### THICKNESS

In this section, you define the thickness of the rib.

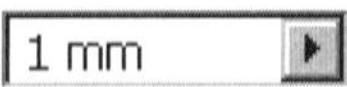


Edit Box — Enter the width of the rib feature using this edit box.

Flip Buttons — Select the flip buttons to specify which side of the profile to apply the thickness value to or to add the same amount of material to both sides of the profile.

### EXTENTS

In this section, you specify a rib or a web.

To Next

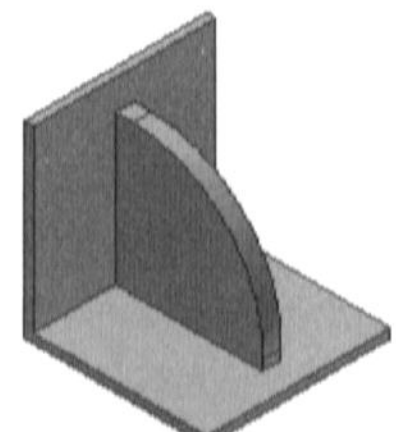

This button will extend the ends of the open profile and the area between the profile and the next available set of faces along the rib direction.

Finite

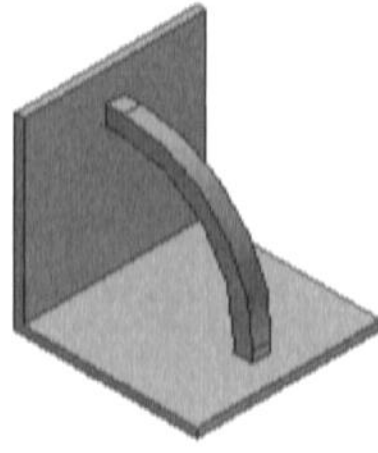

The Finite button enables an edit box in which to specify an offset distance from the sketch geometry. By enabling the Finite button and entering a distance, you can create a web feature. Note the Edit box for Distance is not available with the To Next option.

Extend Profile — Using the Finite option also enables the Extend Profile checkbox. The checkbox specifies whether or not to extend the endpoints of the sketch to the next available face or to leave the ends of the open profile as determined by the end of the rib feature. If checked, the ends are extended; if cleared, the ends cap at the end of the sketched profile. Note the Extend Profile option is not available with the To Next option.

### CREATING RIBS AND WEBS

To create a rib or web, follow these steps:

1 Create an active sketch in the location where the rib or web will be placed.
2 Sketch a single open profile that defines the basic shape of one side of the rib or web.

3 Add constraints and dimensions as needed.
4 Start the Rib tool located on the Part Features Panel Bar.
5 Define the direction of the rib or web. Click the Direction button in the Rib dialog and then in the graphics window position the cursor around the open profile to specify whether the rib or web will extend parallel or perpendicular and in which direction relative to the sketch geometry. Once the correct direction is displayed, click in the graphics window.
6 Enter a value for the rib or web in the Thickness edit box. This will define the width of the rib or web.
7 Also in the Thickness section, use the Flip buttons to specify which side of the profile to apply the thickness value or to add the same amount of material to both sides of the profile.
8 Define the depth of the profile by clicking either the To Next or Finite buttons from the Extents area of the Rib dialog box.
9 If the Finite option is used, enter an offset distance and check the Extend Profile checkbox if the endpoints are to be extended to the next available face.
10 To complete the operation, click the OK button.

### RIB NETWORKS

Rib networks can also be created using the Rib tool. You can use multiple intersecting or nonintersecting sketch objects within a single profile to create a rib network. The process of creation is the same for creating a single rib, except the thickness is applied to all objects within the profile. When you select the profile objects, they will need to be selected individually. If the rib network is to have equal spacing between the objects, use the 2D Rectangular Pattern or 2D Circular Pattern tools to create the profile. The following example shows a part with multiple intersecting lines defined in a single profile, that same profile used to create a rib network, and again as a web network.

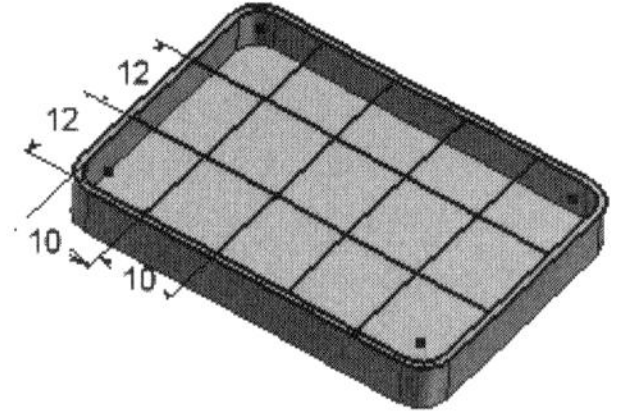

Rib Networks as a Profile

Rib Networks as Ribs

Rib Networks as Webs

**Figure 8-8**

# EXERCISE 8-1 Creating Ribs and Webs

In this exercise, you sketch an open profile and then use the Rib tool to create a rib. Then, you edit the Rib feature to change the rib to a web. You complete the exercise by sketching overlapping lines and creating a rib network. To navigate to the exercise in the *Electronic Student Workbook*, do the following:

1 From the Main TOC page, click Chapter 8.
2 From the TOC for Chapter 8, click Creating Ribs and Webs.

The completed exercise is shown in the following image.

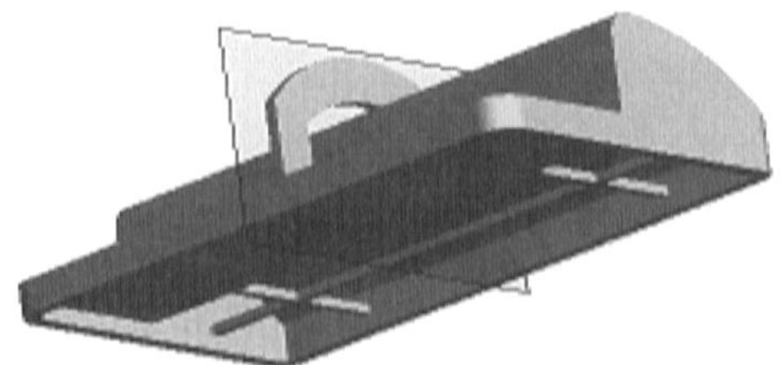

**Figure 8-9 Completed exercise**

## EXTRUDE FEATURE TERMINATION OPTIONS

When extruding a profile with the To extents option and extruding to a closed loop face or surface, it is possible that there are multiple faces at which the profile may terminate, and you can specify a minimum or maximum extrusion solution. The Extrude tool is the only tool that utilizes the minimum or maximum extrusion solution.

To choose between multiple solutions with the Extrude tool, follow these steps:

1 Start the Extrude tool.
2 Select the profile to extrude.
3 Click the To extents option.

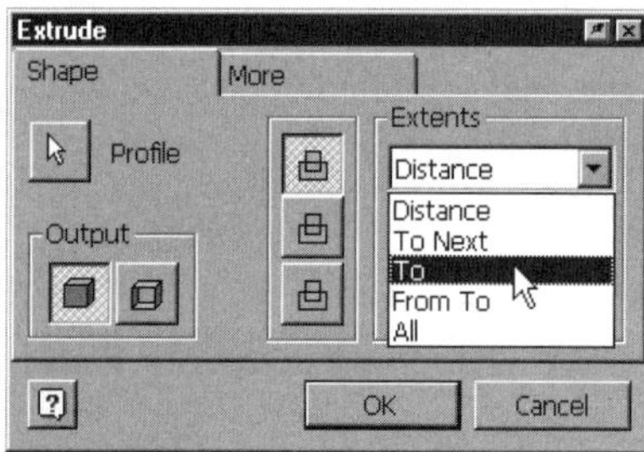


Figure 8-10

4 Select the face at which the profile will terminate.
5 Click the More tab.
6 If needed, reverse the extrusion direction by clicking the direction buttons.
7 Check the Minimum Solution box to have the profile terminate at the closest side of the selected face. If the box is not checked, the profile will terminate to the furthest side.

**NOTE** The Minimum Solution option can also be changed when a feature is edited.

The following images show a profile extruded to the furthest face, and a profile extruded to the closest face with the Minimum Solution option checked in the Extrude dialog box.

Figure 8-11

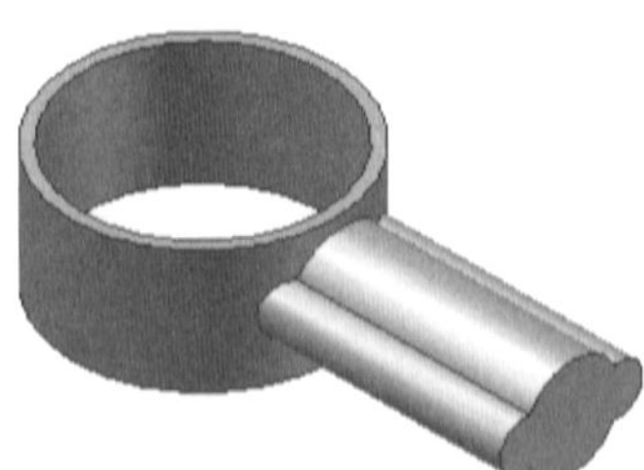

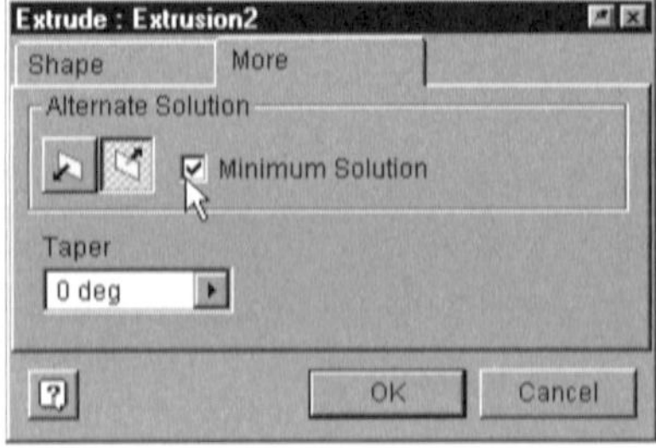


Figure 8-12

## EMBOSS TEXT FEATURES

To better define a part, you may need to have a shape or text either embossed (raised) or engraved (cut) into a model. In this section, you will learn how a closed shape or text can be embossed or engraved onto a planar or curved face. A shape can be defined using the sketch tools from the 2D Sketch Panel Bar. The shape needs to be closed. There are two steps to embossing or engraving: first, you create the shape or text and then you emboss the shape or text onto the part. The steps in embossing text will be covered in the next section.

### STEP 1 – CREATING TEXT

To place text onto a sketch, follow these steps:

1 Make a sketch active.
2 Click the Create Text tool from the 2D Sketch Panel Bar, as shown in the following image.

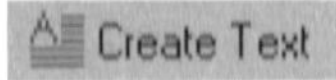


Figure 8-13

3 Click the point where the upper-left corner of the text will begin.
4 From the Format Text dialog box, define the text style and enter the text that will be placed on the sketch. The following image shows the Format Text dialog box with text entered in the bottom pane.

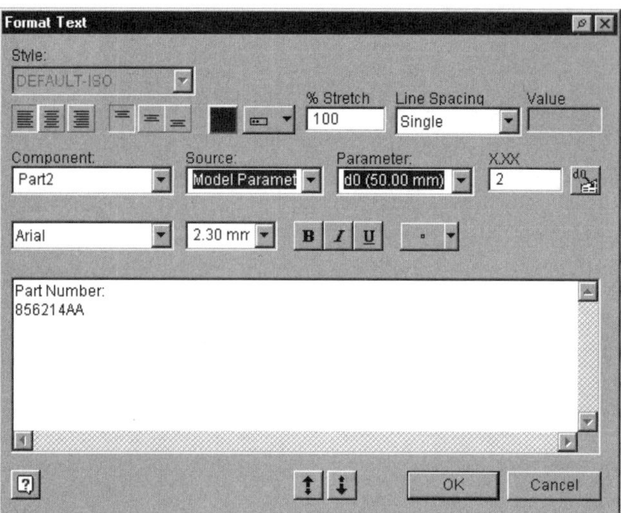


Figure 8-14

5 When you are done entering text, click the OK button.

6 When the text is placed, a rectangular set of construction lines defines the perimeter of the text. You can add dimensions or constraints to these construction lines to refine its position, as shown in the following image.

7 To edit the text, move the cursor over the text, right-click, and select Edit Text from the menu. The same Format Text dialog box will appear that was used to create the text.

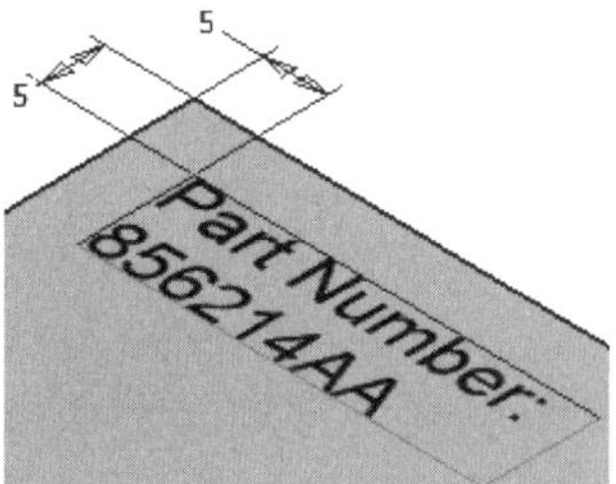


Figure 8-15

## STEP 2 – EMBOSS TEXT

To emboss a closed profile or text, click the Emboss tool from the Part Features Panel Bar, as shown on the left in the following image. The Emboss dialog box will appear as shown on the right. The Emboss dialog has the following options:

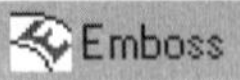


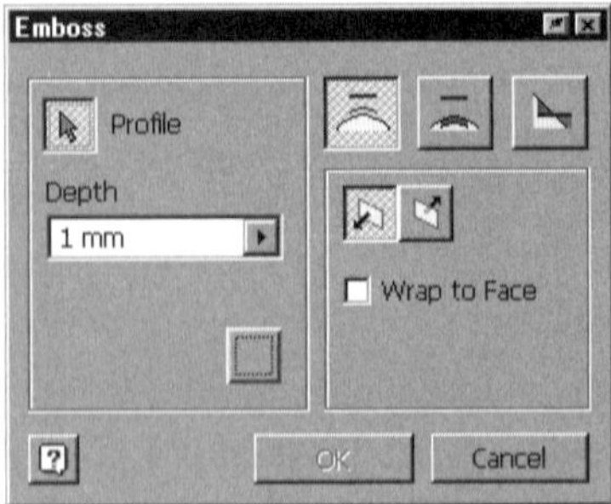


**Figure 8-16**

| | | |
|---|---|---|
| | Profile | Select a profile (closed shape or text) to emboss. You may need to use the Select Other tool to select the text. |
| | Depth | Enter an offset depth to emboss or engrave the profile if the Emboss from Face or Engrave from Face types are selected. |
| | Top Face Color | Select a color from the drop-down list to define the color of the top face of the embossed area (not its lateral sides). |
| | Emboss from Face | Select this option to add material to the part. |
| | Engrave from Face | Select this option to remove material from the part. |
| | Emboss/ Engrave from Plane | Select this option to add and remove material from the part by extruding both directions from the sketch plane. Direction changes at the tangent point of the profile to a curved face. |
| | Flip Direction | Select either of these buttons to define the direction of the feature. |
| Wrap to Face | Wrap to Face | Check this box for Emboss from Face or Engrave from Face types to wrap the profile onto a curved face. Only a single non-seamed face can be selected. The profile will be slightly distorted as it is projected onto the face. The wrap stops when a perpendicular face is encountered. |
| | Face | Click this button after checking the Wrap option, and then select the face around which the profile will be wrapped. |

To emboss a closed shape or text, follow these steps:

1. Click the Emboss tool from the Part Features Panel Bar.
2. Define the profile by selecting a closed shape or text; if needed, use the Select Other tool.
3. Select the type of emboss—Emboss from Face, Engrave from Face, or Emboss/Engrave from Plane.
4. Define the depth, color, direction, and face as needed.

   The embossed feature can be edited like any other feature.

# EXERCISE 8-2
# Creating Text and Emboss Features

In this exercise, you emboss and engrave sketched profile objects on faces of a razor handle model. You then create a sketch text object and engrave it on the handle. To navigate to the exercise in the *Electronic Student Workbook*, do the following:

1 From the Main TOC page, click Chapter 8.

2 From the TOC for Chapter 8, click Creating Text and Emboss Features.

The following image illustrates the opened exercise.

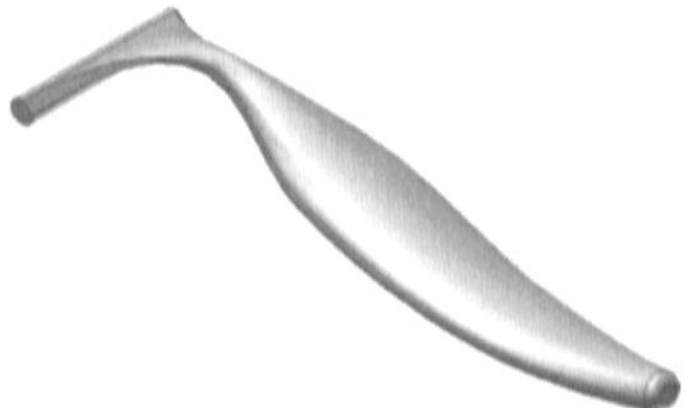

**Figure 8-17 Opened exercise**

The following image illustrates the completed exercise.

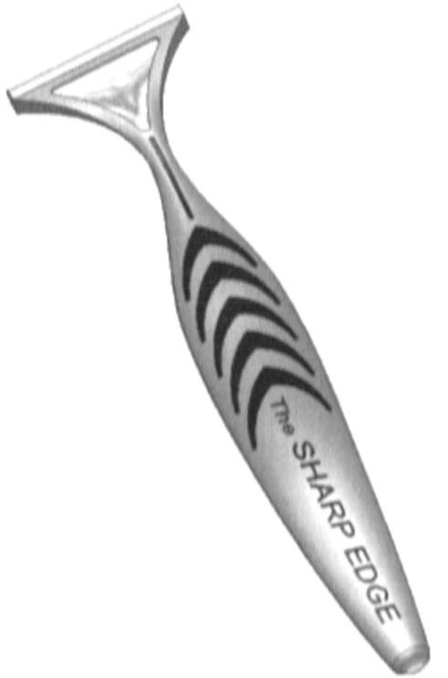


**Figure 8-18 Completed exercise**

## SWEEP FEATURES

A sweep feature is unlike other sketched features. A sweep feature requires two unconsumed sketches—a profile, and a path that the profile will follow. The two sketches cannot lie on the same plane and cannot be parallel. The path can be an irregular shape, or based on a part edge by projecting the edges onto the active sketch. The path can be either open or closed and can lie in a plane or lie in multiple planes. Handles, cabling, and piping are examples of sweep features. A sweep feature can be the base or a secondary feature. To create a sweep feature, use the Sweep tool from the Part Features Panel Bar, as shown on the left in the following image. The Sweep dialog will appear as shown on the right. The following list includes description of the options in the Sweep dialog box.

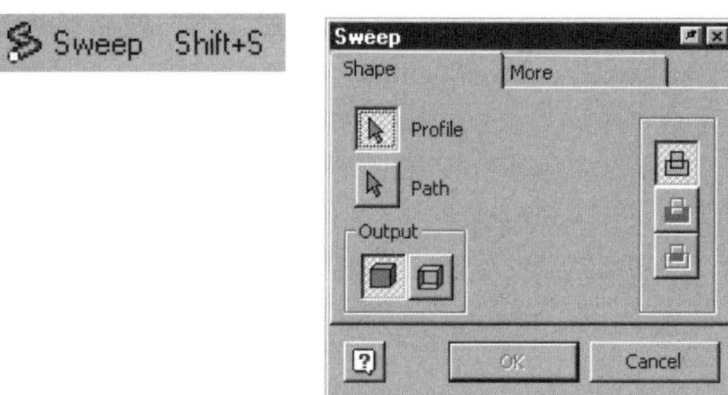


Figure 8-19

### SHAPE

**Profile** Select this button to choose the sketch to sweep. If the Profile button is depressed, it means that a sketch or sketch area needs to be selected. If there are multiple closed profiles, you will need to select the profile that you want to sweep. If there is only one possible profile, Autodesk Inventor will select it for you and you can skip this step. If the wrong profile or sketch area is selected, reselect the Profile button and select the new profile or sketch area.

**Path** Select this button to choose the path along which to sweep the profile. The path can be either open or closed.

**Operation buttons** This is the column of buttons along the right side of the dialog box. By default, the Join operation is selected. Use the operations buttons to add or remove material from the part (using the Join or Cut options), or keep what is common between the existing part and the completed sweep (using the Intersect option).

*Join*—Adds material to the part.

*Cut*—Removes material from the part.

*Intersect*—Creates a new feature from the shared volume of the sweep feature and existing part volume. Material not included in the shared volume is deleted.

**Output Buttons** In this section, click Solid to create the feature as a solid, or Surface to create the feature as a surface.

### MORE

From the More tab, enter a value for the angle from which you want the profile to be drafted. By default, the taper angle is 0°, as shown in the following image.

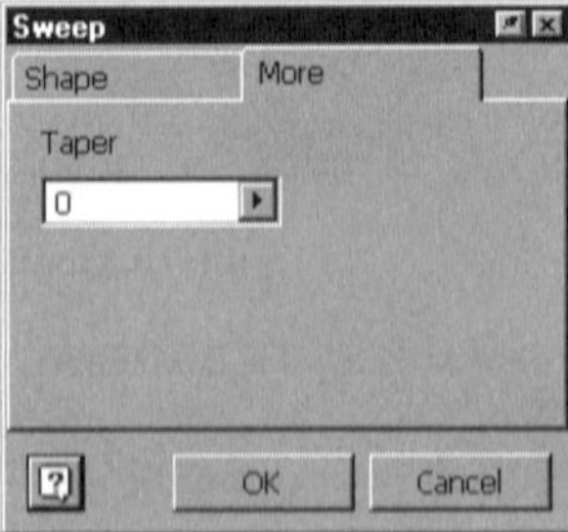


**Figure 8-20**

In the next two sections, you will learn how to create sweep features using both 2D and 3D paths.

### CREATING A SWEEP FEATURE

In this section, you will learn how to create sweep features. You first need to have two unconsumed sketches. One sketch will be swept along the second sketch that represents the path. To create a sweep feature, follow these steps:

1. Create two unconsumed sketches—one for the profile and the other for the path. The profile and path must lie on separate non-parallel planes. Use work planes to place the location of the sketches, if required. The sketch that is used for the path can be open or closed. Add dimensions and constraints to both sketches as needed.
2. Click the Sweep tool from the Part Features Panel Bar.
3. The Sweep dialog box will appear. If two unconsumed sketches do not exist, Autodesk Inventor will notify you that two unconsumed sketches are required.
4. Click the Profile button and then select the sketch that will be swept in the graphics window. If only one closed profile exists, this step is automated for you.
5. If not already depressed, click the Path button and then select the sketch to be used as the path in the graphics window.
6. Determine if the output will be a solid or a surface by clicking either the Solid or Surface button from the Output area.
7. If this is a secondary feature, click the operation that will define whether material will be added, removed, or if what is common between the existing part and the new sweep feature will be kept.

8 If you want the sweep feature to have a taper, click the More tab and enter a value for the taper.
9 Click the OK button to complete the operation.

The drawing on the left of the following image shows two sketches that have dimensions applied to them. The path is lying flat and the profile is drawn on a work plane that is normal to the path. The image on the right shows the completed sweep.

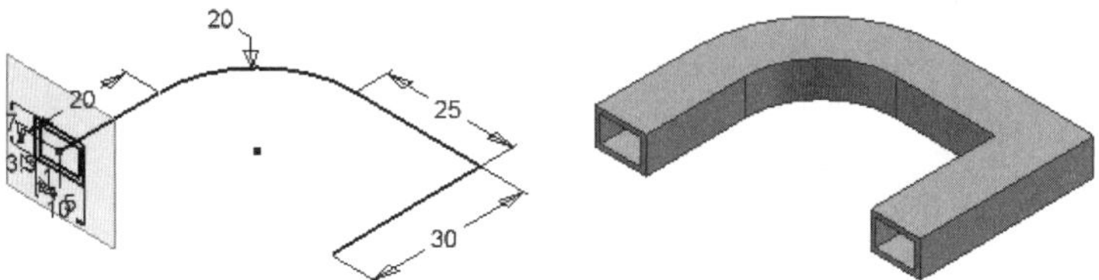


Figure 8-21

## 3D SKETCHING

To create a sweep feature whose path does not lie on a single plane, you need to create a 3D sketch that will be used for the path. You can use a 3D sketch to define the path for a lip or to define the routing path for an assembly component, such as a pipe or duct work that crosses multiple faces on different planes. A 3D sketch needs to be defined in the part environment, and this can be done within an assembly or in its own part file. You can use the Autodesk Inventor adaptive technology during 3D sketch creation to create a path that updates automatically to match changes to referenced assembly components. In this section, you will learn strategies on how to create 3D sketches.

### 3D Sketch Overview

When creating a 3D sketch, you use many of the same sketching techniques that you have already learned with the addition of a few tools. 3D sketches use work points and model edges/vertices to define the shape of the 3D sketch by creating line or spline segments between them. Bends can also be created between line segments. When creating a 3D sketch, you use a combination of lines, splines, fillet features, work features, constraints, and existing edges and vertices.

### 3D Sketch Environment

The 3D sketch environment is used to create 3D curves that are created with a combination of both 2D and 3D curves. Before creating a 3D sketch, change the environment to the 3D sketch environment by clicking the 3D Sketch tool from the Standard toolbar under the 2D Sketch tool, as shown on the left in the following image. The Panel Bar's tools will then change to the 3D Sketch tools, as shown in the center of the following image. The tools that are only available in the 3D Sketch Panel Bar are explained throughout this section. While in the 3D Sketch environment, all features appear in the

Browser with a 3D sketch name. The image on the right shows an example of a 3D Sketch in the Browser. Once a feature uses the 3D sketch, it will be consumed under the new feature in the Browser.

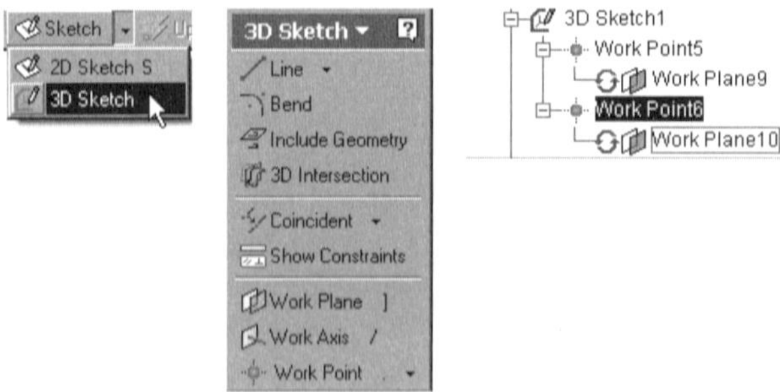


Figure 8-22

## 3D Path From Existing Geometry

The easiest way to create a 3D path is to use existing geometry. If you wanted to create a lip on an existing part, for example, you would use the existing edges to define the path. You can include existing geometry by projecting part edges, vertices, and geometry from visible sketches into a 3D sketch. To use existing geometry to create a 3D path, follow these steps:

1 Change to the 3D Sketch environment by clicking the 3D Sketch tool from the Standard toolbar under the 2D Sketch tool.

2 Start the Include Geometry tool from the 3D Sketch Panel Bar, as shown in the following image. This tool will include existing 2D sketch geometry into a 3D sketch. The projected geometry is updated to reflect changes to the original 2D geometry.

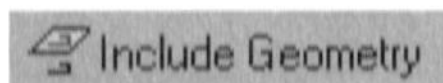


Figure 8-23

3 Click each of the model edges that you want to use for the 3D sketch. When finished, right-click and select Done from the menu, or press the ESC key on the keyboard. If you select a wrong edge, you can manually delete it. The following image shows an example of including outside edges of a part in a 3D sketch.

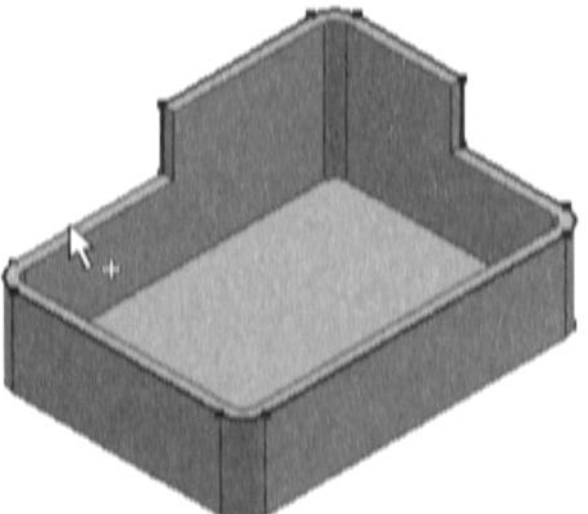

Figure 8-24

4 If a plane does not exist where the profile will be placed, create a work plane that defines the plane on which the profile will be placed.

5 Make the work plane, or the existing plane, the active sketch.

**NOTE** Use the 2D Sketch tool to make the plane the active sketch.

6 Sketch, constrain, and dimension the profile that will be swept. The following image shows a sketch created, constrained, and dimensioned on the work plane.

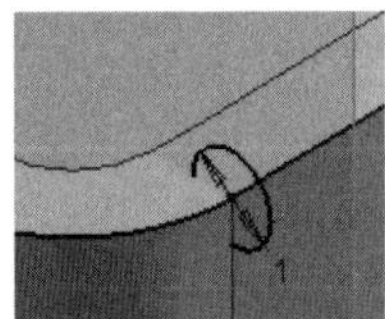

Figure 8-25

7 Click the Sweep tool from the Part Features Panel Bar.

8 If it is not already depressed in the Sweep dialog box, click the Profile button and then select the sketch that will be swept in the graphics window.

9 Click the Path button and then select the 3D sketch to use as the path in the graphics window.

10 Select the operation that will define whether material will be added, removed, or if what is common between the existing part and the new sweep feature will be kept.

11 If you want the sweep feature to have a taper, click the More tab and enter a value for the taper.

12 Click the OK button to complete the operation. The following image shows the completed part.

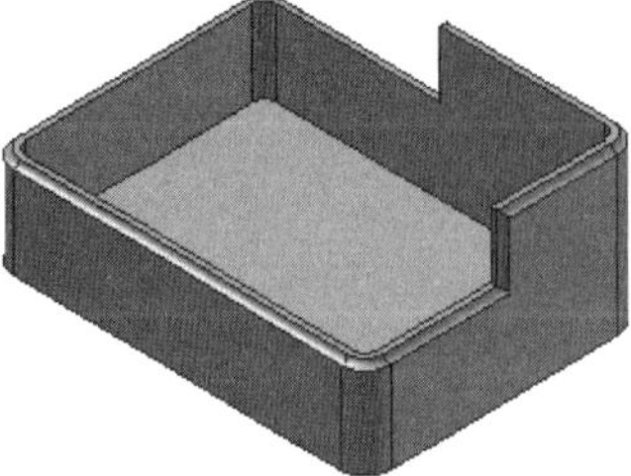

Figure 8-26

### 3D Sketch From Intersection Geometry

Another option to help create a 3D path is to use geometry that intersects with the part. If the intersecting geometry defines the 3D path, it can be used. The intersection can be defined by a combination of the following—a planar or nonplanar part face, a surface face or quilt, or a work plane. To create a 3D path from an intersection, follow these steps:

1. Create the intersecting features.
2. Change to the 3D Sketch environment by clicking the 3D Sketch tool from the Standard toolbar under the 2D Sketch tool.
3. Start the 3D Intersection tool, shown in the following image, from the 3D Sketch Panel Bar.

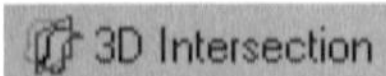


**Figure 8-27**

4. The 3D Intersection Curve dialog box appears, as shown on the right in the following image.
5. Select the two intersecting features.
6. Click the OK button and a 3D path will be created.

The left image shows a cylinder with a work plane that is intersecting it at an angle.

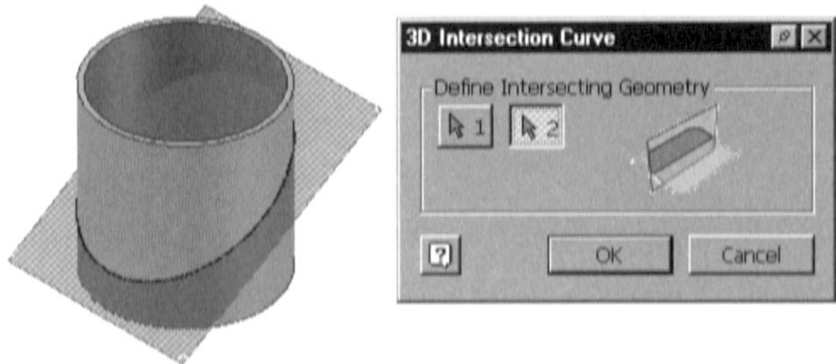


**Figure 8-28**

The following image shows a cylinder with a surface that was extruded to the outside face of the cylinder. A 3D sketch was then created using the 3D Intersection tool.

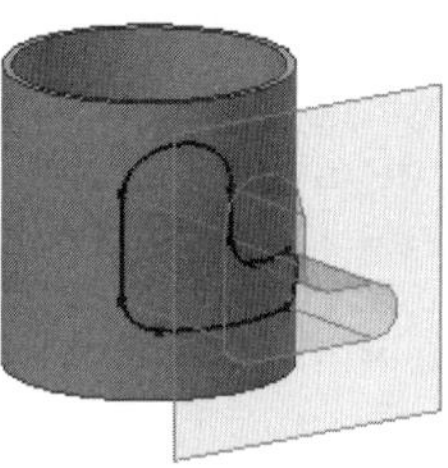

Figure 8-29

## Constructed Paths

Another option for creating 3D paths is to define a path by creating work points at locations where the 3D path will intersect and then connecting the points using a 3D line or spline. Because this method of constructing a 3D path depends on work points, you should review the work point section in Chapter 4 to become comfortable with work points. The following is a review of the methods used when creating a work point:

- Click an endpoint or midpoint of an edge, then click an edge or axis, and a work point is created at the intersection (or theoretical intersection) of the two.
- Click an edge and plane, and a work point is created at the intersection (or theoretical intersection) of the two.
- Click three nonparallel faces or planes, and a work point is created at their intersection (or theoretical intersection).
- A work point is also automatically generated if you select a vertex while creating a 3D line.

Grounded work points can also be used, but they are not associated with the part or any other work features, including the original locating geometry. When surrounding geometry is modified, the grounded work point remains in the specified location.

To create a 3D path using work points and a 3D line, follow these steps:

1. Create work points and grounded work points, as needed, to define the 3D path.
2. Change to the 3D Sketch environment by clicking the 3D Sketch tool from the Standard toolbar under the 2D Sketch tool.
3. If you want the 3D path to automatically place a bend, you can set the size of the bend, click Tools > Document Settings, and change the 3D Sketch Auto-Bend Radius setting from the Sketch tab, as displayed in the following image.

**Figure 8-30**

4 Click the 3D Line tool from the 3D Sketch Panel Bar.

5 Select the work points in the order that the path will follow. By default, a bend is automatically applied between 3D line segments. The automatic bend option can be toggled on and off by right-clicking while in the 3D line command and checking or unchecking Auto-Bend on the menu, as shown in the following image.

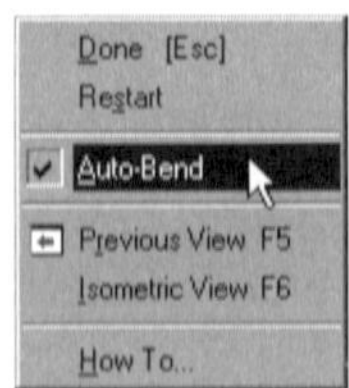


**Figure 8-31**

6 To manually add a bend between two 3D lines, use the Bend tool, shown in the following image, from the 3D Sketch Panel Bar. In the 3D Sketch Bend dialog box, enter a value for the bend and then select two 3D lines. Once the bend is placed, it can be edited by double-clicking on the dimension and entering a new value.

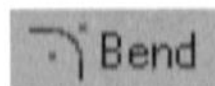


**Figure 8-32**

7 When you are done selecting points, right-click and select Done from the menu.

8 Next, create a profile that will be swept along the 3D path.

9 Start the Sweep tool and create the 3D sweep.

The following images show a part with construction work axes, work planes and work points, the constructed 3D path with bends, and the completed swept part.

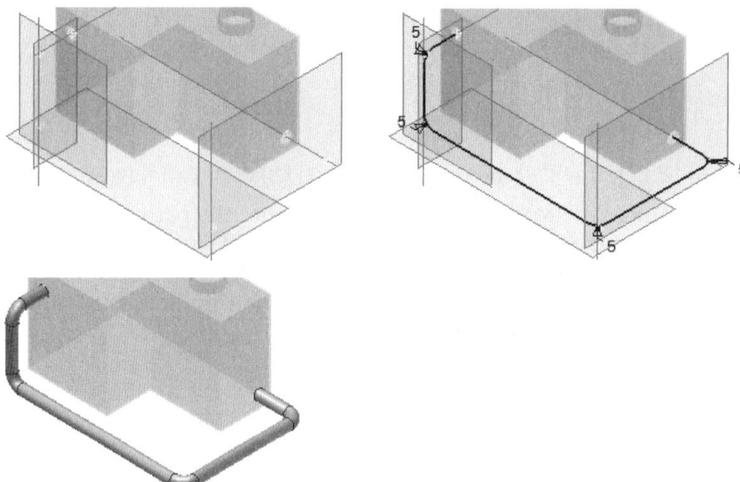

Figure 8-33

## Splines

A spline can also be created between work points or vertices on a part.

To create a 3D path using work points and a 3D line, follow these steps:

1. Create work points and grounded work points, as needed, to define the 3D path.
2. Change to the 3D Sketch environment by clicking the 3D Sketch tool on the Standard toolbar under the 2D Sketch tool.
3. Click the Spline tool on the 3D Sketch Panel Bar as shown in the following image.

Figure 8-34

4. Select the work points in the order that the path will follow, as shown in the following image.

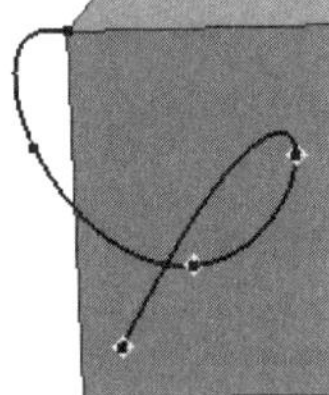

Figure 8-35

5 To exit the spline command, right-click and select Continue from the menu, and then either press the ESC key or right-click and select Done from the menu.

6 If desired, you can add a tangent constraint between the spline and edge of a part by clicking the Tangent tool on the 3D Sketch Panel Bar. The following image shows a tangent constraint being added between the spline and the top edge of the part.

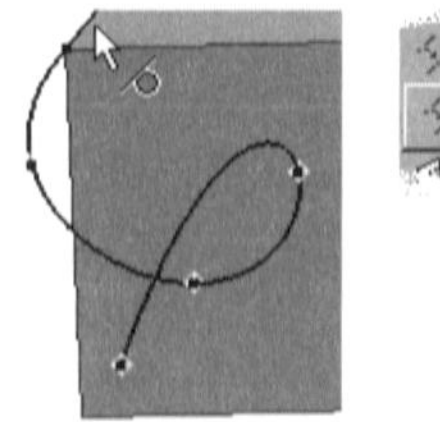

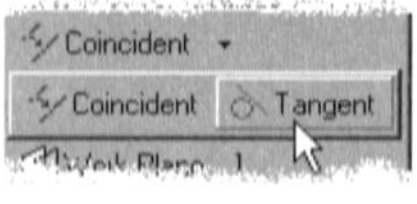


Figure 8-36

## Coincident Constraints

When working with 3D lines, there are two types of 3D coincident constraints that hold the endpoints of the lines and work points together. Line segments created with the Line tool inside a 3D sketch are the only geometry that works with 3D coincident constraints. Each 3D coincident constraint is displayed in the sketch as individual icons. The following is a definition of the two coincident constraints:

A constraint symbol (shown in the following image) near the endpoint of a line indicates a 3D coincident constraint between a line's endpoints and the underlying work point.

Figure 8-37

The following rules apply to this 3D coincident constraint:

1 Deleting the symbol removes the connection between one or more line endpoints and the underlying work point.

2 Deleting the symbol does not delete the constraint between shared line endpoints.

3 Lines joined by a bend have a separate (single) endpoint coincident constraint to the work point.

4 All other line endpoints coincident to the work point share a common coincident constraint.

5 When two lines are connected by a bend, deleting their endpoint coincident constraint does not destroy the bend. The common endpoint can be reattached with a new coincident constraint.

- The 3D coincident constraint symbol (as shown in the following image) near a line's midpoint indicates the constraint between shared line points.

**Figure 8-38**

The following rules apply to this 3D coincident constraint:

1. The symbol will only be displayed if an endpoint is shared by more than one line.
2. Deleting a symbol breaks the sharing of the endpoint.
3. If a bend exists between two line segments, deleting the symbol destroys the bend. The two line endpoints are located at the work point but have different properties.
4. The endpoint of the line with the deleted symbol is not attached to the work point.
5. The endpoint of the other line is constrained to the end point.
6. You can reattach the free endpoint with a new coincident constraint.

The following image shows an example of the 3D constraints shown on 3D lines.

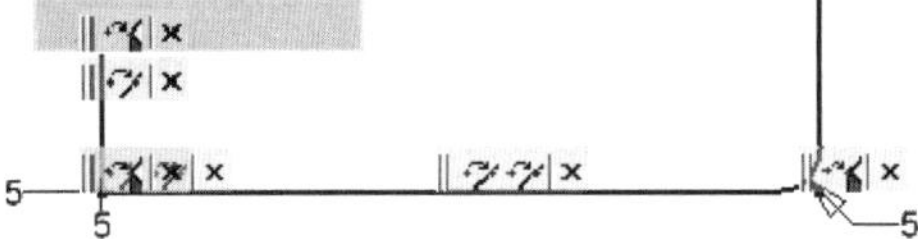

**Figure 8-39**

To add a 3D constraint, click the Coincident tool from the 3D Sketch Panel Bar, as shown in the following image, and follow these guidelines:

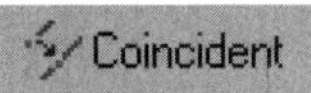


**Figure 8-40**

- You can only add coincident constraints between an unattached line endpoint and an existing work point or vertex.
- You must first delete an existing endpoint or midline (midpoint) constraint symbol to detach a line's end point from an existing work point.
- You must select the line endpoint as the first selection in the coincident constraint.
- A new work point is created if the second selection is a part vertex.

# EXERCISE 8-3 Creating Sweep Features

In this exercise, you construct a 3D path entirely from existing part edges and use the Sweep tool to create a lip. To navigate to the exercise in the *Electronic Student Workbook*, do the following:

1 From the Main TOC page, click Chapter 8.

2 From the TOC for Chapter 8, click Creating Sweep Features.

The following image illustrates the opened exercise.

**Figure 8-41 Opened exercise**

The following image illustrates the completed exercise.

**Figure 8-42 Completed exercise**

## COIL FEATURES

Using the Coil feature, you can easily create many types of helical or coil shapes. You can create many types of springs by selecting different settings in the Coil dialog box. You can also use the Coil feature with the cut operation to remove a helical shape through the outside of a part if you want to represent threads by removing material from a part.

To create a coil, you need to have one unconsumed sketch available in the part. This sketch is the profile (or shape) of the coil. If no unconsumed sketch is available, Autodesk Inventor will prompt you with an error message that states there are "No unconsumed visible sketches on the part." After an unconsumed sketch is available, you can select the Coil tool from the Part Features Panel Bar, as shown in the following image. The Coil dialog box appears. Each of its tabs will be explained in the following sections.

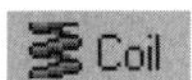


**Figure 8-43**

### COIL SHAPE

The Coil Shape tab, as shown in the following image, allows you to specify the geometry and orientation of the coil.

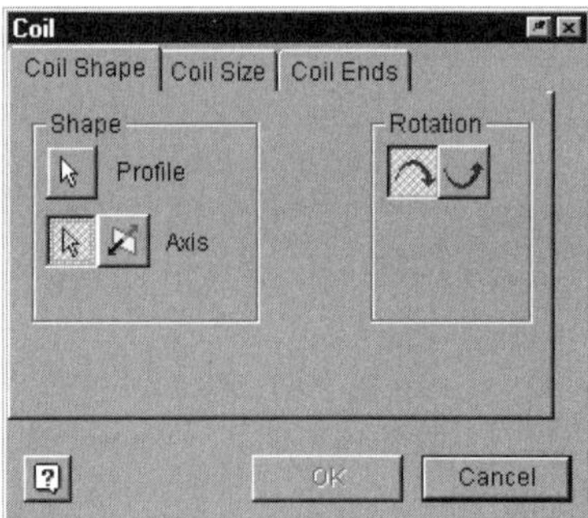


**Figure 8-44**

**Profile** Select this button to select the sketch you will use as the shape of the coil feature. By default, the Profile button is shown depressed—this tells you that a sketch or sketch area needs to be selected. If there are multiple closed profiles, you will need to select the profile that you want to revolve. If there is only one possible profile, Autodesk Inventor will select it for you and you can skip this step. If the wrong profile or sketch area is selected, reselect the Profile button and select a new profile or sketch area. Only one closed profile can be used to create the coil feature.

**Axis** Select this button to select a straight edge, centerline, or axis about which the sketch will be revolved. The edge must be part of the sketch. If a sketched centerline is used as the axis, it will also need to be part of the sketch. If a work axis is used, it cannot intersect the profile.

**Flip** Select this button to change the direction that the coil will be created from the axis. The direction will be changed on either the positive or negative *X*- or *Y*-axis, depending upon the edge or axis that is selected. You will see a preview of the direction on which the coil will be created.

Figure 8-45

**Rotation** Select this button to specify the direction in which the coil will rotate. You can choose to have the coil rotate in either a clockwise or counterclockwise direction. The operation buttons are the column of buttons along the center of the dialog box that will appear if a base feature exists, as shown in the following image. The operation buttons are only available if the Coil feature is not the first feature in the part. By default, the Join operation is selected. You can select the other operations to either add or remove material from the part using the join or cut options, or keep what is common between the existing part volume and the completed coil feature by using the intersect option.

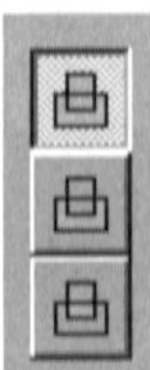

Figure 8-46

*Join*—Adds material to the part.

*Cut*—Removes material from the part.

*Intersect*—Keeps what is common to the part and the sweep feature.

## COIL SIZE

The Coil Size tab, as shown in the following image, allows you to specify how the coil will be created. You are presented with various options for the type of coil that you want to create. Based on the type of coil that you select, the other parameters for Pitch, Height, Revolution, and Taper will become active or inactive. Specify two of the three parameters that are available and Autodesk Inventor will calculate the last field for you.

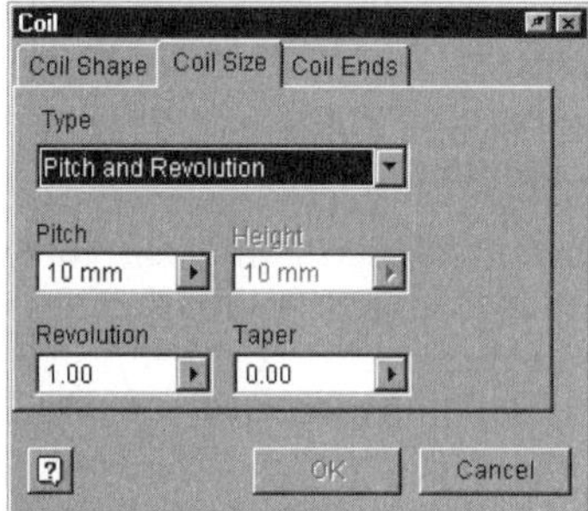


Figure 8-47

**Type** Select the parameters that you want to specify: Pitch and Revolution, Revolution and Height, Pitch and Height, or Spiral.

**Pitch** Type in the value for the height to which you want the helix to elevate with each revolution.

**Revolution** Specify the number of revolutions for the coil. A coil cannot have zero revolutions. Fractions can be used in this field. For example, you can create a coil that contains 2.5 turns. If end conditions are specified (see next section), the end conditions are included in the number of revolutions.

**Height** Specify the height of the coil. This height is measured from the center of the profile at the start to the center of the profile at the end.

**Taper** Type the angle at which you want the coil to be tapered.

**NOTE** A Spiral coil type cannot be tapered.

## COIL ENDS

The Coil Ends tab, as shown in the following image, lets you specify the end conditions for the Start and End of the coil. When selecting the Flat option, the helix—not the profile that you selected for the coil—is flattened. The ends can have end conditions that are not consistent between the start and end of the coil.

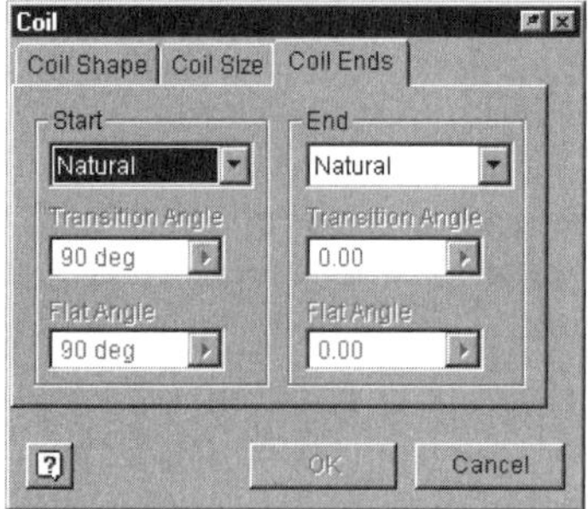


Figure 8-48

**Start** Select either Natural or Flat for the start of the helix. Click the down arrow to change between the two options.

**End** Select either Natural or Flat for the end of the helix. Click the down arrow to change between the two options.

**Transition Angle** This is the distance (specified in degrees) in which the coil achieves the transition. Normally occurs in less than one revolution.

**Flat Angle** This is the distance (specified in degrees) to which the coil extends after transition that does not have a pitch (flat). It provides the transition from the end of the revolved profile into a flattened end.

The following image shows a coil created as the base feature. The image on the left shows the coil in its sketch stage—the rectangle will be used as the profile and the centerline will be used as the axis of rotation. The finished part on the right shows the coil with flat ends.

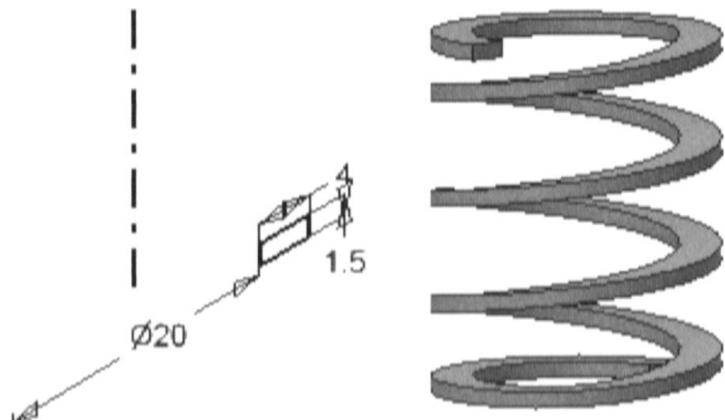


**Figure 8-49**

The following image shows a coil created as a secondary feature. The image on the left shows the coil in its sketch stage. The sketch is drawn on a work plane that is located in the center of the cylinder—the rectangle will be used as the profile and the work axis will be used as the axis of rotation. The finished part on the right shows the coil with flat ends.

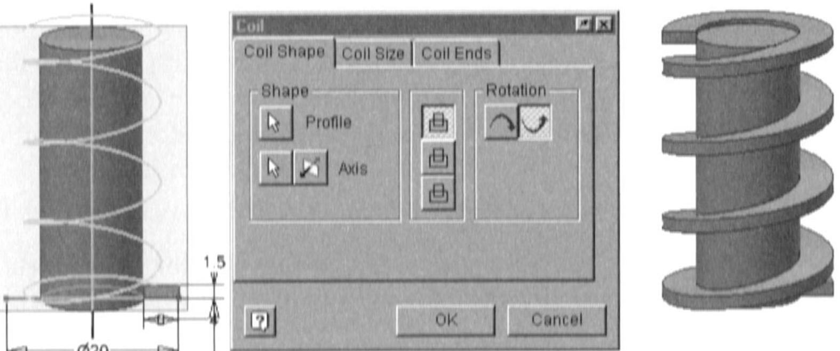


**Figure 8-50**

## LOFT FEATURES

The Loft tool creates a feature that blends a shape between two or more sections (profiles) that have different shapes. Loft features are used frequently in the creation of plastic or molded parts. Many of these types of parts have complex shapes that would be difficult to create using standard modeling features. You can create Loft features that blend between two or more cross-section profiles that reside on different planes. A rail, or multiple rails, can be defined as a path(s) that the loft will follow. There is no limit to the number of sections that can be included in the Loft feature. Three types of geometry are used to create a loft: sections, rails, and points. Each type of geometry is explained here.

**Sections** Define the shape that the loft will blend between. The following rules apply to sections:

1 There is no limit to the number of sections that can be included in the Loft feature.
2 Sections do not have to be sketched on parallel planes.
3 Sketches can be defined by 2D sketches (planar), 3D sketches (nonplanar), or existing faces or edges on a part.
4 All sections must be either open or closed. You cannot mix open and closed profiles.

**Rails** Can be defined by the followin—2D sketches (planar), 3D sketches (nonplanar), or edges on a part. The following rules apply to rails:

1 There is no limit on the number of rails that can be created.
2 Rails must not cross each other and must not cross mapping curves.
3 Points defined by rail curves and mapping sets cannot be coincident on internal sections and rails.
4 Rails affect all of the sections, not just faces or sections they intersect. Section vertices without defined rails are influenced by neighboring rails.
5 All rail curves must be open or closed.
6 Closed rail curves define a closed loft, meaning that the first section is also the last section.
7 No two rails can have identical guide points, even though the curves themselves may be different.
8 Rails can extend beyond the first and last sections. Any part of a rail that comes before the first section or after the last is ignored.
9 If a rail is not an edge of the final body, the surface will be smooth across the rail.

**Points** Can be mapped to help define how the sections will blend to each other. The following rules apply to mapping points:

1 A set of mapping points constitutes a mapping curve.

2 A mapping point can lie anywhere on an edge.
3 The points in a set must either follow the order of the sections (first point on first section, etc.) or follow the reverse of this order (first point on last section, etc.).
4 Mapping curves must not cross other mapping curves or rail curves.
5 Mapping curves can share their first and/or last points.

To create a Loft feature, follow these steps:

1 Create the profiles that will be used as the sections of the loft. Use work features to position the profiles, if required.
2 Click the Loft tool from the Part Features Panel Bar, as shown in the following image.

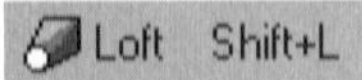


**Figure 8-51**

3 On the Curves tab of the Loft dialog box, the Sections option will be the default option. In the graphics window, click the sketches in the order in which the loft sections will blend.
4 If rails are to be used in the loft, click "Click to add" in the Rails section of the Curves tab and then click the rail or rails.
5 If needed, change the options for the loft from the Conditions and Transition tabs.

The following is an explanation of the options in the Loft dialog box and its tabs.

## CURVES

The Curves tab allows you to select which sketches will be used as sections, whether or not a rail will be used, and determines the output condition (see following image).

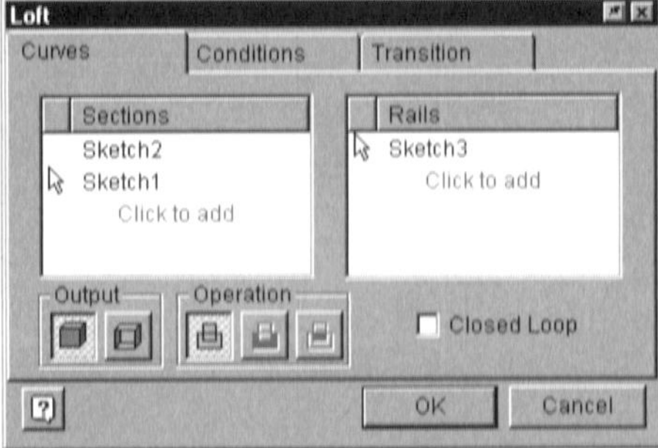


**Figure 8-52**

**Sections** This is where you select the sketches (two or more) that make up the Loft feature. The sketches can be created on any plane and they do not have to be perpendicular to each other.

**Rails** This is where you will select a sketch or sketches to be used as rails.

**Output** Select whether or not the Loft feature will be a solid or a surface.

**Operation** Select the option that will either add or remove material from the part using the Join or Cut options, or keep what is common between the existing part and the completed loft by using the Intersect option. By default, the join operation is selected.

**Closed Loop** When checked, the first and last sections of the Loft feature are joined to create a closed loop.

## CONDITIONS

The Conditions tab, shown in the following image, allows you to control the boundary, angle, and weight condition of the Loft feature.

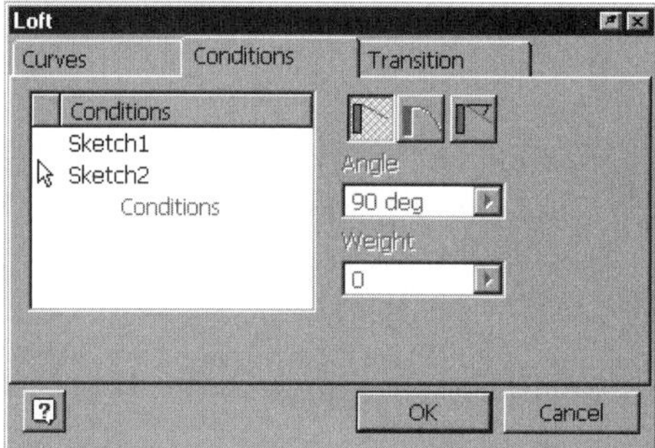


**Figure 8-53**

**Conditions** The column on the left lists the sketches that are defined for the sections. To change a sketch's condition, click on its name and then select a condition option.

**Condition Boundaries** There are three boundary condition buttons available on the top right side of the Conditions tab as shown in the following image.

**Figure 8-54**

*Free*—With this option there is no boundary condition.

*Tangent*—This option is only available when a Loop is selected or the curve is a 2D sketch created from a face. When selected, the loft will be tangent to the selected loop or face.

*Direction*—This option is only available when the curve is a 2D sketch. When selected, you can define the angle.

**Angle** This option is only enabled for a section when the boundary condition is Tangent or Direction. Default is set to 90 degrees and is measured relative to the profile plane. The angle sets the value for an angle formed between the plane that the profile is on and the direction to the next cross section of the loft feature.

**Weight** Default is set to 0. The weight value controls the tangency of the loft shape to the normal of the starting and ending profile. A small value will create an abrupt transition and a large value creates a gradual transition. High weight values could result in twisting the loft and may cause a self-intersecting shape.

## TRANSITION

The Transition tab allows you to define point sets. A point set is used to define how segments blend from one section to the segments of the section before and after. Points are reoriented or added on two adjoining sections.

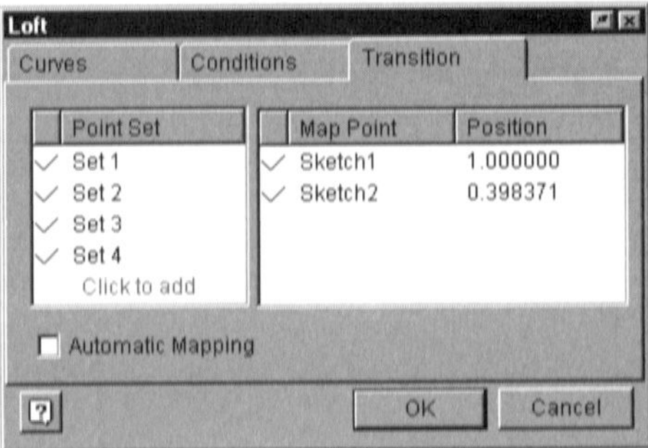


**Figure 8-55**

**Point Set** The name of the point set is displayed here.

**Map Point** The corresponding sketch for the selected Point Set is displayed here.

**Position** Displays the location of the selected Map Point. The position can be modified by either entering a new value or dragging the point to a new location in the graphics window.

To modify the default point sets or to add a point set, follow these steps:

1. Click the Transition tab and clear the Automatic Mapping box. The dialog box will populate the default point data for each section. The list is sorted in the order in which they were specified on the Curve tab.
2. To modify a point's position, first select the point set in which the point is defined. When you select its name, it will also highlight in the graphics window.

3 Click in the Position section of the Map Point that you want to modify and either enter in a new value or drag the point to a new location in the graphics window.

4 To add a point set, click the "Click to add" section in the point set area.

5 Click a point on the profile of two adjoining sections. As the cursor is moved over a valid region of the active section, a green point is displayed. As the points are placed, they are previewed in the graphics window, as shown in the following image.

6 The new point set can be modified similar to the default point set.

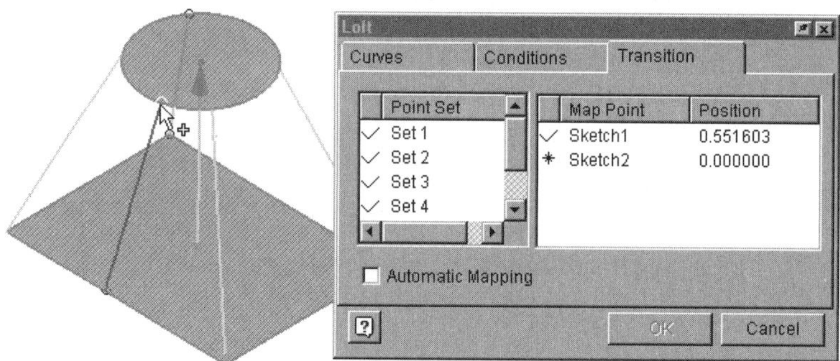


**Figure 8-56**

The following image shows a loft created from two sections and one rail. The image on the left shows two sections and a rail in their sketch stages shown in the top view. The image on the right shows the completed loft.

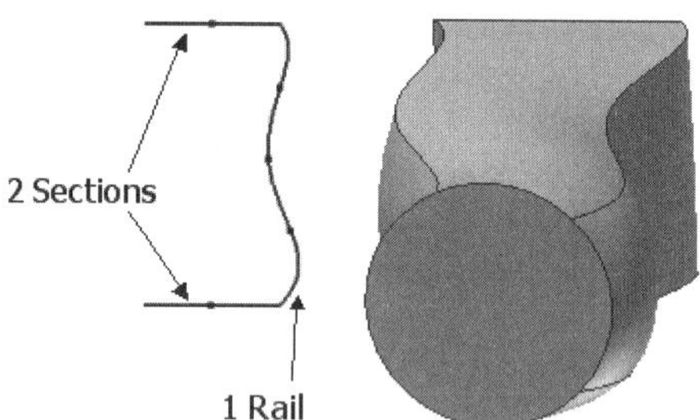


**Figure 8-57**

# EXERCISE 8-4 Creating Loft Features

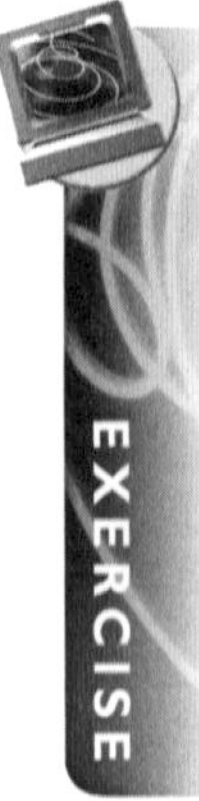


In this exercise, you use Loft options and controls to define the shape of a razor handle. To navigate to the exercise in the *Electronic Student Workbook*, do the following:

1. From the Main TOC page, click Chapter 8.
2. From the TOC for Chapter 8, click Creating Loft Features.

   The following image illustrates the opened exercise.

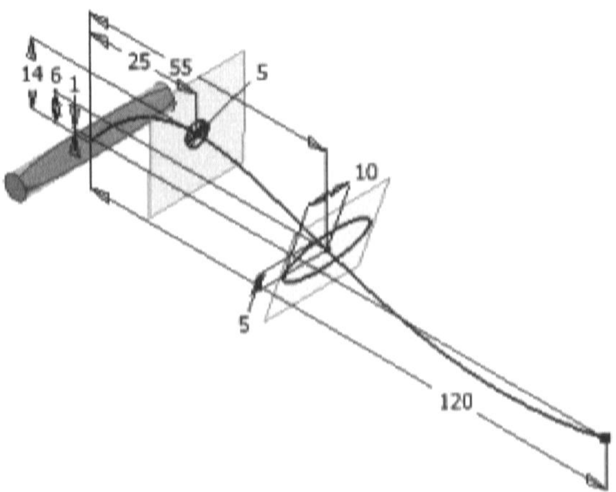


**Figure 8-58 Opened exercise**

The following image illustrates the completed exercise.

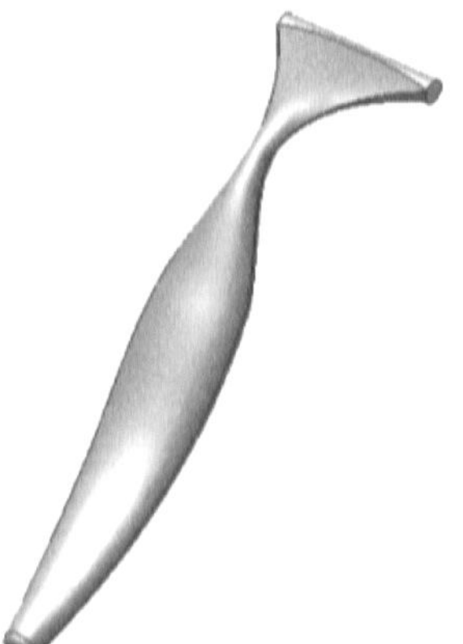

**Figure 8-59 Completed exercise**

## PART SPLIT AND FACE SPLIT

The Split tool allows you to split a part by removing one portion of the part, or split individual faces in order to apply draft to both halves of the part. You can use the Split tool to:

- Split an entire part by sketching a parting line or placing a work plane, and then using it to cut material from the part in the direction you specify. The side that is removed is suppressed rather than deleted. To create a part with the other side removed, edit the split feature and redefine it to keep the other side, then save the other half of the part to its own file using the Save Copy As option.
- Split individual faces by using a surface, sketching a parting line or placing a work plane, and then selecting the faces to split. You can edit the split feature and redefine it to split some or all faces of the part.

The Split tool is located on the Part Features Panel Bar, as shown in the following image. Once selected, the Split dialog box will appear.

**Figure 8-60**

The Split dialog box contains the following sections:

### METHOD

**Split Part** The button on the left of the Method option, as shown in the following image, splits an entire part by selecting a work plane, surface, or sketched geometry to cut (removing material). If you select this option, you are prompted to choose the direction of the material that you want to remove.

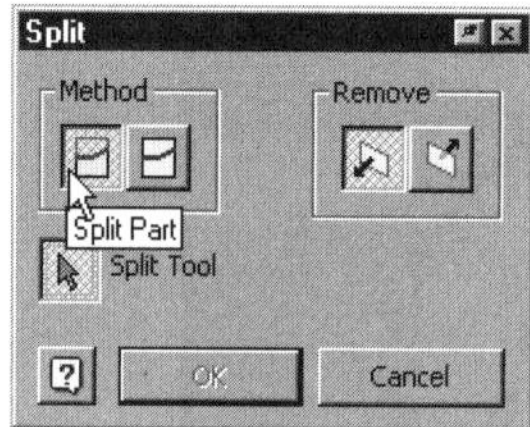


**Figure 8-61**

**Split Face** The button on the right of the Method option, as shown in the following image, splits individual faces of a part by selecting a work plane, surface, or sketched geometry and then selecting the faces to split. The Split face can feature face can split individual faces or all the faces on the part. When the Split Face method is selected, the Remove area in the dialog box will be replaced with the Faces area as shown in the following image on the right.

**Figure 8-62**

**Remove** The option to remove material is only available when the Split Part method is used. After splitting a part, you can retrieve the cut material by editing the split feature and clicking to remove the opposite side, or by deleting the Split feature.

**Figure 8-63**

## FACES

The Faces option is available only when the Split Face method is selected.

**All** This button will select all faces of the part to be split.

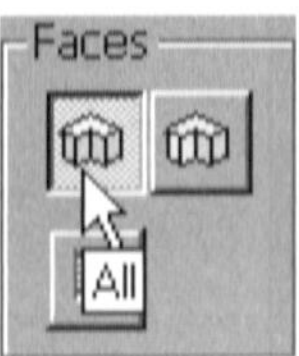


**Figure 8-64**

**Select** By clicking the Select button, you can choose specific faces that you want to split. After clicking the Select button, the Faces to Split tool becomes active.

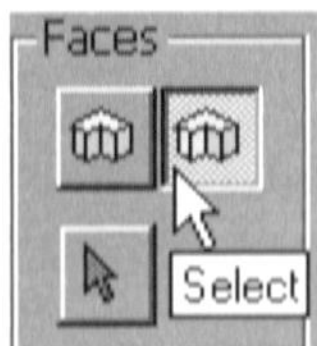


**Figure 8-65**

**Faces to Split** Select this button and then click on the faces of the part that you want to split. Any faces selected will be broken into two halves.

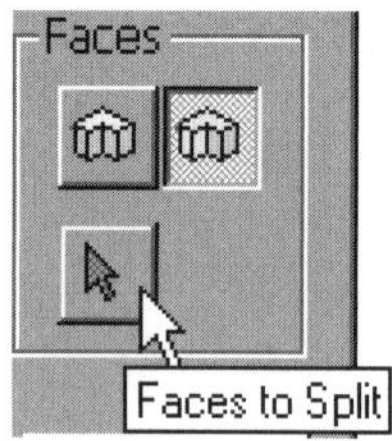


Figure 8-66

**Split Tool** Select this button and then choose either a surface, work plane, or sketch that you want to use to split the part.

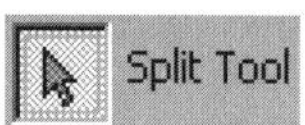


Figure 8-67

# EXERCISE 8-5 Splitting a Part

In this exercise, you use Split features to complete a model of a wireless phone handset. To navigate to the exercise in the *Electronic Student Workbook*, do the following:

1 From the Main TOC page, click Chapter 8.

2 From the TOC for Chapter 8, click Splitting a Part.

The following image illustrates the opened exercise.

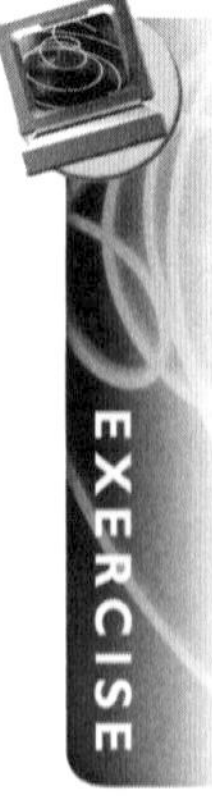

**Figure 8-68 Opened exercise**

The following image illustrates the completed exercise.

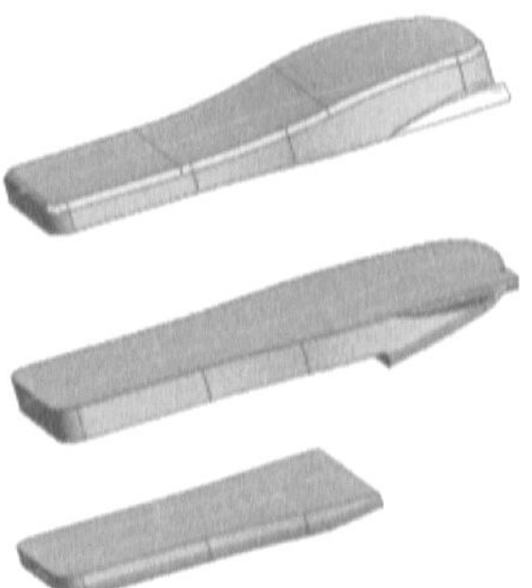

**Figure 8-69 Completed exercise**

## COPYING FEATURES

To increase your productivity, you can copy features from one part to another place on the same part or to a different part. If a feature is copied from one part to another, both files must be open or within the same assembly file. When making a copy of a feature, you have the option to designate whether or not you want the feature, its dependent feature(s), and its parameters to be dependent on or independent of the parent feature(s).

### Dependent

The dimensions of the copied feature(s) will be equal to the dimensions of the parent feature(s). If the parent feature was defined with a value of d4=8 inches, for example, the copied feature would have a parameter similar to d14=d4. The Dependent option is only available when copying features within the same part.

### Independent

The dimensions of the copied feature(s) will contain their own values. Initially, the value will equal the value of the parent feature. If the parent feature was defined with a value of d4=8 inches, for example, the copied feature will have a parameter similar to d14=8 inches. The value of d14 can be modified independent from the value of d4.

The following rules apply when creating a copy of a feature:

- Only selected features are copied by default—not children features.
- Upon pasting the feature, you can choose to copy dependent features.
- When pasting a copied feature, you can specify what plane or orientation the new feature will have.
- The newly pasted feature is completely independent and contains its own sketches and feature definitions within the Browser.

Follow these steps to copy and paste a feature within Autodesk Inventor:

1 Right-click on a feature's name (in the Browser), or change the Select option on the Standard toolbar to Select Features (as shown in the following image) and then move the cursor over the feature that will be copied.

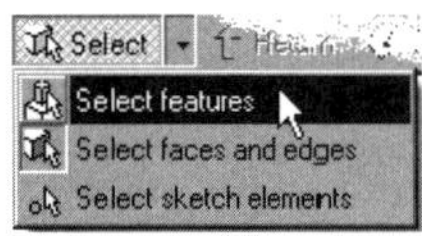


**Figure 8-70**

2 Click Copy from the menu to copy the feature to the clipboard.

3 Start the Paste command by doing one of the following:

a. Right-click and select Paste from the menu.

b. Click Paste from the Edit menu.

c. From the keyboard, press both the CTRL and V keys at the same time.

4 The Paste Features dialog box will appear, as shown in the following image.

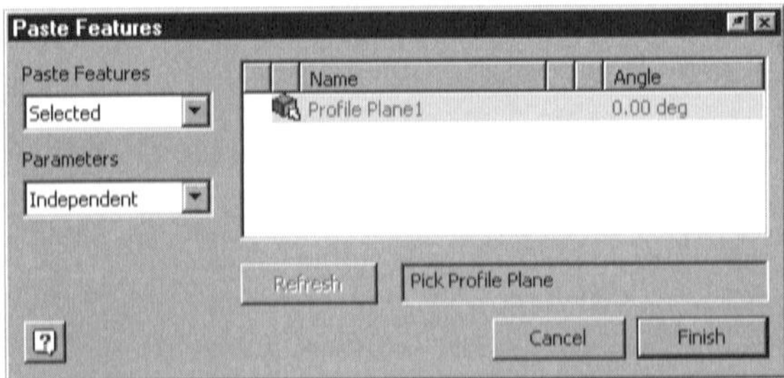


**Figure 8-71**

5 Click a plane on which the feature will be placed. The placement plane can be any planar face on the part or a work plane.

6 The new feature can be roughly moved or rotated by using the compass in the preview image, as displayed in the following image.

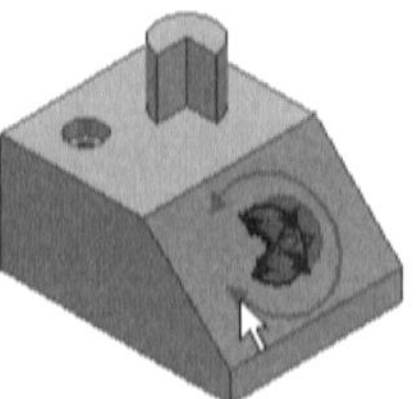

**Figure 8-72**

7 Click the four-headed arrows to move the profile or click the arc around the arrows to rotate the profile. The value for the angle of rotation of the profile can also be entered into the dialog box.

8 To accurately locate the pasted feature, edit its sketch and add constraints and dimensions as needed.

9 From the Paste Features dialog box, adjust the settings as needed.

The following are definitions of the various options in the Paste Feature dialog box:

**Paste Features** Select these options to paste either the Selected feature(s) or the Selected feature(s) and its Dependent feature(s) from the drop-down list as shown in the following image.

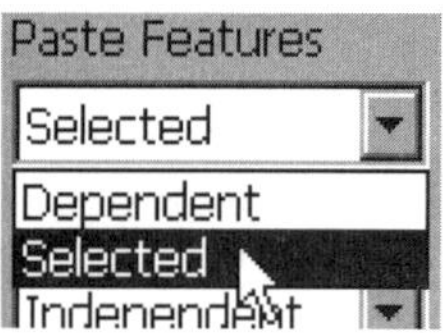


Figure 8-73

*Selected*—This will copy only the features that were selected when initiating the Copy command.

*Dependent*—This will copy the dependent feature(s) of the selected feature(s).

**Parameters** Select to make the parameters of the newly copied feature dependent on the parent feature or independent of the parent feature as shown in the following image.

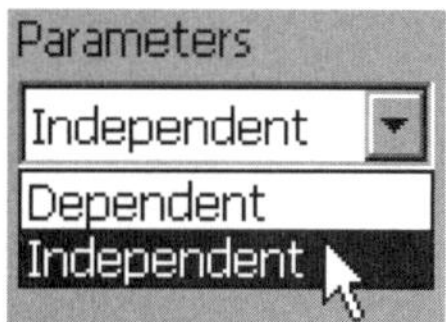


Figure 8-74

*Dependent*—The dimensions of the copied feature(s) will be equal to the dimensions of the parent feature(s). If the parent feature was defined with a value of d4=8 inches, for example, the copied feature would have a parameter similar to d14=d4. The Dependent option is only available when copying features within the same part.

*Independent*—The dimensions of the copied feature(s) will contain their own values. Initially the value will equal the value of the parent feature. If the parent feature was defined with a value of d4=8 inches, for example, the copied feature will have a parameter similar to d14=8 inches. The value of d14 can be modified independent from the value of d4.

**Pick Sketch Plane** Information about the pasted feature will be displayed in this area. A new plane can be selected by clicking on its name in the Pick Sketch Plane area of the Paste Features dialog box and then selecting a new plane in the graphics window. The angle of the pasted feature can be changed by clicking in the Angle section and entering in a new value.

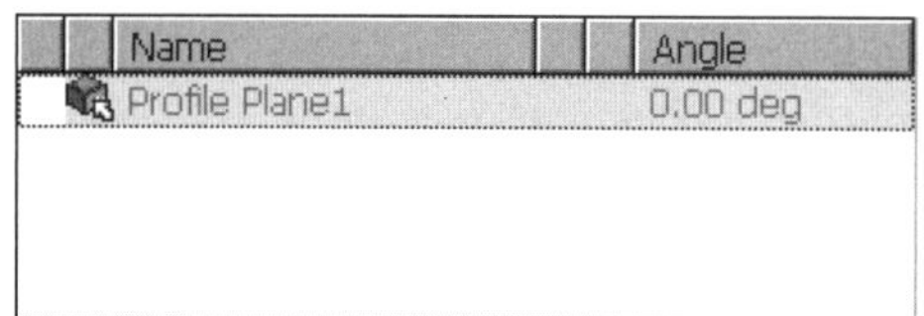


Figure 8-75

## REORDERING FEATURES

Features can be reordered within the Browser. If you have created a Fillet feature using the all fillets or all rounds option, and then created an additional extruded feature (like a boss), you can move the fillet feature below the boss and include the edges of the new feature in the selected edges of the fillet. To reorder features, follow these steps:

1. Hold down the left mouse button (click-and-drag) and move the feature to the desired location in the Browser. A horizontal line will appear in the Browser to show you the feature's relative location while reordering the feature. The image on the left in the following figure shows a hole feature being reordered in the Browser. Click the feature that you want to move by clicking on its icon next to the feature name in the Browser.
2. Release the mouse button, and the model features will be recalculated. The image on the right in the following image shows the Browser and reordered hole feature.

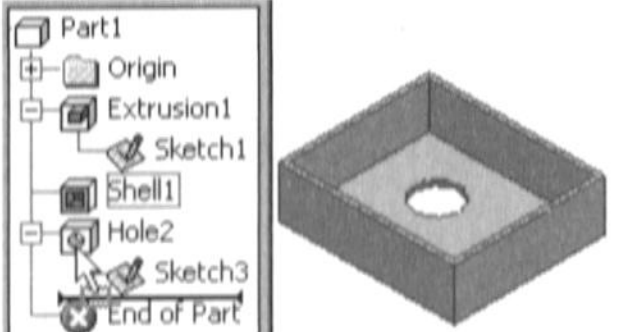


Browser and Part During Feature Reorder

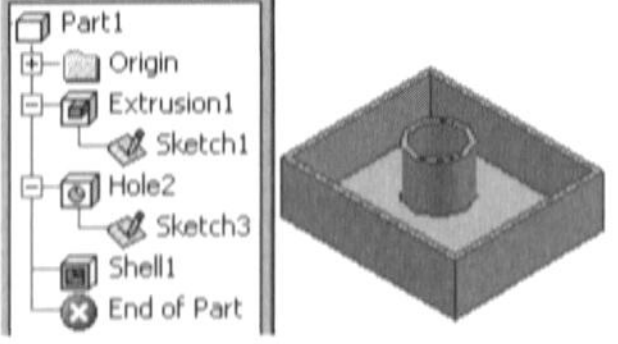


Browser and Part After Feature Reorder

**Figure 8-76**

If the feature cannot be moved due to parent-child relationships with other features, Autodesk Inventor will not allow you to drag the feature to the new position. The cursor will change in the Browser to a NO symbol instead of the horizontal line, as shown in the following image.

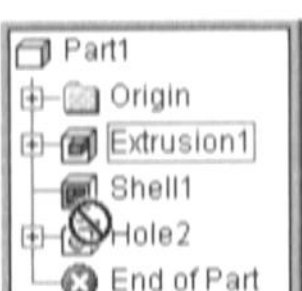


**Figure 8-77**

## MIRRORING FEATURES

When creating a part that has features that are mirror images of one another, you can use the Mirror Feature tool to mirror a feature(s) about a plane instead of recreating each of the features from scratch. Before mirroring a feature, a plane must exist that will be used as the mirror plane. The plane must be located between the existing feature(s) and the new mirrored feature(s), and the plane can be a planar face, a work plane on the part, or an origin plane that has the correct location and orientation. The mirrored feature(s) will be dependent on the parent feature—if the parent feature changes, the resulting mirror feature will also update to reflect the change. To mirror a feature or features, use the Mirror Feature tool from the Part Features Panel Bar as shown on the left in the following image. The Mirror Pattern dialog box will appear, as shown on the right. The options in the Mirror Pattern dialog box are explained next.

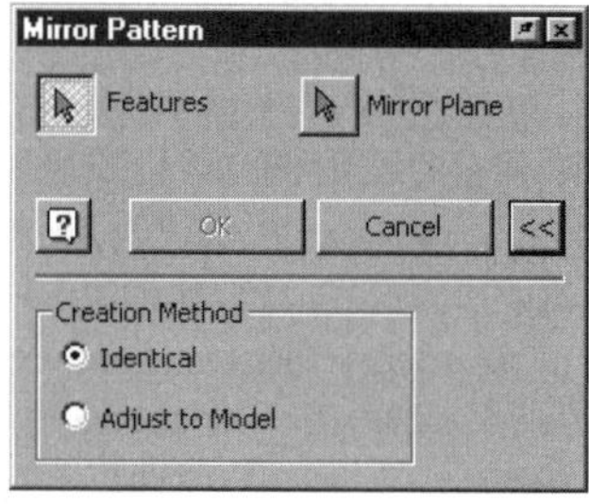


**Figure 8-78**

**Features** Select the feature(s) that you want to mirror.

**Mirror Plane** Chose a planar face or work plane on which to mirror the feature(s).

**Creation Method** Access the Creation Method tools by selecting the More (>>) button from the dialog box.

*Identical*—Create mirrored features that remain identical regardless of where they intersect other features on the part. Use this option when you want to mirror a large number of features.

*Adjust to Model*—Use this option when you want to calculate an independent termination for each feature that is being mirrored. Use this option if the feature(s) being mirrored uses a model face as the termination option.

To mirror a feature or features, follow these steps:

1 Start the Mirror Feature tool from the Part Features Panel Bar, as shown in the previous image. The Mirror Pattern dialog box will appear.
2 Select the feature or features to mirror.
3 Click the Mirror Plane button in the Mirror Pattern dialog box and then select the plane on which the feature will be mirrored.
4 From the More area in the Mirror Pattern dialog box, determine if the mirrored feature(s) will be identical or will adjust to the model.
5 Click the OK button to complete the operation.

The following images show the front features on the cylinder being mirrored about a work plane:

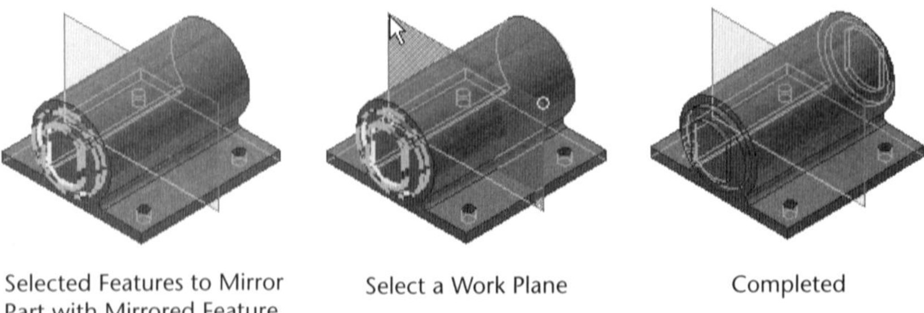

Selected Features to Mirror Part with Mirrored Feature | Select a Work Plane | Completed

**Figure 8-79**

## SUPPRESSING FEATURES

You can suppress a model's features to temporarily turn off their display. Feature suppression can be used to simplify parts, which also increases system performance. This capability also shows the part in different states throughout the manufacturing process and can be used to access faces and edges that you would otherwise not be able to reach. If you need to dimension to a theoretical intersection of an edge that was filleted, for example, you could suppress the fillet and add the dimension, then unsuppress the fillet feature. If the feature you suppress is a parent feature to other features (children), the children's features will also be suppressed. Features that are suppressed appear as a light gray color in the Browser and have a line drawn through them, as shown in the following image. A suppressed feature will remain suppressed until it is unsuppressed, which will also return the Browser display to its normal state. The following image also shows the suppressed children features that were dependent on the extrusions.

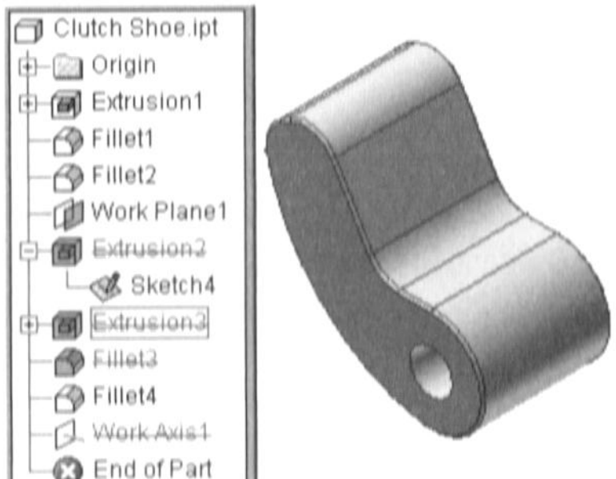


**Figure 8-80**

To suppress a feature on the active part, follow one of these methods.

1 Click on the feature's name that you want to suppress in the Browser.

2 Right-click on the feature and select Suppress Features from the menu, as shown in the following image.

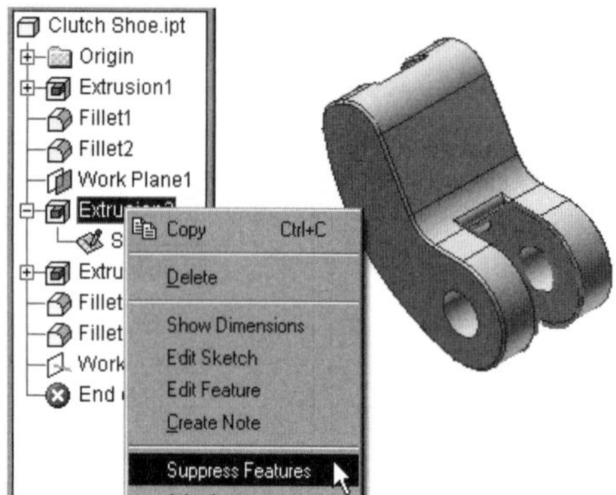


Figure 8-81

or

1 Click the Select Features option from the Standard toolbar, as shown in the following image. This option allows you to select features on the parts in the graphics window.
2 Click the feature that you want to suppress in the graphics window.
3 Right-click to access the Suppress Features option from the menu, as shown in the preceding image.

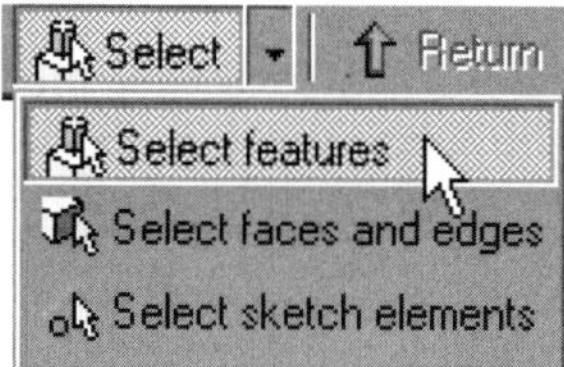


Figure 8-82

To unsuppress a feature, right-click on the suppressed feature's name in the Browser and select Unsuppress Features from the menu.

## FEATURE ROLLBACK

While designing, you may not always place features in the order that your design later needs. Earlier in this chapter, you learned how to reorder features, but reordering features will not always allow you to create the correct design. To solve this problem, Autodesk Inventor allows you to roll back the design to an earlier state and then place the new features. To roll back a design, drag the End of Part marker in the Browser to the location where the new feature will be placed. To move the End of Part marker, follow these steps:

1 Move the cursor over the End of Part marker in the Browser.

2 With the left mouse button pressed down, drag the End of Part marker to the new location in the Browser. While dragging the marker, a line will appear as is shown on the left in the following image.

3 Release the mouse button and the features below the End of Part marker are temporarily removed from the part and the End of Part marker will be moved to its new location in the Browser. The image on the right shows the End of Part marker in its new location.

4 Create new features as needed. The new features will appear *above* the End of Part marker in the Browser.

5 To return the part to its original state (that includes the new features), drag the End of Part marker below the last feature in the Browser.

**Figure 8-83**

## FILE PROPERTIES

You can specify properties for files you create in Autodesk Inventor using the iProperties option from the File menu or by right-clicking on the part's name in the Browser or graphics window and clicking Properties from the menu. After selecting the Properties option, the Properties dialog box for the active file will be opened. The following image shows the Properties dialog box with the Summary tab active.

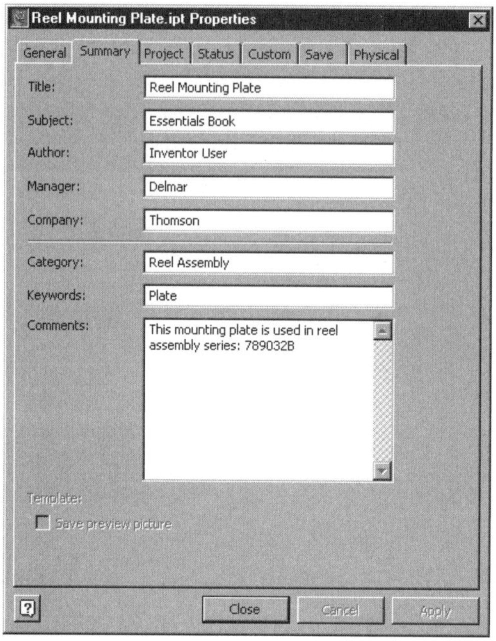

Figure 8-84

The Properties dialog boxes for part and assembly files both contain seven tabs, described below. The Properties dialog boxes for a drawing file contain six tabs—they do not include the Physical tab. The properties of your files can be used to classify and manage your Autodesk Inventor files. You can also use properties that have been populated with data for search criteria, for creating reports, and for automatically updating your title blocks and parts lists in drawings and bills of material. The tabs and the contents of the tabs are listed below.

## GENERAL

This tab lists the file name, location, size, date created, date modified, and last date the file was accessed. The information in this tab cannot be modified via the dialog box.

## SUMMARY

The Summary tab contains fields in which you can enter data for the file: Title, Subject, Author, Manager, Company, Category, Keywords, and Comments. The preceding image shows an example of data entered into the Summary tab.

## PROJECT

The Project tab contains fields in which you can enter data for the file: Location, File Subtype, Part Number, Description, Revision Number, Project, Designer, Engineer, Authority, Cost Center, Estimated Cost, Creation Date, Vendor, and WEB Link.

### STATUS

The Status tab contains fields in which you can enter data or change the status via drop-down lists for the file: Part Number, Status, Design State, Checked By, Checked Date, Eng. Approved By, Eng. Approved Date, Mfg. Approved By, Mfg. Approved Date, File Status, Checked out By, Checked out, Checkout Workgroup, and Checkout Workspace.

### CUSTOM

You can add custom properties to the active file using the Custom tab. The general Microsoft custom properties are also available and can be added to your Autodesk Inventor files.

### SAVE

With the Save tab, you can specify whether you want to save a preview picture of your files for the preview image or you can specify an image file to use as the preview picture.

### PHYSICAL

The Physical tab calculates and displays the physical and inertial properties for a part or assembly file. The Mass, Surface Area, and Volume are automatically calculated. Physical properties of the model change as you create the structure of your files. To get accurate results for the mass, you will need to select a material for each part. If you are in an assembly, this will need to be done by making each part active and then adjusting the material. Selecting the Update button on the Physical tab of the Properties dialog box updates the physical properties based on changes to your models. If the material is changed, the part's color will also reflect the change.

## CENTER OF GRAVITY

As you are designing, it may be helpful to know where the center of gravity is for the active part or assembly. To view a symbolic representation (*X*-,*Y*-, and *Z*-axis arrows) of the center of gravity for the active part or assembly, follow these steps:

1 Click Center of Gravity from the View menu.

2 If the mass properties of the active document are not up-to-date, Autodesk Inventor will prompt you to update them, as shown in the following image.

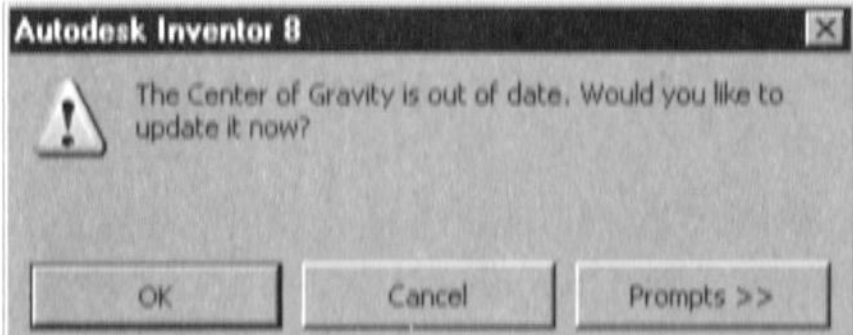


**Figure 8-85**

3 Click OK to update the mass properties and the Center of Gravity symbol will appear at the location of the center of gravity.

4 To see the numeric value of the center of gravity, move the cursor over (or click) the center of the Center of Gravity symbol (you may need to use the Select Other tool) and the *X*-, *Y*-, and *Z*-axis location will appear, as shown on the left in the following image. These values are relative to the origin point for the active document.

5 If the part's physical size changes or parts in an assembly change location, the Center of Gravity symbol will appear colorless—indicating the center of gravity is not up-to-date. The image on the right shows an example of an out-of-date center of gravity since a fillet was added and the holes sizes were edited.

6 Either click Update Mass Properties on the Tools menu or click the Update tool from the Physical tab of the Properties dialog box to update the Center of Gravity to reflect these changes.

7 To turn off the Center of Gravity symbol, click Center of Gravity from the View menu.

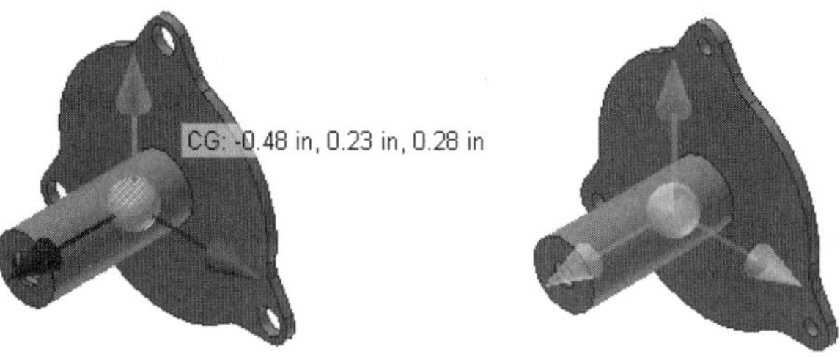


**Figure 8-86**

## OVERRIDE MASS AND VOLUME PROPERTIES

While designing, you may not always draw parts that are 100% complete; for example, you may model only the bounding area and critical features of a purchased part. You still want the mass and volume to be accurately represented in the Properties dialog box, however. Both the mass and volume values of a part or assembly can be overridden by following these steps:

1 Click iProperties from the File menu, or right-click on the part's name in the Browser and click Properties from the menu.

2 The Properties dialog box will appear, as shown on the left in the following image.

3 If a material has not been assigned, select a material from the Material drop-down list.

4 Click the Update button in the Properties dialog box to update the properties if they are displaying N/A. Notice the calculator symbol next to the Mass and Volume values—the symbol shows that Autodesk Inventor has calculated these properties based on the material and physical part.

5 To override a value for either the Mass or Volume, click in their text area and enter a new value. The symbol next to the overridden value will change to a hand to reflect that the value has been overridden, as shown in the image on the right. Notice that the * in the Inertial Properties area states "Values do not reflect user-overridden mass or volume."

6 To change the overridden value back to the default value, delete the information in the text box area of either Mass or Volume area and click the Apply button in the Properties dialog box.

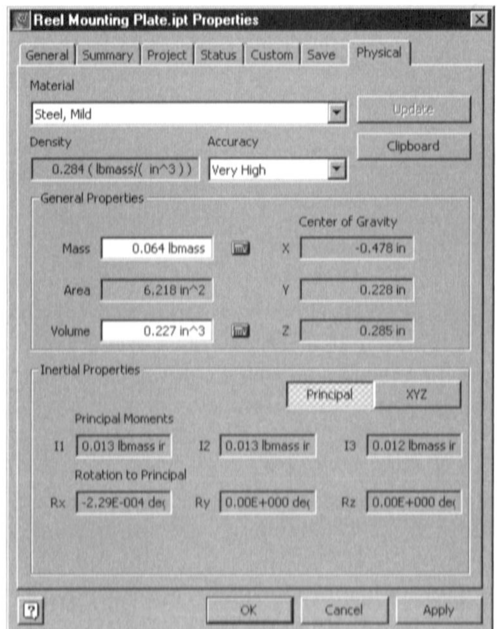


Default Properties

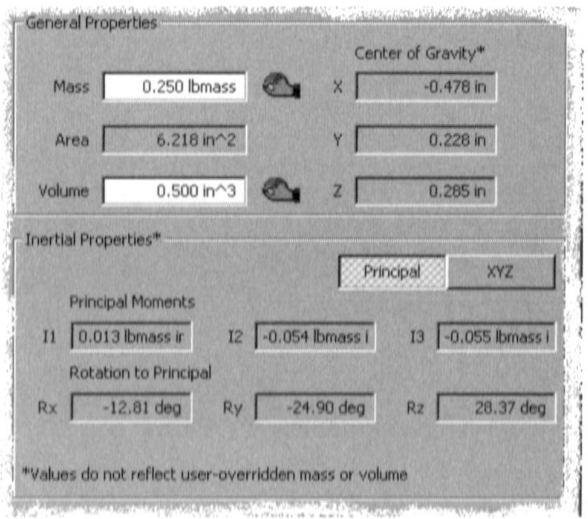


Overridden Mass and Volume

**Figure 8-87**

To copy the information in the Physical tab of the Properties dialog box into another application, such as Microsoft Word, click the Clipboard button in the Physical tab of the Properties dialog box. Next, make the other application active and paste the information into that file. The following image shows the physical property information of a part file when copied into Microsoft Word.

```
Physical Properties for
General Properties:
    Material: {Steel, Mild}
    Density:     7.860E-006 (kg/( mm^3 ) )
    *Volume:     2.5E+003 mm^3
    *Mass:0.2 kg
    Area:  3.970E+003 mm^2
**Center of Gravity:
    X:    -12.146 mm
    Y:    5.801 mm
    Z:    7.129 mm
**Mass Moments of Inertia
    Ixx          3.816 kg mm^2
    Iyx Iyy      1.776E-015 kg mm^2      3.816 kg mm^2
    Izx Izy Izz  3.553E-016 kg mm^2      4.767E-005 kg mm^2      3.436
kg mm^2
**Principal Moments of Inertia
    I1:   3.816 kg mm^2
    I2:   3.816 kg mm^2
    I3:   3.436 kg mm^2
**Rotation from XYZ to Principal
    Rx:   -0.01 deg
    Ry:   0.00E+000 deg
    Rz:   0.00E+000 deg
*Calculations are based on user overwritten values
**Values do not reflect user-overridden mass or volume
```

Figure 8-88

## VISUALIZATION

While working on complex parts, distinguishing between different faces can be difficult. In this section, you learn how to change the colors of individual faces and how to place a bitmap onto a face and apply shadows. Lastly, you will learn how to analyze the part for draft and tangencies.

### FACE COLORS

In Chapter 3, you learned how to change the color of a feature. In this section, you learn how to change the color of a face. When changing a face color, it is important to understand the precedence of colors—the part color is the basic color, a feature color overrides the part color, and a face color overrides the feature color. When changing the color of a face, multiple faces can be changed in one operation. To change the color of a face, follow these steps:

1 Click on one or more faces in the graphics window. To select multiple faces, hold down the CTRL key while clicking on the faces.

2 Right-click and select Properties from the menu, as shown on the left in the following image.

3 Select a color style from the Face Color Style drop-down list in the Face Properties dialog box, as shown on the right in the following image.

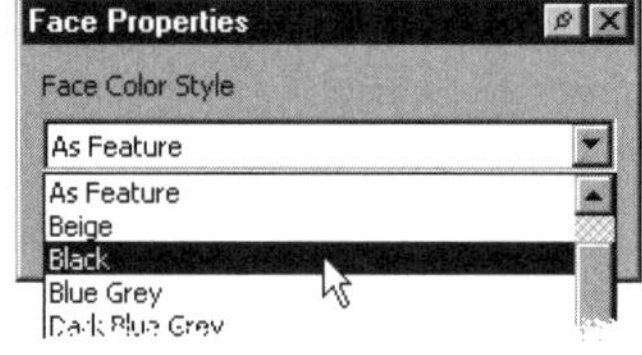


Figure 8-89

4 Click OK in the Face Properties dialog box. All the selected faces will be updated to reflect the new color style, as shown in the following image.

**NOTE** If a feature has a face color that has been overridden, and that feature is patterned, the overridden face color is not reflected in the patterned instances. You will need to also modify the face colors of the pattern.

**Figure 8-90**

## DECALS

You may not always be able to create features that fully describe your design. Raster images or data from a Microsoft Excel or Microsoft Word file may prove to be cumbersome to create using typical 3D modeling techniques. Autodesk Inventor provides the decal tool to make this task easier.

In this section, you learn the two steps that are required to create a decal. First you place an image onto a sketch and then create a decal feature.

### Images In A Sketch

When placing an image onto a sketch, use one of the following file formats: BMP, CALS1, XLS, FLIC, GEOSPOT, GIF, IG4, IGS, JPG, PCX, PCT, PNG, RLC, TGA, TIF, or DOC.

The Microsoft Excel or Microsoft Word data is converted to a raster image. Once this image is placed onto a sketch, it can be rotated, proportionately scaled, or mirrored, and its background can be set to transparent. If a Microsoft Excel file is inserted, the printed area will be used. If a Microsoft Word file is inserted, the contents of the page width are used.

To place a bitmap or the contents of a Microsoft Excel or Microsoft Word file, follow these steps:

1 Make active the sketch on which the image will be placed.

2 Click the Insert Image tool from the 2D Sketch Panel Bar, as shown in the following image.

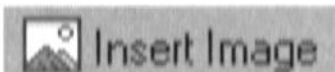


**Figure 8-91**

3 Locate the file that will be inserted from the Open dialog box. If needed, change the file type to Bitmap, Microsoft Excel, or Microsoft Word, as shown

in the following image. To maintain a link to the file, select the Link box in the lower left corner of the Open dialog box. If left unchecked, the image is embedded into the file.

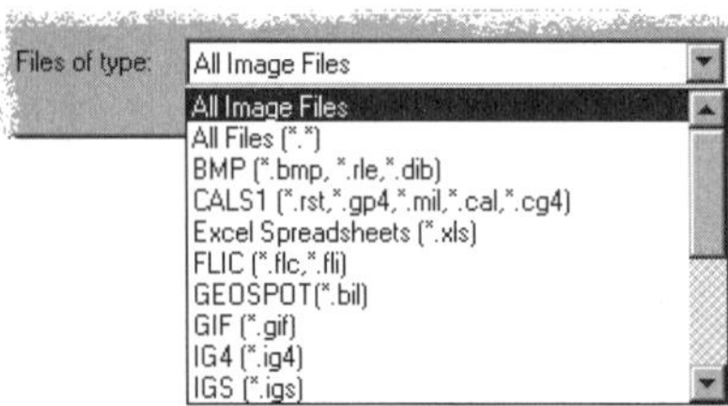


Figure 8-92

4 Place the contents onto the active sketch. A coincident constraint will be added if the cursor is moved over another point. The following image shows a bitmap being inserted on the upper left corner with a coincident constraint being applied.

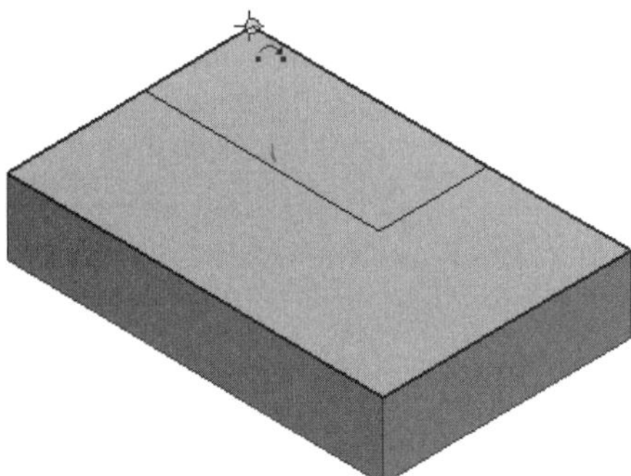

Figure 8-93

5 The inserted image can be edited by right-clicking on its name in the Browser and clicking Properties, as shown on the left in the following image. The Image Properties dialog box is displayed on the right. You can rotate or mirror the image using the rotate or mirror tool in the Image Properties dialog box. Click the Use Mask option to make the image background transparent.

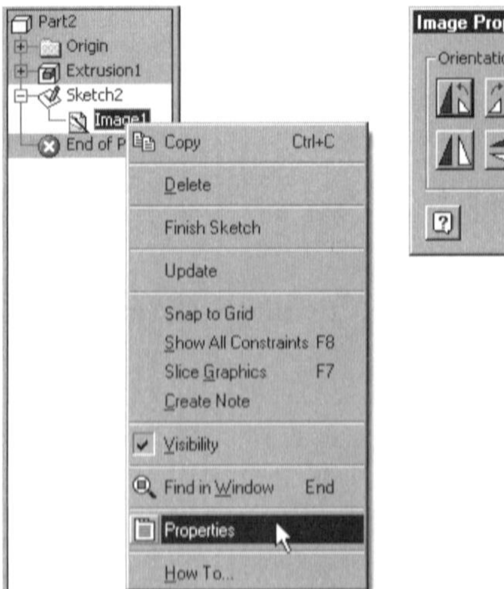


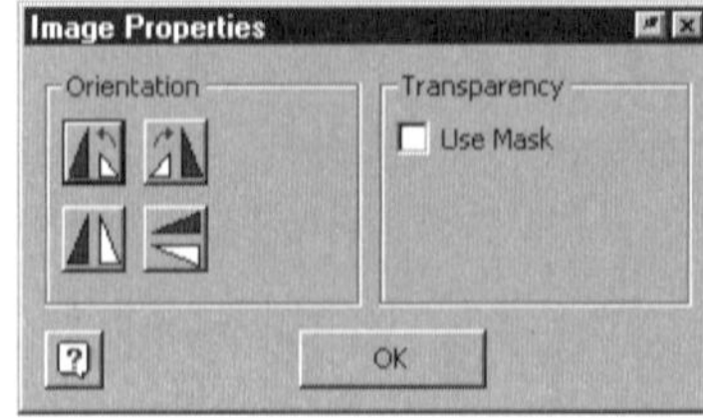


**Figure 8-94**

6 The image can be constrained and dimensioned to the bounding rectangle of the image.

7 You can resize the image by clicking and dragging a handle.

The following image shows an example of a bitmap image representing a nameplate that was placed onto an extruded rectangle.

**Figure 8-95**

## Creating a Decal Feature

After placing an image onto a sketch, the last step is to use the image to create a decal feature. A decal feature will only appear while in the Shaded Display or Hidden Edge mode in the part or assembly environment, and only in a shaded view in the drawing environment. Once the decal feature is created, other features can be added to the part. Additional features can either add or remove material from the decal. To create a decal feature, follow these steps:

1 Click the Decal tool from the Part Features Panel Bar, as shown on the left in the following image.

2 The Decal dialog box will display, as shown on the right in the following image.

3 Click the Image button on the Decal dialog box and then click the image that will be used to create the decal feature.

4 Click the Face button and then click a face in the graphics window to which to apply the decal.

5 Click OK to create the decal feature.

6 To edit the decal feature, right-click on the decal feature in the Browser and select Edit Feature from the menu. The Decal dialog box that you used to create the decal will appear.

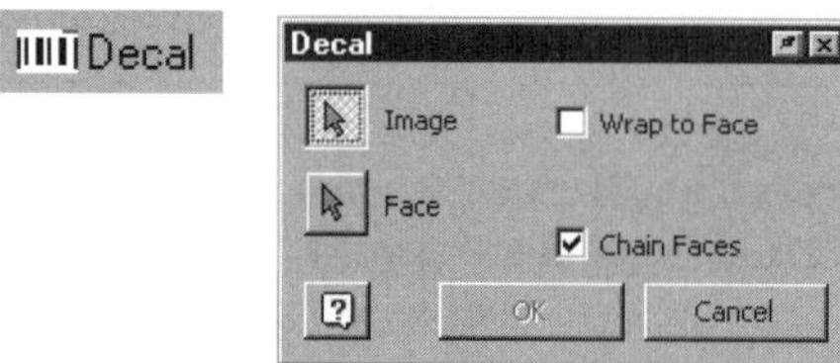


Figure 8-96

The options in the Decal dialog box are described next.

**Image** Select this button to choose the image that will be used as a decal.

**Face** Select this button and then select a face to which the decal will be applied.

**Wrap** Click in the box to wrap the image around a curved face, as shown in the following image.

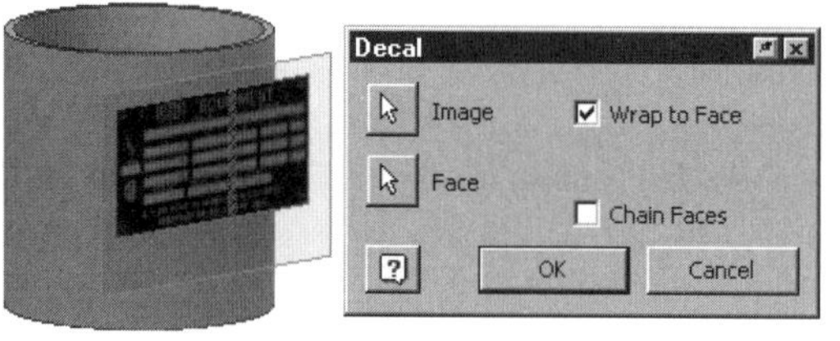


Image Shown on a Work Plane with the Decal Dialog Box's Wrap Option Checked

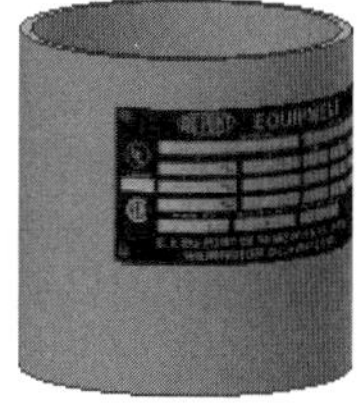

Completed Wrapped Decal Feature

Figure 8-97

**Chain Faces** Check this box if the decal will be applied over adjacent faces that have a gap.

Decal Feature with Two Holes Features

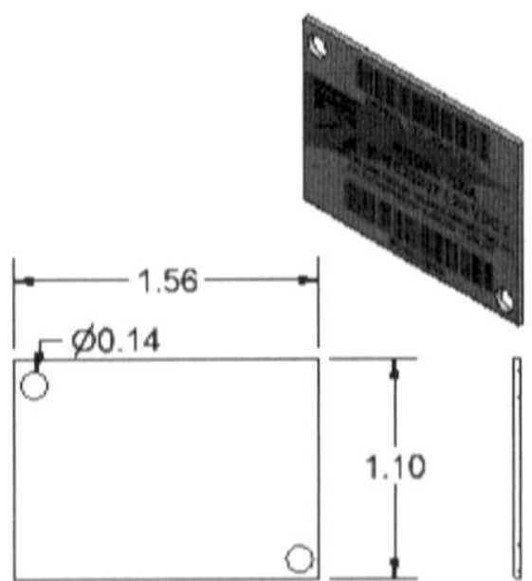


Part with Decal Feature Shown in Drawing Views, the Decal Only shows in the Shaded View

**Figure 8-98**

## PART COLORS

The color of a part can be changed by selecting a color from the Color drop-down list on the Standard toolbar, as shown in the following image. This color will not affect any material that was applied to it in the Properties dialog box.

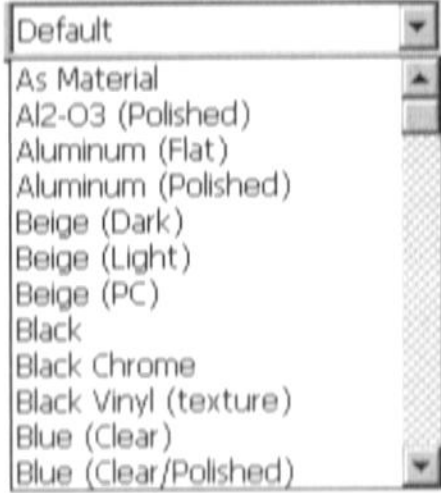


**Figure 8-99**

## TEXTURE MAPPING

Autodesk Inventor provides the ability to define surface finishes with Texture and Opacity Mapping. Textures can be defined by image files using the standard Windows *.bmp* file format. A 16-color, 256-color, or TrueColor *.bmp* file may be used. A default texture library is installed with Autodesk Inventor and can be used in conjunction with user-defined textures in your designs. The default texture library is located in the following directory:

Program Files\Autodesk\Inventor (version number)\Textures\Surfaces

User-defined textures should be located in project directories alongside the parts and assemblies that use them, or in special library project locations. Sample materials are shown in the following image:

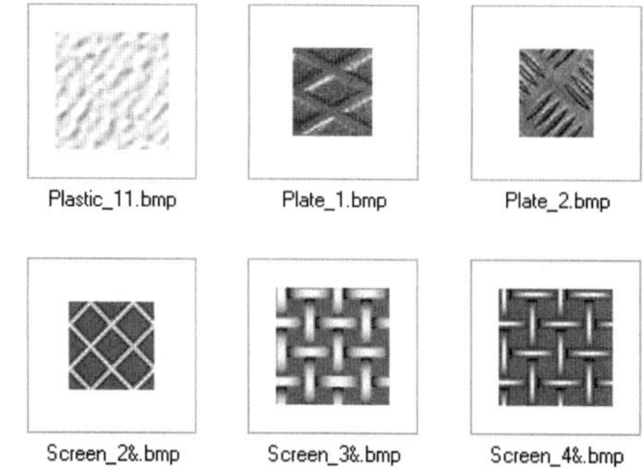


**Figure 8-100**

Textured images whose names end in an ampersand (&) are treated as perforated textures. This means that a particular color will define *voids* in the texture. Perforated textures are defined by making a particular color opaque within the *.bmp* file. The opaque color is defined by the lower right-most pixel of the image. Use perforated textures to simulate materials like grills and screens without having to model all of the voids. The default texture library that Autodesk Inventor supplies uses magenta as the color to define voids. The texture color combines with the part color to produce the finished texture. If you have problems with a texture displaying correctly, set the color's Diffuse, Emissive, Specular, and Ambient properties to white.

To assign and use textures, follow these steps:

1. Click the Colors option from the Format menu and the Colors dialog box will appear, as shown in the following image.
2. If a part is active, its color will appear in the Style Name area. This is the material that will be modified.
3. If you want to create a copy of a color, click the New button, rename and save the color, and the color will be added to the Color Style drop-down list.

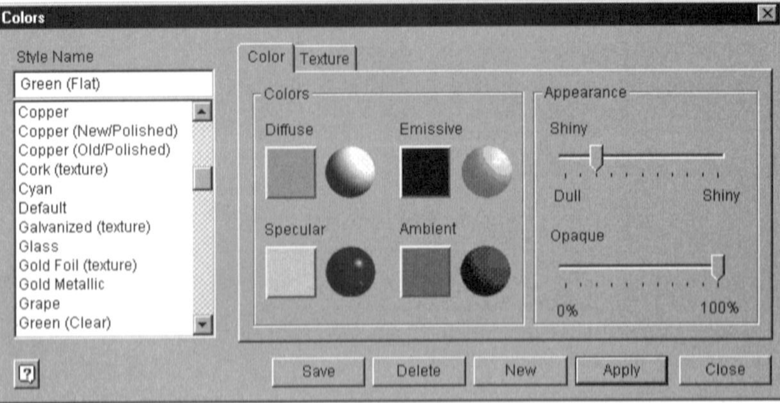


**Figure 8-101**

4 Click the Texture tab from the Colors dialog box, as shown in the following image. The Texture tab shows the currently selected texture, if any, and the scale and rotation settings that control the appearance of the texture.

**% Scale** Scales the texture in a range between 25 to 400. The default value is 100.

**Rotation** Rotates the texture between -180° and +180°. The default value is 0.

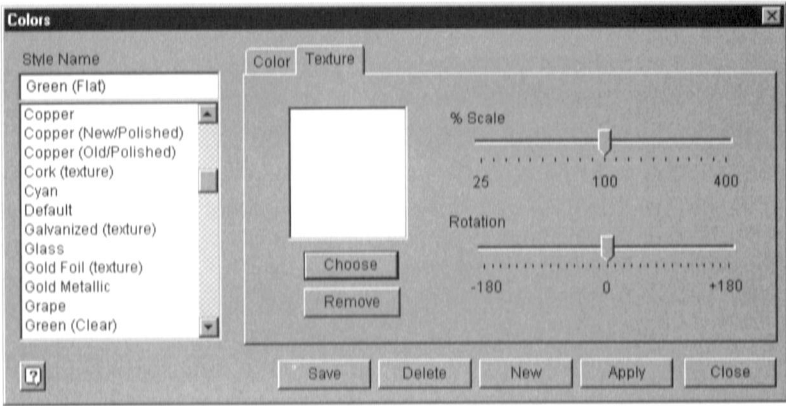


**Figure 8-102**

5 Select a texture by clicking the Choose button. The Texture Chooser dialog box is displayed, as shown in the following image. The currently selected texture is displayed in the area on the right side of the dialog box.

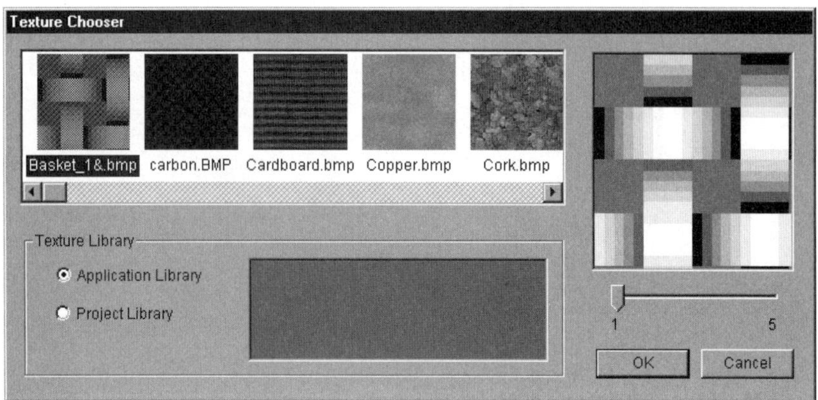


Figure 8-103

6 Use the preview slider on the right side of the dialog box to visually adjust the tiling characteristic (alignment of the pattern) from a value of 1 to 5.

7 In the bottom-left corner of the Texture Chooser dialog box is the Texture Library area. In this area, select an option to choose where the texture will be saved.

You can also specify the X, Y, or Z direction for any texture and how it is mapped based on the & character and the appropriate letter being placed after it (i.e., *FILENAME&X.bmp*).

To make a texture reflective add the & and V after the FILENAME (i.e., *FILENAME&V.bmp*).

**Application Library** This is the default library that provides access to the textures that are installed with Autodesk Inventor.

**Project Library** This option allows you to define a texture library specific to the active project. There are several advantages to using a project-specific texture library: the textures are archived with the project, which eliminates the chance that a future release of Autodesk Inventor will change the look of the model by either changing or removing a texture. This feature also provides the ability to define project-specific textures.

1 After selecting and adjusting the texture, click the OK button.

2 The Colors dialog box will appear with the Texture tab active. Click the apply button to preview the texture.

3 Make changes to the scale, rotation angle, or tile matching, as needed.

4 Click the Close button to complete the operation.

5 To remove a texture from a color, click the Colors option from the Format menu. From the Texture tab of the Colors dialog box, first click the Remove button and then the Close button.

The following images show an example of a part with and without a texture applied to it.

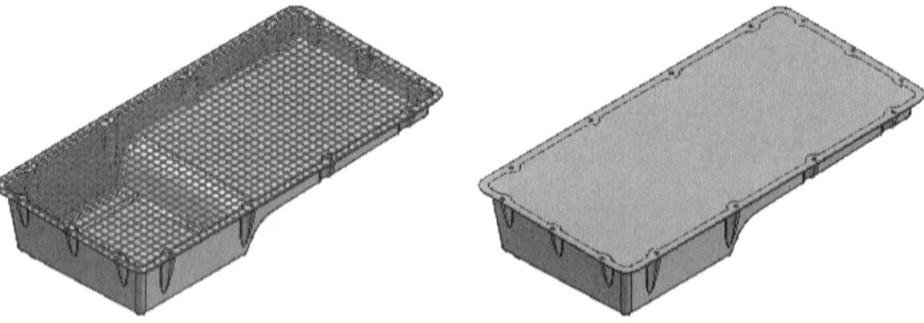

A Screen Texture Applied to the Top Plate    No Texture Applied to the Top Plate

**Figure 8-104**

### Texture Visibility

If the texture does not appear on the part, click Tools> Application Options and click the Show Reflections and Textures option on the Colors tab of the Options dialog box, as shown in the following image:

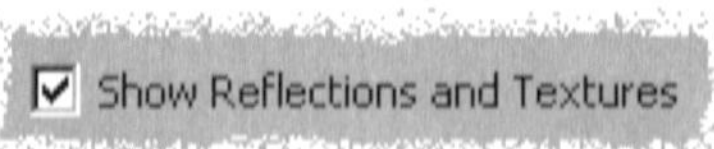


**Figure 8-105**

### Resolving the Texture

If a texture file being referenced in a part does not resolve to a usable texture file, the object is displayed with a special texture that contains the pattern ??, as shown in the following image. This makes it easily distinguishable from legitimate system- and user-supplied textures.

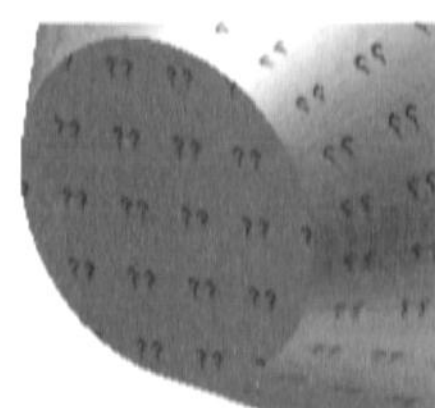

**Figure 8-106**

## PART MATERIALS

To change the part's physical properties and appearance to a specific material, click the iProperties option from the File menu or right-click on the part's name in the Browser and select a material from the material drop-down list from the Physical tab, as shown on the left in the following image. As you learned in the File properties section in this chapter, the properties will reflect the attributes of this material. Click the OK button to complete the operation, and the color of the part should change to reflect the material. If the material was not updated on the part, click the As Material option from the Color area of the Standard toolbar, as shown on the right in the following image. If a color is selected from the Color drop-down list on the Standard toolbar, it will only change the appearance of the part and will not change the physical material or properties of the part.

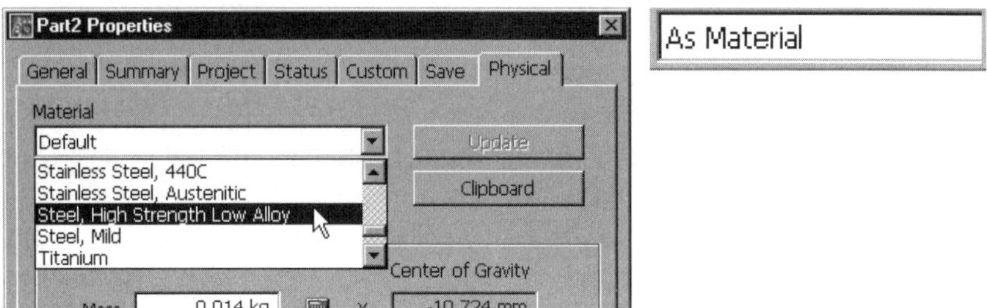


Figure 8-107

## SHADOWS

To give your model(s) a realistic look, or to get a better orientation of the model(s), you can choose to turn on shadows that will be displayed on a plane below the model. When shadows are displayed, they will dynamically update as the model(s) is rotated. From the Standard toolbar, there are three choices: No Shadow, Shadow, and X-Ray Shadow, as shown in the following image.

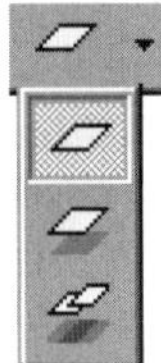

Figure 8-108

**No Shadow** This is the default setting and no shadow is displayed below the model when it is selected.

**Shadow** When this feature is selected, a shadow is displayed on a plane below the model, as shown on the left in the following image. Work geometry, sketches, origin indicators, engineer's notes, and trails in presentation documents do not cast shadows.

**X-Ray Shadow** When this feature is selected, a shadow of the model and all features within the model are cast. Internal features are displayed as a darker shadow, as shown on the right in the following image.

Model Shown with
Shadows Displayed

Model Shown with
X-Ray Shadows Displayed

Figure 8-109

## PERSPECTIVE VIEW

To change the camera view from the default orthographic view to a perspective view, click the Perspective Camera icon from the Standard toolbar, as shown in the following image. The camera position and lens focal length for the perspective view can be adjusted by holding down the SHIFT + CTRL + F3 keys and then either moving the cursor with the left mouse button depressed or spinning the wheel on a wheel mouse. It is recommended that you change back to an orthographic view while working, however. This will help prevent you from seeing the model incorrectly. The following example shows a part in both orthographic camera and perspective camera views.

Perspective
Camera Tool

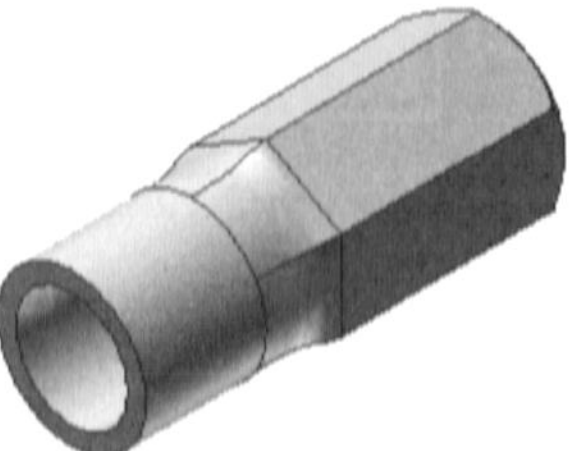

Orthographic Camera View

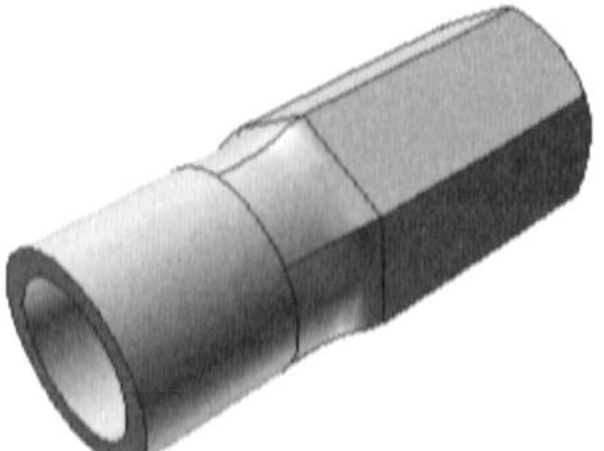

Perspective Camera View

Figure 8-110

## BACKGROUND IMAGE

You can also change the background of the graphics window from a color to an image. The background image needs to be a bitmap and should be located in the following directory: Autodesk\Inventor (version number)\Backgrounds. Once the image has been placed in this directory, you should change Autodesk Inventor's background image by clicking Tools> Application Options and change the Background option to Background Image from the drop-down list on the Colors tab of the Options dialog box, as shown on the left in the following image. From the File name area, you then click the

Browse tool as shown on the right in the following image. The open dialog will be displayed where you can select the image you want to use. Click the OK button to place the image as your background.

**NOTE** The Show Reflections and Textures option on the Colors tab of the Options dialog box must be checked in order for the background image to appear.

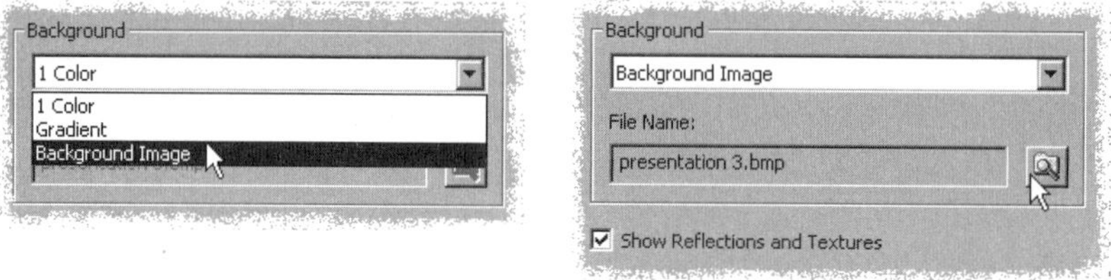


Figure 8-111

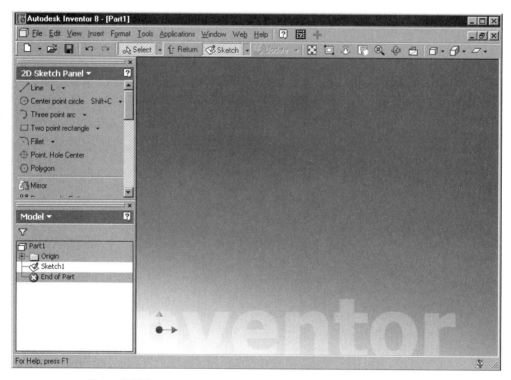

Figure 8-112

# EXERCISE 8-6 Visualization

In this exercise, you create a new material that contains texture mapping and experiment with the Perspective Camera tool. You also control face colors, use the Decal tool, and display ground shadows. To navigate to the exercise in the *Electronic Student Workbook*, do the following:

1 From the Main TOC page, click Chapter 8.

2 From the TOC for Chapter 8, click Visualization.

The following image illustrates the opened exercise.

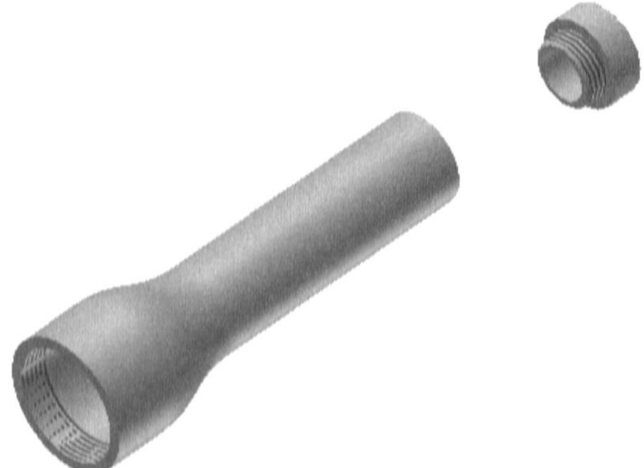

**Figure 8-113 Opened exercise**

The following image illustrates the completed exercise.

**Figure 8-114 Completed exercise**

## CHAPTER SUMMARY

| To | Do This | Tool |
|---|---|---|
| To extrude an Open Profile | Click the Extrude tool, then click the Open Profile and click on the side of the part that will be filled in. | Extrude +E |
| To create a rib or web | Use the Rib tool and select an open profile. | Rib |
| To change the extrusion termination | Start the Extrude tool, click the More tab, and then click the Minimum Solution box. | Extrude +E |
| Place text on a sketch | Click the Create Text tool from the 2D Sketch Panel Bar. | Create Text |
| Emboss text | Click the Emboss tool from the Part Features Panel Bar. | Emboss |
| Create a sweep feature | Click the Sweep tool from the Part Features Panel Bar. | Sweep Shift+S |
| Create a coil feature | Click the Coil tool from the Part Features Panel Bar. | Coil |
| Create a loft feature | Click the Loft tool from the Part Features Panel Bar. | Loft Shift+L |
| Split a face or part | Click the Split tool from the Part Features Panel Bar. | Split |
| Copy a feature | Right-click on a feature's name in the Browser and click Copy from the menu and then Paste the feature. | |
| Reorder a feature | Click on the feature's name in the Browser and, with the left mouse button depressed, drag the feature to the desired location. | |
| Mirror a feature | Click the Mirror Feature tool from the Part Features Panel Bar. | Mirror Feature Shift+M |
| Suppress a feature | Right-click on the feature's name in the Browser and click Suppress Features from the menu. | |
| Rollback features | Click the End of Part marker in the Browser and, with the left mouse button depressed, drag the End of Part marker to the new location. | |
| Adjust a file's properties | Click the iProperties option from the File menu. | |
| View the center of gravity of a part or assembly | Click Center of Gravity from the View menu. | |
| To create a decal feature | Click the Insert Image tool from the 2D Sketch Panel Bar and place an image, then click the Decal tool from the Part Features Panel Bar to make the image a decal feature. | Insert Image<br>Decal |

# APPLYING YOUR SKILLS

## Skill Exercise 8-1

In this exercise, you use complex part modeling techniques to create a housing for an electronic device. To navigate to the exercise in the *Electronic Student Workbook*, do the following:

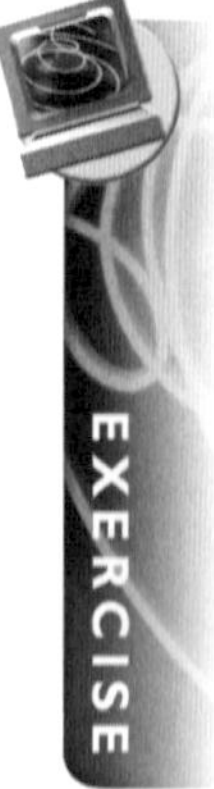


1 From the Main TOC page, click Chapter 8.
2 From the TOC for Chapter 8, click Exercise 1 under Applying Your Skills.

The following image illustrates the completed exercise.

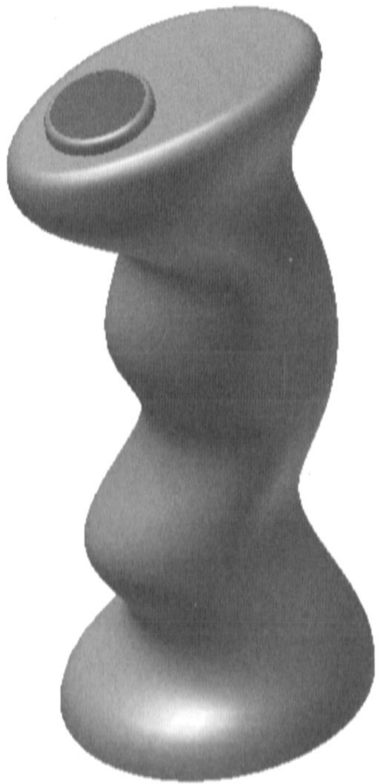

**Figure 8-115 Completed exercise**

## CHECKING YOUR SKILLS

Use these questions to test your knowledge of the material covered in this chapter.

1 **True___ False___** When creating a single rib or a web feature, only a closed profile can be selected as the profile.

2 **True___ False___** Both the Extrude and Revolve tool can use the minimum or maximum extrusion solution.

3 **True___ False___** Embossed text can only be placed on a planar face.

4 **True___ False___** A sweep feature requires three unconsumed sketches.

5 **True___ False___** A 3D curve can be created with a combination of both 2D and 3D curves.

6 Explain how to create a 3D path using geometry that intersects with a part.

________________________________________

________________________________________

________________________________________

7 **True___ False___** The easiest way to create a helical feature is to create a 3D path and then sweep a profile along this path.

8 **True___ False___** You can control the twisting of profiles in a loft by defining point sets.

9 Explain how to save both halves of a part after splitting it.

________________________________________

________________________________________

________________________________________

10 **True___ False___** Features can be copied between parts using the Copy Feature tool from the Part Features Panel Bar or Toolbar.

**11** Explain the difference between suppressing and deleting a feature.

_______________________________________________

_______________________________________________

_______________________________________________

**12** **True___ False___** After mirroring a feature, the mirrored feature is independent from the parent feature. If the parent feature changes, the mirrored feature will not reflect this change.

**13** Explain why you would want to override a part's mass and volume properties.

_______________________________________________

_______________________________________________

_______________________________________________

**14** **True___ False___** When creating a decal feature, the image must be a *.jpg* file.

**15** **True___ False___** After changing a part's physical material properties, the part's color in the graphics window will change to match the material.

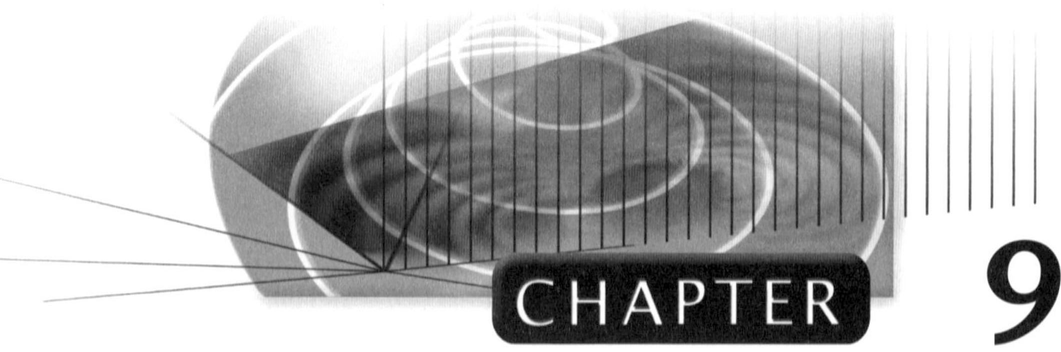


# Complex Drawing View Creation and Editing

In a previous chapter, you learned how to create base and projected views. This chapter will expand the creation of views to include, broken, break-out, draft, and perspective views. You will learn techniques for showing work features and model sketches in drawing views. You will also learn advanced dimensioning techniques, which include the creation of dual, baseline, and ordinate dimensions. The topic of hole tables will be introduced as a method used to create a table that consists of *X*- and *Y*-axis coordinate hole locations. You will learn how to create and insert sketched symbols into your drawings. Drawing edges will be projected and used for specialty situations. Advanced features of using parts lists will also be covered, as will the creation and application of revision blocks and tags.

## CHAPTER OBJECTIVES

**After completing this chapter, you will be able to**

- Create broken views
- Create break-out views
- Create perspective views
- Show and reference work features in drawing views
- Use sketches in drawing views
- Create dual, ordinate, and auto-baseline dimensions
- Create hole tables
- Create and use sketched symbols
- Understand all tools associated with the Parts List dialog box
- Create a revision block

## CREATING DRAWING VIEWS

Initially, a base drawing view is created followed by projected views. Additional views may need to be created to better document an object. These additional views include broken, break-out, and perspective views. These and draft views will be discussed in greater detail in this chapter.

### BROKEN VIEWS

When creating drawing views of long parts, you may want to remove a section or multiple sections from the middle of the part and show just the ends—this type of view is referred to as a *broken* view. You may, for example, want to create a drawing view of a 50 x 50 x 6 angle in the following image that is 1500 mm long and only has the ends chamfered. When a drawing view is created, the detail of the ends is small and difficult to see because the part is so long.

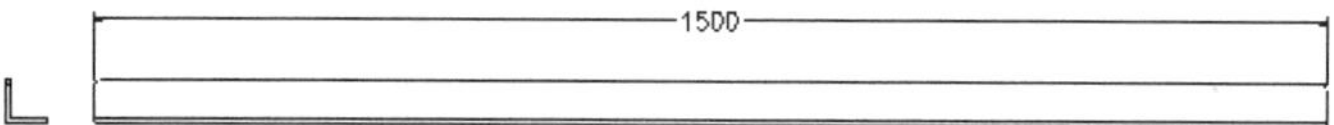


**Figure 9-1**

In this case, you can create a broken view that removes a 1000 mm section from the middle of the angle and leaves 250 mm on each end. When an overall length dimension is placed that spans over the break, it is displayed as 1500 mm and the dimension has a break symbol to note that it is derived from a broken view (see the following image).

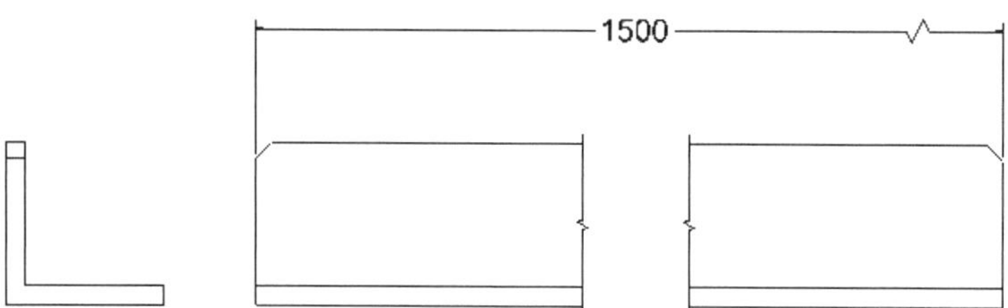


**Figure 9-2**

A broken view is created by adding breaks to an existing drawing view. The view types that can be changed into broken views are: part views, assembly views, projected views, isometric views, auxiliary views, section views, breakout views, and detail views. After creating a broken view, the breaks can be moved dynamically to change what is seen in the broken view.

To create a broken drawing view, follow these steps:

1 First, create a base view (one that will eventually be shown as broken).

2 Select the Broken View tool from the Drawing Views Panel Bar, as shown in the following image. You can also right-click inside the bounding area of an existing view box (shown as a dotted rectangle when the mouse is moved into the view). Click Create View>Broken from the menu.

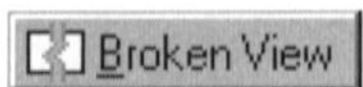


Figure 9-3

3 If the Broken View tool was selected from the Drawing Views Panel Bar, click inside the view that the broken view will be created from.

4 The Broken View dialog box will appear, as shown in the following image. Do not click the OK button at this time, as this will end the operation.

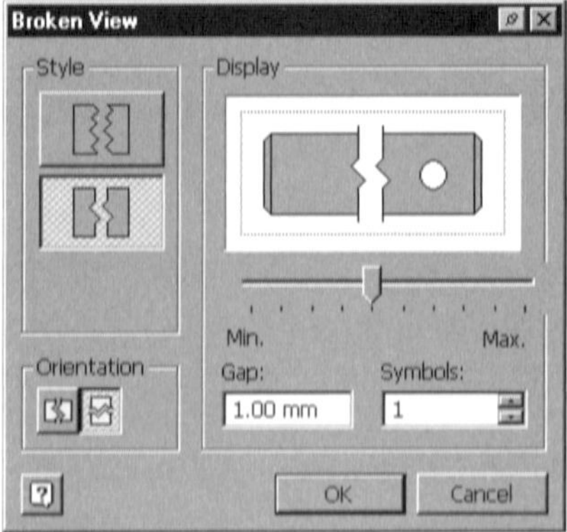


Figure 9-4

5 In the drawing view that will be broken, select a point where the break will begin (see the following image).

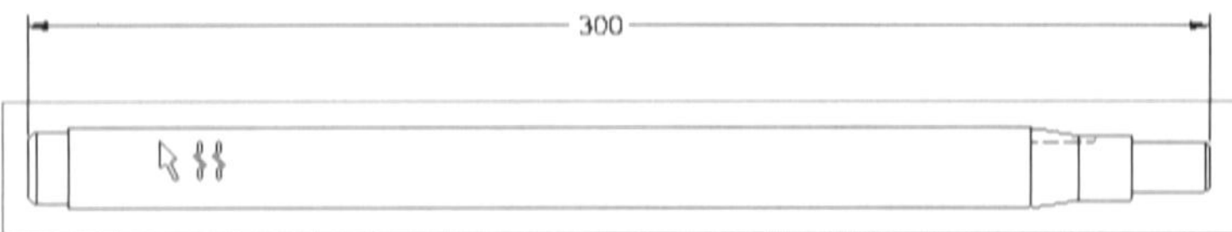


Figure 9-5

6 Then select a second point to locate the second break, as shown in the following image. As you move the mouse, a preview image will appear showing the placement of the second break line.

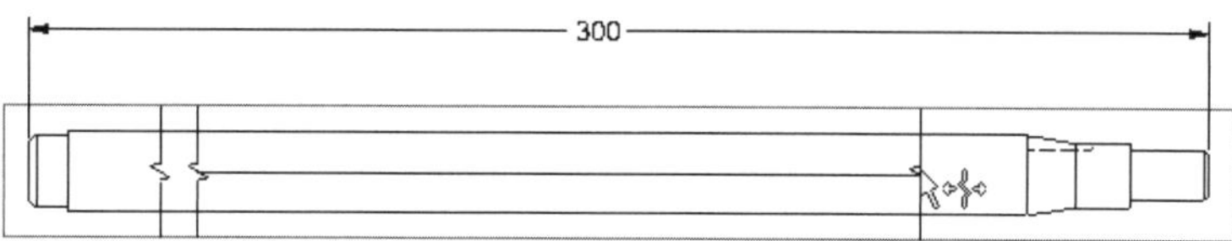


Figure 9-6

7 The results of creating a broken view are illustrated in the following image. Notice the dimension is displayed with a break symbol to signify the view is broken.

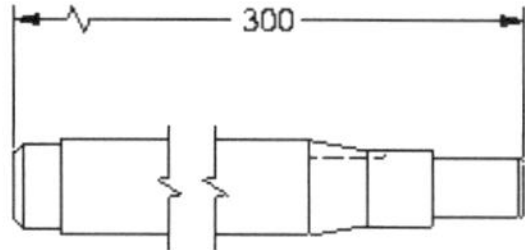


Figure 9-7

8 To edit the properties of the broken view, move the mouse over the break lines and a green circle will appear in the middle of the break. Right-click and select Edit Break from the menu. The same Broken View dialog box will appear. Edit the data as needed and click the OK button to complete the edit.

9 To move the break lines, click on one of the break lines and, with the mouse button pressed down, drag the break line to a new location, as shown in the following image. The other break line will follow to maintain the gap size.

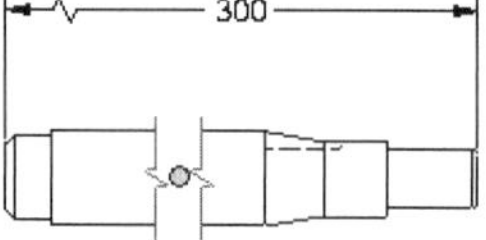


Figure 9-8

### BREAK-OUT VIEWS

For cases when you want to expose internal components or features, a method in Autodesk Inventor is available that allows you to cut or peel away a body and expose those internal parts. The method just described is called a *break-out view.* It is not unusual for assemblies to have housings or covers that hide internal components. Break-out views make it possible to expose these components. You can create break-out views on assemblies as well as part files.

When performing break-out views, you will need to be able to define two items: a closed profile boundary over the area to break, and the depth of material to remove or portion of component to remove in order to see other components. As with broken views, the break-out view is defined and displayed on the same view.

To create a break-out view, choose the Break-Out View tool from the Drawing Views Panel Bar, as shown in the following image.

**Figure 9-9**

After selecting the view in which the break out will occur, the Break-Out View dialog box will display, as shown in the following image. The termination options of From Point, To Sketch, To Hole, and Through Part are available to give you control over how the break-out section is created.

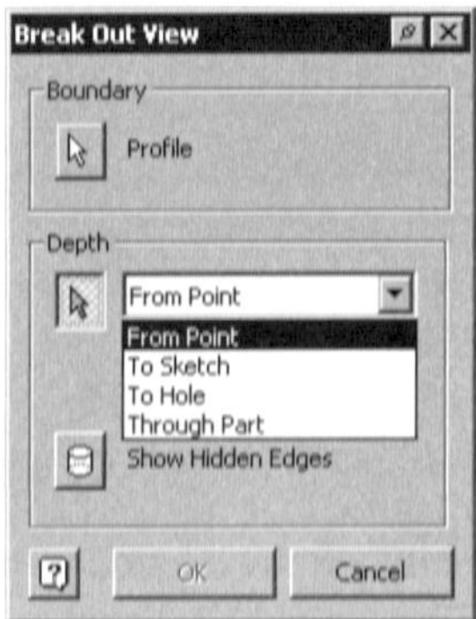


**Figure 9-10**

All four options are explained as follows:

**From Point** The break out will occur at a specified distance from a point located in a view. The point could be located in the base view or in a projected view.

**To Sketch** This option uses sketched geometry associated with another view to define the depth of the break out.

**To Hole** The break out will be based on a hole feature, whose axis determines the termination depth of the break.

**Through Part** Selected components located in the Browser will have drawing content removed from inside a break profile. This termination option is especially useful when you want to reveal internal components or features.

### Creating a Break-Out View Using the From Point Option

Creation of the break-out view using the From Point option consists of the following steps:

1. Click on the view in which the break-out section will be created.
2. Click on the Sketch button on the Standard toolbar and begin sketching a closed profile over the area you want broken out. When done, right-click and select Finish Sketch from the menu.

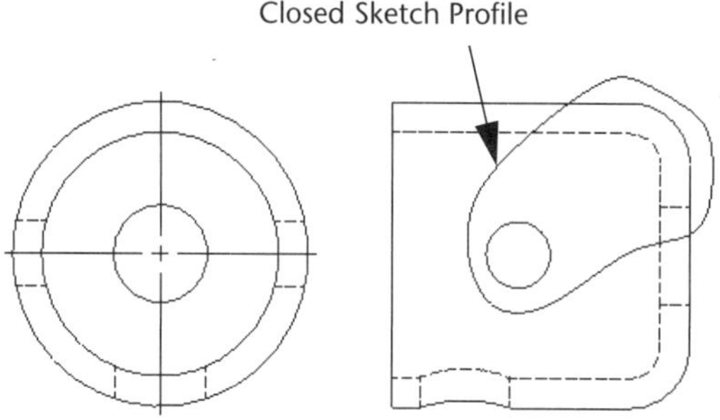


**Figure 9-11**

3. Click the Break-Out View button; this will activate the Break-Out View dialog box. Select the view in which the break out will occur and select the defined boundary. Select the sketch you just created as the boundary in the view that will contain the break-out.
4. Click on the From Point option in the dialog box, then click on the Depth arrow and select a point in the adjacent projected view. This point will be used to calculate the depth of cut, based on the distance (see the following image).

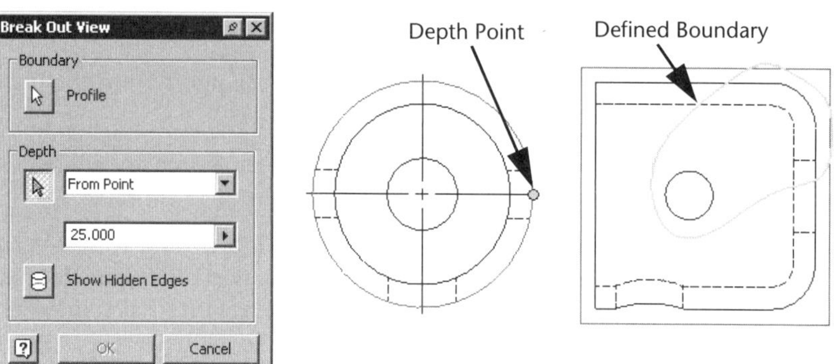


**Figure 9-12**

5 The results of using the From Point option for creating a break out are illustrated in the following image. The depth of 25 units cuts the view at the middle of the circular features, thus displaying the wall thickness of the part.

**NOTE** You can also select a point at the center of the circle for the depth of the cut. In this case, the distance value would change to 0 because the depth of the cut was specified by a point.

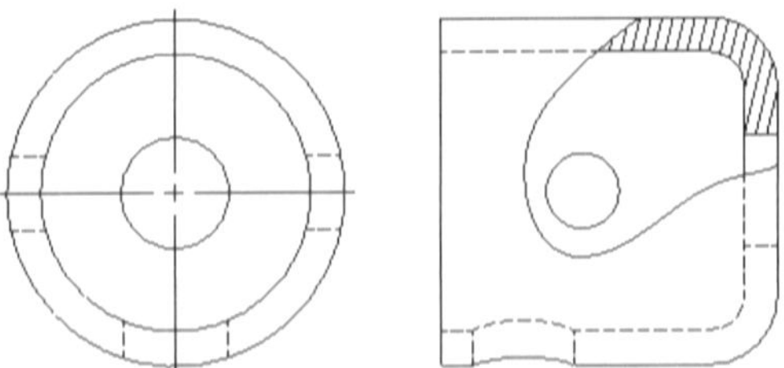

Figure 9-13

## Creating a Break-Out View Using the To Sketch Option

Creating a break-out view using the To Sketch option consists of the following steps:

1 Activate the view in which the break out will occur and sketch a closed profile over the area you want broken out. In the following image, two closed profiles will be used to create the break-out view. When finished with this operation, right-click and select Finish Sketch.

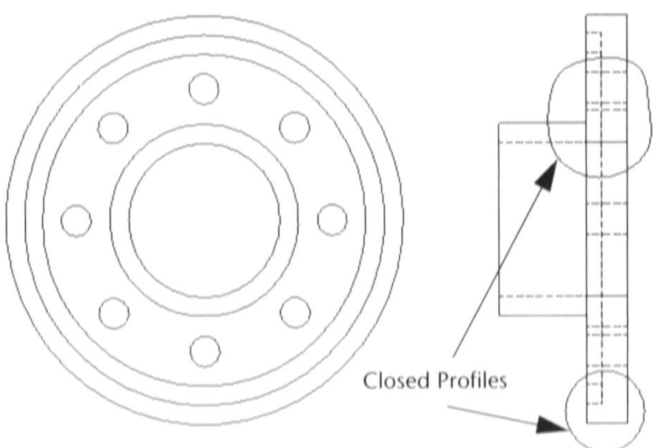


Figure 9-14

2 Activate the adjacent projected view and create another sketch. This sketch will be used to determine the depth of cut for the break out. When finished with this operation, right-click and select Finish Sketch.

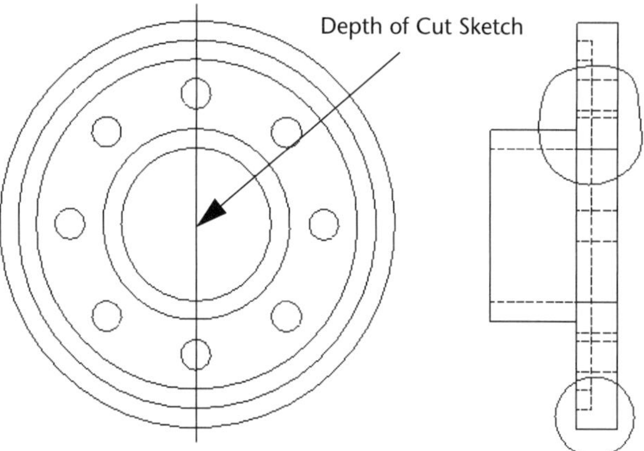


Figure 9-15

3 Select the Break-Out View tool and select a view as the base for the break-out view.

4 When the Break-Out View dialog box appears, select the initial sketch or sketches to use as the boundary profile. In the Depth area of the dialog box, change the option to Sketch. In the adjacent projected view, select the sketch that will determine the depth of cut for creating the break-out view (see the following image).

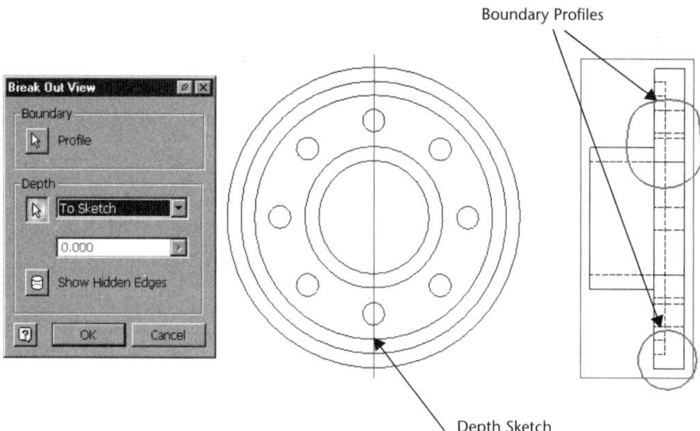


Figure 9-16

5 The break-out view is created in the following image based on the first group of sketched boundaries and the second sketch, which determined the depth of cut.

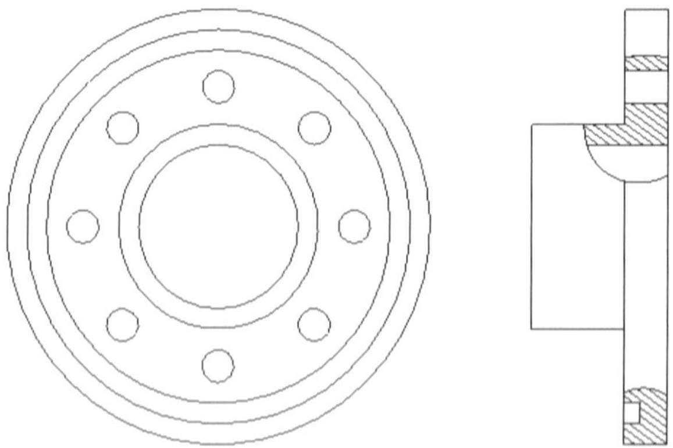

Figure 9-17

### Creating a Break-Out View Using the Hole Option

The Hole option of creating break-out views is similar to the Sketch option. Instead of basing the break out on a sketch, a hole will be used instead. It is the axis of the hole feature that determines the termination depth.

1 Click the Break-Out View button on the Drawing View toolbar.

2 In the graphics area, click to select the view. When the Break-Out View dialog box appears, click to select the defined boundary. The boundary profile must be on a sketch associated to the selected view. In the Break-Out View dialog box, click the arrow next to the Depth type box and select the To Hole option, as shown in the following image.

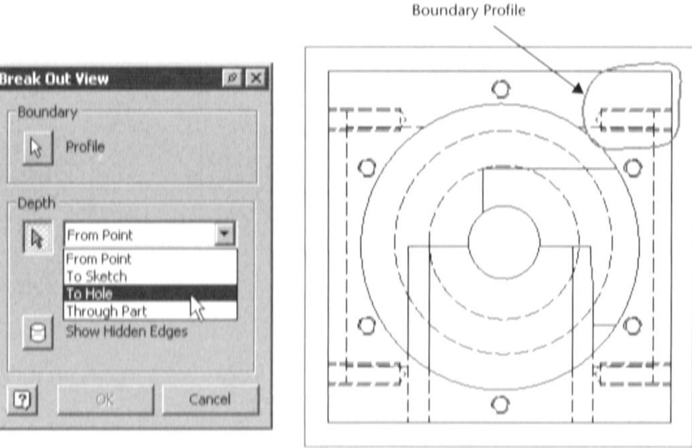


Figure 9-18

3 Click the select arrow, then click to select the hole feature in the graphic window, as shown in the following image. The depth is defined by the axis of the hole.

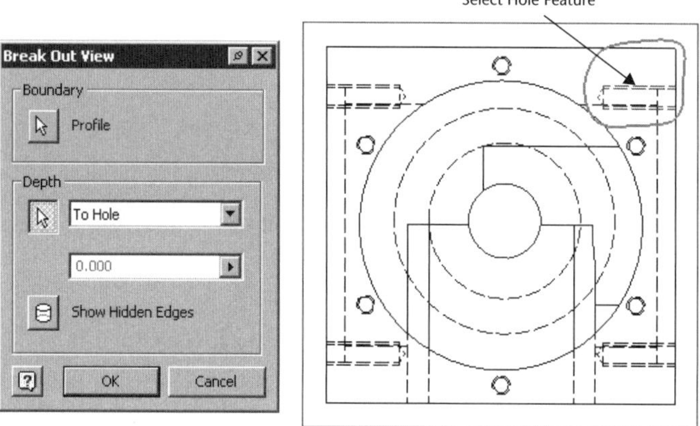


Figure 9-19

**NOTE** If the hole feature is hidden, click the Show Hidden Edges button to temporarily show it.

4 When the view is fully defined, click OK to create the view (as shown in the following image).

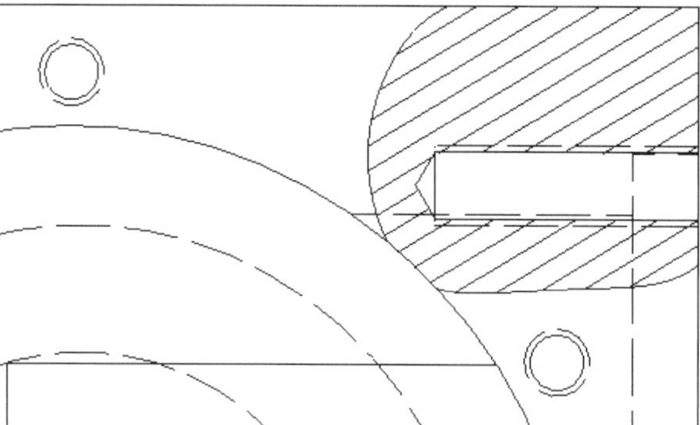

Figure 9-20

## Creating a Break-Out View Using the Through Part Option

In the following image, a housing blocks out internal details of an assembly. A segment of the housing can be cut away to expose obscured parts or features using the Through Part option of the Break-Out View tool.

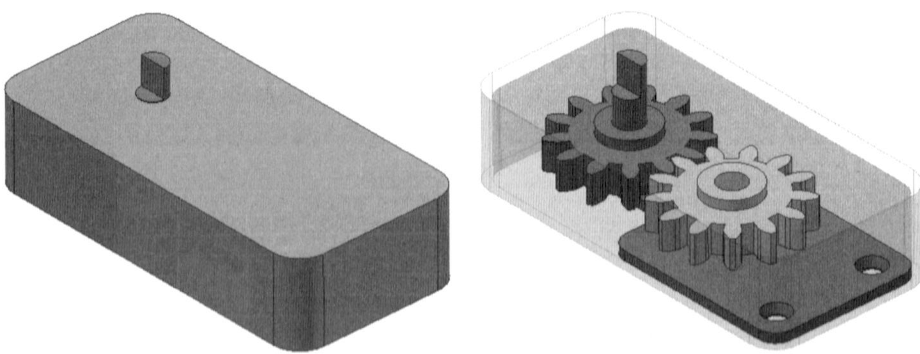

Figure 9-21

Creating a break-out view with the Through Part option consists of the following steps:

1 Activate the view you wish to cut through and click on the Sketch button in the Standard toolbar.
2 Next, sketch a closed profile (as shown in the following image) as the area you wish to cut through. When finished, right-click and select Finish Sketch.

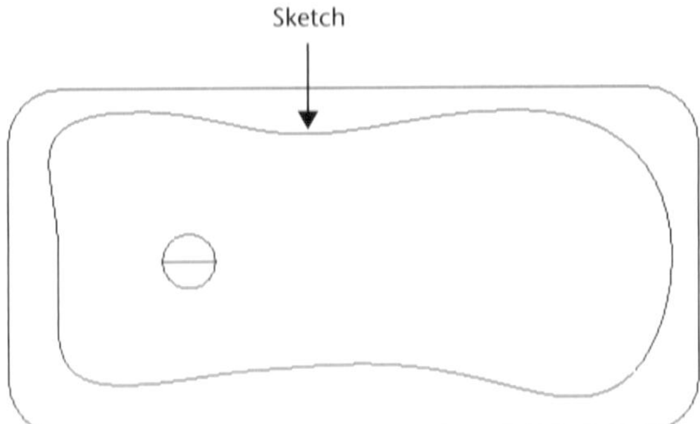


Figure 9-22

3 Click on the Break-Out View tool. Selecting the view to perform the operation on will activate the Break-Out View dialog box. The boundary profile consisting of the previous sketch will automatically select. In the Depth area of the dialog box, click on the Through Part option (see the following image).

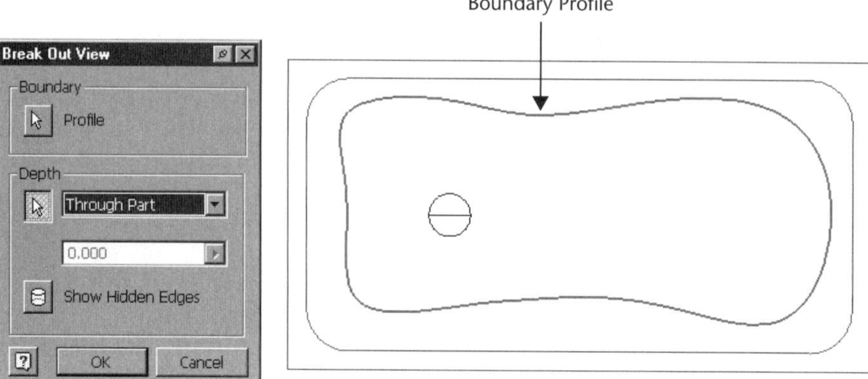


Figure 9-23

4 Move over to the Browser and select the part on which to perform the penetration operation—in this case, the Rotation_Gear Box part. When finished, click the OK button in the Break-Out View dialog box.

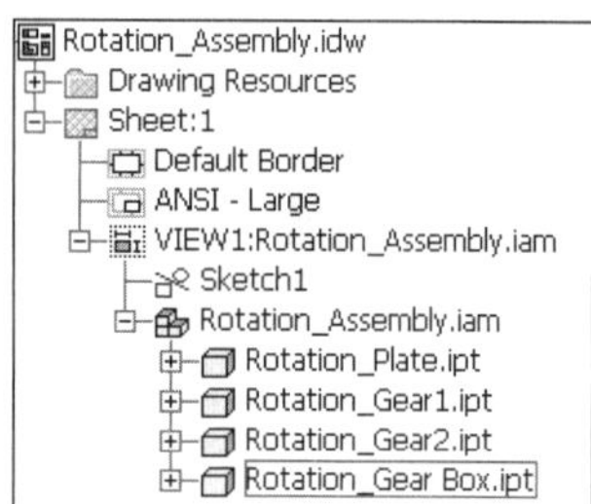


Figure 9-24

5 The results of the Through Part option are illustrated below. The gear cover has been sliced away based on the sketched boundary to expose the internal workings of the gear mechanism.

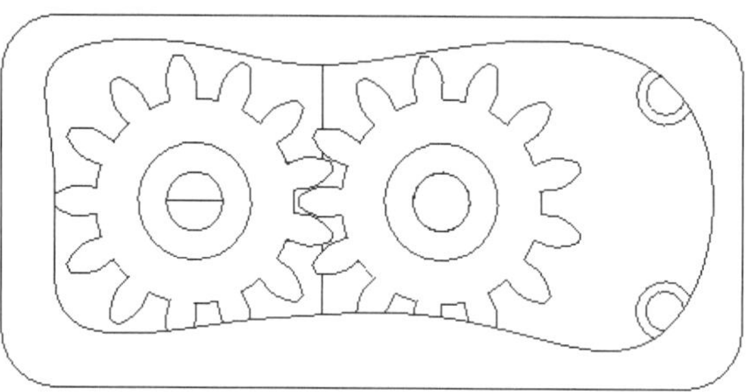

Figure 9-25

The Through Part option can be very effective when applied to an isometric view. In the following image, a spline sketch was created on an isometric view. Notice how the sketch covers the top and sides of the isometric. The completed break-out view is illustrated in the image on the right. All sides surrounded by the spline sketch will be broken out to display the internal components.

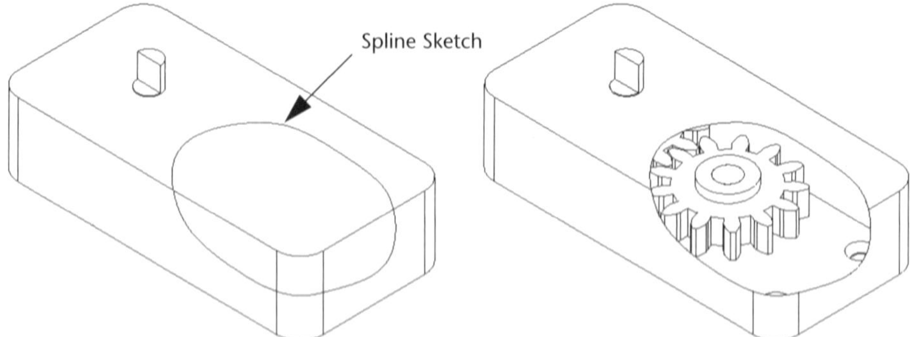


**Figure 9-26**

## DRAWING VIEW OPTIONS

Additional drawing view options are available to make your drawing views more dynamic and realistic. Autodesk Inventor has always had the ability to create high quality isometric views in drawing mode. You can also easily create perspective views in drawing mode, which will improve the realism of your drawing view designs. You can display work features such as work axes, planes, and points. These work features are used for reference purposes as a means of further documenting designs in drawing mode. You can also display unconsumed sketches in part files in drawing views for the purpose of viewing conceptual design information.

### Creating Perspective Views

In addition to creating isometric drawing views, you can create perspective drawing views based on part, assembly, presentation, and custom view information. Perspective views are used in providing a more natural or realistic view of an assembly or component.

Because perspective views have a very narrow and specific use, they are not treated as normal views because other views are not expected to be projected or developed from them. This means that when a perspective view is selected, view creation commands that require a base view will not be enabled. You will also not be able to apply dimensions to perspective views.

To create a perspective view, use the following steps:

1 While in drawing mode, click on the Base View button located in the Drawing Views Panel Bar.

2 While in the Drawing View dialog box, select the model file to be used for the perspective view. Then click on the Custom View button, as shown in the following image (this button is labeled "Change view orientation").

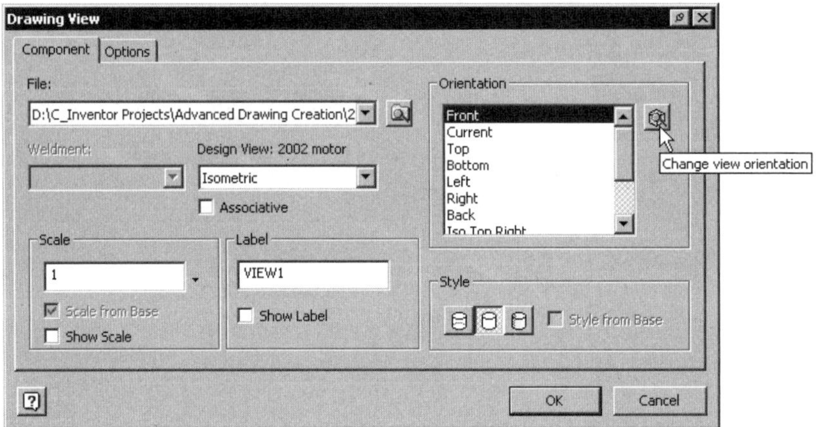


Figure 9-27

3 The Custom View window containing the current model image will appear. Use Pan, Zoom, Rotate, or other viewing tools from the Standard toolbar to position the view, as shown in the following image. The following image also illustrates the Incremental View Rotate dialog box, which is used for fine tuning the position of the view.

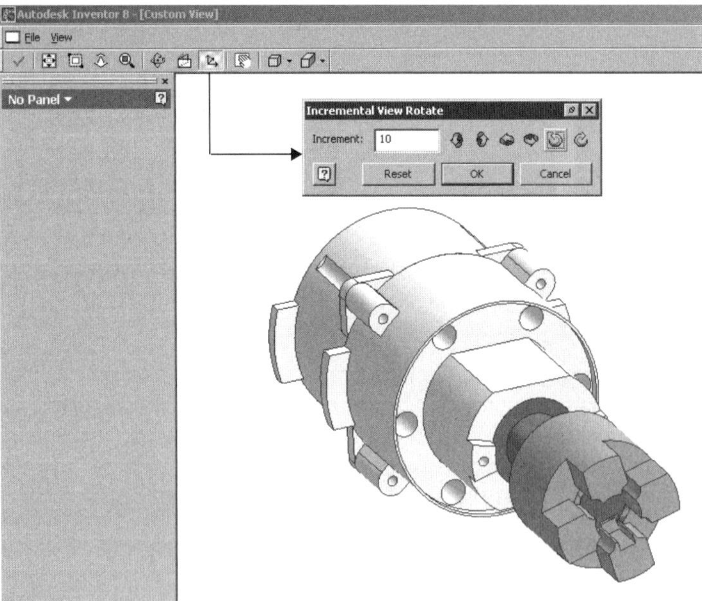


Figure 9-28

4 Use the Camera Perspective tool, as shown in the following image, to switch the view to a perspective.

**NOTE** You can resize the Custom View panel by double-clicking on its title bar.

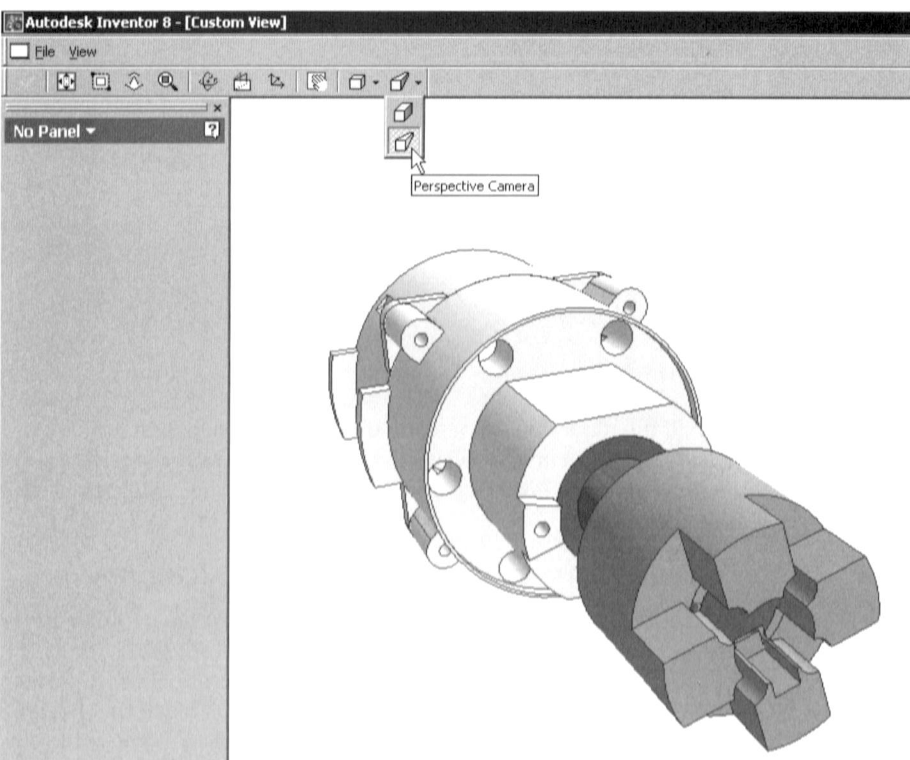


Figure 9-29

5 When you are pleased with the view displayed in the Custom View window, click on the check mark located in the Standard toolbar to accept this view and close the window, as shown in the following image. While back in the Drawing View dialog box, make any additional changes—such as scale—before placing the view inside the drawing border.

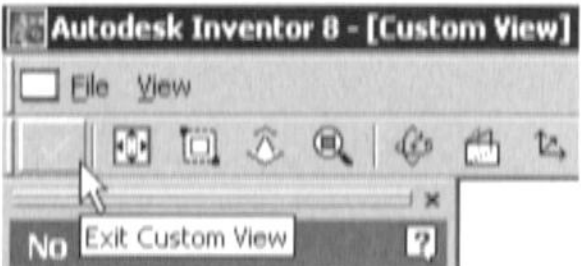


Figure 9-30

The perspective drawing view is created as shown in the following image.

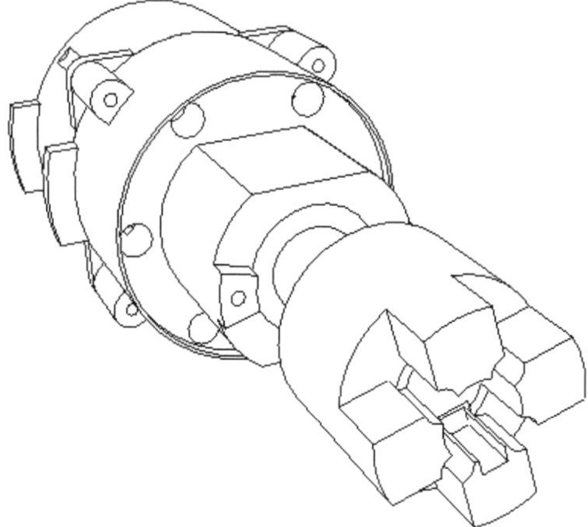

Figure 9-31

**NOTE** If the model file is open and the current display is set to Camera Perspective, you can set up the perspective view in the drawing by selecting Current from the Orientation list of the Drawing View dialog box, and then placing the view in the drawing.

## CREATING DRAWING VIEWS FROM DESIGN VIEWS

A design view is created in an assembly file and can have information such as component visibility and viewing angle saved in it under a unique name. For example, you can turn off the visibility of a number of components and save these changes under a Design View. When you recall this Design View name, the components you turned off are not displayed. The actual creation process of Design Views will be explained in greater detail in Chapter 10. However, Design Views are discussed in this chapter as they relate to drawing views.

To generate a Drawing View based on a Design View, activate the Drawing View dialog box and click in the edit box to display all valid design views, as shown in the following image. Projected views will be based on the base view created from the Design View as is also shown in the following image.

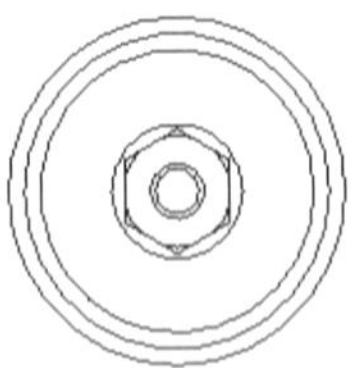

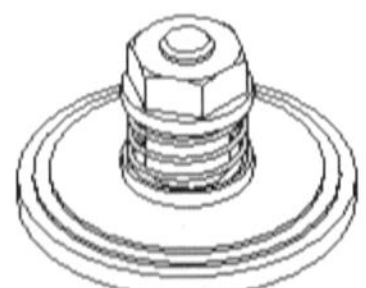

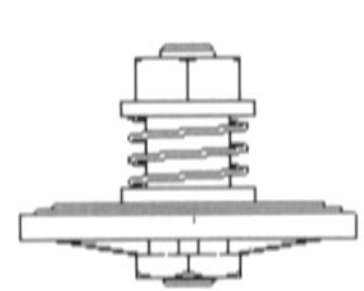

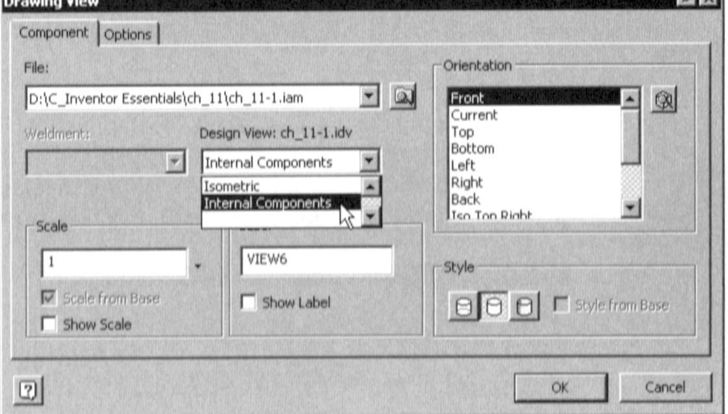


Figure 9-32

**NOTE** Design Views can also be made associative with the drawing view. This means that when a change is made to the Design View contained in an assembly file, the drawing view associated with this Design View will be updated automatically. Design Views that are identified by a *.default* extension cannot be made associative.

Once a drawing view is generated based on a Design View, the drawing view or views can be changed if a different Design View is selected. To perform this operation, first select the drawing view and then right-click. Then select Apply Design View from the menu as shown in the following image.

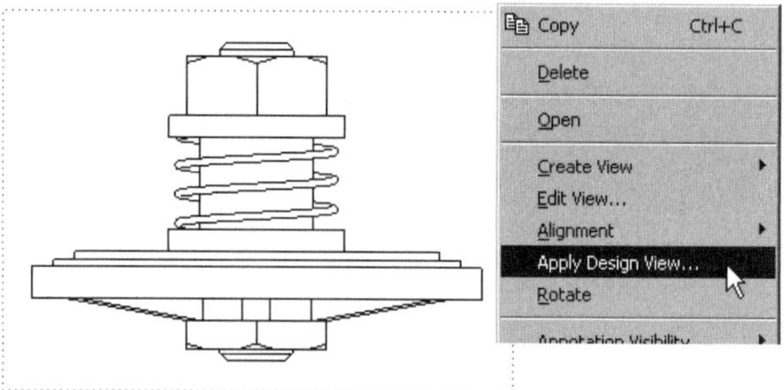


Figure 9-33

When the Apply Design View dialog box appears, click in the edit box and select a new Design View from the list, as shown in the following image. If multiple views of the assembly appear on the sheet, set the Apply status for each view. Also set the Associative status by changing the value from No to Yes.

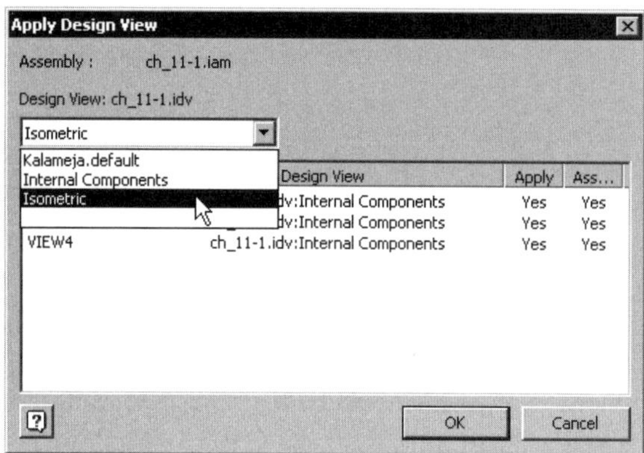


Figure 9-34

The results are illustrated in the following image. Originally, a Design View called “Internal Components” was used to generate the first set of drawing views. After a different Design View was selected, the drawing views get updated to reflect the new Design View.

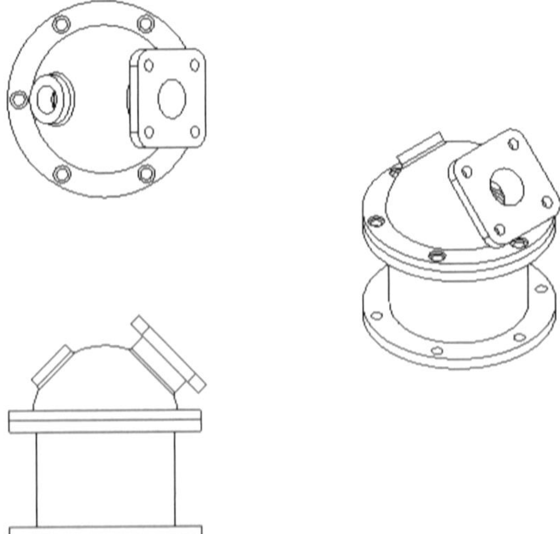

Figure 9-35

## OPENING A MODEL FROM A DRAWING TO EDIT

While working inside of a complex drawing view, you may need to make changes to a part or assembly file. Rather than close down the drawing and open the individual part file, you can open a part or assembly file directly from the drawing manager. Use the following steps to accomplish this task:

1 Select the part or assembly file from the graphics window or the Browser.

2 Right-click on the part or assembly name in the Browser to display the menu in the following image and then click Open to launch the assembly file.

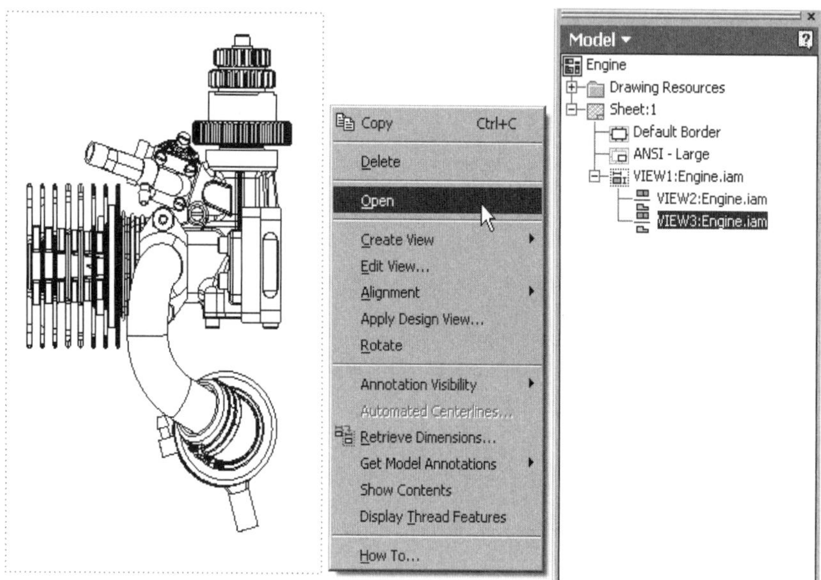


Figure 9-36

If you want to open individual part files from the assembly drawing view, first show the contents of the drawing view by right-clicking on the view in the Browser and picking Show Contents from the Browser, as shown in the following image.

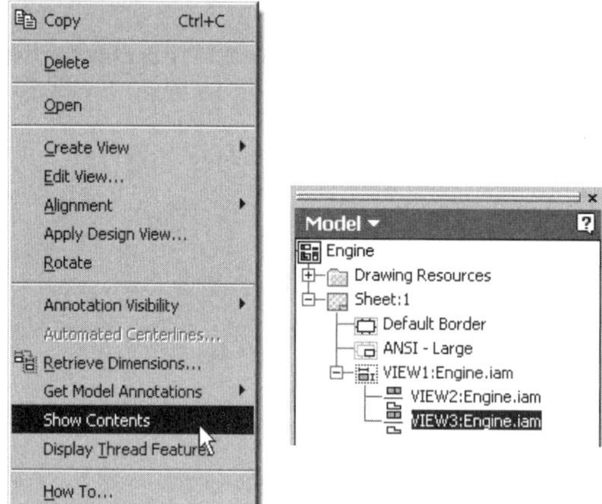


Figure 9-37

This will expand the drawing view to include all assembly and part files. Right-click on the desired part file from the Browser and select Open from the menu, as shown in the following image.

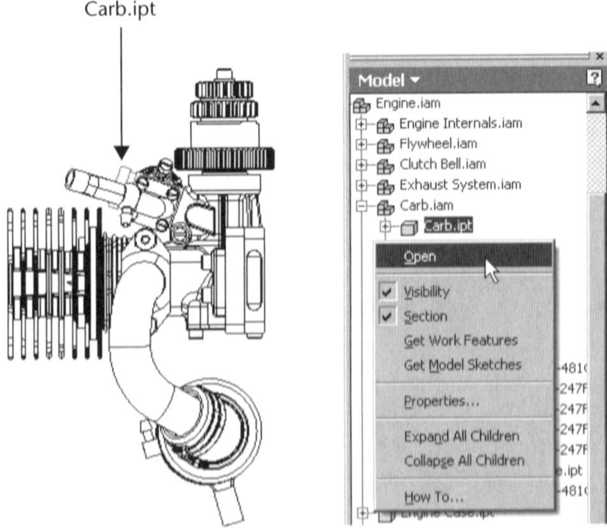


Figure 9-38

**NOTE** Another technique used to open a part file from a drawing is to first change the selection filter to Part. Then right-click on the part in the graphics screen and pick Open from the menu, as shown in the following image. Doing so will open the source part file.

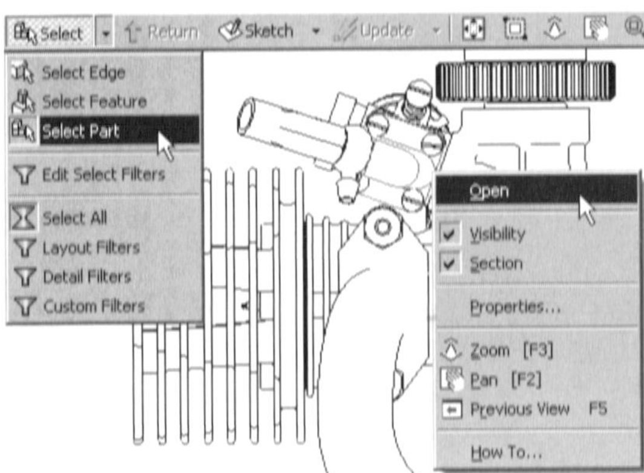


Figure 9-39

## Showing and Referencing Work Features in Drawing Views

Part and assembly models have work features (planes, axes, points, and surfaces) that aid in accurately defining and capturing the design. These features have purpose in the drawing (e.g., axes and planes can serve as center lines or define datums for critical dimensions or features). Points can be datums as well, such as in the intersection point (*X0*, *Y0*) for aerospace, where critical dimensions are taken from that one point.

These work features can be displayed in drawing views. Annotations in the form of dimensions can be made to these work features. You also have the full benefit of the dimensions being associative to the work features. Changes in the model or assembly will be reflected in the drawing views.

If work features are visible in the model, they will be visible in the drawing when the view is created. Work features turned off in the model will not initially display in the drawing view. You will have to manually turn on the visibility of the work feature through the Browser.

The command used to control the display of work features in the drawing view is Show Contents. This will display the part file and its features in the Browser. Right-clicking on the part file will allow you to select the Get Work Features command. This will populate the Browser with the work features from the part file. You control the visibility of these work features through the Browser.

To display work features in a drawing view, follow these steps:

1 Identify the drawing view in the Browser to display work features and right-click. In the menu, click the Show Contents command.

2 When the contents of the view are displayed, click on the part file name in the following image and right-click.

3 Click the Get Work Features command.

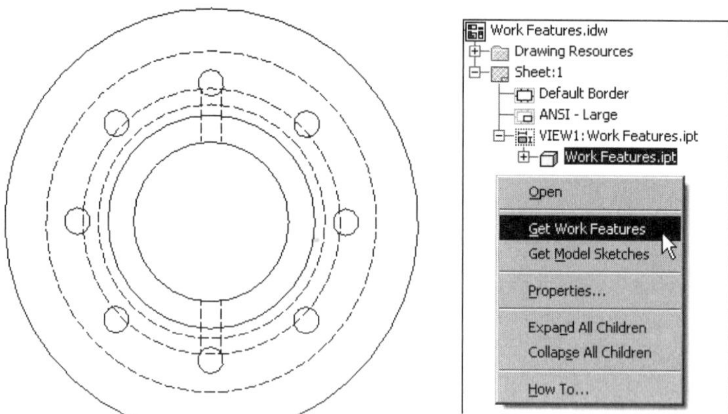


**Figure 9-40**

4 In the following image, notice that the work features are visible in the given view. In this example, the work features were already visible in the part file. As a result, the work features automatically display in the drawing view. You may have to stretch the work features as a means of lengthening them. This is accomplished by dragging the green dots until you achieve the desired results.

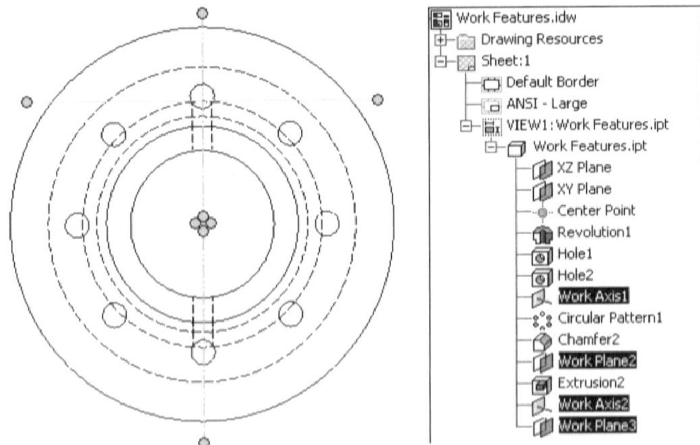


**Figure 9-41**

5 If work features are turned off in the model, they will not display in the drawing. To display a work feature, right-click on the feature, as shown in the following image, and check the Visibility box. This will turn the work feature on in the drawing view.

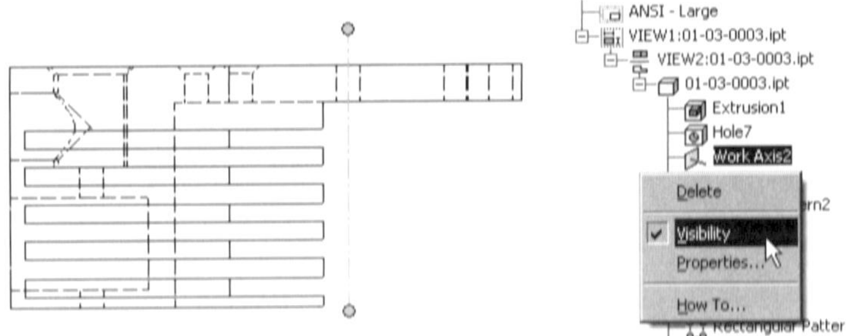


**Figure 9-42**

Another method of displaying work features is through the Drawing View dialog box, as shown in the following image. As you create a drawing view, you can check the box next to Work Features under the Options tab of the View dialog box. The work features will display as the view is created.

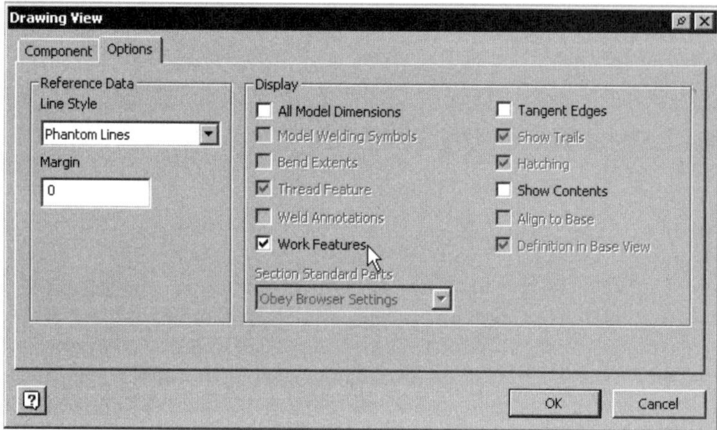


Figure 9-43

## Using Sketches in Drawing Views

You have the ability of easily sharing conceptual model sketches in a drawing view. Typical applications could include being able to simulate printed circuit board traces, sharing a screen printing image on a component, or having a model sketch display in an already created drawing view.

If you want model sketches to be displayed in drawing views, the sketches you wish to share must be unconsumed. Start this process by right-clicking on the view to share the unconsumed model sketch. Then select Show Contents in the menu, as illustrated in the following image.

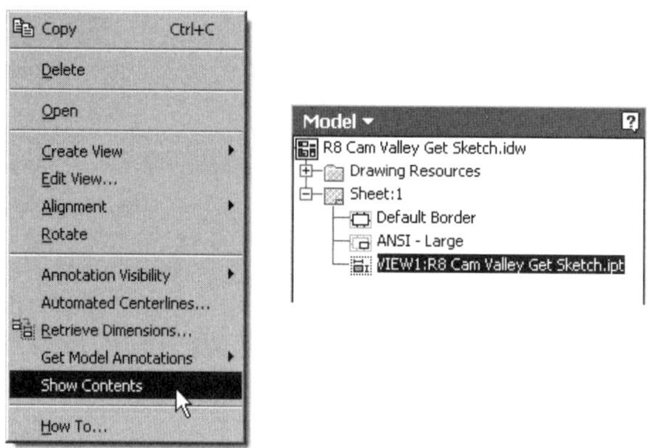


Figure 9-44

For the purpose of clarity, the original drawing view is illustrated in the following image. Once the contents of the drawing view have been expanded, click in the part icon in the Browser to display the menu, also shown in the following image. Then, click on Get Model Sketches.

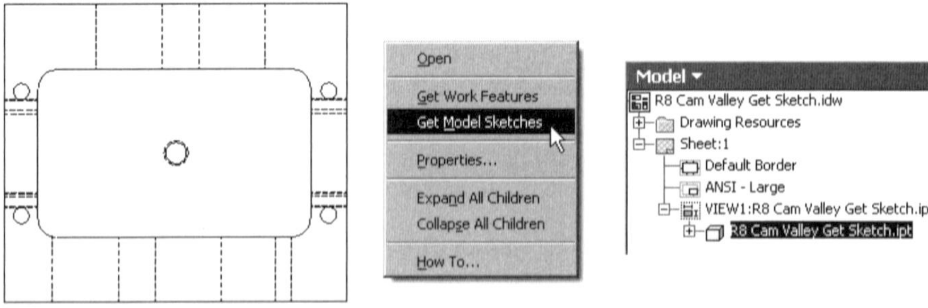


**Figure 9-45**

The results are illustrated in the following image with the model sketch displaying in the drawing view.

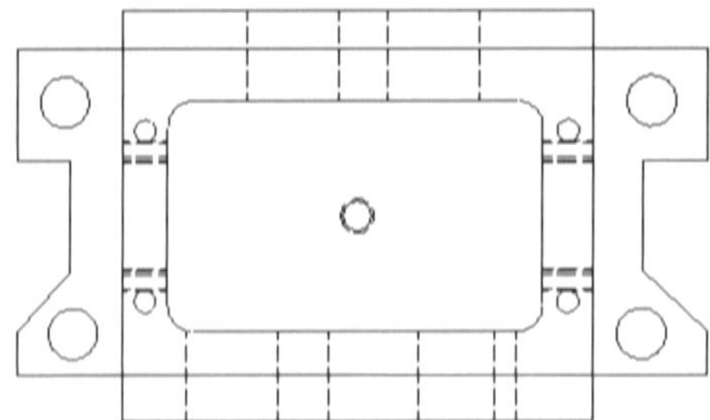

**Figure 9-46**

## CREATING DRAFT VIEWS

Another type of drawing view is a draft view. A draft view is not created from a 3D part, but contains one or more associated 2D sketches. A draft view can be created on its own sheet, in its own Autodesk Inventor drawing file, or can be used in an existing sheet with other drawing views to provide detail that is missing in a model. From the Drawing Sketch Panel Bar you can add geometry, constraints, parametric dimensions, text, and other annotations in the draft view. All these objects will be placed in the draft view and, if the draft view moves, so will all the objects associated with it. When an AutoCAD 2D file is imported into an Autodesk Inventor drawing, the geometry is placed in a draft view.

To create a draft view, follow these steps:

1 Select the Draft View tool from the Drawing Views Panel Bar, as shown in the following image.

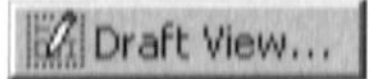


**Figure 9-47**

2 The Draft View dialog box will appear, as shown in the following image. Fill in information for the Label and Scale of the draft view as you want them to appear in the drawing view. Then, click the OK button to continue. The Scale setting that you set will be the scale to which you are drafting the objects.

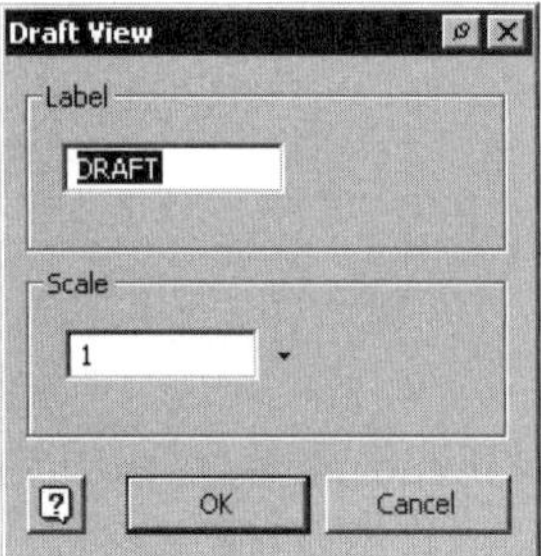


Figure 9-48

3 Using the tools from the Drawing Sketch Panel Bar, create 2D geometry and add constraints, dimensions, and annotations.

4 When you have finished creating the geometry that will make up the draft view, either right-click and click Finish Sketch from the menu, as shown in the following image, or click the Return button from the Standard toolbar. The draft view will now act like a regular drawing view that can be edited.

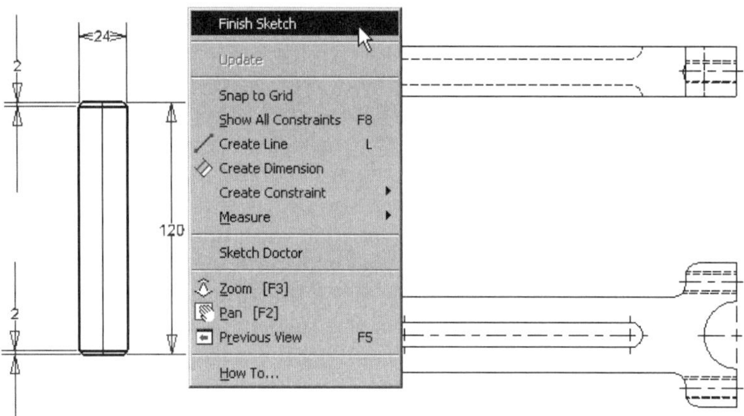


Figure 9-49

# EXERCISE 9-1
# Complex Drawing View Techniques

In this exercise, you create a variety of complex drawing views from a model of a cover. To navigate to the exercise in the *Electronic Student Workbook*, do the following:

1 From the main TOC page, click Chapter 9.

2 From the TOC for Chapter 9, click Complex Drawing View Techniques.

The completed exercise is shown in the following image.

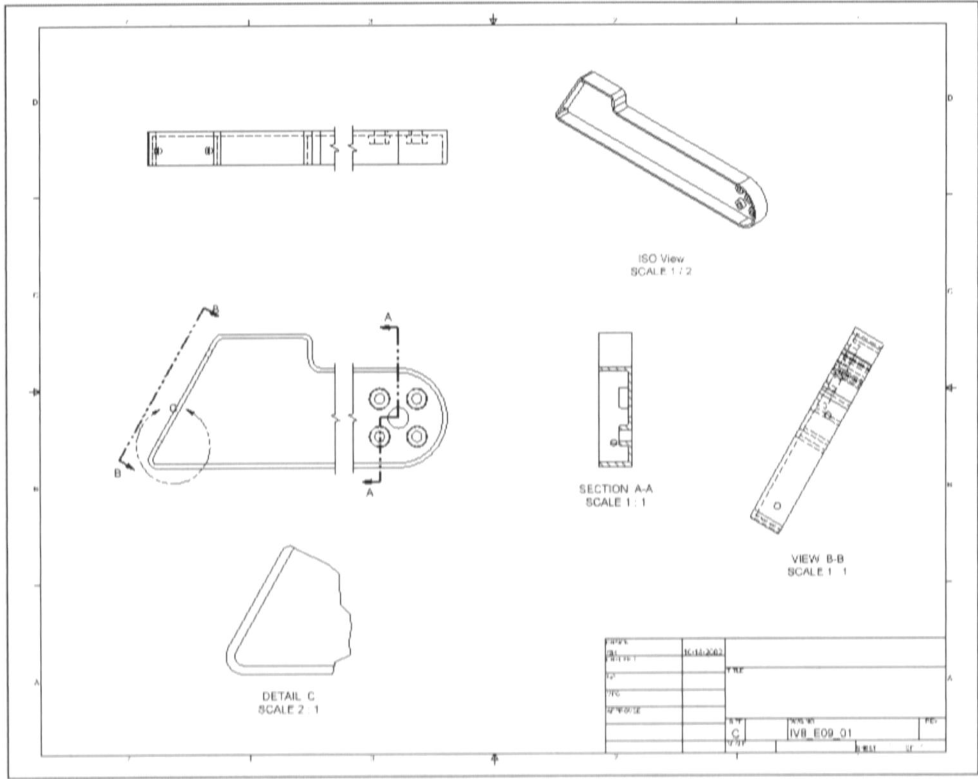

**Figure 9-50 Completed exercise**

EXERCISE

## MANAGING SHEETS

An Autodesk Inventor drawing file can have more than one sheet. A multiple sheet drawing would typically be used where the model cannot be represented on a single sheet. Each sheet is numbered, and a count is maintained of the total number of sheets. This information can be displayed in the title block. The title block, for example, can indicate that the current sheet is Sheet:2 of a 5-sheet drawing.

### CREATING MULTIPLE SHEETS

You can add sheets to a drawing in three ways:

1 Create a new sheet using Panel Bar, Menu, or Browser tools.

- From the Drawing Views Panel Bar, click New Sheet, as shown in the following image.

**Figure 9-51**

- From the Insert menu, select Sheet, as shown in the following image.

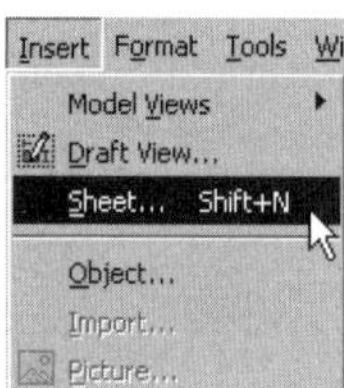


**Figure 9-52**

- Right-click the background and select New Sheet from the menu, as shown in the following image.

**Figure 9-53**

- Right-click the Browser and select New Sheet from the menu, as shown in the following image.

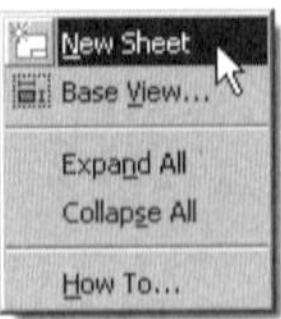


**Figure 9-54**

2 Create a new sheet with a predefined layout. In the Browser, expand Drawing Resources > Sheet Formats to list the defined formats. Pick a format, right-click, and select New Sheet, as shown in the following image.

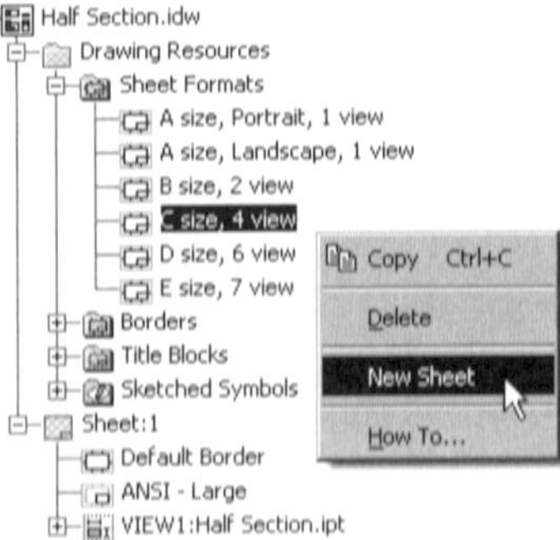


**Figure 9-55**

3 Create a new sheet with an existing layout from another drawing. Right-click a sheet border in the graphics window or the Browser and select Copy, as shown in the following image.

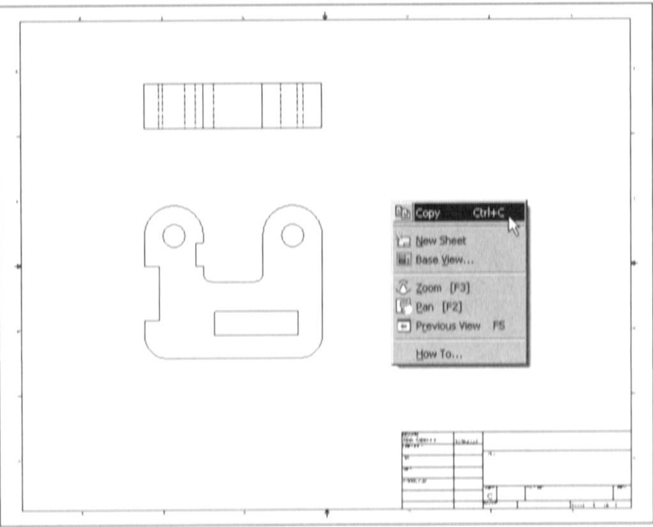


**Figure 9-56**

Switch to another open drawing, right-click the drawing name in the Browser, and select Paste (as shown in the following image). A copied sheet contains the same views as the source.

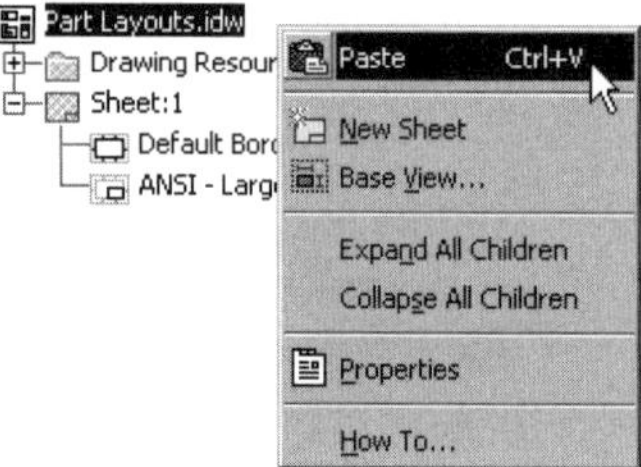


**Figure 9-57**

**NOTE** You cannot copy and paste a sheet within the same drawing.

While numerous drawing sheets can be created, only one sheet can be active at a time. In the following image, Sheet:2 is currently the active sheet while Sheet:1 and Sheet:3 are inactive, as identified by the gray background. To make another sheet active, double-click on the sheet number or right-click on the sheet in the Browser and select Activate.

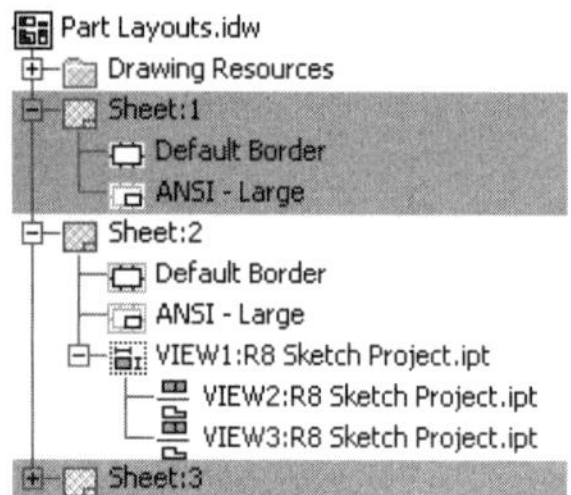


**Figure 9-58**

## CREATING SHEET FORMATS

A sheet format contains a predefined layout that can set the sheet size, border, title block, and views for a new sheet.

**NOTE** Auxiliary, broken, break-out, detail, and section views cannot be predefined.

To set up a sheet format, you create a sheet layout as described in Chapter 5, complete with border and title block. Create the views you want as default views in a new sheet. Right-click the sheet and select Create Sheet Format from the menu, as shown in the following image.

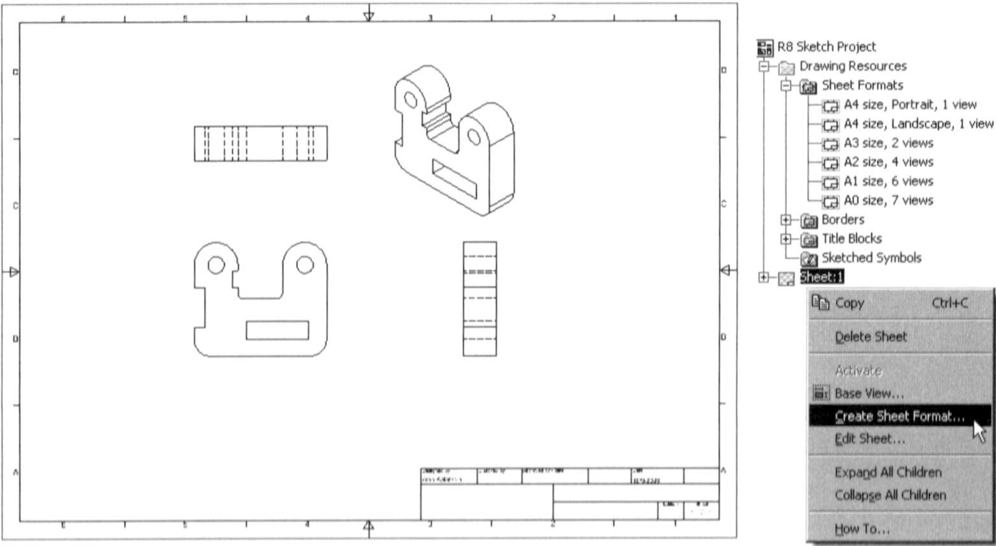


**Figure 9-59**

The Create Sheet Format dialog box is displayed. Enter a descriptive name for the sheet and click OK. The new format is added to Drawing Resources, as shown in the following image.

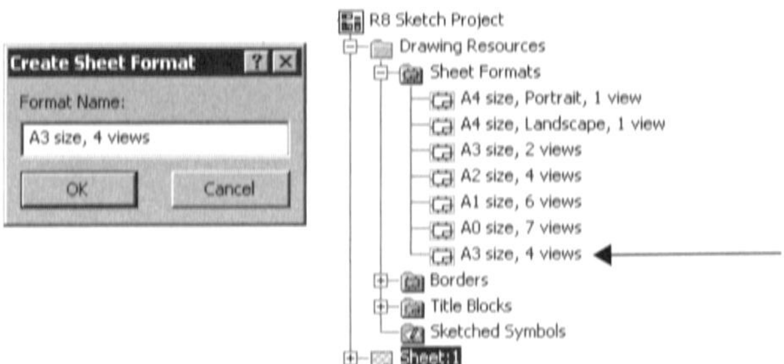


**Figure 9-60**

**TIP** Set up a template file that contains formats for the standard sheet sizes.

## COPYING VIEWS BETWEEN SHEETS

When you want to use the same view in two different layouts, you can copy views between sheets in the same or different drawings.

To copy a view between sheets, you expand the sheet listing for the sheet containing the source view in the Browser, and then right-click the view and select Copy, as shown in the following image.

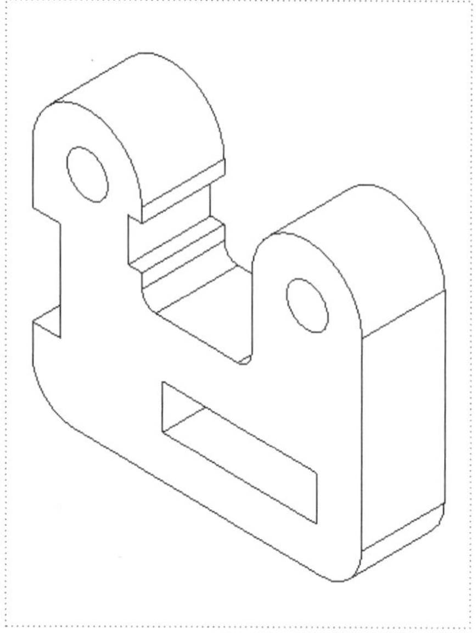

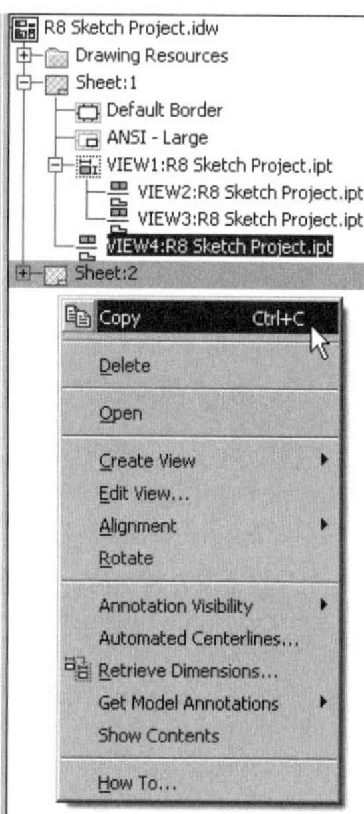


Figure 9-61

Select the destination sheet, right-click, and then select Paste as shown in the following image. On the sheet, position and edit the view as required.

Figure 9-62

## MOVING VIEWS BETWEEN SHEETS

To move a series of views from one sheet to another, activate the sheet that contains the views. In the Browser, select the views to move. If multiple views will be moved, press and hold down the CTRL key and select the views. In the following image, View3 and View4 are selected and will be moved into Sheet:2. After selecting the views to move, release the CTRL key. In the Browser, drag the selected views to Sheet:2 and release the mouse button when the horizontal bar appears below Sheet:2.

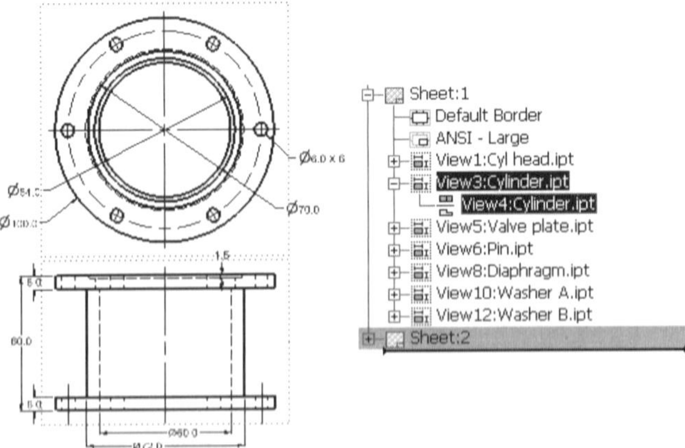


**Figure 9-63**

You can move a single view from one sheet to another within the same drawing by dragging the view icon in the Browser. If the view is related to another view in the drawing (either as a dependent view or a base view), a shortcut icon appears in the Browser as shown in the following image. The shortcut icon indicates that the base or dependent view for this view resides on another sheet.

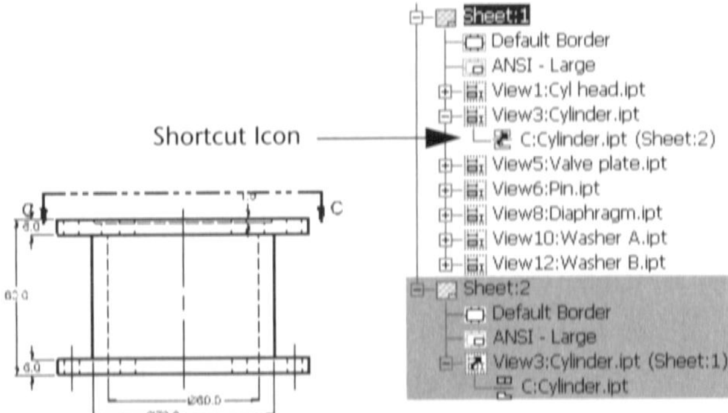


**Figure 9-64**

To switch to the sheet with the moved view, right-click the shortcut icon and select Go To, as shown in the following image.

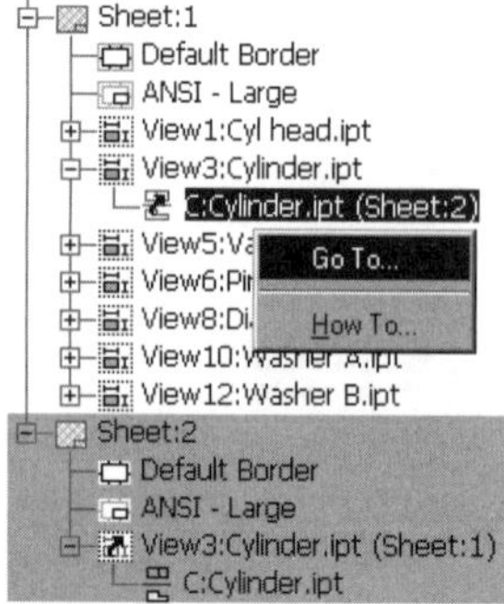


Figure 9-65

**NOTE** If you move a base view to a different sheet, the dependent views remain on the original sheet. If you move a dependent view, the base view remains. In both cases, the Browser displays a shortcut icon to remind you of the dependency between the base view and its dependents.

## ADVANCED DIMENSIONING TECHNIQUES

Additional tools are available to assist you with annotating drawings in Autodesk Inventor. When the dimension standard calls for distances to be called out in metric and English units, dual dimensions are used. The Auto Baseline tool is used to add multiple dimensions to a drawing view in one operation. Ordinate dimensions utilize a common origin on the drawing view. All dimensions reference this origin. Ordinate dimensions also do not utilize dimension lines or arrows. This allows a complex drawing to remain uncluttered while communicating the required drawing information. Dimensions can be easily rearranged to improve readability within Autodesk Inventor.

## DUAL DIMENSIONING

Dual or alternate dimensions are useful when you need to display dimensions in a drawing using two different units of measure. For instance, you should use alternate unit dimensions when the same drawing will be used in different countries that use different units (such as inches and millimeters for length). Autodesk Inventor provides six different dual-dimensioning formats, as illustrated in the following image.

30 / 1.2 in    [30] 1.2 in    30 [1.2 in]    30<br>1.2 in    [30]<br>1.2 in    30<br>[1.2 in]

Figure 9-66

To create a new dimension style consisting of alternate dimensions, follow these steps:

1 Open the drawing or template file and begin the process of creating a new dimension style that will utilize alternate unit dimensions. Activate the Dimension Style dialog box by clicking on the Format menu, followed by Dimension Style.

2 From the Standard list, shown in the following image, select a standard on which to base the new dimension style. Then, select the New button. A new style name will appear in the Style Name text box. Replace this with a more meaningful style name.

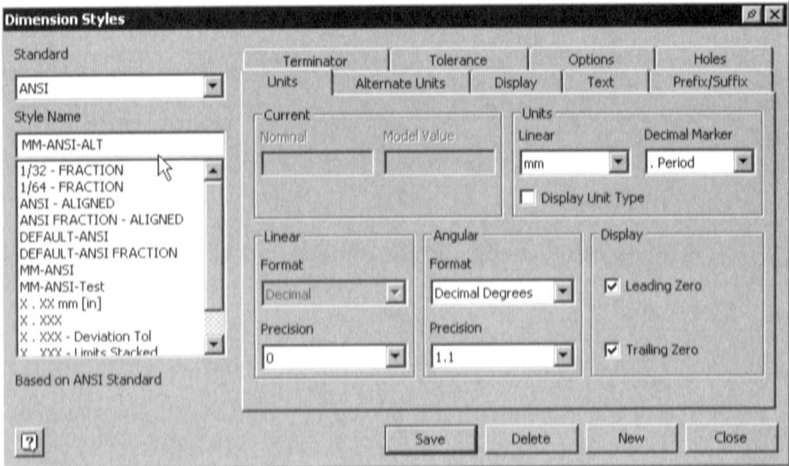


**Figure 9-67**

3 Select the Alternate Units tab, then check the box next to the Display Alternate Units option, as shown in the following image. You may also want to check the box next to Display Unit Type and select the desired type of units. In this example, inches has been selected. You should also notice the various Display Format styles that are available. Click on the desired style to display alternate units.

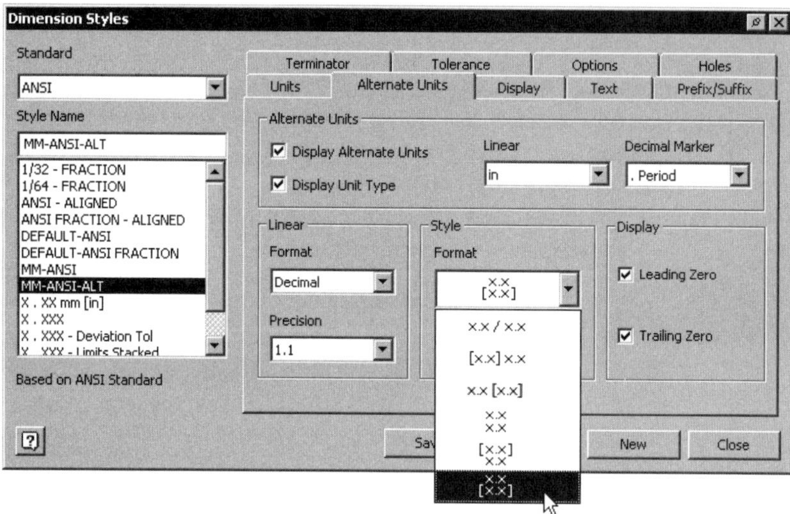


Figure 9-68

4 Make other changes to the dimension settings as desired. When finished, click Save to save the dimension style and then click Close to close the dialog box.

To add alternate unit dimensions to your drawing, follow these steps:

1 Click the General Dimension button from the Drawing Annotation Panel Bar.

2 From the Style list on the Standard toolbar, select a dimension style that uses the alternate unit dimensions shown in the following image. In this example, the dimension style in use is MM-ANSI-ALT.

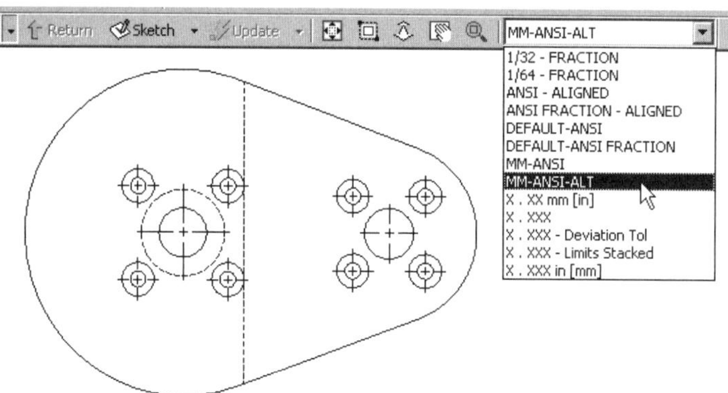


Figure 9-69

3 In the graphics window, select the geometry and drag to display the dimension.

4 Click to place the dimension in the desired location. The alternate unit dimension appears in the format you specified, as shown in the following image.

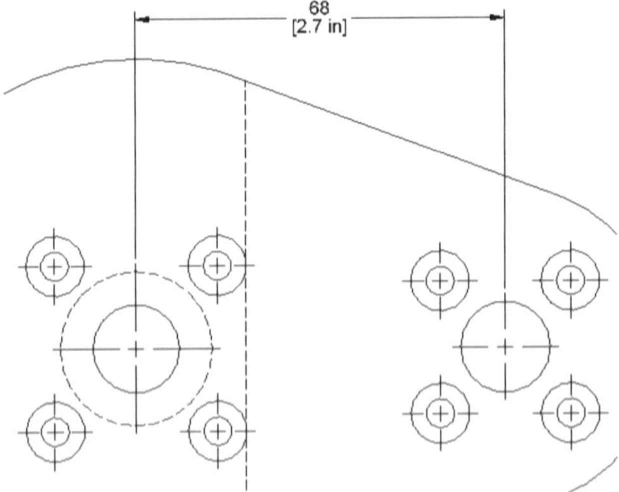


Figure 9-70

**NOTE** Changes to a custom dimension style apply only to the open drawing. To automatically assign a modified dimension style to all new drawing files, make the modifications in the template file.

## AUTO BASELINE DIMENSIONS

To add multiple drawing (reference) dimensions to a drawing view in a single operation, use the Baseline Dimension tool, which will place a baseline dimension about the selected geometry. The dimensions can be either horizontal or vertical. Once the dimensions are placed, they are grouped together into a set and members can be added and deleted from the set, a new origin can be placed, and the dimensions can be automatically rearranged to reflect the change.

To create automatic baseline drawing dimensions, follow these steps:

1 Select the Baseline Dimension tool from the Drawing Annotation Panel Bar, as shown in the following image.

Figure 9-71

2 Define the origin line by selecting an edge of the object to be dimensioned. This edge will act as the baseline for other dimensions.

3 Draw a box around the geometry that you want to dimension. Move the mouse into position where the first point of the box will be. Press and hold down the left mouse button and move the mouse so the preview box encompasses the geometry that you want to dimension, as shown in the following image, and then release the mouse button.

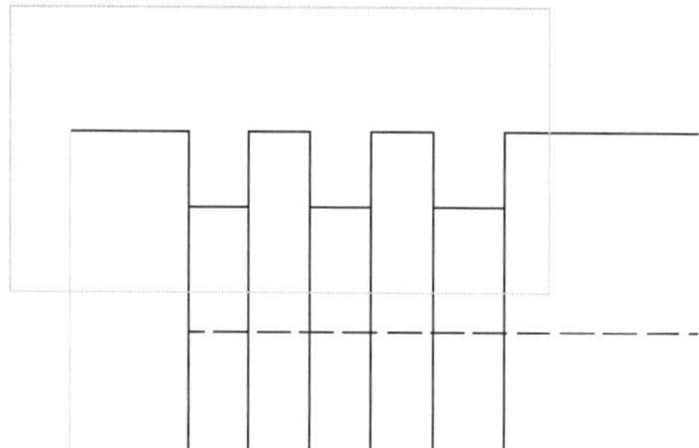

**Figure 9-72**

4 When you are finished selecting geometry, right-click and select Continue from the menu.

5 Move the mouse to a position where the dimensions will be placed, as shown in the following image. When the dimensions are in the correct place, click the point with the left mouse button to anchor the dimensions.

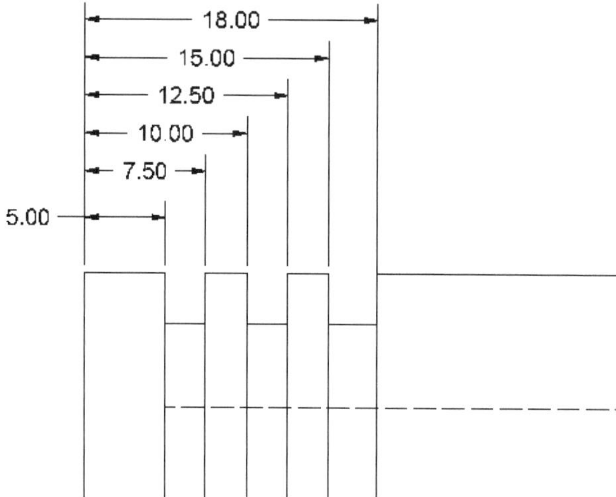


**Figure 9-73**

6 To complete the operation, either press the ESC key or right-click and select Done from the menu.

After creating dimensions using the Baseline Dimension tool, there are various edit operations that can be performed beyond the basic dimension options. To perform these edits, move the mouse over the baseline dimension and small green circles will appear on the dimensions. Right-clicking will display the menu, as shown in the following image.

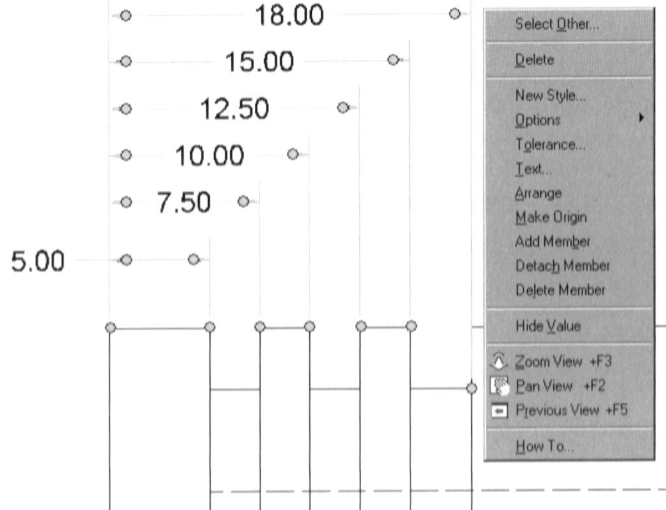


**Figure 9-74**

A number of options for editing baseline dimensions are described as follows:

**Delete** This option will delete all the baseline dimensions that are in the set. With the green circles on the screen, right-click and select Delete from the menu.

**Arrange** This will rearrange the dimensions to the spacing that is determined through the Dimension Style.

**Make Origin** This option will change the origin of the baseline dimensions. With the green circles on the screen, select an edge that will define the new baseline, then right-click and select Make Origin from the menu.

**Add Member** This option will add a drawing dimension to the baseline dimensions. After creating a drawing dimension, move the mouse over the baseline dimensions and with the green circles on the screen, right-click and select Add Member from the menu. Now select the drawing dimension. The drawing dimension will be added to the set of baseline dimensions and rearranged in the proper order.

**Detach Member** This option will remove a dimension from the set of baseline dimensions. With the green circles on the screen, right-click on the green circle on the dimension that you want to detach and then right-click and select Detach Member from the menu. The detached dimension can then be moved and edited like a normal drawing dimension.

**Delete Member** This option will erase a dimension from the set of baseline dimensions. With the green circles on the screen, right-click on the green circle on the dimension that you want to erase and then select Delete Member from the menu. The dimension will be deleted.

## ORDINATE DIMENSIONS

Ordinate dimensions are used to indicate the location of a particular point along the *X*- or *Y*-axis from a common origin point. This type of dimensioning is especially suited for describing part geometry for numerical control tooling operations. There are two methods for creating ordinate dimensions: in ordinate dimension sets and with the Ordinate Dimension tool.

In an ordinate dimension set, all the ordinate dimensions that are created in a single operation will be grouped together but can be edited individually or as a set. When creating an ordinate dimension set, the first dimension created will be used as the origin. The origin dimension needs to be a member of the set and can later be changed to a different dimension. If the location of the origin or the origin member changes, the other members will be updated to reflect the new location.

Ordinate dimensions created with the Ordinate Dimension tool are recognized as individual objects and an origin indicator will be created as part of the operation. If the origin indicator location is moved, the other ordinate dimensions will be updated to reflect the change.

Both methods will create drawing dimensions that reference the geometry and will be updated to reflect any changes in the geometry to which they are dimensioned. Ordinate dimensions can be placed on any point, center point, or straight edge. Before adding ordinate dimensions to a drawing view, create or edit a dimension style to reflect your standards. When you place ordinate dimensions, they will automatically be aligned to avoid interfering with other ordinate dimensions.

To create an ordinate dimension set, follow these steps:

1 Create a drawing view.

2 Select the Ordinate Dimension Set tool from the Drawing Annotation Panel Bar, as shown in the following image.

**Figure 9-75**

3 Select the desired dimension style from the Style list in the Standard toolbar.

4 Select a point to set the origin for the dimensions.

5 Select a point on the drawing view to locate the dimension and then move the mouse to place the dimension in a horizontal or vertical orientation. Click to place the dimension.

6 Continue selecting points and placing dimensions that will make up a dimension set, as shown in the following image.

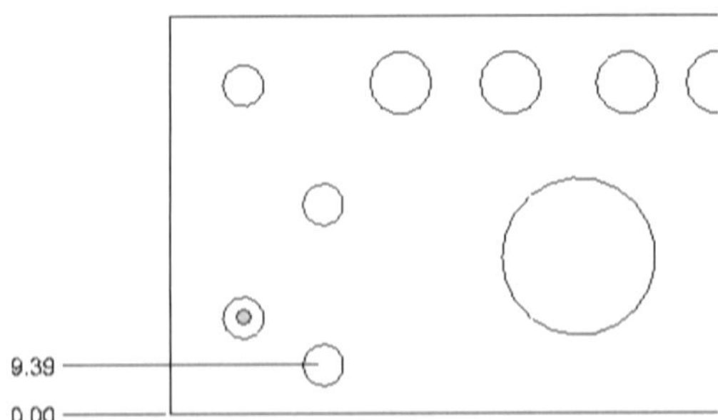


**Figure 9-76**

7 To change options for the dimension set, right-click at any time and then select Options, as shown in the following image.

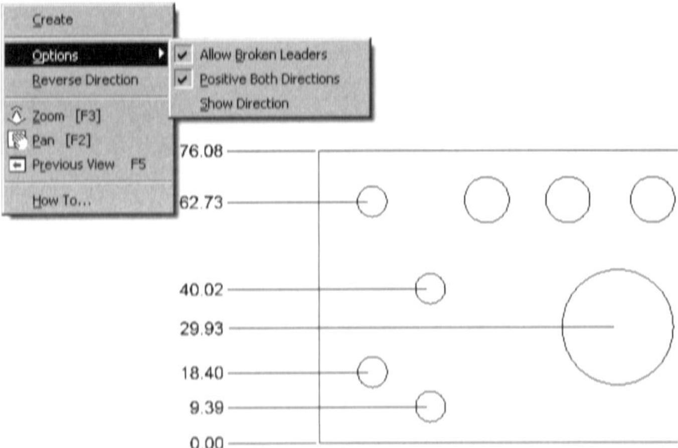


**Figure 9-77**

8 To create the dimensions, right-click and click Create from the menu.

9 To edit the origin, right-click on the dimension that will be the origin and click Make Origin from the menu (see following image). The other dimension values will be updated to reflect the new origin.

10 To edit a dimension set, right-click on a dimension in the set and select the desired option from the menu, as shown in the following image.

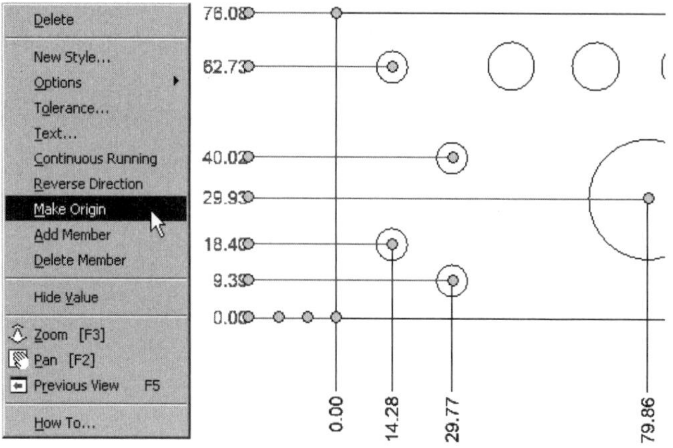


**Figure 9-78**

To create an ordinate dimension using the Ordinate Dimension tool, follow these steps:

1 Create a drawing view.
2 Click the Ordinate Dimension tool from the Drawing Annotation Panel Bar, as shown in the following image.

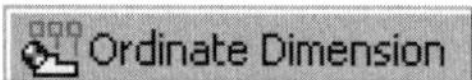


**Figure 9-79**

3 Select the desired dimension style from the Style list in the Standard toolbar.
4 In the graphics window, select the view to dimension.
5 Select a point to set the origin indicator for the dimensions, as shown in the following image.

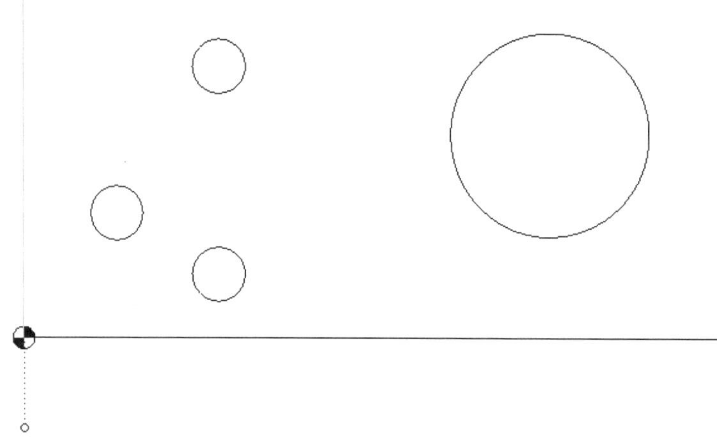

**Figure 9-80**

6 Select a point to locate the dimension and then move the mouse to place the dimension in a horizontal or vertical orientation. Click to place the dimension.

7 Continue selecting points and placing dimensions.

8 To create the dimensions, right-click and select Create from the menu, as shown in the following image. You can then continue placing ordinate dimensions in another orientation.

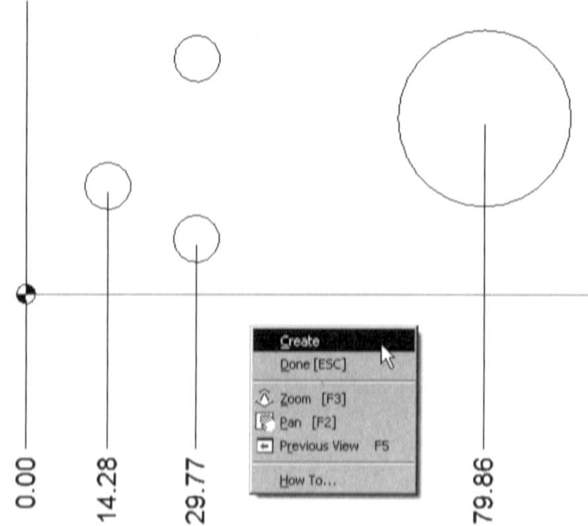


**Figure 9-81**

9 To create the dimensions and end the operation, right-click and select Done from the menu.

10 To edit a dimension, right-click on a dimension and select the desired option from the menu, as shown in the following image.

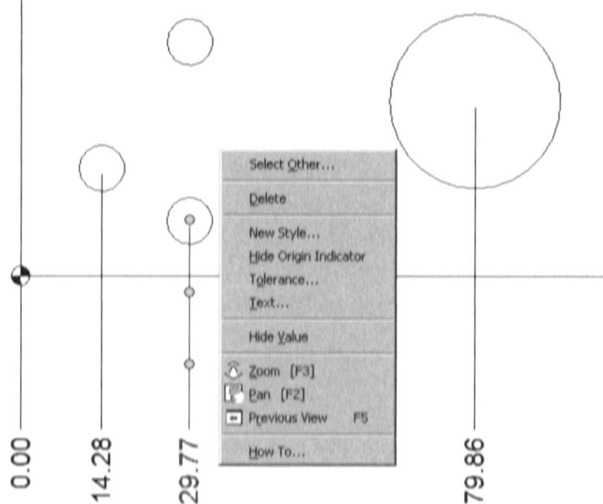


**Figure 9-82**

11 To edit an ordinate dimension leader, grab an anchor point (green circle) and drag it to the desired location.

12 To edit the origin indicator, do one of the following:

- Grab the origin indicator and drag it to the desired location.
- Double-click the origin indicator and enter the precise location in the Origin Indicator dialog box.

### MOVING DIMENSIONS

To move a dimension by either lengthening or shortening the extension lines or moving the text, position the mouse over the dimension until the dimension highlights, as shown in the following image. An icon consisting of four diagonal arrows will be attached to your cursor.

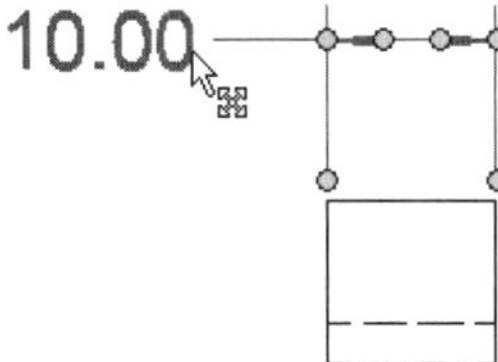


Figure 9-83

Press and hold down the left mouse button and move the dimension to a new location. Depending upon where the dimension has been moved, the text or the text and extension lines will move and stretch. If the text in a vertical dimension is moved up and to the right, for example, the text will move up and the extension lines will stretch to the right. To move or reattach the starting point for an extension line, move the mouse over the dimension's text until the four arrows appear and then select the green circle at the end of the extension line that will be moved. With the mouse button pressed, move the point to a new location. If the point is moved off the geometry, a dialog box will appear warning you of the problem. An example of when you would want to reattach a dimension is in a case where the extrusion dimension appears in the middle of a part and the extension line should begin at the end of the part.

**NOTE** You can only change the extension line placement for dimensions that were created in the drawing. Dimensions on the drawing due to the use of the Retrieve Dimensions command cannot have their extension lines moved because they are tied to the model geometry; however, their text can be moved and the extension lines either lengthened or shorted from the edges of the drawing view.

## HOLE TABLES

If the drawing view that you are dimensioning contains holes, you can locate them by placing individual dimensions or create a hole table that will list the location and size of all the holes (or just the selected holes) in a view. The hole location will be listed in both the *X*- and *Y*-axis coordinates with respect to a hole datum that will be placed before creating the hole table.

**NOTE** If a part contains a hole that was created by extruding a circle, it will not be placed in the hole table; it must be a hole feature.

After placing a hole table, if you add, delete, or move a hole to a part, the hole table will automatically be updated to reflect the change. Each hole in the table is automatically given an alphanumeric tag as its name. It can be edited by either double-clicking on the tag in the drawing view or in the hole table, or by right-clicking on the tag in the drawing view or in the hole table and selecting Edit Tag from the menu. Type in a new tag name, and the change will appear in the drawing view and hole table. There are three tools for creating a hole table:

- Hole Table - Selection
- Hole Table - View
- Hole Table - Selected Type

The tools are located in the Drawing Annotation Panel Bar, as shown in the following image.

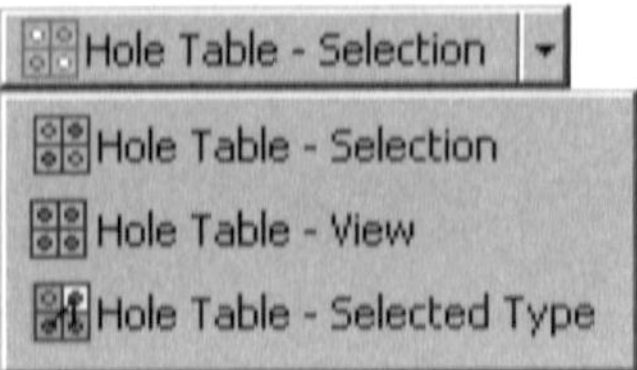


**Figure 9-84**

When using the Hole Table - Selection mode, a hole table will be created from only the selected holes, as shown in the following image. You control the holes that are selected.

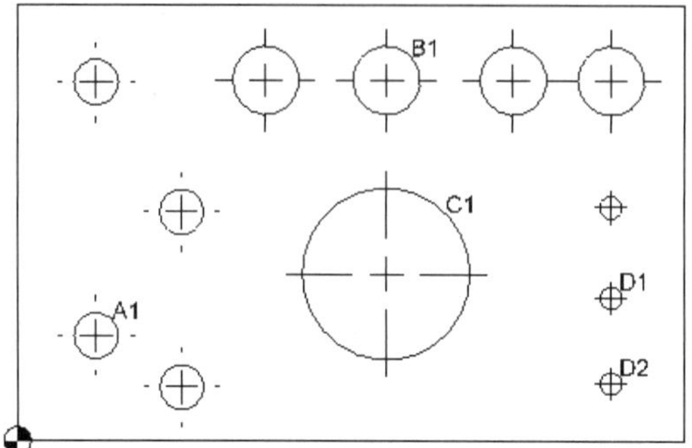


| Hole Table | | | |
|---|---|---|---|
| LOC | XDIM | YDIM | SIZE |
| A1 | 14.28 | 18.40 | Ø8.00 THRU |
| B1 | 66.53 | 63.09 | Ø12.00 THRU |
| C1 | 66.53 | 29.23 | Ø30.00 THRU |
| D1 | 106.89 | 25.12 | Ø4.00 THRU |
| D2 | 106.89 | 10.11 | Ø4.00 THRU |

**Figure 9-85**

When using the Hole Table - View mode, a hole table will be created based on all the holes in a selected view, as shown in the following image. Notice that all holes are given an alphanumeric identifier, which locates each hole by *X*- and *Y*-axis coordinates.

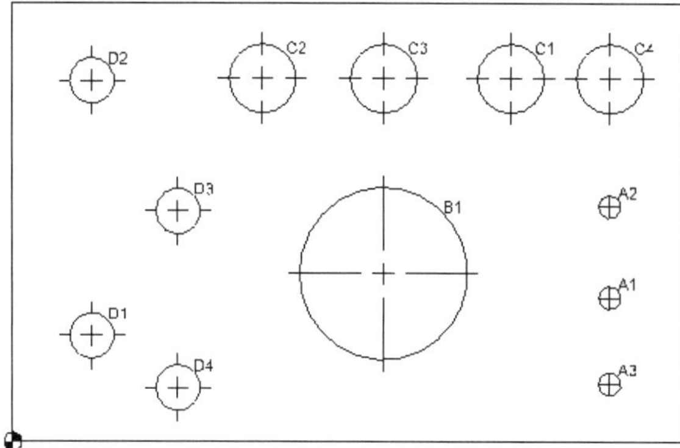


| Hole Table | | | |
|---|---|---|---|
| LOC | XDIM | YDIM | SIZE |
| A1 | 106.89 | 25.12 | Ø4.00 THRU |
| A2 | 106.89 | 40.39 | Ø4.00 THRU |
| A3 | 106.89 | 10.11 | Ø4.00 THRU |
| B1 | 66.53 | 29.23 | Ø30.00 THRU |
| C1 | 89.23 | 63.09 | Ø12.00 THRU |
| C2 | 44.91 | 63.09 | Ø12.00 THRU |
| C3 | 66.53 | 63.09 | Ø12.00 THRU |
| C4 | 106.89 | 63.09 | Ø12.00 THRU |
| D1 | 14.28 | 18.40 | Ø8.00 THRU |
| D2 | 14.28 | 62.73 | Ø8.00 THRU |
| D3 | 29.77 | 40.02 | Ø8.00 THRU |
| D4 | 29.77 | 9.39 | Ø8.00 THRU |

**Figure 9-86**

When using the Hole Table - Selected Type mode, the hole table will be created based on a single grouping of holes. In the following image, one of the larger holes located in the upper part of the object is selected using the Hole Table - Selected Type tool.

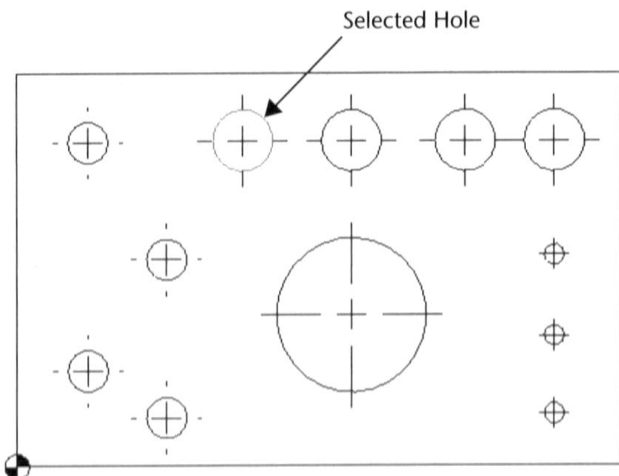


**Figure 9-87**

When the hole table is created, as shown in the following image, all holes that share the same type or diameter will be added to the hole table.

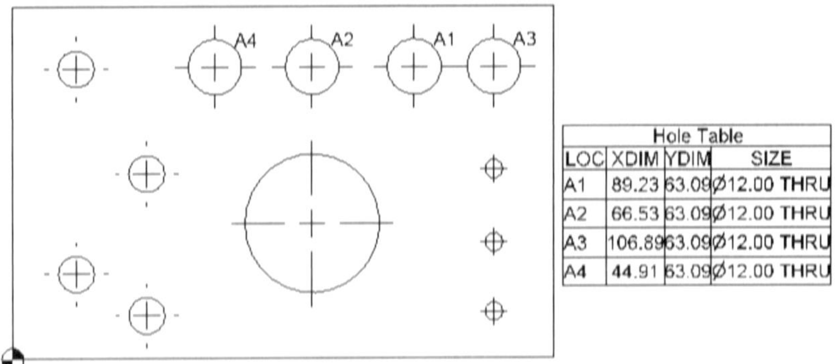


| Hole Table | | | |
|---|---|---|---|
| LOC | XDIM | YDIM | SIZE |
| A1 | 89.23 | 63.09 | Ø12.00 THRU |
| A2 | 66.53 | 63.09 | Ø12.00 THRU |
| A3 | 106.89 | 63.09 | Ø12.00 THRU |
| A4 | 44.91 | 63.09 | Ø12.00 THRU |

**Figure 9-88**

To create a hole table based on an existing drawing view that shows holes in a plan view, follow these steps:

1 Select the desired Hole Table tool from the Drawing Annotation Panel Bar.

2 Select the view on which the hole table will be based.

3 Select a point to locate the origin. Typical origins include the corners of rectangular objects and the centers of drill holes used for datums.

4 If the Hole Table – Selection tool is used, select the holes to include in the hole table. The holes can be selected individually or you can window around the holes. If the Hole Table - Selected Type is used, select the individual hole or holes that will act as references for all other holes of the same type.

5 Select a point to locate the table.

6 The contents of the hole table can be edited by either right-clicking on the hole table, the tag name in the hole table, or on the tag name in the drawing view. Next, choose the desired option from the menu, as shown in the following image.

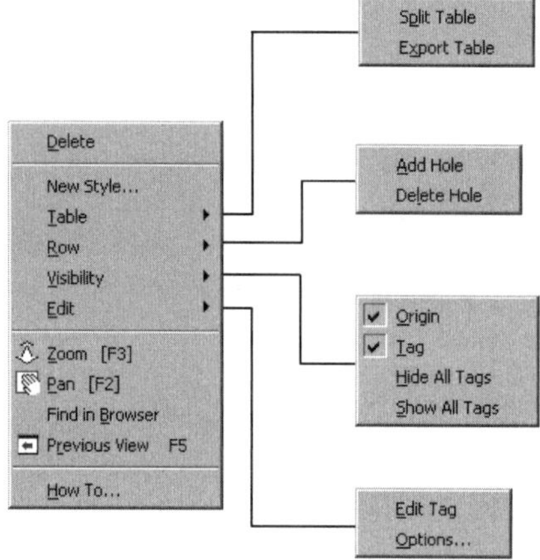


**Figure 9-89**

Clicking on Options displays the Edit Hole Table dialog box, as shown in the following image. Use this dialog box to change such items as:

- Location of the Hole Table Title (either above or below the coordinate information).
- The types of hole properties used in the table. You can add or remove properties from a hole table. You can also create new fields that can be added to the hole table.
- The hole properties of Combine Notes, Numbering, and Rollup.

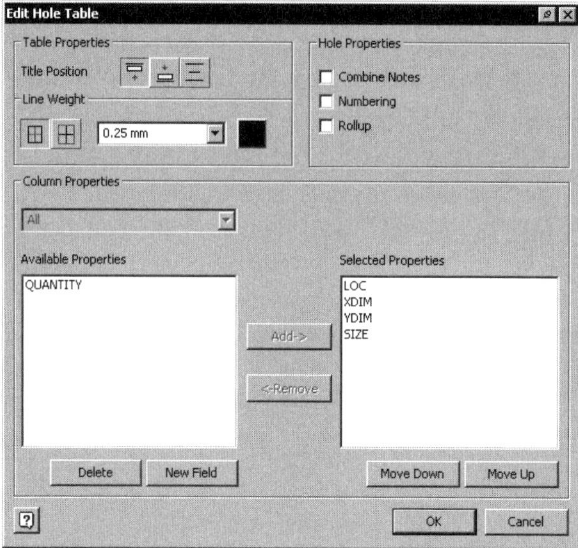


**Figure 9-90**

# EXERCISE 9-2
# Dimensioning Drawing Views

In this exercise, you create a drawing for a shim plate, and then quickly add annotations using baseline dimensions. You then open another shim plate model, and create a hole table. You modify the model to add new holes and update the hole table accordingly. To navigate to the exercise in the *Electronic Student Workbook*, do the following:

1. From the Main TOC page, click Chapter 9.
2. From the TOC for Chapter 9, click Dimensioning Drawing Views.

The following image illustrates the opened exercise.

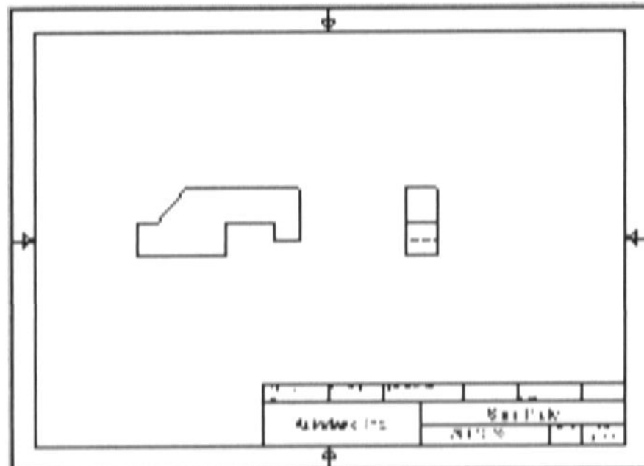

**Figure 9-91 Opened exercise**

The following image illustrates the completed exercise.

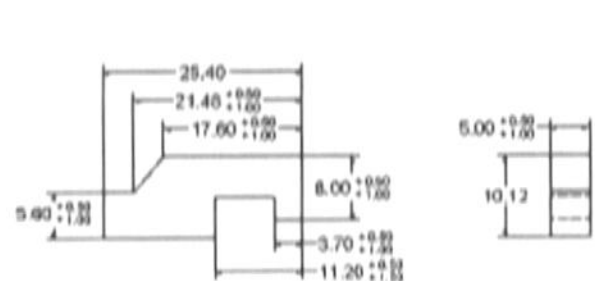

| Hole Table | | | |
|---|---|---|---|
| LOC | XDIM | YDIM | SIZE |
| A1 | 6,00 | 6,00 | Ø2,50 -8,00 DEEP |
| A2 | 6,00 | 10,00 | Ø2,50 -8,00 DEEP |
| A3 | 10,00 | 6,00 | Ø2,50 -8,00 DEEP |
| A4 | 10,00 | 10,00 | Ø2,50 -8,00 DEEP |
| A5 | 14,00 | 10,00 | Ø2,50 -8,00 DEEP |
| A6 | 2,00 | 10,00 | Ø2,50 -8,00 DEEP |
| A7 | 2,00 | 6,00 | Ø2,50 -8,00 DEEP |
| [illegible] | [illegible] | [illegible] | Ø2,50 -8,00 [illegible] |

**Figure 9-92 Completed exercise**

EXERCISE

## CREATING AND USING SKETCHED SYMBOLS

To include a customized symbol on the drawing sheet, such as safety symbols or company logos, you will need to define a new symbol. To create a symbol, right-click on Sketched Symbols under Drawing Resources in the Browser and click Define New Symbol from the menu, as shown in the following image.

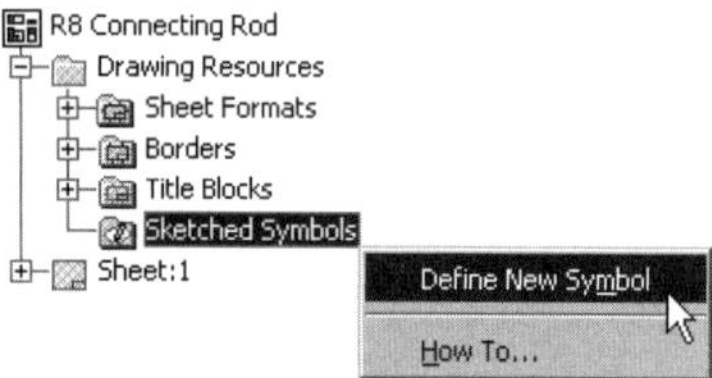


**Figure 9-93**

This action will activate the Sketch Panel Bar. Use the tools from this Panel Bar to create the symbol. To save the symbol, right-click either in the graphics area or select Drawing Resources from the Browser, as shown in the following image. Right-click on Sketched Symbols and click Save Sketched Symbol from the menu.

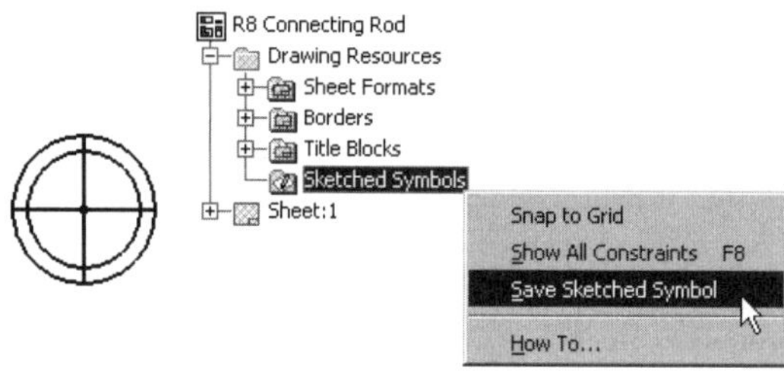


**Figure 9-94**

After selecting the Save Sketched Symbol option, the Sketched Symbol dialog box will display, as shown in the following image. Give the Sketched Symbol a name and click Save in this dialog box.

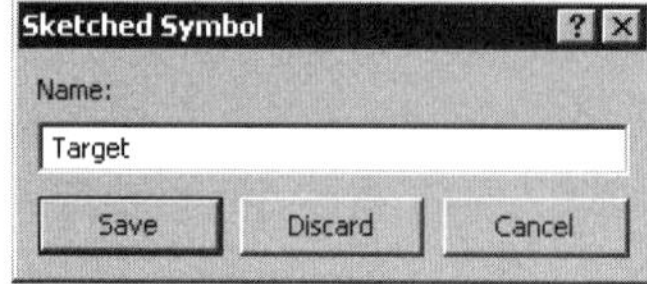


**Figure 9-95**

**NOTE** The new symbol will not be available to new drawings unless it is saved in a template file.

To insert the symbol in the drawing sheet, either double-click on the symbol's name in the Browser under Drawing Resources>Sketched Symbols, as shown in the following image, or right-click on the symbol's name and click Insert from the menu.

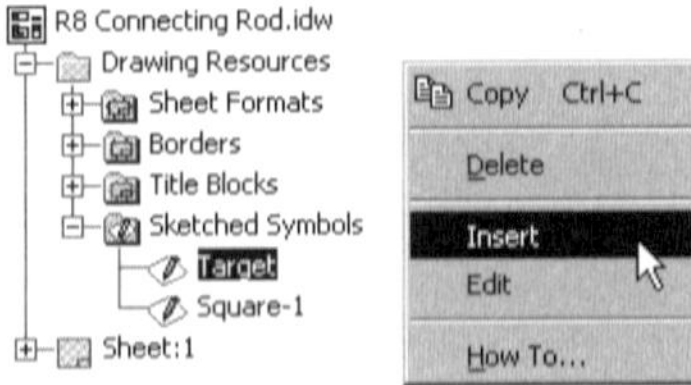


**Figure 9-96**

Once a sketched symbol has been created, an additional method is available to insert the symbol into the drawing through the Symbols dialog box. This dialog box is activated by selecting the Symbols tool from the Drawing Annotation Panel Bar in the following image.

**Figure 9-97**

The Symbols dialog box, as shown in the following image, will display a list of symbols defined in the drawing, along with options for the scaling and rotating of the symbol. A check box is also available to control whether or not a leader is to be applied during the initial placement.

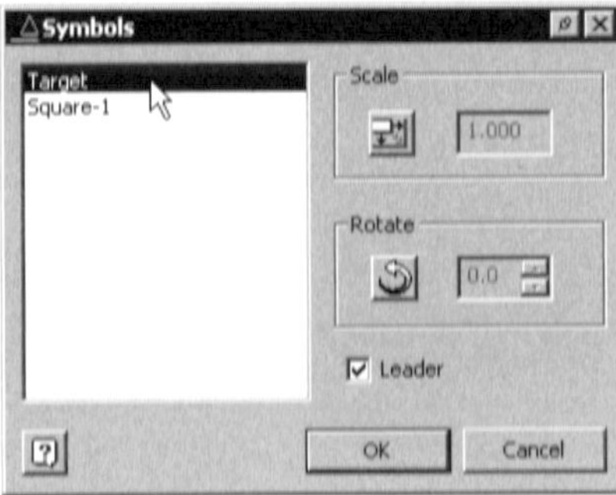


**Figure 9-98**

Insertion points are optional when creating symbols. When an insertion point is not defined in a symbol, scaling and rotating operations occur relative to the geometric center of the symbol. The following image illustrates two symbols placed on a drawing view.

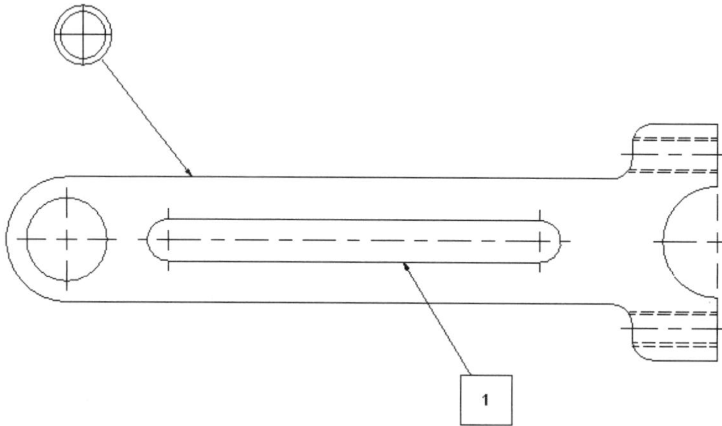


Figure 9-99

## SKETCH OPTIONS

When you sketch in a drawing view, you can also project edges to assist in the creation of a special effect such as a crosshatching operation. To do this, first select the drawing view in which the sketch will be performed. With the view box highlighted, click on the Sketch button in the Standard toolbar. To use existing drawing geometry as a sketch, click on the Project tool, as shown in the following image, and select the edges to sketch. These edges will change color to distinguish them from regular drawing edges.

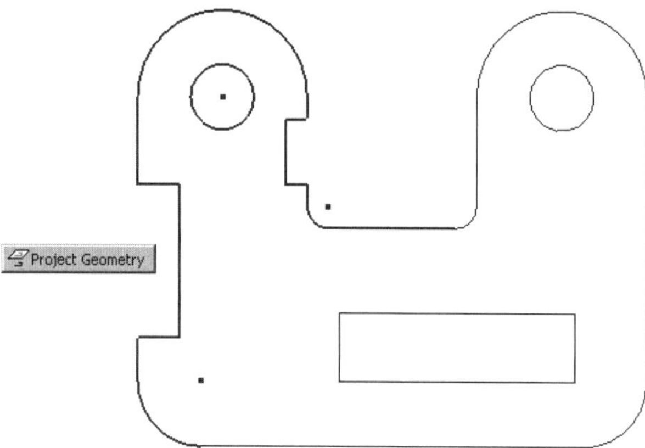


Figure 9-100

In the following image, a spline was created to connect one edge of the sketch with the other. This also formed a closed profile. Next, the Fill/Hatch Sketch Region tool is used to fill in the closed profile with a desired pattern. Right-clicking displays a menu that allows you to select Finish Sketch.

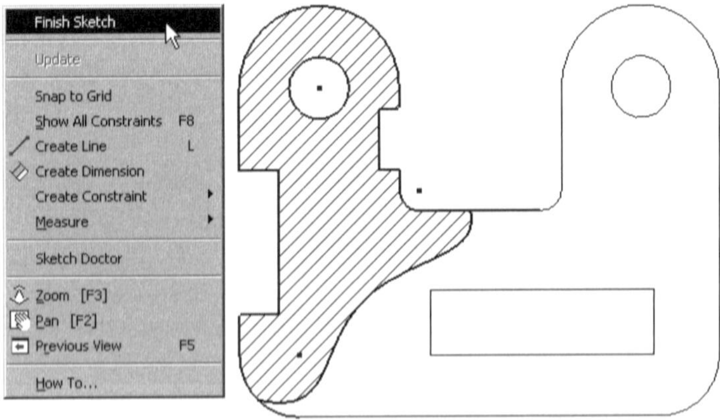


**Figure 9-101**

The results are illustrated in the following image. When the sketch changes back to the drawing view, the line appearance of projected edges, sketched features, and hatch pattern matches that of the current drawing view.

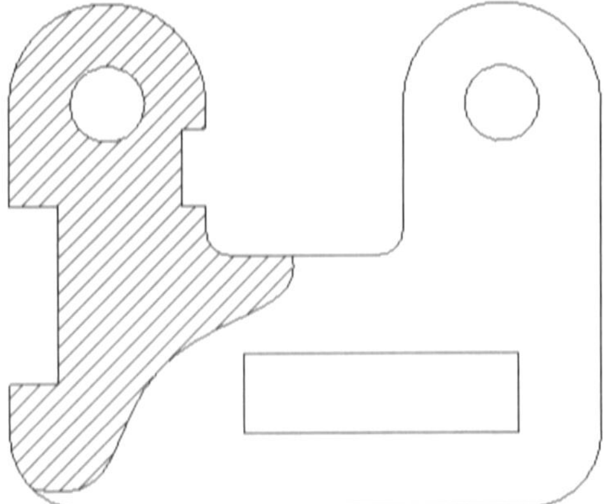

**Figure 9-102**

## LINEAR SYMMETRICAL ARROWHEAD

Based on global standards, you may need to place a dimension in the form of a linear symmetrical arrowhead. This dimension is anchored at one end of the feature being dimensioned. The other end of the dimension consists of double arrowheads that are not attached to an extension line. It is implied that the dimension value is being calculated from one end of the feature to another.

To place a linear symmetric dimension, follow these steps:

1 Click on the General Dimension button located on the Drawing Annotation Panel Bar and then select an edge of the object to be the first extension line location of the dimension.

2 Select another edge to be the second extension line location of the dimension.

3 Move the dimension to a desired location. Instead of placing it at that location, however, right-click to display the menu illustrated in the following image. Select Dimension Type followed by Linear Symmetric.

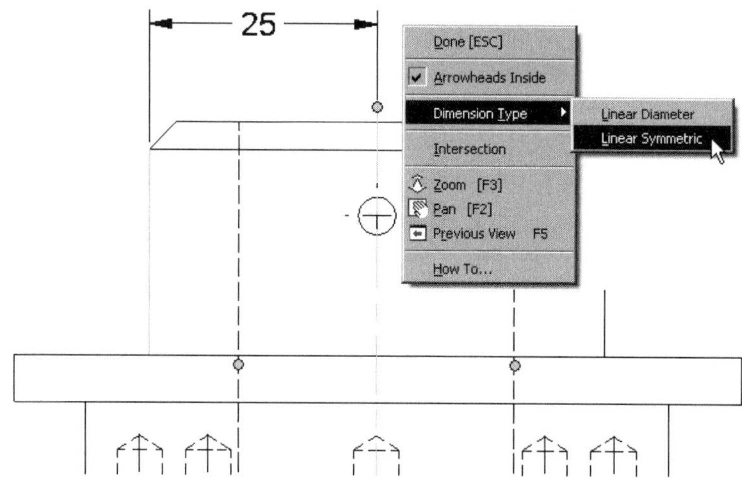


**Figure 9-103**

The results are illustrated in the following image. The two arrows denote the linear symmetric function.

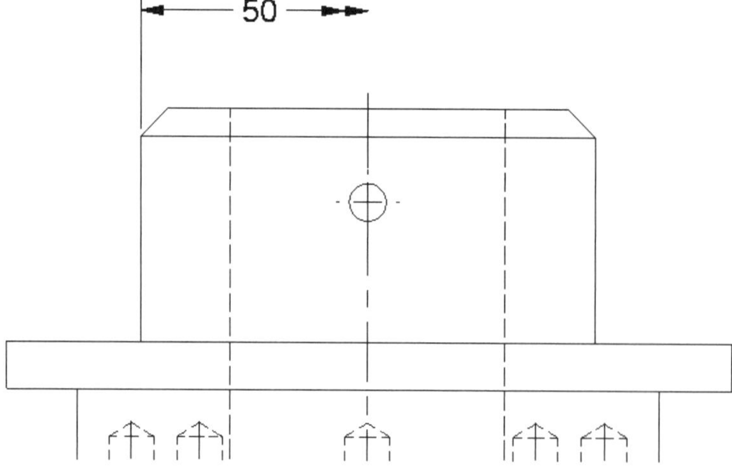

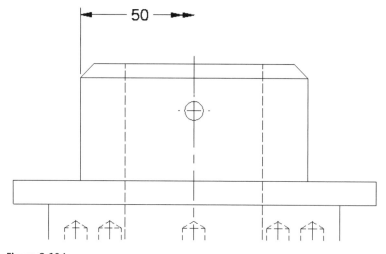

**Figure 9-104**

Selecting the Linear Diameter option in the menu displays the results in the following image. The correct dimension is displayed without the second set of arrows.

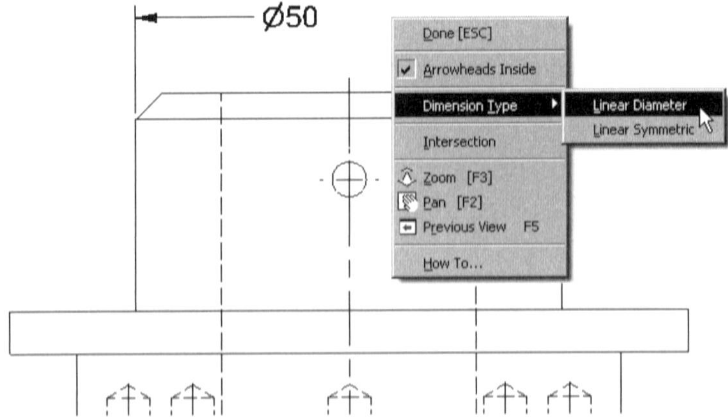


**Figure 9-105**

## Hiding Edges

Depending on the complexity of a view, it may be necessary to hide lines that represent visible edges, hidden edges, or a combination of both. To perform this operation, first select the edges to hide. You can accomplish this by picking the edges individually or by using a window or crossing box. In the following image, a crossing box is used to select a number of hidden edges.

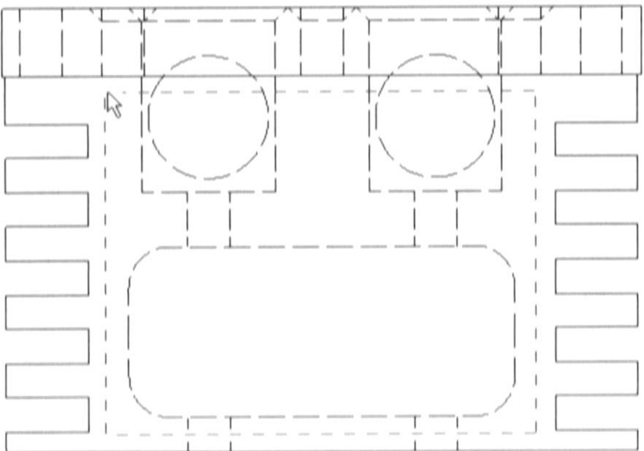

**Figure 9-106**

When these edges are selected, they will turn the color green by default. At this point, right-click to display the menu as shown in the following image. Click on Visibility to remove the check from the box.

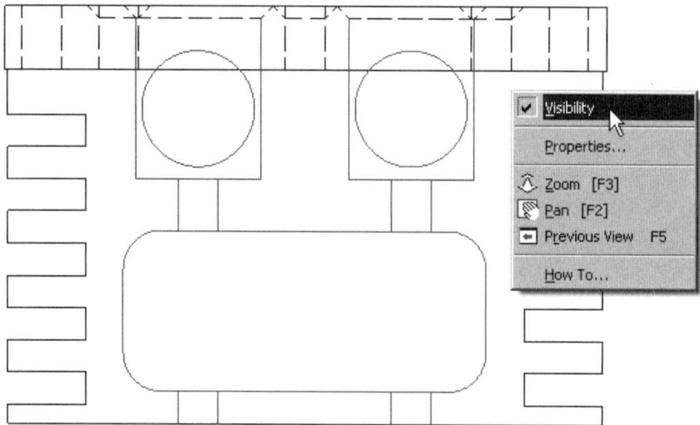


Figure 9-107

The results are illustrated in the following image. All edges that were identified by the green color in the previous image are now turned off or invisible. In the event you wish to reverse this process by turning some or all edges back on, select the view, right-click to display the menu as shown in the following image, and select Show Hidden Edges.

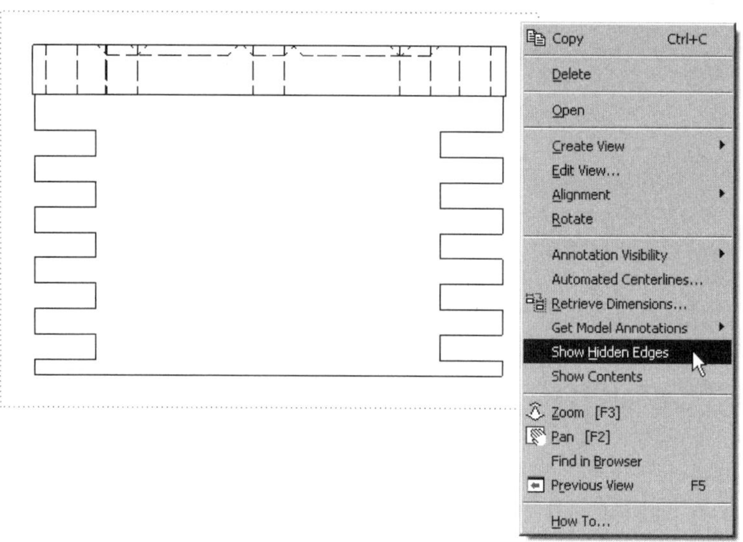


Figure 9-108

All edges previously turned off are now displayed. However, this is only a temporary display. You must pick the edges that you wish to redisplay. You can accomplish this by picking the edges individually or by a window or crossing box. In the following image, the edges located in the upper half of the view are selected; the edges in the lower half of the view remain unselected.

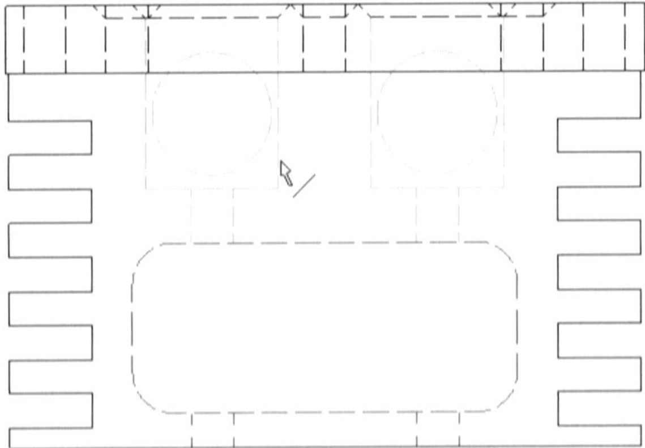

**Figure 9-109**

When you are finished selecting edges to redisplay, right-click and select Done from the menu as shown in the following image.

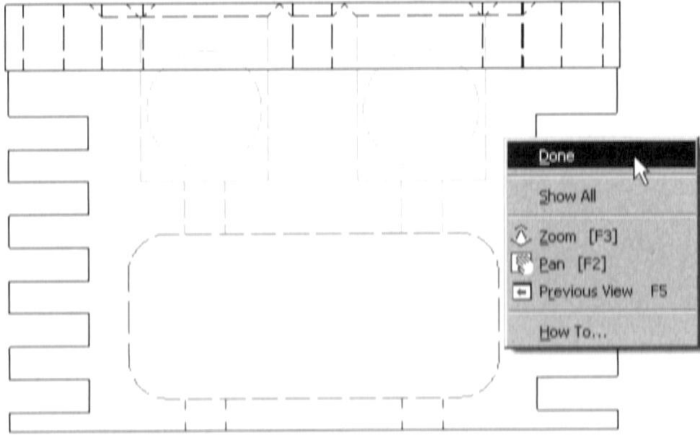


**Figure 9-110**

The results are shown in the following image, with only the selected edges redisplayed. You could also have selected Show all from the menu in the previous image. This would have redisplayed all edges that were initially turned off.

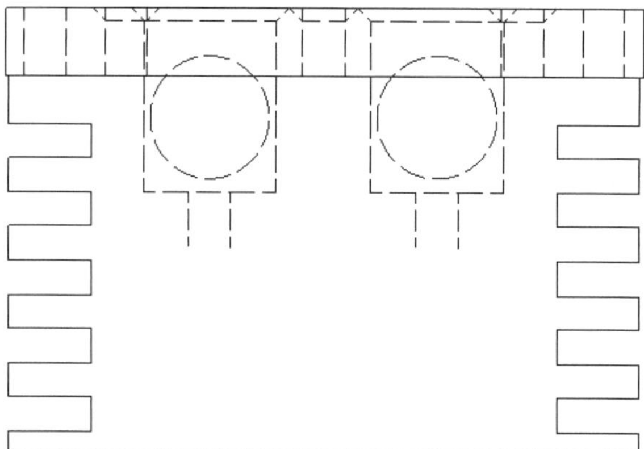

Figure 9-111

## PARTS LIST TOOLS

The information supplied in Chapter 5 developed the basis of inserting and manipulating a parts list. This section will expand the capabilities of the parts list to include the following items:

- A detailed account of the top icons found in the Edit Parts List dialog box.
- The function of the spreadsheet view.
- Information dealing with additional sections of the Edit Parts List dialog box.
- Expanded capabilities of the Edit Parts List dialog box.
- Nested parts lists.

### PARTS LIST OPTIONS (TOP ICONS)

Once a parts list has been placed in the drawing, various controls are available to assist in the modifying of the parts list. With the parts list placed in a drawing, right-click on it to display the menu illustrated in the following image, and click Edit Parts List.

| Parts List | | | |
| --- | --- | --- | --- |
| ITEM | QTY | PART NUMBER | DESCRIPTION |
| 1 | 1 | Main Shaft 1 | Crankshaft Main Shaft |
| 2 | 2 | Shaft End Wheel | Crankshaft End Wheel |
| 3 | 1 | Shaft Center Wheel | Crankshaft Center Wheel |
| 4 | 2 | Shaft Bearing Journals | Crankshaft Journals |
| 5 | 1 | Main Shaft 2 | Crankshaft Aux Shaft |

Figure 9-112

This displays the Edit Parts List dialog box shown in the following image. The top icons of this dialog box consist of numerous features that allow you to change the parts list.

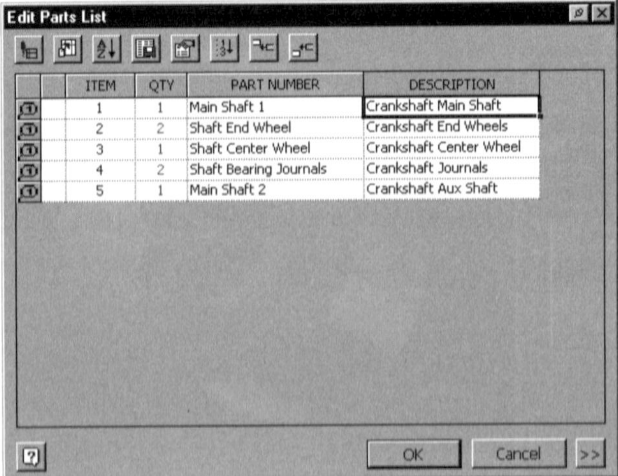


**Figure 9-113**

The following table lists a detailed explanation of all parts list icons:

| | |
|---|---|
| Compare | Compares the parts list to the current properties for the parts. Values that are inconsistent are highlighted in the spreadsheet view. To resolve, right-click on a highlighted cell and select the desired resolution from the menu. |
| Column Chooser | Opens the Parts List Column Chooser dialog box, where you can add, remove, or change the order of the columns for the selected parts list. Data for these columns is populated from the properties of the component files. |
| Row Merge Settings | Opens the Row Merge Settings dialog box, where you can arrange parts, assemblies, and subassemblies that have the same components into a single row. |
| Sort | Opens the Sort Parts List dialog box, where you can change the sort order for items in the selected parts list. |
| Export | Opens the Export Parts List dialog box so that you can save the selected parts list to an external file of the file type of your choice. |
| Table Layout | Opens the Parts List Table Layout dialog box where you can change the title text or heading location for the selected parts list. |
| Renumber | Used to change item numbers of parts in the parts list. |
| Add Custom Parts Below | Adds a new row below a selected row in the parts list so you can then enter the custom parts information. |
| Add Custom Parts Above | Adds a new row above a selected row in the parts list so you can then enter the custom parts information. |

**NOTE** Custom parts are useful for displaying parts in the parts list that are not components or graphical data such as paint or a finishing process.

To remove a custom row, right-click within the row and select Remove Custom Part from the menu.

## SPREADSHEET VIEW

This is the middle portion of the Edit Parts List dialog box where the contents of the selected parts list is shown. As you make changes to the parts list properties, the changes are reflected in the spreadsheet. Items that have corresponding balloons are marked with a balloon symbol on the left hand column.

| ① | +/- | ITEM | QTY | PART NUMBER |
|---|---|---|---|---|
| ① | | 1 | 1 | 2002 motor house bottom |
| ① | | 2 | 1 | 2002 motor house top |
| ① | | 3 | 1 | 2002 Motor Logo |
| ① | | 4 | 1 | 2002 Motor Cover |
| ① | | 5 | 2 | 2002 Motor Clamp |
| ① | | 6 | 2 | motor house screw |
| ① | | 7 | 1 | 2002 motor Transmit 1 |
| ① | | 8 | 1 | 2002 motor Transmit2 |
| ① | | 9 | 1 | 2002 motor Transmit shaft |
| ① | | 10 | 1 | coupler2 |
| | | 11 | 1 | 2002 Transmission Key |
| ① | | 12 | 1 | 2002 motor |
| ① | | 13 | 1 | 2002 motor bushing |
| ① | | 14 | 1 | 2002 motor shaft |
| ① | | 15 | 1 | 2002 motor gear |

Figure 9-114

### Nested Parts Lists

An added function of a parts list is to distinguish individual parts from subassemblies. In the case of subassemblies, a visual representation in the form of a +/- button will be found along the left column. Clicking on a + button will expand the subassembly and display a list of its component parts. In the example shown in the following image, the Crankshaft Assembly is first identified with a + and a 1.1 item number. When expanding the Crankshaft Assembly, the individual components of the nested items are displayed with an additional number, such as the Main Shaft 1 part, which is listed as item 1.1.

| ① | +/- | ITEM | QTY | PART NUMBER |
|---|---|---|---|---|
| | + | 1 | 1 | Crankshaft Assembly |
| | | 2 | 1 | Engine Block |
| | | 3 | 2 | Cylinder |
| | | 4 | 2 | Cylinder Sleeve |
| | | 5 | 2 | Cylinder Head |
| | | 6 | 2 | Valve Head |
| | | 7 | 2 | Piston |
| | | 8 | 2 | Wrist Pin |
| | | 9 | 2 | Connecting Rod |
| | | 10 | 2 | Rod Cap |
| | | 11 | 4 | Rod Cap Screw |

| ① | +/- | ITEM | QTY | PART NUMBER |
|---|---|---|---|---|
| | - | 1 | 1 | Crankshaft Assembly |
| ① | | 1.1 | 1 | Main Shaft 1 |
| ① | | 1.2 | 2 | Shaft End Wheel |
| ① | | 1.3 | 1 | Shaft Center Wheel |
| ① | | 1.4 | 2 | Shaft Bearing Journals |
| ① | | 1.5 | 1 | Main Shaft 2 |
| | | 2 | 1 | Engine Block |
| | | 3 | 2 | Cylinder |
| | | 4 | 2 | Cylinder Sleeve |
| | | 5 | 2 | Cylinder Head |
| | | 6 | 2 | Valve Head |
| | | 7 | 2 | Piston |
| | | 8 | 2 | Wrist Pin |
| | | 9 | 2 | Connecting Rod |
| | | 10 | 2 | Rod Cap |
| | | 11 | 4 | Rod Cap Screw |

Figure 9-115

If balloons are applied to a drawing view, as shown in the following image, the balloons will expand to hold the nested parts list items.

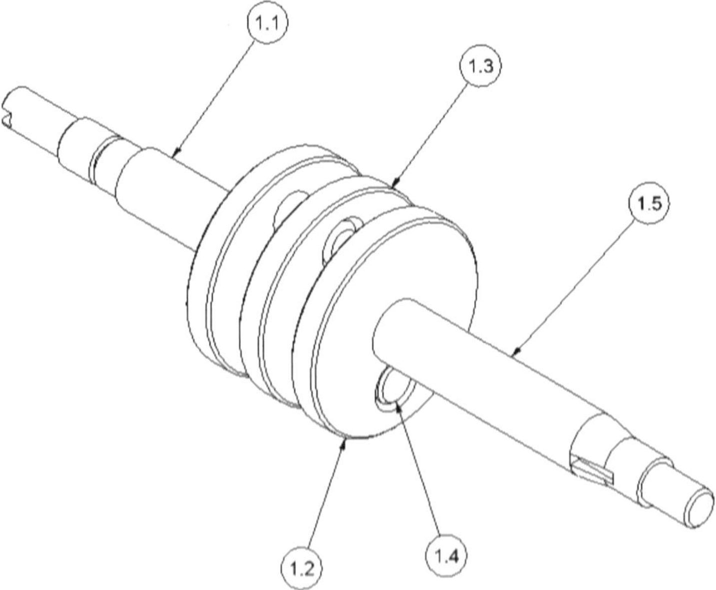


Figure 9-116

**NOTE** If a parts list is in need of being updated, the icon associated with the parts list in the Browser will be identified by a red lightening bolt as a visual reminder. Right-click on the parts list in the Browser and select Update from the menu to make the parts list up-to-date with the changes of the drawing.

Selecting the Parts List tab of the Standards dialog box (shown in the following image) will allow for more controls to be applied to a parts list. These controls are listed as follows:

- Specifying a text style.
- Controlling the position of the heading.
- Adjusting the row and columns.
- Providing a nested list delimiter.
- Activating an inheritance option that expands parent part changes to nested child parts.

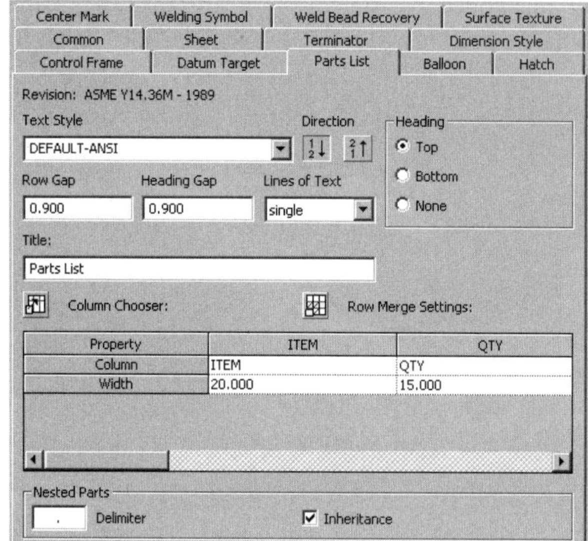


**Figure 9-117**

## Parts List Column Settings

When you right-click on a column heading in the Edit Parts List dialog box, a menu is displayed as shown in the following image. Use this menu to control the table layout, column format, and column width of the parts list. Each of these items will be discussed on the next several pages.

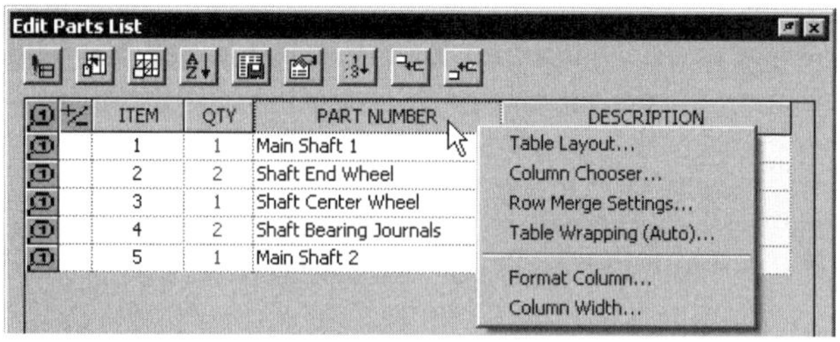

**Figure 9-118**

## Table Layout

While the title and title location are already set through the active drafting standard, you can change this setting for the parts in which you are currently working. To make these changes, use the following steps:

1 In the Edit Parts List dialog box, right-click on a column heading and then select Table Layout from the menu as shown in the following image. You could also click the Table Layout button located in the Edit Parts List dialog box.

2 When the Parts List Table Layout dialog box appears, make changes to the Part List Title if necessary. You can change the placement of the parts list title. You can also change the ordering direction and the spacing in between rows through this dialog box.

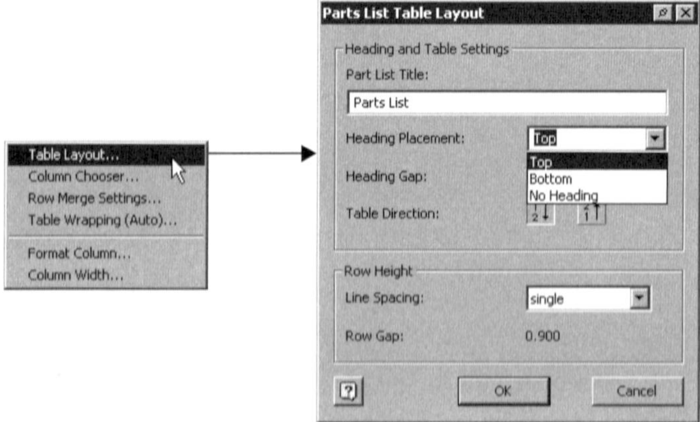


Figure 9-119

## Column Chooser

The Parts List Column Chooser dialog box illustrated in the following image allows you to add, remove, and reorder the columns within the parts list. Numerous properties are available for you to choose. Simply click on a property and click the Add button to add the item to the Selected Properties list. You can also remove items from the Selected Properties list.

1 In the Edit Parts List dialog box, right-click on a column heading and pick Column Chooser from the menu. Alternatively, you could also pick the Column Chooser button from the Edit Parts List dialog box.

2 In the Parts List Column Chooser dialog box, add or remove the desired columns. You can also change the order of the items in the Selected Properties edit box.

3 You can also create additional properties by clicking on the New Property button. After you enter a new property in the Define New Properties dialog box, this entry is added to the Selected Property list.

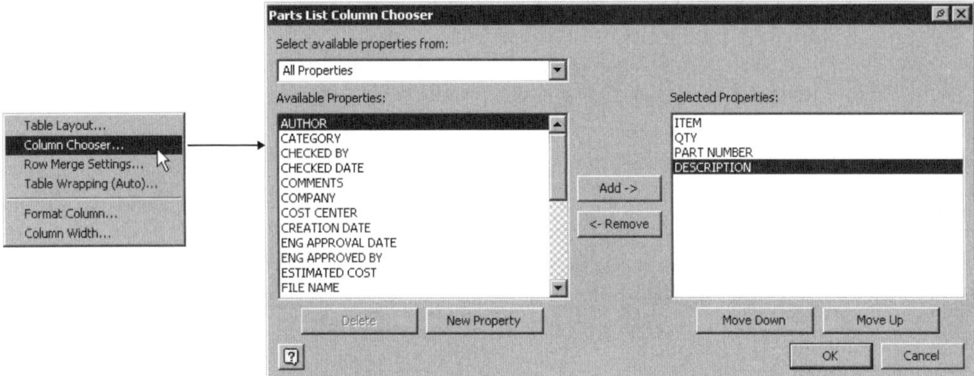


Figure 9-120

## Row Merge Settings

Parts, assemblies, and subassemblies that have the same components can be arranged in a single row in the Parts List. Use the following steps to accomplish this task:

1 In the Edit Parts List dialog box, right-click on a column heading and select Row Merge Settings. Alternatively, you could also click the Merge Row Settings button from the Edit Parts List dialog box.

2 When the Merge Row Settings dialog box is displayed, select Merge Similar Components as shown in the following image.

3 Click the arrow to select the Component Types to Merge in the edit box. Component types include Parts only, Parts and Assemblies Separately, and Parts and Assemblies Together.

4 Select a property to merge in the First Key box.

5 Select another property in the Second Key box if further filtering is necessary.

6 Select another property in the Third Key box if more filtering is needed.

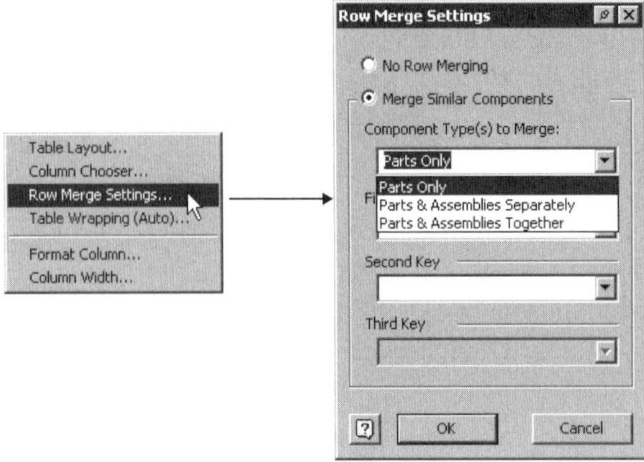


Figure 9-121

## Table Wrapping (Auto)

If a parts list is too long, you can divide it into sections extended to the left or right of the main table:

1. Select the parts list, right-click, and then select Edit Part List.
2. Right-click on a column heading and then select Table Wrapping (Auto).
3. Set the number of sections and the placement direction relative to the first column. Select Wrap Automatically to cause the table to be split equally.

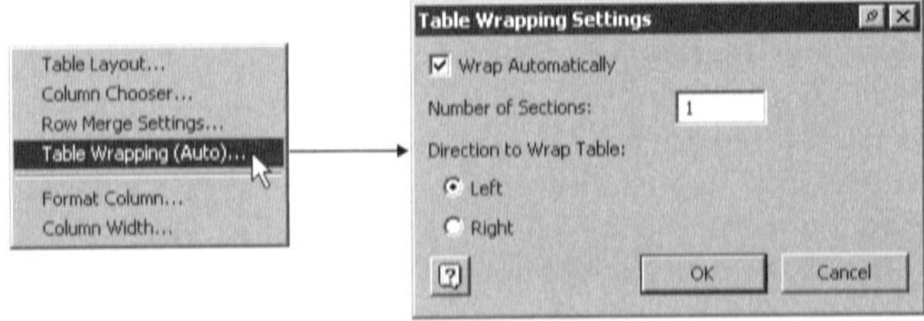


Figure 9-122

## Format Column

You can change the default column formatting properties and create substitution values for the selected parts lists.

While in the Column Format tab, shown in the following image, you can set the formatting and alignment properties for the column selected in the Edit Parts List dialog box.

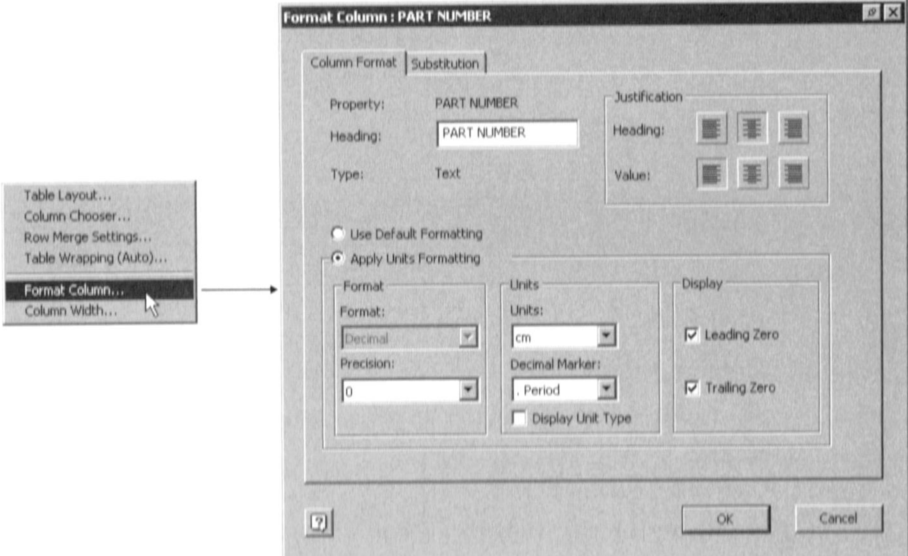


Figure 9-123

The Substitution tab shown in the following image is used to substitute the values of the selected column with the new values from another column. To activate the various fields under this tab, first select Enable Value Substituting. You will need to make changes to both of the following fields.

### When exists, use value of:

The name in this edit box represents the property to be substituted for the selected column. When you click on the arrow, you will be able to browse for a property from the supplied list. You can also create and add a new property to the list by clicking on the New Property button.

### When rows are merged, value used is:

This edit box contains two settings that control how the substituted value is calculated when the rows are merged. Setting this edit box to First Row will calculate only the first row. Setting this edit box to Sum of Values will calculate all rows. This option adds the values of the cells of the substituted row and shows the sum in the property fields of the selected column.

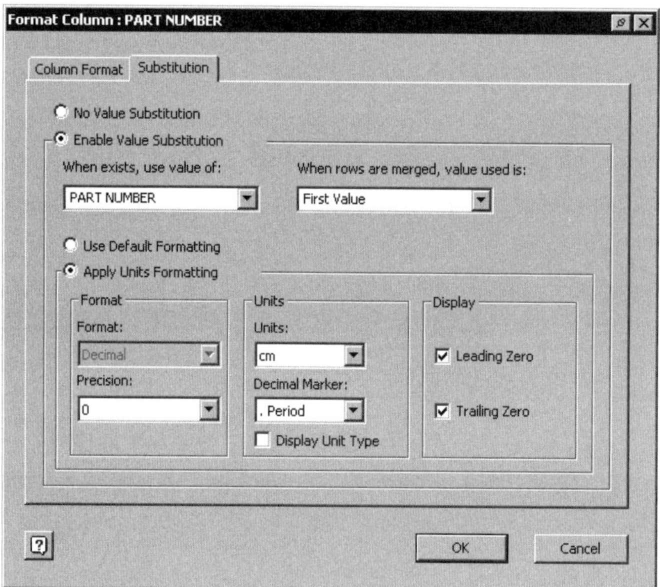


Figure 9-124

## Column Width

The width of any column can be easily changed using the following steps:

1 While in the Edit Parts List dialog box, right-click on a column heading, and then select Column Width.

When the Column Width dialog box is displayed, as shown in the following image, enter a new value for the column width.

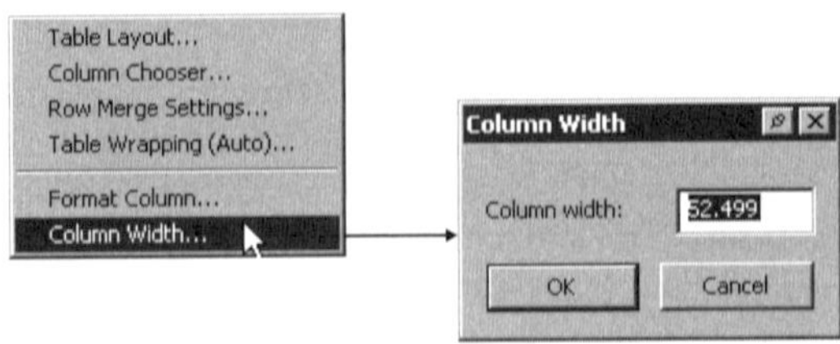


Figure 9-125

## Parts List Cell Settings

A special menu is also available when you right-click on a cell in the Edit Parts List dialog box. Options designed to affect the cell include Insert Custom Part, Remove Custom Part, Visibility, Wrap Table at Row, Update All, Update Values, and Freeze All (as shown in the following image).

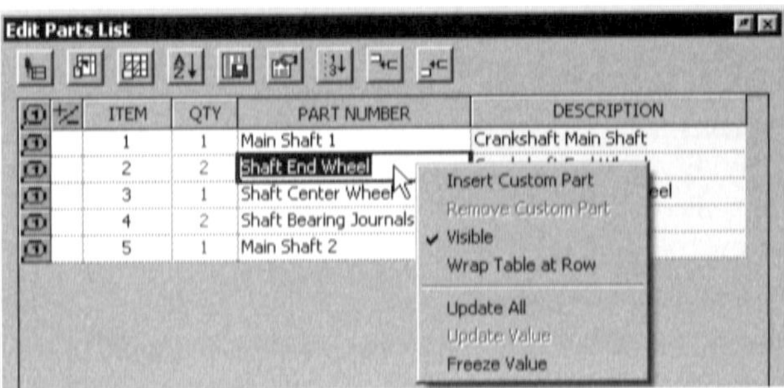


Figure 9-126

## Insert Custom Part

Custom parts can be added to the parts list. While in the Edit Parts List dialog box, right-click on any value in a cell and then choose Insert Custom Part from the menu. You can also create custom parts using two buttons located in the Edit Parts List dialog box. These buttons will insert a custom part either before or after the selected cell.

## Remove Custom Part

If you need to remove the entry of a custom part from the parts list, right-click on any value in a cell and then choose Remove Custom Part from the menu. This will delete the custom part from the parts list

## Visible

This option allows you to display or hide rows or columns in a parts list. While in the Edit Parts List dialog box, select one or more rows or columns in the table. Right-clicking in any selected cell will display a menu. Choose Visibility to hide cell values. These values will now be grey.

### Wrap Table at Row

A parts list that is very long can be divided into sections that extend to the left or right of the main table. To accomplish this task, first select a row or cell in the parts list. Then right-click and select Wrap Table at Row from the menu. Rows that wrap from the selected row or cell are highlighted in the color orange. Finally, click OK to apply the wrapping operation.

**NOTE** You can also specify settings for wrapping the table into equal columns by right-clicking on a column heading, and then selecting Table Wrapping (Auto). The cell color indicates the state of the row.

Purple: Automatic Table Wrap is active. Automatic is selected in the Table Wrapping Settings dialog box. The table wraps into equal columns as specified in the Table Wrapping Settings dialog box.

Orange: Custom Table Wrap is active. The table wraps at the selected row.

### Update Value/ Update All

Parts list cells will be displayed in a yellow color in the Edit Parts List dialog box when changes are made to the model that affect the parts list values. Parts list cells that have override values are also shown in yellow. Follow the steps below to update the values in the cells for them to be in sync with the model.

1 While in the Edit Parts List dialog box, select a cell with the yellow highlighting and right-click.
2 Choose Update Value to update the value in the cell to the same value contained in the model.
3 Choose Update All to update all values in cells highlighted in yellow to the values contained in the model.

**NOTE** You can also elect to keep the values and remove the yellow highlighting from the cells

### Freeze Value

In order to protect the values found in parts list cells from being updated, you may elect to freeze the values. This will prevent any updates from occurring in the selected cells. You cannot edit values in these cells until the Freeze condition is changed.

To freeze cells in a parts list, select a table row containing the cell to freeze in the Edit Parts List dialog box. Then right-click in the selected sell and choose Freeze Value.

**NOTE** To signify that a cell is frozen, a blue highlight appears.

## Using Part and Assembly Model Properties

You can provide additional functionality to your drawing in the form of properties. Activating the Drawing tab of the Document Settings dialog box will display the following image. In this dialog box, you may specify a Custom Property Source. This custom property source is any Autodesk Inventor file or template.

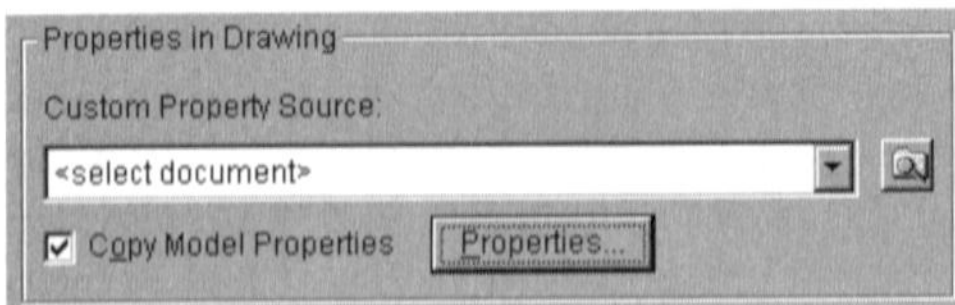


**Figure 9-127**

Checking the box next to Copy Model Properties activates the Properties button. Clicking this button will display the Properties dialog box, as shown in the following image. If a custom property source has been specified, this list will include all custom properties from the source. If no custom property source is identified, the list will contain only the standard model properties.

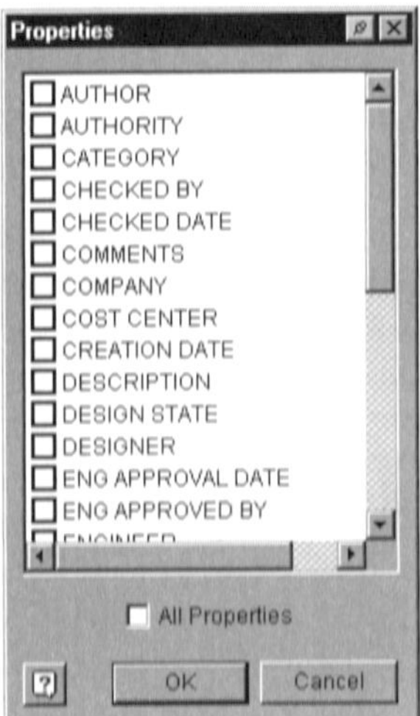


**Figure 9-128**

## Default Drawing File Name

The default drawing file name can be derived from the model's file name. When you first enter a new drawing file, the file name is listed as *Drawing1.idw*, as shown in the following image.

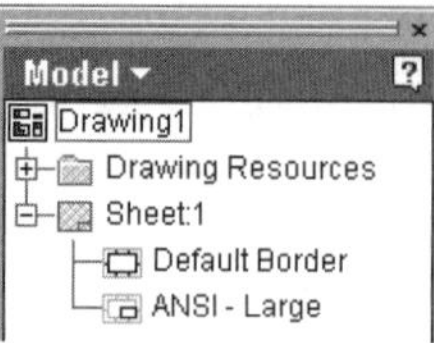


Figure 9-129

Once the first drawing view is placed, this action derives a suggested file name. In the following image, the file name *01-01-0010* is derived from the model's file name and is only a suggestion. When the drawing is saved for the first time, this name is presented for you to accept or override with a different name.

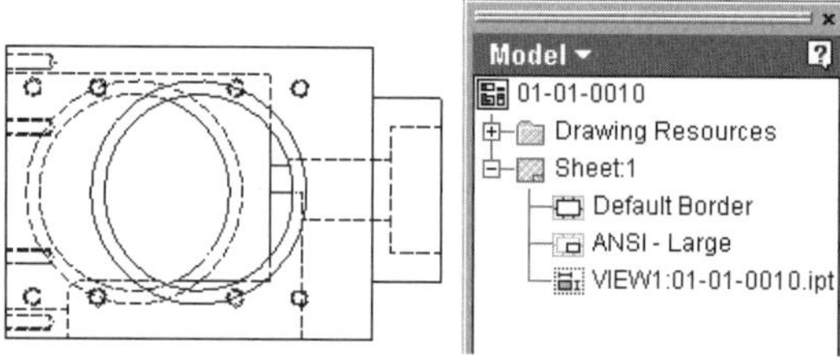


Figure 9-130

## Defer Drawing Update

You can turn off the automatic updating of drawings. This option is set through the Drawing tab of the Document Settings dialog box as shown in the following image. Checking the Defer Updates box will place the drawing in a deferred state. In this state, the drawing will not automatically update to changes in the part, assembly, or presentation files.

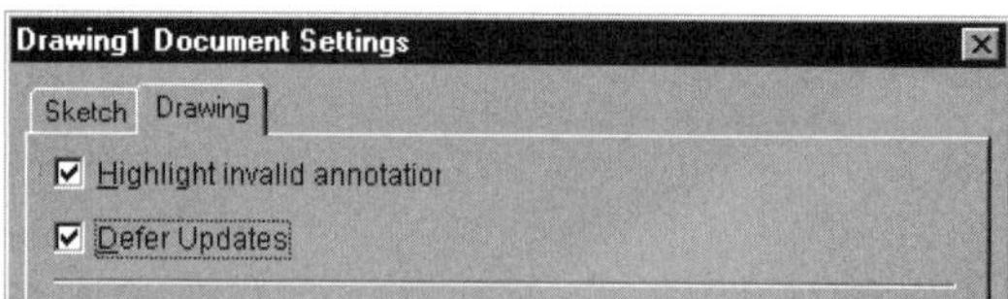


Figure 9-131

After you save a deferred drawing and re-enter it at a later time, the following dialog box will display alerting you to the fact that you are opening a drawing file where updates are deferred.

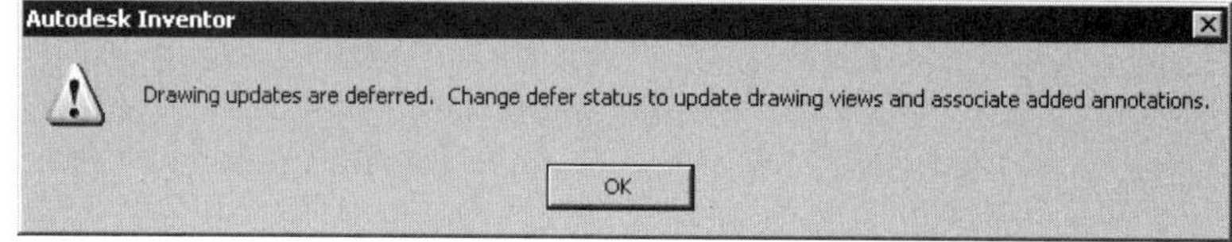


Figure 9-132

The drawing icon in the Browser turns blue-green to show that it has been set to a deferred state. This makes it easier recognize drawings that will not update.

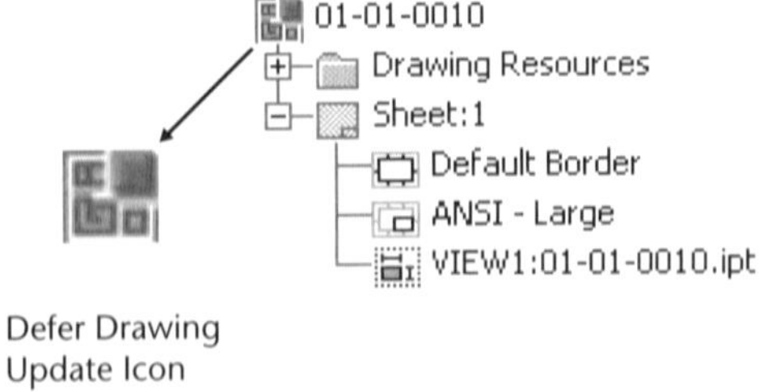


**Figure 9-133**

## DEFER DRAWING UPDATE ON OPEN

In the previous example of deferring a drawing update, you first had to load the drawing file before setting the controls to defer the drawing updates. You can also make changes to defer a drawing update by first locating the idw file in the Open File dialog box. Selecting this drawing file and then clicking on the Options button shown in the following image will launch the File Open Options dialog box. It is here that you can set the defer drawing update control without opening up the drawing file. This is especially beneficial when you are dealing with very large drawing files that you do not wish to load.

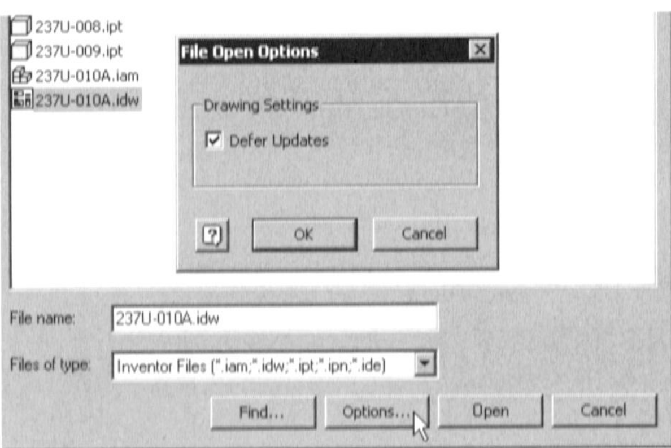


**Figure 9-134**

## CREATING A REVISION BLOCK

During the course of documenting mechanical designs, the drawing document typically undergoes numerous changes. For legal, technical, and other reasons, companies have adopted the practice of maintaining records of these changes (also known as revisions). Revision records come in the form of Engineering Change Orders (ECO) and Engineering Change Notices (ECN). To formally display and keep track of these changes on the drawing, a revision block is inserted. The following tools are available in Autodesk Inventor for tracking revisions in a drawing: the Revision Table tool and the Revision Tag tool.

The Revision Table tool in the following image is found under the Drawing Annotation Panel bar.

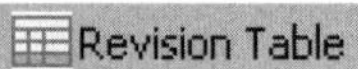


**Figure 9-135**

Once the Revision Table button is selected, a revision block attaches to your cursor for placement in a drawing. The default revision block is illustrated in the following image. A drawing can only have one revision block. If more than one is used, all occurrences will reflect the same information. As with the placement of a parts list, the revision block will initially have a series of green dots displayed at its corners. These green dots will allow the revision block to be accurately placed in any corner of the title block. All examples of revision blocks for this chapter will be placed in the upper right corner of the title block.

| REVISION HISTORY | | | | |
|---|---|---|---|---|
| ZONE | REV | DESCRIPTION | DATE | APPROVED |
| 1 | 1 | Value | 7/26/2002 | Name |

**Figure 9-136**

Once you have placed the revision block in the desired location, you must edit the contents of the various fields in order to reflect the most current changes in the drawing. To do this, double-click on the text within the revision block. This will launch the Format Text dialog box, as shown in the following image. Add the appropriate text to the dialog box and, when finished, click the OK button. This operation should be reflected in the specific field of the revision block.

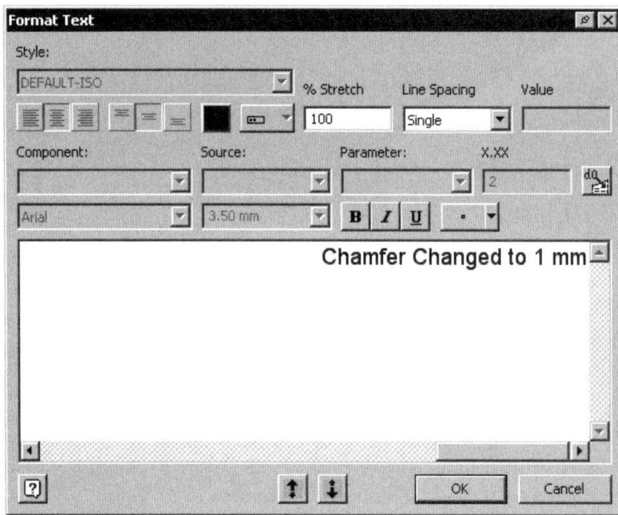


**Figure 9-137**

To edit the format of the revision block, right-click on the revision block and choose the Edit option. This will launch the Edit Revision Table dialog box, as shown in the following image. Revision block column properties can be added or removed depending on the desired information you want to display.

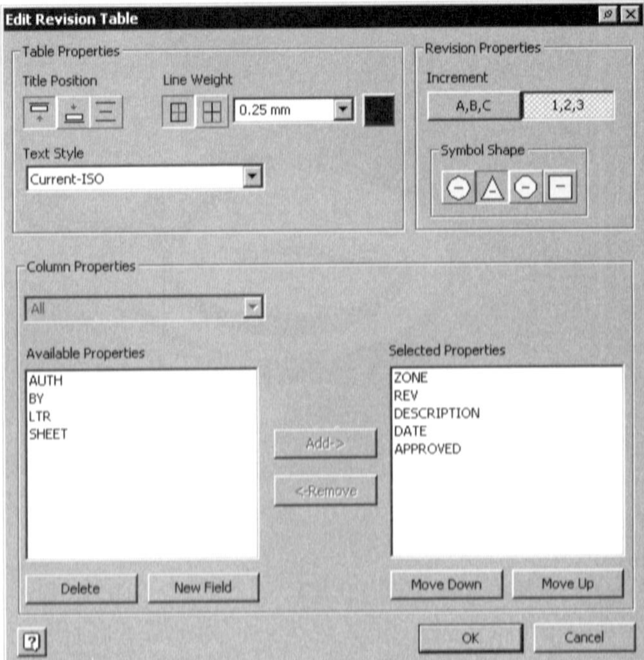


**Figure 9-138**

To lend further support for the revision block, a revision tag is also available. The Revision Tag tool, illustrated in the following image, is accessed through the Drawing Annotation Panel Bar. This tool allows you to tag an item on the drawing. The tag information is formatted according to the options selected in the Revision Tag dialog box. Revision numbers or letters will automatically increment.

**Figure 9-139**

The following image illustrates a revision block and tag applied to an object in a drawing file.

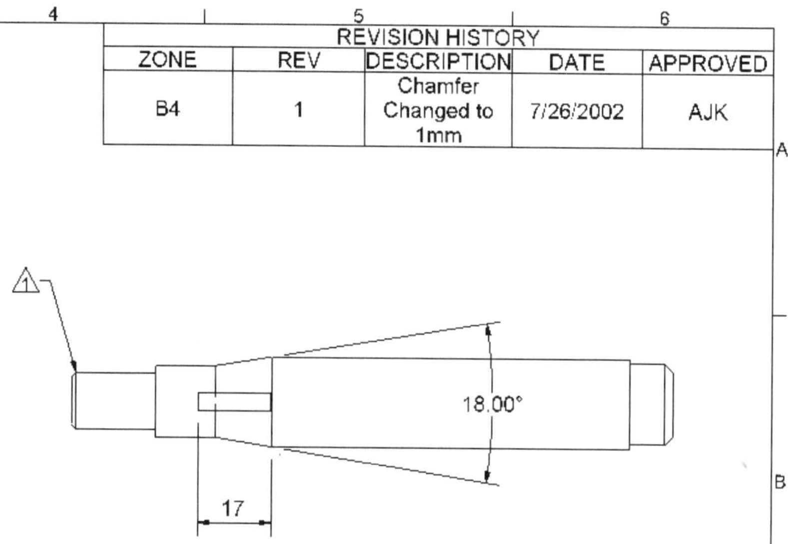


Figure 9-140

## PRINTING DRAWING FILES

When the time comes to print or plot drawing sheets, click Plot located under the File menu to activate the Plot dialog box as shown in the following image.

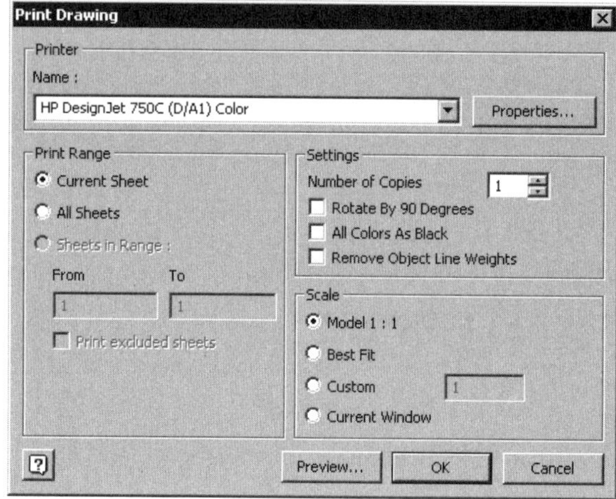


Figure 9-141

The various areas of this dialog box are explained as follows:

**Name:** Click the arrow and then select a valid printer or plotter from the list.

**Properties Button** Clicking on this button will open the Print Setup dialog box. It is here that you can set the paper size and orientation of the plot.

**Print Range** Use this area to specify the sheets to print. If you click on Current Sheet, only the active drawing sheet will be printed. Clicking on All Sheets will print all drawing sheets. Checking the box next to Exclude from Printing will allow you to identify the drawing sheets to exclude from the printing process. If you click on the Sheets in Range button, you will be able to specify the range of sheets to print in the From and To boxes.

**Settings** Use this area of the Print Drawing dialog box to set overrides for color, lineweight, and rotation. Print settings are applied at print time. Settings options are described in greater detail in the table that follows.

| | |
|---|---|
| Number of Copies | Enter the number of copies to print in the edit box provided. |
| Rotate By 90 Degrees | Checking this box will rotate the drawing by 90 degrees on the paper. |
| All Colors As Black | This will force all drawing colors to be printed in black and white. |
| Remove Object Line Weight | Checking this box will print all lines with the same width regardless of how the line weight is set in the drawing. |

**Scale** This area is used to set the scale between the drawing sheet and the paper size defined by the printer or plotter. Scale options are described in greater detail in the table that follows.

| | |
|---|---|
| Model 1:1 | This setting sets the drawing sheet and paper to the same scale. |
| Best Fit | This setting forces the drawing sheet to fit the paper size. |
| Custom (Model:Paper) | Use this setting to enter a custom scale. |
| Current Window | This setting will scale the entire drawing to fit the paper size. |

# EXERCISE 9-3 Annotations

This exercise is a continuation from a previous exercise originally performed in Chapter 6. On a different drawing sheet, you place a hole note, thread note, and insert a sketched symbol into the drawing title block. On another drawing sheet, you add a revision table and revision tag to keep track of changes made to the drawing. To navigate to the exercise in the *Electronic Student Workbook*, do the following:

1 From the Main TOC page, click Chapter 9.

2 From the TOC for Chapter 9, click Annotations.

The following image illustrates the opened exercise.

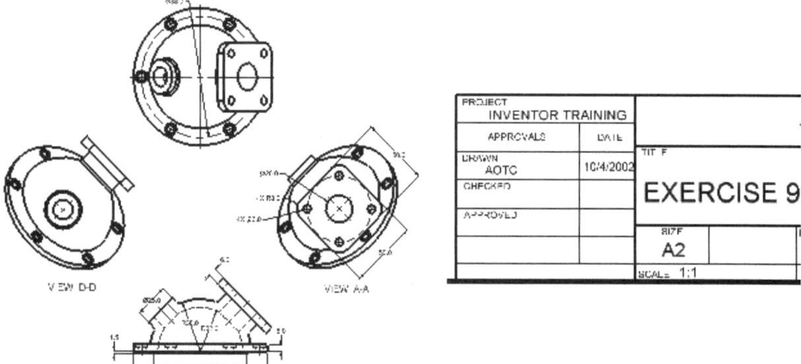


**Figure 9-142 Opened exercise**

The following image illustrates the partially completed exercise.

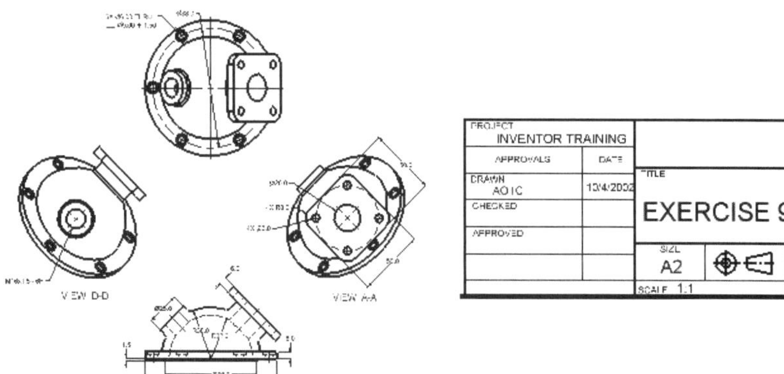


**Figure 9-143 Partially completed exercise**

## CHAPTER SUMMARY

| To | Do This | Tool |
|---|---|---|
| Create a broken view | Click the Broken View tool from the Drawing Views Panel Bar. |  |
| Create a break-out view | Click the Break-Out View tool from the Drawing Views Panel Bar. |  |
| Create a draft view | Click the Draft View tool from the Drawing Views Panel Bar. |  |
| Create dual dimensions | Create a new dimension style that includes alternate dimensions. | |
| Create auto baseline dimensions | Click the Baseline Dimension tool from the Drawing Annotation Panel Bar. |  |
| Create an ordinate dimension set | Click the Ordinate Dimension Set tool from the Drawing Annotation Panel Bar. |  |
| Create ordinate dimensions | Click the Ordinate Dimension tool from the Drawing Annotation Panel Bar. |  |
| Create a hole table by selection | Click on the Hole Table - Selection tool from the Drawing Annotation Panel Bar. |  |
| Create a hole table by view | Click the Hole Table - View tool from the Drawing Annotation Panel Bar. |  |
| Create a hole table by selected type | Click the Hole Table - Selected Type tool from the Drawing Annotation Panel Bar. |  |
| Create a sketched symbol | Click on the Sketched Symbol listing in the Browser. |  |
| Create a revision block | Click on the Revision Table tool from the Drawing Annotation Panel Bar. |  |
| Create a revision tag | Click on the Revision Tag tool from the Drawing Annotation Panel Bar. |  |

# APPLYING YOUR SKILLS

## Skill Exercise 9-1

In this exercise, you open an existing drawing and then use complex drawing view creation techniques to document a part. You also create a revision block and tag to complete the drawing. To navigate to the exercise in the *Electronic Student Workbook*, do the following:

1 From the Main TOC page, click Chapter 9.

2 From the TOC for Chapter 9, click Exercise 1 under Applying Your Skills.

The following image illustrates the completed exercise.

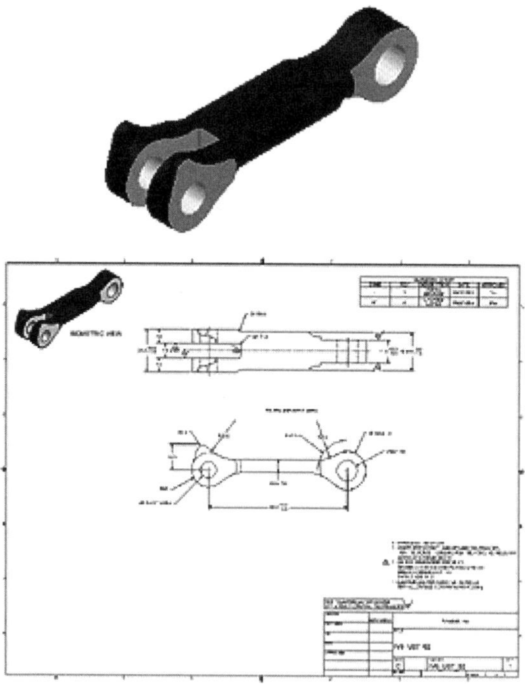

**Figure 9-144 Completed exercise**

## CHECKING YOUR SKILLS

1 **True___ False___** A broken view can only be derived from a base view.

2 **True___ False___** Auxiliary views typically show internal features of a part that have been cut.

3 **True___ False___** Imported AutoCAD 2D information is shared with Autodesk Inventor as a draft view.

4 **True___ False___** A work feature will display in a drawing view even if the same work feature in the part model is turned off.

5 Explain how to create a perspective drawing view.

______________________________________________________________

______________________________________________________________

______________________________________________________________

6 **True___ False___** When using model sketches in drawing views, be sure the model sketch is completely constrained and dimensioned.

7 **True___ False___** Ordinate dimensions are for reference only and cannot parametrically change the part's size.

8 **True___ False___** Hole tables can be created from extruded circles.

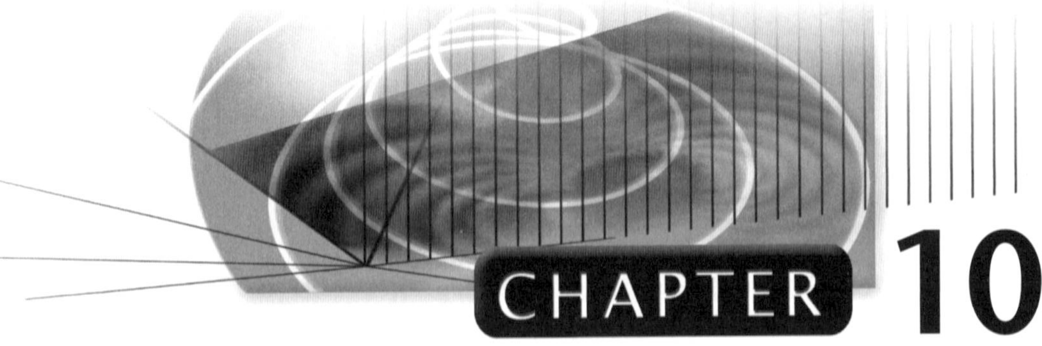


# Complex Assembly Modeling

This chapter is a continuation of Chapter 6, Creating and Documenting Assemblies. Topics covered in this chapter include the following: the use of Design Views as a means of displaying an assembly in different states; creating predefined assembly constraints called iMates; animating the assembly model through the use of drive constraints; replacing one component with another; the sharing of standard items such as nuts, bolts, and washers through the Content Library; the creation of rectangular and circular component patterns; the ability to add features while in an assembly file without changing the individual part components; and creating a 2D design layout in an assembly to test the design for functionality. Adaptive design techniques will also be expanded on to include sketches and features at the subassembly level.

## CHAPTER OBJECTIVES

### After completing this chapter, you will be able to

- Create Design Views to show the different states and viewing directions of an assembly
- Use assembly selection tools to narrow the number of parts that you can work with
- Create iMates
- Drive assembly constraints to simulate motion in an assembly model
- Check moving components for freedom using the contact detection solver
- Replace a component with another component
- Insert parts from the content library
- Set up a new content library
- Add a new definition to the content library
- Create rectangular and circular patterns of components
- Mirror an assembly
- Create assembly work features
- Create assembly features
- Create a 2D design layout
- Create adaptive sketches
- Create adaptive features
- Create an adaptive subassembly
- Create Assembly Section Views

## DESIGN VIEWS

While working in an assembly, you may want to save configurations that show the assembly in different states and from different viewing positions. Design Views can store the following information:

- Component visibility (visible or not visible)
- Component selection status (enabled or not enabled)
- Color settings and style characteristics applied in the assembly
- Zoom magnification
- Viewing angle

The information in the Design View is saved to an associated file that has the IDV file extension. The Design View file is saved with the same name and in the same directory as the assembly file. There can be multiple Design Views saved in the same IDV file. Design Views can be used while working on the assembly and when creating presentation views or drawing views. Once the screen orientation and part visibility is set, you can create a Design View by clicking the arrow next to the Design Views icon from the top of the Browser and then either clicking Other, as shown in the following image, or clicking Design Views from the View menu.

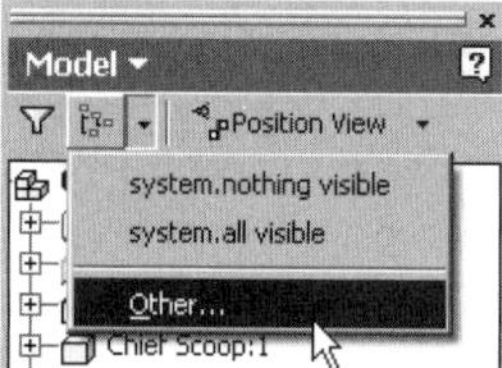


**Figure 10-1**

The Design Views dialog box will appear, as shown in the following image. Select a file name, enter a name for the new Design View, and then click the Save button. To make a Design View current, select it from the drop-down list where you selected the Other option (or after clicking Other), select its name from the list, and click the Apply button. To delete a Design View from the Design View dialog box, select the Design View's name from the list and then click the Delete button.

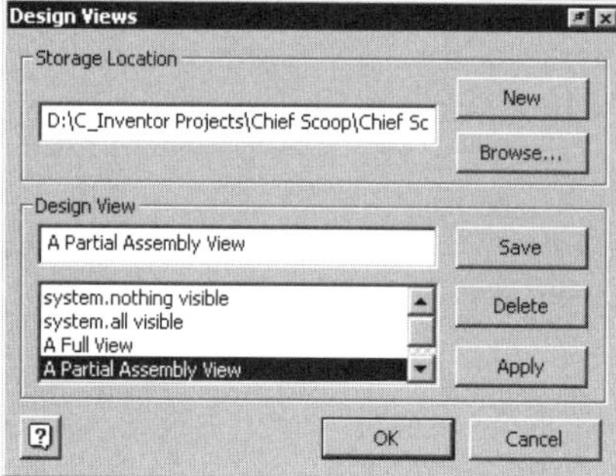


**Figure 10-2**

### VISIBILITY OVERRIDES

In addition to storing component visibility, zoom magnification, and viewing angle, Design Views can also store the display of origin work planes, origin work axes, origin work points, user work planes, user work axis, user work points, and sketches made within an assembly.

### INCREASED PERFORMANCE THROUGH DESIGN VIEWS

Through Design Views, additional control of the visibility of each subassembly is possible, allowing you to turn the visibility of unimportant components off to increase performance. Two Design Views are available:

**System.Nothing Visible** When this Design View is selected, NONE of the components in the assembly will be visible.

**System.All Visible** When this Design View is selected, all of the components in the assembly will be visible.

**NOTE** You can even import a Design View for a subassembly.

These default Design Views will override any existing visibility setting in the subassembly. These Design Views cannot be altered and saved with the same name. You can select these Design Views when you place an assembly into an existing assembly, and you can change any existing subassembly in an assembly to a different Design View by importing a Design View that resides in the associated IDV file. The Design Views in a subassembly are not associated with the Design Views created in the original subassembly. After a subassembly is placed in an assembly, any changes or additional Design Views are not automatically added to the placed subassembly. You must re-import the Design Views to see the new Design View or modifications made to a Design View in the originating assembly.

To place an assembly into another assembly with a selected Design View, follow these steps:

1 Start the Place Component tool from the Assembly Panel Bar (as shown in the following image) and then press the shortcut key P, or right-click and select Place Component from the menu.

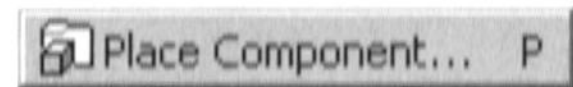


Figure 10-3

The Open dialog box is displayed.

2 Navigate to and select the assembly that will be placed.

3 While still in the Open dialog box, click the Options button as shown in the following image.

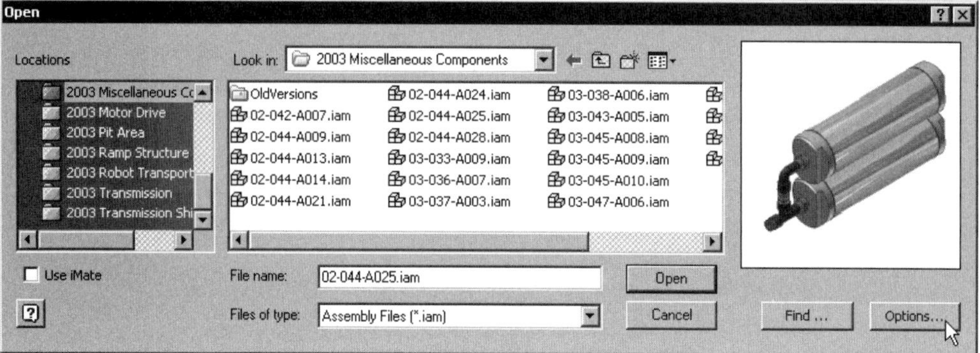


**Figure 10-4**

The File Open Options dialog box opens, showing the available Design Views.

4 Select the Design View that will be applied to the subassembly.

**Figure 10-5**

5 Click the OK button in the File Open Options dialog box.

6 In the Open dialog box, click the Open button.

7 In the graphics window, place the subassembly. If the system.nothing visible Design View is selected, a wireframe box is displayed in the graphics window, as shown in the following image. To add assembly constraints you will need to turn on the visibility of some components or change to a Design View that displays the necessary components. See the next section for how to change from one Design View to another Design View.

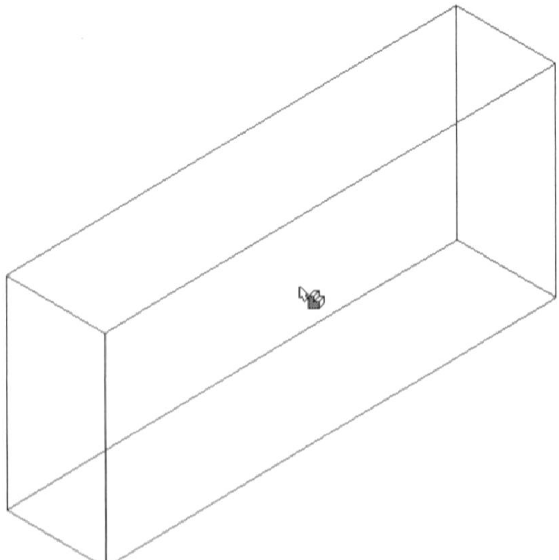

Figure 10-6

After an assembly has been placed or created as a subassembly in a parent assembly, you can change the Design View to a different Design View by following these steps:

1 In the Browser, right-click on the assembly's name that you want to change; or in the graphics window, move the cursor over the assembly that you want to change and then right-click. In the menu, click Import Design View as shown in the following image.

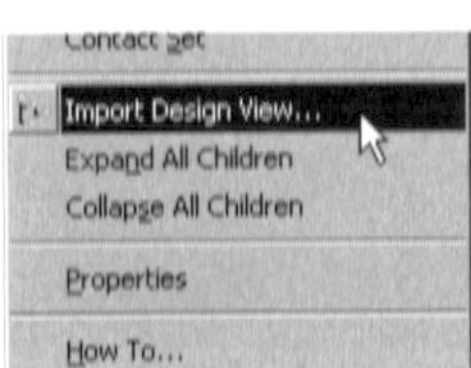


Figure 10-7

The Import dialog box is displayed as shown in the following image.

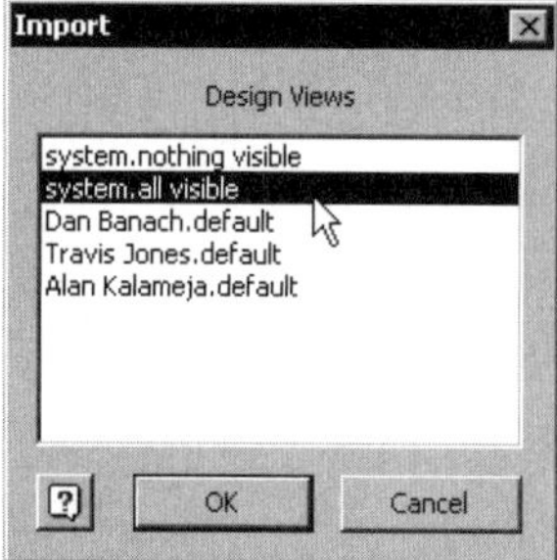


Figure 10-8

2 From the list in the dialog box, select the needed Design View.
3 To complete the operation, click the OK button.

### Associate Design Views

Drawing views generated from a Design View can be made associative. In the following image, a Design View called Front View, which shows the engine on the right, was saved.

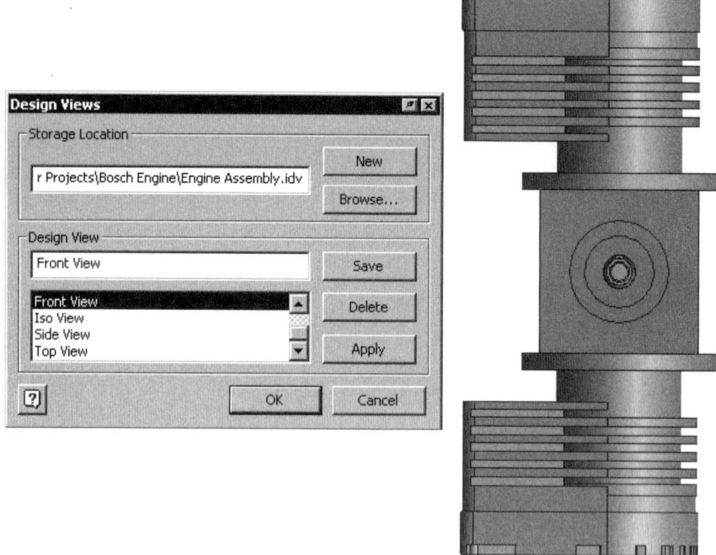


Figure 10-9

When you generate drawing views based on a Design View, you first select the Design View in the Drawing View dialog box, as shown in the following image. You can also check the box next to Associative. This will make the drawing view generated by the Design View associated with the same Design

View found in the assembly file. Any changes made to the Design View in the assembly will update the drawing view.

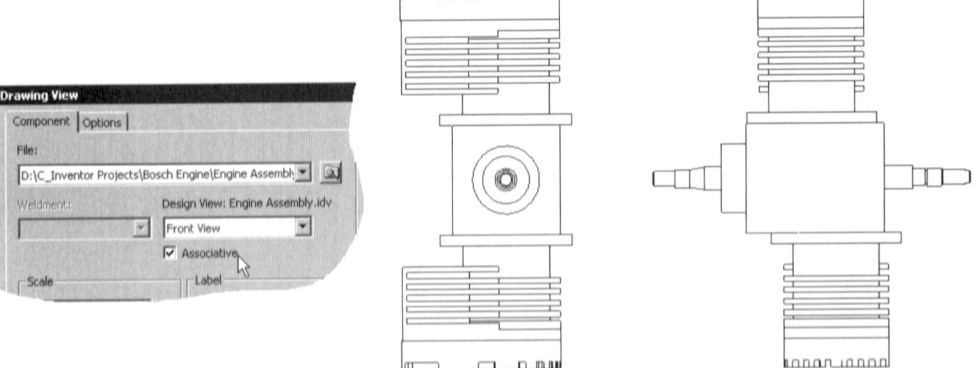


Figure 10-10

To see this in operation, the assembly file of the engine is opened up. The visibility of the engine block is turned off in order to expose the internal connecting arms. These changes are then saved to the Front View Design View, as shown in the following image.

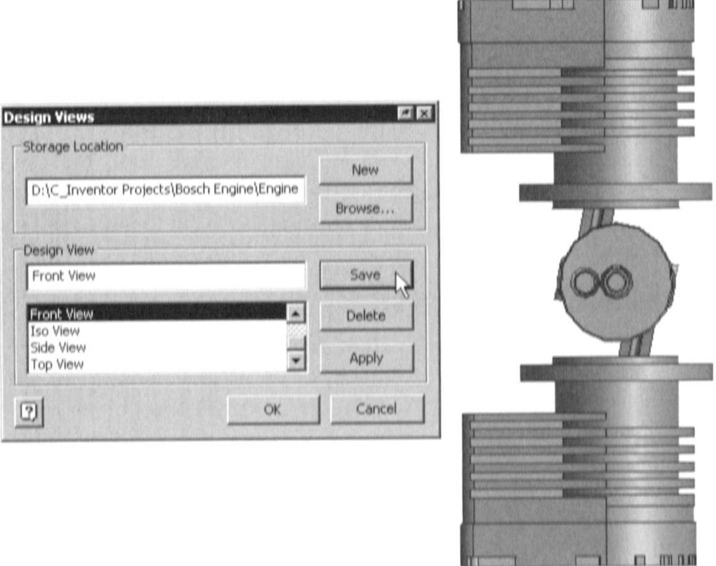


Figure 10-11

When the drawing file of the assembly is opened again, the engine block does not appear in any drawing views because this part was turned off back in the assembly file.

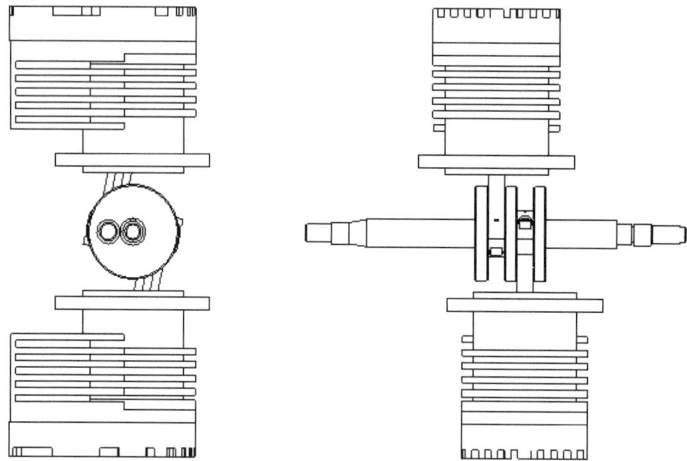

Figure 10-12

**NOTE** Design Views identified by the *.default* extension cannot be associated with drawing views.

## ASSEMBLY BROWSER TOOLS

Additional tools are available through the Assembly Browser as a means of better controlling and managing data in an assembly file. These tools include In-Place Activation, Visibility Control, Assembly Reorder, Restructuring an Assembly, Demoting and Promoting assembly components, and the use of Browser filters.

### IN-PLACE ACTIVATION

The level of the assembly that is currently active determines whether components or features can be edited. Some actions can only be taken in the active assembly and its first-level children, while other operations are valid at all levels of the active assembly.

Double-click any subassembly or component occurrence in the Browser to activate it, or right-click the occurrence in the Browser and select Edit. All components not associated with the active component are shaded in the Browser, as shown in the following image.

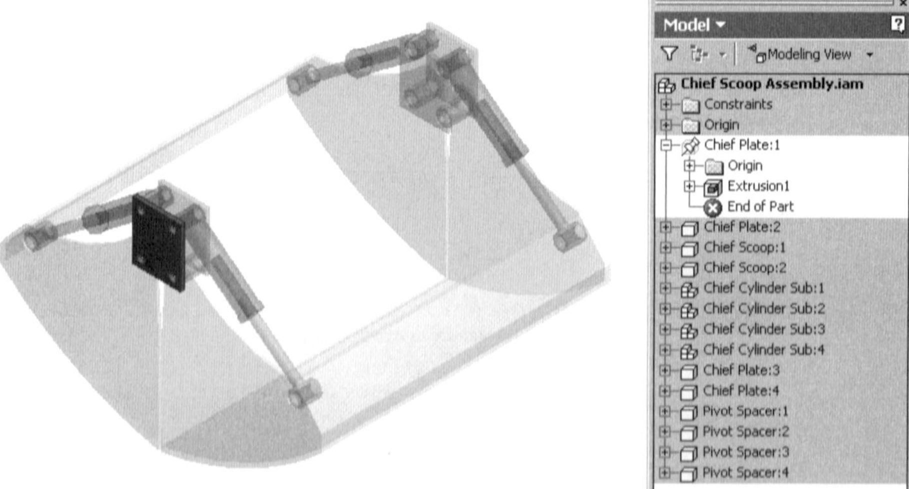


**Figure 10-13**

If you are working with a shaded display, the active component appears shaded in the graphics window and all other components appear translucent. If you are working with a wireframe display, the active component appears in a contrasting color.

The following actions can be performed on the first-level children of the active assembly:

- Deleting a component.
- Displaying the degrees of freedom of a component.
- Designating a component as adaptive.
- Designating a component as grounded.
- Editing or deleting the assembly constraints between first-level components.

The features of an activated part can be edited in the assembly environment. The Panel Bar and toolbars change to reflect the part environment when a part is activated.

**NOTE** Double-click a parent or top-level assembly in the Browser to reactivate it.

## VISIBILITY CONTROL

Controlling the visibility of components is critical to managing large assemblies. You may need some components only for context, or the part you need may be obscured by other components. Assembly files open and update faster when the visibility of nonessential components is turned off.

The visibility of any component in the active assembly can be changed, even if the component is nested many layers deep in the assembly hierarchy. To change the visibility of a component, expand the Browser until the component occurrence is visible, right-click the occurrence, and click Visibility as shown in the following image.

**NOTE** You can also right-click on a component in the graphics window and select Visibility.

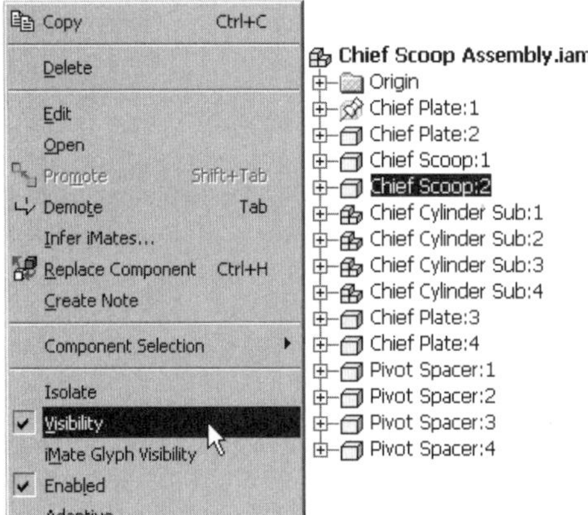


Figure 10-14

## ASSEMBLY SELECTION TOOLS

Assembly selection tools allow you to easily select subassemblies, individual components, features that make up a component, faces and edges, and even the elements that make up a sketch. This tool is especially useful when you attempt to select items in a large assembly. All assembly-selection tools are found by clicking on the Select button located on the Standard toolbar, as shown in the following image.

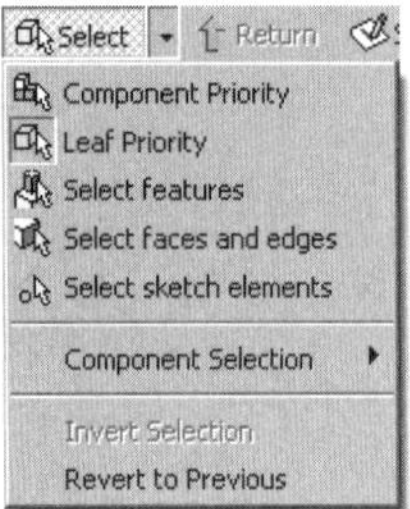


Figure 10-15

The following table lists each assembly selection tool and the function it performs:

| Tool | Title | Purpose |
|---|---|---|
| | Component Priority | Use this tool to select components in an assembly. This can include an individual component or even a subassembly. A component that is part of a subassembly cannot be selected using this tool. |
| | Leaf Priority | This selection tool allows you to select either individual components or components that belong to a subassembly. This tool does not allow you to select features or sketch geometry for a component. |
| | Select Features | Use this tool to select individual features that make up a component. These features can include work features. |
| | Select Faces and Edges | Use this tool to select faces on a component while inside an assembly file. Individual curves or edges that define these faces can also be selected. |
| | Select Sketch Elements | This selection tool allows you to select the sketch geometry used to create a feature in a component. |

## COMPONENT SELECTION

Additional selecting capabilities are available through the series of Component Selection tools found under the Select button located in the Standard toolbar as shown in the following image. These tools can select an entire set of components in order to perform a given task.

No matter what Component Selection tool is used, it is considered good practice to set the selection mode to Leaf Priority if you want to select individual components or components that belong to a subassembly. If you want to select the first level components for an editing session, then set the selection mode to Component Priority.

The four Component Selection tools described in greater detail include Constrained To, Component Size, Component Offset, and Component Sphere.

### Constrained To

This Component Selection tool will highlight all components constrained to one or more preselected components. In the following image, the gripper lift mechanism is selected as the Constrained To component.

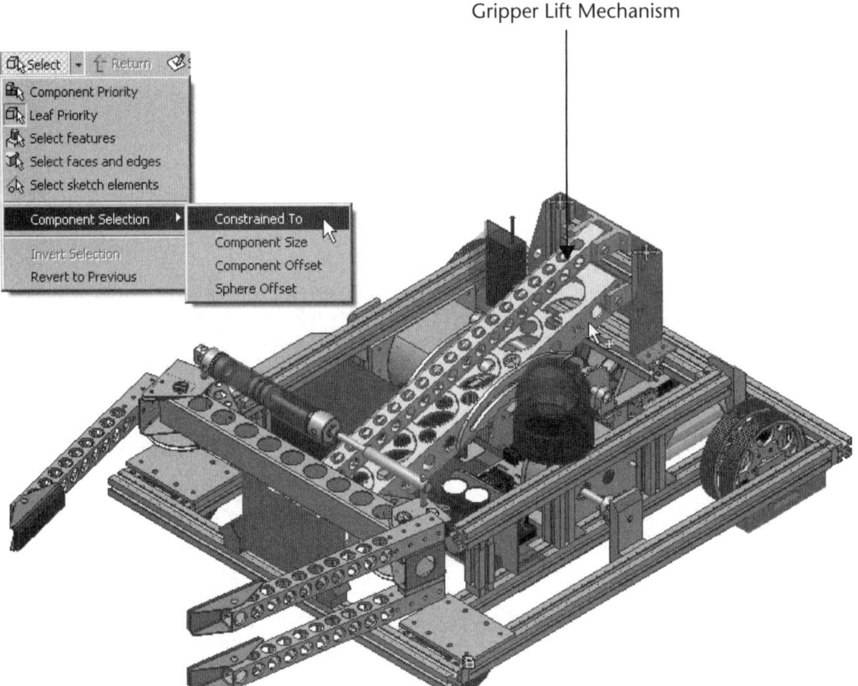


**Figure 10-16** **Robot images courtesy of US FIRST Team 342—Robert Bosch Corporation, Trident Technical College, Fort Dorchester High School, Summerville High School, Charleston, SC.*

Selecting the gripper lift mechanism in the previous image will highlight all components that come in contact through assembly constraints with this mechanism. The selected components will take on a highlighted color for identification purposes. Once these components are selected, right-click and choose Isolate from the menu shown in the following image.

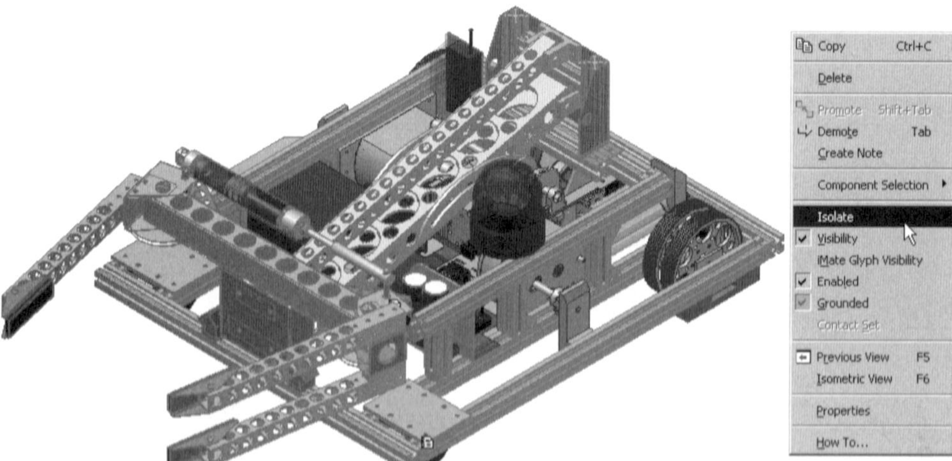


**Figure 10-17**

The results of the Constraint To selection process are illustrated below. All components that are tied to the gripper lift mechanism through assembly constraints are displayed. The visibility of all other components is turned off.

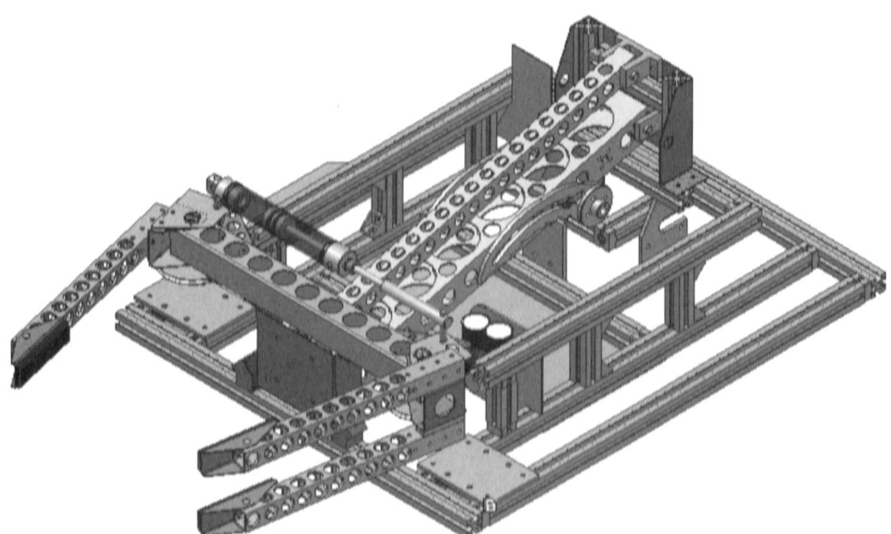

**Figure 10-18**

The purpose of creating selection sets in this manner is to have Autodesk Inventor select the desired components for editing. The time will come when you will have completed an editing operation and wish to display the entire assembly. Rather than right-click on each individual component in the Browser and turn its visibility on, it would be much more efficient to display the Design Views dialog box and apply the system.all visible Design View to the assembly as shown in the following image. This will turn on all components in the assembly.

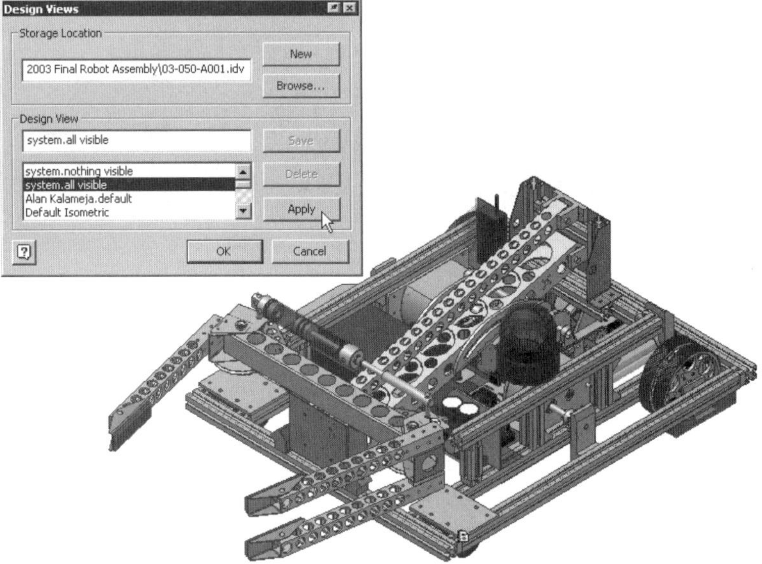


Figure 10-19

## Component Size

This Component Selection tool will highlight all components based on the selected component size. Choose this command under the Select button, as shown in the following image.

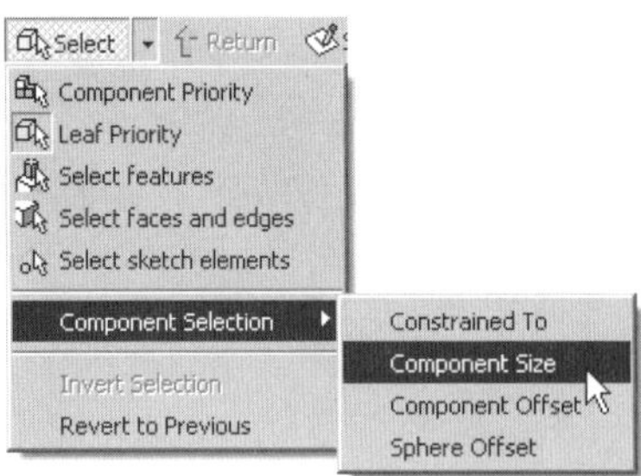


Figure 10-20

Selecting a component will display the Select by Size dialog box. The value in the dialog box represents the current volume of the part. You can select all parts either greater than or less than the displayed value. In the following image, all parts less than the value entered in the dialog box will be selected.

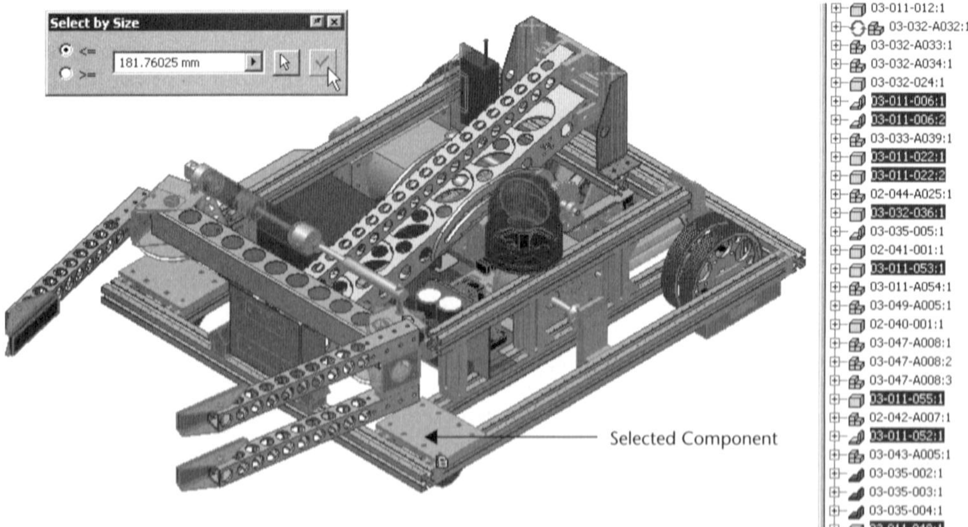


**Figure 10-21**

Right-clicking as the components are selected and then picking Isolate from the menu will display the results of selecting by component size, as shown in the following image. To redisplay the entire assembly, activate the Design View dialog box and select system.all visible.

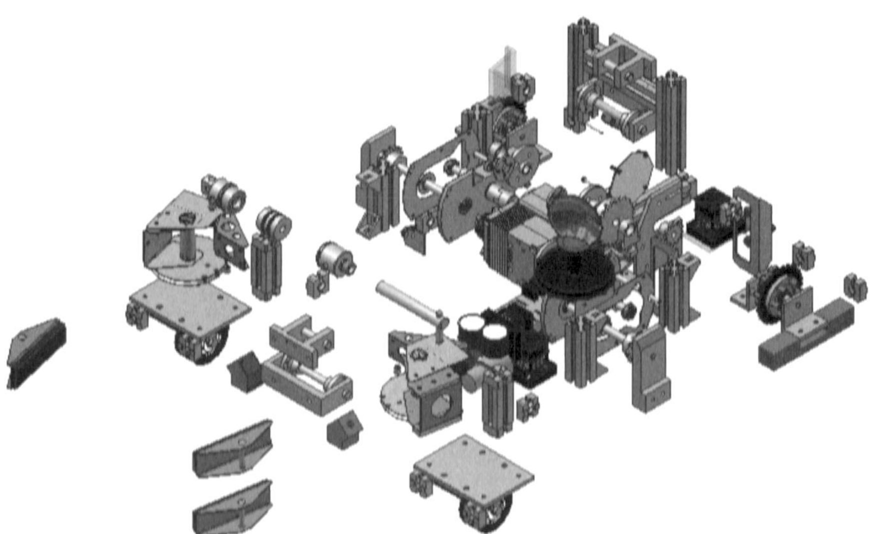

**Figure 10-22**

## Component Offset

This Component Selection tool will highlight all components contained within the area of a bounding box plus an offset distance. Choose this command under the Select button, as shown in the following image.

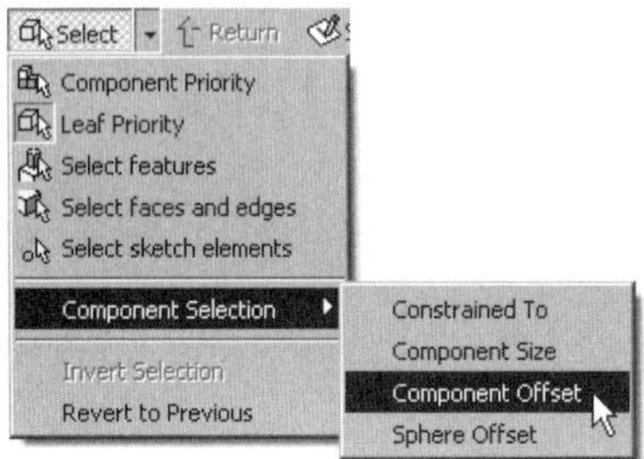


**Figure 10-23**

In the following image, a controller is first selected. A transparent gray cube appears. You can press and drag the corners of this cube in order to establish an offset distance, which will display its value in the Select by Offset dialog box. If you want to select objects partially contained in the cube, place a check in the box next to Include partially contained.

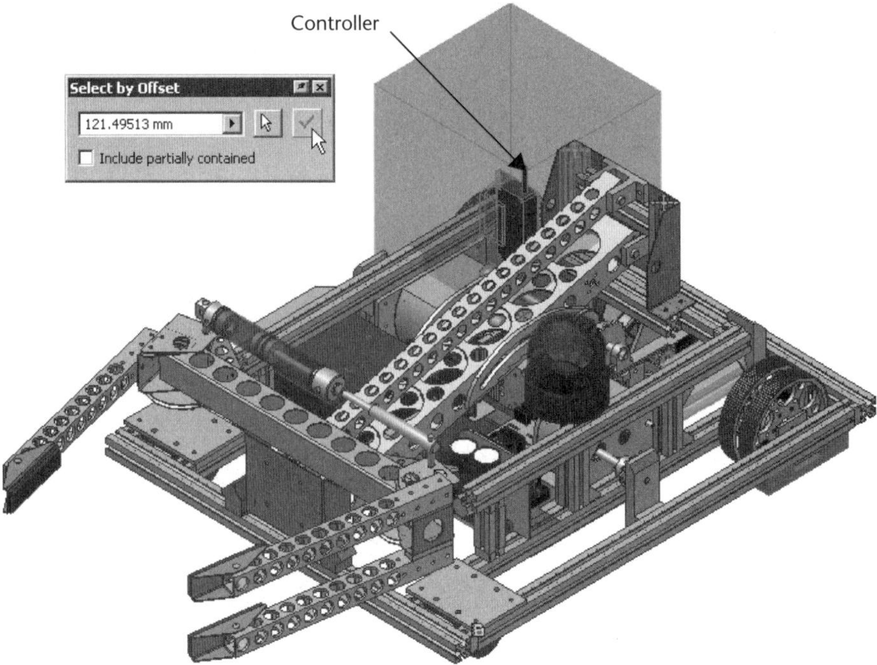


**Figure 10-24**

The results of performing a selection by component offset are illustrated in the following image. Only those components completely inside of the selection cube will be selected. To redisplay the entire assembly, activate the Design View dialog box and select system.all visible.

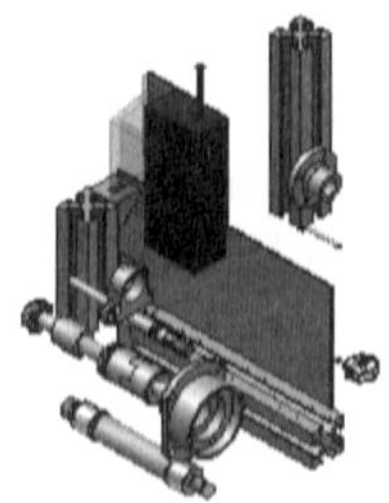

**Figure 10-25**

## Sphere Offset

This Component Selection mode will highlight components contained within a sphere. Choose this command under the Select button, as shown in the following image.

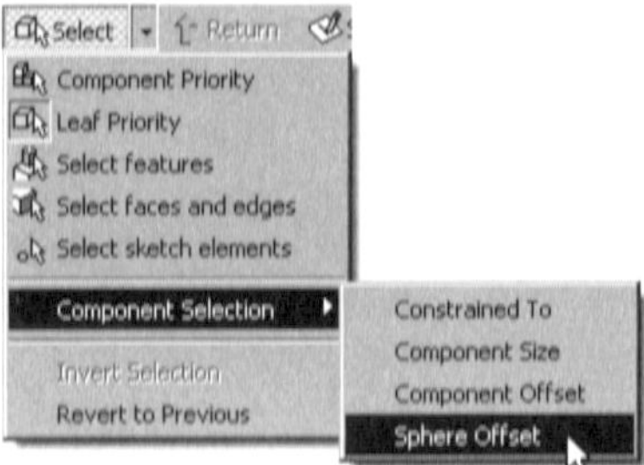


**Figure 10-26**

As shown in the following image, you want to select all objects contained within a sphere. The center of the sphere is based on the light globe. Selecting the globe will display the sphere, whose size you can increase or decrease by dragging its edge. The size of the sphere is displayed in the Select by Sphere dialog box. If you want to select objects partially contained in the sphere, place a check in the box next to Include partially contained.

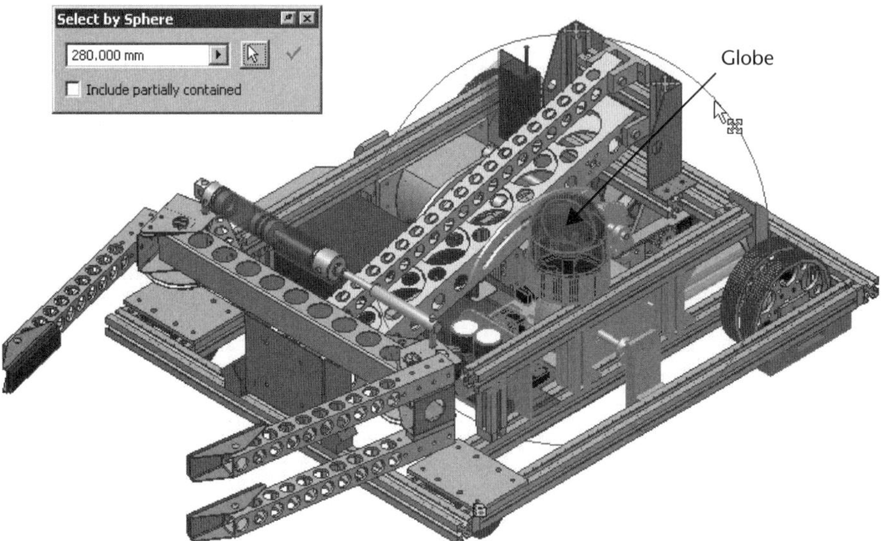


Figure 10-27

The results of performing a selection by component sphere are illustrated in the following image. Only those components completely inside of the selection sphere will be selected. To redisplay the entire assembly, activate the Design View dialog box and select system.all visible.

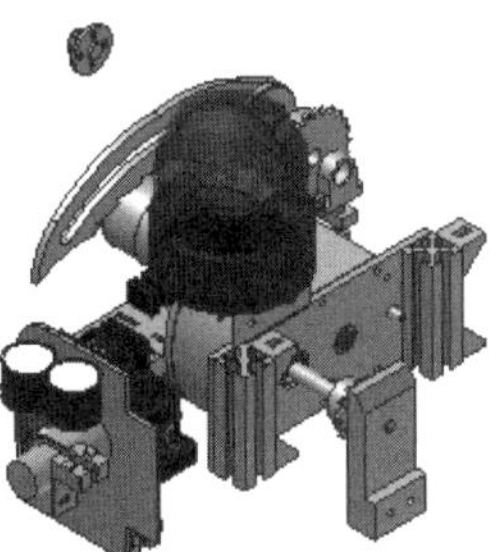

Figure 10-28

## ADDITIONAL SELECTION OPTIONS

### Invert Selection

This item will be highlighted when a valid selection set of components is created. This tool reverses the selection process. All current selections will be deselected; all previously unselected items will now be selected and highlighted.

### Revert to Previous

Clicking on this item will return you to the previous selection set. While this action may be similar to performing an Undo, you cannot repeat this command to back up through a number of selection sets.

## ASSEMBLY REORDER

The order of components can be changed in the Browser based on design intent. To change or reorder the position of a component in the Browser, use the following steps:

1 Click a component in the Browser.

In the following image, the example on the right shows the component Shaft Bearing Journals:1 selected.

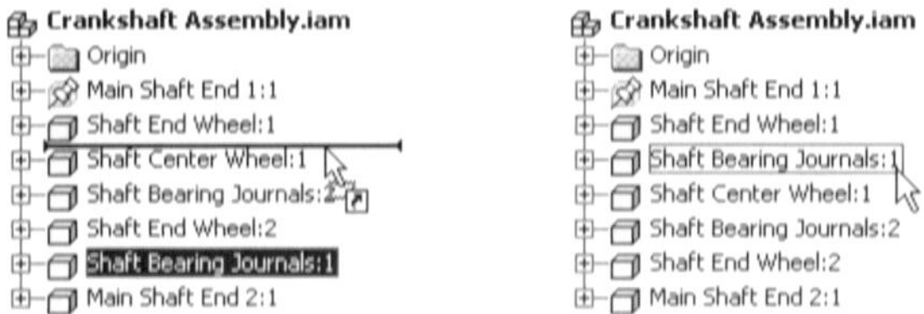


**Figure 10-29**

2 While in the Browser, drag the Shaft Bearing Journals:1 component to its new location under the Shaft End Wheel:1 component.

The results are illustrated in the previous image in the example on the right. When you reorder a component from the same parent subassembly, the assembly constraints are retained.

## ASSEMBLY RESTRUCTURE

Your assembly hierarchy may need to change as your design progresses. The need to combine related components into a subassembly or move components between assemblies may not be apparent early in the design process. With Autodesk Inventor, you can restructure the assembly hierarchy by dragging selected components to a different assembly level in the Browser. In the following image, two components at the top of the assembly hierarchy are dragged to the crankshaft subassembly level using the restructuring process.

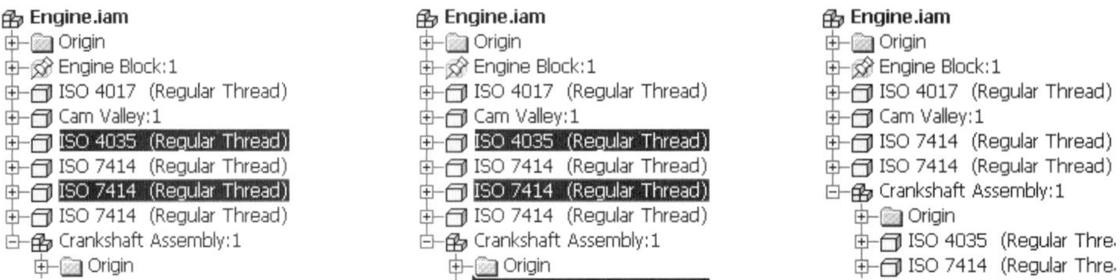


**Figure 10-30**

To create a new subassembly containing selected components, right-click and select Demote from the menu, as shown in the following image.

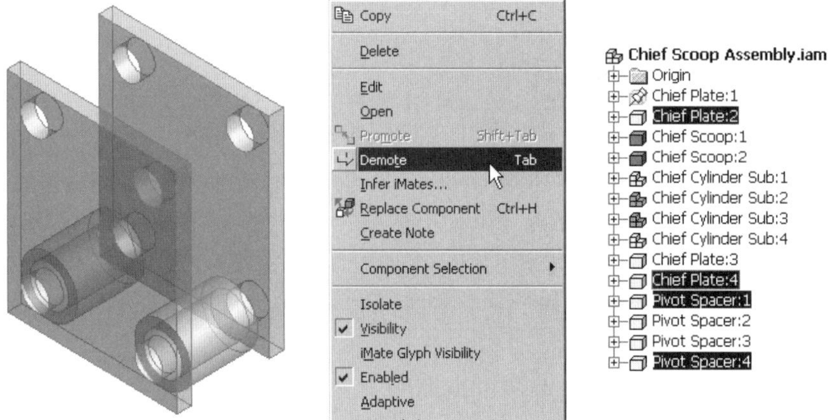


**Figure 10-31**

A new subassembly is created and populated with the selected components.

To move selected components from a subassembly to the parent assembly level, select the component or components from the subassembly and select Promote from the menu in the following image.

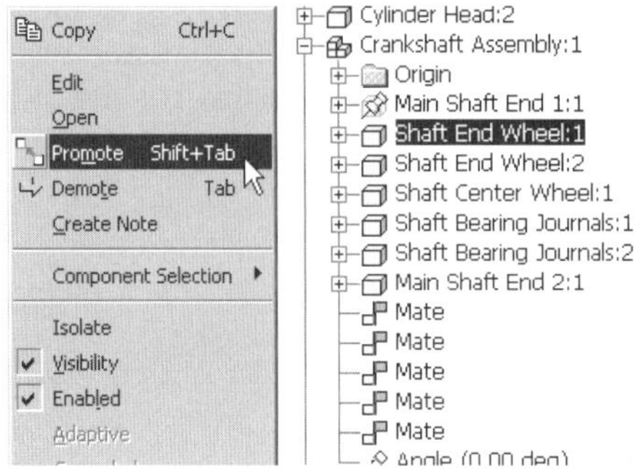


**Figure 10-32**

Assembly constraints are retained between restructured components that originate from the same assembly. Assembly constraints between two parts that are first-level children of the top-level assembly would be maintained if, for example, both parts were moved to a subassembly. Components moved to a different assembly lose assembly constraints with components outside the new assembly.

## BROWSER FILTERS

Adjacent to the Design Views tool located in the Browser are two buttons used to control the display of information in the Browser when in an assembly. The first button, identified by the funnel button in the following image, is the Browser Filters tool.

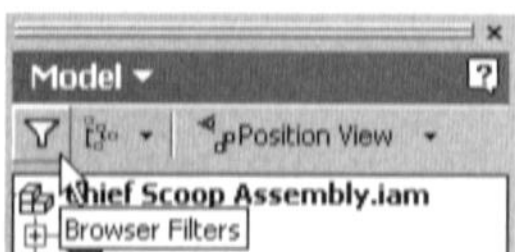


**Figure 10-33**

This tool consists of multiple filters that can be applied to the Browser at the same time. Clicking on the Browser Filters button activates the menu shown in the following image:

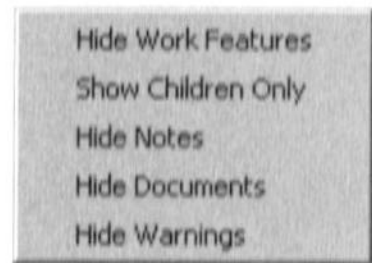


**Figure 10-34**

A brief description of each filter type is described as follows:

**Hide Work Features** Turns off the display of all work planes, work axes, and work points in the Browser, including the default work planes and origin for all components.

**Show Children Only** Causes the Browser to show only the first-level children of the active assembly.

**Hide Notes** Turns off the display of all notes in the Browser.

**Hide Documents** Hides any external objects such as spreadsheets or text files. These objects appear under 3$^{rd}$ Party near the top of the Browser when the Hide Documents filter is off.

**Hide Warnings** Turns warning symbols attached to constraints in the Browser off or on.

The second button represents the Position View button. Clicking on this button displays the menu shown in the following image.

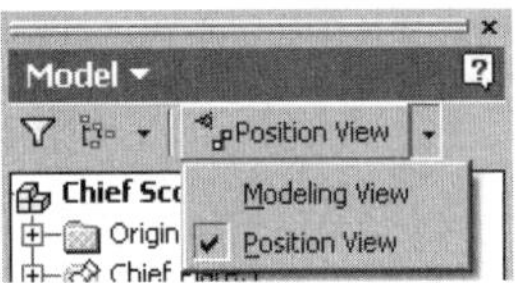


Figure 10-35

The two items in the Browser menu, Modeling View and Position View, define the organization of information in the Browser. Selecting one of these filters disables the other. Each item is described in detail as follows:

**Modeling View** Causes each part occurrence in the Browser to list its defining features directly under the occurrence name. All assembly constraints are displayed as a single group labeled Constraints, under the top-level assembly (see the following image on left).

**Position View** Causes components to display information pertaining to assembly tasks. Part feature information such as sketches and extrusions is suppressed. Assembly constraints are listed below each component (see the following image on right).

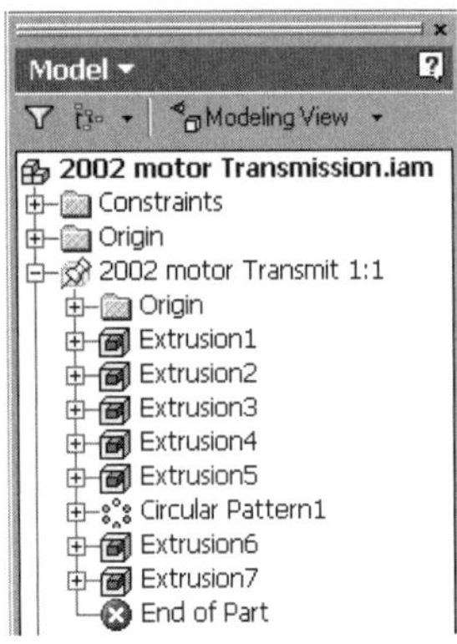


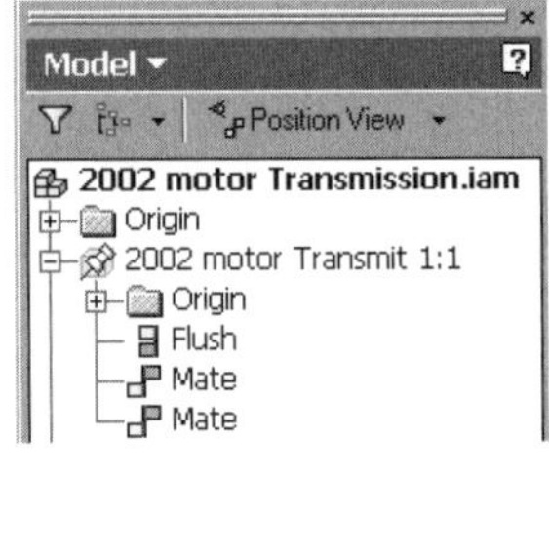


Figure 10-36

## IMATES

Another way to apply assembly constraints is to create iMates. An iMate holds information in the component or subassembly file on how the component or subassembly is to be assembled. An iMate needs to have the same name on both of the components or subassemblies that are being constrained together. Each component or subassembly holds half of the iMate information and when put together they form a pair, or a complete iMate. iMates are useful when similar components are switched in an assembly. You may, for example, have different pins that go into an assembly. The same iMate name can be assigned to each pin or corresponding hole. When the pin is placed into the assembly, it will be automatically constrained to the corresponding hole. An iMate can only be used once in an assembly—once used, it is consumed.

To create an iMate, follow these steps:

3 Click the Create iMate tool on the Tools menu as shown in the following image. (You can also click the iMate tool from the Part Features Panel Bar.)

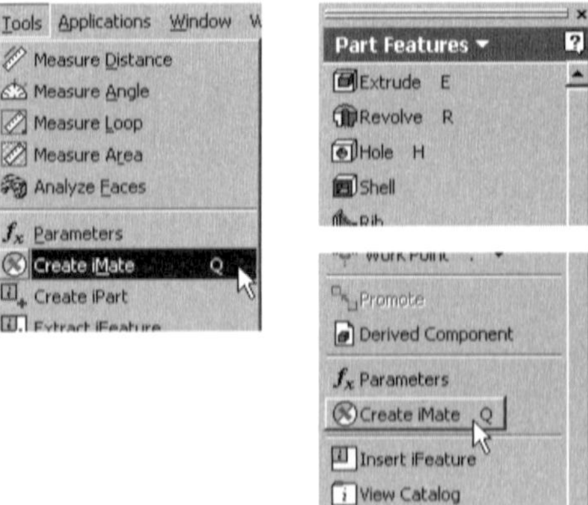


Figure 10-37

4 In the Create iMate dialog box illustrated in the following image, click the type of constraint to apply: mate, angle, tangent, or insert on the Assembly tab.

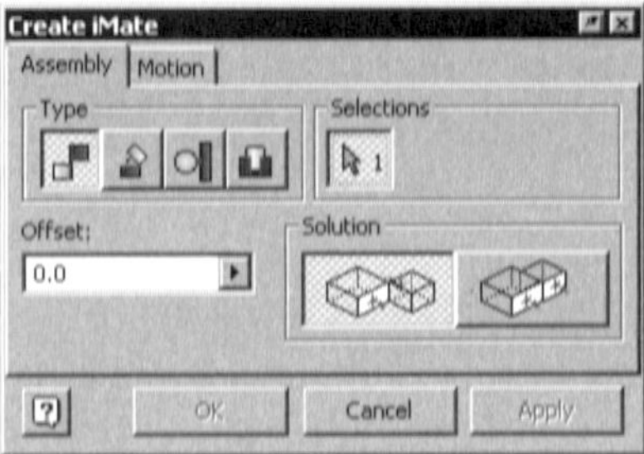


Figure 10-38

5 In the graphics window, select the geometry you want to use as the primary position geometry. A mate-center constraint is being applied to the part, as shown in the following image.

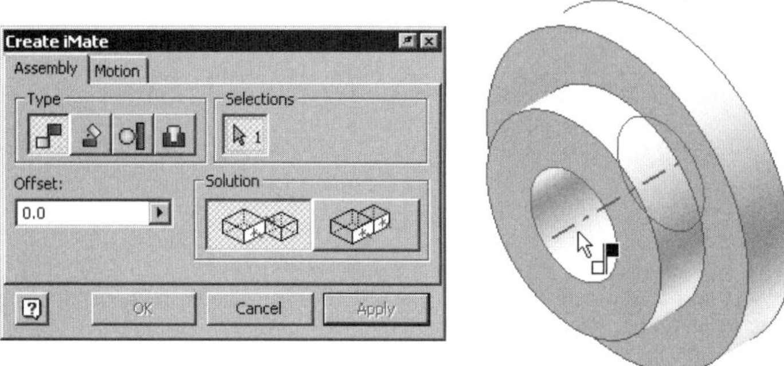


Figure 10-39

6 Click Apply. An iMate symbol will appear on the component and in the Browser, as shown in the following image.

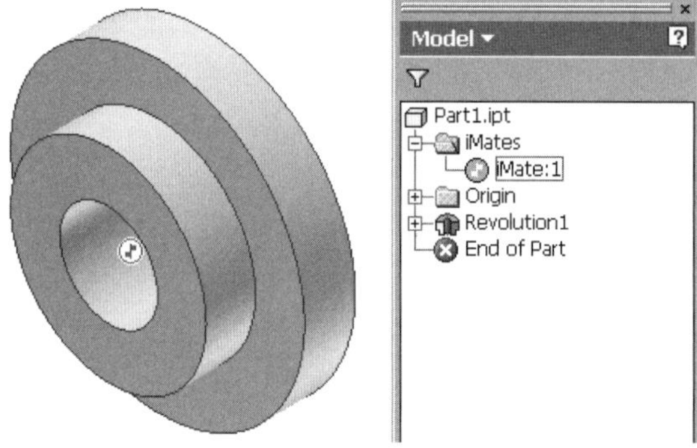


Figure 10-40

7 Continue to create iMates. Click Apply after each one.

The symbol shows the type and state of the iMate. When an iMate is created, it is given a default name such as *iInsert:1* or *iAngle:1*. The iMates can, however, be renamed to better reflect the condition that they represent. Slowly double-click on the iMate name in the Browser and type in a new name.

Right-click in the Browser, or on the iMate symbol in the graphics window, and select Properties from the menu, as shown in the following image. Type in a new name.

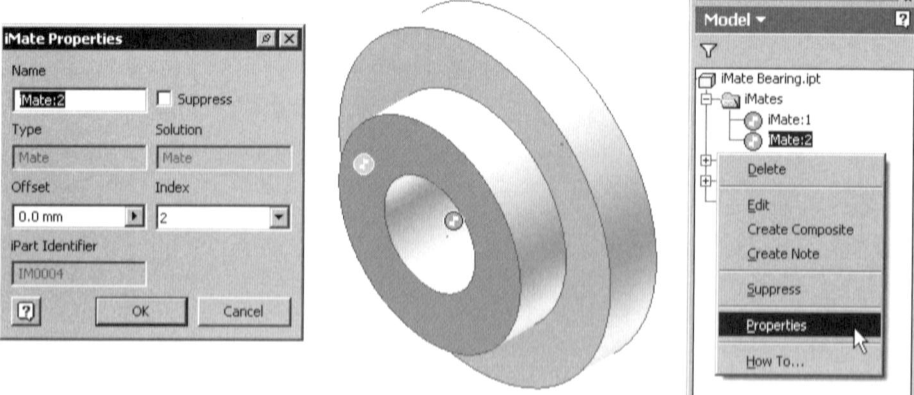


**Figure 10-41**

There are two methods for assembling components that exist in the same assembly and have corresponding iMates:

1 Use the Place Constraint tool on the Assembly Panel Bar.
2 Click an iMate symbol on a component, then select a matching iMate symbol on another component or drag the selected iMate symbol over a matching iMate symbol on another component.
3 When the second iMate symbol is highlighted, click to position the components.
4 Click the Apply button to create the constraint.

A component that has an iMate can automatically be constrained to another component that has the same iMate name. Use the following steps to perform this operation:

1 Click the Place Component tool on the Assembly Panel Bar.
2 Select the component to place and check the box next to Use iMate in the lower-left corner of the Open dialog box, as shown in the following image.
3 Click the Open button.

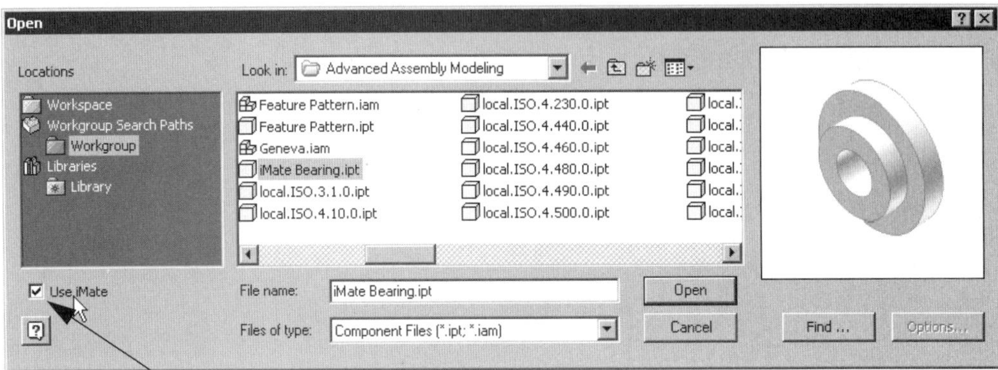


**Figure 10-42**

If a matching iMate on another component exists, it will be placed and a consumed iMate symbol will be shown in the Browser. If not, you should repeat the sequence with the second component making sure to select Use iMate in the Open dialog box. The second component will be automatically constrained to the first.

## Composite iMates

You can also group multiple iMates into a single, composite iMate. In the following image of the standard drill transmission housing, the three separate iMates can be combined into a single, composite iMate.

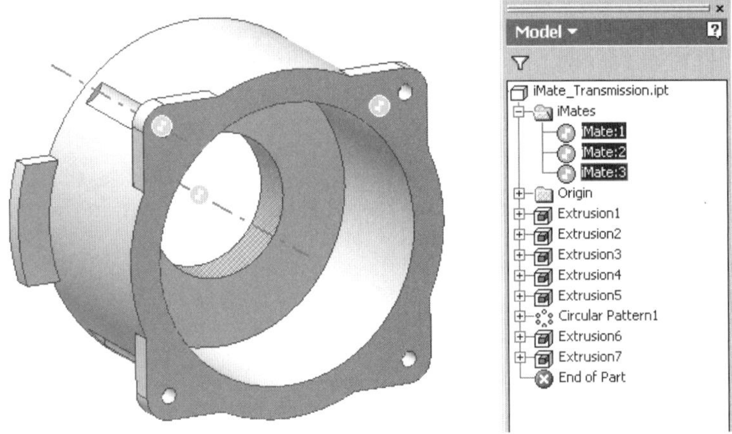


**Figure 10-43**

You can then completely position the motor assembly in other assemblies in one easy step by Alt+Dragging (an operation that is covered later in this chapter) the composite iMate rather than using individual iMates in multiple steps.

To create a composite iMate for any component, first create the individual iMates. Once created, press the CTRL key while selecting multiple iMates in the Browser. Right-click on any iMate and select Create Composite from the menu, as shown in the following image.

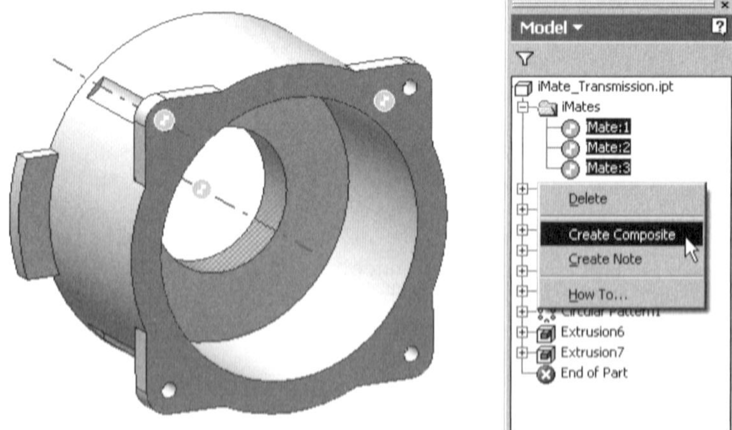


**Figure 10-44**

The results are shown in the following image. The three iMate glyphs that were originally created on the part are now combined into a single glyph. The Browser also displays the newly created composite iMate. Expanding the composite iMate will display the three original iMates.

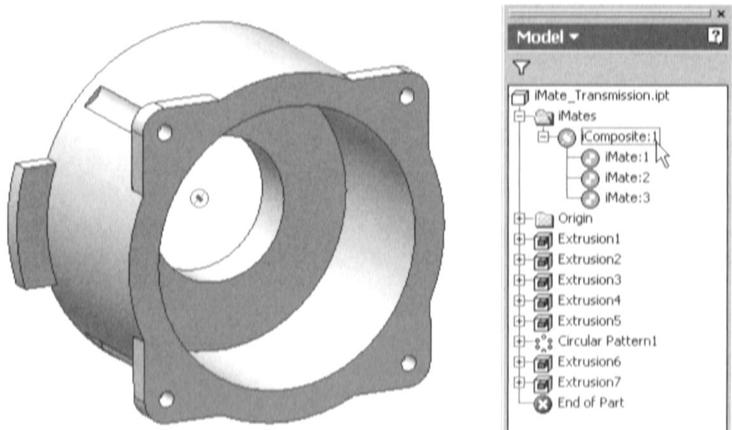


**Figure 10-45**

To revise a composite iMate from the Browser, use the following steps:

1 Rename the composite iMate or individual iMate members.

2 Drag and reorder individual iMate members.

3 Delete individual iMate members.
4 Remove individual iMate members from the composite.
5 Delete the entire composite.

### Using Composite iMates

**NOTE** You cannot reorder individual iMates into existing composite iMates.

There are two ways to orient components in your assemblies using composite iMates:

- Select the Use iMate option in the Place Component dialog box to automatically search for matching component iMates.
- Use the ALT+ Drag shortcut to manually match composite iMates between two components.

To ensure a successful match between any two iMates in an assembly,

- The iMate type and values must match.
- The iMate names are used for pairing if multiple matches exist for a given type and value.
- The iMates are paired by sequence if multiple matches cannot be paired using names .

For composite iMates, the entire set of iMate members must match these same requirements to be a valid match.

When you place a component in an assembly using the Use iMate option in the Place Component dialog box,

- Autodesk Inventor searches the assembly for a valid composite iMate with an identical name.
- Autodesk Inventor searches for other valid composite iMates if an identical name is not found.
- The component is placed in the assembly but not positioned if no valid matches are found.

**NOTE** iMate names must match on the placed component and the unconsumed iMate in the assembly.

When the two matching iMates join in the assembly, a single consumed iMate is created. Because the relationship is specific to two components with matching iMate halves, multiple occurrences cannot be placed.

The solution that is selected for iMates (mate and flush, inside and outside, or opposed and aligned) must be the same for the matching iMates.

## IMATE OPTIONS

### iMate Publish

You can extract existing constraint data from a part and convert this information into iMates. This process allows for iMates and Composite iMates to be generated from only the constraints on the one part or from all constraints of all occurrences of the selected part in the active assembly. This method of converting existing constraints to iMates is especially helpful if you will be reusing parts along with their exact constraints.

Follow the next series of steps for publishing iMates:

1 While in an assembly file, the three highlighted Mate constraints will be converted into iMates, as shown in the following image.

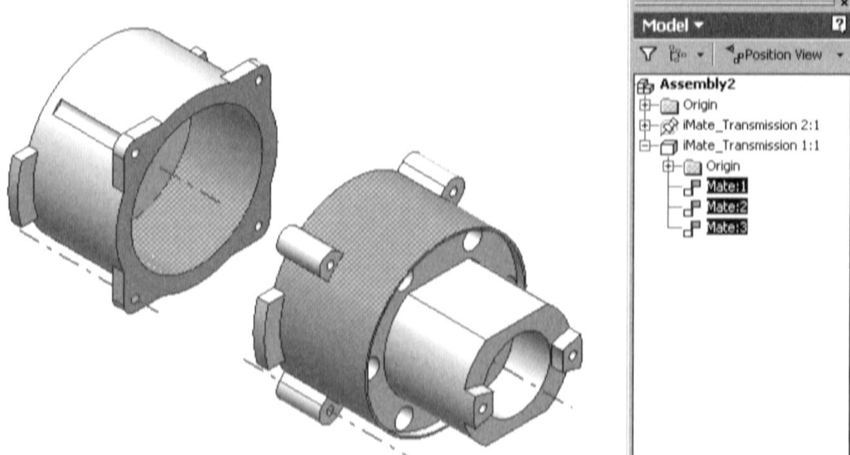


**Figure 10-46**

2 With the desired part selected, right-click to display the menu and select Infer iMates, as shown in the following image.

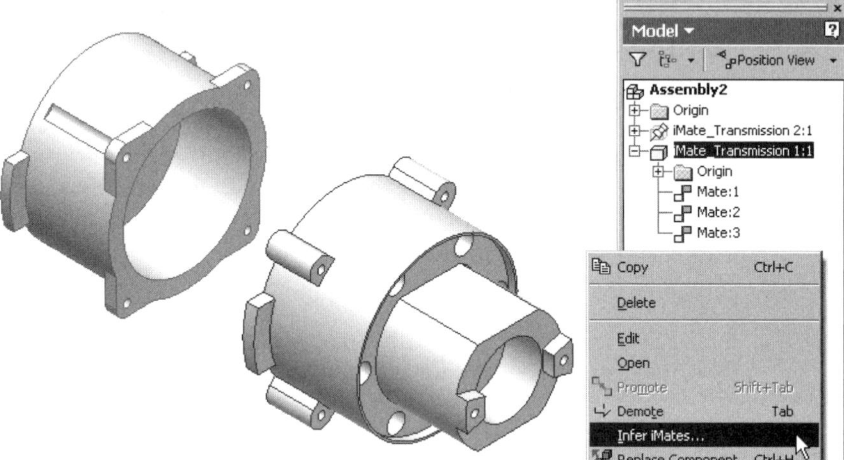


**Figure 10-47**

3 Selecting Infer iMates displays the Infer iMates dialog box, as shown in the following image.

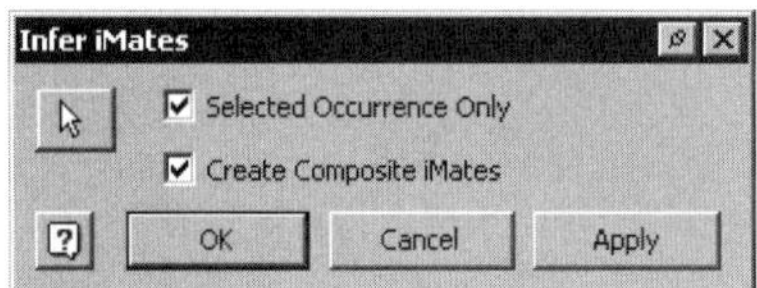


**Figure 10-48**

Based on the current settings in the dialog box, a single composite iMate definition will be created from each of the three constraints. Had there been two placements of the component in the current assembly, the dialog box default setting would have specified that only constraints on the selected occurrence would be converted to iMates. Clearing the checkbox for this option would allow constraints placed to all occurrences to be converted to iMate definitions.

Once you click the OK button to dismiss the Infer iMates dialog box, you will continue to remain in the assembly. It may seem as if nothing has changed in the assembly; however, the iMates were just created in the individual part model file. In other words, this process does not replace the existing assembly constraints with iMates in the active assembly.

## ALT+Drag iMate Behavior

When using the ALT+Drag feature of constraining iMates on a component part, the other component part that contains the matching iMate solution will display the iMate glyph (as shown in the following image).

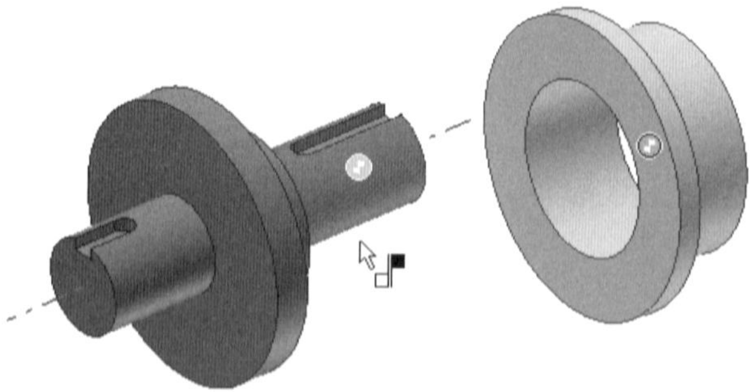

Figure 10-49

**NOTE** The above behavior differs from existing behavior as follows: iMate glyphs that are currently displayed will not display during an ALT+Drag if they don't match the selected iMate. This behavior relies upon glyphs being turned off by default and requires the matching glyphs to display.

## iMate Visibility Controls

Visibility controls are available for displaying iMates. First, select the component with iMates from the Browser while in an assembly file, then right-click to display the menu as shown in the following image.

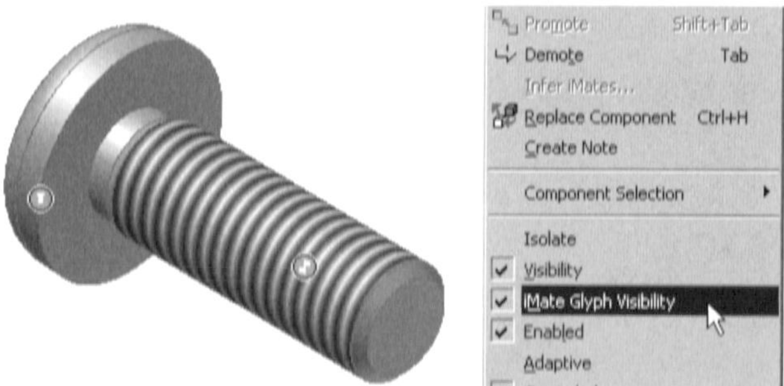


Figure 10-50

When checked in the menu, any iMate glyphs within the selected component shall be displayed in the assembly. When the check is not visible, the iMate glyphs will not be displayed.

# EXERCISE 10-1 Creating and Using iMates

In this exercise, you add iMate halves to existing parts and assemblies. You then use the iMates to place components in an assembly. You create a composite iMate from the original iMates, which you use to place components in another assembly. You complete the exercise by replacing a component in the assembly. To navigate to the exercise in the *Electronic Student Workbook*, do the following:

1 From the Main TOC page, click Chapter 10.
2 From the TOC for Chapter 10, click Creating and Using iMates.

The following image illustrates the opened exercise.

**Figure 10-51 Opened exercise**

The following image illustrates the completed exercise.

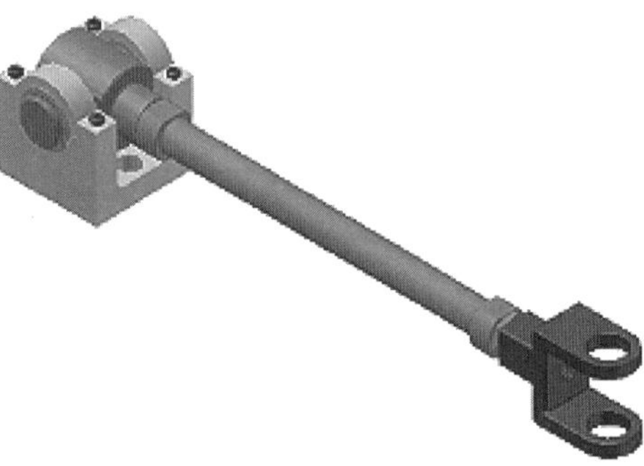

**Figure 10-52 Completed exercise**

## DRIVING CONSTRAINTS

Mechanical motion can be simulated (driven) using the Drive Constraint tool. To simulate motion, either an angle, mate, tangent, or insert assembly constraint must exist. Only one assembly constraint can be driven at a time, but equations can be used to create relationships to drive multiple assembly constraints. To drive a constraint, right-click on the constraint in the Browser, as shown in the following image, and select Drive Constraint from the menu.

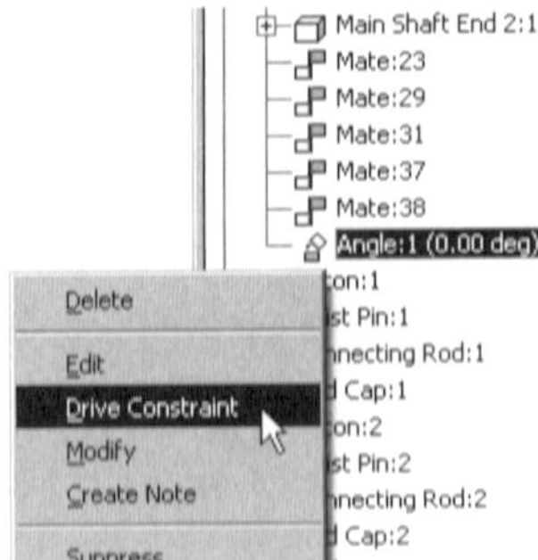


**Figure 10-53**

The Drive Constraint dialog box will appear, as shown in the following image. Depending upon the constraint that you are driving, the units may be different. Enter a Start value—the default value is the angle or offset for the constraint. Enter a value for the End and a Pause Delay if you want time between the steps. In the top half of the dialog box, you can also choose to create an animation (AVI) file that will show the motion. The AVI file can be replayed without Autodesk Inventor being installed.

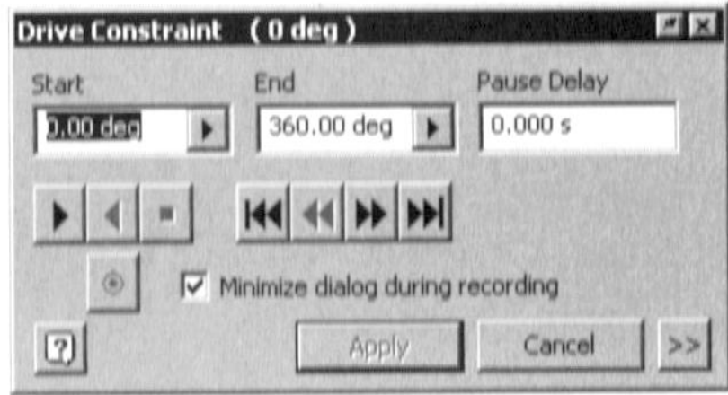


**Figure 10-54**

The Start, End, and Pause Delay controls are described as follows:

**Start** Sets the start position of the offset or angle.

**End** Sets the end position of the offset or angle.

**Pause Delay** Sets the delay between steps.

Use the following Motion and AVI control buttons to control the motion and to create an AVI file:

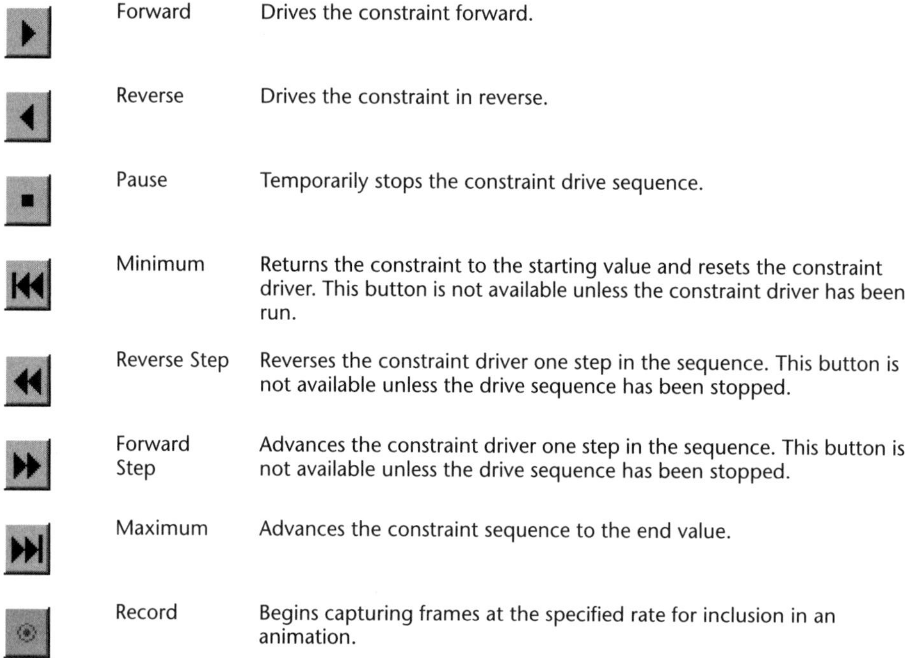

| Button | Description |
|---|---|
| Forward | Drives the constraint forward. |
| Reverse | Drives the constraint in reverse. |
| Pause | Temporarily stops the constraint drive sequence. |
| Minimum | Returns the constraint to the starting value and resets the constraint driver. This button is not available unless the constraint driver has been run. |
| Reverse Step | Reverses the constraint driver one step in the sequence. This button is not available unless the drive sequence has been stopped. |
| Forward Step | Advances the constraint driver one step in the sequence. This button is not available unless the drive sequence has been stopped. |
| Maximum | Advances the constraint sequence to the end value. |
| Record | Begins capturing frames at the specified rate for inclusion in an animation. |

To set more details on how the motion will behave, select the More (>>) button. This will display the expanded Drive Constraint dialog box, as shown in the following image. Each option of the dialog box will be explained as follows:

**Figure 10-55**

**Drive Adaptivity** When selected, the component will adapt while the constraint is driven.

**Collision Detection** When selected, the constraint will be driven until collision is detected. When interference is detected, the drive constraint will stop and the components where the collision occurs will be highlighted. Also shows the constraint value for the collision.

**Increment** This area describes the value that the constraint will be incremented during the animation.

**Amount of Value** Drives the constraint in this number of increments.

**Total # of Steps** Drives the constraint equally per number of steps.

**Repetitions** Sets how the driven constraint will act when it completes a cycle and how many cycles there will be.

**Start/End** Drives the constraint from the start value to the end value and resets at the start value.

**Start/End/Start** Drives the constraint from the start value to the end value, and then in reverse to the start value.

**AVI Rate** Specifies how many frames are skipped before a screen capture is taken of the motion that will become a frame in the completed AVI file.

**NOTE** If you try to drive a constraint and it fails, you may need to suppress or delete another assembly constraint. To reduce the size of an AVI file, reduce the screen size before creating the file and use a solid background color in the graphics window.

# EXERCISE 10-2 Driving Constraints

In this exercise, you drive a constraint and use collision detection to identify the point of interference. To navigate to the exercise in the *Electronic Student Workbook*, do the following.



1 From the Main TOC page, click Chapter 10.
2 From the TOC for Chapter 10, click Driving Constraints.

The following image illustrates the opened exercise.

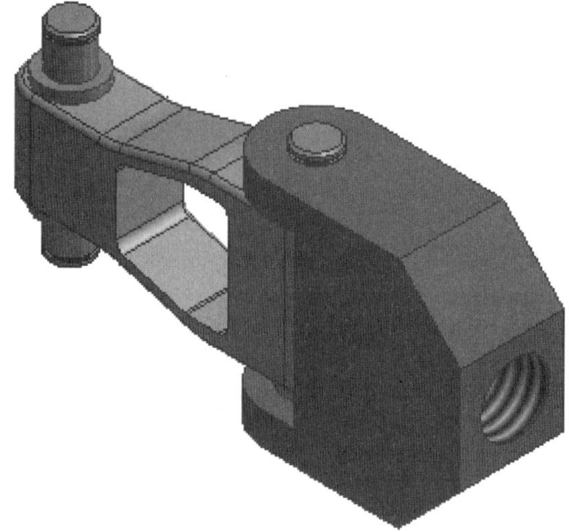

**Figure 10-56 Opened exercise**

The following image illustrates the completed exercise.

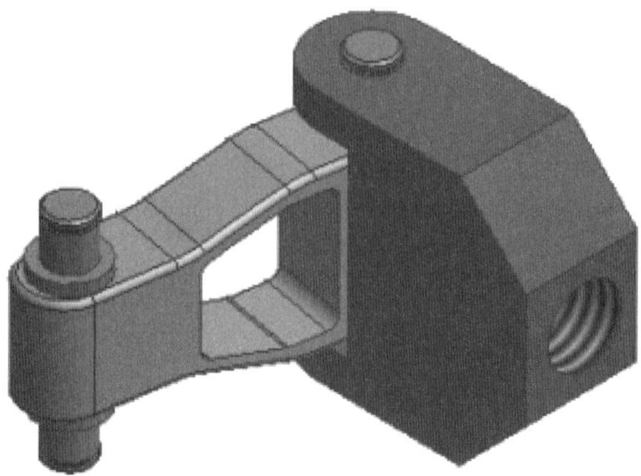

**Figure 10-57 Completed exercise**

## CONTACT DETECTION SOLVER

The contact solver determines how assembled components behave when a mechanical motion is applied. To get a better idea of what happens with the contact solver, study the three components illustrated in the following image. The base component, #1, consists of two slots cut into its underside. The second component, #2, has two pins in addition to a single slot. The two pins are designed to run inside of slots found on component #1. The third component, #3, has a single pin designed to run inside of the slot found on component #2.

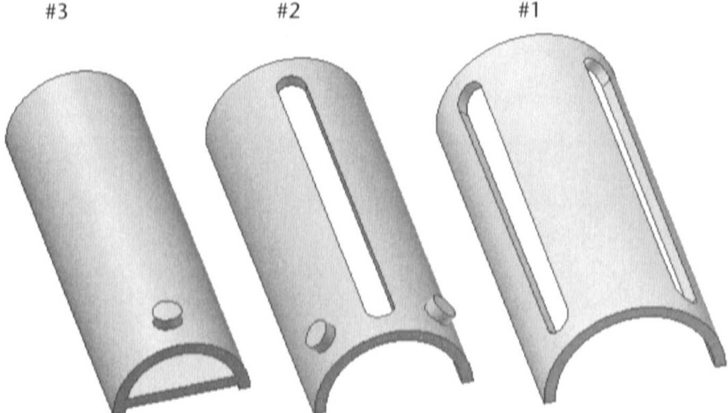


Figure 10-58

In the image below, all three components are assembled with the freedom to move out. However, as the pins should stop inside of the end of a slot, the components continue to drag beyond the slots. The contact solver works based on your designation of certain components to be included in a contact set. It is this contact set that will limit the motion of the objects so they stop when the pin detects the end of a slot.

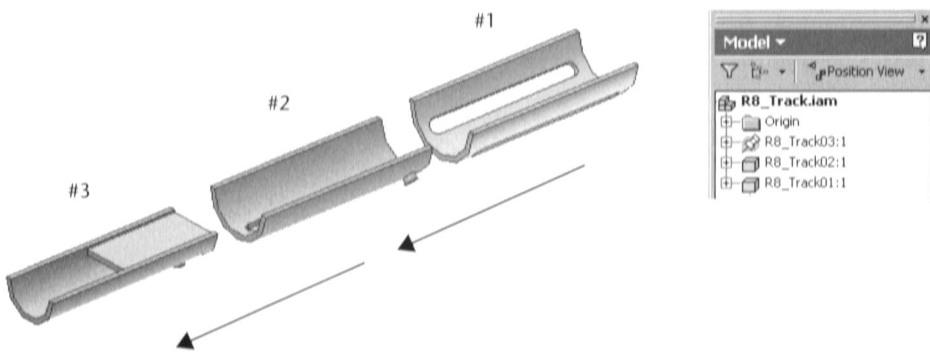


Figure 10-59

To designate components as part of a contact set and test a mechanism, follow these steps:

1 Click on Tools on the Standard toolbar and then click on Application Options to launch the Options dialog box. Click on the Assembly tab and place a check in the box next to Activate Contact Solver, as shown in the following image. By default, the Contact Set Only option is selected. You could change the option to All Components and have the content solver act on all components in the assembly.

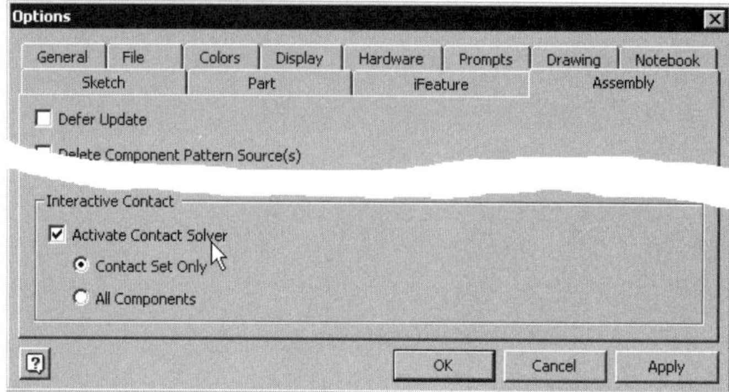


Figure 10-60

**NOTE** Another way of activating the contact solver is by picking Tools on the Standard toolbar and then picking Contact Solver, as shown in the following image. This action will also turn on the contact solver back under the Assembly tab of the Options dialog box. The undersides of all three components are shown for better clarity in this example of the contact solver.

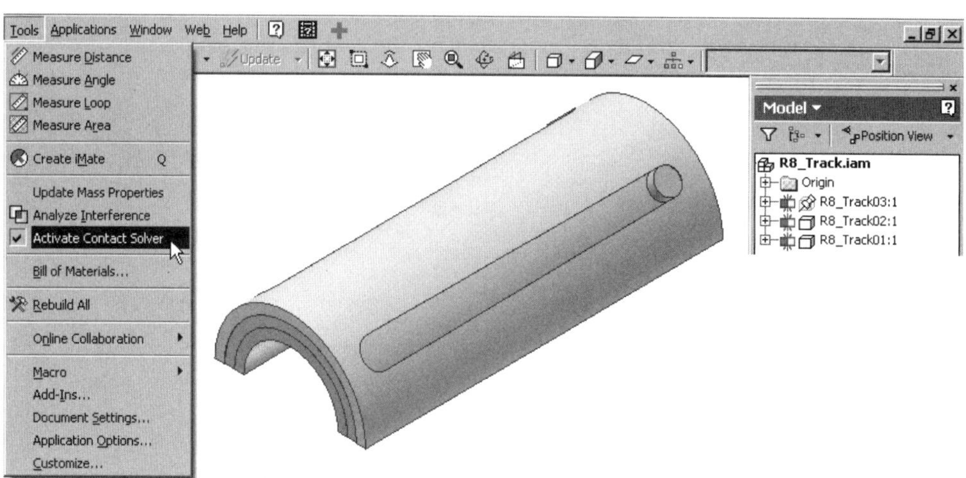


Figure 10-61

2 Once the contact solver is activated, you need to identify those components that will act upon each other. This is called a contact set. To create a contact set, right-click on one or more components in the Browser and select Contact Set from the menu as shown in the following image.

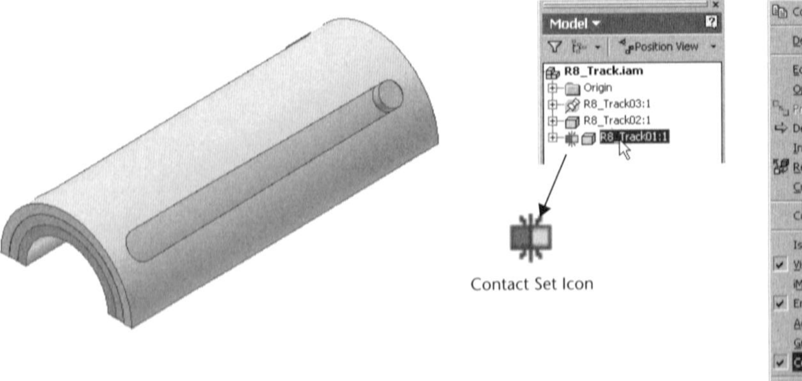


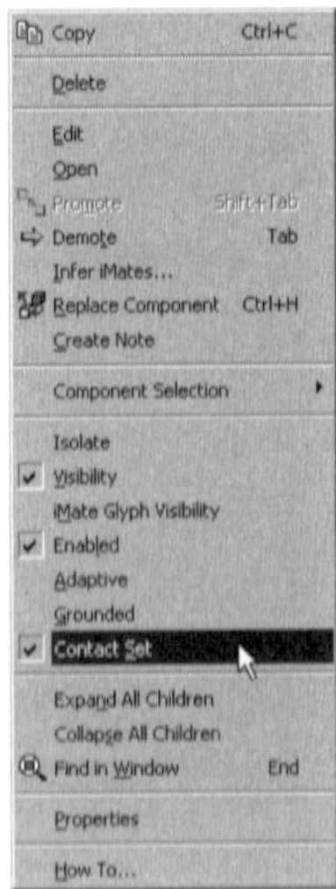


Figure 10-62

**NOTE** Another way of creating a contact set is to right-click on one or more components and select Properties from the menu. When the Properties dialog box appears, click on the Occurrence tab and select Contact Set from this dialog box. Click the OK button to create the contact set, as shown in the following image.

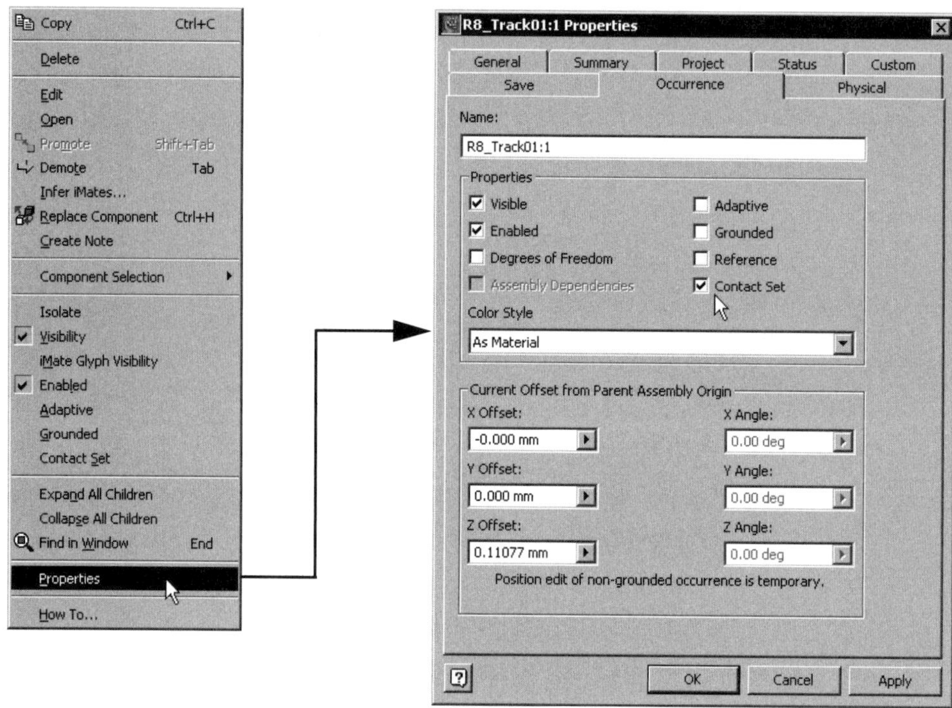


**Figure 10-63**

3 You can now test the mechanism based on the contact solver. In the example illustrated in the following image, begin dragging components through their intended motion. The pins should now stop when they contact the ends of the slots. You may have to adjust the component positions and repeat the motion if the pins did not stop.

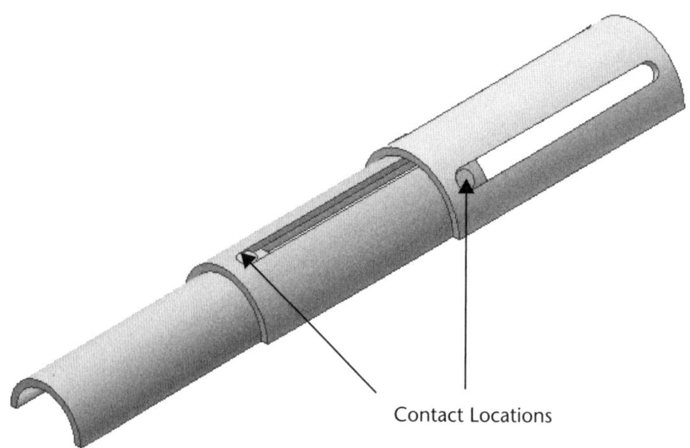


**Figure 10-64**

Drive constraints can also be used to produce intended motion. In the following image, a cam illustrates the motion of a Geneva wheel as the pin comes in contact with a slot. The use of the contact solver for these types of mechanisms usually produces dramatic results.

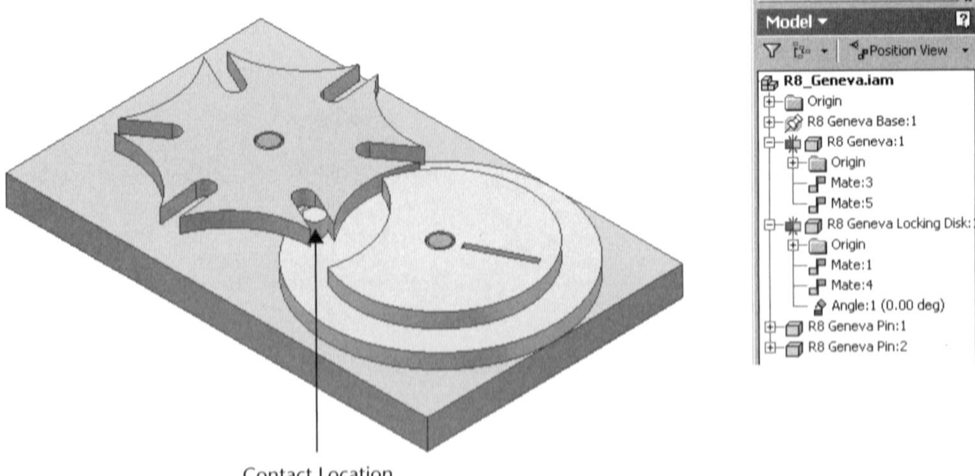


Figure 10-65

## PRESENTATION HIGHLIGHTING

When activating the More (>>) button of the Animation dialog box, as shown in the following image, the sequence scheme that controls the order of the animation is displayed. As the animation plays, each sequence will highlight in the dialog box to match the animation playback in the graphics window. This highlighting makes it easier to match the sequence number and name with the animation currently playing in the graphics window.

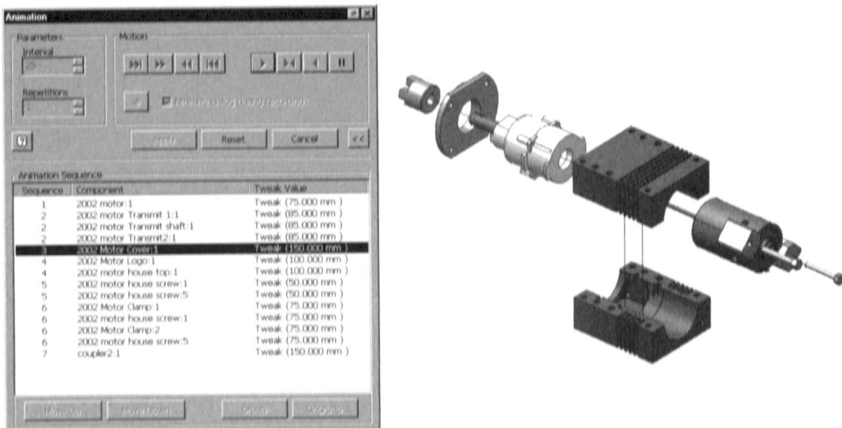

Figure 10-66

## SETTING UNITS IN A PRESENTATION FILE

The units of measure can be set for use in a presentation file. You can therefore set the units in which tweaks will be created. The specified units can be different than the units that are set in the parent assembly. To set the units in a presentation file, create a presentation view, click Tools > Document Settings, and, on the Units tab, set the unit under the Length drop-down list as shown in the following image.

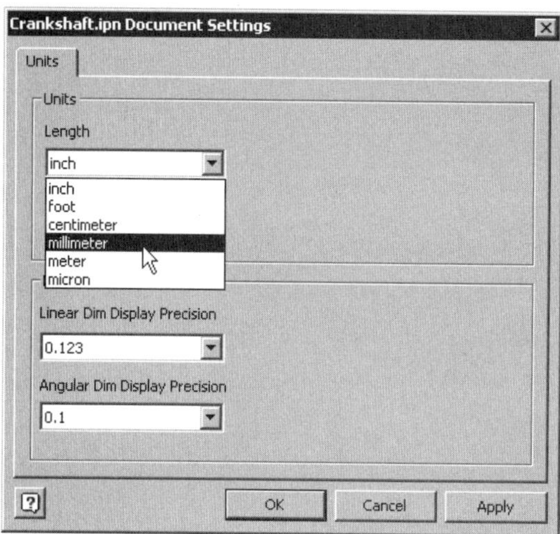


**Figure 10-67**

You can also edit an existing tweak and apply a unit other than the unit set in the Document Settings dialog box. To edit a tweak's value, double-click on the name of the tweak that you want to edit in the Browser, enter a new value in the tweak edit box (displayed at the bottom of the Browser), and press the ENTER key. The following image shows a tweak with a value of 30.000 mm being edited to 1 in.

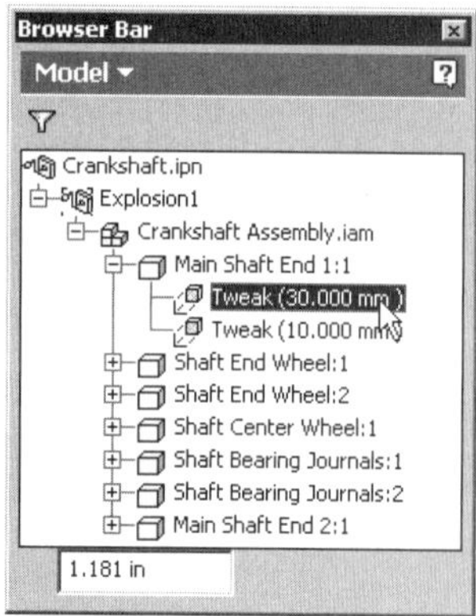


**Figure 10-68**

Another method used to edit a tweak is to move the cursor over the tweak in the graphics window until it is highlighted, right-click, and select Edit from the menu as shown on the left side of the following image. The Tweak Component dialog box is displayed. Enter a new value in the Transformations area of the dialog box, as shown on the right side of the following image.

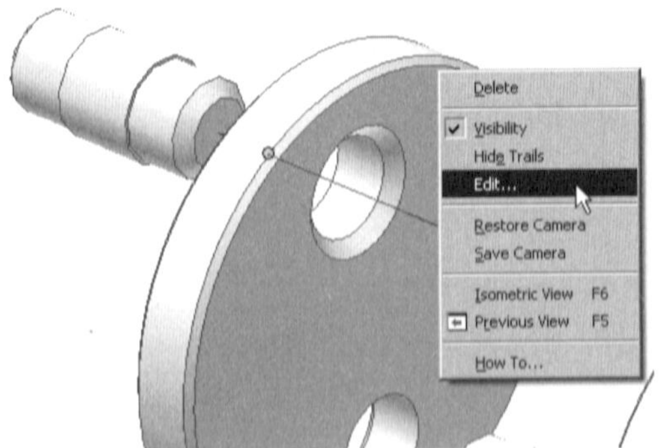


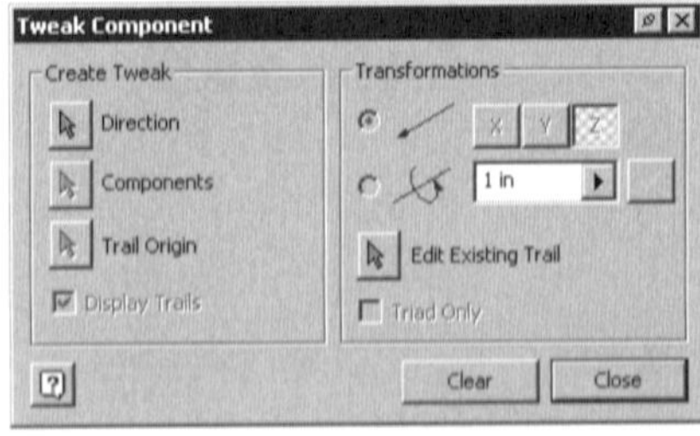


Figure 10-69

## CONSTRAINT TOOLS

Additional tools are available to manipulate and edit assembly constraints. These tools include the Find Other Half tool, the Constraint Tool Tip tool, and the Constraint Offset Value Modification dialog box. Each of these tools is described in the sections that follow.

### FIND OTHER HALF

The Find Other Half tool can be used to find the matching part that participates in a constraint placed in a large assembly. As you add additional parts over time, you may wish to highlight an assembly constraint and find the part that matches this constraint. In the following image, a Mate constraint has been highlighted. Half of this constraint has been applied to a part called Engine Block. To view the part sharing this common constraint, right-click on the constraint in the Browser and choose Other Half from the menu. The Browser will expand and expose the second constraint. In the case of this example, the other half of the Mate constraint was made to a part called Cylinder.

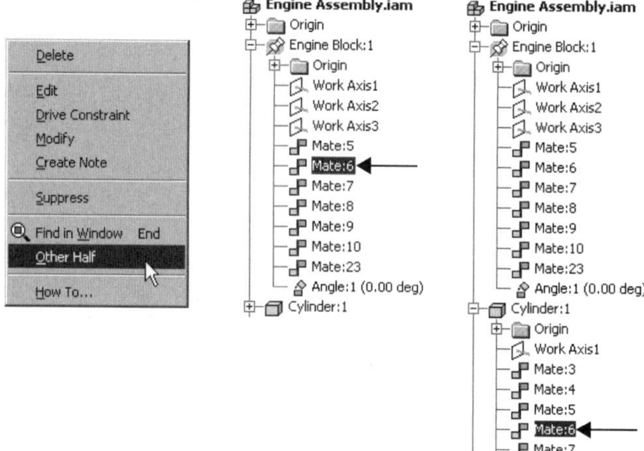


**Figure 10-70**

## CONSTRAINT TOOL TIP

In order to display all property information for a specific constraint, move your cursor over the constraint icon and a tool tip will appear, as shown in the following image.

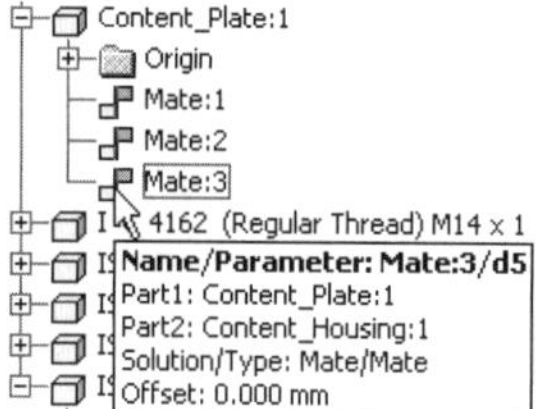


**Figure 10-71**

The following information is displayed in the tool tip:

- Constraint name and Parameter name (for offset and angle parameters).
- Constrained components (the two part names from the Browser).
- Constraint solution and type.
- Constraint offset or angle value.

**NOTE** Although the constraint name is highlighted in the previous image, you must hover your cursor over the constraint icon to view the tool tip information.

## CONSTRAINT OFFSET VALUE MODIFICATION

When editing a constraint offset value, use the standard value edit control. This process is similar to editing work plane offsets and sketch dimensions and will allow you to measure while editing constraint offset values. Right-click on the constraint to edit in the Browser and select Modify from the menu, as shown in the following image. The Edit Dimension dialog box will appear, allowing you to edit the offset value.

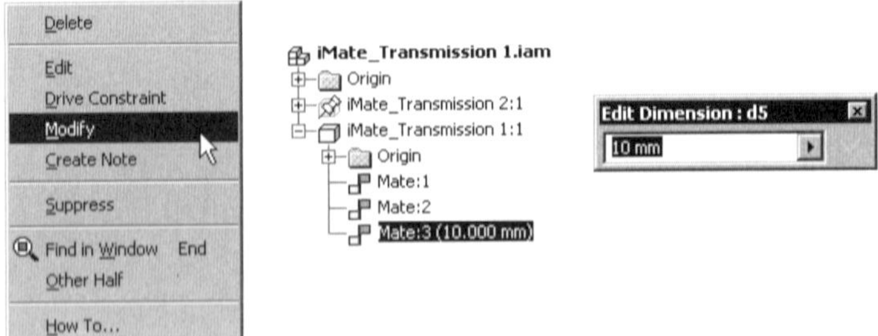


**Figure 10-72**

## REPLACING COMPONENTS

While designing, you may need to replace one component in an assembly with another component. A single component or all occurrences of the component can be replaced. The new component(s) will be placed in the same location if the origin of the replaced component is coincident with the origin of the placed component. During this replacement process, some assembly constraints may need to be recreated to correctly position the component. In the example of the hinge assembly, shown in the following image, a simpler design of the hinge pivot has been replaced with a more sophisticated design.

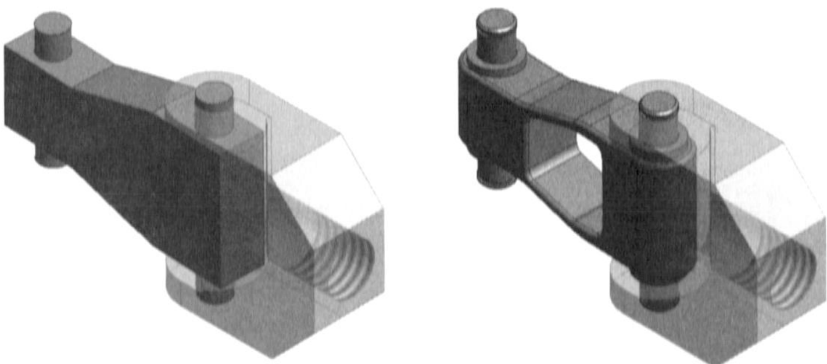

**Figure 10-73**

To replace a component, use the following steps:

1 Select the component to replace from the Browser or the graphics window or click the down arrow on the Replace Component tool on the Assembly Panel Bar. Then, choose either the Replace Component or Replace All tool, as shown in the following image.

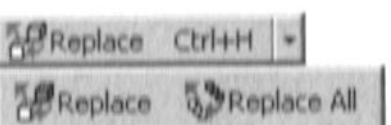


**Figure 10-74**

2 In the Open dialog box, select the component that will replace the existing component(s) and click Open.

3 A message notifies you that constraints and iMates will be retained, if possible. Click OK to continue or Cancel to not replace a component (see the following image).

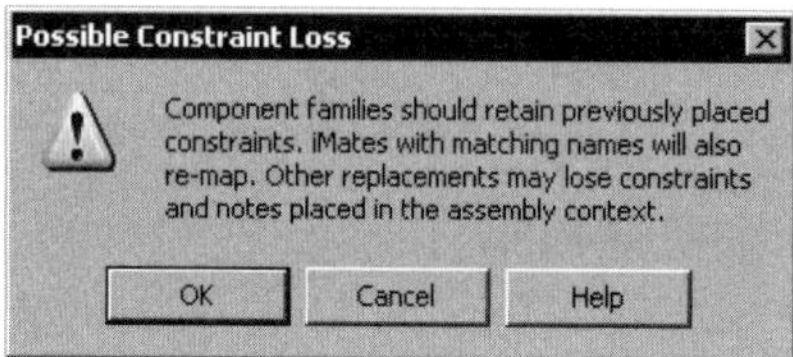


**Figure 10-75**

## CONTENT LIBRARY

A Content Library consisting of screws, nuts, washers, and other standard components is supplied with Autodesk Inventor. This library will be automatically installed if you choose to perform a complete installation. Use it to share predefined part files with assemblies. This library will reduce the need for you to draw standard items from scratch.

### CONTENT LIBRARY OPTIONS AND I-DROP

To display the Content Library, select the Model button from the Browser, and then select Library, as shown in the following image.

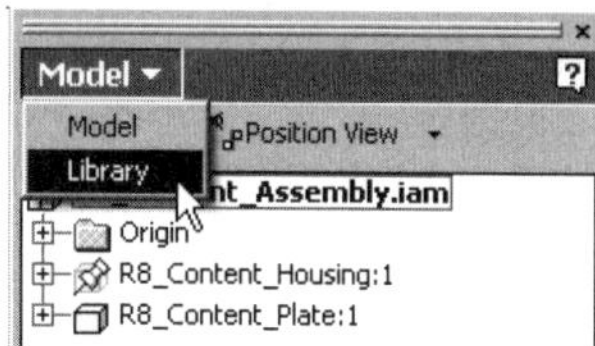


**Figure 10-76**

The Browser changes to display the Catalogs page. The Content Library supports 18 international standards and up to 12 different classes of standard parts in the form of screws, nuts, pins, rivets, and even structural steel shapes. To view a list of standards, click on the Standards filter button as shown in the following image. To deactivate a standard, remove the check from the box adjacent to the standard. This is a popular method of filtering out the standards that you do not use on a regular basis. Double-clicking on Standard Parts will change the Browser to reflect the three categories that house content: namely Fasteners, Steel Shapes, and Shaft Parts (as shown in the following image).

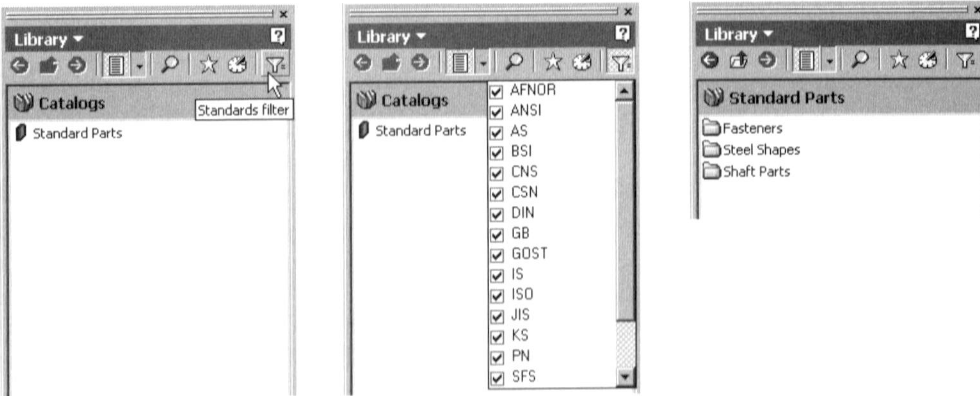


Figure 10-77

**NOTE** To return to the display of Model part and subassembly components, click on the Library button at the top of the Browser and select Model.

## The List View Option

By default, all content libraries and component categories, as shown in the following image, are displayed as a folder and title. This is referred to as a *list view*. Changing the view type to Icon View will display a pictorial representation of the Content Library or even the individual part category.

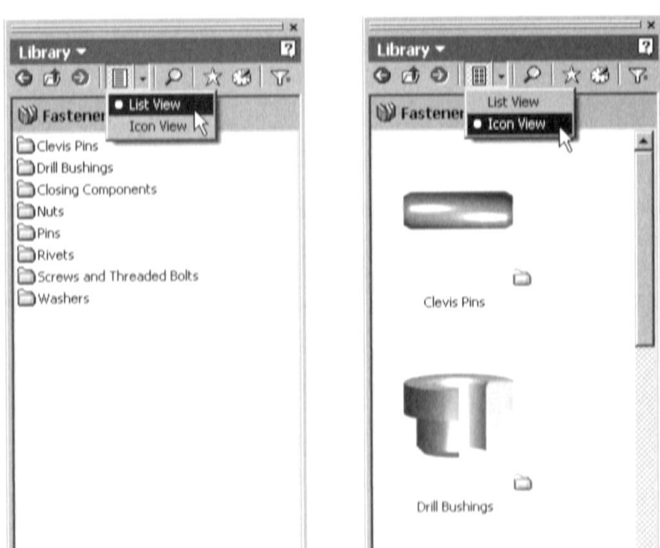


Figure 10-78

## The Search Option

When working with the Content Library, you can perform simple or advanced searches on components contained in the Content Library. Clicking on the magnifying glass button in the Browser displays the Search Browser, as shown on the left in the following image.

In this example, a simple search will be performed on the word "Woodruff" in all 18 standard catalogs. Clicking the Search button will perform this search and display the results shown on the right in the following image.

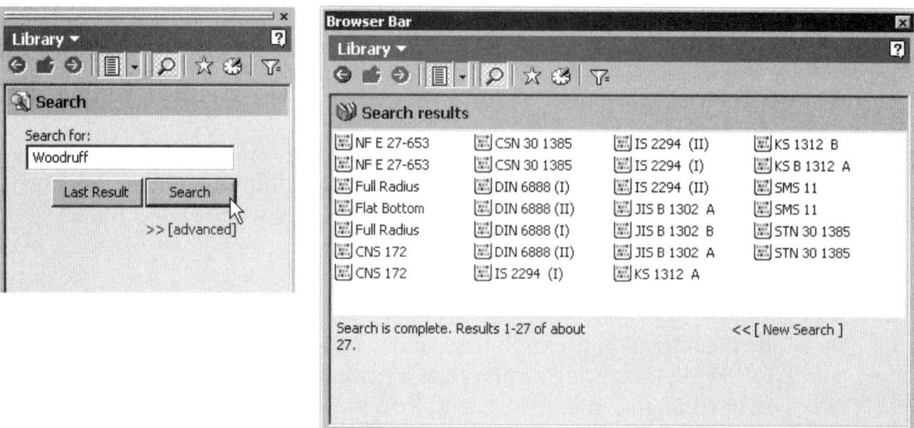


**Figure 10-79**

When performing searches, you can elect to select all 18 catalogs to include in the search or be more selective and remove the checks from boxes on catalogs you do not wish to be included in the search. Use the Standards filter button to reduce the number of catalogs when performing a search. In the following image, an advanced search will be performed only in the ISO standard. You then identify the Part Types as a means of narrowing down your search. In this example, a search will be performed in the ISO standard for all Fasteners that include nuts, screws and threaded bolts, and washers (see the following image).

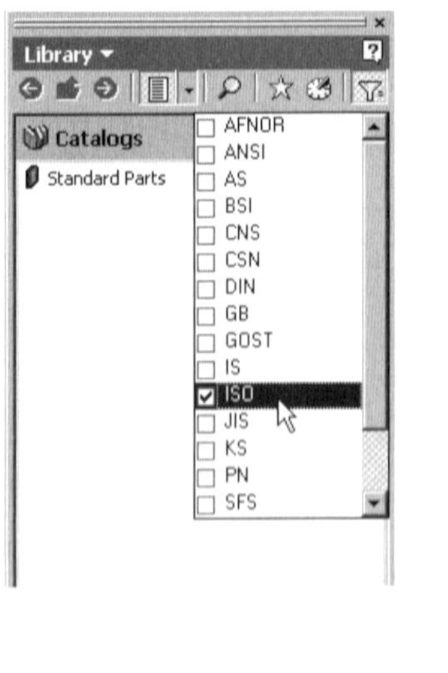


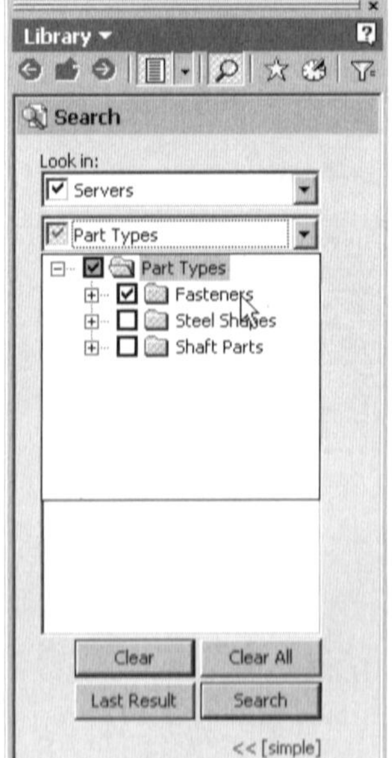


**Figure 10-80**

With the Property type set to Nominal Diameter, enter a value of 14 as the search value, as shown in the following image. When finished, click the Search button at the bottom of the Browser. The results are displayed in the following image. All nuts, screws, threaded bolts, and washers that share a nominal diameter of 14 are displayed in the Browser.

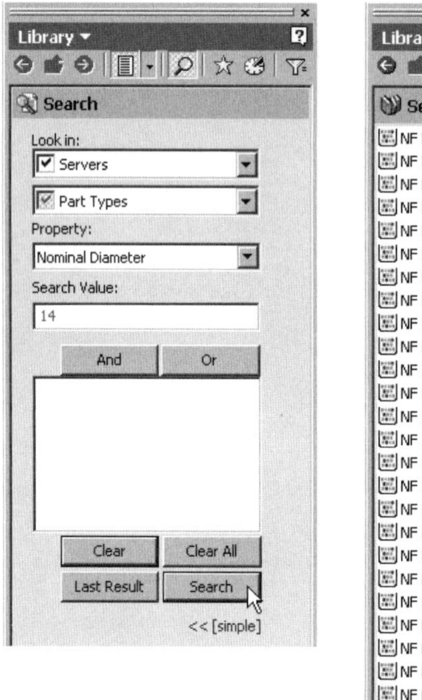


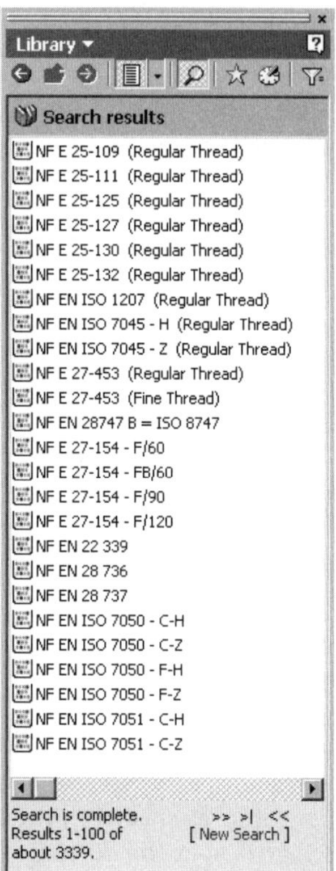


**Figure 10-81**

## The Add to Favorites Option

For those commonly used content items, you can select the item from any catalog and designate it as a favorite. You designate an item as a favorite through the Add to Favorite option in the menu, as shown in the following image. Any level of catalog, folder, or individual catalog component can be added as a favorite. Subfolders can also be created under Favorites, which will allow you to drag and drop components into the subfolder of your choice. Any item within the Favorites Browser can be deleted or renamed through the menu.

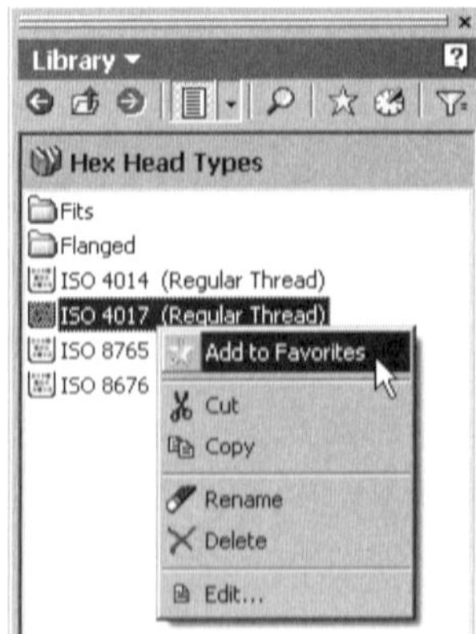


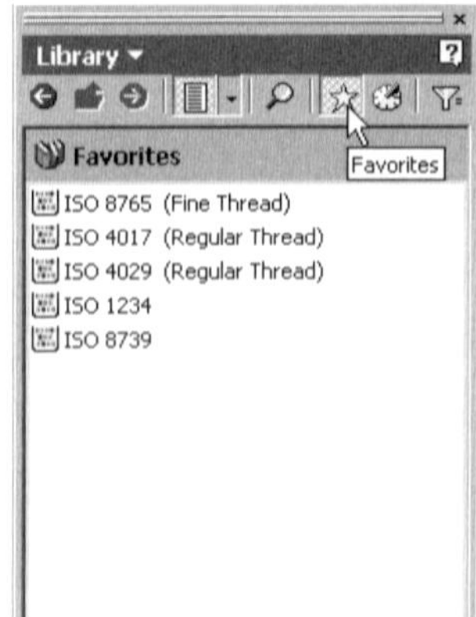


Figure 10-82

## The History Option

As in the following image, you can obtain a list of the content used most recently with the History tool.

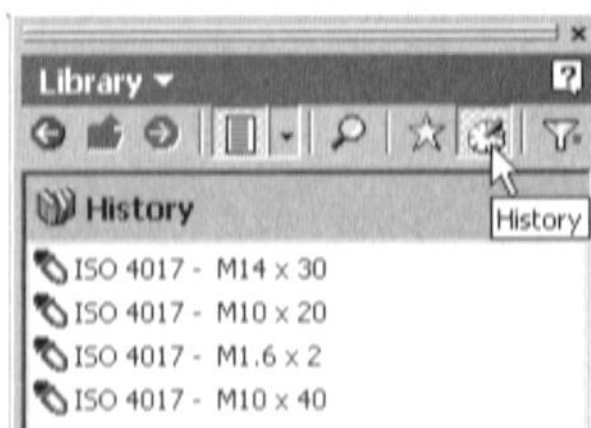


Figure 10-83

### LIBRARY SETUP

It has already been stated that Autodesk Inventor ships with a Content Library. Web or network server-based catalogs can be added to this library list by clicking on the Configure button shown in the following image.

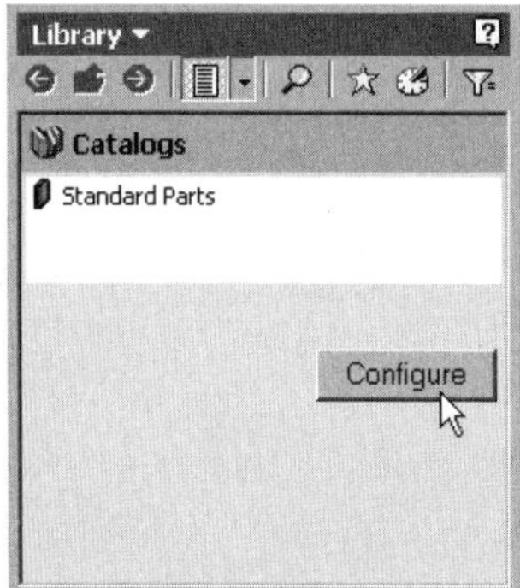


**Figure 10-84**

This activates the Configure servers dialog box, as shown in the following image. The current server name and address is listed in the edit box. Click on the Add button to define any new Web or network server-based standard parts catalogs. If you want to change the configuration of the current server, double-click on its name. In the case of the following image, double-click on the *local* server name.

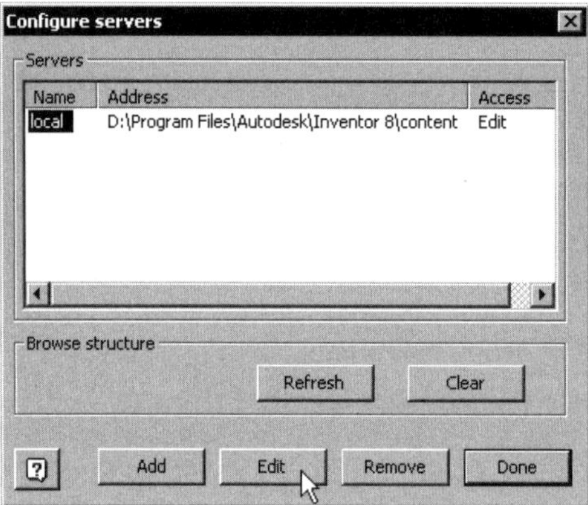


**Figure 10-85**

This action will launch the Configure server local dialog box, as shown in the following image. Use this dialog box to change the display of catalog list information. You can filter out those components that you do not use on a regular basis. In the following image, the Rivets and the Steel Shapes categories have been filtered out. You can also select a new storage location for the components generated from the server. By default, the location is in your Parent Assembly (see Storage Location in the following image). When finished, click the Save button to save the changes made in the Catalog List. This will return you back to the Configure server dialog box where you will click the Done button.

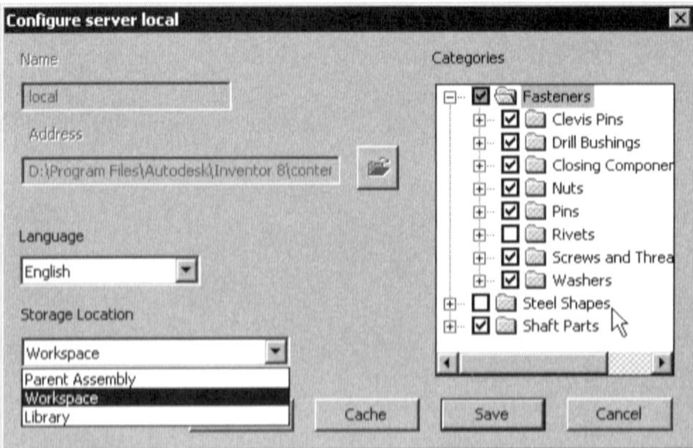


**Figure 10-86**

The results of filtering out certain component categories are illustrated in the following image. Notice that Steel Shapes is not listed under Standard Parts. Also notice in the image on the right that Rivets is not listed under the Fasteners category. This allows you to reduce the amount of information when performing searches on library components.

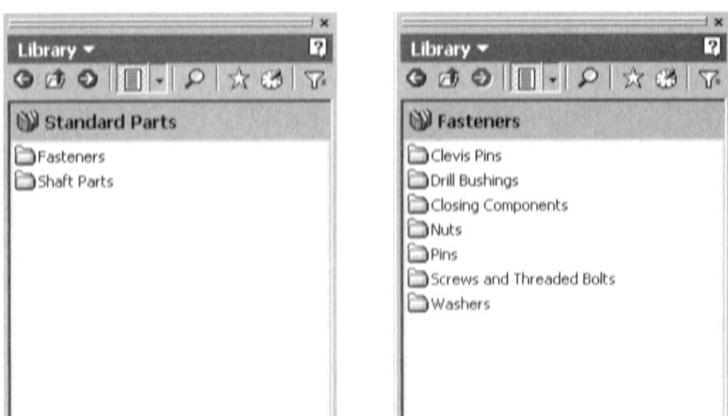


**Figure 10-87**

## PLACING CONTENT

Once you have identified a library part to add to an assembly, you can easily drag and drop this information into an assembly from the Browser. In the following image, the hex head fastener ISO 4017 will be shared with an assembly. Double-clicking on the item under the Hex Head Types area of the Browser displays the fastener where you can change the Nominal Diameter and Nominal Length. The eyedropper icon in the lower-left corner of the symbol box is a reminder that you can drag the symbol icon and drop the fastener into your assembly.

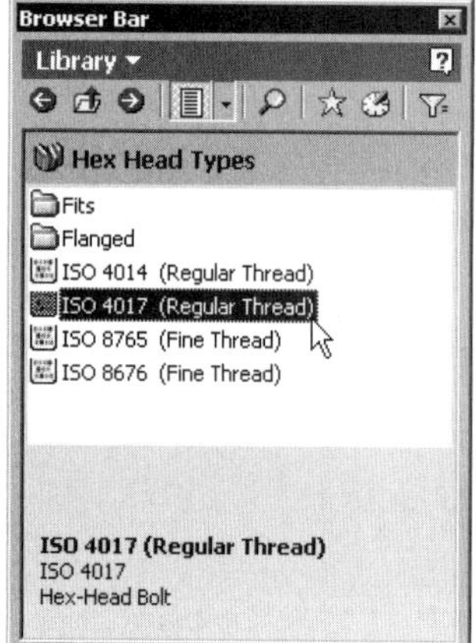


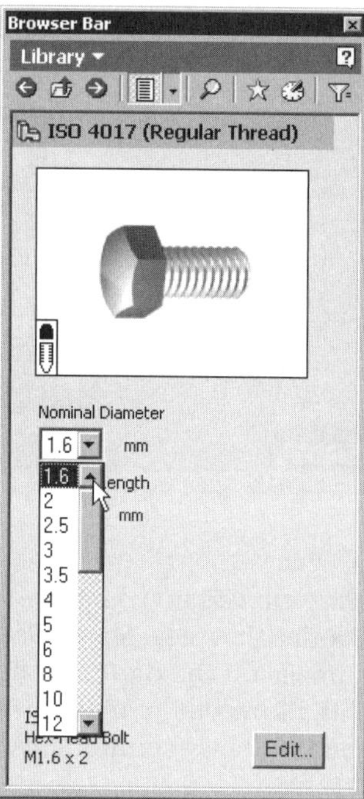


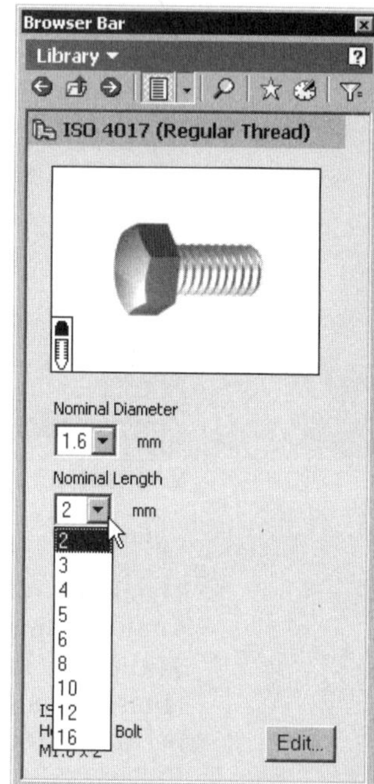


**Figure 10-88**

Hover the cursor over the image bit map associated with the part. In the following image, the cursor changes to the empty eyedropper as shown on the left. Click and hold down the left mouse button. The cursor changes to the full eyedropper as shown in the image on the right. Drag the part and drop it into the current assembly file.

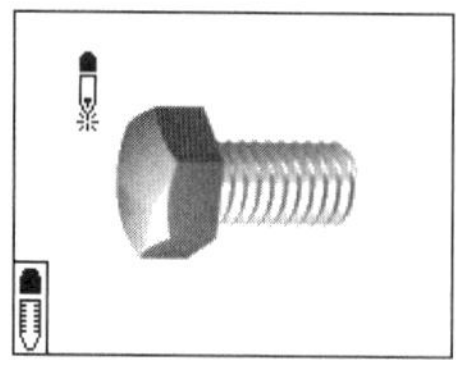
Empty Eyedropper

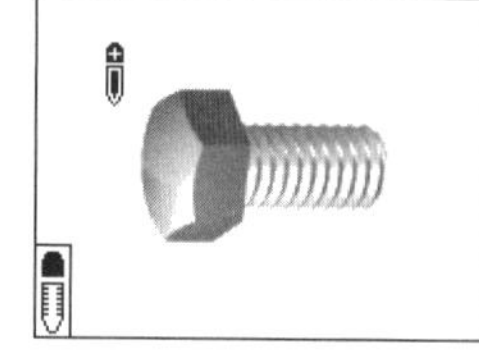
Full Eyedropper

**Figure 10-89**

The following image shows a typical hex head fastener. Clicking on the part displays pre-defined iMate glyphs that can be used to easily place the standard part in your assembly.

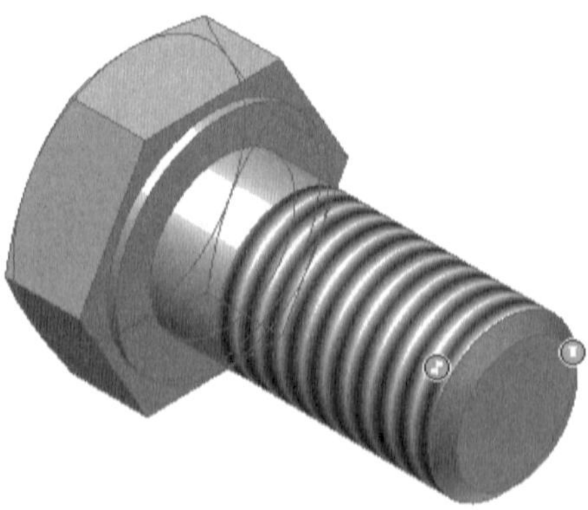

Figure 10-90

## Replacing Content

A single component or multiple instances of the same component can easily be replaced in an assembly. Use the following steps to replace Autodesk Inventor Content Library parts in the assembly:

1 In the assembly, select the part you want to find in Library Browser, and then right-click. From the menu, select the Find in Catalog option, as shown in the following image.

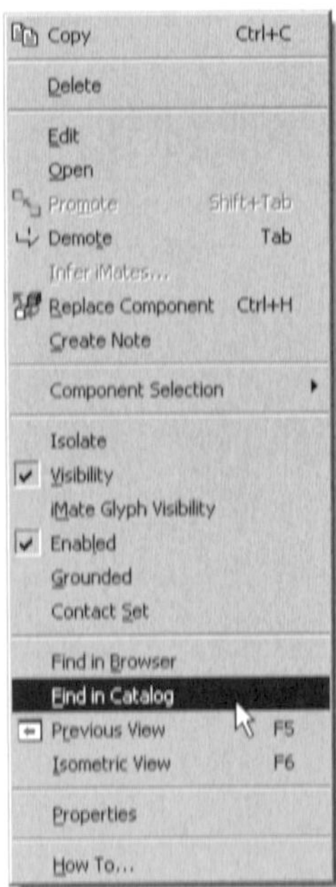


Figure 10-91

2 This action will find the catalog page with the part displayed in Content Library Browser, as shown in the following image.

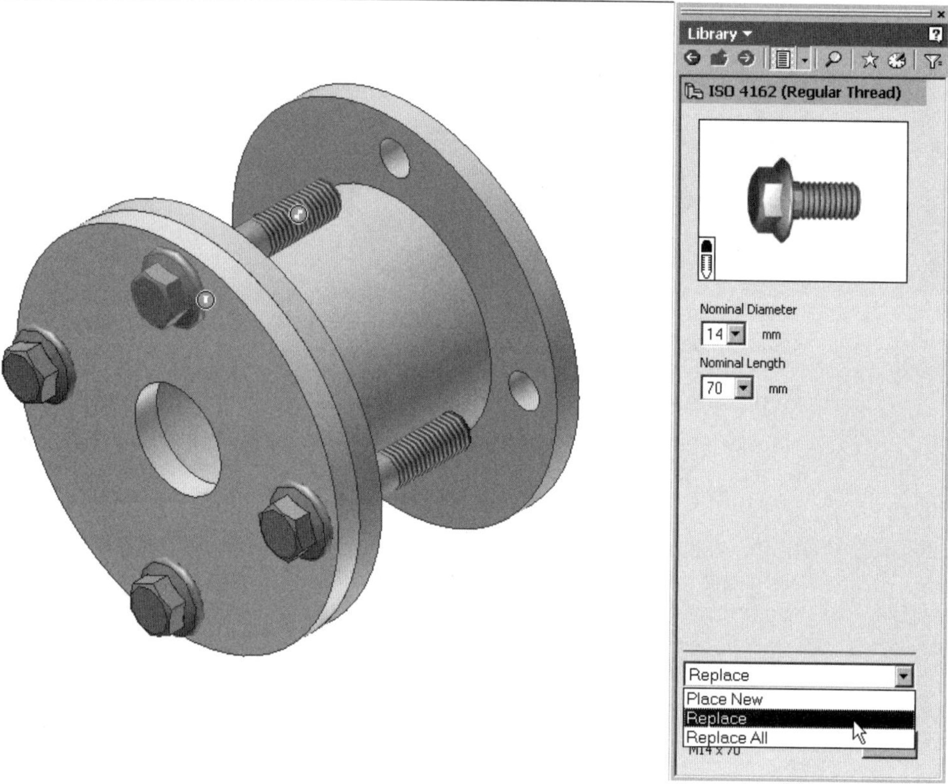


Figure 10-92

3 On the part page, use the list boxes and edit boxes to set new parameters for the part. In the following example, the Nominal Length of 70 mm has been changed to 130 mm. In the drop-down list at the bottom of the page, you can perform the following functions:

- Select Place New to insert a new instance of the part in the assembly.
- Select Replace to replace the selected part in the assembly with the part with new parameters.
- Select Replace All to replace all instances of the selected part.

In the following image, the Replace function is used to replace one of the ISO 4162 fasteners.

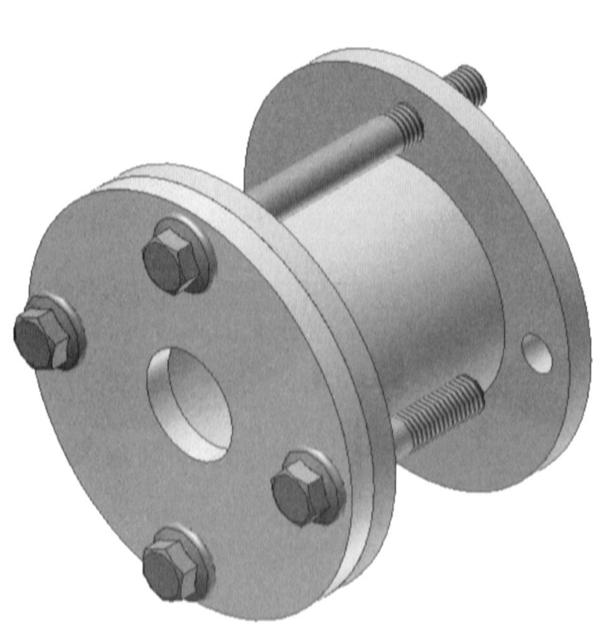

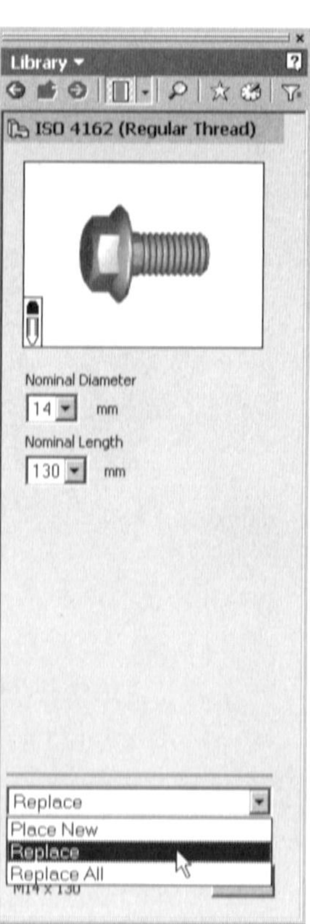


Figure 10-93

4 To perform the replacement of all ISO 4162 fasteners, drag the fastener from the image at the top of the Library Browser and drop it into the graphics window. This will replace all ISO 4162 fasteners in the assembly with a new length of 130 mm, as shown in the following image.

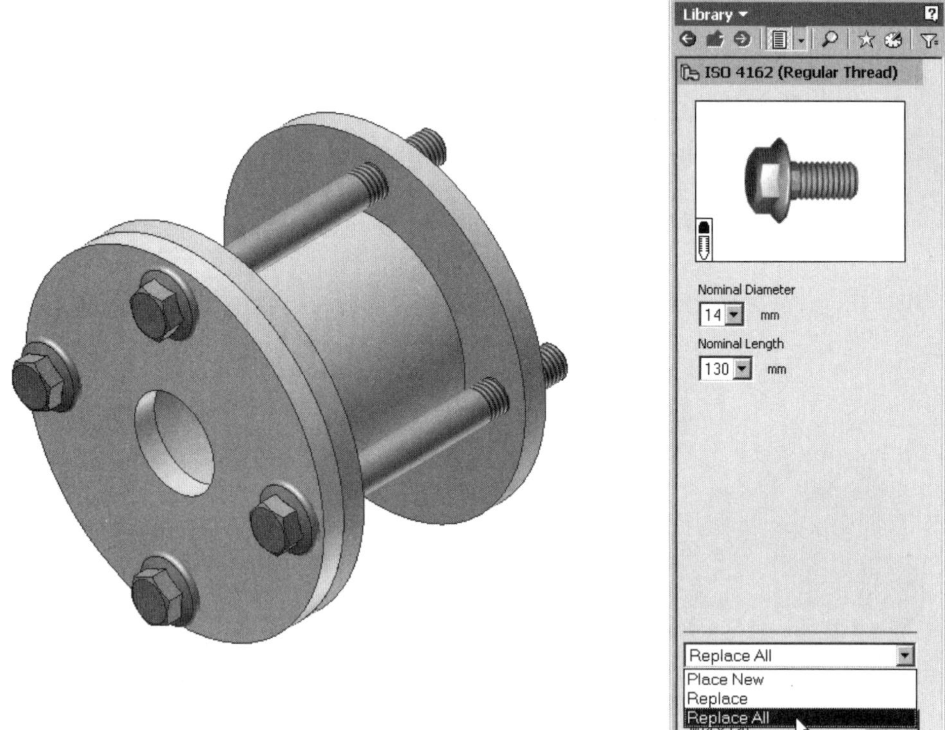


Figure 10-94

## CONTENT LIBRARY EDIT

A Content Library dialog box is available and it allows you to edit the data fields saved in the Content Library. To launch the dialog box, display the part in the Content Library Browser and click the Edit button at the bottom of the Browser as shown in the following image. A number of dialog box areas will be explained in the pages that follow.

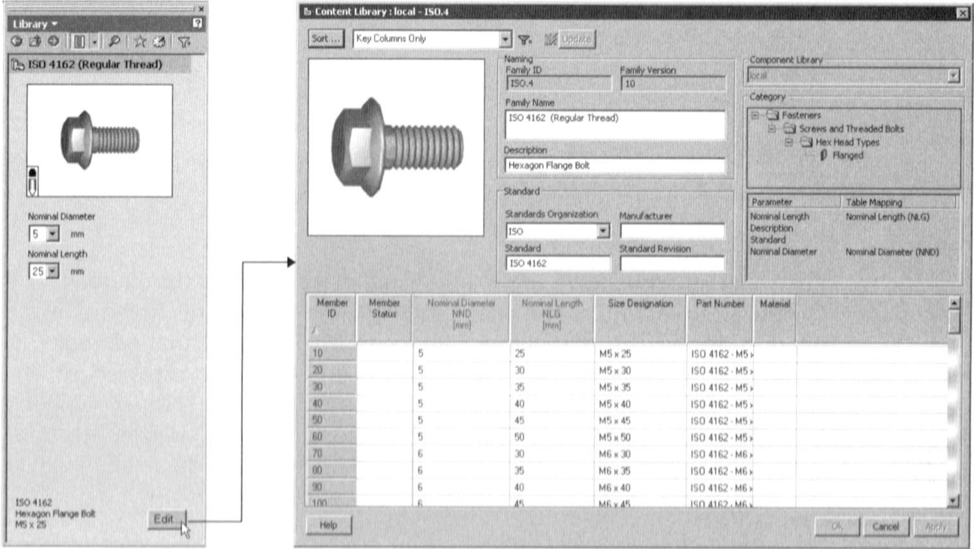


**Figure 10-95**

The Sort button shown in the image below displays the Sort dialog box. Information located in the edit box can be displayed in ascending or descending order through this dialog box. The Key Columns Only parameter applies to the database information listed in the edit box area of the Content Library dialog box. Each listing can be used to reduce or filter the amount of information viewed at any one time. To display all database information, click on All Columns. The Update button will activate whenever changes are made to the content database. Click this button to recalculate the database.

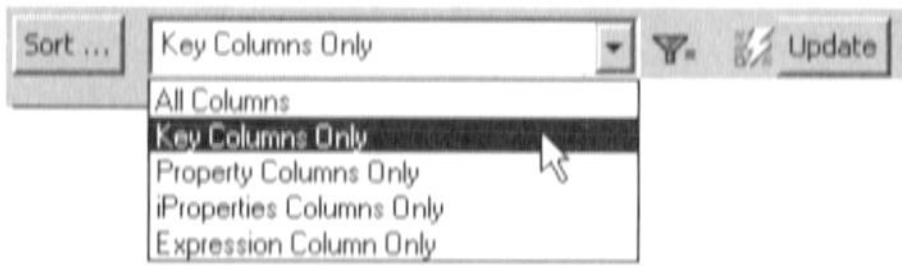


**Figure 10-96**

The Naming area shown in the following image displays a Family ID, which identifies the database family type. The Family Version displays the version number of the part family. This number is automatically changed whenever the part family is edited and saved in the database. The Family Name displays the name of the part being edited. The Description displays a general description of the part family.

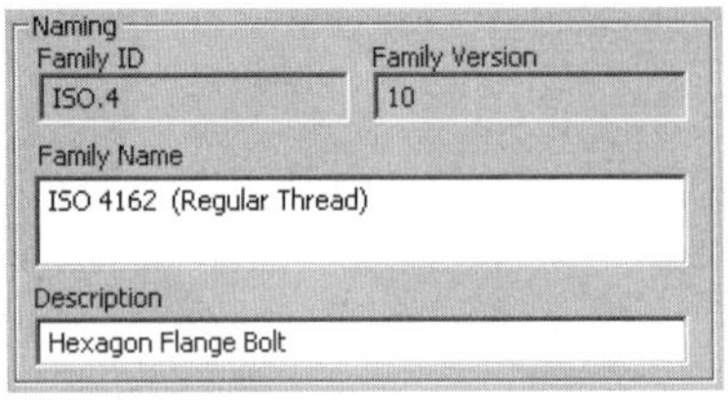

Figure 10-97

The Standard area shown in the following image contains the Standards Organization. An arrow is available to select other standards from the list provided in the Content Library dialog box. The Manufacturer edit box allows you to enter text for a manufacturer. When the part is placed into an assembly, this manufacturing information is stored in the iProperty dialog box under Manufacturer. The Standard edit box displays the current standard on which the part family is based. The Standard Revision edit box allows you to enter a revision value that can be specific to each part family.

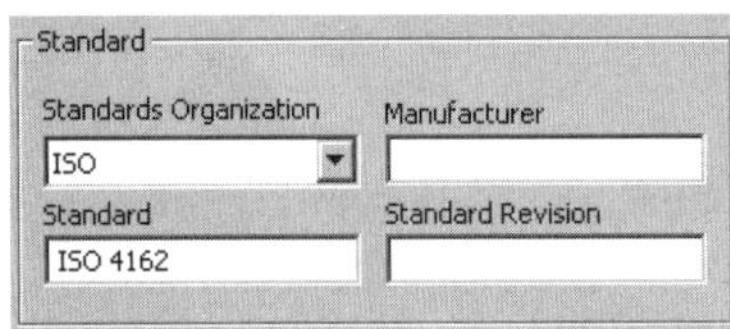


Figure 10-98

The Component Library area shown below displays the main database in which the part library is stored. The Category displays a tree view that is used to navigate to the part family.

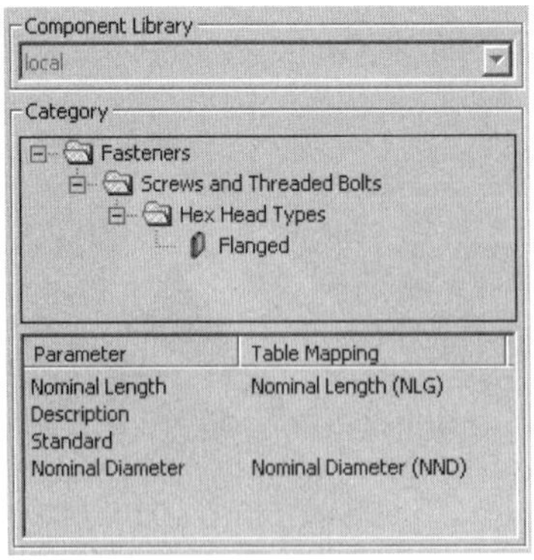

Figure 10-99

## Adding a Part to the Part Family

The following examples illustrate the creation of a new part based on the fastener type ISO 4162. The longest fastener in the M14 group is 140 mm. A new part file will be created with a length of 160 mm. Follow the steps below for adding this new part to the part family.

1 Select a row where you want to insert the new row in the editing table of the Content Library dialog box, right-click in the row header, and then select Insert from the menu, as shown in the following image.

| | | | | |
|---|---|---|---|---|
| 540 | 14 | 130 | ISO 4162 - M14 | M1 |
| 550 | 14 | 140 | ISO 4162 - M14 | M1 |
| 560 | | 55 | ISO 4162 - M16 | M1 |
| 570 | | 60 | ISO 4162 - M16 | M1 |
| 580 | | 65 | ISO 4162 - M16 | M1 |
| 590 | | 70 | ISO 4162 - M16 | M1 |

Copy
Paste
Insert

**Figure 10-100**

2 The new row is added directly above the row you just selected. Select a row that best matches the new part, right-click, and then select Copy, as shown in the following image. It is recommended that you copy the data from an existing row to the new row, and then edit the parameters for the new part.

| | | | |
|---|---|---|---|
| 520 | 14 | 110 | 5.6 |
| 530 | 14 | 120 | 5.6 |
| | 14 | 130 | 5.6 |
| | | | |
| | 14 | 140 | 5.6 |
| | 16 | 55 | 6.7 |

Copy
Paste
Insert
Delete

**Figure 10-101**

3 Select the new row, right-click, and then select Paste to create a duplicate row of information, as shown in the following image.

| | | | | | |
|---|---|---|---|---|---|
| 520 | | 14 | 110 | 5.6 | 1 |
| 530 | | 14 | 120 | 5.6 | 1 |
| 540 | | 14 | 130 | 5.6 | 1 |
| | Update Req. | 14 | 130 | 5.6 | 1 |
| 550 | | 14 | 140 | 5.6 | 1 |

**Figure 10-102**

4 Edit all parameters for the new part to reflect the new length of 160 mm, as shown in the following image. You will have to scroll horizontally to change all parameters of this new part.

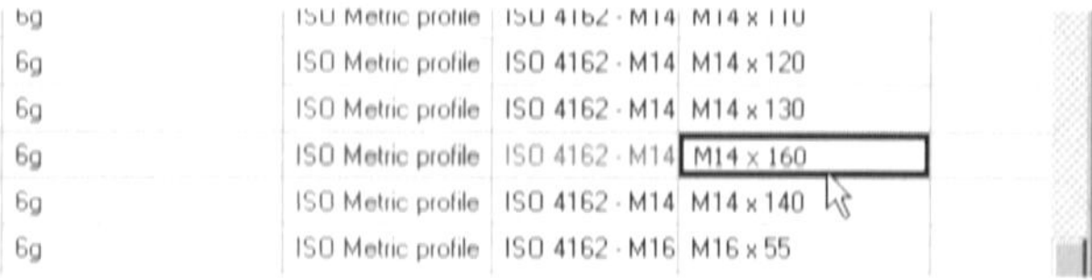

| | | | |
|---|---|---|---|
| 6g | ISO Metric profile | ISO 4162 - M14 | M14 x 110 |
| 6g | ISO Metric profile | ISO 4162 - M14 | M14 x 120 |
| 6g | ISO Metric profile | ISO 4162 - M14 | M14 x 130 |
| 6g | ISO Metric profile | ISO 4162 - M14 | M14 x 160 |
| 6g | ISO Metric profile | ISO 4162 - M14 | M14 x 140 |
| 6g | ISO Metric profile | ISO 4162 - M16 | M16 x 55 |

**Figure 10-103**

5 When you have finished editing the new part, click the Update button as shown in the following image to recalculate the new part.

Figure 10-104

6 Click the Apply button located in the lower-right corner of the Content Library dialog box to save the change and publish the new part to the Content Library. The new part is added at the end of the database.

| | | | | | |
|---|---|---|---|---|---|
| 680 | | 16 | 160 | 6.7 | 15.1 |
| | | 14 | 160 | 5.6 | 12.9 |
| | | | | | |

Figure 10-105

## Deleting a Part from the Part Library

1 To delete a part from the Content Library dialog box, select the row that contains the information on the part you want to delete, right-click, and then select Delete from the menu as shown in the following image.

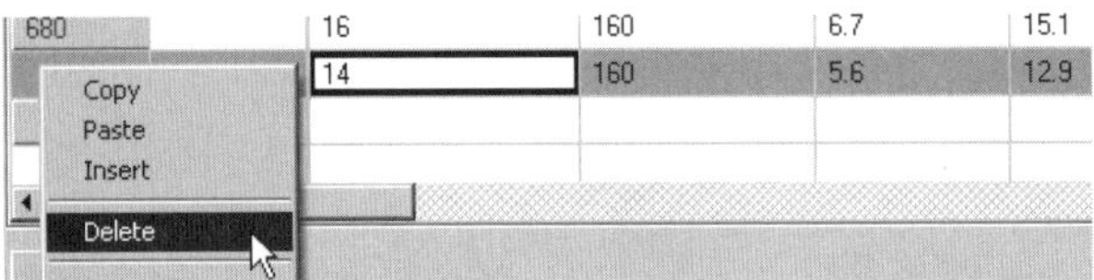


Figure 10-106

2 Click on the Update button and then click the Apply button to save this change and publish the results to the Content Library.

### STANDARD CONTENT SECTION ATTRIBUTE

You can now control whether standard parts in drawing views will be sectioned or not.

From the Assembly tab in the Options dialog box, there is a new option "Section Standard Parts." When this option is checked, Autodesk Inventor will section all standard parts in a section view. This option is NOT checked by default.

There are three options for the Section Standard Parts option:

**Always** Standard parts are always sectioned.

**Never** Standard parts are never sectioned.

**Obey Browser Settings** Sectioning defaults to how the part's visibility is set in the Browser.

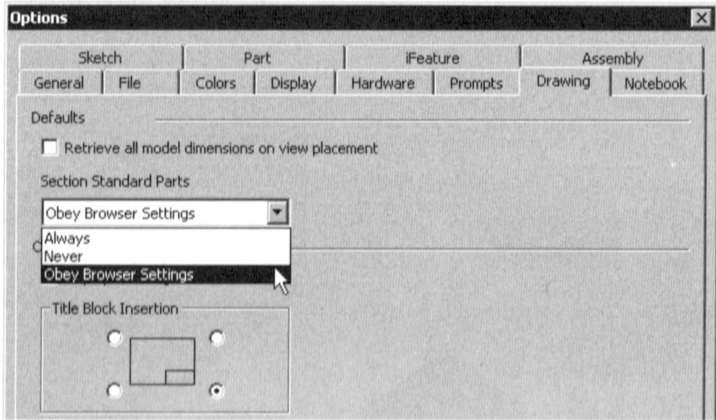


**Figure 10-107**

The following image illustrates the results of sectioning and not sectioning standard parts.

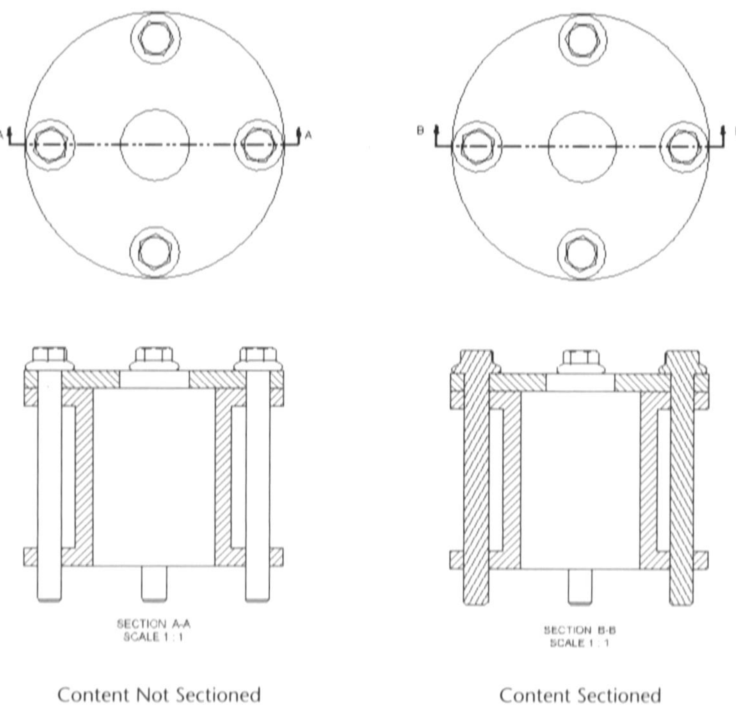


**Figure 10-108**

After a section view is placed, you can modify how standard part(s) are sectioned by right-clicking in the parent view and selecting Edit View from the menu. When the Drawing View dialog box is displayed, click the Options tab and select the desired option under the Section Standard Parts heading, as shown in the following image.

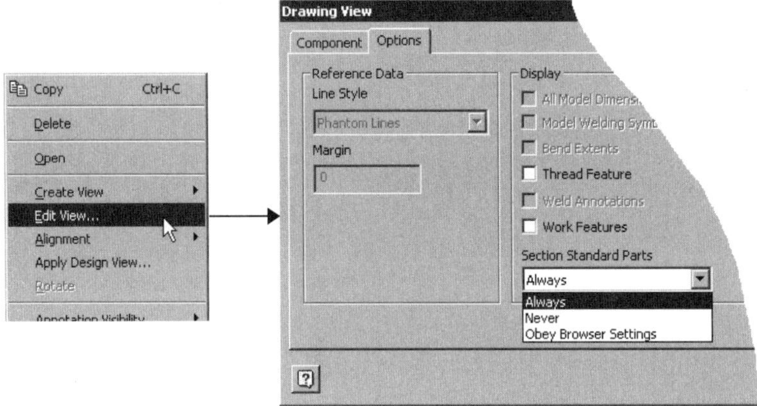


Figure 10-109

# EXERCISE 10-3 Fasteners and Parts Library

In this exercise, you place standard bearings from the Autodesk Inventor content library in a shaft assembly. You then edit the shaft diameter and replace the bearings to match the new diameter. Finally, you override the mass properties of the bearing to reflect the actual weight of the bearing and then you infer iMates. To navigate to the exercise in the *Electronic Student Workbook*, do the following:

1 From the Main TOC page, click Chapter 10.

2 From the TOC for Chapter 10, click Fasteners and Parts Library.

The following image illustrates the opened exercise.

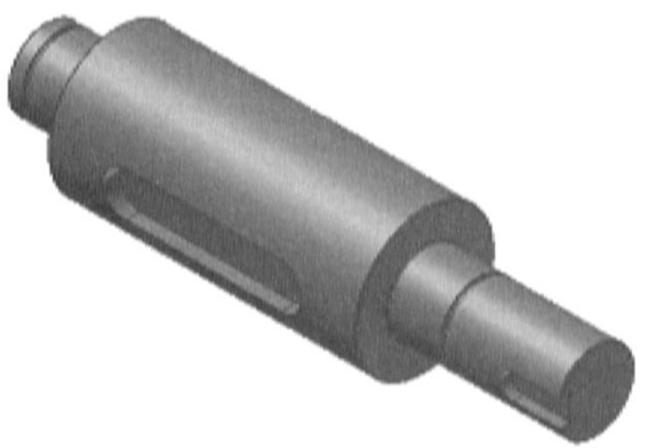

**Figure 10-110 Opened exercise**

The following image illustrates the completed exercise.

**Figure 10-111 Completed exercise**

## PATTERNING COMPONENTS

You can use the Pattern Component tool from the Assembly Panel Bar shown in the following image when you are placing multiple occurrences of the same component or subassembly that match a feature pattern on another part (component pattern) or have a set of circular or rectangular part patterns in an assembly (assembly pattern). Both the component and assembly patterns will be described in the next section.

Figure 10-112

### COMPONENT PATTERNS

A component pattern will maintain a relationship to the feature pattern that is selected. A bolt is component-patterned, for example, to a bolt-hole pattern that consists of 4 holes. If the feature pattern (the bolt-hole) changes to 6 holes, the bolts will move to the new locations and two new bolts will be added for the two new holes. To create a component pattern, there must be a feature-based component pattern, and the part that will be patterned should be constrained to the parent feature (that is, the original feature that was patterned). Issue the Pattern Component tool from the Assembly Panel Bar, and the Pattern Component dialog box will appear, as shown in the following image.

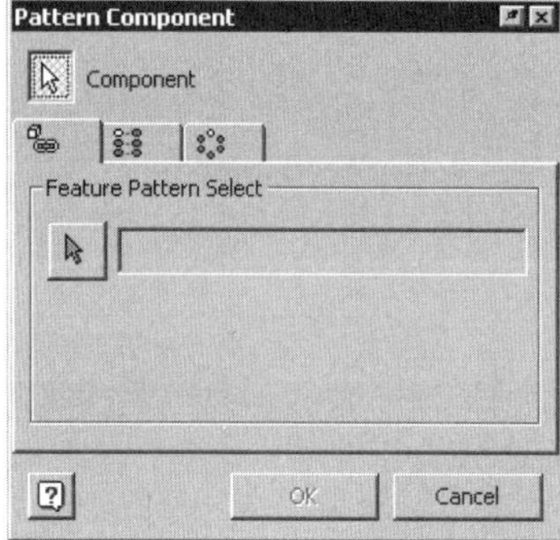


Figure 10-113

There are three tabs: Associative, Rectangular, and Circular (see preceding image). The Associative tab is the default and the tab that will be used to create a component pattern. By default, the Component selection option is active. Select the component (such as the screw) or components that will be patterned. Next, select the Feature Pattern Select button in the dialog box and select a feature (such as a hole) that is part of the feature pattern. Do not select the original feature. After selecting the pattern, it will highlight on the part, and the pattern name will appear in the dialog box. When done, select OK to create the component pattern and exit the operation (see the following image).

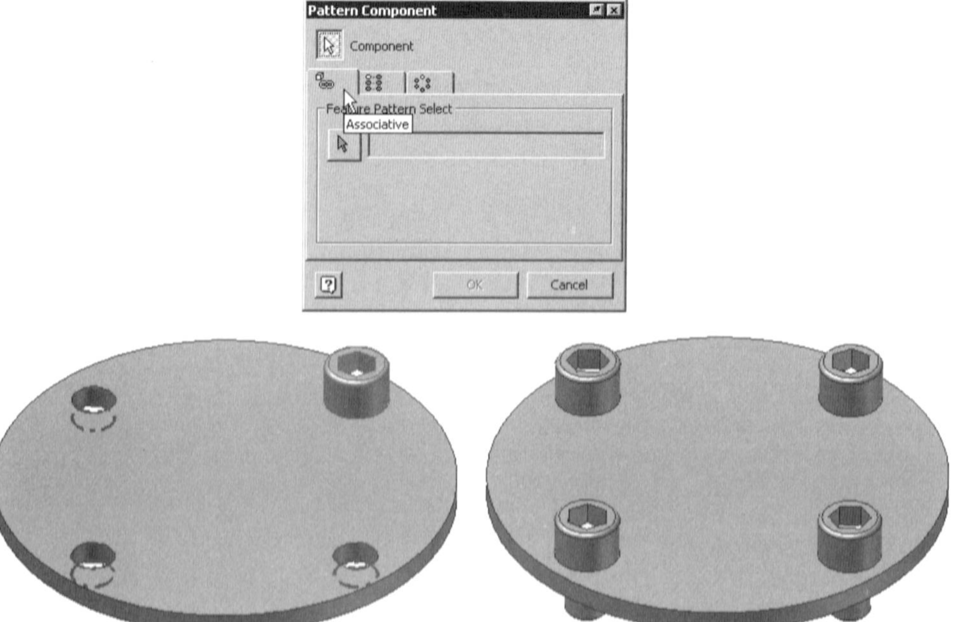


Figure 10-114

The pattern can be edited by either selecting the pattern in the graphics window, or by right-clicking in the Browser and choosing Edit from the menu. In the Browser, the component that was patterned will be consumed into a Component Pattern and the part occurrences will appear as elements, which can be expanded to see the part. You can suppress an individual part by right-clicking on it and selecting Suppress from the menu, as shown on the right in the following image.

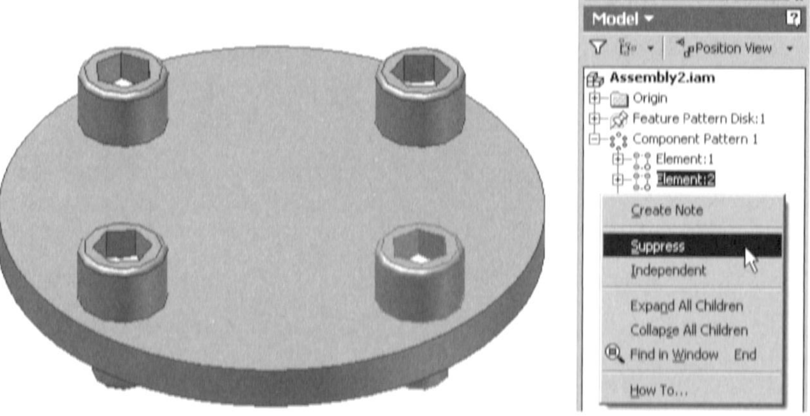


Figure 10-115

You can also break an individual part out of the pattern by right-clicking on it and selecting Independent from the menu, as shown on the right in the following image. Once a part is independent, it no longer has a relationship with the pattern.

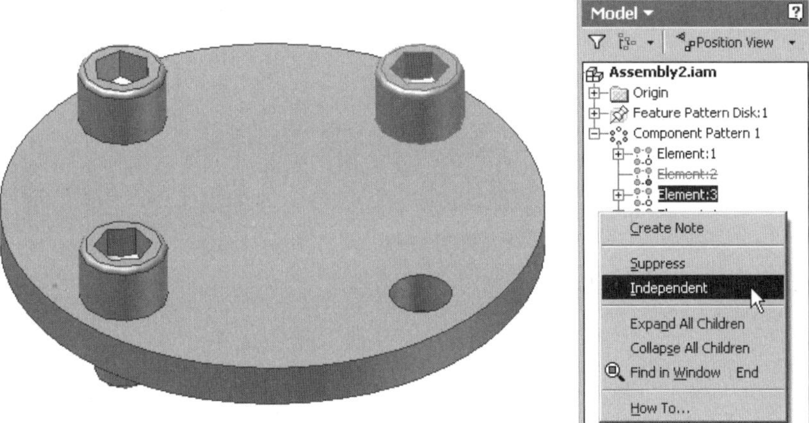


Figure 10-116

## Deleting a Pattern Component

To delete a pattern, select the pattern in the graphics window or in the Browser, and then either right-click and select Delete from the menu or press the DELETE key on the keyboard. If the Delete Component Pattern Source(s) option from the Options dialog box (under the Assembly tab) is checked, the source component will be automatically deleted. If the source component should not be deleted when the pattern elements are deleted, you should uncheck this option as shown in the following image.

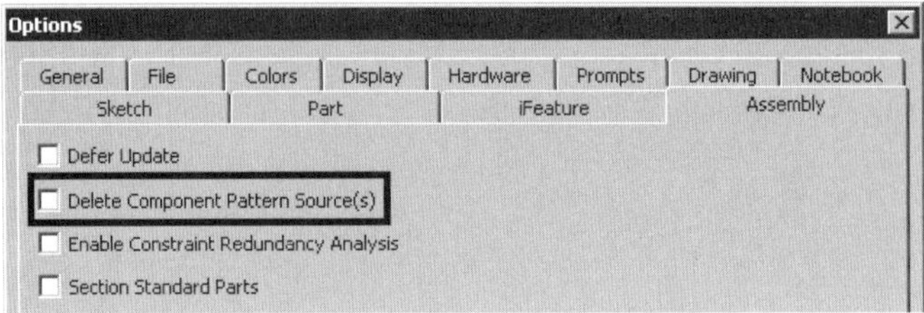


Figure 10-117

### ASSEMBLY PATTERNS

A component in an assembly can also be patterned in either a rectangular or circular fashion. The resulting pattern acts like a feature pattern. After creating it, you can edit it to change its numbers, spacing, and so on. To pattern a component, you must first create it or place it in an assembly. Then, issue the Pattern Component tool from the Assembly Panel Bar and the Pattern

Component dialog box will appear. Click the Rectangular or Circular tab and enter the placement values as needed.

Examples of Rectangular and Circular patterns are shown in the following image.

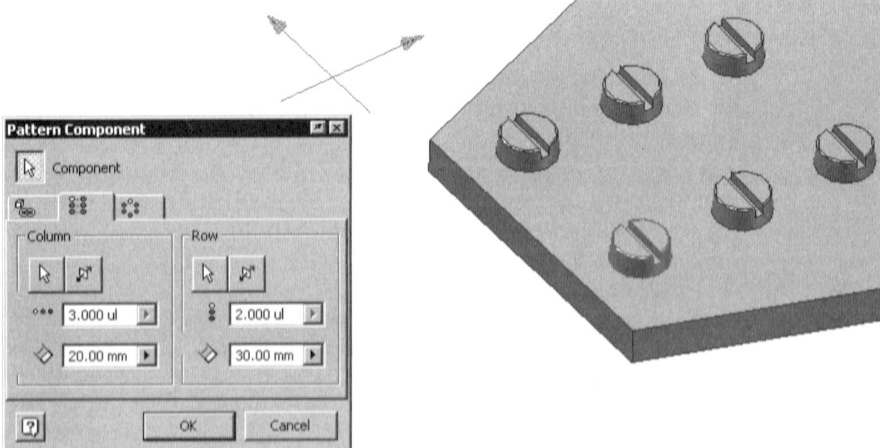


**Figure 10-118**

**NOTE** In the previous image of the rectangular pattern component, work axes are used to define the direction of the pattern in the *X* and *Y* directions.

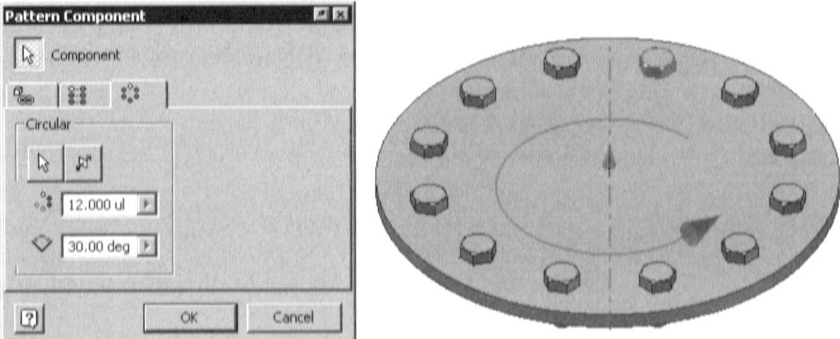


**Figure 10-119**

The completed pattern then acts as a single part. If one part moves, they all move. The component that was patterned will be consumed into a Component Pattern in the Browser and each of the part occurrences will also appear as elements that can be expanded. The pattern can be edited by either selecting the pattern in the graphics window and right-clicking, or right-clicking on the pattern's name in the Browser and selecting Edit from the menu.

## ADDITIONAL COMPONENT PATTERN OPTIONS

Additional options that are available when editing and manipulating component patterns include replacing all occurrences in an assembly pattern in a single step, restructuring a component pattern, and controlling the component pattern visibility. All of these features are discussed below.

### Component Pattern Replace

When replacing a component in an assembly pattern, all occurrences in the selected assembly pattern can be replaced with a newly selected component. Expand one of the elements in the Browser, as shown in the following image, and select the component to replace. With this item highlighted in the Browser, right-click and select Replace Component from the menu. The Open dialog box will appear and enable you to select the replacement component.

**Figure 10-120**

After selecting the replacement component from the Open dialog box, an Alert dialog box will appear, as shown in the following image. Component families should retain their previously placed constraints. Click OK to dismiss this dialog box.

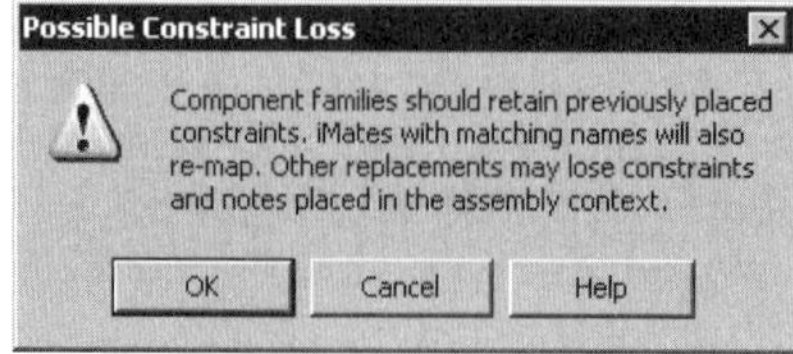


**Figure 10-121**

The result of performing the replace pattern component operation is illustrated in the following image. Since the use of component patterns allows for better capture of design intent, replacement of all occurrences maintains this design intent without manually replacing each component in the pattern. Overall ease of assembly use is improved and component patterns can be completed more quickly.

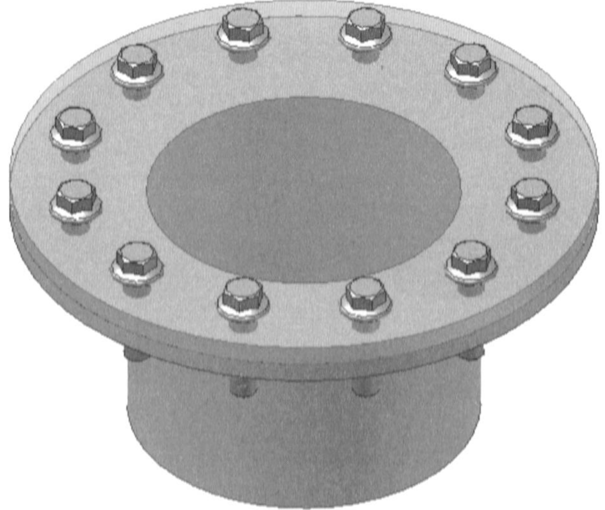

Figure 10-122

## Component Pattern Restructure

Component patterns can be restructured into or out of a subassembly. These restructuring tools are found on the menu when you are in the assembly design environment. First, select the component pattern from the Browser to restructure. Next, right-click and select Demote from the menu. A dialog box stating that restructuring may remove assembly constraints will appear. Click OK to complete the component pattern restructuring process.

## Component Pattern Visibility

Instead of expanding a pattern in the Browser and selecting each pattern instance on which to toggle off the visibility of patterned components, the visibility of the entire component pattern can be toggled off in one easy step. Select the component pattern in the Browser and right-click the menu, as shown in the following image. Clicking Visibility will remove the check to turn the component pattern off.

Figure 10-123

## MIRRORING AN ASSEMBLY

To create right- and left-handed versions of assemblies and components, use the Mirror Components tool. Mirrored components can be created using one of two methods:

1 Create a new assembly:

A new assembly can be created by selecting the components of a source assembly. A mirror copy is generated from the original assembly. The new assembly is created relative to a mirror plane.

2 Create instances of components:

Individual components can be selected in a current assembly to create mirrored instances or copies. Each new mirrored component is considered a new file. The individual components are mirrored based on a mirror plane.

To begin, open the assembly you want to mirror. The following image illustrates a portion of a hold-down clamping mechanism. Since the image represents half of the total assembly, the Mirror Components tool will be used to generate the other half of the clamp.

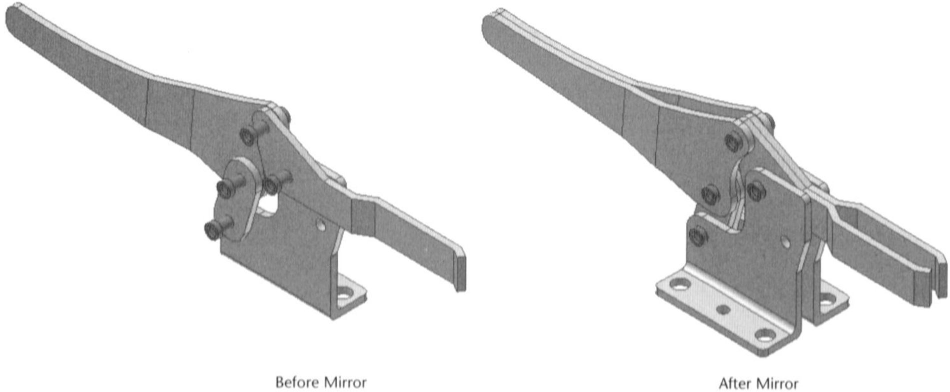


**Figure 10-124**

Follow the next series of steps to create a mirrored group of components in an assembly:

1 Click the Mirror Components button located on the Assembly Panel Bar as shown in the following image.

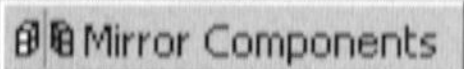


**Figure 10-125**

2 The Mirror Components dialog box will appear. If it is not already selected, click on the Mirror Plane button and select a work plane or planar face on the assembly, as shown in the following image.

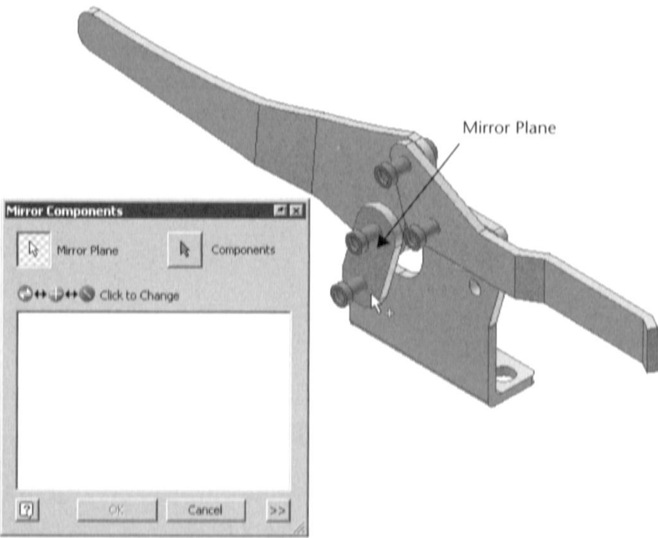


**Figure 10-126**

3 If it is not already selected, click on the Components button. Select all components from the graphics screen or from the Browser. As each component is selected, it is added to the Mirror Components dialog box list. You will also see a preview of the mirrored components on the graphics screen, as shown in the following image.

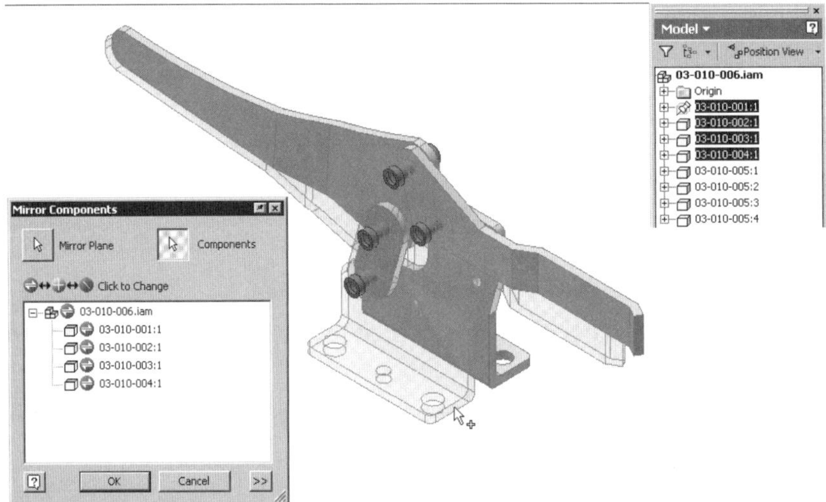


Figure 10-127

4 The following status buttons are available in the Mirror Components dialog box. Click the status button next to a component to change its status based on the following information:

| Symbol | Title | Meaning |
|---|---|---|
| (mirrored symbol) | Mirrored | Clicking on this button will create a mirrored copy (instance) of the component. |
| (reused symbol) | Reused | Clicking on this button will create a new copy (instance) in the current or new assembly file. |
| (excluded symbol) | Excluded | Click on this button excludes a component or subassembly in the mirror operation. |
| (mixed symbol) | Mixed Reused/Excluded | This button indicates that a subassembly contains components with reused and excluded status. It could also mean that the reused subassembly is not complete. |

Figure 10-128

5 Click on the More (>>) button to display previewing options and controls for how content library components are handled.

### Reuse Content Library Components

Placing a check in this box will restrict the mirrored state for library components. Instances of the library part are created in the current or new assembly file instead.

### Preview Components

Place a check in each box to preview the mirrored components on the graphics screen.

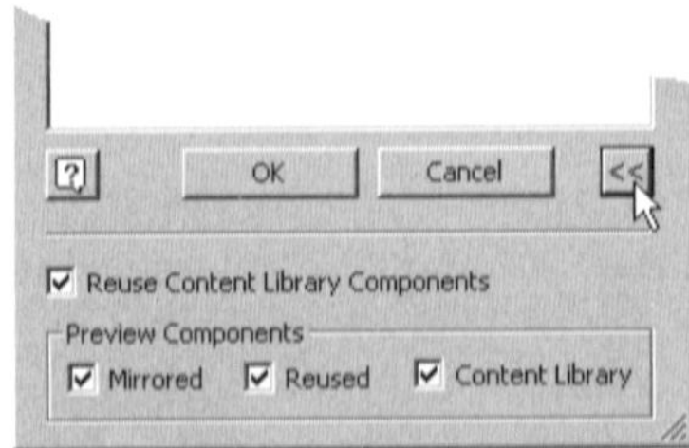


**Figure 10-129**

6 When you have finished making changes to settings, click the OK button to continue.

7 Clicking the OK button in the previous step does not complete the mirror operation. Instead, the Mirror Copy: File Names dialog box appears. Use this dialog box to change the names of the mirrored components or to keep the default names, as shown in the following image.

To change a part name, simply click on a current name under the New Name heading.

You could also change the file location by right-clicking on Source Path under the File Location heading. However, it is considered good practice to keep the default file location in order for the mirrored components to be located with reopening the assembly.

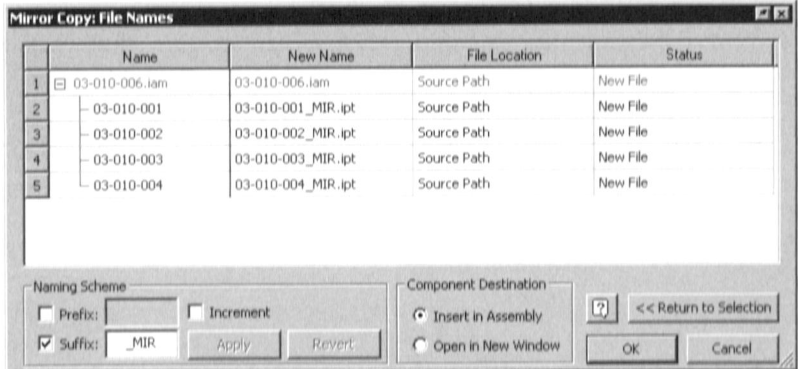


**Figure 10-130**

8 You can also control the listing of a prefix or suffix in the part name in the Naming Scheme area of the Mirror Copy: File Names dialog box, as shown in the following image. Changing the MIR listing to a different name activates the Apply button. Clicking on the Apply button will make the change to all components under the New Name heading. This action will also activate the Revert button if you want to return to the original values. Placing a check in the box next to Increment will increase all file name numbers.

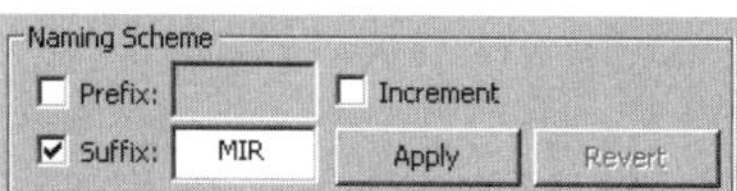


**Figure 10-131**

9 You can also make changes in the Component Destination area as shown in the following image. Clicking the Insert in Assembly button will place the mirrored components in the current assembly. Clicking the Open in New Window button will place the mirrored components in a new assembly file.

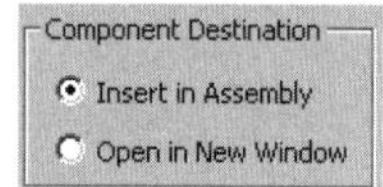


**Figure 10-132**

10 Clicking on the Return to Selection button in the following image will allow you to change the status of the mirrored components or select new components. When you have finished making changes, click the OK button.

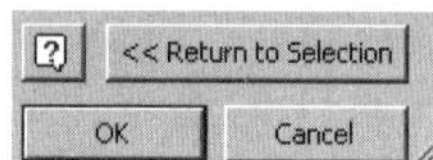


**Figure 10-133**

The results of the mirror operation are illustrated in the following image. All selected components were mirrored about the selected plane. The Browser was also updated to reflect the new part additions. In this image, the default _MIR suffix was added to the end of each part name.

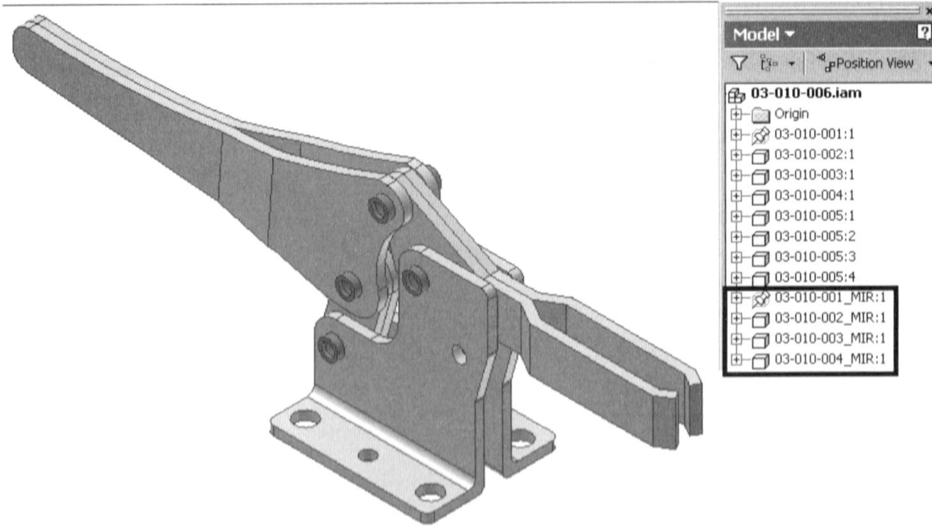


Figure 10-134

In the following image, each component name was changed under the New Name heading. This method ensures a different, autonomous part number for each mirrored component.

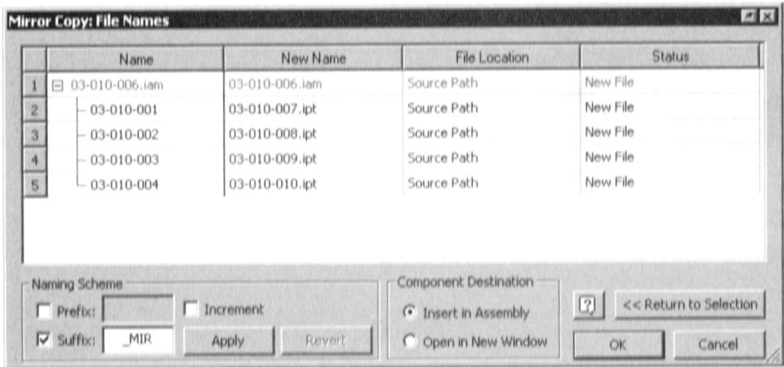


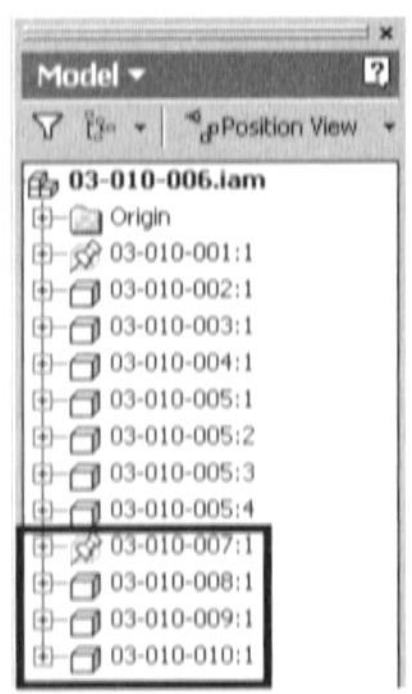


Figure 10-135

EXERCISE

# EXERCISE 10-4 Patterning Components

In this exercise, you create a fastener component pattern to match an existing hole pattern. You then replace the bolt in the pattern. You complete the exercise by demoting components to create a subassembly with the cap plate and patterned fasteners.

This exercise is available in the *Electronic Student Workbook*. To navigate to the exercise in the *Electronic Student Workbook*, do the following:

1 From the Main TOC page, click Chapter 10.
2 From the TOC for Chapter 10, click Patterning Components.
3 Open *IV8_E10_04.iam*.

The completed exercise is shown in the following image.

**Figure 10-136 Completed exercise**

## ASSEMBLY WORK FEATURES

In the assembly environment, you can create work features to help you construct, position, and assemble components. You can create work planes and axes between parts in an assembly by selecting an edge or point on each part. These work features remain tied to each part and adjust accordingly as the assembly is modified. You can use assembly work features to parametrically position new components, check for clearance in an assembly, and as construction aids. You can also use work planes to help you generate section views of your assemblies.

Autodesk Inventor also provides a means for you to globally turn off the visibility of work features. This is important in the assembly environment, where the display of work features from individual parts can quickly clutter the graphics window. Options for turning off work feature visibility are available by selecting the View > Object Visibility menu, as shown in the following image.

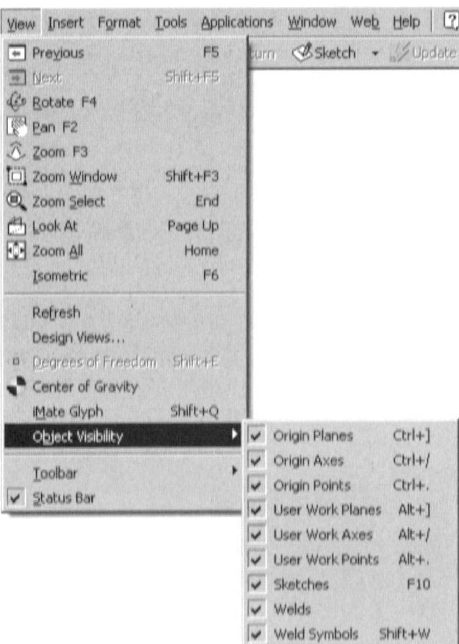


**Figure 10-137**

You can use these controls to turn off the visibility of work features by type. By default, all types of work geometry are initially selected for display. Thus, any work feature with its individual visibility turned on in the Browser is visible in the assembly file.

To globally turn off the visibility of a particular type of work geometry, select the type from the menu, as shown in the following image, and clear the check mark next to it. This overrides the visibility setting for individual work features of that type in the assembly and in each part in the assembly. Although the work features' visibility in the assembly is suppressed, their individual visibility control remains turned on.

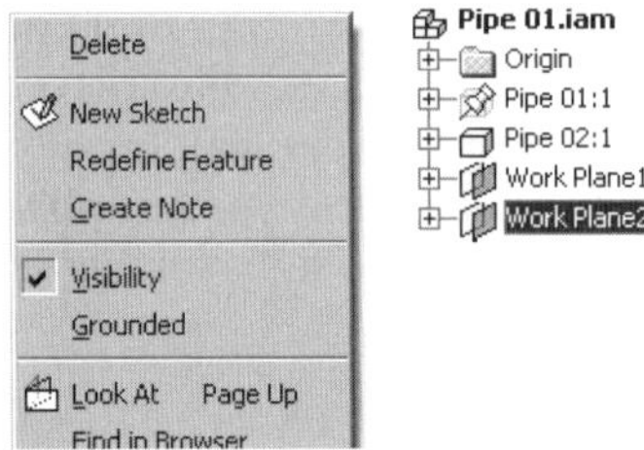


Figure 10-138

## ASSEMBLY FEATURES

Assembly features are features that are defined in an assembly. They only affect a part when the part is viewed in the context of the assembly. Assembly features allow you to remove material from various components after they have been assembled. Examples of material removal processes include match-drilling operations and post-weld machining operations. Typical operations involving assembly features include cutting extrusions, drilling holes, and cutting chamfered edges.

In the following image of a post-machining operation, three blocks have been assembled through the use of mate-mate and mate-flush constraints. With all three items assembled, a feature in the form of two slots needs to be cut through the lower and upper faces of the assembly. A sketch plane is created on the front face of one of the parts, geometry is sketched and dimensioned, and the assembly feature is cut using an extrusion operation. Notice in the following image that, in the completed part, a chamfer was also applied to one end of the assembly using a chamfer operation.

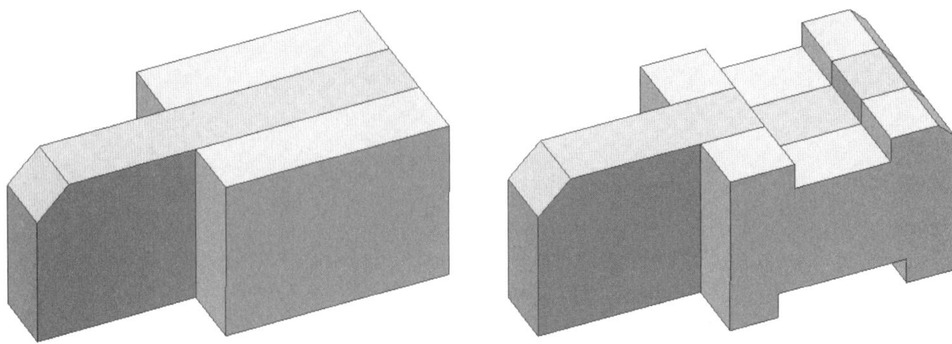

Figure 10-139

A number of assembly practices involve the addition of a set of features (often including parts) following the assembly of the *as designed* piece parts. A common example, as shown in the following image of the timing cam assembly, would be to "drill and pin the following assembly." Parts are placed that need to be precision-located following some assembly adjustments, making the prior location of the holes impractical from a cost efficiency perspective.

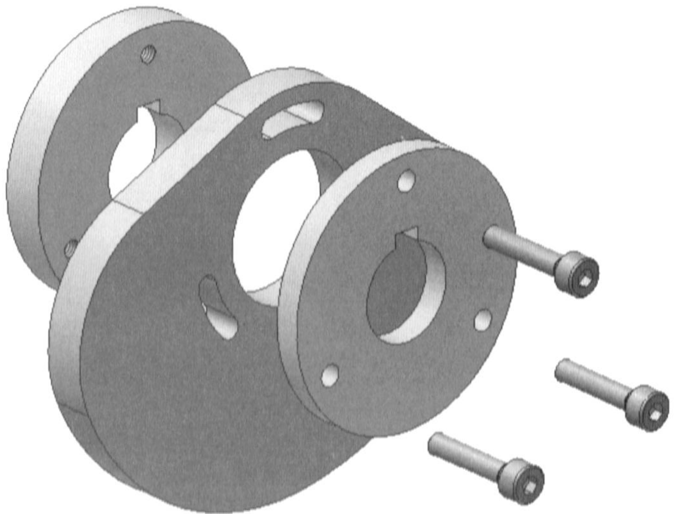

**Figure 10-140**

In the following image, an assembly has been created consisting of the timing cam, front and back hubs, and bolts. The cam has three radial slots, which allow the cam to be advanced and retracted following installation by loosening the bolts and twisting the cam about the main axis.

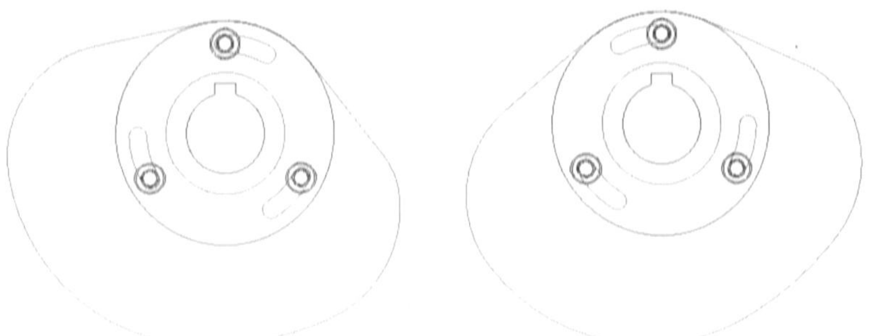

**Figure 10-141**

Once the correct timing position is set relative to the key seat and other parts in the drive train, the cam is pinned in position, as shown in the following image. A hole, which does not exist in the hubs or cam plate, is drilled and the pin is pressed in place.

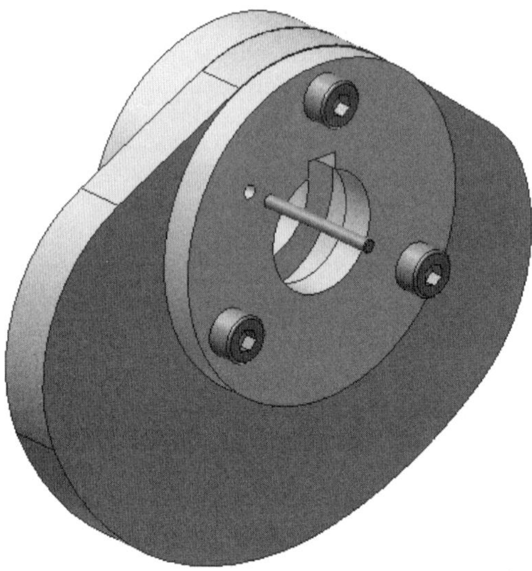

Figure 10-142

In the following image, three parts have all been constrained to form an assembly model. A rectangular hole with filleted corners needs to be cut in the middle of all three component parts. This needs to be performed without the original part files being affected.

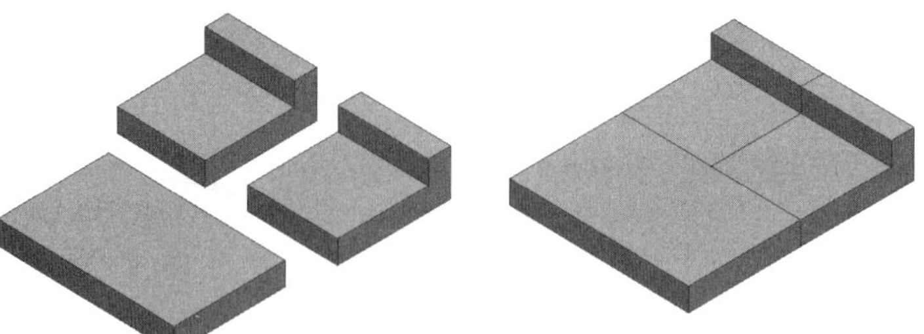

Figure 10-143

## ASSEMBLY SKETCHES

Geometry can be added in an assembly as sketches. You can sketch on a part's face, a part's work plane, or an assembly work plane. Features can then be created from these sketches. As with creating features in part-modeling mode, you create a new sketch plane on the top of the assembly. It does not matter which assembled part is selected for the sketch plane. This will activate the 2D Sketch Panel Bar.

When you create a sketch profile, geometry is projected from various parts to the assembly sketch and constraints and dimensions are added, as shown in the following image.

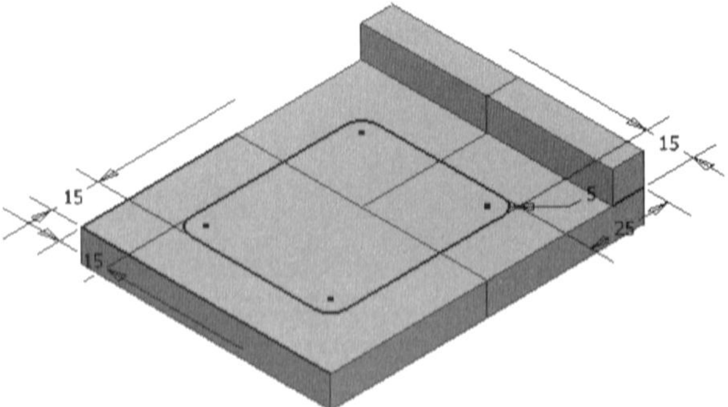


Figure 10-144

**NOTE** When creating assembly sketches, iMates cannot be added and 3D sketches are not supported.

## ASSEMBLY FEATURES

Once you have finished creating the desired sketch, return to the assembly mode. In the following image, three tools are available for creating Assembly Features: Extrude, Hole, and Chamfer. These tools join the existing Work Plane, Work Axis, and Work Point tools that function inside of assembly models.

**NOTE** Assembly features exist only at the assembly level. They do not affect the individual part files.

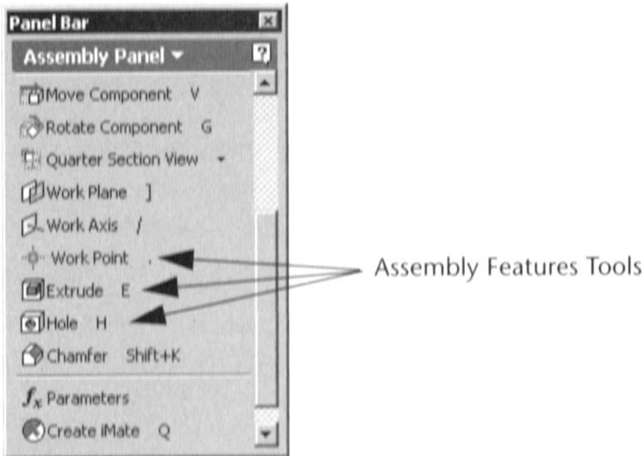


Figure 10-145

Use one of the three feature tools on the sketch geometry to create the assembly feature. In the following image, an extrusion operation is being performed by cutting the rectangular slot through the entire base of the assembly.

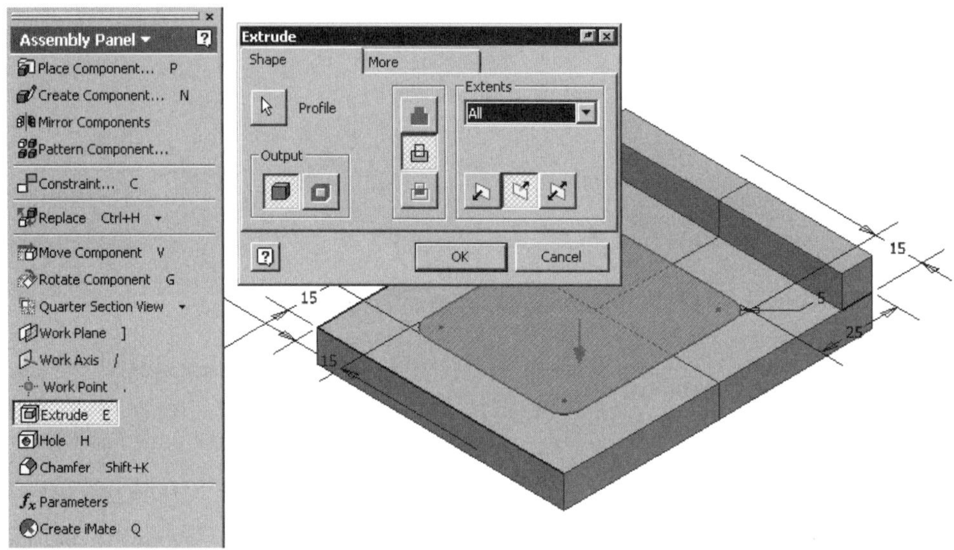


Figure 10-146

The result of the extrude cut operation on the assembly model is illustrated in the following image. Assembly features affect a part or component only when the part is viewed in the context of the assembly. The original part file remains unaffected.

**NOTE** When extruding an assembly feature, only the cut operation is available.

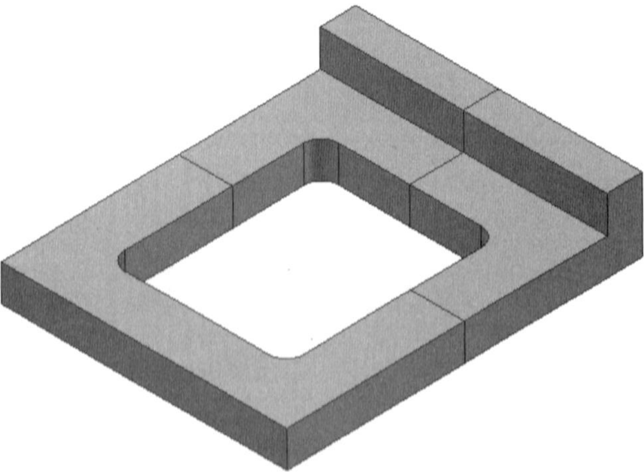

Figure 10-147

A different node is present in the Browser to separate assembly features from assembly components. In the following image, various assembly components are displayed in their usual format. When an assembly feature is created, this feature is added above all assembly components. In the following image, the assembly feature named Extrusion 1 was added to the Browser after the cut was made in the assembly. You can find all parts affected by the created feature under the assembly feature. These affected parts are called *participants*, as they are all members of the group of components being cut by the assembly feature. They are shown as *children* of the feature in the Browser.

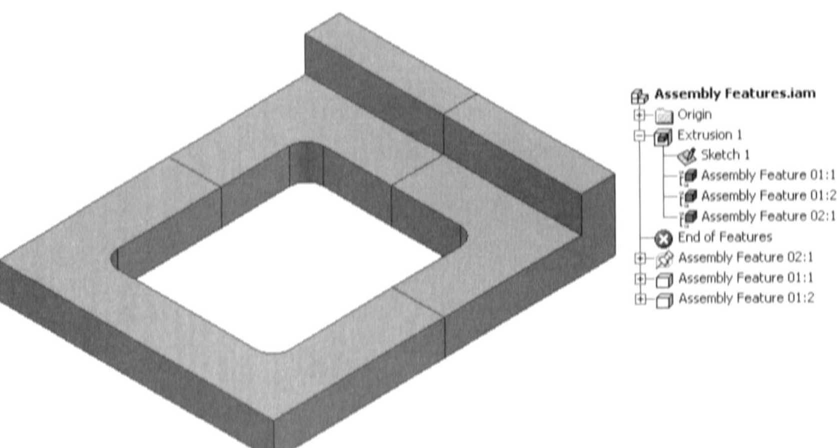


Figure 10-148

## Removing and Adding Participants

When working with assembly features, it is possible to add and remove participants that have been affected by the assembly feature. To remove a participant, right-click on the feature and select Remove Participant from the menu, as shown in the following image.

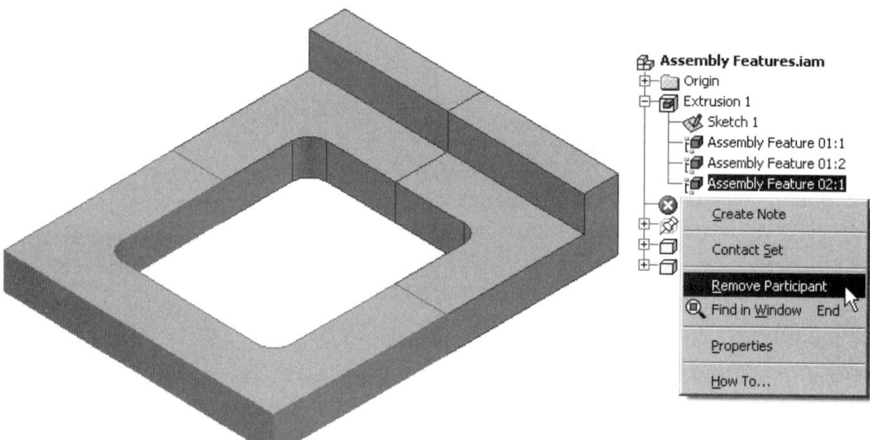


Figure 10-149

The result of removing a participant is shown in the following image. The assembly feature no longer affects the component being removed.

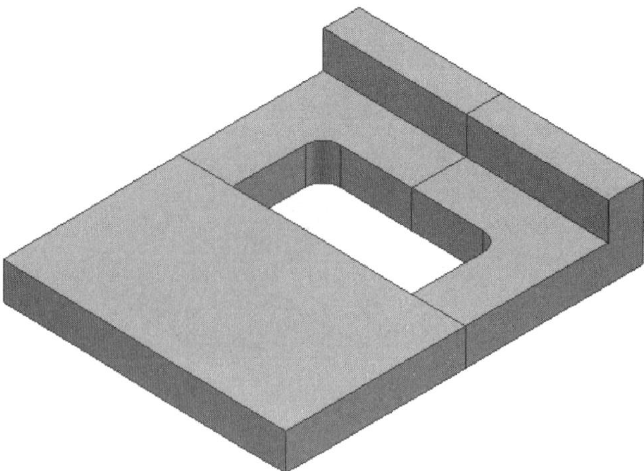

Figure 10-150

To add a participant to an assembly group, right-click on the feature in the Browser and select Add Participant from the menu, as shown in the following image. You can then select the part to add from the graphic screen or the Browser.

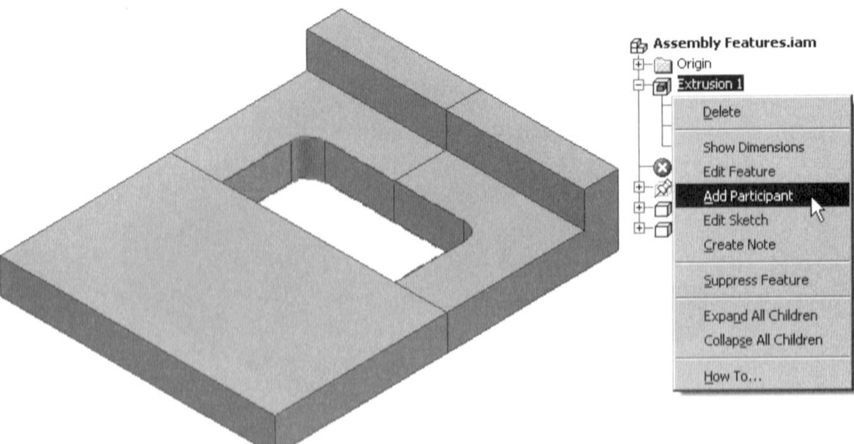


**Figure 10-151**

Assembly features can be suppressed by dragging the End of Features icon to a new position before the actual assembly feature, which is represented by *Extrusion 1* in the following image.

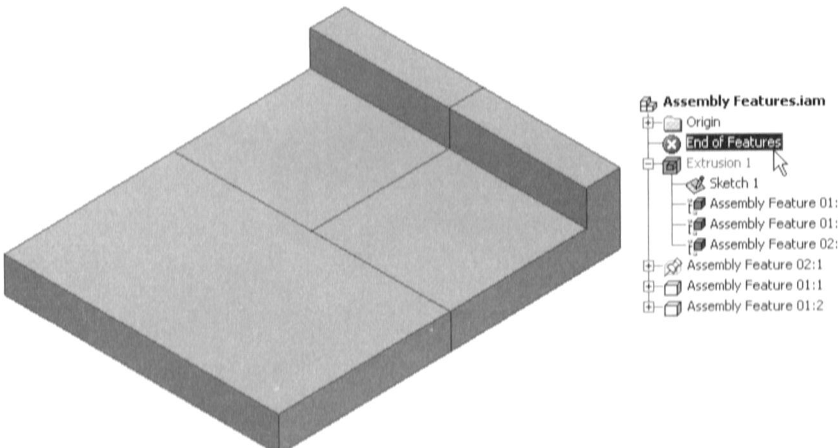


**Figure 10-152**

When editing the constraints of the assembled blocks, an offset value has been applied to the mate constraints of one part, as shown in the following image. The results show the part offset from the main assembly. The assembly feature, however, has not shifted from its original location. The offset part appears different compared to the same part that remains mated to the main assembly.

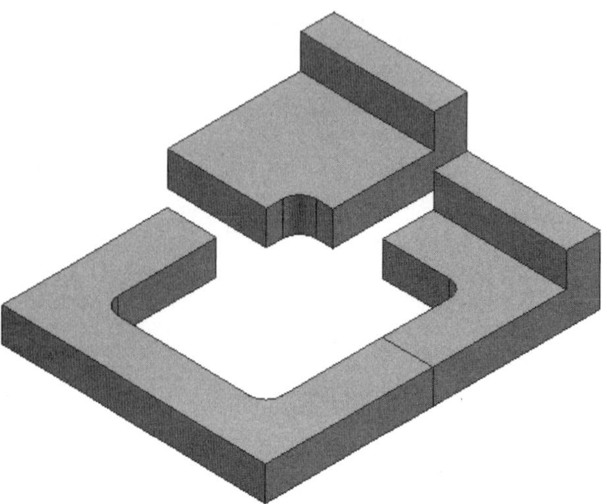

**Figure 10-153**

In the following image, the component is relocated using the Move Component tool. This component will snap back to its original location whenever an assembly update is performed. Editing the component in-place will also snap the component back to its original constrained position.

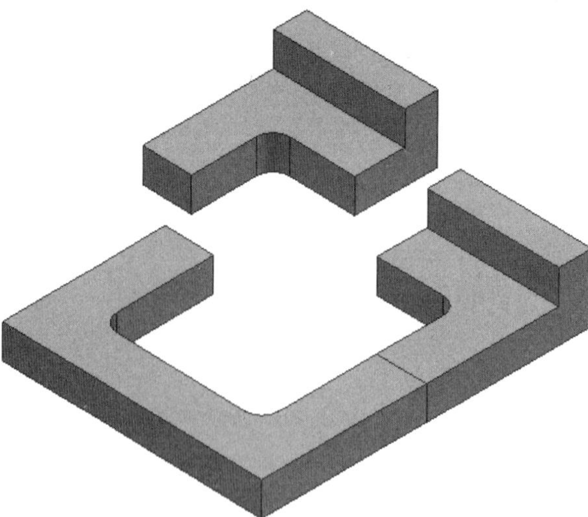

**Figure 10-154**

# EXERCISE 10-5 Assembly Features

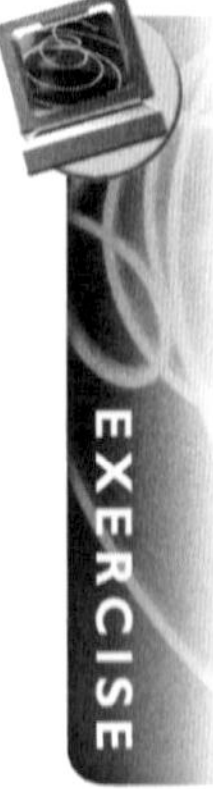


In this exercise, you add a sketched assembly feature to an existing assembly. The assembly feature creates an alignment feature spanning two assembly components. You add an assembly hole feature, and then change the components that participate in the hole feature. To navigate to the exercise in the *Electronic Student Workbook*, do the following:

1 From the Main TOC page, click Chapter 10.

2 From the TOC for Chapter 10, click Assembly Features.

The following image illustrates the opened exercise.

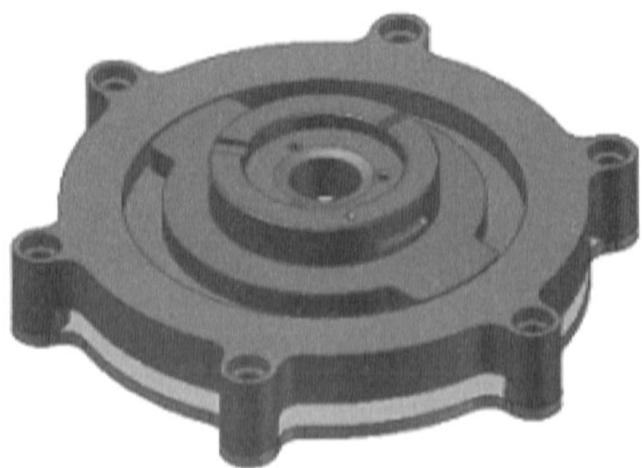

**Figure 10-155 Opened exercise**

The following image illustrates the completed exercise.

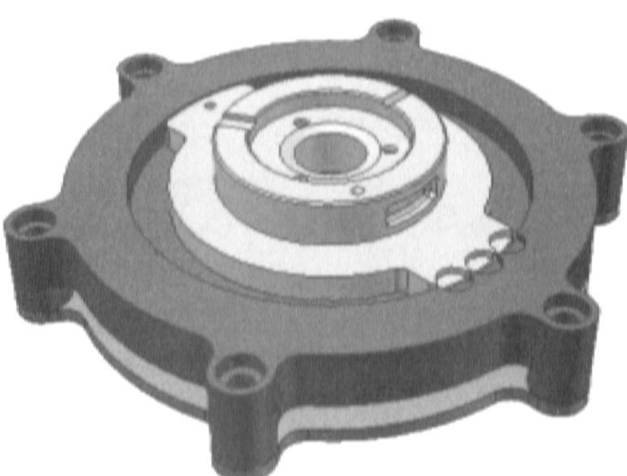

**Figure 10-156 Completed exercise**

## 2D DESIGN LAYOUT

When you create a new design, you often begin with design criteria and create components that meet those criteria. From a list of known parameters, you may create an engineering layout (a 2D design that evolves throughout the design process) and then build components that reference this layout.

The following image illustrates an assembly that consists of an arm designed to move a slide. The link between the arm and the slide has not yet been created. Instead of creating the link as a solid model part, a 2D sketch will be used as a layout to test for motion.

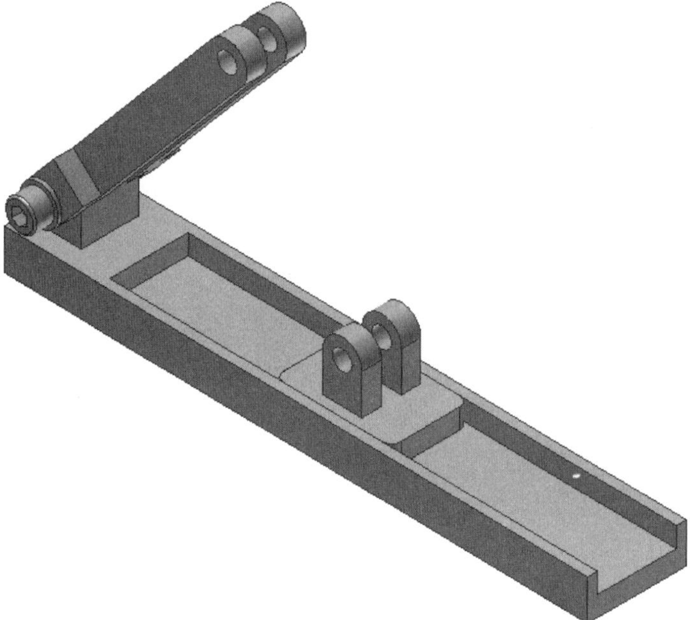

**Figure 10-157**

Before constructing the 2D design layout of the link, a work plane is added to the slide part and a sketch plane is created from this work plane (see the following image).

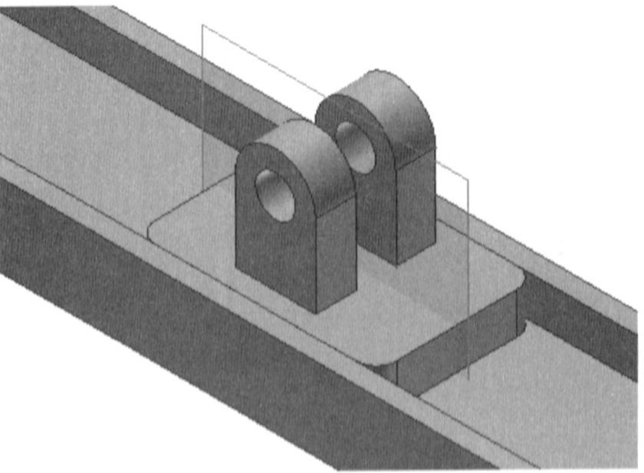

Figure 10-158

With the work plane as the current sketch plane, the link is sketched using a combination of sketch, constraint, and dimension tools, as shown in the following image.

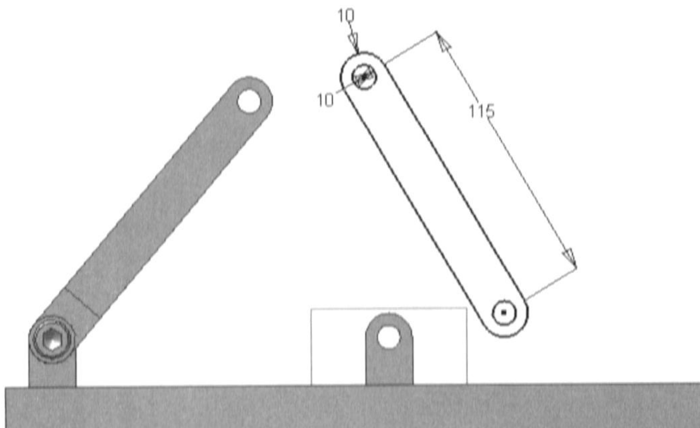


Figure 10-159

After the sketch is finished and you return to assembly mode, assembly constraints are added between the sketch geometry and other parts.

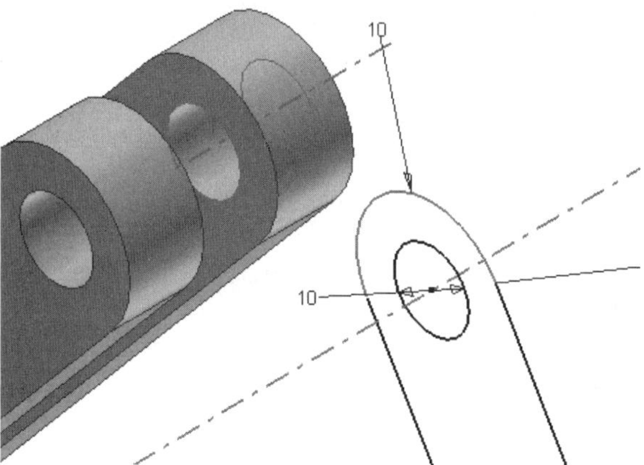


Figure 10-160

The completed sketch and assembly layout is shown in the following image. By using a top-down layout and constraining the components to the assembly, multiple design configurations can be evaluated by making simple changes to the layout parameters. You can quickly examine minimum and maximum values for component parameters, enabling you to determine if the working envelope of the mechanism meets the design criteria.

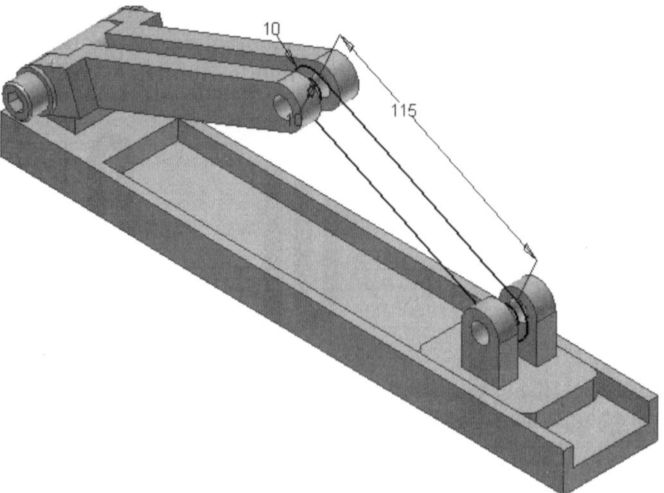


Figure 10-161

# EXERCISE 10-6 2D Design Layout

In this exercise, you create two simple 2D links and apply assembly constraints to verify the links sizes. You then modify one of the 2D links and turn it into a 3D part. To navigate to the exercise in the *Electronic Student Workbook*, do the following:

1 From the Main TOC page, click Chapter 10.

2 From the TOC for Chapter 10, click 2D Design Layout.

The following image illustrates the opened exercise.

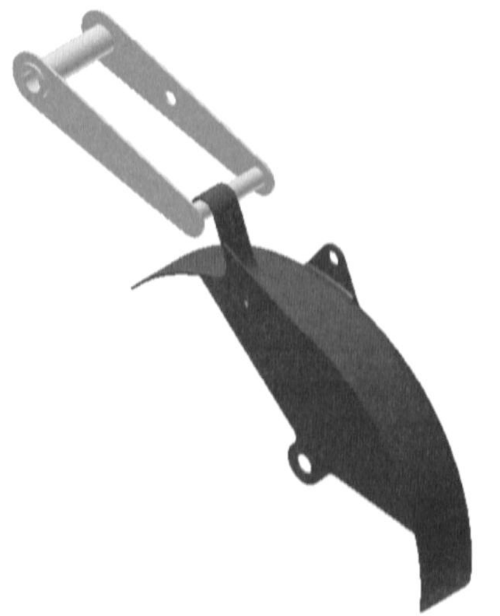

**Figure 10-162 Opened exercise**

The following image illustrates the completed exercise.

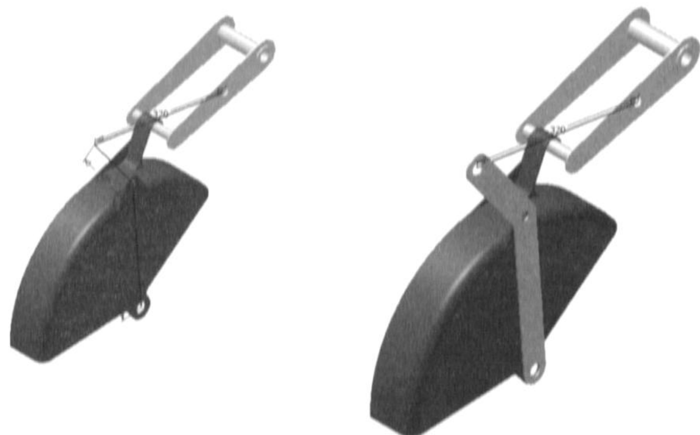

**Figure 10-163 Completed exercise**

## ADAPTIVE DESIGN TECHNIQUES

Adaptivity is the functionality in Autodesk Inventor that allows the size of a part to be determined by setting a relationship between the part and another part in the assembly. Adaptivity allows under-constrained sketches, features that have undefined angles or extents, hole features, and subassemblies that contain parts that have adaptive sketches or features to be adaptive. The adaptivity relationship is acquired by applying assembly constraints between an adaptive sketch or feature and another part. If a sketch is fully constrained, it cannot be made adaptive. The extruded length or revolved angle, however, can be. A part can only be adaptive in one assembly at a time. In an assembly that has multiple placements of the same part, only one occurrence can be adaptive. The other occurrences will reflect the size of the adaptive part. An example of adaptivity would be to have the diameter of a pin get its size from a hole. In the same example, the hole could get its diameter from the pin. Adaptivity can be turned on and off as needed. Once a part's size is determined through adaptivity, you may want to turn its adaptivity off. If you want to create adaptive features, there are options within Autodesk Inventor that will speed the process of creating them. From the Tools menu, select Application Options and then click on the Assembly tab. There are three areas that relate to adaptivity as shown in the following image. Each option will be described.

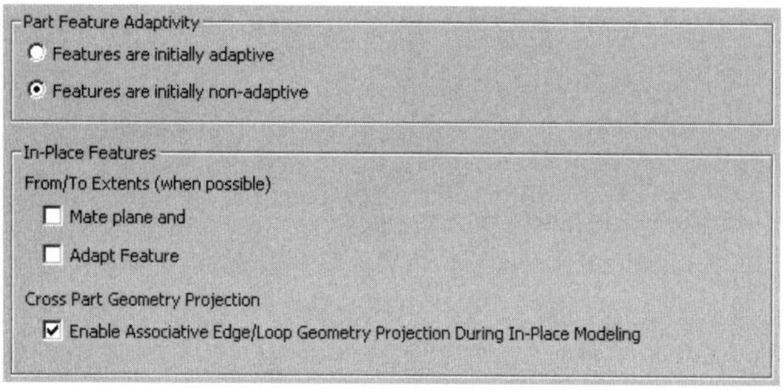

Part Feature Adaptivity
Features are initially adaptive
Features are initially non-adaptive
In-Place Features
From/To Extents (when possible)
Mate plane and
Adapt Feature
Cross Part Geometry Projection
Enable Associative Edge/Loop Geometry Projection During In-Place Modeling

**Figure 10-164**

**Part Feature Adaptivity** This determines the adaptability or non-adaptability of a feature.

**Features are initially adaptive** When checked, features will be adaptive when they are created. Select this option if you know you will be creating many adaptive features.

**Features are initially non-adaptive** When checked, features will not be adaptive when they are created. Select this option if you know you will not be creating many adaptive features.

**In-Place Features From/To Extent (when possible)** This determines whether or not a feature will be adaptive when the To or From/To option is selected for the Extent. If both options are selected, Autodesk Inventor will try to make the feature adaptive—if it cannot, it will terminate at the selected face.

**Mate plane and** When checked, you can create a new component and have a mate constraint applied to the plane on which it was constructed without it being adaptive.

**Adapt Feature** When checked, you will create a new component that adapts to the plane on which it was constructed.

**Cross Part Geometry Projection: Enable Associative Edge/Loop Geometry Projection During In-Place Modeling** When checked and geometry is projected from another part onto the active sketch, the projected geometry is associative (sketch associativity) and will update when changes are made to the parent part. Projected geometry can be used to create a sketched feature.

## SKETCHES

In Autodesk Inventor, you can use reference sketches to create new parts that automatically update to match changes in existing parts of the assembly.

### Using Reference Sketches

To create a new part in an assembly using a reference sketch, use the following steps:

1 Create or place the part that the new adaptive part is intended to match. The container body shown in the following image will be used for this example.

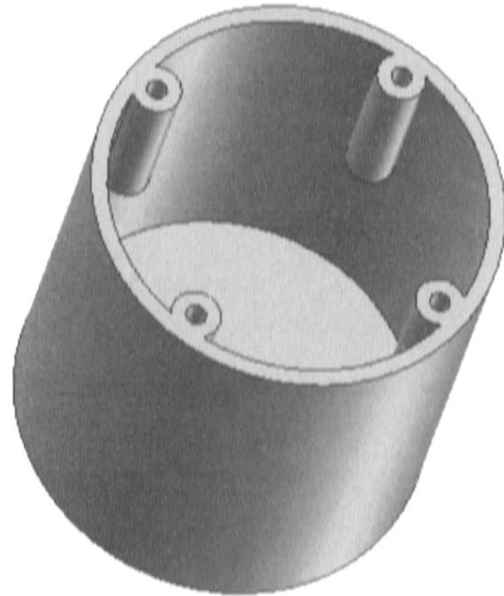

**Figure 10-165**

2 Next, create a new part in-place in the assembly, as shown in the following image. When prompted to place the new sketch, select either the face on the existing part that you want the new part to match or an offset work plane. In the example of the container body, the top face is selected as the new sketch plane.

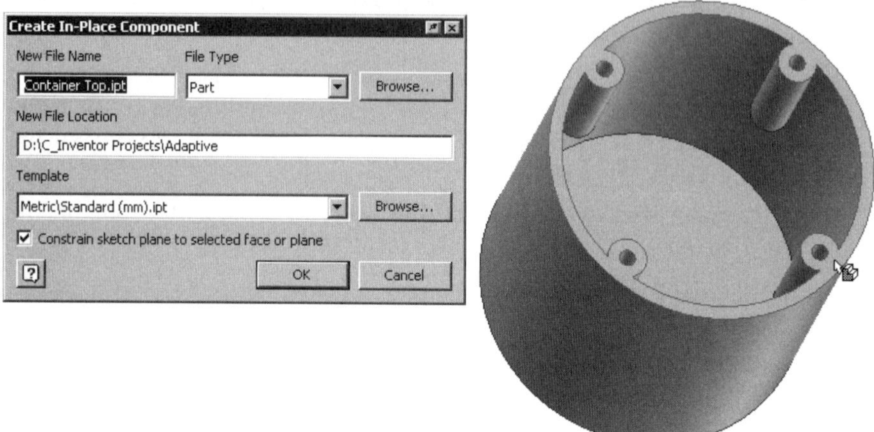


**Figure 10-166**

3 From the 2D Sketch Panel Bar, select Project Geometry. Then select the face or other geometry on the existing part that you want to project. In the following image, the top face of the container body is selected as the face to project. When the edges project, the adaptive property is automatically applied to the sketch and the part located in the Browser, as shown in the following image.

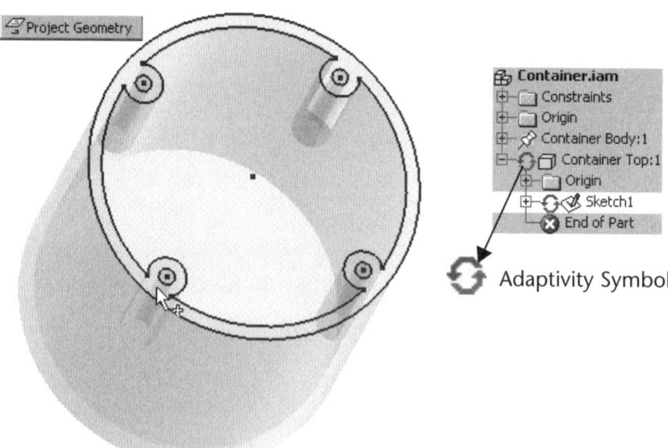


**Figure 10-167**

4 When the geometry has been projected onto the new sketch, exit the sketch environment and create the base feature of the new part. In the following image, the highlighted face of the container body will be extruded a distance of 10 mm.

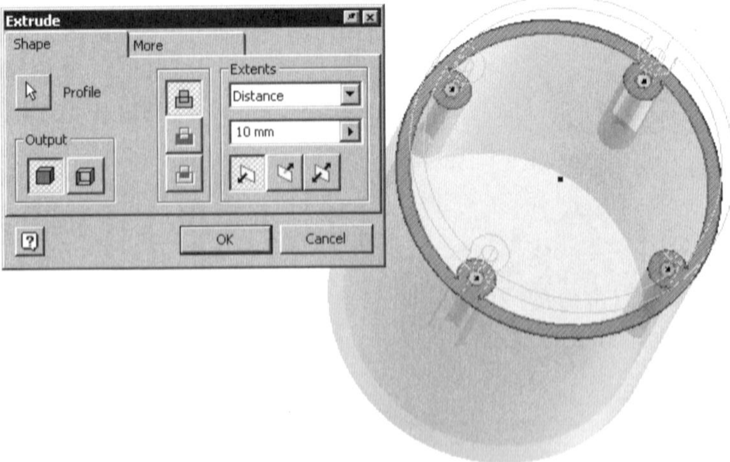


Figure 10-168

5 In the following image, the adaptive icon is displayed in the Browser next to the sketch, the feature created from it, and the new part. The associative geometry projected onto the sketch is displayed as the reference sketch below the Reference1 icon. Notice that the reference sketch has its own icon. When you modify the original part, the associated part created with the reference sketch automatically adapts to reflect the changes.

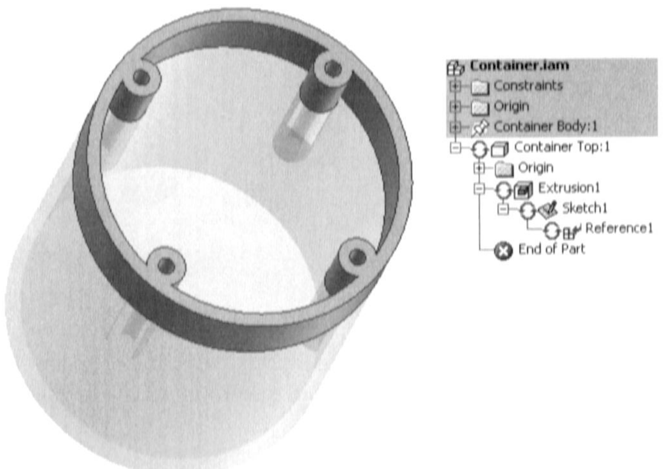


Figure 10-169

6 Complete the cover by extruding a 2 mm top and then return to the assembly modeling environment. The new part (container top) will be assembled to the container bottom. In the following image, the container top was rotated to view its underside details.

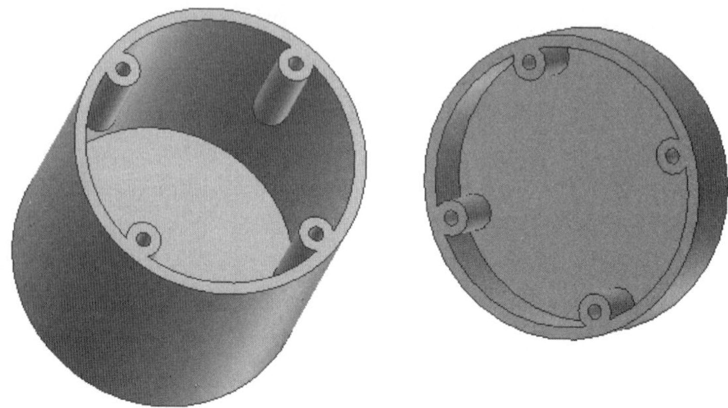

Figure 10-170

In the following image, the diameter of the container body was changed from 50 mm to 90 mm, and the container top was updated to reflect these new diameter values.

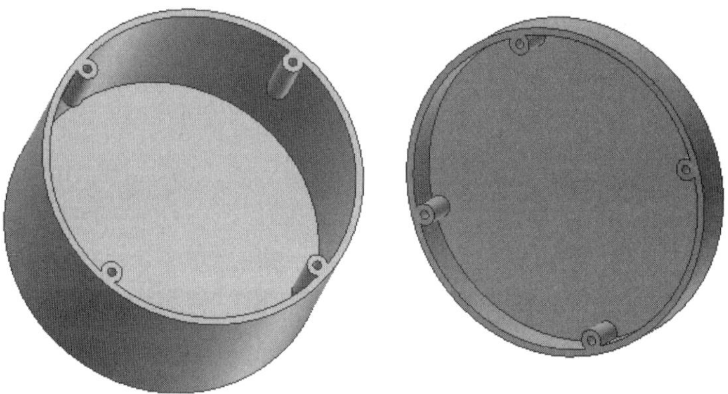

Figure 10-171

**NOTE** If you place additional assembly constraints between the original part and the adaptive part created from it, the Browser may display some constraints as redundant.

## Controlling Reference Sketch Adaptivity

When you use a reference sketch to create a part, the reference sketch, the sketch, the feature created from the sketch, and the part are all automatically made adaptive. You can turn off the adaptivity of any of these by right-clicking them in the Browser and using one of the following two options:

- To turn off adaptivity for every sketch and reference sketch under a feature in the part, right-click the feature and clear the check mark displayed next to the Adaptive option, as shown in the following image. The adaptivity of the sketch and reference sketch is turned off, and the adaptive part remains in its last adapted state.

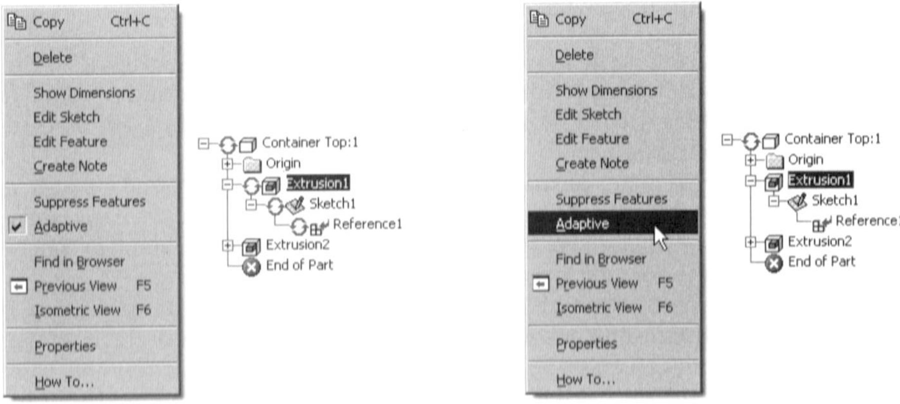


**Figure 10-172**

- To turn off adaptivity for a particular part, right-click the part and clear the check mark next to the Adaptive option.

**NOTE** To make a sketch or a feature adaptive, it must exist in an assembly that will be adaptive.

## Tips and Considerations for Using Adaptivity

Follow these guidelines to make sure adaptive parts created from reference sketches update predictably:

- You can project as many reference sketches (from different source parts) as necessary to obtain the desired geometry on a single sketch. If you do this, make sure the assembly constraint scheme maintains proper positional relationships between parts.
- You can use projected reference sketch geometry as construction geometry (without changing its style).
- Although it is helpful in most cases when you create adaptive base features, the Constrain Sketch Plane to Selected Face or Plane option in the Create In-Place Component dialog box is not required to create reference sketches.
- Create reference sketches for mating faces with patterned features when the design is stable. Otherwise, you may want to control patterned features separately (not by using reference sketches).

- If you project geometry from part faces in another assembly level or from 2D sketches at any assembly level, the projected geometry will not be adaptive reference sketches. You must project 3D features from the same assembly level in order to create adaptive reference sketch geometry.

Follow these guidelines to make sure under-constrained adaptive features and parts update predictably:

- A zero offset mate constraint between two planes or two lines is the most reliable assembly constraint for resizing adaptive features. Avoid non-zero offsets when applying constraints between two points, two lines, or a point and a line.
- Avoid mate constraints between two points, a point and a plane, a point and a line, and a line and a plane. Also avoid tangency constraints between a sphere and a plane, a sphere and a cone, and two spheres. Use only one tangency constraint per revolved feature.
- Place the assembly constraint on the adaptive parent feature. The parent feature won't adapt if you try to constrain a child feature. If you place a work axis on a hole, for example, you can not use an assembly constraint on the work axis to reposition the hole. Most errors for updating adaptive parts are a result of trying to constrain child features.
- You may need to use a different modeling approach when you create adaptive features. As an example, you may include a circle in a sketch instead of making a subsequent hole feature. You can then adapt the rest of the sketch by the location of the circle.
- Whenever possible, you should drag unconstrained sketch geometry to preview how the adaptive feature will react to assembly constraints.
- You can also turn on visibility of a feature's sketch and apply assembly constraints to the sketch geometry. This is particularly useful when you want to use a 2D layout to control an assembly.

Follow these guidelines to make sure adaptive features and parts inside subassemblies update predictably:

- Carefully choose the grounded component in a subassembly to prevent undesirable movement. Whenever possible, you should instance adaptive parts last in the assembly. For cases where this is not possible, make sure you apply assembly constraints to parts that will affect the adaptive features, instead of applying constraints on parts that are dependent on adaptive features.
- If your subassembly does not update correctly, you may need to place additional temporary or permanent constraints on other members of the subassembly before you place constraints on the parts that affect adaptive features.
- For best results in your own designs, you should remove adaptivity (and suppress constraints as necessary) for moving parts after you stabilize their design. This prevents inadvertent changes to individual parts. In this

manner, you use adaptive parts and subassemblies as a temporary tool to design moving mechanisms.

- If your parts and subassemblies are not used in a moving mechanism, you can leave them adaptive to change their size, shape, or position.

## ADAPTIVE FEATURE PROPERTIES

A part can resize to meet assembly constraints if one or more features of the part are defined as adaptive. Extruded or revolved features, hole features, and work planes can all be defined as adaptive.

You can make a feature adaptive in one of two ways. Right-click the feature in the Browser and choose one of the following methods:

- Select Adaptive from the menu to make all available parameters of the feature adaptive.
- Select Properties from the menu and select one or more parameters to be adaptive in the Feature Properties dialog box.

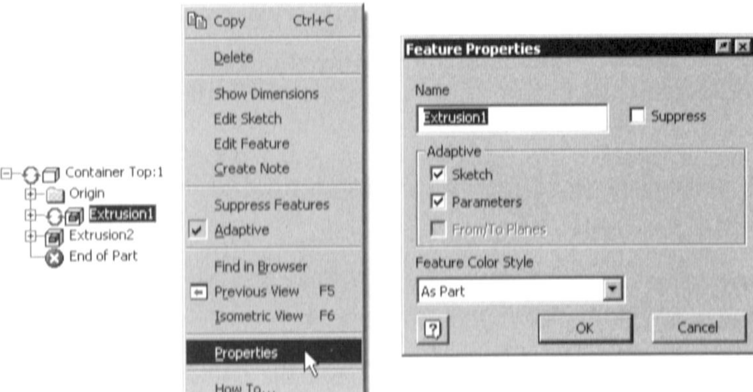


**Figure 10-173**

**NOTE** You can set the default status of newly created features from the Assembly tab of the Options dialog box. Select the Features are Initially Adaptive option to give features adaptive status at the time of their creation.

The available adaptive parameters for each feature type are described in the following sections.

### Extruded Features

Selecting an extrusion feature from the Browser activates the Feature Properties dialog box, as shown in the following image.

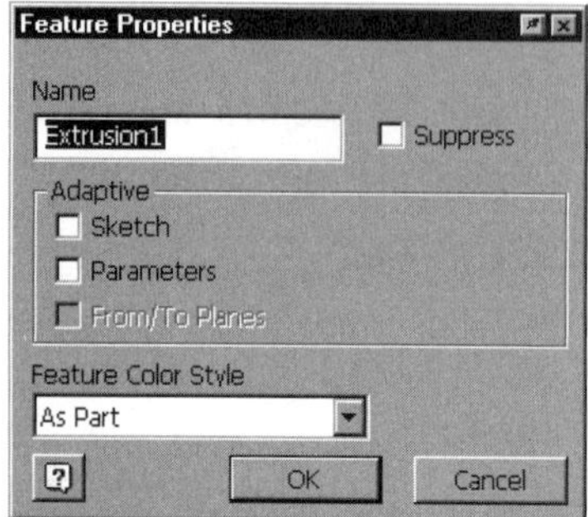


**Figure 10-174**

This dialog box sets the adaptive status of sketched features. Select the appropriate check box to indicate adaptive status of underconstrained geometry. Clear the check box to remove adaptive status. A brief description of each option in this dialog box follows:

**Suppress** Suppresses the feature in the Browser and the graphics window.

**Sketch** You can intentionally leave out specific geometric constraints or dimensions on a sketch and then make it adaptive. An undimensioned line in a sketch, for example, allows the length of the face it defiles to adapt to meet assembly constraints. Removal of parallel or perpendicular constraints may allow an angle between faces to adapt to meet assembly constraints.

**Parameters** The extrusion distance, originally defined as a numeric value, becomes adaptive when selected.

**From/To Planes** You can use an adaptive work plane as the termination of an extrusion. If you place an assembly constraint between the adaptive plane and fixed geometry, then the termination face of the extrusion extends and tilts to satisfy the applied constraint.

Depending on the termination specified for the extrusion, either the Parameters option or the From/To Planes option is available. Options that do not apply to the feature appear as shaded in the dialog box.

## Revolved Features

Selecting a revolved feature from the Browser activates the Feature Properties dialog box, as shown in the following image.

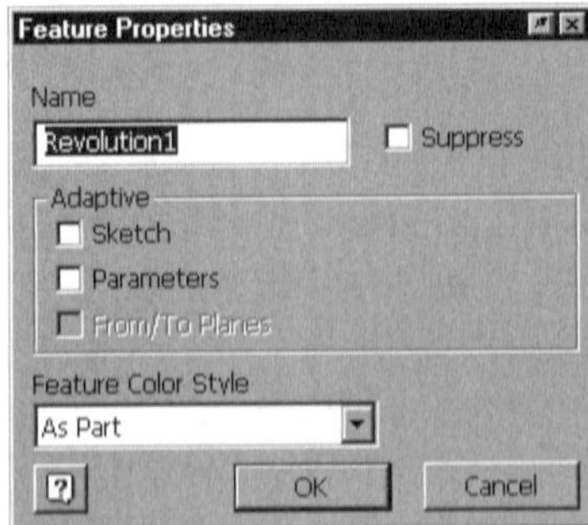


**Figure 10-175**

As with an extruded feature, this dialog box sets adaptive status of sketched features. Select the appropriate check box to indicate adaptive status of underconstrained geometry. Clear the check box to remove adaptive status. A brief description of each option of this dialog box follows.

**Sketch** You can intentionally leave out specific geometric constraints or dimensions on a sketch and then make it adaptive. An undimensioned distance between the revolution centerline and a parallel sketch line, for example, would allow the radius of a feature to adapt, given a suitable assembly constraint.

**Parameters** The angle of revolution, originally defined as a numeric value, becomes adaptive when selected.

## Hole Features

Selecting a hole feature from the Browser activates the Feature Properties dialog box, as shown in the following image.

**Figure 10-176**

Use this dialog box to set adaptive status to parameters of hole features. A brief description of each option of this dialog box follows:

**Suppress** Suppresses feature in the Browser and the graphics window.

**Sketch** The position of the sketch point defining the hole center becomes adaptive when selected. This point must be underconstrained—meaning one or more located dimensions are not specified.

**Hole Depth** Applies to blind termination holes. Specify a Flat Drill Point type in the Holes dialog box so that the bottom of the hole can adapt with a mate or flush constraint.

**Nominal Diameter** The diameter of the hole becomes adaptive when selected.

**Counterbore Diameter** Applies to counterbored holes only. The diameter of the hole counterbore becomes adaptive when selected.

**Counterbore Depth** Applies to counterbored holes only. The counterbore depth becomes adaptive when selected.

### Work Planes

The Properties option is not available for work planes, but the work plane itself can be specified as adaptive or not adaptive. The offset value for a work plane or the angle between a work plane and a planer face or another work plane can be adaptive.

**NOTE** To fix a feature at its current size and shape, right-click the feature in the Browser, and then clear the Adaptive check mark.

## ADAPTIVE SUBASSEMBLIES

In Autodesk Inventor, you can use adaptive subassemblies in your models to control assembly constraints for moving parts inside any subassembly nesting level. When you define a subassembly occurrence to be adaptive, parts inside the subassembly can automatically (and independently) adjust their size or position to fit changing conditions in a higher level of the assembly.

Subassemblies when merged into assembly files are typically defined as rigid bodies. Drag constraints are used to work on underconstrained subassembly components. A typical example of this concept in action is an air cylinder. All parts of the assembly are fully dimensioned; however, the rod can translate along the axis of the cylinder.

In the following image of the industrial shovel, the air cylinders are constrained to the shovel. Unfortunately, the air cylinder motion is restricted due to the rigid nature of the subassemblies.

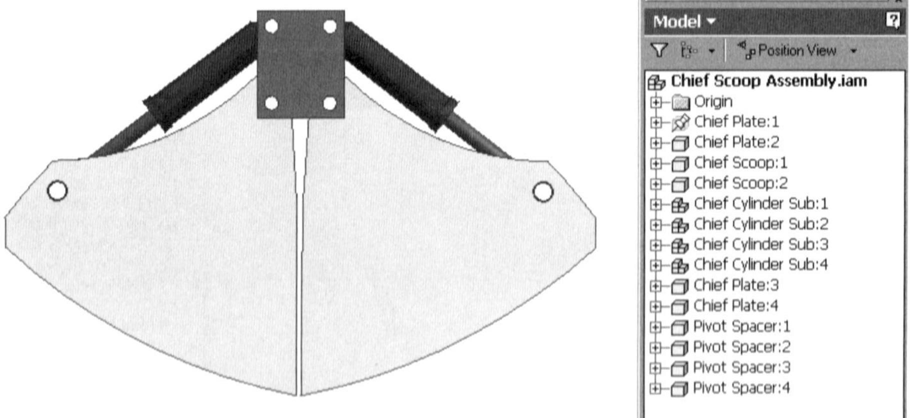


**Figure 10-177**

In the following image, one of the air cylinders is toggled to adaptive. Notice the appearance of the adaptive symbol next to the subassembly in the Browser. Where multiple occurrences of the same subassembly are present in the assembly, it is not necessary to turn adaptivity on for each subassembly.

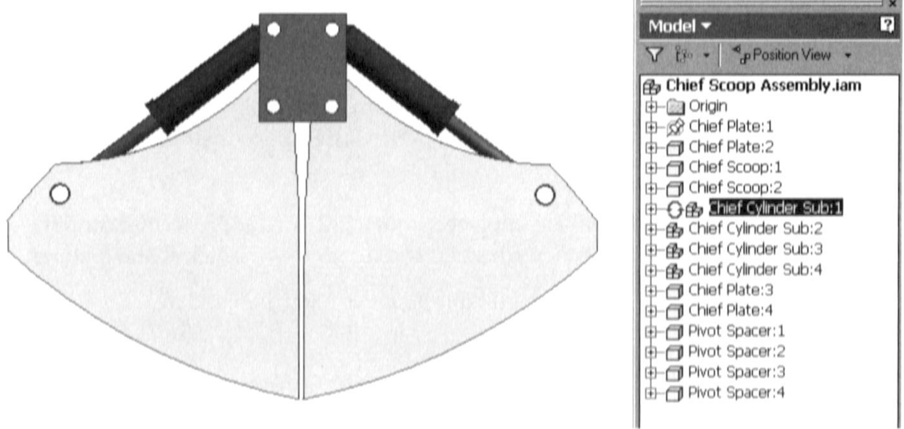


**Figure 10-178**

With the air cylinder toggled to adaptive, the underconstrained rod of the air cylinder can now move along the cylinder axis and affect the other shovel components of the assembly, as shown in the following image.

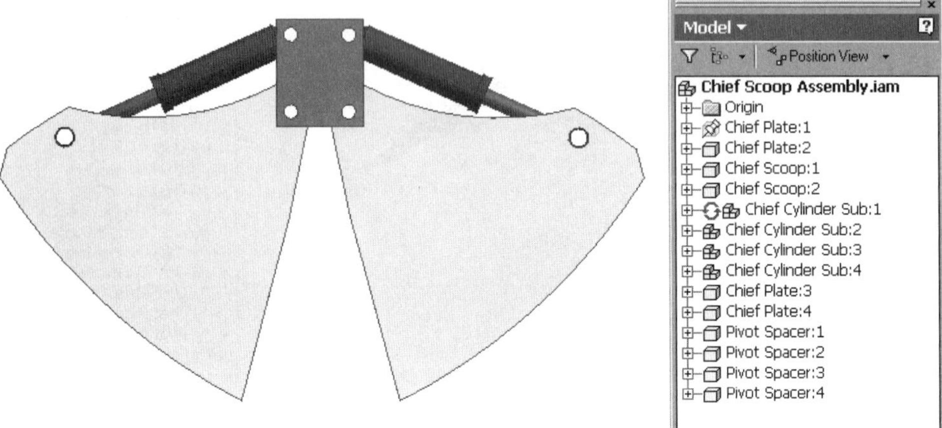


**Figure 10-179**

**NOTE** To turn off the adaptivity for sketches, features, and subassemblies, right-click on the sketch, feature, or subassembly and uncheck Adaptivity from the menu or from the Feature Properties dialog box.

### ADAPTING THE SKETCH OR FEATURE

After making a sketch or feature adaptive, the part itself must be made adaptive. To make a part adaptive, make the top level of the assembly (where the part exists) active. Right-click on the part's name and select Adaptive from the menu. Apply assembly constraints that will define the relationship for the adaptive sketch or feature. As the assembly constraints are being applied, degrees of freedom are being removed.

**NOTE** Parts that are imported from a SAT or STEP file format cannot be made adaptive because they are static and do not have sketches and features.

In assemblies that have multiple adaptive parts, two updates may be required to solve correctly.

For revolved features, use only one tangency constraint.

Avoid offsets when applying constraints between two points, two lines, or a point and a line. Incorrect results may occur.

## Assembly Section Views

Assembly section views can be used to visualize portions of an assembly especially where components are hidden or obscured within chambers. While the assembly is sectioned, part and assembly tools can be used to create or modify parts within the assembly.

To begin, open an assembly file containing one or more components as shown in the following image. Depending on the type of assembly section, the presence of work planes could play a key role by assisting in the creation process.

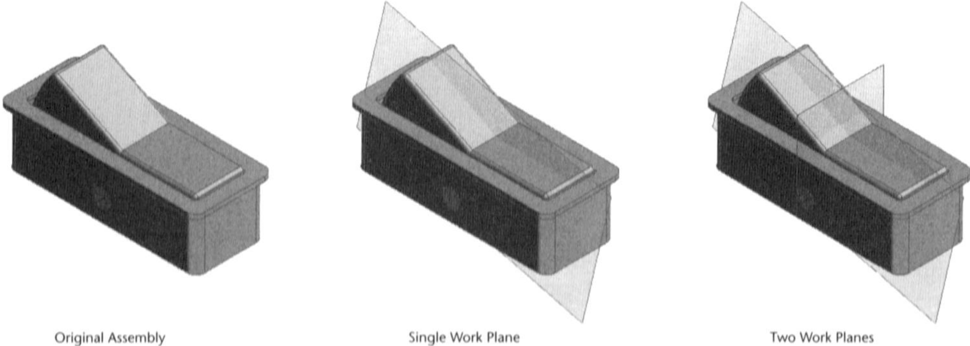


**Figure 10-180**

Four Section View tools are located on the Assembly Panel Bar. To establish the cutting plane, select a planar face or a work plane. All four Section View tools are explained as follows:

- Quarter Section View will display ¼ or 25% of the assembly. In the image below, two work planes were selected to define the Quarter Section. When this view is created, you do have the option to right-click and cycle through all of the other valid quarter sections.

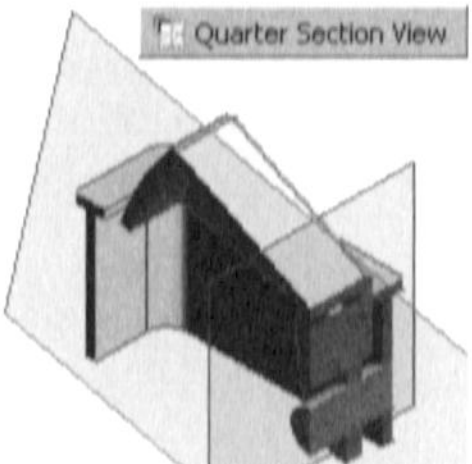


**Figure 10-181**

- Half Section View cuts the assembly based on a single plane or face. In the image below, a single work plane is used to generate the half section view. In order for the assembly to be cut exactly in half, the work plane must be positioned accordingly.

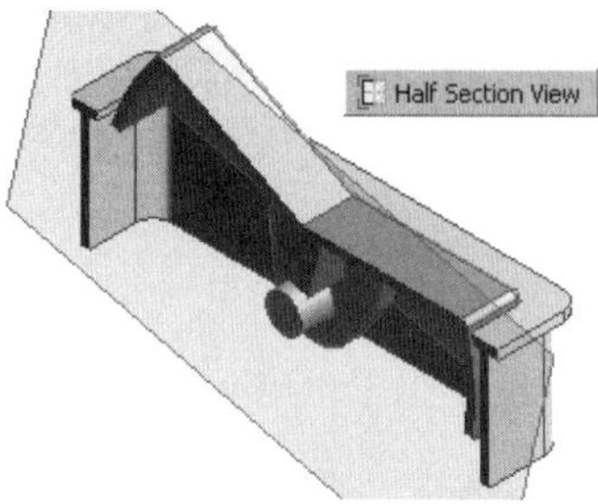


Figure 10-182

- Three-Quarter Section View represents ¾ or 75% of the assembly. As with the Quarter Section View, two work planes were used to define the section as shown in the following image. You can also cycle through other Three-Quarter Sections by right-clicking and selecting Flip from the menu.

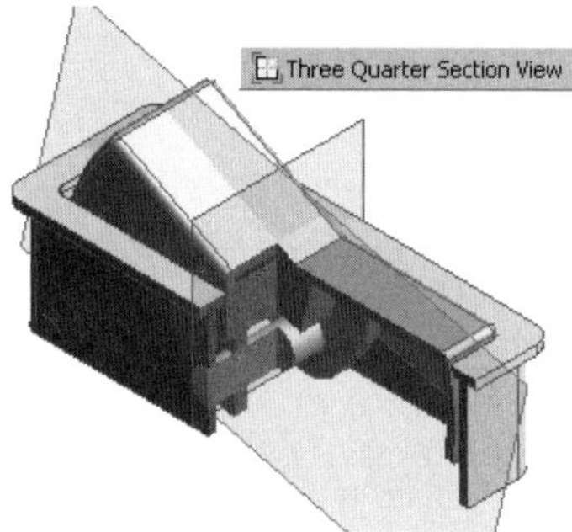


Figure 10-183

- After you have finished sketching or creating a new part in the context of an assembly, it will be necessary to return the sectioned assembly back to its full viewing position. Clicking on the End Section View tool will perform this operation, as shown in the following image.

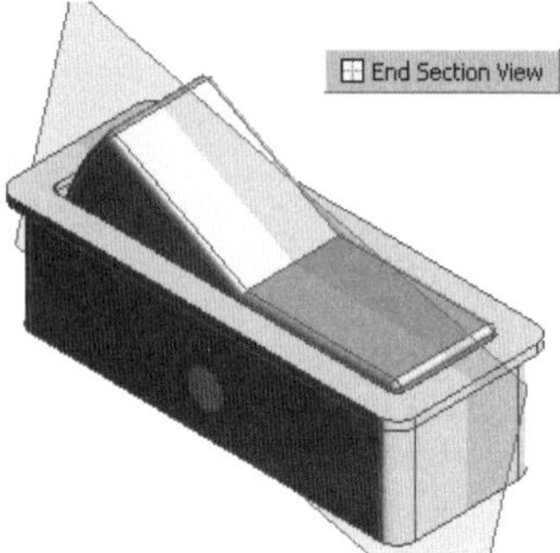


Figure 10-184

As you create an assembly section view, you may accidentally generate the wrong half of the section. If this occurs, right-click and select Flip Section from the menu as shown in the following image. This should display the desired view of the section.

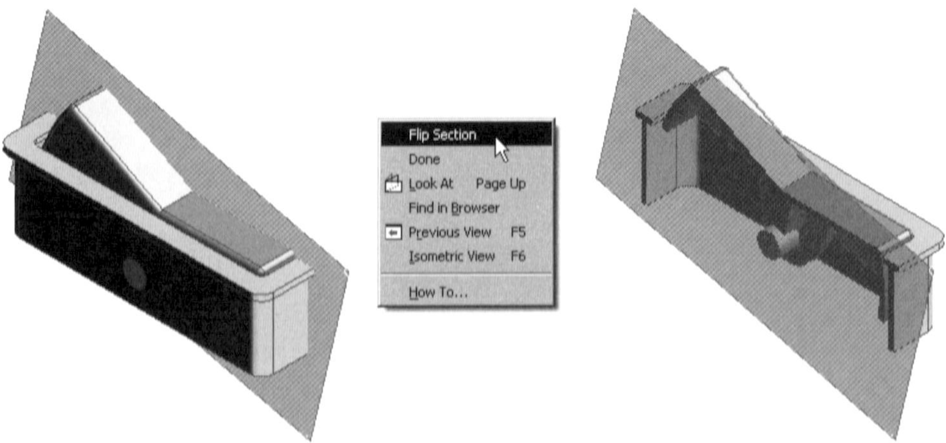


**Figure 10-185**

Activate the part to edit by double-clicking on its name in the Browser. Create a sketch plane (the work plane in the following image), right-click, and then select Slice Graphics to view objects at the sketch plane.

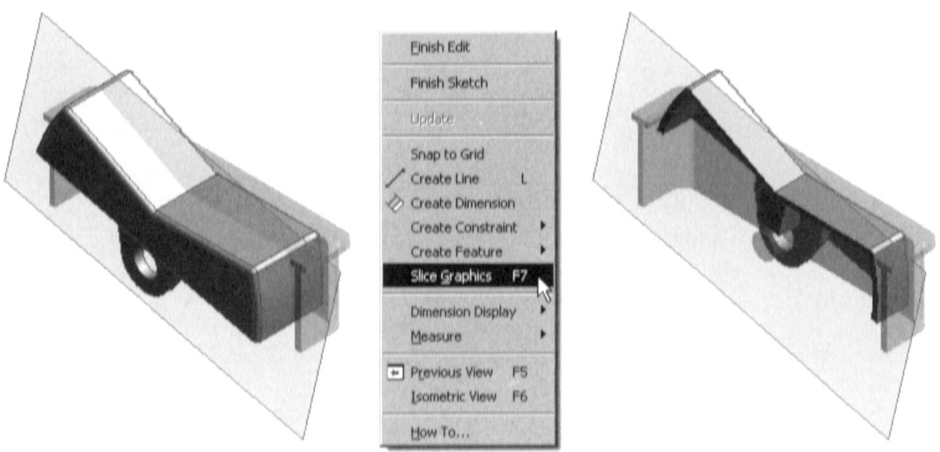


**Figure 10-186**

If desired, use the Project Cut Edges tool on the 2D Sketch Panel Bar to project edges of a part sliced by the cutting plane onto the current sketch plane as shown in the following image.

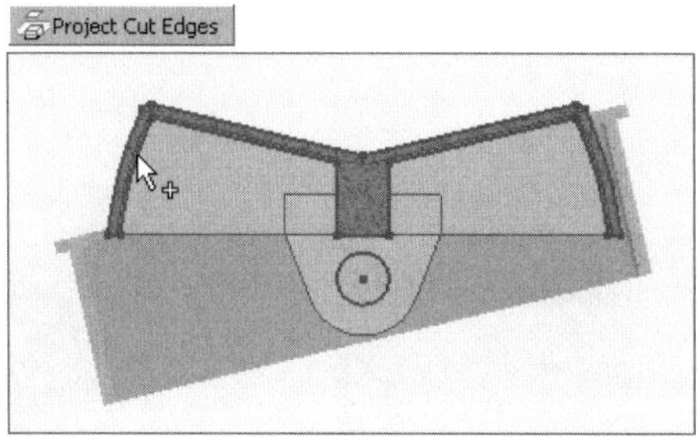


**Figure 10-187**

Use traditional sketch tools to define the new shape and use feature tools such as Extrude to create new geometry as shown in the following image.

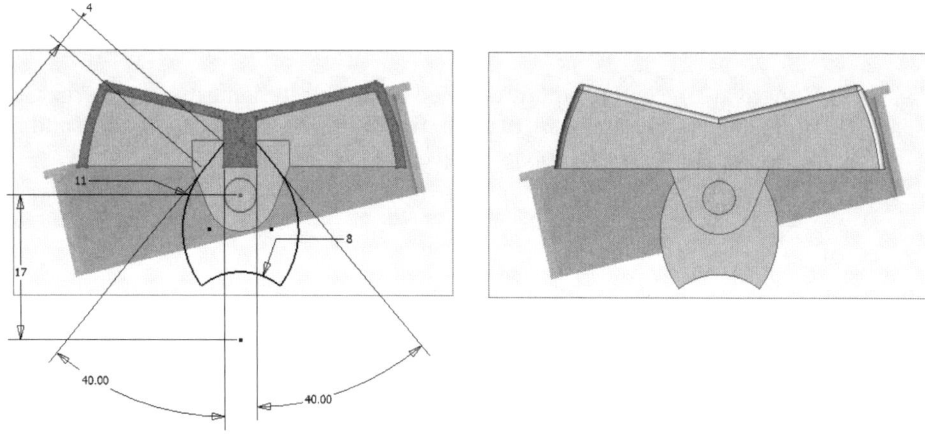


**Figure 10-188**

When you have finished editing the part, right-click and pick Finish Edit from the menu. This will return you to the assembly model still converted to a section view. To turn off the current section mode, use the End Section View tool. Doing so will display the assembly in its entirety as shown in the following image.

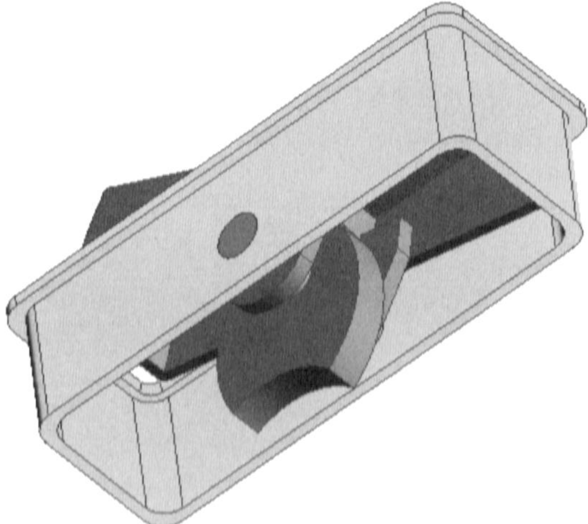

**Figure 10-189**

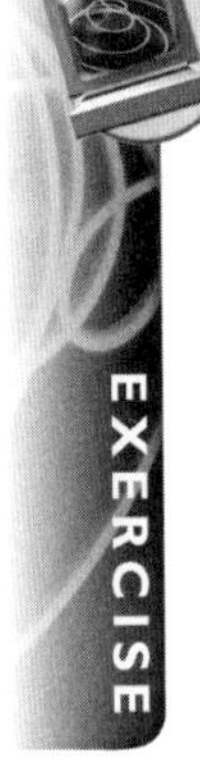


# EXERCISE 10-7
# Adaptive Design Techniques

In this exercise, you create a variety of adaptive features to complete an assembly model of a kettle. To navigate to the exercise in the *Electronic Student Workbook*, do the following:

1 From the Main TOC page, click Chapter 10.
2 From the TOC for Chapter 10, click Adaptive Design Techniques.

The following image illustrates the opened exercise.

**Figure 10-190 Opened exercise**

The following image illustrates the completed exercise.

**Figure 10-191 Completed exercise**

## CHAPTER SUMMARY

| To | Do This | Tool |
|---|---|---|
| Create Design Views | Click the Design Views button located above the Browser. | |
| Create iMates | Click the iMate tool located in the Assembly Panel Bar or from the Tools toolbar. | |
| Drive constraints | Right-click on an assembly constraint and choose Drive Constraint from the menu. | Drive Constraint |
| Replace a component | Click the Replace Component tool located in the Assembly Panel Bar. | |
| Insert components from the standard parts library | Click the arrow next to Model in the Browser and select Library. | Model ▾ Model Library |
| Create component patterns | Click the Pattern Components tool located in the Assembly Panel Bar. | |
| Create mirrored components in an assembly | Click on the Mirror Components tool located in the Assembly Panel Bar. | |
| Apply assembly work features | Click the Work Plane, Work Axis, or Work Point tools located in the Assembly Panel Bar. | |
| Create assembly features | Click the Extrude, Hole, or Chamfer tools located in the Assembly Panel Bar. | |

# APPLYING YOUR SKILLS

## Skills Exercise 10-1

In this exercise, you create a lid that will adapt to fit the hole in the top of the kettle. To navigate to the exercise in the *Electronic Student Workbook*, do the following:

1 From the Main TOC page, click Chapter 10.
2 From the TOC for Chapter 10, click Exercise 1 under Applying Your Skills.

The following image illustrates the completed exercise.

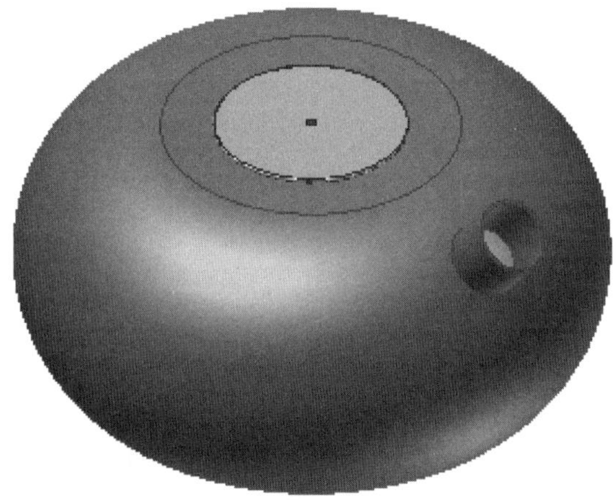

**Figure 10-192 Completed exercise**

## CHECKING YOUR SKILLS

1 **True___ False___** A Design View can control the display style (shaded or wireframe) of an assembly model.

2 **True___ False___** iMates are created while inside of a part file.

3 **True___ False___** The creation of an AVI file is one of the functions that can be accessed when driving constraints.

4 **True___ False___** When replacing components in an assembly, assembly constraints will automatically be applied to the replacement component.

5 **True___ False___** The visibility of an entire set of patterned components can be turned off in a single operation.

6 **True___ False___** Of the total number of standard libraries supplied with Autodesk Inventor, you can narrow the number down to the most commonly used libraries for your applications.

7 **True___ False___** Creating features in the context of an assembly will automatically update the individual parts that make up the assembly.

8 **True___ False___** A 2D design layout consists of a series of 2D sketches that are constrained to 3D parts in an assembly file.

9 **True___ False___** Hole features cannot be made adaptive.

10 **True___ False___** A sketch must be fully constrained to adapt.

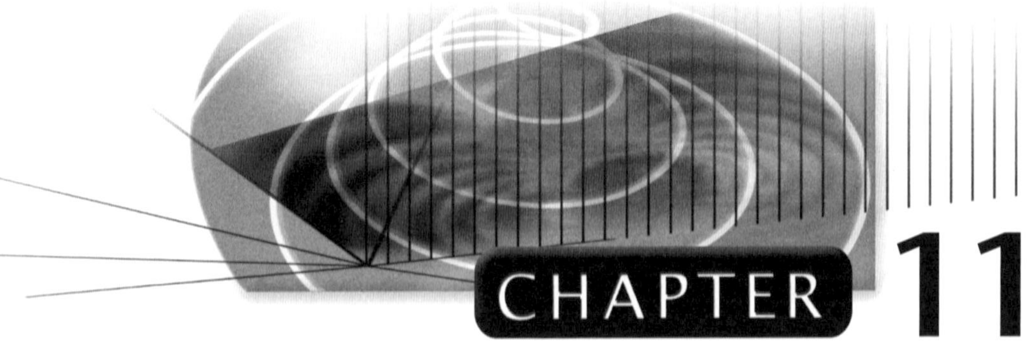


# Sheet Metal Design

In this chapter, you will learn how to create sheet metal parts in Autodesk Inventor. Assemblies often require components that are manufactured by bending flat metal stock to form brackets or enclosures. Cutouts, holes, and notches are cut or punched from the flat sheet, and 3D deformations such as dimples or louvers are often formed into the flat sheet. The punched sheet is then bent at specific locations using a press brake or other forming tools to create a finished part. The sheet metal environment in Autodesk Inventor presents specialized tools for creating models in both the folded and unfolded states.

## CHAPTER OBJECTIVES

**After completing this chapter, you will be able to**

- Start the Autodesk Inventor sheet metal environment
- Modify settings for sheet metal design
- Create sheet metal parts
- Modify sheet metal parts to match design requirements
- Create sheet metal flat patterns
- Create drawing views of a sheet metal part

## INTRODUCTION TO SHEET METAL DESIGN

A metal blank folded into a finished shape is a common component in a mechanical assembly. Examples include enclosures, guards, simple-to-complex bracketry, and structural members. Although the term *sheet metal* is often associated with these components, heavy plates (1"+) can also be formed using similar methods.

Common sheet metal components include electronic and consumer product chassis and enclosures, lighting fixtures, support brackets, frame components, and drive guards.

Sheet metal fabrication can include a number of processes:

- Drawing
- Stamping
- Punching and cutting
- Braking
- Rolling
- Other more complex operations

The tools in the Autodesk Inventor sheet metal environment enable you to create press brake-formed models. You can also create rolled shapes including cones and cylinders. In addition, sheet metal parts can include formed features such as nail holes, lances, and louvers. Using lofts and surfacing tools, you can create parts that are fabricated through deformation processes such as drawing or stamping, but Autodesk Inventor cannot unfold these parts to determine the shape of the flat blank.

### SHEET METAL FABRICATION

Brake-formed parts must be described in a minimum of two states; the flat blank shape prior to bending, and the finished part in its folded form. Manufacturing processes are often performed on the flat sheet before folding the part into a finished shape. Holes and other openings are punched into the flat stock, and deformations such as dimples are stamped into the blank with forming tools and dies. Preparing the flat stock can be done manually, or more commonly with CNC punching machines. High volume sheet metal parts are often created on progressive die lines where a continuous feed of flat stock (from a coil) is passed through a series of forming dies that remove material or form the shape of the component. For lower volume components, manual or CNC press brakes are the primary tools used to bend the finished blank into its bend up form.

In most sheet metal designs, the folded shape of the part is known. You can create the folded model using various techniques, the most basic being a process of adding individual faces, flanges, and other sheet metal features to build the final state of the folded part. A key face is the first feature added to the part and additional "as bent" features are added, and joined to open edges of the part. During feature creation, a bend is automatically created at the intersecting edge of the new feature and the existing part. You can also add bends between disjointed faces, a technique that is very helpful when designing a sheet metal component in the context of an assembly. When the folded design is complete, Autodesk Inventor can create a flattened version of the model, commonly called a flat pattern. The flat pattern locates the position of bend centerlines used to form the part in a press brake or other sheet metal forming tools. The flat pattern also contains any holes, cutouts, and other features placed on faces of the folded part. The size of the flat pattern is determined by the settings in the current sheet metal style.

When a metal blank or sheet is bent in a press brake, the metal in the bend area deforms. Material on the inside of the bend compresses, and material on the outside of the bend is stretched. This deformation must be taken into account when calculating the flattened state of the folded model. The amount of deformation is dependent on the material, the radius and angle of the bend, and the process and equipment used to create the bend. Within the sheet there is a plane where the material neither compresses nor stretches. If the location of this plane is known, the length of the flat sheet can be calculated. For many materials and processes, the location of this neutral plane is known and can be expressed as a percentage of the thickness of the part, as measured from the surface on the inside of a bend. This offset value is referred to as a kFactor. The following image shows the neutral plane of a sheet metal bend. The location of the neutral plane is kFactor * Thickness, where 0 < kFactor < 1. The default sheet metal style uses a kFactor of 0.44. This value is appropriate for many common materials and processes.

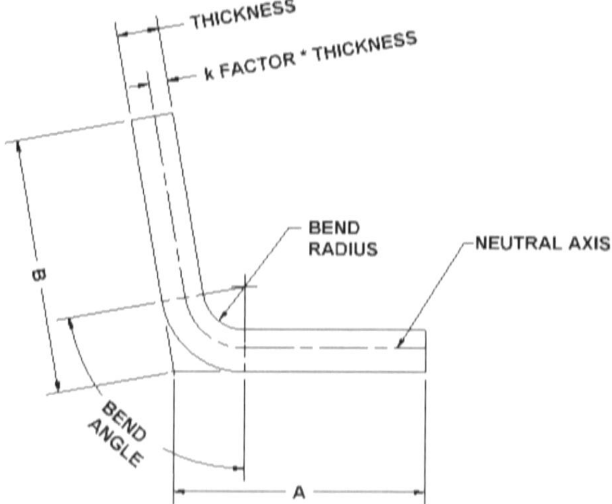


**Figure 11-1**

You can edit the default sheet metal style, or add additional styles. You store sheet metal styles in a template, and apply the appropriate style for the design. The current sheet metal style applies to all sheet metal features in the part. Although you cannot apply separate styles to different sheet metal features in a part, you can override default values during feature creation or edit. For example, some materials have different bending properties dependent on the orientation of the bend relative to the grain of the material. As the sheet is manufactured, the material microstructure aligns with the rolling direction. This results in a sheet with bend properties related to its orientation in the press brake. Stainless steel sheets often display this anisotropic behavior.

The use of a constant kFactor is not always appropriate. For some materials and processes, when formed parts require precise tolerances, a bend table can be used in place of the kFactor. A bend table uses empirically derived information to apply length adjustments to bends. With overrides, you can apply a separate kFactor or bend table to each bend on a sheet metal part.

## Bend Tables

In the calculation of the unfolded length of a bend, a bend table replaces the kFactor with a set of known adjustments to the unfolded length. A bend table can provide greater precision of unfolded dimension since the values are derived from your measurements of bend length of a particular material on a specific machine. When a bend table is used to calculate an unfolded length, the following formula applies:

$$L = A + B - x$$

Referring to the previous equation, the variables in the formula are:

L Unfolded length
A Length of folded face 1
B Length of folded face 2
x Adjustment from bend table

The value of x is dependant on the sheet thickness, the angle of the bend, and the inner radius of the bend.

Note that the measurements of A and B are to the intersection of the extended outer faces on either side of the bend. This intersection is used when the angle of the bend is less than or equal to 90 degrees. The following image is used when the bend angle is greater than 90 degrees. The measurements are parallel to the face and tangent to the outer surface of the bend. The same formula is used to determine the unfolded length of the part.

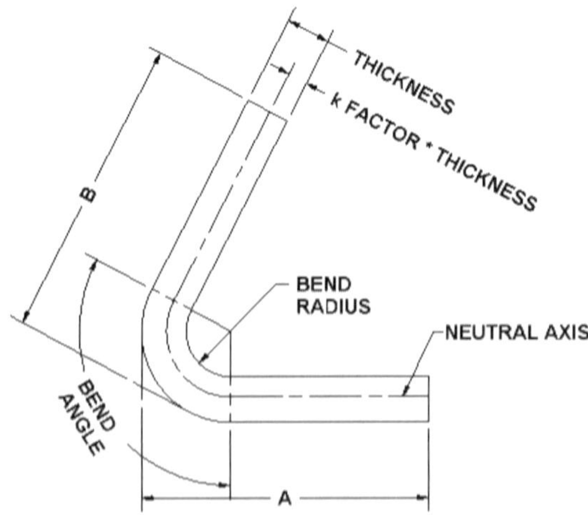


**Figure 11-2**

Autodesk Inventor can use a bend table that is a plain text file (*.txt* extension), but you can create or edit a spreadsheet version of the bend table using Microsoft Excel and then save the table in text format.

Autodesk Inventor includes both metric (mm) and English unit (in) bend tables that you can modify. Each bend table is supplied in both *.txt* and *.xls* formats. The following image illustrates the angle information required in a bend table. A flat sheet bent at an angle of 110 degrees will have an opening angle of 70 degrees. The opening angles are entered in the bend table.

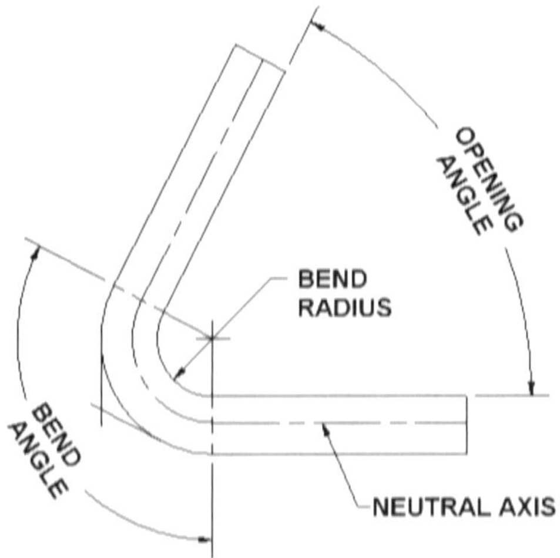


**Figure 11-3**

## SHEET METAL PARTS

Sheet metal parts can be designed separately or in the context of an assembly. Since sheet metal parts are often used as supports for, or to enclose, other components, designing in the assembly environment can be advantageous.

### SHEET METAL DESIGN METHODS

Sheet metal parts are most often created in the folded state. The model is then unfolded into a flat sheet using the Flat Pattern tool. To create the folded model, the first sketch is extruded the thickness of the sheet to create a face. Additional faces or flanges are added to open edges of the part and bends are automatically added between the features. Cuts and other special sheet metal features are added to the part as required.

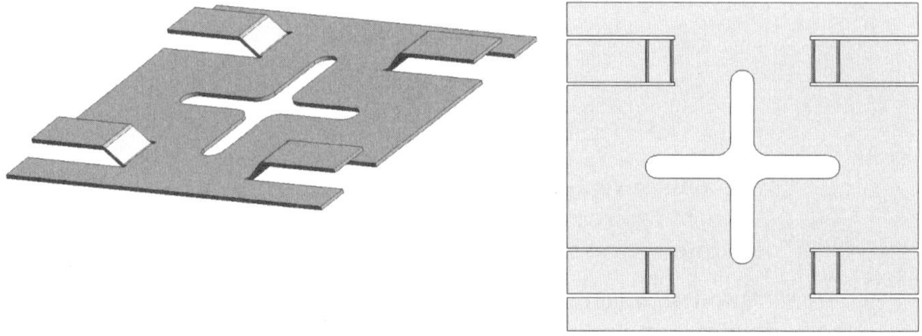

**Figure 11-4**

Autodesk Inventor's ability to create models with disjointed solids (unconnected features) is very helpful when building sheet metal parts in an assembly. Separated faces of a sheet metal part can be built quickly by referencing faces on other parts in the assembly. Additional sheet metal features can then join the distinct faces.

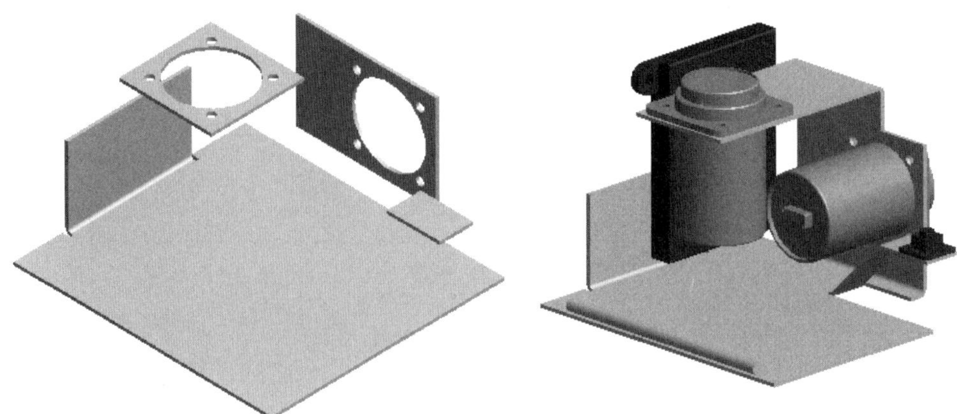

**Figure 11-5**

Sheet metal parts can also be created from standard parts that have been shelled. All walls of the shelled part must be the same thickness, and match the thickness of the flat sheet. The solid corners of the shelled part are ripped open to enable the model to unfold, and appropriate bends are added along the edges between faces.

## CREATING A SHEET METAL PART

The first step in creating a sheet metal part is either to select the New icon from the What To Do section or to select New from the File menu on the Standard toolbar. You can then select the *Sheet Metal.ipt* icon from the Default tab, as shown in the following image.

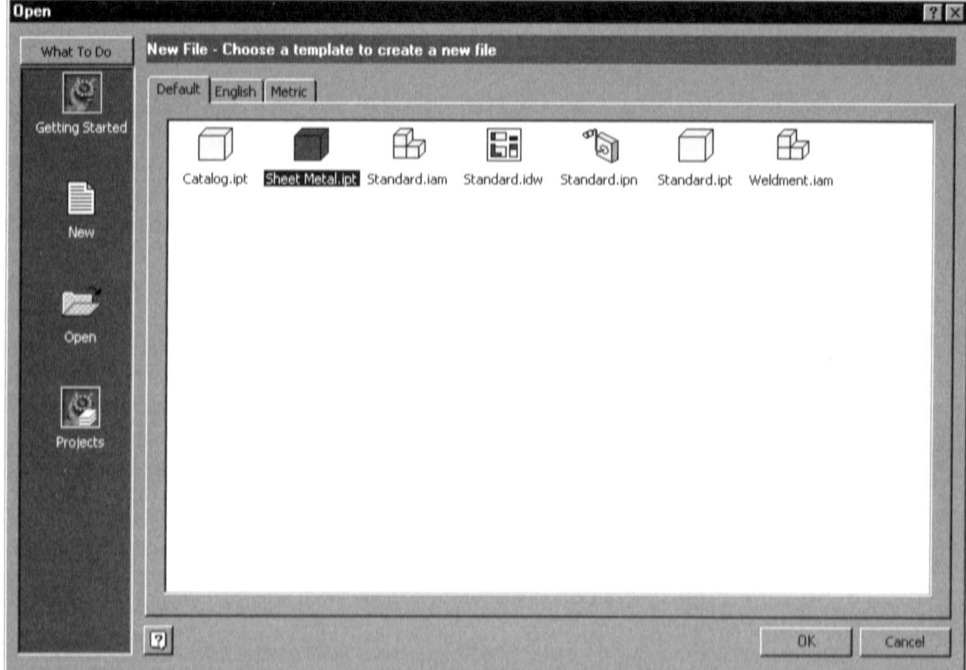


**Figure 11-6**

You can switch between the standard modeling environment and the sheet metal environment at any time by selecting Sheet Metal from the Applications menu, as shown in the following image. Autodesk Inventor's general modeling tools can be used to add features to a sheet metal part, even when the Sheet Metal application is active.

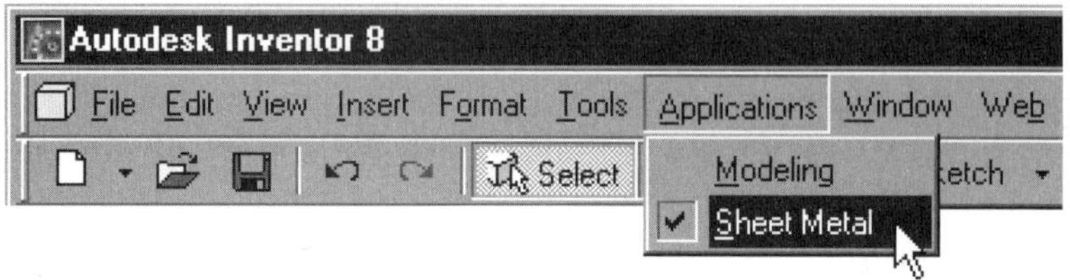


Figure 11-7

The creation of a sheet metal part usually begins by specifying the sheet metal style or settings, such as sheet thickness, material, and bend radius. You then use sheet metal-specific tools to create sheet metal parts by either adding faces at existing edges or connecting disjointed faces with bends. Use the optimized sheet metal tools to add, cut, and clean up features to the part. Sheet metal faces and parts can be adaptive and their size adjusted to meet design rules specified by assembly constraints.

## SHEET METAL TOOLS

After creating the new sheet metal document from a sheet metal template file, Autodesk Inventor's screen changes to reflect the sheet metal environment. As with standard parts, a sketch is created and becomes active. The sketch tools are common to both sheet metal and standard parts. The first sketch of a sheet metal part must be either a closed profile that is extruded the sheet thickness to create a sheet metal face or an open profile that is thickened and extruded as a contour flange. Additional tools are used to add sheet metal features to the base feature. The following is a description of the tools used to create sheet metal features. You can also add standard part features to a sheet metal part, but these features may not unfold when a flat pattern is generated from the folded model.

Use sheet metal tools to:

- Build a sheet metal part by adding faces along edges of existing faces.
- Create individual key faces of the sheet metal part and then connect these disjointed faces by adding bends between them.
- Automatically extend faces to create corner seams where unjoined edges meet.
- Cut shapes from faces with tools enhanced for sheet metal design.
- Add standard features, such as chamfers and fillets, using tools that are optimized for working with thin sheets.
- Create a flat pattern model of the part with a single button click. This flat pattern model is updated automatically as features are added, removed, or edited.
- Create drawing views of the folded part and flat pattern to document your design for manufacturing.

## SHEET METAL STYLES

The sheet metal-specific parameters of a part are stored in a sheet metal style. These include the thickness and material of the sheet metal stock, a bend allowance factor to account for metal stretching during the creation of bends, and various parameters dealing with sheet metal bends and corners. You can create additional sheet metal styles to account for various materials and manufacturing processes or material types. If you create sheet metal styles in a template file, they are available in all sheet metal parts based on that template.

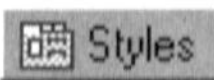


**Figure 11-8**

The thickness of the sheet metal stock is the key parameter in a sheet metal part. Sheet metal tools such as Face, Flange, and Cut automatically use the Thickness parameter to ensure that all features are the same wall thickness, a requirement for unfolding a model. In the default sheet metal style, all other sheet metal parameters, such as bend radius, are based on the Thickness value.

To edit or create a sheet metal style, follow these steps:

1. Click the Styles tool from the Sheet Metal Features Panel Bar. The Sheet Metal Styles dialog box is displayed, as shown in the following image.
2. To create a new sheet metal style, click the New button in the Sheet Metal Styles dialog box. A copy of the style that is currently displayed in the dialog box is created.
3. Edit the values on the Sheet, Bend, and Corner tabs to define the feature properties of parts created with this sheet metal style.
4. Rename the style and click the Save button.
5. When multiple sheet metal styles exist in a part, set the active style by selecting it from the Active Style list.

Changing to a different sheet metal style or making changes to the active style updates the sheet metal part to match the new settings.

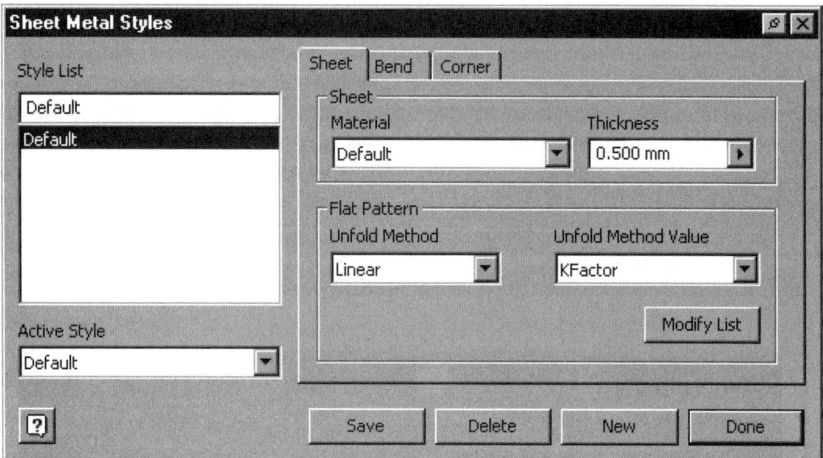


**Figure 11-9**

The following is a description of the settings in a sheet metal style. The parameters with numeric values, such as Thickness and Bend Radius, are saved as model parameters. The parameter values are updated when a sheet metal style is modified or a new sheet metal style is activated. Other model parameters and user-defined parameters can reference these parameters in equations.

## Sheet Tab

The Sheet tab contains settings for the sheet metal material and the folding method. Multiple styles can be created that contain different materials and thicknesses to be applied to a sheet metal model. The Sheet tab is shown in the previous image.

**Material** List of defined materials in the document. The material color setting is applied to the part.

**Thickness** Thickness of the flat stock that will be used to create the sheet metal part.

**Unfold Method** Method used to calculate bend allowance. Bend allowance accounts for material stretching during bending. Options are Linear or a Bend Table for more complex requirements.

**Unfold Method Value** Default value used for calculating bend allowances. A kFactor is a value between 0 and 1 that indicates the relative distance from the inside of the bend to the neutral axis of the bend. A kFactor of 0.5 specifies the neutral axis lies at the center of the material thickness. The Unfold Method Value is combined with the Unfold Method.

**Modify List** A list of kFactors and named Bend Tables included in the sheet metal style. Each bend can have the default bend allowance overridden by a value in this list.

## Bend Tab

The Bend tab contains settings for sheet metal bends. Most bend settings are typically defined as a function of the sheet thickness. The Bend Radius is the inside radius of the completed bend. This setting is the default for all bends but can be overridden during the creation of any bend.

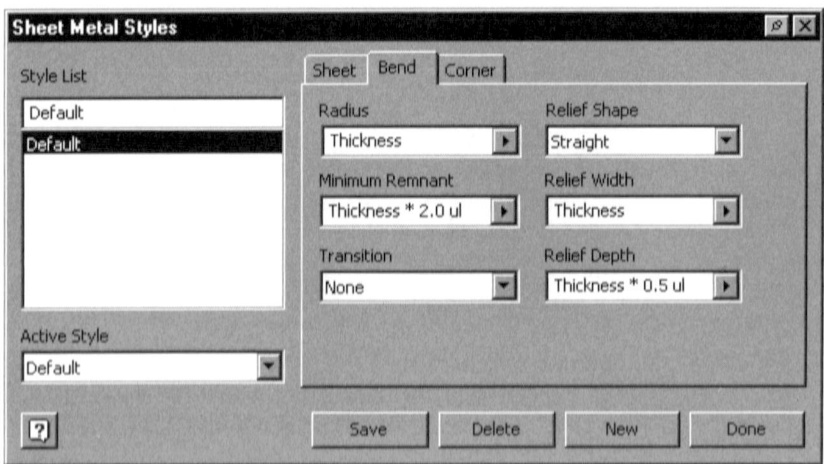


**Figure 11-10**

Bend reliefs can be specified when a bend zone (the area deformed during a bend) does not completely cross a face (see the following image).

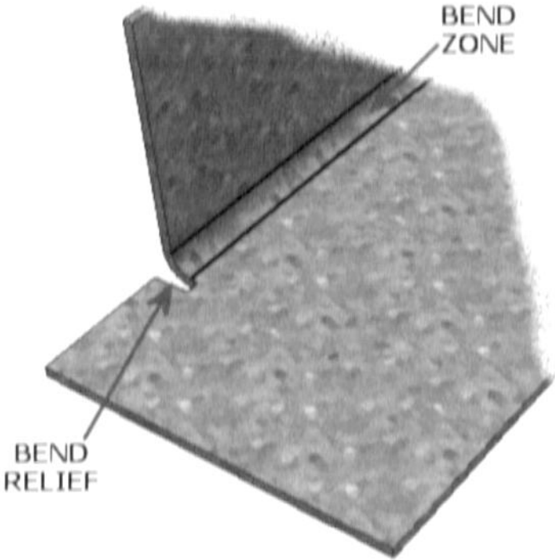


**Figure 11-11**

If bend reliefs are used, they are incorporated in the flat blank prior to folding. You can generate bend reliefs by punching or with laser, water-jet, or other cutting methods. The creation of bend reliefs increase production costs. It is common practice with thin, deformable materials such as mild steel to add bends without bend reliefs. The material is allowed to tear where the bend zone intersects with the adjacent face, as shown in the following image.

**Figure 11-12**

Descriptions of the settings on the Bend tab follows:

**Radius** Determines the inside bend radius between adjacent, connected faces.

**Minimum Remnant** Specifies the distance from the edge at which, if a bend relief cut is made within this distance, the small tab of remaining material is also removed.

**Transition** Controls the intersection of edges across a bend in the flattened sheet. For bends without bend relief, the unfolded shape is a complex surface. Transition settings simplify the results creating straight lines or arcs, which can be cut in the flat sheet before bending. Four of the five transition types are shown in the following image. Trim to Bend, which is not shown, does not create a transition.

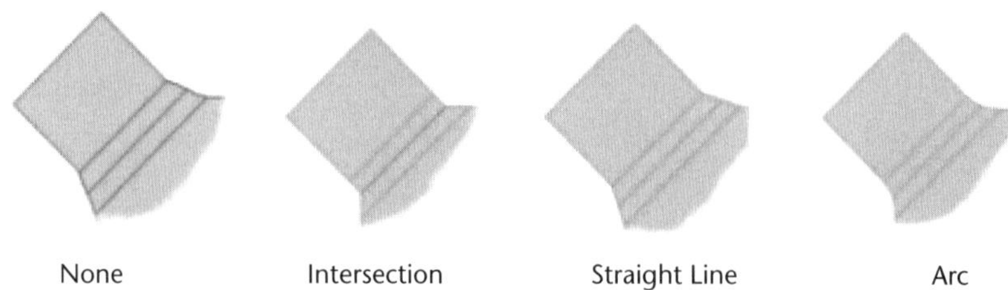


**Figure 11-13**

**Relief Shape** If a bend does not extend the full width of an edge, a small notch is cut next to the end of the edge to keep the metal from tearing at the edge of the bend. You can select from Straight, Round, or None for the shape of the relief. The Straight and Round shapes are shown in the following image.

**Relief Width** Width of the bend relief.

**Relief Depth** Setback amount from the edge. Round relief shapes require this be at least one half of the Relief Width.

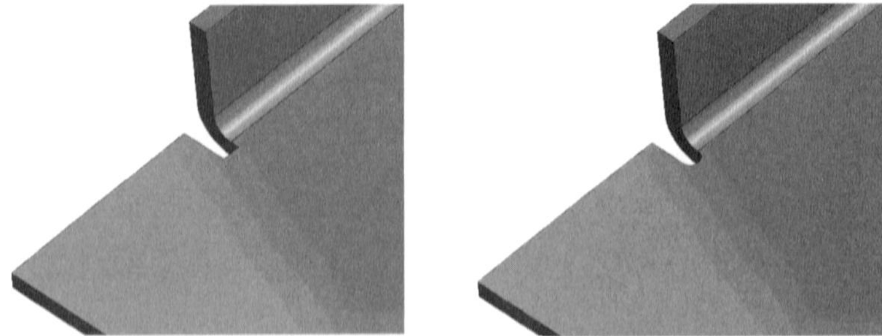

**Figure 11-14**

## Corner Tab

A corner occurs where three faces meet. The corner seam feature controls the gap between the open faces and the relief shape at the intersection. As with bend reliefs, corner reliefs are added to the flat sheet prior to bending. You can choose from four corner relief options, as shown in the following image. The shape of the corner relief can be viewed in the flat pattern of the model.

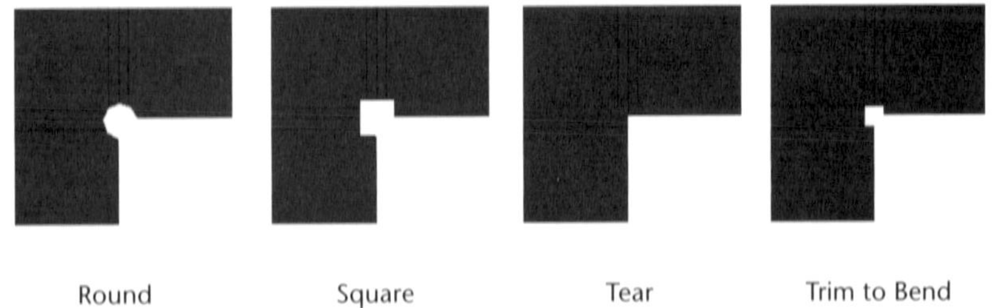


**Figure 11-15**

The Corner tab, shown in the following image, is where you set how corner relief will be applied to the model. A corner relief size and shape can be designated.

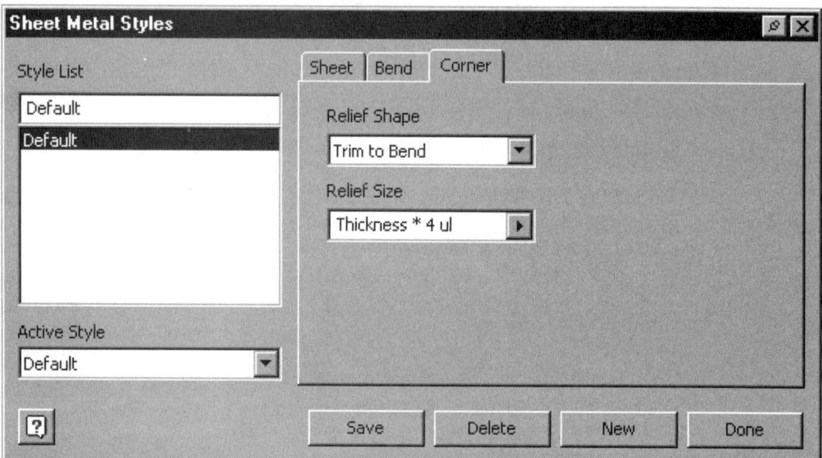


**Figure 11-16**

**Relief Shape** Specifies the shape of the corner relief. The five shapes available are Round, Square, Tear, Trim to Bend, and Linear Weld.

**Relief Size** Sets the size of the corner relief.

## FACE

The Face tool, shown in the following image, extrudes a closed profile for a distance equal to the sheet metal thickness. If the face is the first feature in a sheet metal part, you can only flip the direction of the extrusion. When a face is created later in the design, an adjacent face can be connected to the new face with a bend. If the sketch shares an edge with an existing feature, the bend is automatically added. As an option, you can select a parallel edge on a disjointed face. This will extend or trim the attached face to meet the new face, with a bend created between the two faces.

**Figure 11-17**

To create a sheet metal face, follow these steps:

1 Create a sketch with a single closed profile or a closed profile containing islands. The sketch is most often on a work plane created at either a specific orientation to other part features or by selecting a face on another part in an assembly.

**NOTE** You can create a single face from multiple closed profiles.

2 Select the Face tool from the Sheet Metal Features Panel Bar. The Face dialog box is displayed, as shown in the following image.

If a single closed profile is available, it is selected. If multiple closed profiles are available, you must click inside one to define the face area.

3 If required, flip the side to extrude the profile.

4 If the face is not the first feature and the sketch does not share an edge with an existing feature, click the Edges button and select a parallel edge. The two faces are extended or trimmed as required, meeting at a bend. The bend is listed as a child of the new face in the Browser. If the face attached to the selected edge is parallel but not coplanar with the new face, a set of double bend options are presented. An additional face is added to connect the two parallel faces. The orientation and shape of this face is determined by the selected double bend option.

The Unfold Options and Relief Options tabs contain settings for overriding the default values in the sheet metal style. Overrides are applied to individual features.

5 Click OK.

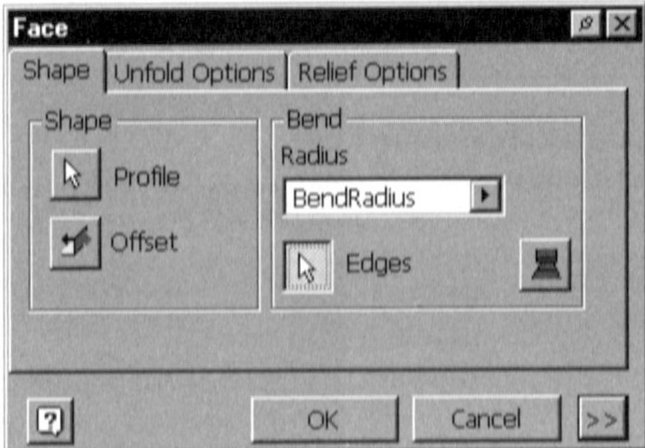


**Figure 11-18**

## Shape tab

The Shape tab is where you select the profile and direction of the face. You can also override the bend radius that is specified in the sheet metal style and select edges of other existing faces or edges that exist in the sheet metal part to apply a bend feature upon creation of the face. The following options are available on the Shape tab:

| | | |
|---|---|---|
|   | Profile | Allows you to select a profile(s) to extrude a distance equal to the Thickness parameter. |
|  | Offset | Toggles the direction for the creation of the feature. |

| | |
|---|---|
| Radius | Displays the default bend radius. If the face will be attached to another edge of the part, you can specify the value for the bend radius to be used. The default value can be modified on a *per feature* basis. |
| Edges | Allows you to select an edge to include in the bend. When selected, the edge is extended to match the edge of the face. A bend feature is created between the two faces. |
|  Extend Bend Aligned to Side Faces | Extends material along the faces of the edges connected by the bend instead of perpendicularly to the bend axis. An image of a part before this setting is applied is shown below on the left, and an example of the part when this option is active is shown on the right. |

**Figure 11-19**

There are options for creating a double bend or full-radius bend that can be accessed by clicking the More (>>) button. These options are discussed in the section on bends later in the chapter.

## Unfold Options

The Unfold Options tab is available so you can override the values set in the Sheet Metal Style dialog box, if necessary. This tab also exists when working with other sheet metal tools (i.e., Flange, Contour Flange, Hem, Bend, and so on). The following image exhibits the Unfold Options tab, and is followed by descriptions of its components.

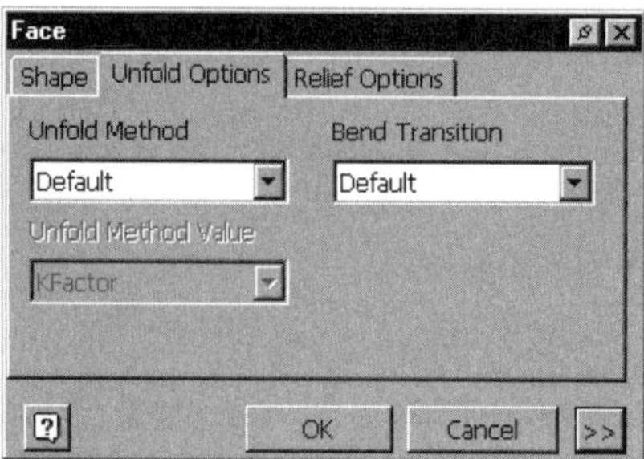


**Figure 11-20**

**Unfold Method** Sets the unfold method. Available methods are Default (reverts to the setting in the Sheet Metal Style), Linear, and Bend Table. (Linear and Bend Table give you the ability to choose a different Unfold Method if one has been defined in the Sheet Metal Style.)

**Unfold Method Value** Specifies a value for the Unfold Method. The list provides access to the values defined in the active Sheet Metal Style.

**Bend Transition** Sets the transition type for the sides of a bend. The following six types are available:

*Default*—Bend is created using the value in the active style.
*None*—A spline is used to join the tangencies.
*Intersection*—Adjacent edges join in the bend zone.
*Straight Line*—A straight line is used to join the tangencies.
*Arc*—An arc is used to join the tangencies.
*Trim to Bend*—No transition is created.

## Relief Options

The Relief Options tab, shown in the following image, is available so you can override the values set in the Sheet Metal Style, if necessary. This tab also exists when working with a number of the sheet metal tools (e.g., Flange, Contour Flange, Hem, Bend, and so on). These settings allow you to override how bends and bend reliefs are created on a per feature basis.

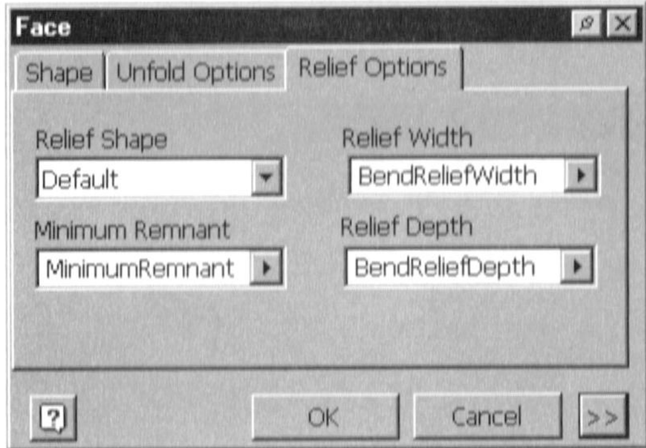


**Figure 11-21**

**Relief Shape** Overrides the relief shape specified by the active sheet metal style. Available shapes are:

*Default*—Uses the setting in the active sheet metal style.
*Straight*—Relief with square corners is created.
*Round*—Relief with full radius corners is created.
*None*—No bend relief is created.

**Minimum Remnant** Specifies a value for the minimum remnant. Any value entered will override the value specified in the active sheet metal style.

**Relief Width** Specifies a value for the Relief Width. Any value entered will override the value specified in the active sheet metal style.

**Relief Depth** Specifies a value for the Relief Depth. Any value entered will override the value specified in the active sheet metal style.

## CONTOUR FLANGE

A contour flange (see following image) is created from an open sketch profile. The sketch is extruded perpendicular to the sketch plane and the open profile is thickened to match the sheet metal thickness. The profile does not require a sketched radius between line segments—bends are added at sharp intersections. Arc or spline segments are offset by the sheet metal thickness. A contour flange can be the first feature in a sheet metal part, or it can be added to existing features.

**Figure 11-22**

To create a contour flange, follow these steps:

1 Create a sketch with a single open profile. The sketch can contain line, arc, and spline segments, and it can be constrained to projected reference geometry to define a common edge between the contour flange and an existing face.

2 Select the Contour Flange tool from the Sheet Metal Features Panel Bar. The Contour Flange dialog box is displayed, as shown in the following image.

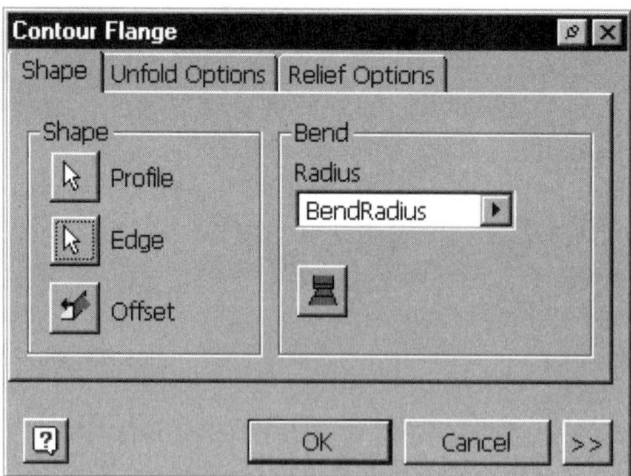


**Figure 11-23**

3 Select the open sketch profile to define the shape of the contour flange.

4 If required, flip the side to offset the profile.

5 If the contour flange is the first feature in the part, enter an extrusion distance and direction. If the contour flange is not the first feature, an edge perpendicular to the sketch plane can be selected to define the extrusion extents. If the sketch is attached to an existing edge, select that edge. If the

sketch is not attached to any projected geometry, the contour flange is extended or trimmed to meet the selected edge. The selected edge is typically the closest edge to the end of the sketch.

6 Apply any unfold and bend overrides on the Unfold Options and Bend Relief Options tabs.

7 If an edge is selected to define the extrusion extents, you can select from three options to further refine the length of the contour flange.

8 Click OK.

## Shape Tab

The Shape tab is where you select the profile, edge, and offset direction of the contour flange. You can also override the bend radius that is specified in the sheet metal style. The following options are available on the Shape tab:

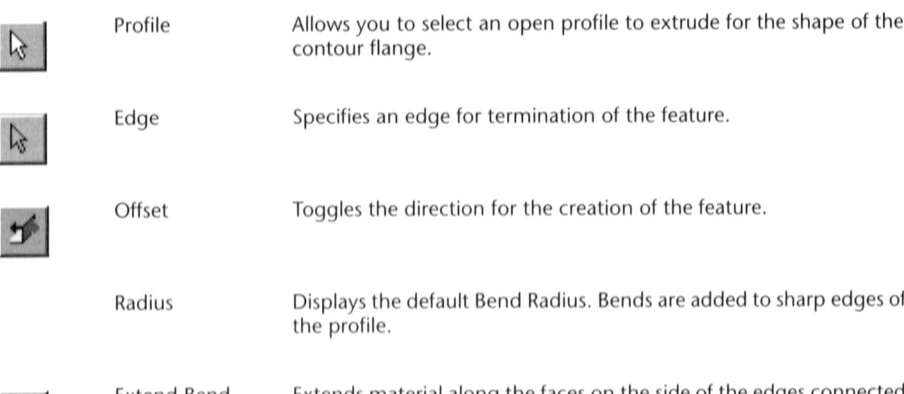

| | | |
|---|---|---|
| | Profile | Allows you to select an open profile to extrude for the shape of the contour flange. |
| | Edge | Specifies an edge for termination of the feature. |
| | Offset | Toggles the direction for the creation of the feature. |
| | Radius | Displays the default Bend Radius. Bends are added to sharp edges of the profile. |
| | Extend Bend Aligned to Side Faces | Extends material along the faces on the side of the edges connected by the contour flange. |

## More Button (>>)

The More button provides access to specify the type of extents for the feature, as shown in the following image.

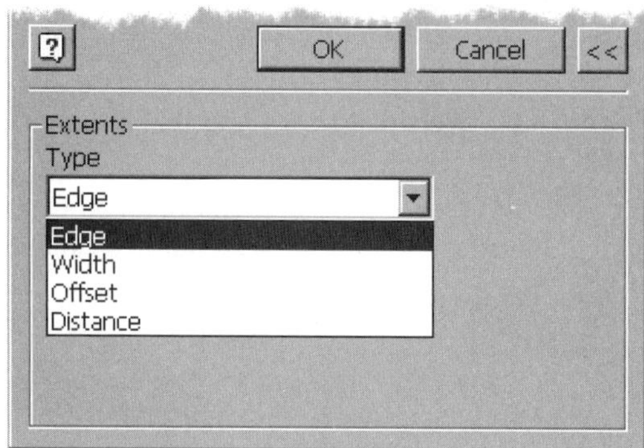


**Figure 11-24**

These extent types are also available for Flange and Hem features. The Flange and Hem features do not have the Distance option because the Distance is specified in the main body of the dialog box. A definition of the four options follows.

**Edge** The contour flange extends the full length of the selected edge.

**Width** A point defines one extent of the contour flange. The flange can be offset from this point and extends a fixed distance. The starting point is defined by selecting an endpoint on the selected edge, a work point on a line defined by the edge, or a work plane perpendicular to the selected edge.

**Offset** Similar to the Width option, two points are selected that define both extents of the flange. An offset distance from each point can be specified.

**Distance** You can enter a distance over which you want the contour flange to be extruded, as shown in the following image.

**Figure 11-25**

Samples of the Edge, Offset, and Width extent types are shown in the following image.

**Figure 11-26**

## FLANGE

A sheet metal flange is a simple rectangular face created from an existing face. A sketch is not required when creating a flange. The flange can extend the full length of the selected edge. The Flange tool, shown in the following image, creates a sheet metal face and a bend to an existing face. Flange length and the angle relative to the adjacent face are set in the Flange dialog box. The selected edge defines the bend location between the two faces.

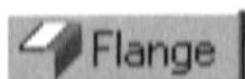


**Figure 11-27**

**NOTE** A minimum of one sheet metal face must exist before creating a flange.

To create a Flange, follow these steps:

1. Select the Flange tool from the Sheet Metal Features Panel Bar. The Flange dialog box is displayed, as shown in the following image.
2. Select an edge on an existing face.
3. Enter the distance and angle for the flange in the Flange dialog box. The flange preview updates to match the current values.
4. If required, flip the direction for the offset side of the flange.
5. Expand the dialog box and select the appropriate extent type. See Contour Flange for additional information on extents.
6. Apply the Flange tool to continue creating flanges, or click OK to complete the flange and exit the dialog box.

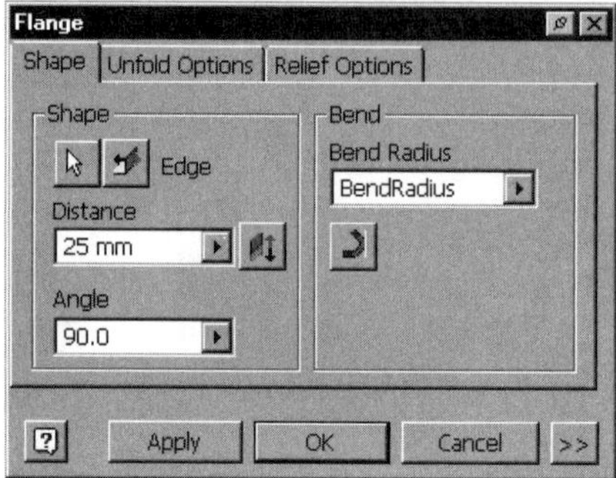


**Figure 11-28**

## Shape Tab

The Shape tab is where you select the edge and offset direction for the flange. You can also override the bend radius that is specified in the sheet metal style. The following options are available on the Shape tab:

| | |
|---|---|
| Select Edge | Allows you to select an edge for the flange to be created along. |
| Flip Offset | Toggles the flange creation to be on the inside or outside of the face. |
| Distance | Allows you to enter a distance for the depth of the flange. |
| Flip Direction | Toggles the side of the face used to create the flange. |
| Angle | Allows you to enter an angle for the flange. The value must be less than 180 degrees. |
| Bend Radius | Allows you to override the default bend radius set by the active sheet metal style. |
| Bend Tangent to Side Face | Creates a flange that is tangent to the side face. |

The options available on the Unfold Options and Relief Options tabs are covered in the previous section on creating faces. The More button provides access to the Edge, Width, and Offset extent types discussed in the previous Contour Flange section.

# EXERCISE 11-1 Creating Sheet Metal Parts

In this exercise, you create sheet metal faces, flanges, and seams to build a small enclosure. To navigate to the exercise in the *Electronic Student Workbook*, follow these steps:

1 From the Main TOC page, click Chapter 11.
2 From the TOC for Chapter 11, click Creating Sheet Metal Parts.

The following image illustrates the opened exercise.

**Figure 11-29 Opened file**

The following image illustrates the completed exercise.

**Figure 11-30 Completed exercise**

## HEM

Hems are used to eliminate sharp edges or strengthen an open edge of a face. Material is folded back over the face with a small gap between the face and the hem. A hem does not change the length of the sheet metal part; the face is trimmed so the hem is tangent to the original length of the face. Hems are created using the Hem tool (see following image).

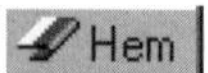


**Figure 11-31**

**NOTE** A minimum of one sheet metal face must exist before creating a hem.

To create a hem, follow these steps:

1 Select the Hem tool from the Sheet Metal Features Panel Bar. The Hem dialog box is displayed, as shown in the image following this list.

2 Select an open edge on a sheet metal face.

3 Select the hem type (examples shown in the following image):

- Single—A 180° flange
- Double—Single hem folded 180° resulting in a double thickness hem
- Teardrop—A single hem in a teardrop shape
- Rolled—A cylindrical hem

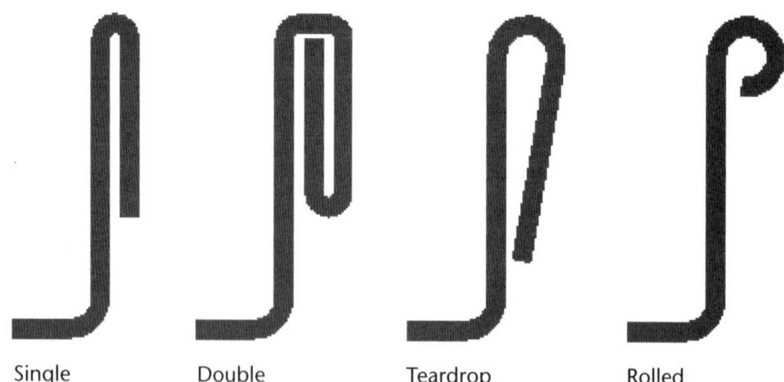


**Figure 11-32**

4 Enter values for the hem. Teardrop and rolled hems require radius and angle values, while single and double hems require gap and length values. The hem preview changes to match the current values.

5 Expand the dialog box by clicking the More (>>) button and select Edge, Width, or Offset for the hem extents.

6 Apply the Hem tool to continue creating hems, or click OK to complete the hem and exit the dialog box.

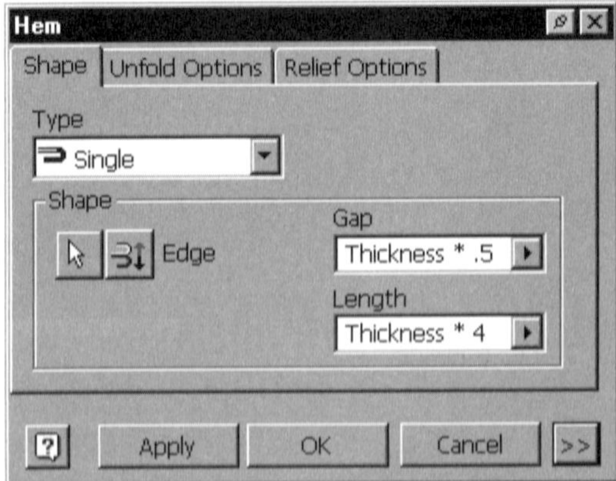


**Figure 11-33**

## Shape Tab

The Shape tab is where you select the edge and type of the hem to be created. Based on the type of hem that you want to create, different options will become active in the Hem dialog box. The following options are available on the Shape tab:

| | |
|---|---|
| Type | Selects one of the four hem types. |
| Select Edge | Selects the edge along which the hem will be created. |
| Flip Direction | Toggles the direction that the hem will be created. |
| Gap | Specifies the distance between the inside faces of the hem. Available when the single or double hem types are selected. |
| Length | Specifies the length of the hem. Available when the single or double hem types are selected. |
| Radius | Specifies the bend radius to apply at the bend. Available when the rolled or teardrop hem types are selected. |
| Angle | Specifies the angle applied to the hem. Available when the rolled or teardrop hem types are selected. |

The options available on the Unfold Options and Relief Options tabs are covered in the previous section on creating faces. The More (>>) button provides access to the Edge, Width, and Offset extent types discussed in the Contour Flange section.

# EXERCISE 11-2 Hems

In this exercise, you create two hem features. To navigate to the exercise in the *Electronic Student Workbook*, do the following:

1 From the Main TOC page, click Chapter 11.

2 From the TOC for Chapter 11, click Hems.

The following image illustrates the opened exercise.

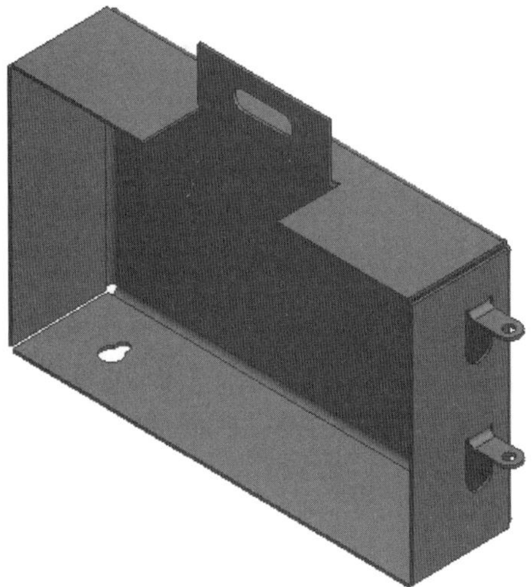

**Figure 11-34 Opened exercise**

The following image illustrates the completed exercise.

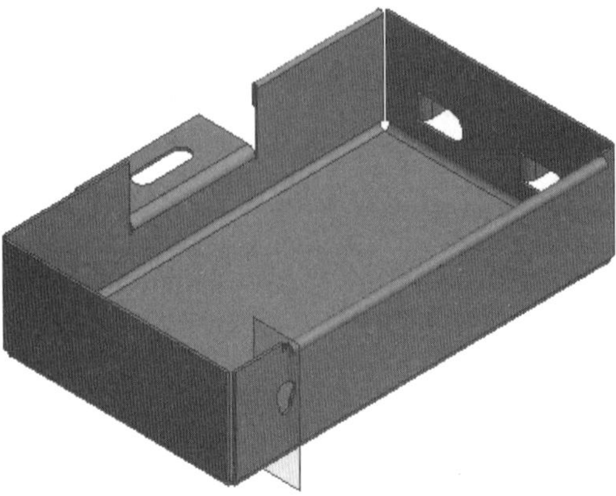

**Figure 11-35 Completed exercise**

## FOLD

An alternate method for creating sheet metal features is to start with a known flat pattern shape and then add folds at sketched lines on a face. The Fold tool, shown in the following image, can create sheet metal shapes that are difficult to create using Face or Flange tools.

**Figure 11-36**

To create a fold, follow these steps:

1 Create a sketch on an existing face. Sketch a line between two open edges on the face.

**NOTE** The line endpoints must be coincident to the face edge.

2 Select the Fold tool from the Sheet Metal Features Panel Bar. The Fold dialog box is displayed, as shown in the following image.

3 Select the sketch or bend line. The fold direction and angle are previewed in the graphics window. The fold arrows extend from the face that will remain fixed. The face on the other side of the bend line will fold around the bend line.

4 If required, flip the fold direction and side.

5 Enter the angle of the fold.

6 Select the positioning of the fold with respect to the sketched line. The line can define the start, centerline, or end of the bend. The fold preview updates to match the current settings.

7 Make any changes to Unfold or Bend Relief options.

8 Apply the Fold tool to continue creating folds (an unconsumed sketch with valid geometry must be available), or click OK to complete the fold and exit.

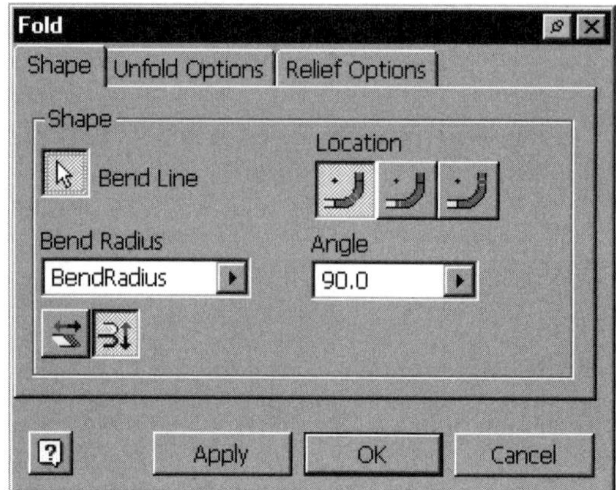


**Figure 11-37**

## Shape Tab

The Shape tab is where you select the bend line for the fold to be created. You can set the location of the selected bend line relative to the fold feature that determines the folded shape. The angle for the fold and direction are also specified on the Shape tab. You can also override the bend radius that is specified in the sheet metal style. The following options are available on the Shape tab:

| Option | Description |
|---|---|
| Bend Line | Selects a sketch line to use as the fold line. |
| Bend Radius | Overrides the active sheet metal style used by default. |
| Flip Side | Flips the side used for the angle of the fold. |
| Flip Direction | Toggles the direction that the fold will be created. |
| Centerline of Bend | Determines the centerline of the fold from the selected sketch line. |
| Start of Bend | Determines the start of the fold from the selected sketch line. |
| End of Bend | Determines the end of the fold from the selected sketch line. |
| Angle | Specifies the angle that is applied to the fold. |

The options available on the Unfold Options and Relief Options tabs are covered in the previous section on creating faces.

## BEND

Bends are created with the Bend tool (see following image) as child objects of other features when the feature connects two faces. Bends can also be created as independent objects between disjointed faces.

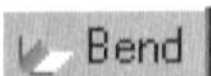


**Figure 11-38**

A preview of a bend is shown in the following image, and the faces that are being joined are either trimmed or extended to connect the faces.

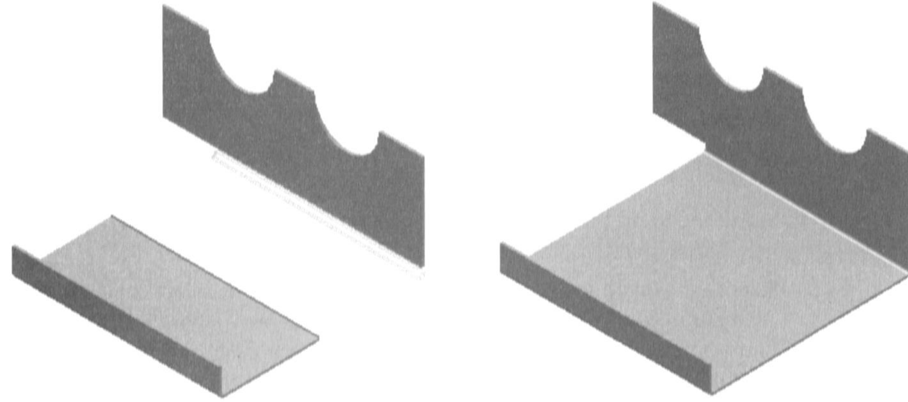

**Figure 11-39**

When parallel faces are selected for the bend feature, a jog or z-bend will be created, depending on the edge location. Jogs are often used to allow overlapping material. A sample of a jog feature is shown in the image below (left). Double bends can also be created when the two faces are parallel and the selected edges face the same direction.

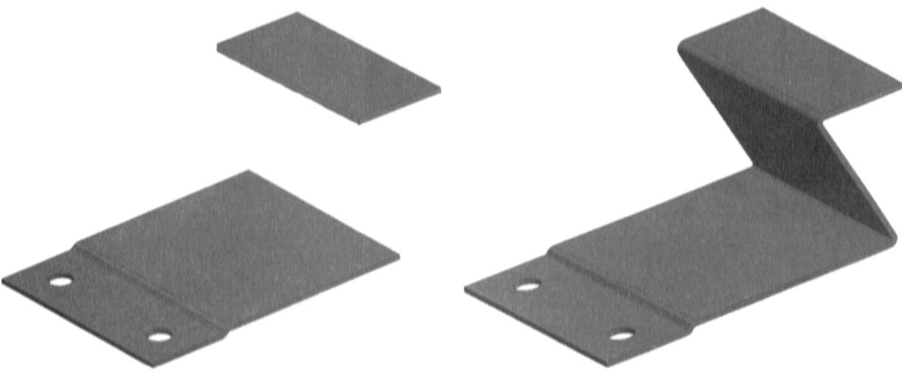

**Figure 11-40**

You can create the bend to be two 90° bends with a tangent face between the bends or a full radius bend between the two faces, as shown in the following image.

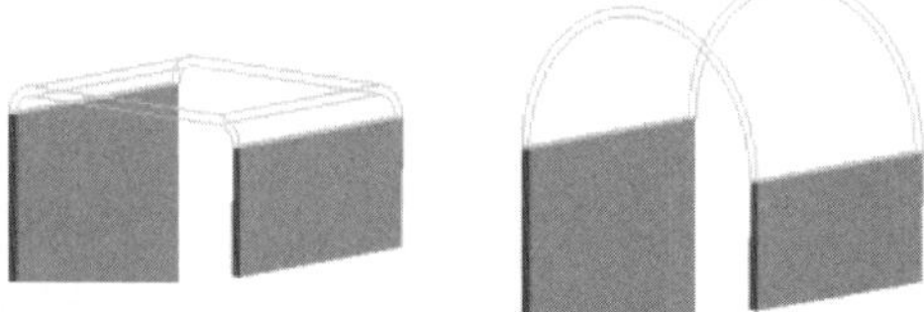

Figure 11-41

**NOTE** The sheet metal part must have two disjointed or intersecting faces. An example of intersecting faces is a shelled box that is being changed into a sheet metal part. The intersection of the box base and a wall is an edge that can be changed to a bend.

To create a bend, follow these steps:

1 Select the Bend tool from the Sheet Metal Features Panel Bar. The Bend dialog box is displayed, as shown in the following image.

2 Select the common edge of two intersecting faces, or select two parallel edges on disjointed, non-coplanar faces. If the two faces are parallel, a set of Double Bend options is presented. Depending on the position of the two faces, the Double Bend options will be either 45° and Fixed Edges or 90° and Full Radius.

3 Make any changes to Unfold or Bend Relief options.

4 Apply the Bend tool to continue creating bends, or click OK to complete the bend and exit the dialog box.

Figure 11-42

### Shape Tab

The Shape tab is where you select the edges for the bend to be created between and specify the type of bend that you want to create. You can also override the bend radius that is specified in the sheet metal style. The following options are available on the Shape tab:

| | |
|---|---|
| Edges | Selects the edges where the bend will be applied. The selected edges will be trimmed or extended as needed to create the bend feature. |
| Radius | Allows you to override the default Bend Radius set by the active sheet metal style. |
|  Extend Bend Aligned to Side Edges | Extends material along the faces of the edges connected by the bend instead of perpendicularly to the bend axis. |
| Fix Edges | Creates equal bends to the selected sheet metal edges. |
| 45 Degree | Creates 45° bends on the selected edges. |
| Full Radius | Creates a semi-circular face between the selected edges. |
| 90 Degree | Creates 90° bends between the selected edges. |
|  Flip Fixed Edge | Reverses the order of the edges being fixed. Normally, the first edge selected is fixed by default, and the second edge will be trimmed or extended. This button reverses the order. |

The options available on the Unfold Options and Bend Relief Options tabs are covered in the section on creating faces.

A bend that is not listed under a face in the Browser can be edited at any time, allowing you to reselect the edges that define the bend. The bend can even be edited to connect two faces that were not initially joined at the bend.

# EXERCISE 11-3
## Modifying Sheet Metal Parts



In this exercise, you complete the design of a sheet metal bracket within the context of an assembly and then modify the sheet metal part to eliminate component interference. You add bends to connect disjointed sheet metal faces and then modify one bend to eliminate interference with a component in the assembly. A hinge feature is then added to the sheet metal part using the general modeling tools. To navigate to the exercise in the *Electronic Student Workbook*, do the following:

1 From the Main TOC page, click Chapter 11.

2 From the TOC for Chapter 11, click Modifying Sheet Metal Parts.

The following image illustrates the opened exercise.

**Figure 11-43 Opened exercise**

The following image illustrates the completed exercise.

**Figure 11-44 Completed exercise**

## CUT

The Cut tool, shown in the following image, is a sheet metal specific implementation of the standard Extrude tool. The Cut tool always performs an extrude cut—join and intersection are not available. The default extents are a blind cut at a distance equal to the sheet metal Thickness parameter. This ensures that the cut only extends through the face containing the sketch and not through other faces that may be folded under the sketch face.

**Figure 11-45**

Most cuts are manufactured on the flat sheet stock before the sheet is bent to form the folded part. The cuts often cross bend lines. Because the part is modeled in the folded state, representing cuts that cross bend boundaries requires the bend be unfolded to represent the flat sheet. When the bend is refolded, the cut deforms around the bend to ensure that the extrusion remains perpendicular to the sheet metal faces on both sides of the bend. When sketching a cut profile that will cross a bend, the Project Flat Pattern tool projects the unfolded flat pattern geometry onto the sketch face, as shown in the following images.

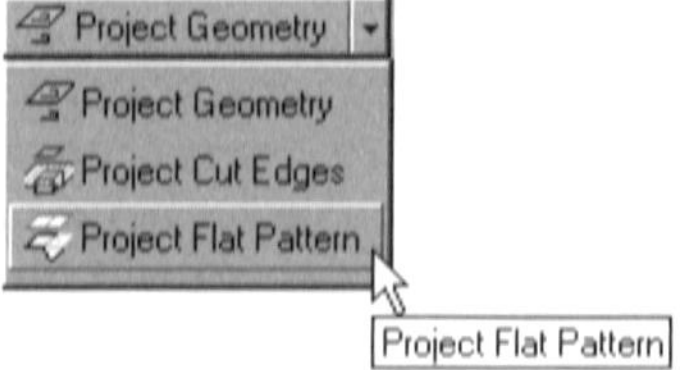


**Figure 11-46**

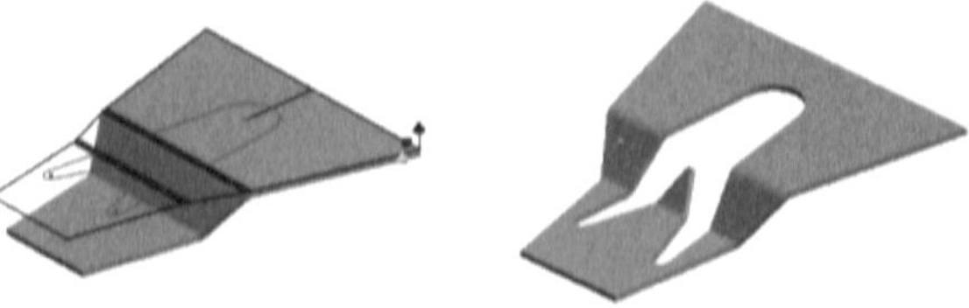

**Figure 11-47**

To create a cut feature, follow these steps:

1 Create a sketch on a sheet metal face that includes one or more closed profiles representing the area(s) to cut. If required, use the Project Flat Pattern tool to project the unfolded geometry of connected faces onto the sketch.

2 Select the Cut tool from the Sheet Metal Features Panel Bar. The Cut dialog box is displayed, as shown in the following image.

3 Select the profiles to cut.

4 If a profile crosses a bend, check the Cut Across Bend option in the Cut dialog box.

5 Click OK to complete the cut and exit the dialog box.

**Figure 11-48**

The following items are available in the Cut dialog box:

| | |
|---|---|
| Profile | Allows you to select the profile(s) to be cut into the sheet metal part. |
| Cut Across Bend | Projects the profile across faces that are bent in the sheet metal part if the box is checked. |
| Extents | Allows you to choose one of the typical extrusion options: Distance, To Next, To, From To, All |
| Direction | Specifies the direction for the cut to be created in the part. |

# EXERCISE 11-4 Cut Across Bend

In this exercise, you complete a sheet metal bracket using a fold and double bend. You then use the Project Flat Pattern tool to create a cut across bends. To navigate to the exercise in the *Electronic Student Workbook*, do the following:

1 From the Main TOC page, click Chapter 11.
2 From the TOC for Chapter 11, click Cut Across Bend.

The following image illustrates the opened exercise.

**Figure 11-49 Opened exercise**

The following image illustrates the completed exercise.

**Figure 11-50 Completed exercise**

## CORNER SEAM

When three faces meet in a sheet metal part, a gap is required between two of the faces to enable unfolding. Using a box as an example, the walls are connected to the floor, and gaps between the walls enable the box to be unfolded. The gap between adjacent faces is a corner seam, and this gap is created using the Corner Seam tool, shown in the following image.

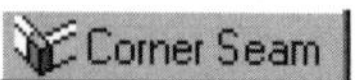


**Figure 11-51**

Corner seams can also be used to create mitered gaps between coplanar faces, as shown in the following image. Finally, a corner seam can be created by ripping open a corner at an intersection between faces.

**Figure 11-52**

To create a corner seam, follow these steps:

1 Select the Corner Seam tool from the Sheet Metal Features Panel Bar. The Corner Seam dialog box is displayed, as shown in the following image.

2 Select face edges that meet one of the following criteria:

- Two open edges on faces that share a common connected edge (two walls that share a connection to a floor).
- Two nonparallel edges on coplanar faces (creates a mitered joint with a gap between the extended faces).
- A single edge at the intersection of two faces (the wall intersections of a shelled box).

3 Select the Seam option in the Corner Seam dialog box.

4 Enter a value for the corner seam gap.

5 Make any changes to the corner options.

6 Apply the Corner Seam tool to continue creating corner seams, or click OK to complete the corner seam and exit the dialog box.

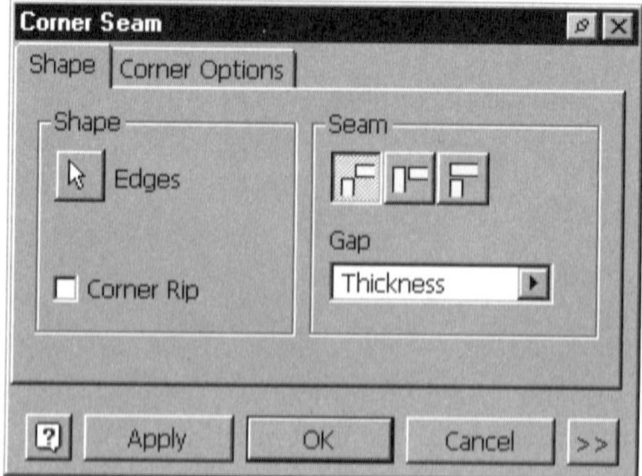


**Figure 11-53**

## Shape Tab

The Shape tab is where you select the edges where the corner seam will be created. The orientation of the seam and if a corner should be ripped can also be specified. The following options are available on the Shape tab:

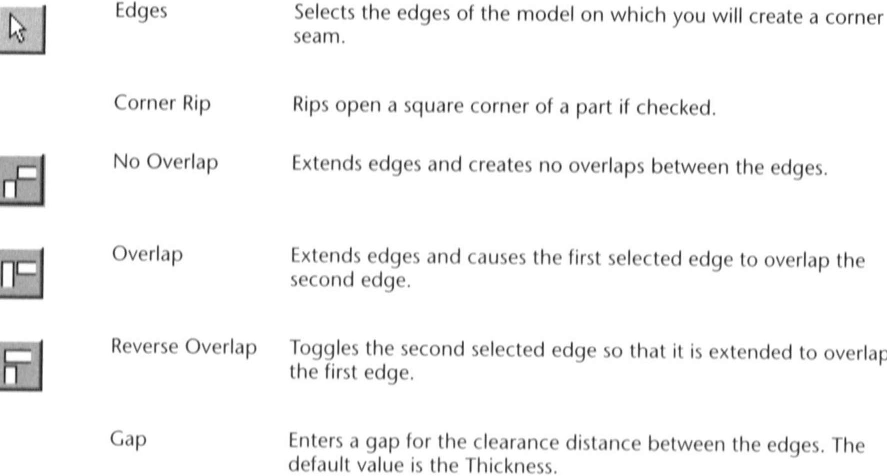

| Option | Description |
|---|---|
| Edges | Selects the edges of the model on which you will create a corner seam. |
| Corner Rip | Rips open a square corner of a part if checked. |
| No Overlap | Extends edges and creates no overlaps between the edges. |
| Overlap | Extends edges and causes the first selected edge to overlap the second edge. |
| Reverse Overlap | Toggles the second selected edge so that it is extended to overlap the first edge. |
| Gap | Enters a gap for the clearance distance between the edges. The default value is the Thickness. |

## Corner Options

The Corner Options tab gives you access to override the settings for relief and bend transition on a per feature basis. By default, the settings specified in the sheet metal style are used. The following options are available:

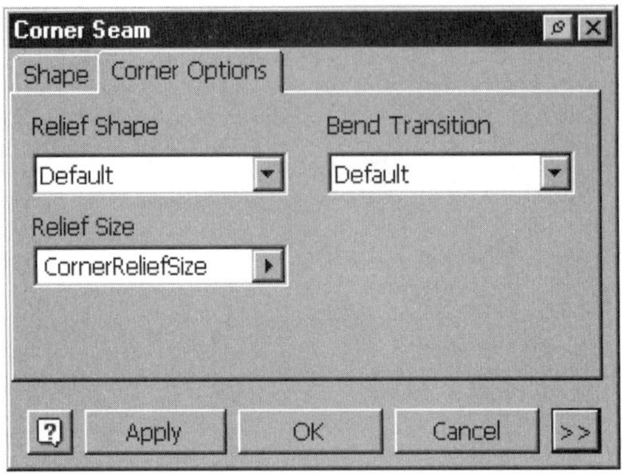

Corner Seam
Shape | Corner Options
Relief Shape: Default
Bend Transition: Default
Relief Size: CornerReliefSize
Apply | OK | Cancel | >>

Figure 11-54

**Relief Shape** Specifies the shape of the relief generated by the corner seam. There are six available shapes:

*Default*—Uses the setting in the active sheet metal style.
*Round*—Creates a circular corner relief.
*Square*—Creates a square corner relief.
*Tear*—Creates a torn relief.
*Trim to Bend*—Creates no transition.
*Linear Weld*—Creates a narrow gap for corners that are to be welded.

**Relief Size** Overrides the Corner Relief Size set by the active sheet metal style.

**Bend Transition** Specifies the Transition type for the bend. Five types are available:

*Default*—Creates bends using the value in the active style.
*None*—Uses a spline to join the tangencies.
*Intersection*—Joins adjacent edges in the bend zone.
*Straight Line*—Uses a straight line to join the tangencies.
*Arc*—Uses an arc to join the tangencies.

## More Button

There are additional options available when the More (>>) button is clicked as shown in the following image. The options are described below.

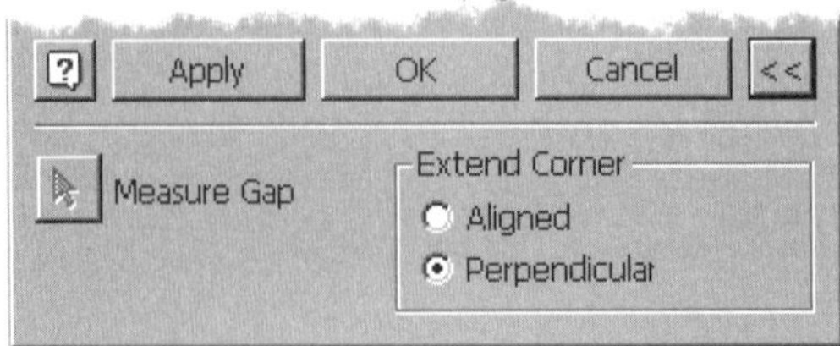


Figure 11-55

**Measure Gap** Allows you to select two edges to measure the distance that exists between them.

**Aligned** Makes the first face align to the second face that is selected.

**Perpendicular** Makes the first face perpendicular to the second face that is selected.

EXERCISE

# EXERCISE 11-5
# Corner Seams (Shelled Solids)

In this exercise, you rip open appropriate edges to create an unfoldable part and add bends to other sharp edges. You also create a flat pattern to confirm that the part can unfold. To navigate to the exercise in the *Electronic Student Workbook*, do the following:

1 From the Main TOC page, click Chapter 11.
2 From the TOC for Chapter 11, click Corner Seams (Shelled Solids).

The following image illustrates the opened exercise.

**Figure 11-56 Opened exercise**

The following image illustrates the completed exercise.

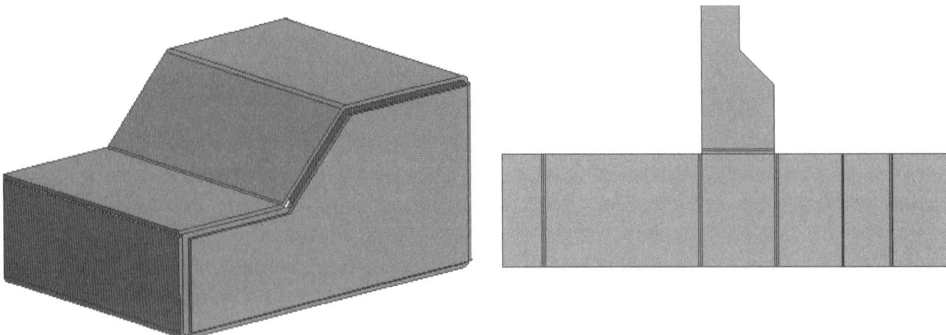

**Figure 11-57 Completed exercise**

## CORNER ROUND

The Corner Round tool shown in the following image is available when no sketch is active. It is a sheet metal-specific fillet tool. All edges other than the ones at open corners of faces (all these edges have a Length = Thickness) are filtered out. This enables you to easily select these small edges without zooming in on the part.

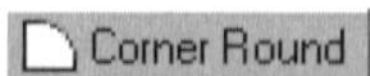


**Figure 11-58**

To create a corner round, follow these steps:

1. Select the Corner Round tool from the Sheet Metal Features Panel Bar. The Corner Round dialog box is displayed, as shown in the following image.
2. Enter a radius for the corner round.
3. Select the Corner or Feature Select Mode in the Corner Round dialog box.
4. Select the corners or features to include.
5. Add additional corner rounds with different radii, and select corners or features for the additional corner rounds.
6. Click OK to complete the corner round and exit the dialog box.

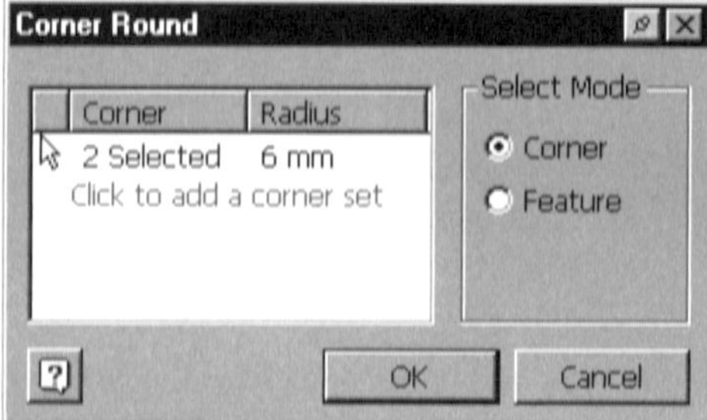


**Figure 11-59**

### Select Mode

Selecting edges to which you apply a corner round is similar to selecting edges for the fillet feature. There are two select modes that enhance the feature for use in the sheet metal environment: Corner and Feature.

**Corner** Selects individual thickness edges to be rounded or filleted.

**Feature** Selects all thickness edges of a feature automatically.

## CORNER CHAMFER

The Corner Chamfer tool, shown in the following image, is a sheet metal-specific chamfer tool. As with the Corner Round tool, all edges other than the ones at open corners of faces are filtered out. Because the edges are always discontinuous, the Edge Chain and Setback settings available with the standard Chamfer tool are not available.

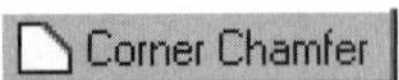


**Figure 11-60**

To create a corner chamfer, follow these steps:

1. Select the Corner Chamfer tool from the Sheet Metal Features Panel Bar. The Corner Chamfer dialog is displayed, as shown in the following image.
2. Select the chamfer style: One Distance, Distance and Angle, or Two Distances (see the following list for explanations of these styles).
3. Select the corners (or edge and corner for Distance and Angle) you wish to include.
4. Enter chamfer values.
5. Click OK to complete the corner chamfer and exit the dialog box.

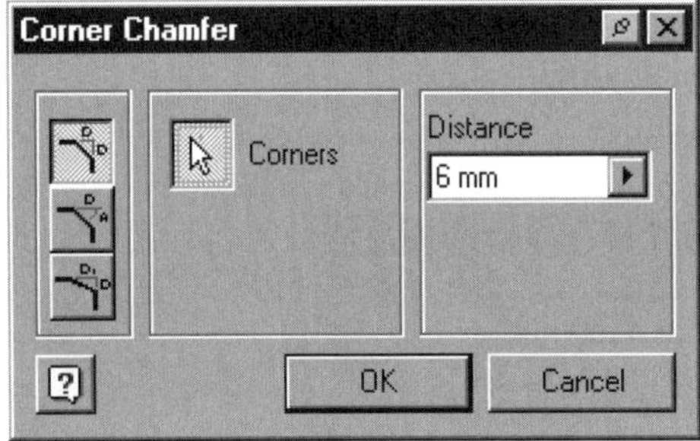


**Figure 11-61**

| | | |
|---|---|---|
| | Distance | Creates a 45° chamfer on the selected edge. The size of the chamfer is determined by typing a distance in the dialog box. The value is offset from the two common faces. |
|  | Distance and Angle | Creates a chamfer offset from a selected edge on a specified face, at an angle from the number of degrees specified. Enter an angle and distance for the chamfer in the dialog box, then click the face that the angle is based on and specify an edge to be chamfered. |
|  | Two Distances | Creates a chamfer offset from two faces, each the amount that you specify. First, click an edge and enter a value for Distance1 and Distance2. A preview image of the chamfer is displayed. To reverse the direction of the distances, pick the Flip button. |
|  | Corners | Selects individual corners to chamfer. |
|  | Edges | Selects an edge for chamfers when using the Distance and Angle option. |

| | | |
|---|---|---|
|  | Flip | Flips the direction for the chamfer distance when using the Two Distances option. |
| | Distance | Enters a distance to be used for the chamfer feature. |
| | Angle | Enters an angle for the chamfer feature when using the Distance and Angle option. |

## PUNCHTOOL

Cuts and 3D deformation features, such as dimples and louvers, are usually added to the flat sheet metal stock in a turret punch before the sheet is bent into the folded part. Turret punch tools are positioned on the flat sheet by a center point corresponding to the tool center, at an angle to a fixed coordinate system. The PunchTool tool, shown in the following image, places specially designed iFeatures on sketch points that define the center point of the tool. A streamlined interface simplifies the selection and placement of the iFeatures.

**Figure 11-62**

A default Punch folder is installed under the top-level Catalog folder when Autodesk Inventor is installed. A selection of punch tools is included in the folder. PunchTool lists all iFeatures in the designated punch folder. You can set the default Punch folder on the iFeature tab of the Application Options dialog box, as shown in the following image.

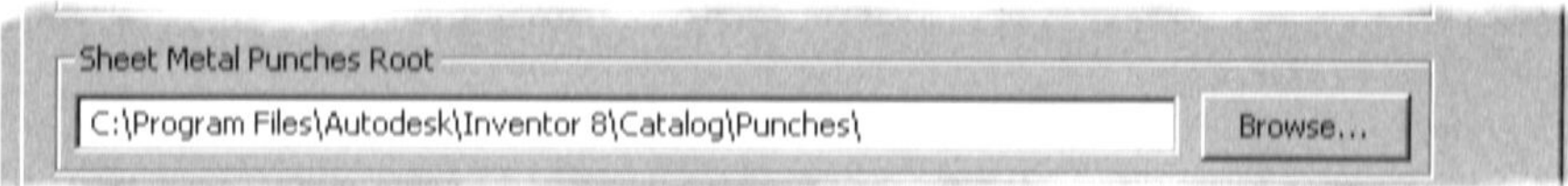


**Figure 11-63**

To qualify as a PunchTool, a saved iFeature must have a single Hole Center point in the sketch of the first feature included in the iFeature. The point corresponds to the center of the tool. Asymmetrical shapes must have the center point position controlled by geometric relationships or equations to ensure that it remains centered when the iFeature parameters are changed. Create the iFeature in a sheet metal part to ensure that sheet metal parameters such as Thickness are saved with the iFeature.

To place a PunchTool, follow these steps:

1 Create a sketch on a sheet metal face. Place at least one sketch point in the sketch. Hole Center points are automatically selected as punch centers during PunchTool placement. Sketch points, line or curve endpoints, and arc centers can be manually added as additional punch centers.

2 Select PunchTool from the Sheet Metal Features Panel Bar. The PunchTool dialog box is displayed, as shown in the following image.
3 Select the desired PunchTool from the list in the PunchTool dialog box. Click Next.
4 All Hole Centers are automatically selected. Hold down the SHIFT key and select Hole Centers to exclude them, and select other center points as required.
5 Select a rotation angle for all occurrences of the punch. At zero degrees rotation, the *X*-axis of the first feature in the saved iFeature is aligned with the *X*-axis of the current sketch. Click Next.
6 Enter values for the PunchTool parameters.
7 Click OK to complete the PunchTool and exit the dialog box.

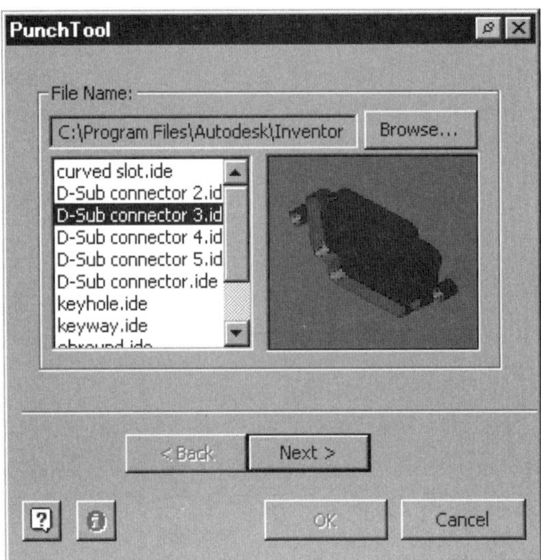


Figure 11-64

# EXERCISE 11-6 Punch Tool

In this exercise, you create an iFeature that can be used as a punch, and then use the PunchTool tool in another sheet metal part. To navigate to the exercise in the *Electronic Student Workbook*, do the following:

1 From the Main TOC page, click Chapter 11.

2 From the TOC for Chapter 11, click Punch Tool.

The following image illustrates the opened exercise.

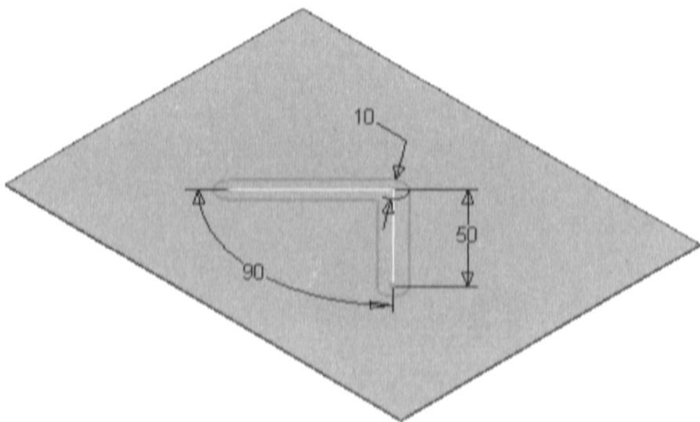


**Figure 11-65 Opened exercise**

The following image illustrates the completed exercise.

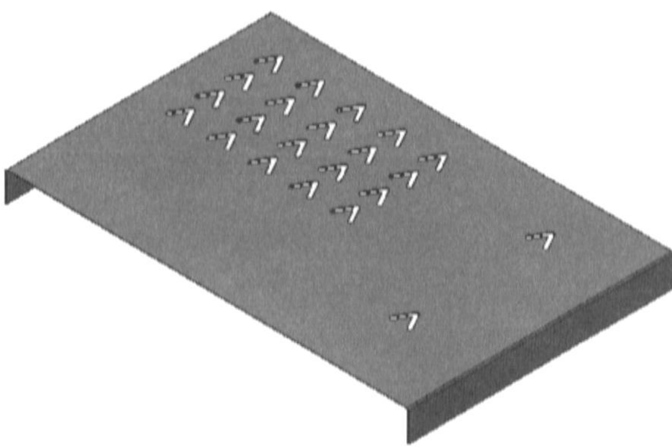

**Figure 11-66 Completed exercise**

## FLAT PATTERN

The final step is the creation of a sheet metal flat pattern that unfolds all of the sheet metal features. The flat pattern represents the starting point for the manufacturing of the sheet metal part. The flat pattern can be displayed in a 2D drawing view, complete with lines indicating bend centerlines. The Flat Pattern tool shown in the following image creates a 3D model of the unfolded part.

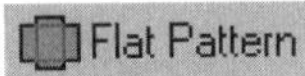


**Figure 11-67**

You can also directly export the flat pattern in 3D (SAT file) or 2D (DXF or DWG) formats, to be used to create machine tool programming for flat pattern punching or cutting. The following image shows a sheet metal part in its formed state on the left and the same part in its flat state on the right.

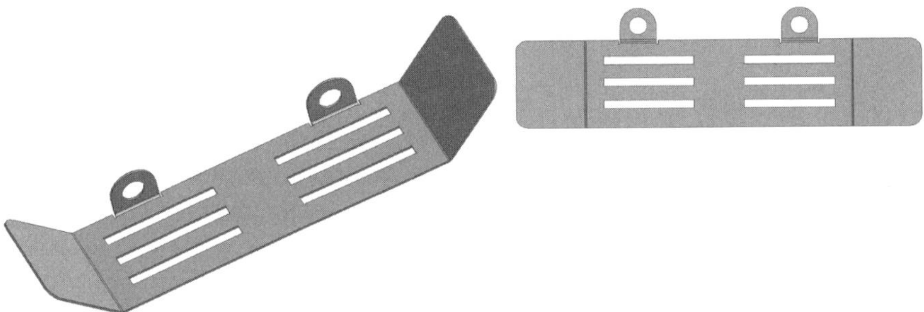

**Figure 11-68**

Manually modeling a flat pattern of a sheet metal part is straightforward, but can take considerable effort for a complicated shape. Autodesk Inventor can quickly create a 3D unfolded model of your sheet metal part and update this model automatically as you modify the features that make up the part. You can create the flat pattern model at any time during the sheet metal modeling process.

The flat pattern model and icon are shown under the part name in the Browser, as shown in the following image. The flat pattern model is viewed in a separate window. Both the model and flat pattern windows can be open and visible at the same time. Select Arrange All from the Window menu to display both windows. Close this window or select the model window from the Window menu to return to the window displaying the folded part. The flat pattern model is automatically updated as you modify sheet metal features. Right-click the flat pattern in the Browser and select Open Window to view the current flat pattern shape. If the flat pattern window has already been opened but is not the active window in Autodesk Inventor, the option is displayed as Show Window.

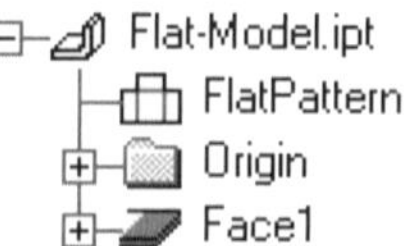


Figure 11-69

**NOTE** If you delete the flat pattern model (from the Browser), you will not be able to create a drawing view of the flat pattern. If you delete the flat pattern model after the creation of a flat pattern drawing view, the drawing view will be deleted.

Right-click the flat pattern in the Browser and select Extents to display the dimensions of the smallest rectangle that encloses the flat pattern, as shown in the following image. Use this information for stock selection and multiple flat pattern layouts on large sheets.

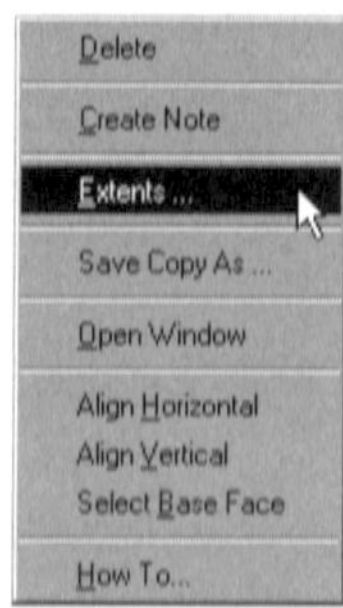


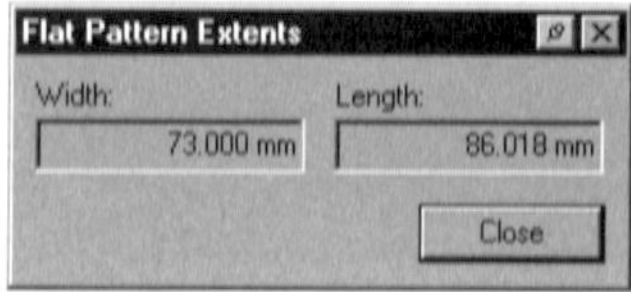


Figure 11-70

To create a flat pattern, follow these steps:

1 Select a face that you expect will not be removed in future edits. The selected face will remain fixed and all other faces will unfold from this face.

**NOTE** This is not a strict requirement, but selecting a persistent face before creating a flat pattern is good practice.

2 Select the Flat Pattern tool from the Sheet Metal Features Panel Bar. The flat pattern model is displayed in a new window. The flat pattern model is saved in the same document as the folded model. A Flat Pattern node is also added to the Browser.

3 Right-click Flat Pattern at the top of the Browser. Menu items are available for saving the flat pattern, aligning the flat pattern to a model edge, and displaying the overall dimensions (extents) of the flat pattern.

4 To document a flat pattern, start a new drawing. Place a drawing view and select Flat Pattern from the Sheet Metal View list in the Drawing View dialog box, as shown in the following image.

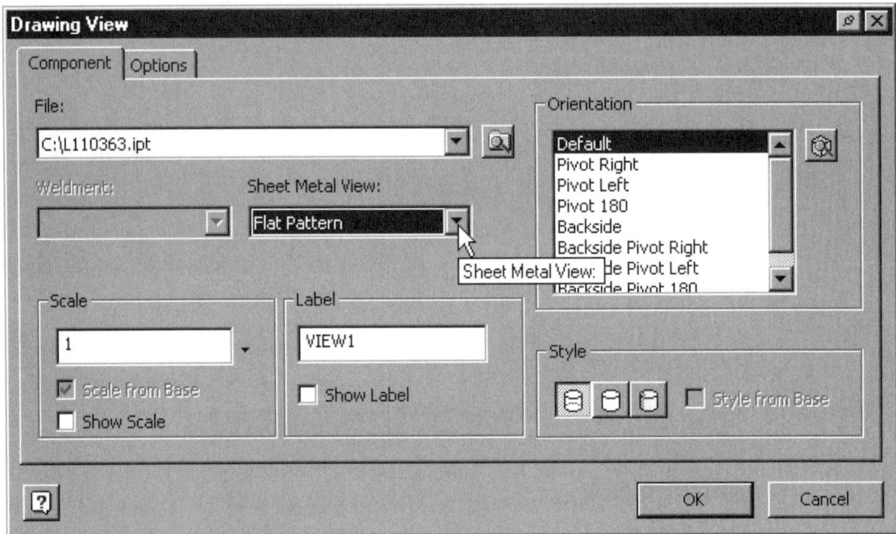


Figure 11-71

## COMMON TOOLS

A number of tools on the Sheet Metal Features Panel Bar are common to sheet metal and standard parts. Following is a short description of how to use these tools in the sheet metal environment.

**Work Features** Work planes are often used to define sketch planes for disjointed faces.

**Holes** The hole feature is common to both the part modeling and sheet metal environments. You can enter sheet metal parameters such as Thickness in numeric fields such as hole depth.

**Catalog Tools** iFeatures can be saved and placed in the sheet metal environment. See the PunchTool section for information on special iFeatures for sheet metal.

**Mirror and Feature Patterns** Sheet metal features can be mirrored or patterned like any other modeled feature. Holes, cuts, iFeatures, and features created with the standard part feature tools can also be patterned or mirrored.

**Promote** The Promote tool allows IGES files to be imported in the sheet metal environment.

**Derived Component** The Derived Component tool can be used to reference another part or assembly (sheet metal or not). Parameters from other files can also be referenced and used in a sheet metal part using the Derived Component tool.

**Parameters** The Parameters tool gives you access to model parameters and equations. You can add user parameters, link to an Excel spreadsheet, or reference exported parameters from other models. These are the sheet metal specific parameters that are added to the parameters dialog box when the sheet metal environment is activated: Thickness, BendRadius, BendReliefWidth, BendReliefDepth, CornerReliefSize, MinimumRemnant, and TransitionRadius.

**Create iMate** iMates can be applied to sheet metal components prior to placement in an assembly.

## DETAILING SHEET METAL DESIGNS

Drawing views of both the 3D model and flat pattern can be created when detailing a sheet metal part. The flat pattern drawing view enables all bend locations, bend and corner reliefs, and cutouts to be located and sized with dimensions. The 3D model views describe the folded shape of the part and allow the forming tool operator to validate the folded part.

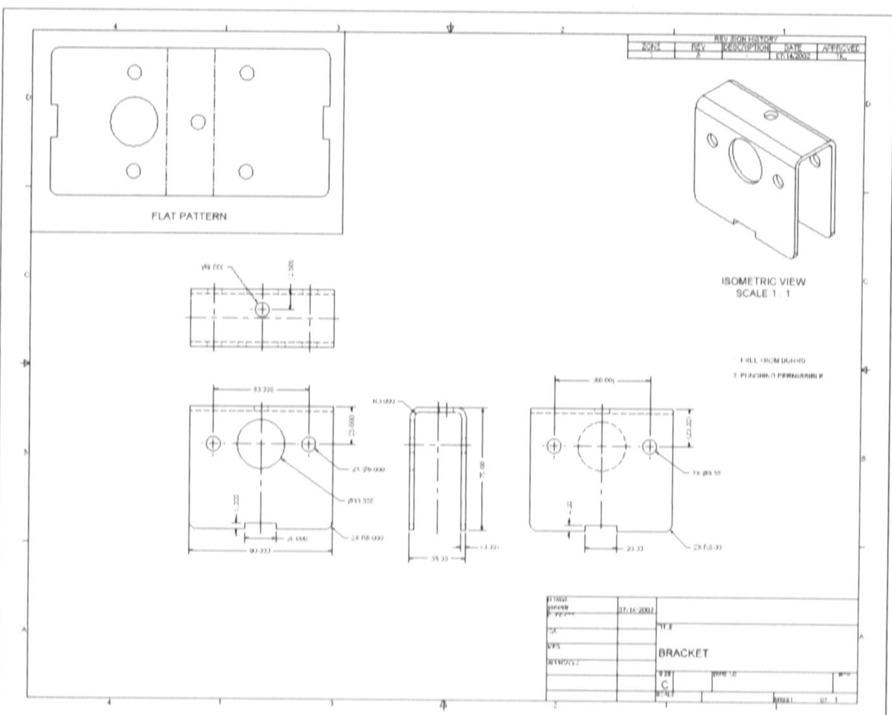


**Figure 11-72**

# EXERCISE 11-7
# Documenting Sheet Metal Designs

In this exercise, you create a flat pattern model of a sheet metal part and observe the live nature of the flat pattern as features are added or modified. You obtain the flat pattern extents and create model and flat pattern drawing views of the sheet metal part. To navigate to the exercise in the *Electronic Student Workbook*, do the following:

1 From the Main TOC page, click Chapter 11.

2 From the TOC for Chapter 11, click Documenting Sheet Metal Designs.

The following image illustrates the opened exercise.

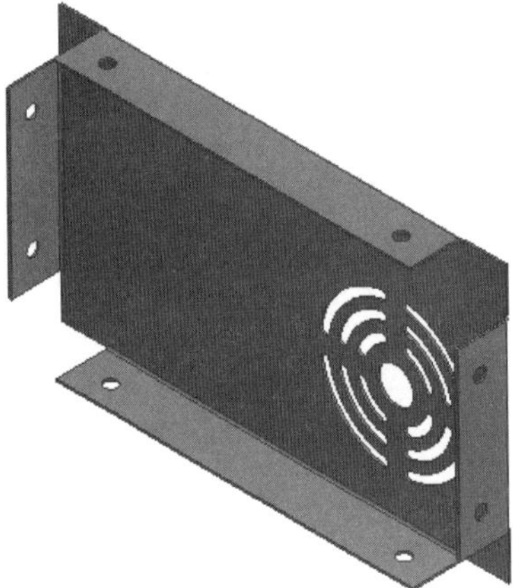

**Figure 11-73 Opened exercise**

The following image illustrates the completed exercise.

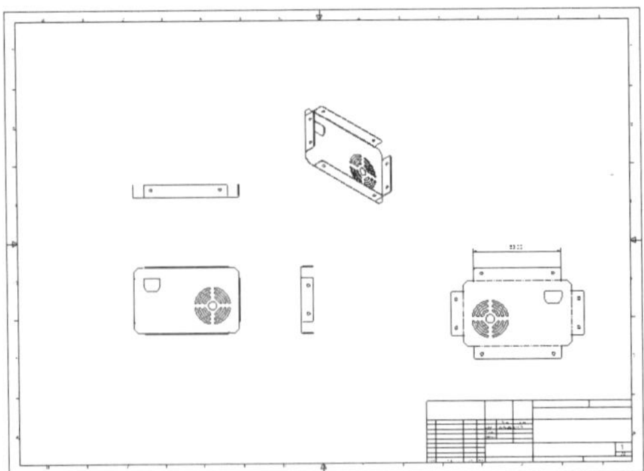

**Figure 11-74 Completed exercise**

## CHAPTER SUMMARY

| To | Do This | Tool |
|---|---|---|
| Create a sheet metal part | Select Sheet Metal from the Applications menu or use the *Sheet Metal.ipt* template file. | |
| Edit sheet metal styles | Click the Styles tool in the Panel Bar or from the Sheet Metal Features toolbar. |  |
| Create a sheet metal face | Click the Face tool in the Panel Bar or from the Sheet Metal Features toolbar. |  |
| Create a sheet metal flange | Click the Flange tool in the Panel Bar or from the Sheet Metal Features toolbar. |  |
| Create a sheet metal bend | Click the Bend tool in the Panel Bar or from the Sheet Metal Features toolbar. |  |
| Create a sheet metal corner seam | Click the Corner Seam tool in the Panel Bar or from the Sheet Metal Features toolbar. |  |
| Create a sheet metal flat pattern model | Click the Flat Pattern tool in the Panel Bar or from the Sheet Metal Features toolbar. |  |
| Create a flat pattern drawing view | Click the Base View tool in the Panel Bar, or from the Drawing Views toolbar. | 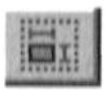 |
| Create a PunchTool | Select Extract iFeature from the Tools menu. | |

# APPLYING YOUR SKILLS

## Skill Exercise 11-1

In this exercise, you use sheet metal tools to create sheet metal and flat pattern models of a mounting bracket. To navigate to the exercise in the *Electronic Student Workbook*, do the following:

1 From the Main TOC page, click Chapter 11.

2 From the TOC for Chapter 11, click Exercise 1 under Applying Your Skills.

The following image illustrates the completed exercise.

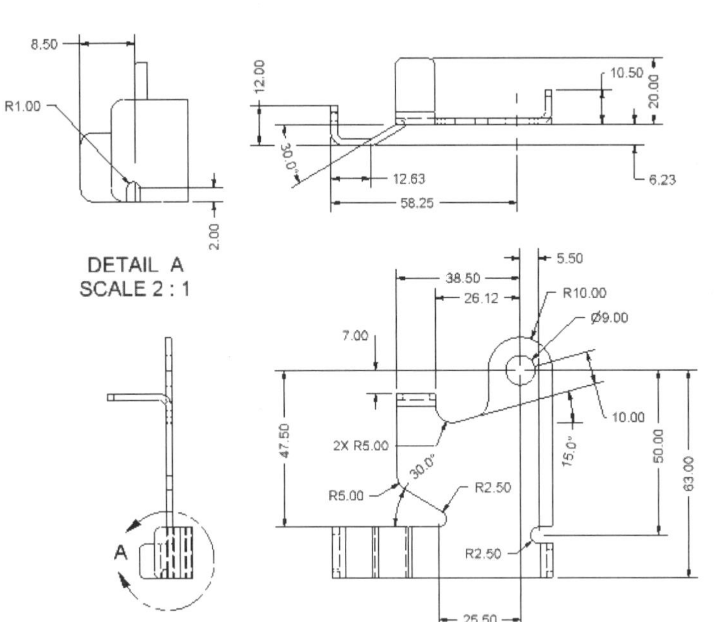


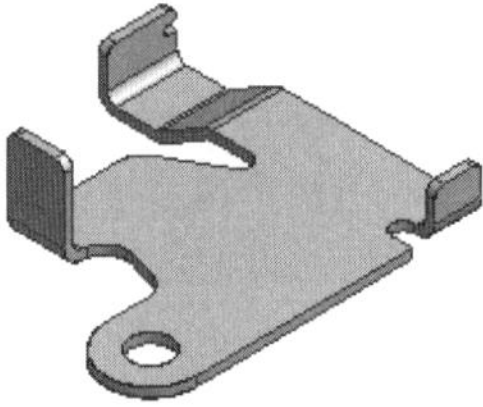

NOTES:

1. PART IS 2.00 mm. THICK.
2. ALL ROUNDS ARE 2.00 mm. UNLESS OTHERWISE SPECIFIED.
3. INSIDE BEND RADII 2.00 mm.

**Figure 11-75 Completed exercise**

## CHECKING YOUR SKILLS

Use these questions to test your knowledge of the material covered in this chapter.

1 The base feature of a sheet metal part is most often a:

a. Revolve

b. Face

c. Extrude

d. Flange

2 What is the procedure to change the edges connected by a bend feature?

a. Suppress the existing bend and add a new bend.

b. Delete the existing bend and add a new bend.

c. Edit the bend and select the new edges.

d. Create a corner seam between the desired faces.

3 Which tool would you use to create a full-length rectangular face off of an existing face edge?

a. Flange

b. Extrude

c. Face

d. Bend

4 What is required to update a flat pattern model?

a. Right-click on the flat pattern in the Browser and select Update.

b. Erase the existing flat pattern and recreate it.

c. Create a new flat pattern drawing view.

d. The flat pattern is updated automatically.

5 **True___ False___** Sheet metal parts can contain features created with Autodesk Inventor modeling tools.

6 **True___ False___** Sheet Metal Style settings cannot be overridden; a new Style must be created for different settings.

7 **True___ False___** During the creation of a sheet metal face, it can extend to meet another face and connect to it with a bend.

# Design Automation Techniques

This chapter introduces the importance of welding operations in an assembly model. Weldments can be applied between component parts using a number of welding operations. These operations, in addition to a number of basic welding operations, are explained in the beginning of this chapter. Component parts can be prepared before a weld bead is placed using the Chamfer, Extrude, and Hole tools. Once prepared, a fillet weld can be placed between parts through the Weld Feature dialog box. A cosmetic weld, which represents a simplified form of a weld, will also be discussed in this chapter. Once welds have been placed, machining operations can be performed on the welded assembly model. These machining operations include chamfering edges, extruding cuts, and drilling holes. Once the weldment process has been completed in the assembly, drawing views can be generated from that assembly.

You will also learn how to automate and reuse the intelligence that is captured while designing with Autodesk Inventor. You will build on the concepts learned in previous chapters by creating iParts and reusing model intelligence with iFeatures. You will also examine how derived parts and assemblies can be used to explore additional designs and work with specific parameters or reference data between models.

## CHAPTER OBJECTIVES

**After completing this chapter, you will be able to**

- Understand the importance of creating weldments in the context of an assembly model
- Prepare component parts for welding by chamfering edges
- Apply cosmetic and fillet welds between component parts
- Machine weldments using extrusion and drilling (hole) operations
- Convert an existing assembly to a weldment
- Document weldments in the drawing environment
- Add weldment annotations such as caterpillars, end treatments, and welding symbols to drawing views
- Create and place standard iParts
- Create and place custom iParts
- Create iFeatures
- Create table-driven iFeatures
- Use derived parts to work with existing design criteria

## WELDMENTS

The weld bead is possibly the most common assembly feature. It is so common that it is generally not considered an assembly feature. A weldment typically consists of specific lengths of stock material and detailed models assembled into their correct locations using existing assembly techniques. Operations such as weld prep, adding cosmetic or fillet welds, and post-weld machining operations can be performed on the assembled work pieces. Once assembled, all parts, features, and weld information are saved to a single assembly file and called a weldment. The weldment can also be used as a subassembly in other upper-level assemblies.

The following image shows a spool hand truck that has been assembled using typical assembly constraints with the assistance of various work planes. As shown in the following image, the individual components that make up the spool hand truck are listed in the Browser, and the weldment environment contains the new Preparations, Welds, and Machining groups.

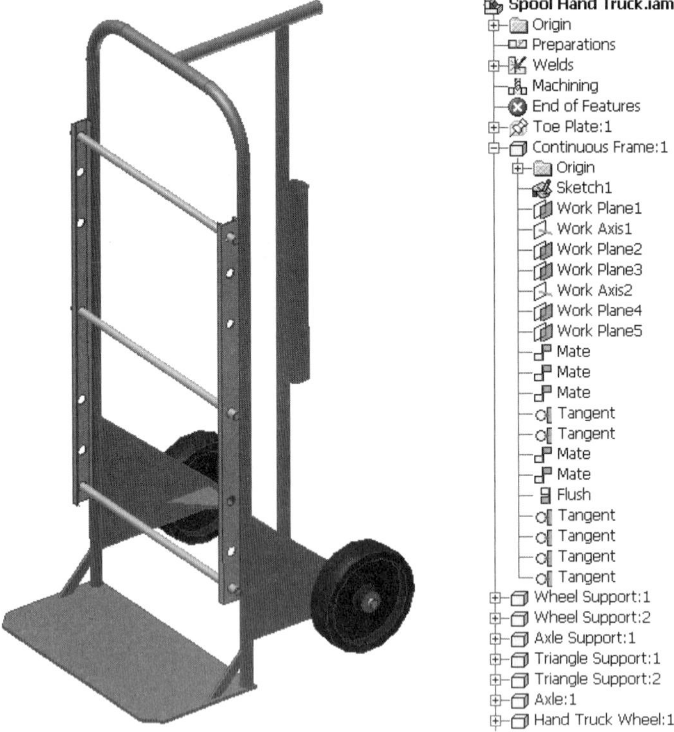


Figure 12-1

In the following image, welds are not present in the assembly example on the left. The parts are assembled correctly, but, without fastener information, it is difficult to determine how the parts are to be held together. The example on the right shows the presence of weld beads, which answers this assembly question.

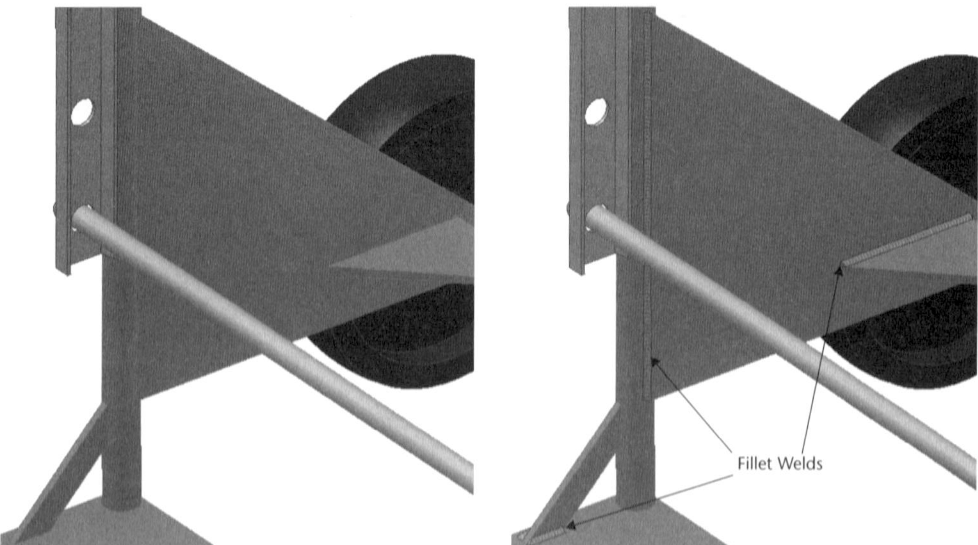


Figure 12-2

In the following image, the rectangular bar is fastened to the round bar stock using a method called *intermittent welds*. Here the weld is created using the width of the weld and the weld pitch (the center-to-center distance between welds).

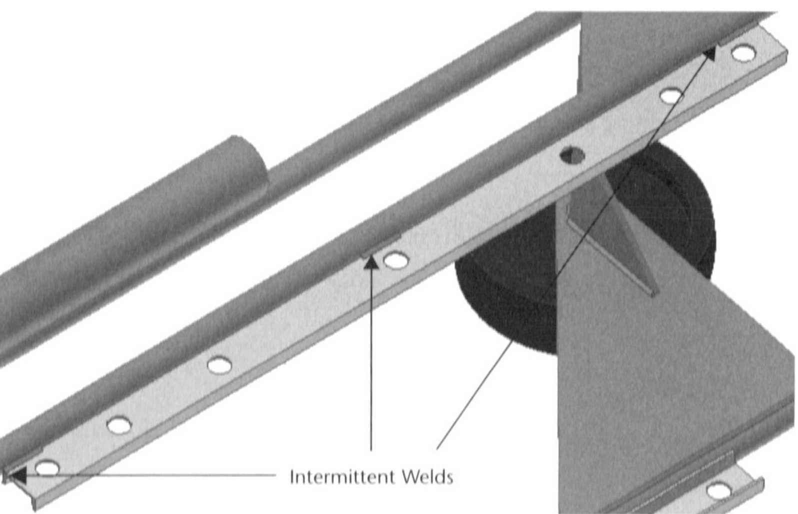


Figure 12-3

## CREATING A NEW WELDMENT

To create a weldment, click the New icon from the What To Do section of the Open dialog box and then click the desired tab. In the following image, the Metric tab is selected. You can choose from a number of weldment templates, including ANSI-mm, BSI, DIN, GB, ISO, and JIS. Notice that all weldment documents are identified by the IAM file type.

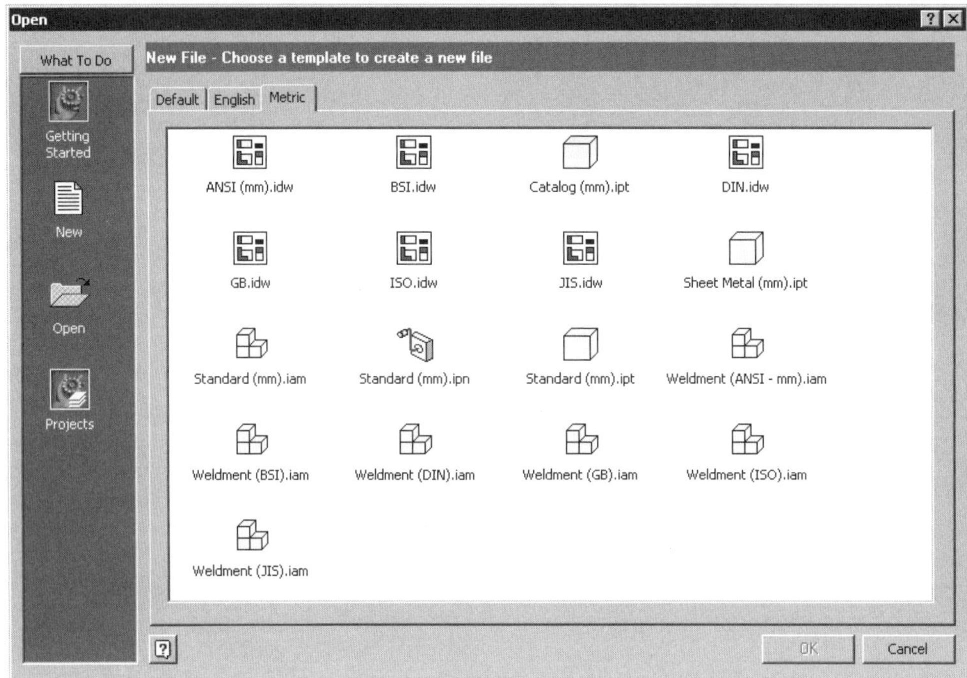


Figure 12-4

### CONVERTING AN EXISTING ASSEMBLY TO A WELDMENT

Any assembly can be easily converted to a weldment. To begin this process, open the existing assembly file and make sure that the assembly has been updated and is not in a rolled-back state. Then, select the Weldment command, which is located on the Applications menu as shown in the following image.

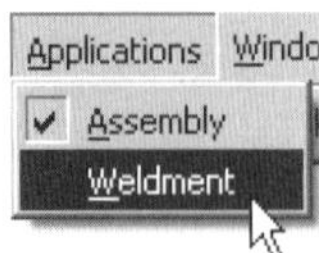


Figure 12-5

An alert box will appear, as shown in the following image, warning you that once the assembly has been converted to a weldment, you cannot convert it back to the assembly again. Clicking on the Yes button will continue with the process; clicking on the No button will cancel this operation.

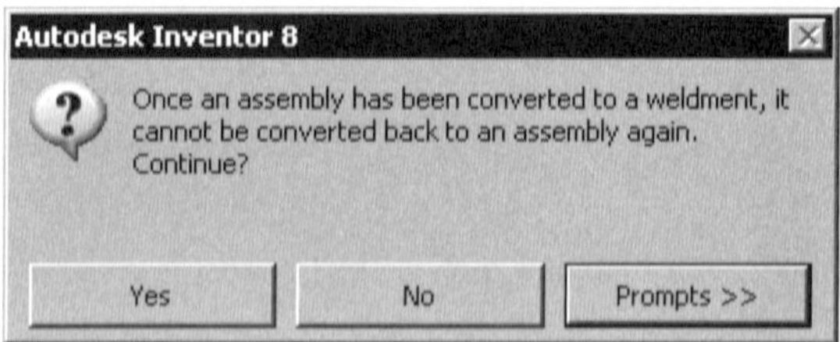


**Figure 12-6**

Clicking the Yes button in the previous alert box will activate the Convert to Weldment dialog box shown in the following image. Click the desired standard to which you want to convert the assembly. If the assembly contains features such as extrusions, holes, or chamfers, you must determine in the Feature Conversion area of the dialog box whether the features will be converted to the Weld preparations group or in the Machining features group. You can also specify the type of Weld Bead Material from the drop-down list. Clicking the OK button converts the assembly into a weldment.

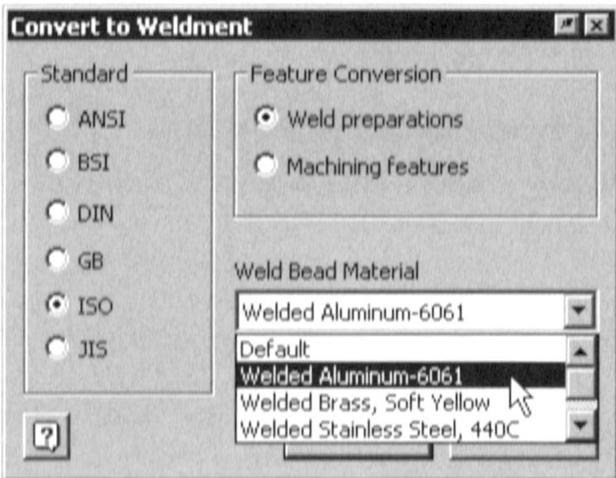


**Figure 12-7**

**NOTE** The Applications menu can be used to convert an assembly model to a weldment, but not vice versa.

## Weldment Standards

In order to create completed weld symbols in the weldment environment and associate this weld specification with either a weld bead or a set of edges, a weldment standard must first be established. Clicking on Tools in the Standard menu, followed by Document Settings, will display the Document Settings dialog box for the given weldment. Selecting the Weldment tab will display information similar to that shown in the following image. Six different predefined standards are available.

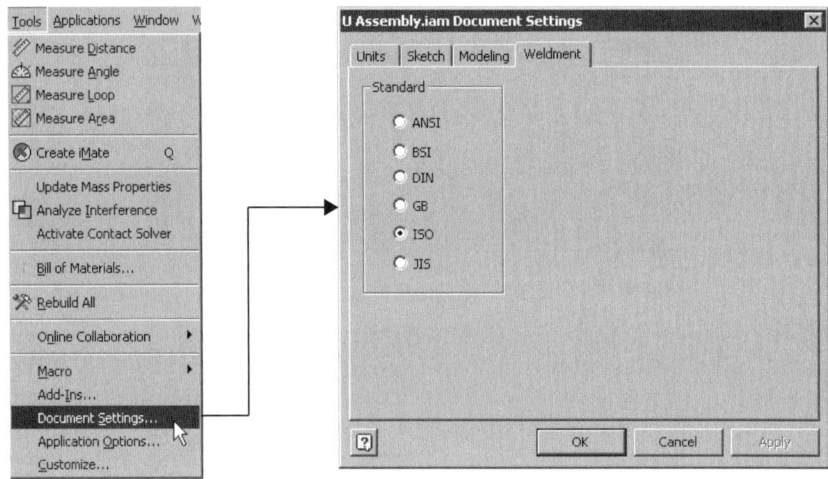


Figure 12-8

**NOTE** If you change the active weldment standard in any given document, preexisting features or welding symbols will not change to the new standard but will retain the standard with which they were created.

### The Weldment Browser and Panel Bar

Once you are inside the weldment environment, the Browser reflects a series of new icons that are operational only when you are producing weldments, as shown in the following image. The icons represent three unique groups referred to as the weld feature groups, one each for weld preparations, weld beads, and post-weld machining features. In order for you to add features, the weld feature groups must be in place and activated by double-clicking on the group name. Once a group has been activated, Autodesk Inventor will automatically enable the appropriate feature commands in the Panel Bar.

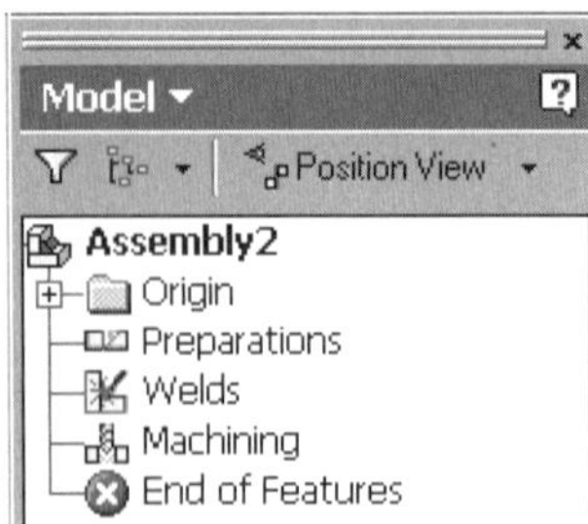


Figure 12-9

In the following image, various panel bars are present. The Panel Bar that appears depends on the type of weld group you have activated. For instance, when you first create a weldment, the Panel Bar is similar to the Assembly Panel Bar. You build a new weldment the same way you would build an assembly model. This means placing components and constraining these components to each other. When you double-click on the Preparations group in the Browser, the Weldment Assembly Panel Bar switches to Weldment Features. You prepare weldments by performing extrusion, hole, and chamfer operations. Notice that these same commands are available when you double-click on the Machining group in the Browser. Double-clicking on Welds in the Browser activates another Weldment Features Panel Bar. In addition to the Work Plane, Work Axis, and Work Point tools, the Weld command is activated. This tool is used for applying welds to component parts in a weldment assembly model.

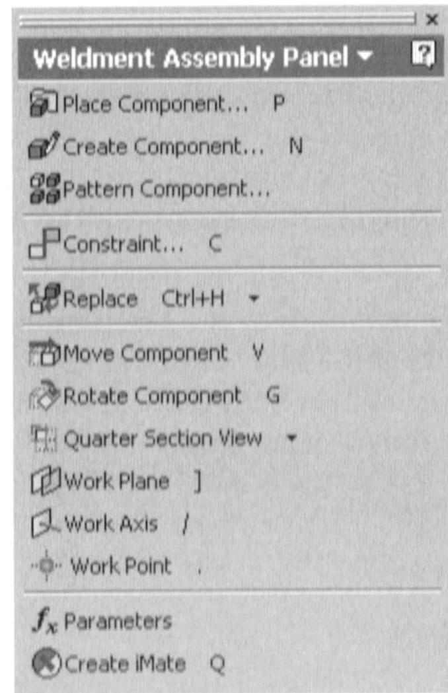


Assembly

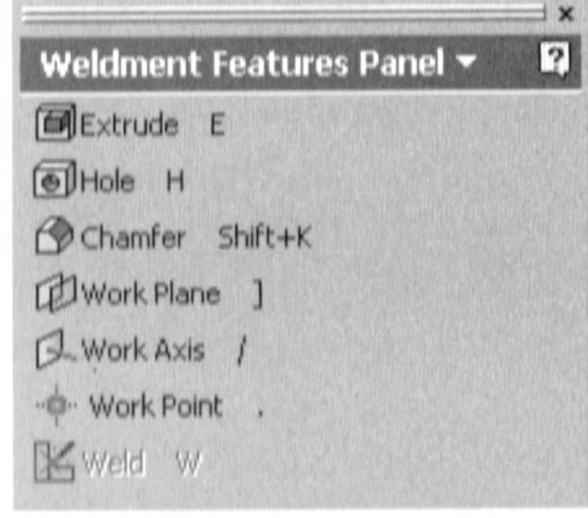


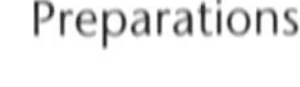

Preparations

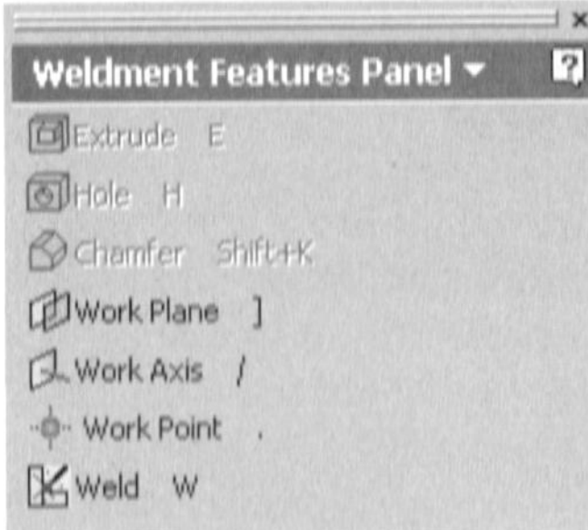


Welds

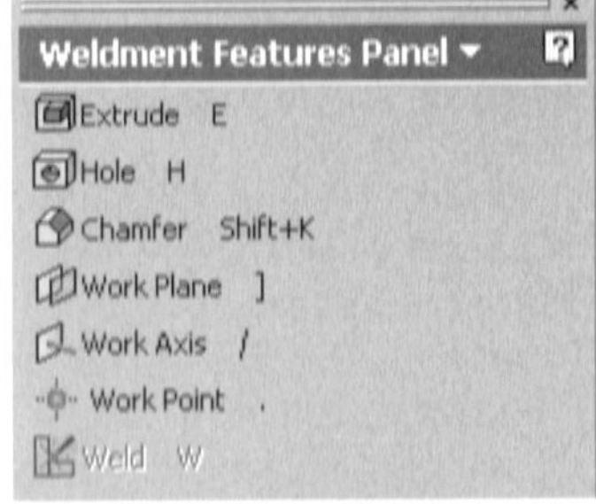


Machining

**Figure 12-10**

Additional support may be available through the Browser, depending on the weld group you are affecting. Right-clicking on the Welds group in the Browser will display the menu shown in the following image.

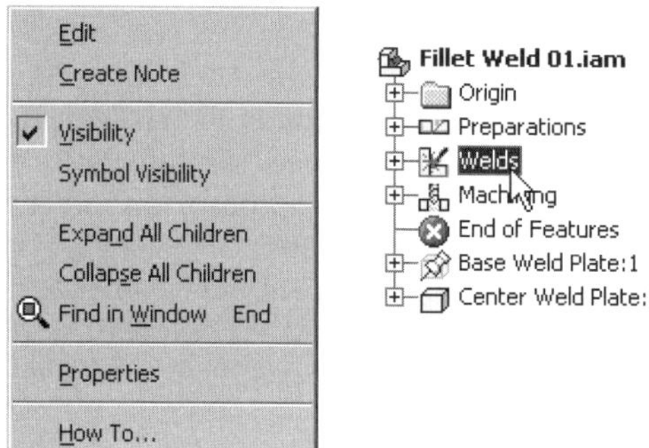


Figure 12-11

The Visibility checkbox controls the appearance of the actual weld. Removing the check from this box will turn off the weld bead. Symbol Visibility controls the appearance of a weld symbol automatically applied to a weld feature placed in a weldment. Removing the check from this box turns off the weld symbol display.

If you right-click on either the Preparations or the Machining group in the Browser, the menu displayed in the following image will appear. Clicking on Edit from the menu will allow you to make changes to any assembly features created under the Preparations or Machining group.

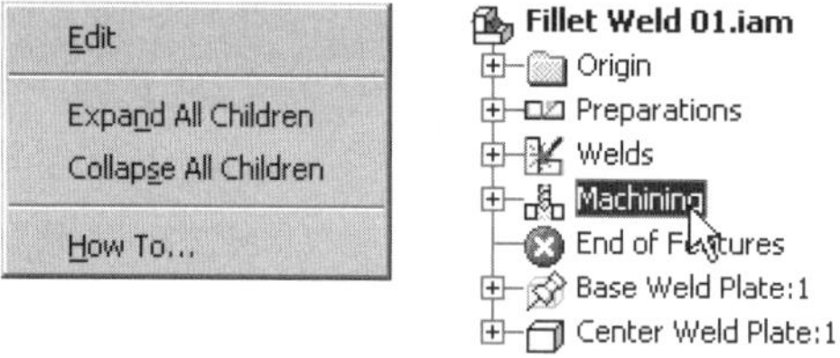


Figure 12-12

## WELDMENT CREATION OVERVIEW

As an overview of the complete weldment process, the following workflow illustrates how the Browser groups work together to prepare weldments, create a weld bead, and perform a post-machining operation. The details of applying the features will be discussed in later sections of this chapter. All Browser listings are displayed as Modeling views. You begin a weldment by creating a new weldment assembly. As with typical assembly modeling steps, you place a number of components and apply assembly constraints. Notice the appearance of the Weldment Assembly Panel Bar and the Weldment Assembly Browser in the following image.

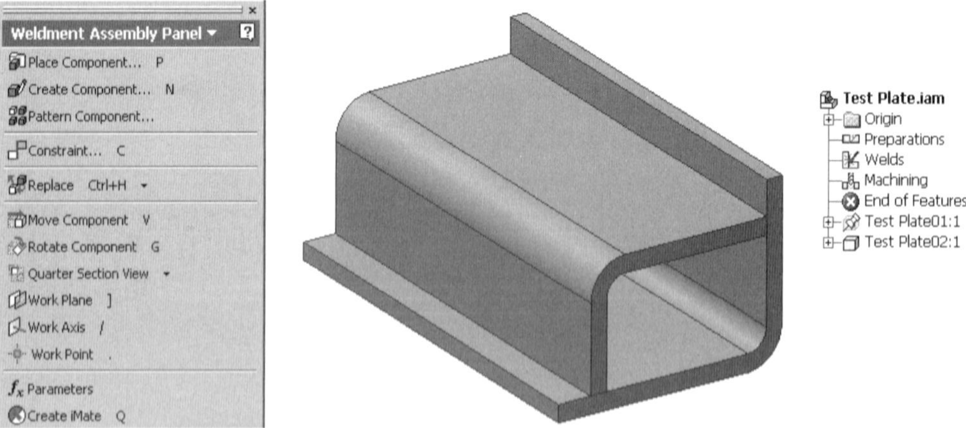


**Figure 12-13**

Double-clicking on Preparations in the Browser activates the Weldment Features Panel Bar, as shown in the following image. Only those assembly feature commands capable of creating welding preparation features are enabled in the Panel Bar.

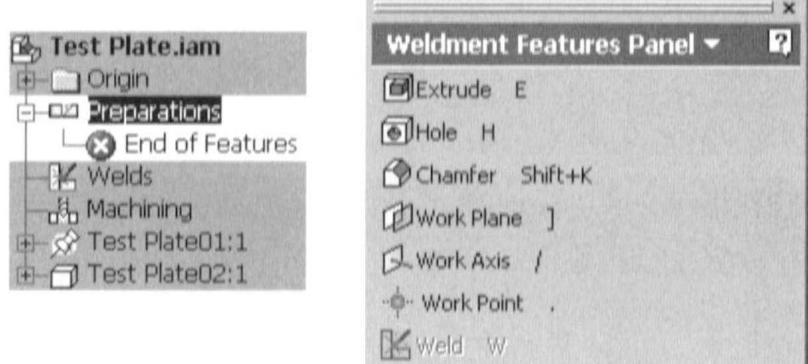


**Figure 12-14**

You need to create two beveled edges that will be used to accept the weld bead in a later operation. The Chamfer tool is used to create the two assembly features, as shown in the following image.

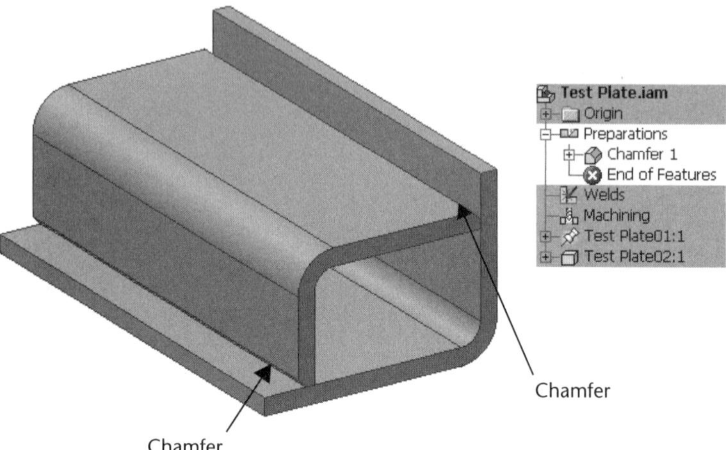


**Figure 12-15**

To place a fillet weld, double-click on the Welds group located in the Browser. This will activate the Weld Features Panel Bar, as shown in the following image. Notice that in addition to the Weld feature command being enabled, the three work features (Work Plane, Work Axis, and Work Point) are also enabled for you to use. Since there are no welds present in the model, it has not changed.

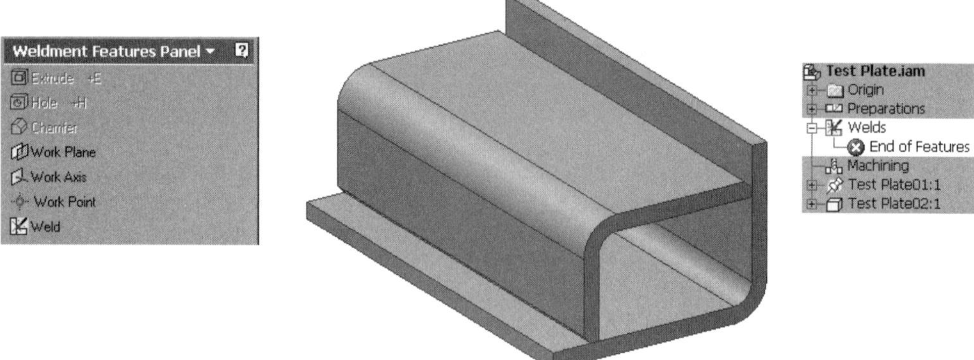


**Figure 12-16**

The Weld Feature dialog box is used to add a fillet weld to the object, as shown in the following image. This topic will be covered in greater detail later in this chapter. As the weld is placed, a weld symbol identifies the type and size of the weld. Also, the weld feature (in this case, Weld Bead 1) is added under the Welds group that is located in the Browser.

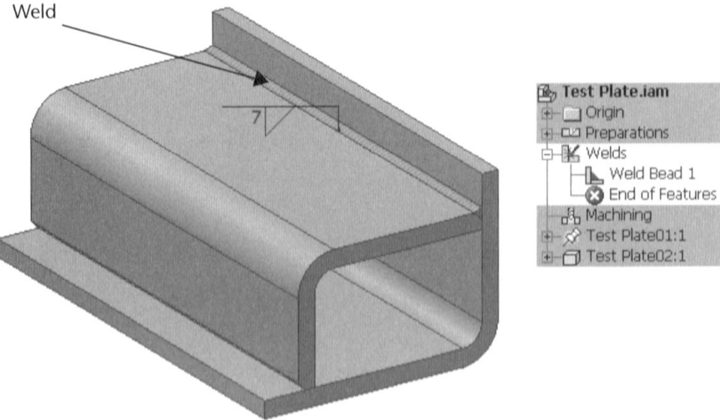


**Figure 12-17**

Once the weld has been placed, double-click on Machining in the Browser to activate this group if you need to perform a post-weld machining operation. Notice in the following image how the Panel Bar changes to display the three assembly features (Extrude, Hole, and Chamfer) and work feature commands (Work Plane, Work Axis, and Work Point).

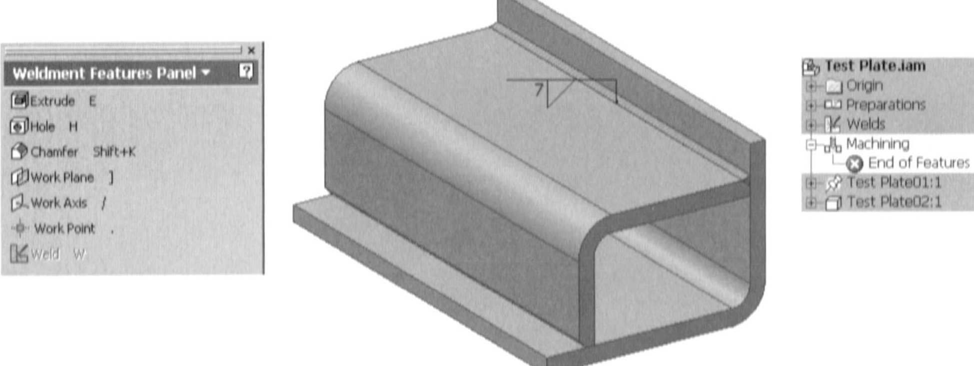


**Figure 12-18**

Once the Machining group is active, typical sketch operations are performed on the object. In the following example, a sketch has been placed on the top of the object. Edges are projected and a Center point is placed and dimensioned. The sketch is also displayed under the Machining group in the Browser.

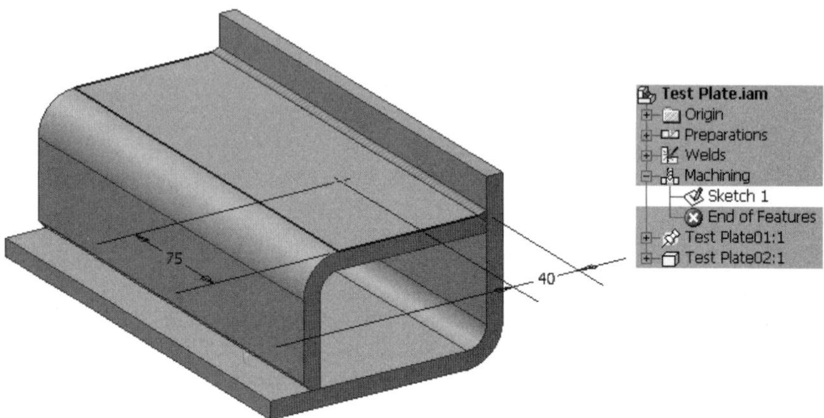


Figure 12-19

After the sketch has been completed, a hole feature is drilled through the top and bottom plates, as shown in the following image. Notice that the hole feature has been added under the Machining group.

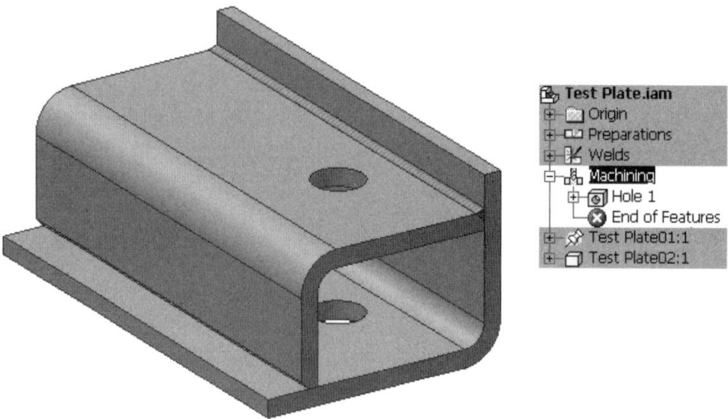


Figure 12-20

With the hole already created, you may elect to create an extrusion through the top and bottom plates. Create a new sketch on the top surface, construct and dimension a rectangle, and cut the extrusion through the assembly. This new feature is also added under the Machining group.

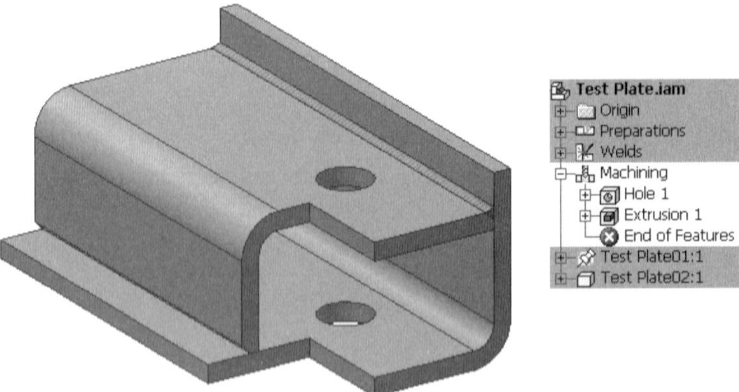


Figure 12-21

In a previous example, a single weld was placed. You now want to add a second weld to the other area prepared with a chamfer. Double-clicking on Welds in the Browser will roll the model back to the last weld bead, as shown in the following image. Notice that the two machining features also roll back; the weld preparations, however, remain unaffected.

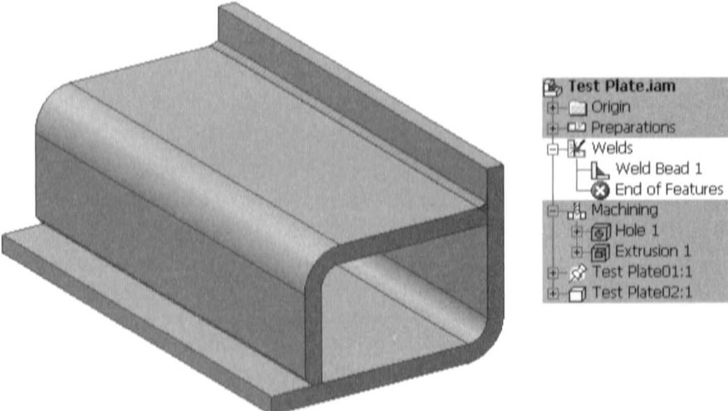


Figure 12-22

A second fillet weld is added to the chamfered area, as shown in the following image. A second weld symbol is also added to this weld. This new fillet weld is added under the Welds group located in the Browser.

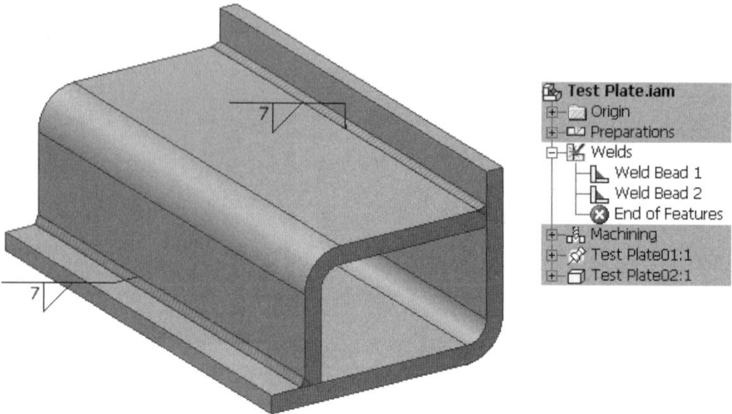


Figure 12-23

Double-clicking on the top-level weldment located in the Browser (*Test Plate.iam*) exits the weld group and displays the model, as shown in the following image. Notice the reappearance of the Weldment Assembly Panel Bar. Even though the second weld bead was placed along the whole edge of the plate, the model rolls forward to complete all features. This causes the second weld bead to be cut by the Extrude1 machining feature.

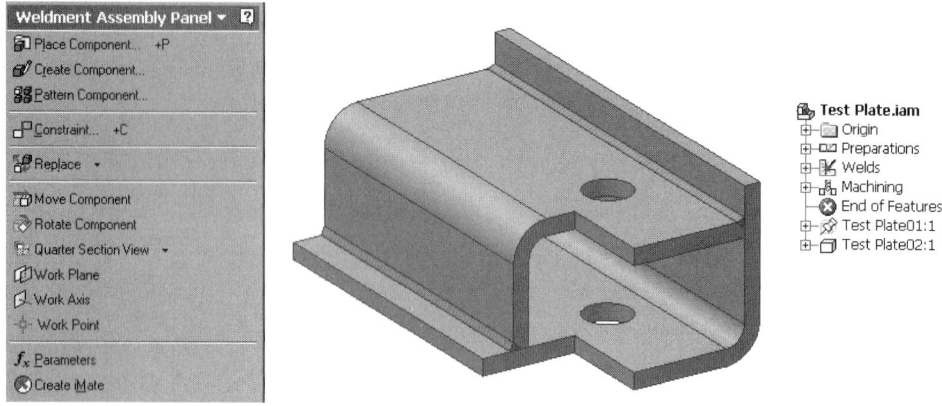


Figure 12-24

## PREPARING WELDMENTS

You prepare work pieces for welding by removing material along edges where the welds are to be applied. The amount of material removed and the shape of the edge after removal can be very important to some designers, since weld strength is affected by the preparation. This step is known as weld prep.

To create preparation features in the weldment, activate the Preparations group and add the needed assembly features. If weld bead and post-weld machining features exist when you activate the Preparations group, the weldment assembly in the graphics window temporarily reverts back to the state that existed before weld and machining features were added.

You can add assembly features can be added to the Preparations group at any time, even after adding machining features. To prepare work pieces for weld preparations, follow these steps:

1 Double-click the Preparations feature group.
2 Create assembly features as needed.
3 To return to the weldment assembly, click the Return tool on the Standard Toolbar.
4 To activate a different weld feature group, double-click on the feature group.

The metal removed in the weld preparation process is typically filled back in by the weld bead.

### WELD PREPARATION BY EXTRUSION

Following is an example of preparing a weld with an extrusion. A rectangular slot is cut into the vertical plate. Notice that in the Browser, the Preparations group is active and contains the various features.

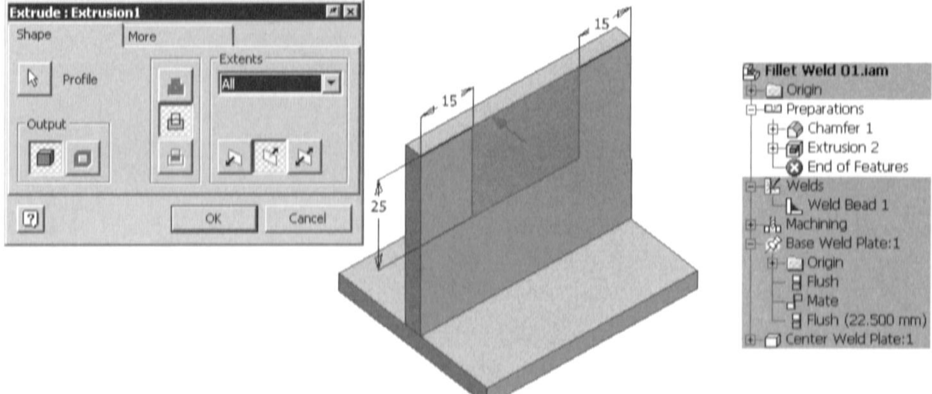


**Figure 12-25**

The result is illustrated in the following image. A U-shaped trough is assembled to the rectangular slot using typical assembly constraints. This trough will be welded to the vertical support member.

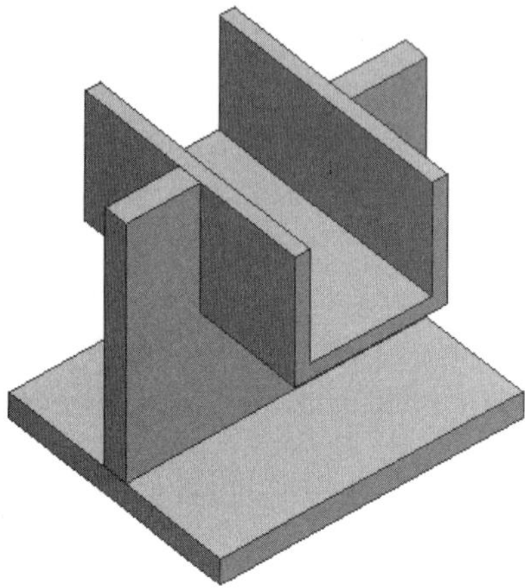

Figure 12-26

## WELD PREPARATION BY HOLE

The following image is an exmaple of a hole assembly feature is to be created in a rectangular plate. A series of four holes are to be made.

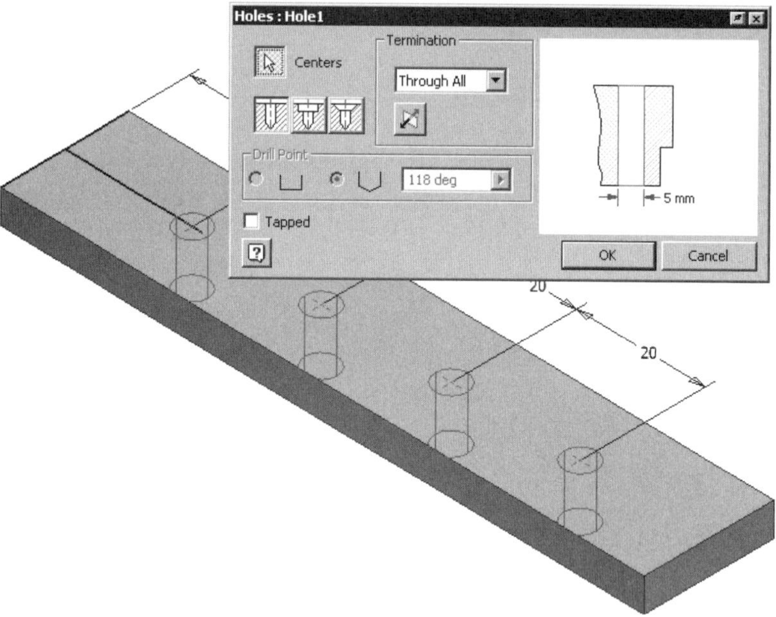


Figure 12-27

The results are shown in the following image. A series of posts were assembled and constrained to the rectangular base, and a fillet weld was applied to all four posts.

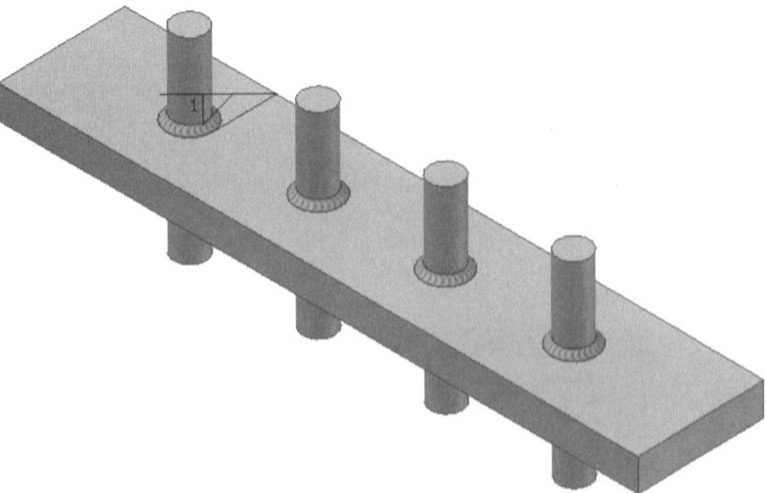

Figure 12-28

## WELD PREPARATION BY CHAMFER

Chamfer operations are the most popular preparation methods that can be performed on items that will be welded together. On the left in the following image, a single chamfer is placed at the base of the vertical plate. This configuration is called a Single Bevel T groove, and it will be welded using a fillet weld. The result is illustrated on the right in the following image.

Figure 12-29

## CREATING WELDS

The Weld Feature dialog box is the centerpiece of the weldment environment, and you activate it by clicking on the Weld button shown in the following image.

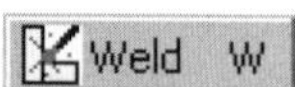


Figure 12-30

This action activates the Weld Feature dialog box, which is shown in the following image. Components of this dialog box are divided into three areas: Type, Orientation, and Main. Each of these areas is described as follows.

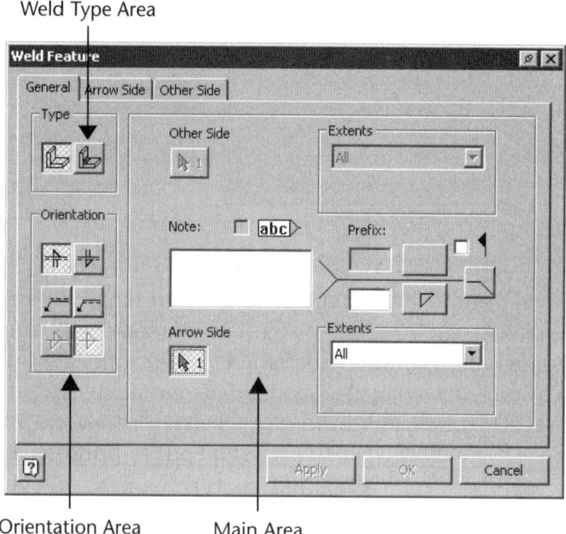


Figure 12-31

### Type Area

Choose the type of weld to create from this area. Click the button shown in the following image to set the desired weld feature type. You have a choice between a cosmetic weld and a fillet weld.

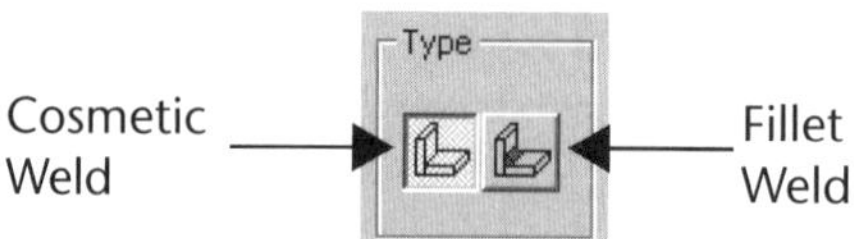


Figure 12-32

The Cosmetic Weld button creates a weld feature that annotates edges in the weldment but does not add weld bead geometry. The Fillet Weld button creates a fillet weld feature with 3D geometry that represents a weld bead.

## Orientation Area

Choose the orientation for the components of the weld symbol from this area. Click the button to set the desired orientation for each component, as shown in the following image.

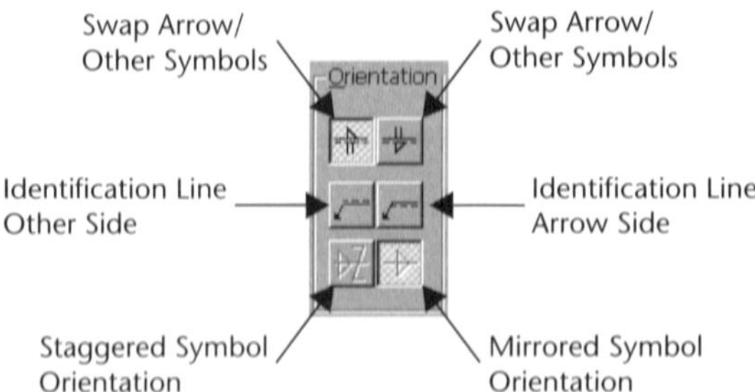


**Figure 12-33**

An explanation of each orientation tool follows:

**Swap Arrow Side** These two buttons reverse the Arrow Side and Other Side for the selected reference line on cosmetic weld features. On 3D weld features, it reverses Arrow Side and Other Side for the selected reference line and moves the symbol to the Other Side on the model.

**Identification Line** These buttons are available for the ISO and DIN standards only. These buttons set the location of the identification line to the Arrow Side or the Other Side for the selected reference line.

**Symbol Orientation** These buttons set the location of the Arrow Side and Other Side symbols on the reference line as either staggered or mirrored.

## Main Area

This area allows you to select model geometry and choose the type of weld for the Arrow Side and Other Side. This area contains other formatting options, as shown in the following image.

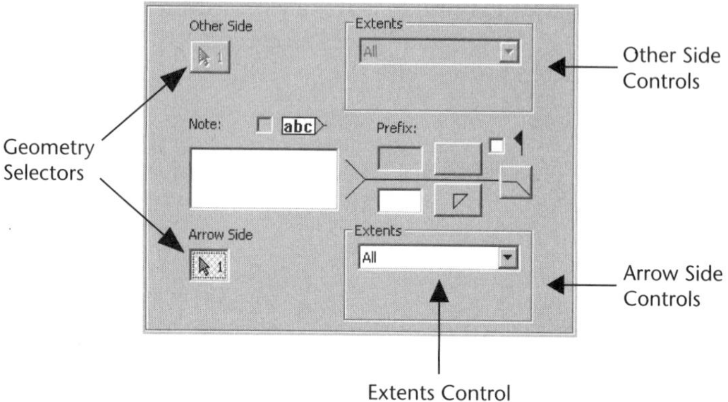


**Figure 12-34**

The selectors and controls shown are defined as follows:

**Geometry Selectors** These buttons are used to select faces on a weldment when placing a weld bead. The Other Side set of selectors will activate only if a weld symbol is selected first.

**Weld Controls** This tool allows you to choose different weld symbols. The default symbol is the fillet weld. You can click on a button to designate a *weld all-around* operation or place a check next to the flag to perform a field weld. You can also add text, in the form of a welding note, to the selected reference line. Enter the text in the box provided.

**Extents Control** This determines the method for ending a weld feature. Weld features can terminate on a work plane or extend across all selected geometry for a full-length weld. Click the arrow to list the extent methods and then select one. The All option creates the weld on all features and sketches in the specified direction. The From-To option enables you to select beginning and ending faces or planes on which to terminate the weld feature. In a weldment, the faces or planes may be on other parts but must be parallel.

## WELD TYPES

The Weld Feature dialog box can create two kinds of welds—Cosmetic and Fillet. Cosmetic welds, as shown in the following image, annotate edges in the weldment but do not create weld bead geometry. Cosmetic welds will have a lightweight appearance. Essentially, the selected edges will have an altered color and be slightly thicker than a normal edge. The example shown in the following image illustrates a U-Groove weld and a width of 4 mm applied to the work pieces.

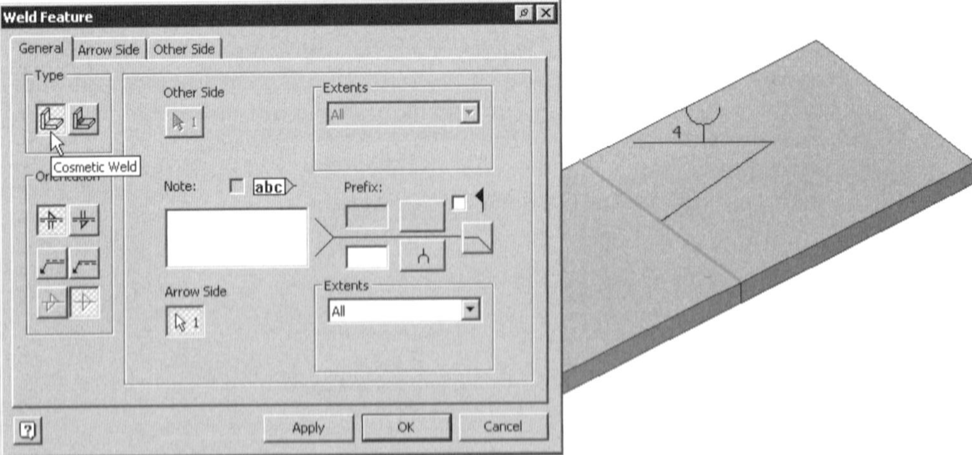


**Figure 12-35**

Fillet welds create actual 3D geometry that represents the weld bead, as shown in the following image. The appearance of 3D welds will be defined by their geometry. A caterpillar texture map will be automatically applied to the faces of a weld, with the exception of the end caps.

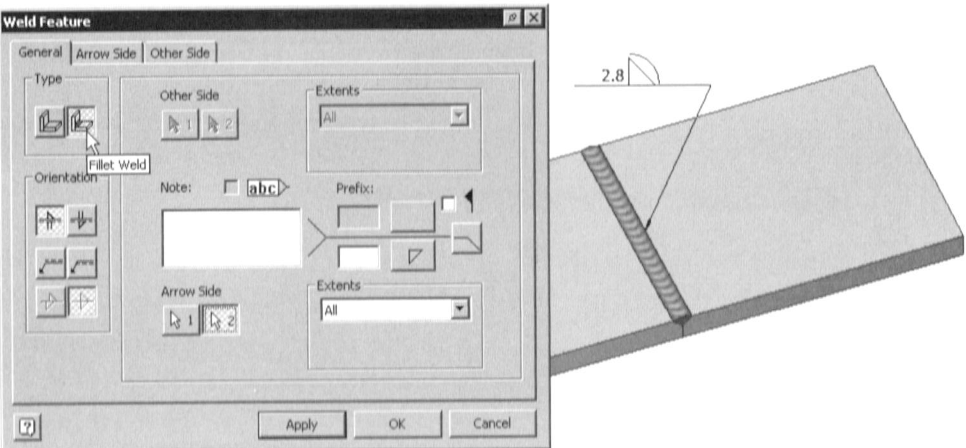


**Figure 12-36**

**NOTE** Both cosmetic welds and weld beads contain and display a welding symbol, which can be recovered in drawing views.

## Weld Feature Type

As stated above, the Weld Feature dialog box can create two types of features—a fillet weld, which captures a weld specification and uses that information to create a 3D representation of the weld bead; and a welding symbol (or a cosmetic weld), which captures the weld specification but only creates a welding symbol. Cosmetic welds change the appearance of the edges to indicate they have been welded, but no weld bead geometry is created.

Each of these feature types has their own unique representation in the Browser, as shown in the following image.

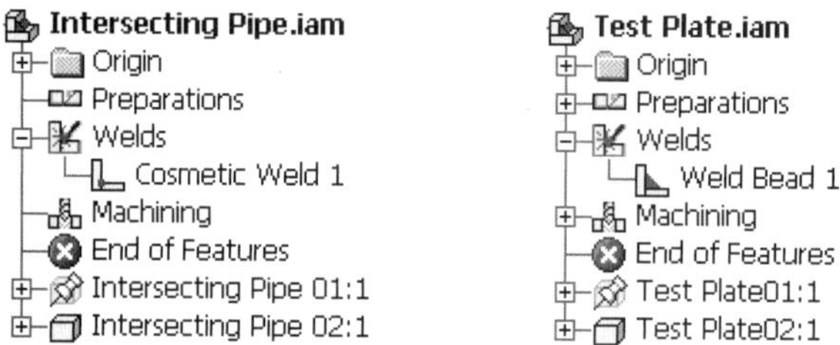


**Figure 12-37**

## Cosmetic Welds

When you create or edit a cosmetic weld, the Weld Feature dialog will appear, as shown in the following image. (The ISO standard is presently active.) The fillet weld symbol is displayed as the default symbol. To change to a different welding symbol, click on the Arrow Side icon to display all available weld symbols.

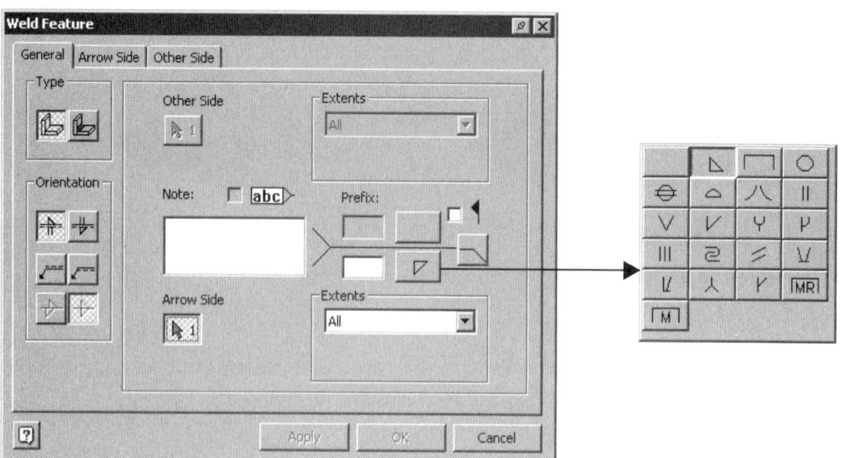


**Figure 12-38**

Note that the Extents controls are hidden, and you are offered a single Select button for each side. These Select buttons allow edges to be selected.

Clicking on the Arrow Side tab of the Weld Feature dialog box will display the following image. Use this tab to set the depth and size of the weld. In the following image, the depth is set to 4 mm and the size is left blank. When this blank field is detected, Autodesk Inventor will create an equal leg length weld.

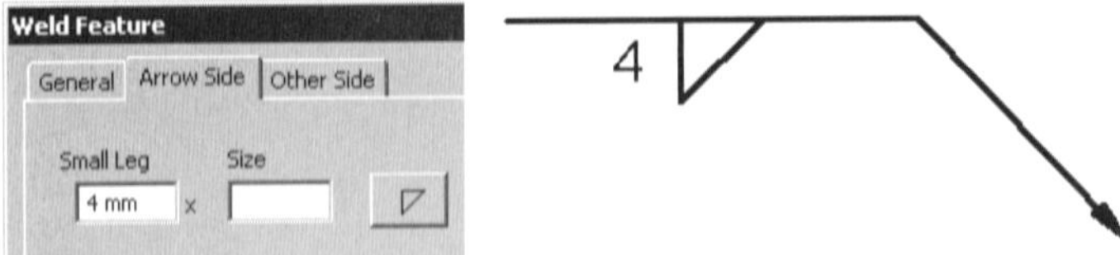


**Figure 12-39**

If you enter a value for the Small Leg (Leg1) and you enter a different value for Size (Leg2), the result will be a welding symbol created to specify an unequal length weld, as shown in the following image. Note that in the welding symbol, the *x* character separates the two different leg sizes. Also, the units of measure are omitted from the welding symbol.

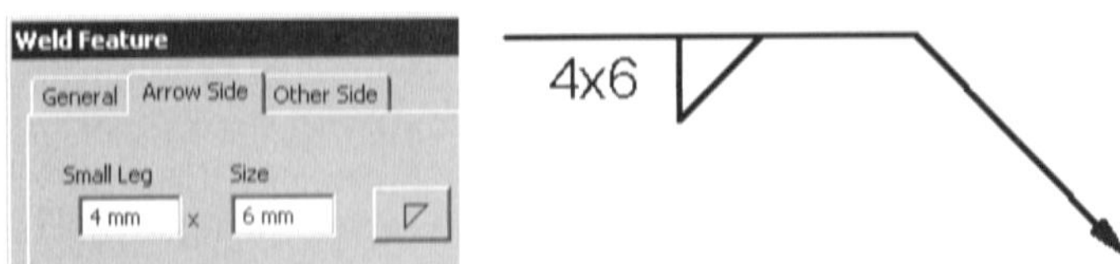


**Figure 12-40**

## Constructing Cosmetic Welds

Each cosmetic weld will be associated with one or more model edges. You have to create separate selection sets of edges when constructing cosmetic welds along the Arrow Side and Other Side. In the following image, the example on the left shows two contiguous edges selected as the first selection set for the first cosmetic weld. The example on the right represents a second selection set before the cosmetic weld is placed. Both examples illustrate the two cosmetic weld features needed to place the weld.

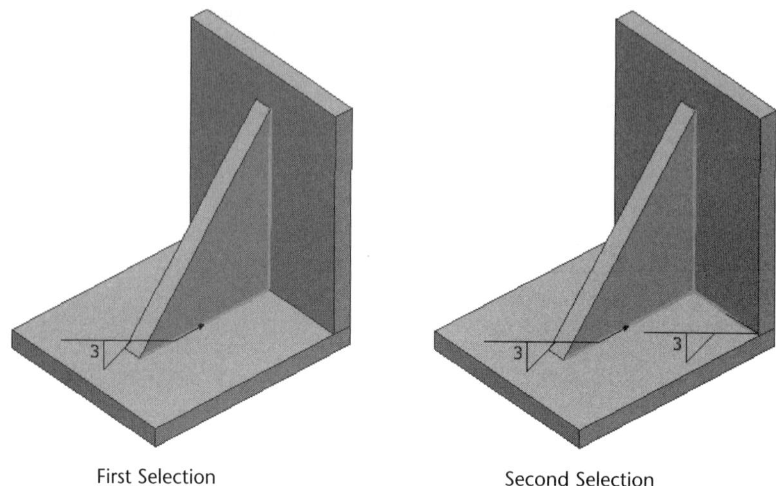


**Figure 12-41**

## Arrow Side versus Other Side

You must select at least one edge using the Arrow Side selection tool in order for the cosmetic weld to be valid. The user will be able to define a weld specification on the Other Side, however, even if no edges have been selected with the Other Side selection tool. This allows the user to create a typical weld specification, where the leader points to one side of the welded part but calls out the weld for both sides, as shown in the following image.

Other Side Selection Tool

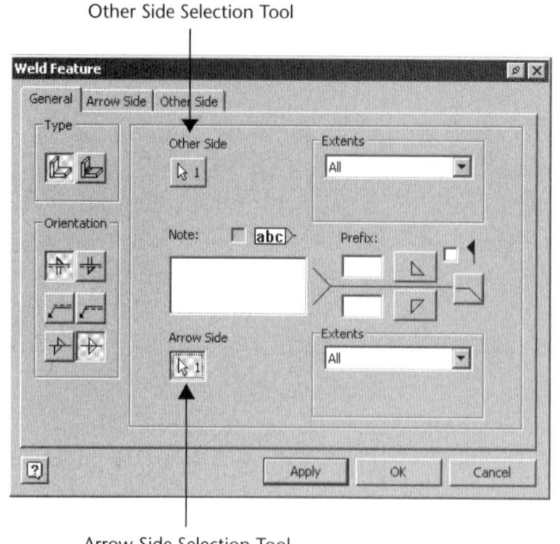


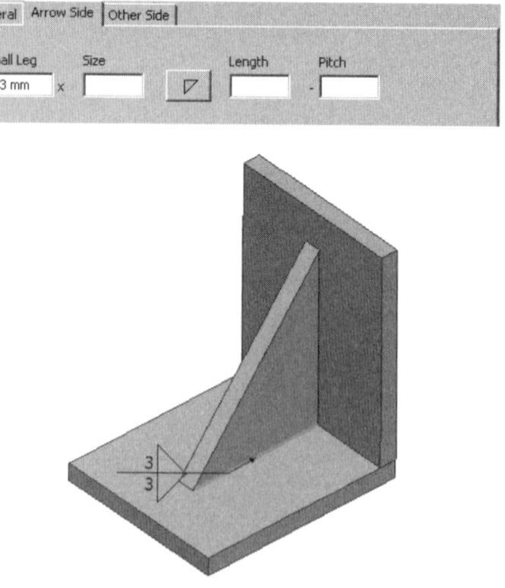


Arrow Side Selection Tool

**Figure 12-42**

## Cosmetic Weld Menu

Right-clicking on the Welds group in the Browser activates the menu, as shown in the following image. You can turn off all weld symbols by removing the check from the box next to Symbol Visibility. Removing the check from the box next to Symbol Visibility will turn off all welding symbols and the highlighted cosmetic weld edge.

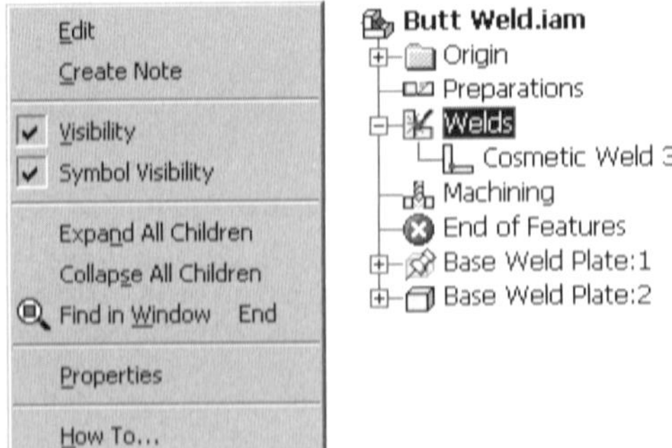


**Figure 12-43**

If you want to selectively control a single welding symbol, you can right-click on the welding symbol listed in the Browser under the Welds group, as shown in the following image. You can control whether the welding symbol is turned on or off through the Symbol Visibility listing in the menu.

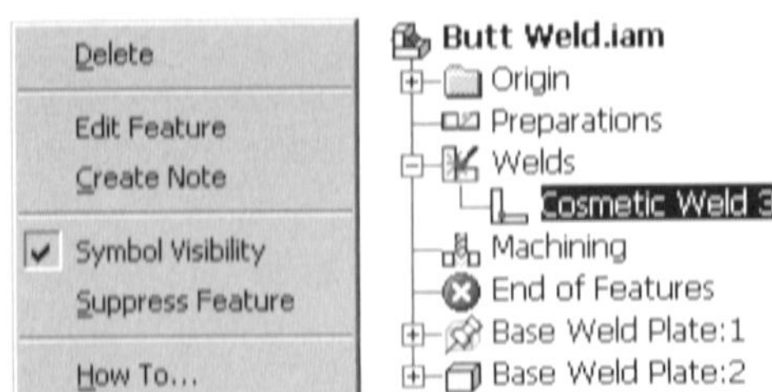


**Figure 12-44**

If you click on the Suppress Feature listing in the previous image, an Autodesk Inventor Alert dialog box will appear, as shown in the following image. This dialog box warns you that suppressing this assembly feature will cause any dependent constraints to also be suppressed, and it asks if you want to continue. Clicking Yes will turn off the welding symbol in addition to the cosmetic weld highlighted edge. The cosmetic weld listing in the Browser will also take on a gray appearance with a line struck through it, signifying its suppressed state. To unsuppress this weld feature, right-click on the cosmetic weld in the Browser and remove the check from the Suppress Feature listing in the menu.

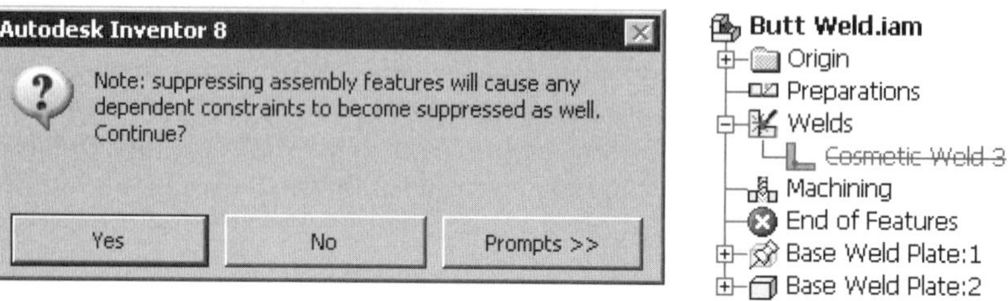


Figure 12-45

## FILLET WELDS

Use the solid Fillet Weld tool in the Weld Feature dialog box to add fillet weld bead features. Fillet welds build up corners by adding material between one or more faces of a single part or two parts. Weld beads contribute to mass property calculations and can take part in interference analyses.

When the Weld Feature dialog is set to Fillet Weld, the fillet weld is the only weld type that is available. The remaining weld types are not present unless you are creating a cosmetic weld bead.

### Creating Fillet Welds

Clicking on the Fillet Weld button located under the General tab of the Weld Feature dialog box activates two selection buttons when you place a weld on the Arrow Side. Click on the Arrow Side 1 button and select a face on the model to define the first leg of the weld. Then click on the Arrow Side 2 button and select a face to defile the second leg of the weld. You can use the **SHIFT** key to deselect unwanted faces. A previewed image of the weld will appear as shown in the following image. Notice in this example that the weld appears large in size.

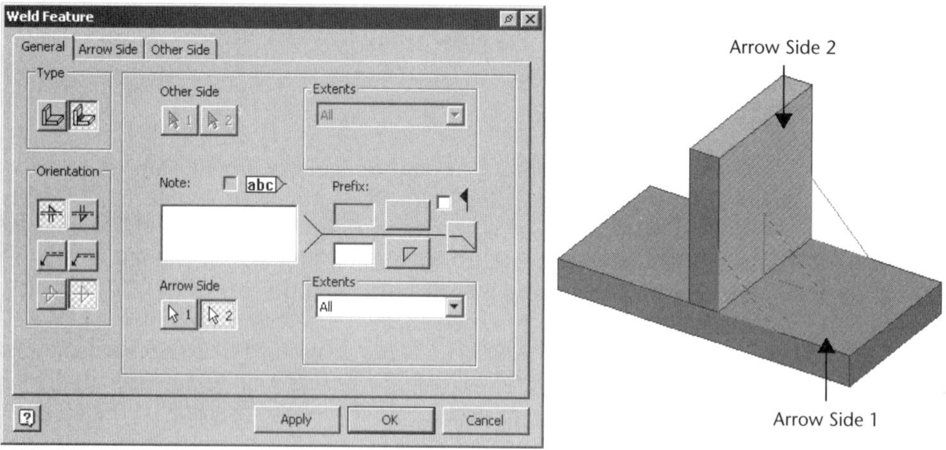


Figure 12-46

Pick the Arrow Side tab to change the depth and/or size of the weld. In the following image, the weld depth has been changed from its default value of 10 mm to a new value of 3 mm. The results are shown in the model; the previewed weld is updated to reflect the new depth value.

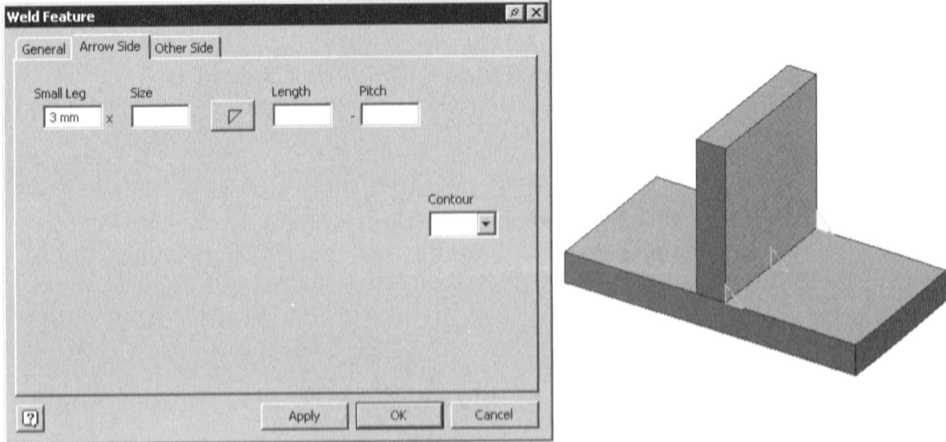


**Figure 12-47**

To have a weld placed on the Other Side, continue in the Weld Feature dialog box by clicking on the General tab and selecting the Fillet Weld symbol above the image of the reference line. This will activate the Other Side buttons. Select two faces on the model in the same way you selected the Arrow Side faces. Notice that the Other Side weld will take on the same value as the Arrow Side unless you are placing a weld with two different depth values.

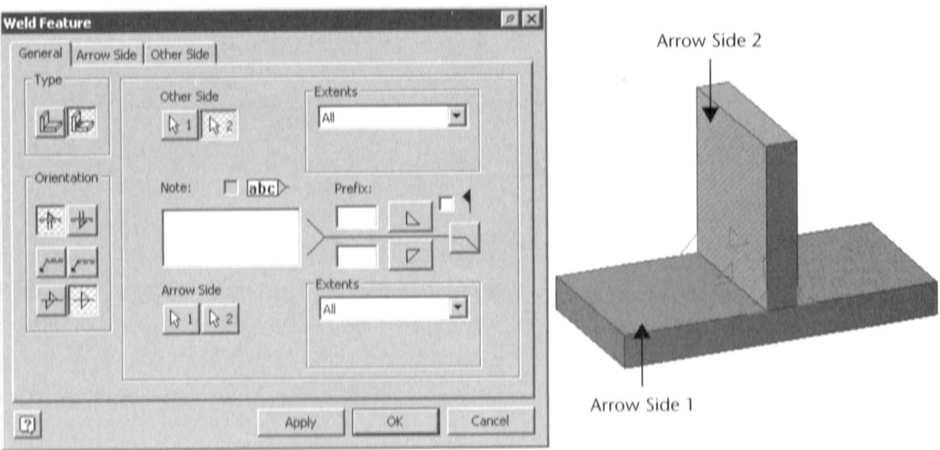


**Figure 12-48**

When both Arrow Side and Other Sides are selected, you can also elect to make changes to the following items:

- Attributes for the welding symbol.
- The Notes input box (to add notes to the welding symbol).
- Orientation for welding symbol components (click on Orientation tools to set the orientation for the welding symbol).
- The Weld Symbol tool (can select a new type if desired).
- The checkbox (select or deselect to include or exclude the flag [field weld] or weld all-around symbol).

Clicking the Apply button will create the Fillet weld, as shown in the following image, with the weld bead placed on both the Arrow and Other Sides. Notice also the appearance of the fillet welding symbol pointing to the Arrow Side and calling out a depth of 3 mm for both sides.

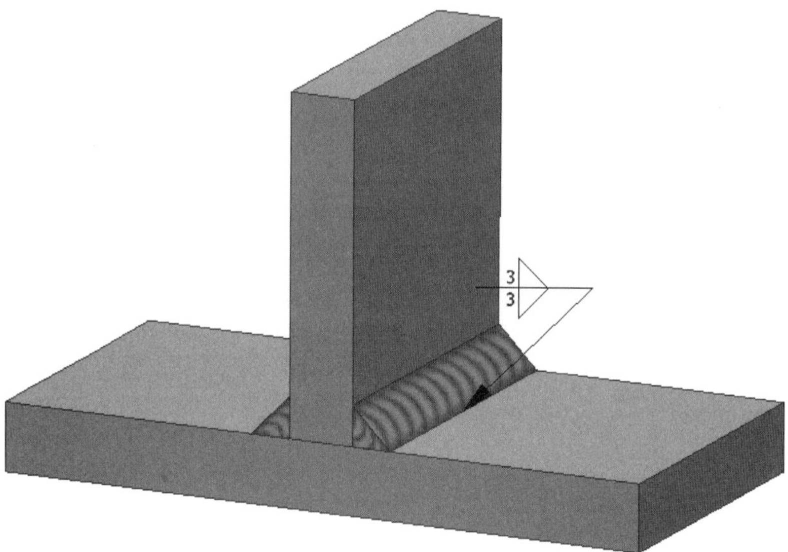


Figure 12-49

## Using the Weld From-To Option

By default, the extents of a Fillet weld is set to All, which means the bead will run the entire length based on the two Arrow Side surfaces selected. Use the From-To option to control the length of a weld bead. First select Arrow Side 1 and Arrow Side 2, as indicated in the following image. Change to the From-To option and select a face for the From option and another face for the To option. The results are illustrated on the right in the following image, with a 3 mm fillet weld placed from the edge of the object to the face of the middle triangular web support.

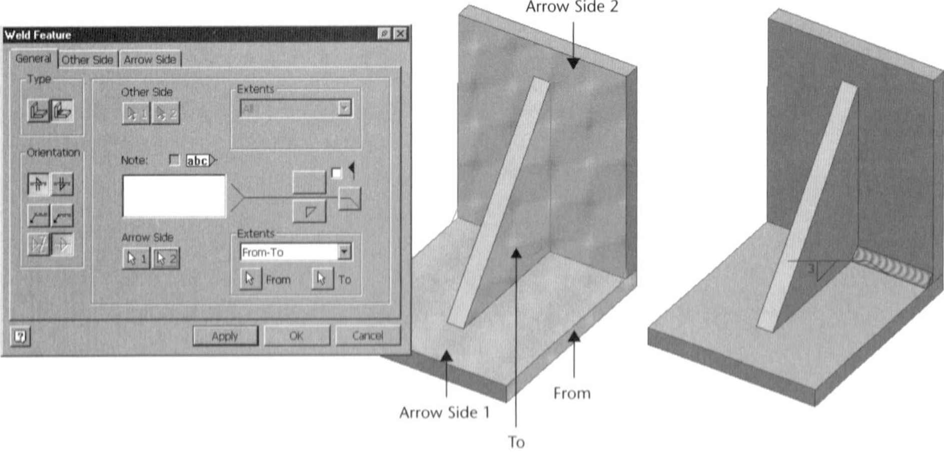


Figure 12-50

## Creating Groove Weld Beads

You can use a fillet weld to create a weld bead for groove applications. In the following image, a groove has been created by chamfering the ends of two plates. Both plates are assembled using traditional assembly constraints. You can use the Weld Feature dialog box to set the weld type to Fillet weld and then select the two chamfered faces of the groove. Set the Arrow Side weld size based on the length of the chamfered edge. The results are illustrated on the right in the following image. Although the weld bead is correct, the corresponding welding symbol specifies the fillet weld and is not correct.

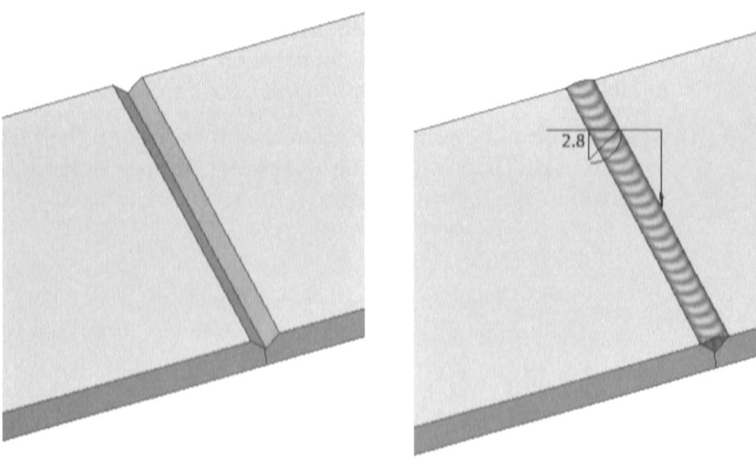


Figure 12-51 Figure 13-80

**NOTE** If the appearance of a fillet weld bead was not important, a cosmetic groove weld could have been applied to capture the true weld design intent in the model.

When documenting the weldment in drawing mode, a groove welding symbol can be manually placed in the drawing to call out the correct weld operation (as shown in the following image).

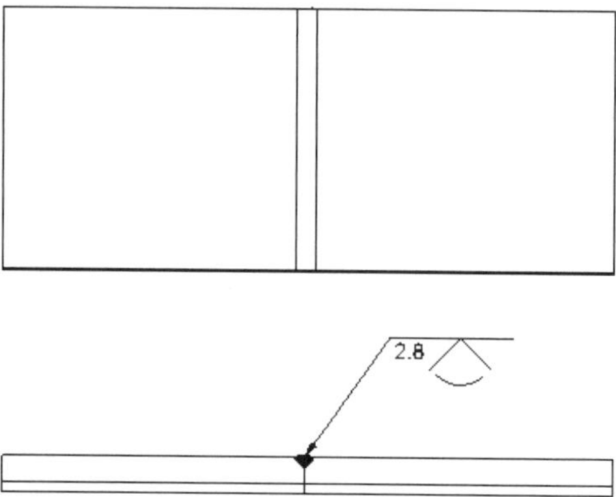


**Figure 12-52**

## Weld Feature Roll Back

There are times when you may need to inspect a weldment feature by feature. You can accomplish this by dragging the End of Features (EOF) symbol up the feature tree in the Browser to roll back the state of an individual weldment group. When in roll-back state, features under the EOF symbol are temporarily disabled and not visible in the model. When the EOF symbol is dragged back down the tree in the weldment group, the features are added back to the model. Use the following steps to roll back a weldment feature:

1 Open an assembly weldment file and double-click on the Machining group to roll it back. Then expand the Machining group to expose the EOF symbol, as shown in the following image.

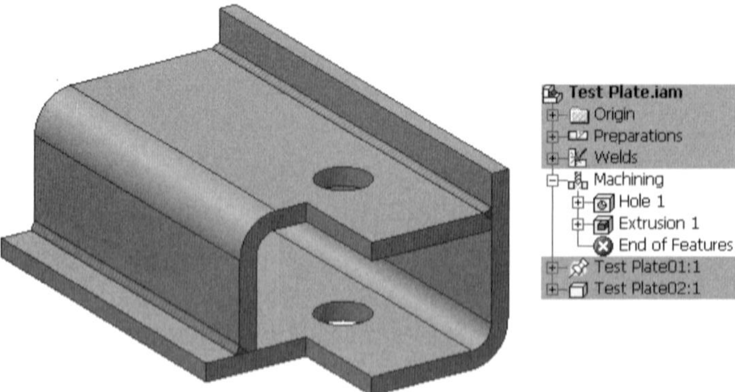


**Figure 12-53**

2 Press and drag the EOF symbol up the Browser tree to a location in that group and examine the results, as shown in the following image. Keep repeating this procedure for each feature in the group.

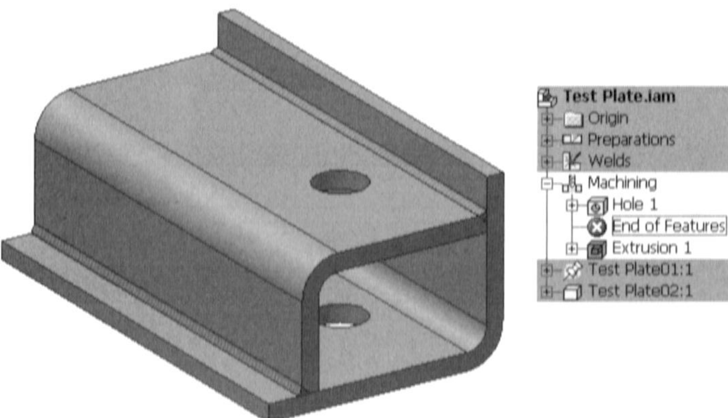


**Figure 12-54**

3 When you have finished examining the weldment, press and drag the EOF symbol back down the tree to restore the features in the weldment.

## Repositioning a Welding Symbol

Welding symbols are attached to a model edge by default. If the welding symbol would be better viewed in a location other than its default position, you can select the welding symbol and drag it to a new location. While being dragged, the symbol remains normal to the screen.

You can move a welding symbol by lengthening or shortening the leader segment of a symbol leader line, or by moving the attachment point or vertex. During the move, the valid edges in the associated weld feature are highlighted.

To begin the process of repositioning a welding symbol, click the arrow next to the Select tool on the Standard toolbar, and then click the Select features mode from the list, as shown in the following image.

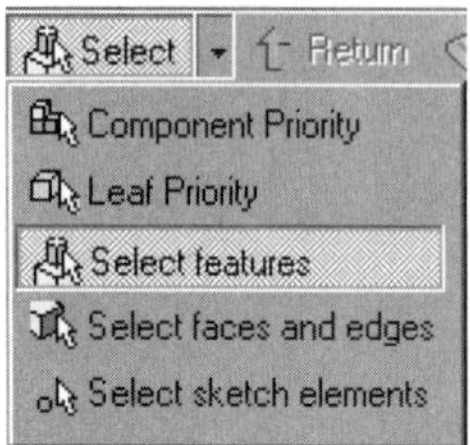


Figure 12-55

In the graphics window, move the cursor over the welding symbol or leader to view the grip points, as shown in the following image. At this stage, you can perform the following functions:

- To move the horizontal reference segment to a new location, click and drag the grip point closest to it.
- To change the attachment point, click and drag the base grip to reposition the leader.
- For a cosmetic weld, drag the grip point to any Arrow Side edge in the feature's selection set.
- For a fillet bead weld, drag the grip point to any Arrow Side edge generated by the fillet weld bead.

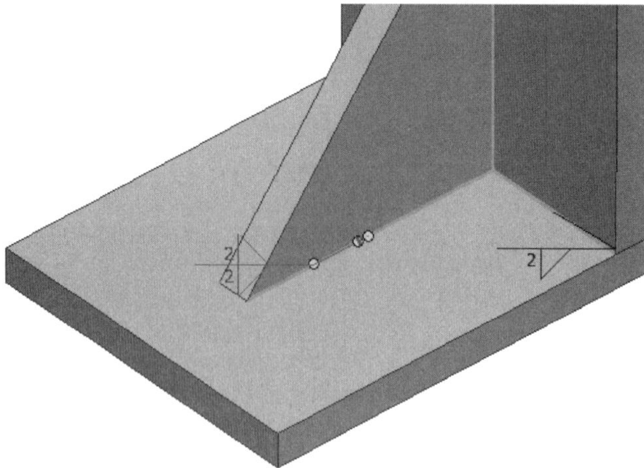

Figure 12-56

Release the mouse button when you are satisfied with the new location, as shown in the following image.

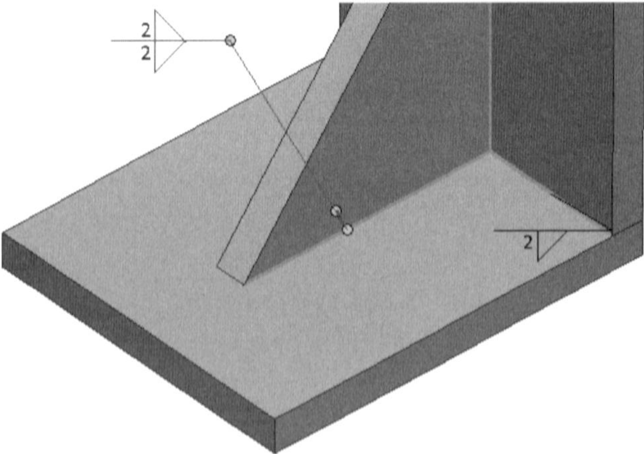


Figure 12-57

The following image shows the results when all three welding symbols have been moved to better viewing locations. Regardless of the viewing angle of your model, the weld symbols will remain normal to the display screen.

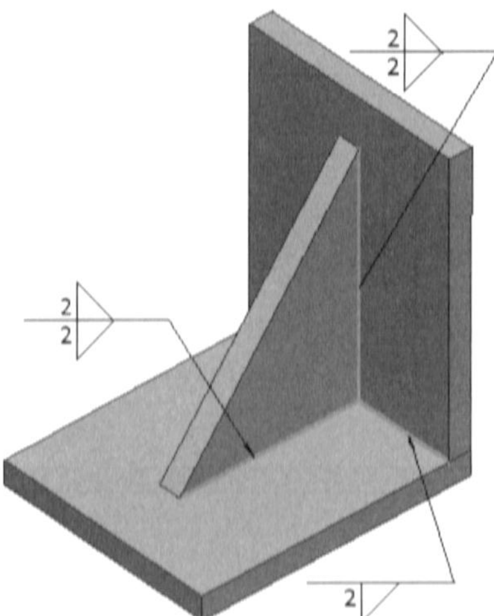


Figure 12-58

## Editing Values or Elements of a Welding Symbol

As you reposition your welding symbols, you may want to update the welding symbols' layout and included components. You can also adjust the values of your weld.

To edit the values or elements of a welding symbol, expand the Welds group in the Browser to display the weld features. Right-click on the feature to display the menu in the following image, and then select Edit Feature.

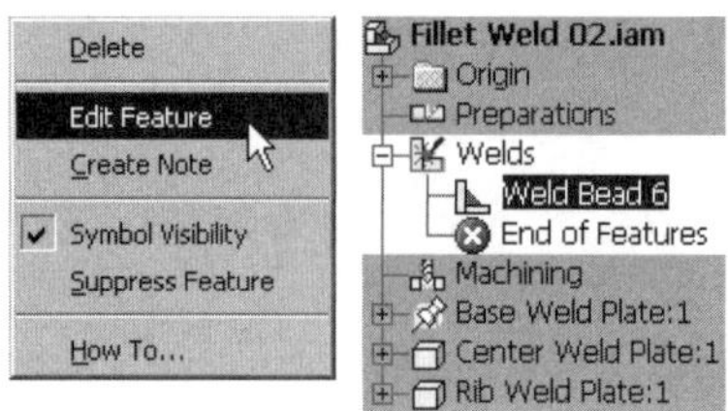


Figure 12-59

In the Weld Feature dialog box, as shown in the following image, change any element of the weld symbol and leader. You can also click on the Arrow Side and Other Side tabs and adjust their values.

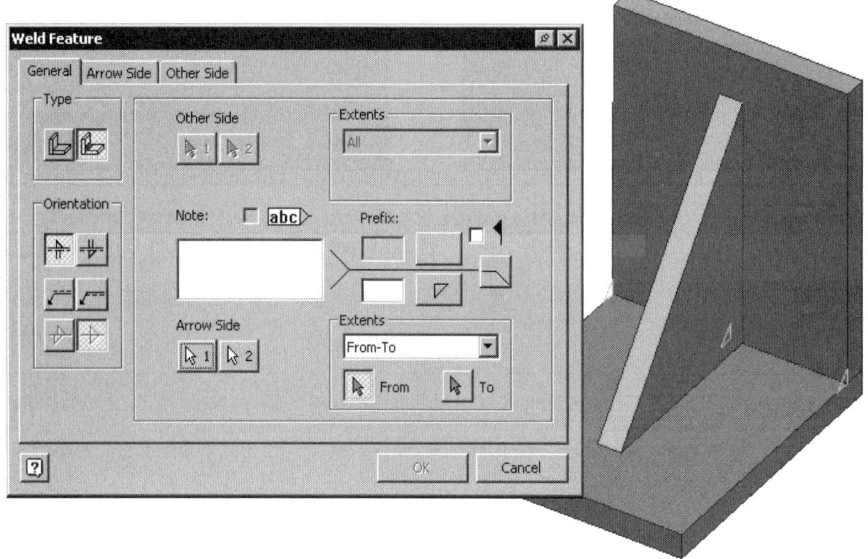


Figure 12-60

During this editing process, you can add notes, change the welding symbol orientation, or change to a different weld type.

## Weld Contours

Various contours are available to display a more realistic example of a weld. In the following image of the weldment, for example, the weld appears to curve outward, signifying a convex weld. To produce this effect, the convex symbol is chosen from the Contour list located under the Arrow Side of the Weld Feature dialog box. An offset value has also been set to control the amount of bulge in the weld bead.

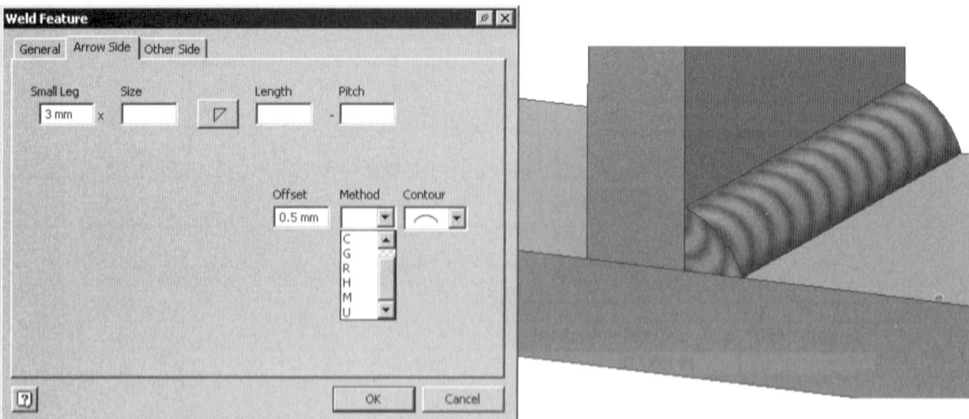


**Figure 12-61**

The example in the following image is that of a concave weld contour. The Concave effect shows the weld bead curving inward. The Concave effect is one of the contour options found under the Arrow Side tab of the Weld Feature dialog box.

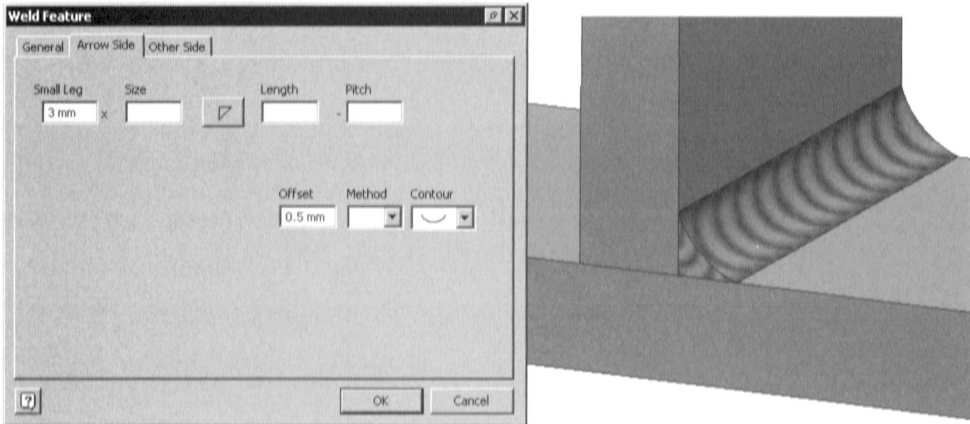


**Figure 12-62**

The following image shows the effect that the offset value has on the type of contour shapes used on a weld bead. The first example shows a straight or flat weld. The contour offset value is not active for this application. For convex or concave contours, the offset value defines the amount of bulge in the weld.

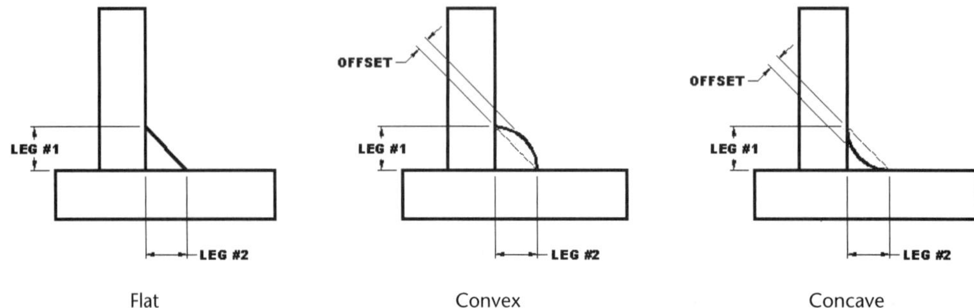


Figure 12-63

## Intermittent Weld Beads

You create an intermittent weld bead by specifying the length of each segment and the distance between the center of each segment, which is called pitch or spacing. The required values vary depending on the current standard. For instance, all standards but the ANSI standard allow you to specify the number of beads in the weld.

**NOTE** Intermittent welds cannot be applied to spline edges. If a spline edge exists in the selection set, it will be welded continuously.

To begin the process of creating an intermittent weld, activate the Welds group in the Browser and click the Weld tool in the Weldment Features toolbar. Specify the first and second Arrow Side faces for the selected weld bead feature, as shown in the following image. Click the tab for the weld side you are defining—either Arrow Side or Other Side.

**NOTE** If the number of beads does not fit on the selection set, they are truncated at the end of the set.

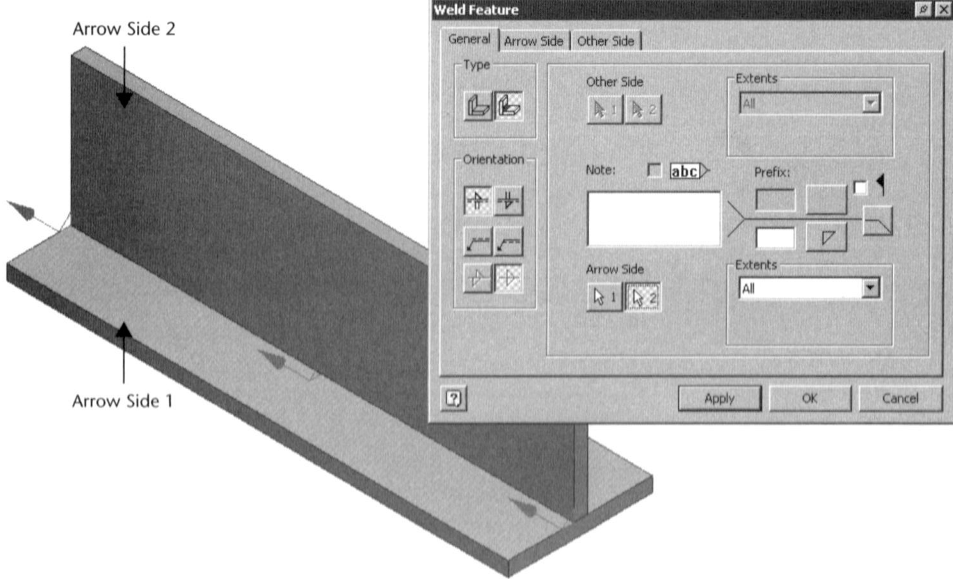


**Figure 12-64**

Once inside the Arrow Side tab, set a value for the weld depth. Then continue setting values for the required intermittent weld fields: Length, and Pitch (the center-to-center distance between welds). When finished, click the Apply button to create the weld, as shown in the following image. Click OK to dismiss the Weld Feature dialog box.

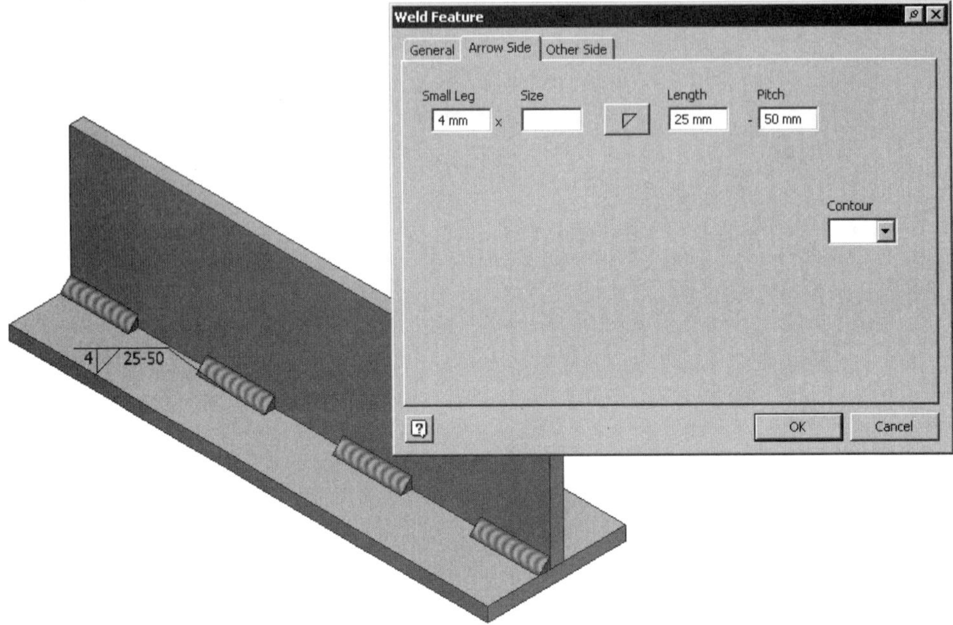


Figure 12-65

**NOTE** The above example illustrates the creation of an intermittent weld using the Metric (ANSI) Standard. When you use the Metric BSI, DIN, GB, ISO, and JIS Standards for creating intermittent welds, the Arrow Side field changes to Number, Length, and Spacing, as shown in the following image.

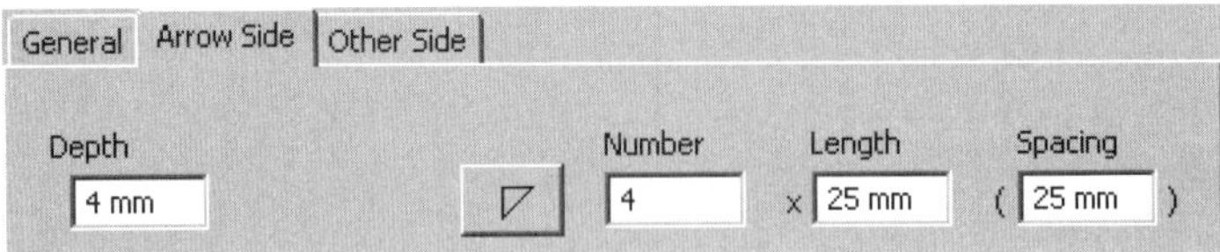


Figure 12-66

## MACHINING WELDMENTS

Post-weld machining operations are performed on the weldment assembly after the preparations and weld beads have been applied to the weldment. Holes and extrude cuts are typical post-weld machining features.

To create machining features in the weldment, activate the Machining group in the Browser and add the needed assembly features. When you activate the Machining group, you can add machining features to the weldment. You can add assembly features to the Machining group at any time. The following steps are used to produce a machining feature in a weldment:

1 Double-click the Machining feature group in the Browser.
2 Create assembly features as needed (extrusions, holes, and/or chamfers).
3 Exit the Machining feature group.

Edges and weld features that are affected by a post-weld machining feature, such as a cut, are updated to show the result. If the machining feature affects the geometry to which the weld symbol is attached, the symbol is automatically updated to accommodate the change.

## Post-Weld Machining Operation—Extrude

The following image illustrates the beginning of an extrusion operation on a weldment. In this example, a sketch plane is created on the top face of the base plate. A rectangle is sketched and extruded a total distance of 10 mm. The extrusion is being performed by a midplane.

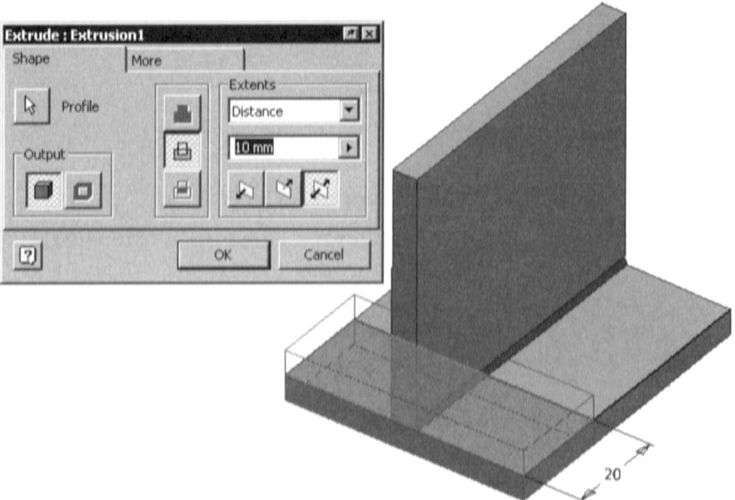


**Figure 12-67**

The results are shown in the following image. The extrusion operation cut a distance of 10 mm through the base and vertical plates. Notice the addition of the Extrusion 1 feature, the sketch (Sketch 1), and the parts affected by the extrusion operation (Base Weld Plate and Center Weld Plate) in the Browser.

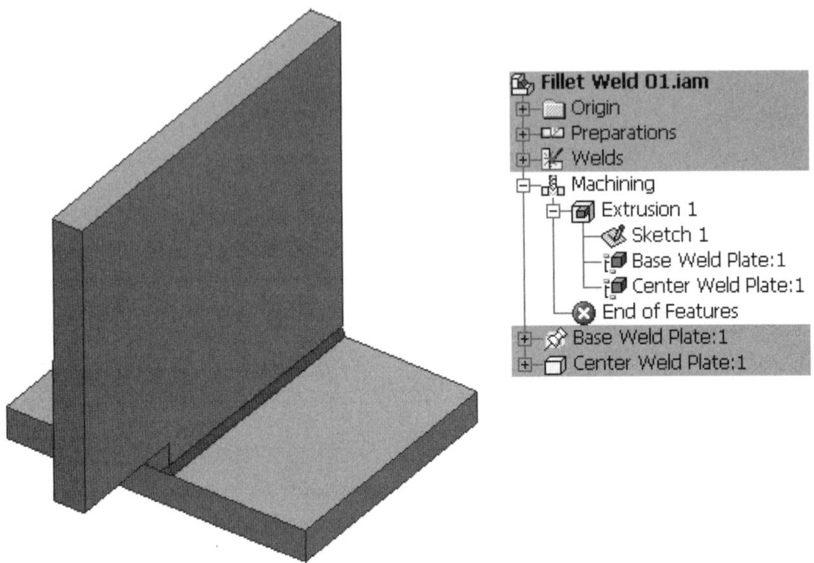


Figure 12-68

## Post-Weld Machining Operation—Hole

The following image illustrates the effects of cutting a series of holes into a weldment. The top face of the base plate is selected as the current sketch plane. A series of center tool points are sketched and dimensioned on the top plate. The Hole tool is used to drill a hole diameter of 5 mm through the part in both center point locations.

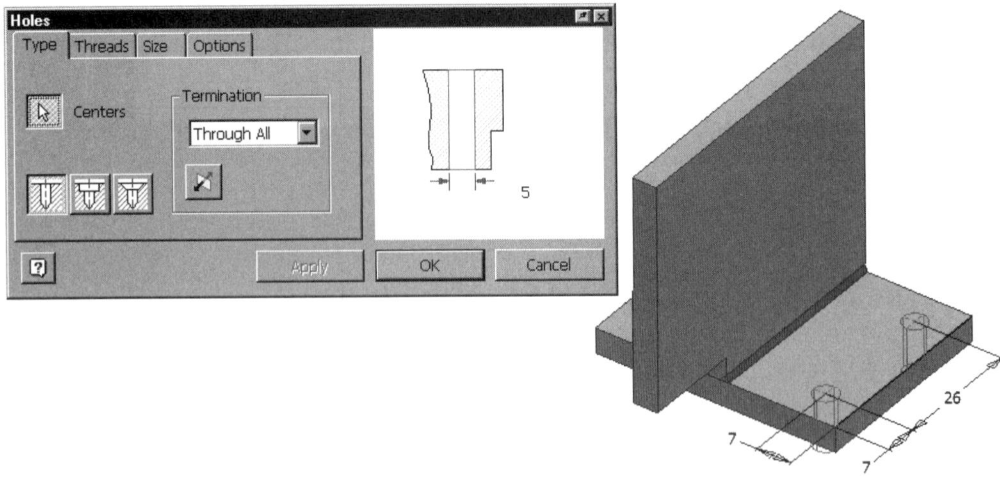


Figure 12-69

The results of this operation are illustrated in the following image. The Browser reflects the contents after you perform a hole operation. Because the holes are cut in the Base Weld Plate, this is the only part listed in the Browser under this feature.

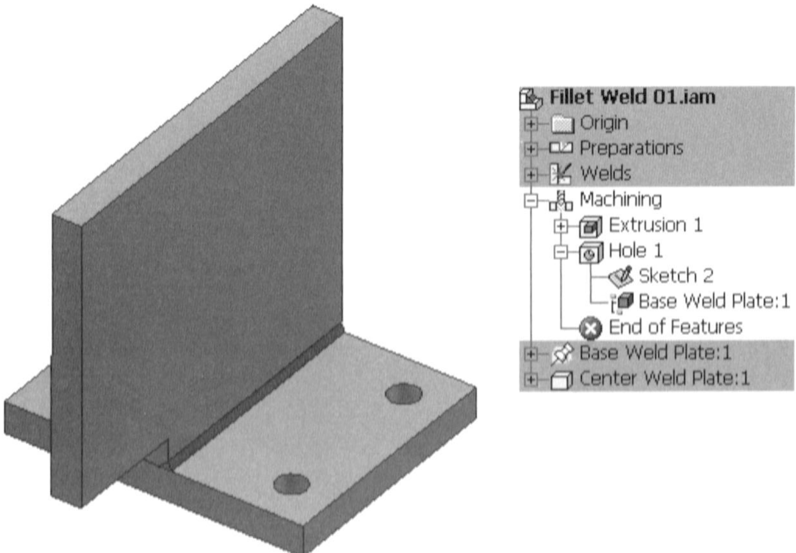


Figure 12-70

## Post-Weld Machining Operation—Chamfer

Once a weldment has been created, you can also cut a chamfer into one of the work pieces. In the following image, a chamfer of 10 mm is being cut into the corner of the vertical plate.

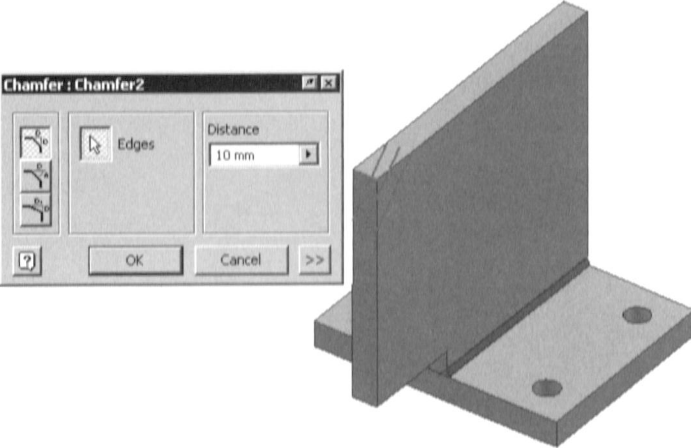


Figure 12-71

The results are illustrated in the following image. The Browser reflects the contents of the new created feature (Chamfer 2) along with the part affected by the feature (Center Weld Plate).

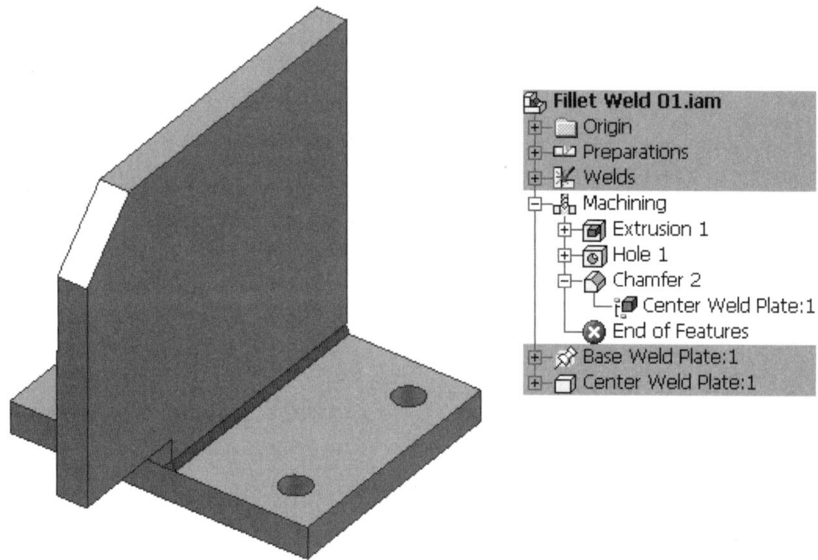


Figure 12-72

## Post-Weld Machining Effects on a Welding Symbol

Welding symbol leader start points are automatically updated when model edges are affected by post-weld machining features. If the welding symbol attachment point is cut away, as shown in the following image, the symbol will move to the closest attachment point located on the same edge. In this example, a post-weld extrusion operation cuts through the welding symbol attachment point. After you perform the extrusion operation, the welding symbol is updated to a new attachment point. The leader length and orientation, however, will not change as a result of this move.

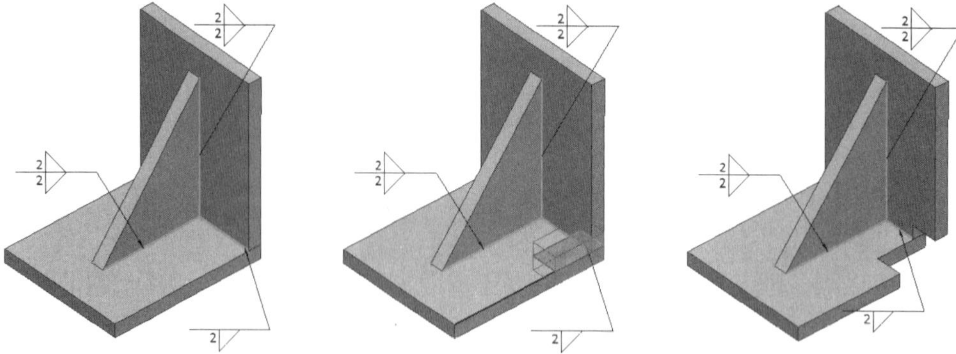

Figure 12-73

# EXERCISE 12-1 Creating Weld Beads and Machining Weldments

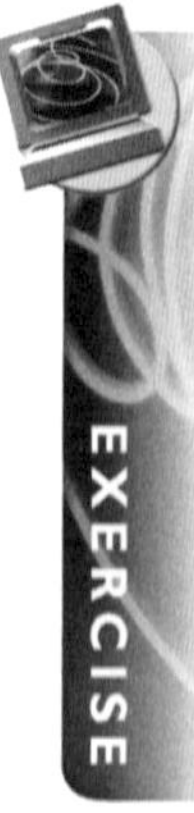


In this exercise, you will build a weldment. You will create weld preparation features, use cosmetic and 3D welds, and add post-weld machining features. To navigate to the exercise in the *Electronic Student Workbook*, do the following:

1 From the Main TOC page, click Chapter 12.

2 From the TOC for Chapter 12, click Creating Weld Beads and Machining Weldments.

The following image illustrates the opened exercise.

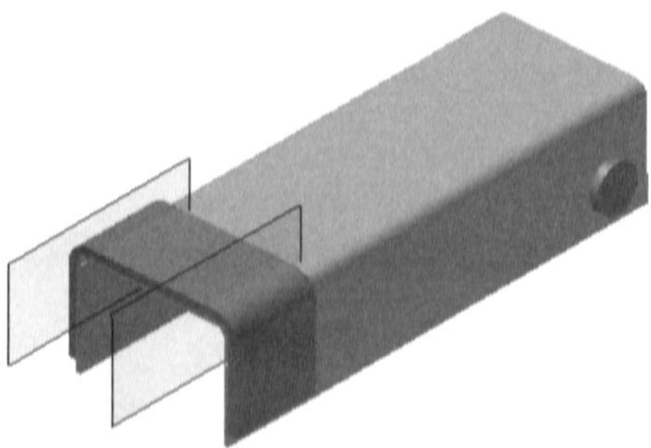

**Figure 12-74 Opened exercise**

The following image illustrates the completed exercise.

**Figure 12-75 Completed exercise**

## DOCUMENTING WELDMENTS

Weldments are a special type of assembly model that track four states of the assembly: assembly, preparations, welds, and machining. In a drawing, you can create views for any of the following states:

- Views of the assembly state show the model without defined weld preparations or welds.
- Views of the preparations state show the model with defined weld preparations.
- Views of the welds state show the model with weld preparations and welds. You can use the model welding symbols in the drawing. For solid body fillet welds, you can also use the model information to automatically generate the caterpillars and end treatments.
- Views of the machining state show the model with weld preparations, welds, and defined post-weld machining features.

You have the opportunity to determine which of the three weld features groups (in addition to the main assembly) are active for any given view when you create the view, or when you edit the view with the Drawing View dialog box shown in the following image.

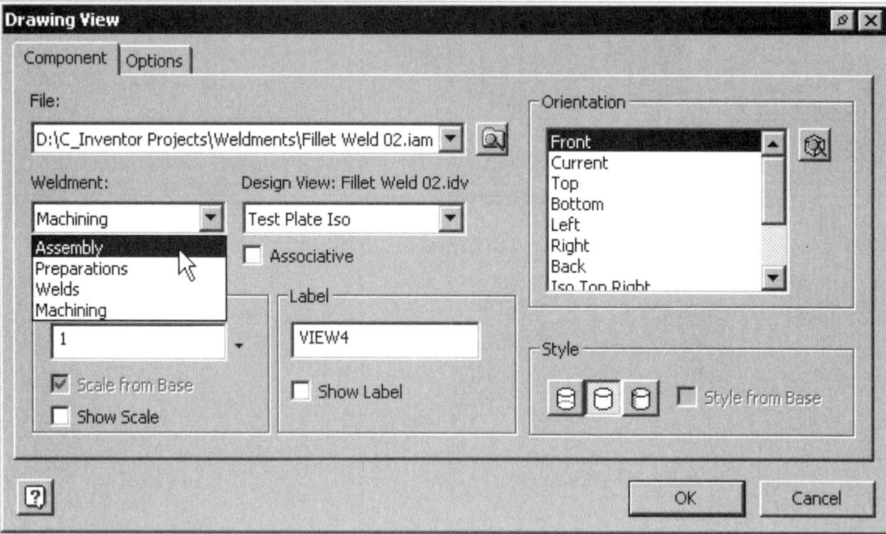


**Figure 12-76**

Views of the weldment can be generated that represent the various stages of fabrication. Most commonly, these views are the following:

- Assembly Only (shows the model without any preparations or welds).
- Preparations (shows the model with chamfers, extrusions, and/or holes).
- Welds (shows the model with preparations and welds).

- Machining (shows the model with preparations, welds, and post-weld machining features).

These stages can be selected from the Drawing View dialog box during the creation of a base view. In the 3D weldment environment, these various stages would appear, as shown in the following image.

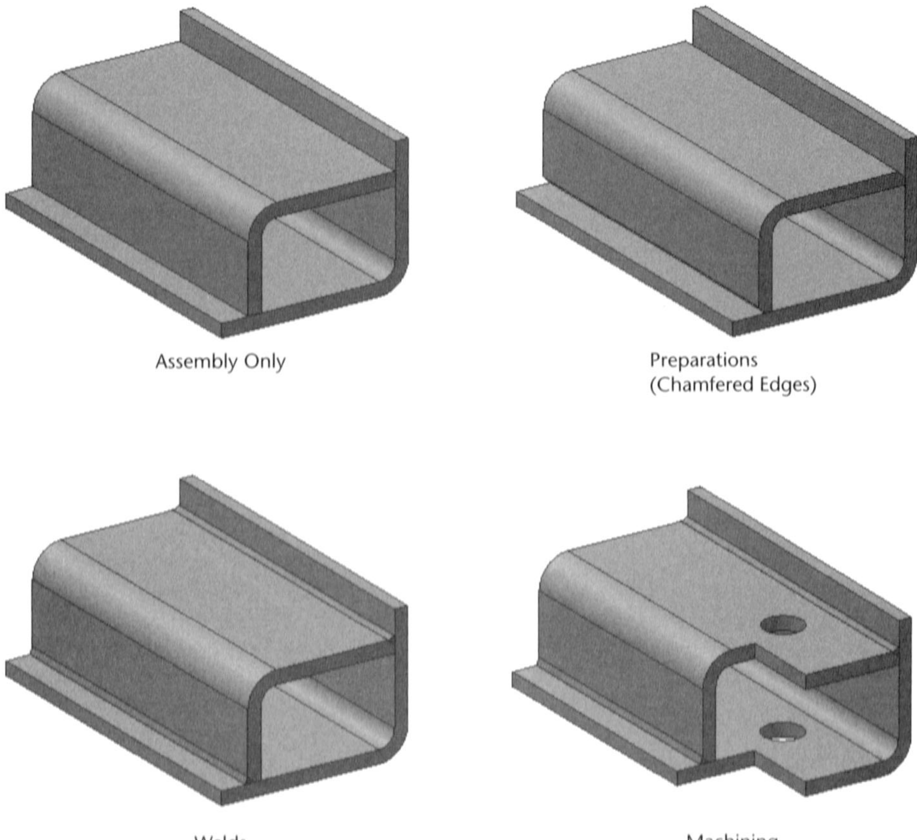


**Figure 12-77**

When base and projected drawing views of these various states are created, the result of each appears, as shown in the following image.

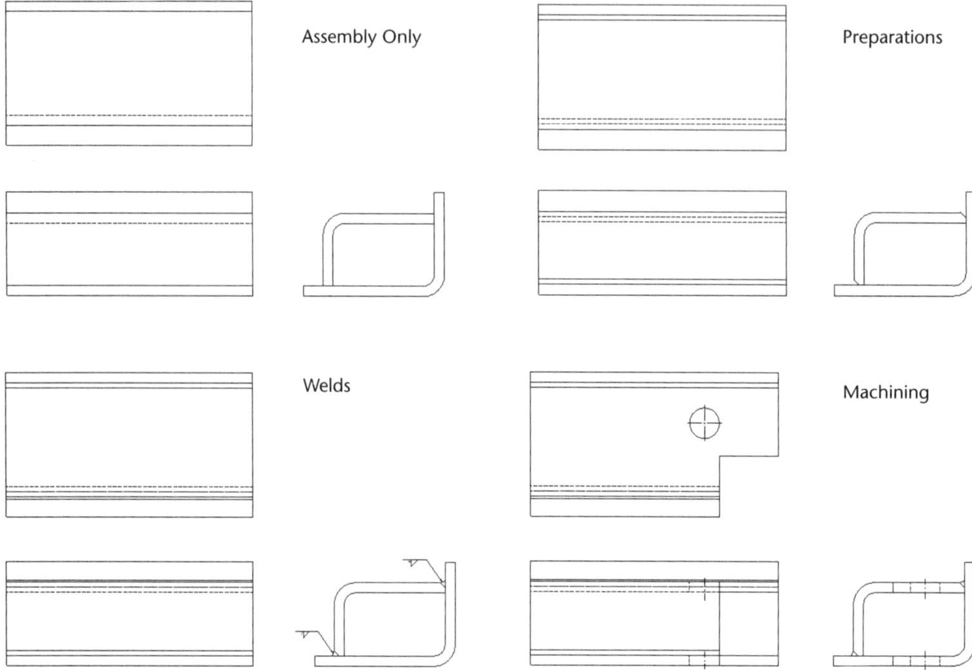


**Figure 12-78**

## Using Model Weld Annotations in Drawings

When you use model weld annotations and symbols in a drawing view, they are associated with the model and updated when the drawing changes.

You should adhere to the following guidelines for using model weld annotations:

- The model in the drawing view must be a weldment assembly.
- You cannot recover weld annotations if any of the feature groups are currently active in the model.
- Only solid body fillet welds will generate caterpillars and end treatments in the drawing. Cosmetic welds must be manually annotated.
- You can apply manual caterpillars, end treatments, and welding symbols to any drawing view.
- Any given welding symbol will appear in only one view. You can drag welding symbols between views.
- You can change the format and display attributes of model weld annotations, but you cannot change the values from the drawing.

To retrieve the model welding symbols for a drawing view that has already been created, right-click on the drawing view and select Edit View from the menu. This will display the Drawing View dialog box shown in the following image.

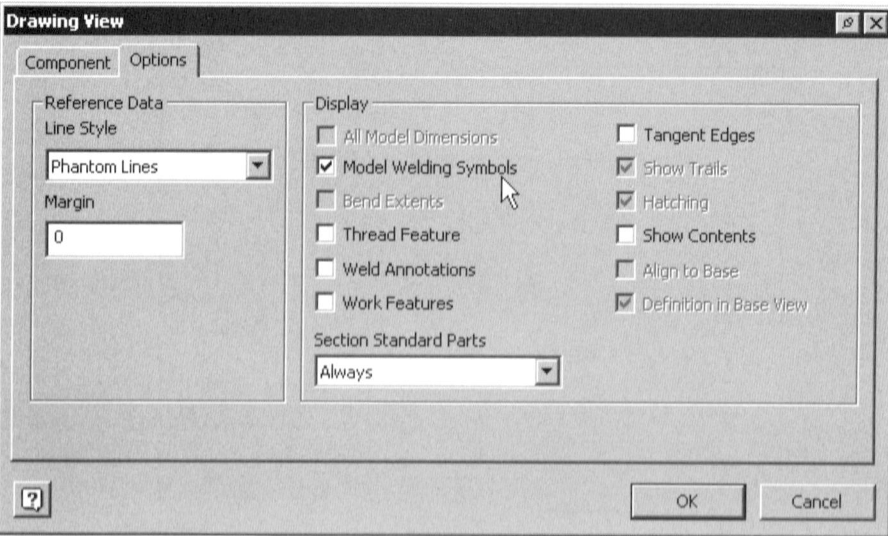


**Figure 12-79**

Select the Options tab to expose the display controls used for weldments. Placing a check in the box next to Model Welding Symbols will generate welding symbols directly from the model in which they were originally created. Checking the box next to Weld Annotations will automatically generate end treatments and caterpillars. End treatments and caterpillars will be discussed later in this chapter.

Another method of extracting welding symbols is to right-click on the view and select Get Model Annotations from the menu, as shown in the following image. Selecting Get Welding Symbols will retrieve the welding symbols from the drawing, while selecting Get Weld Annotations will add end treatments and caterpillars to the view.

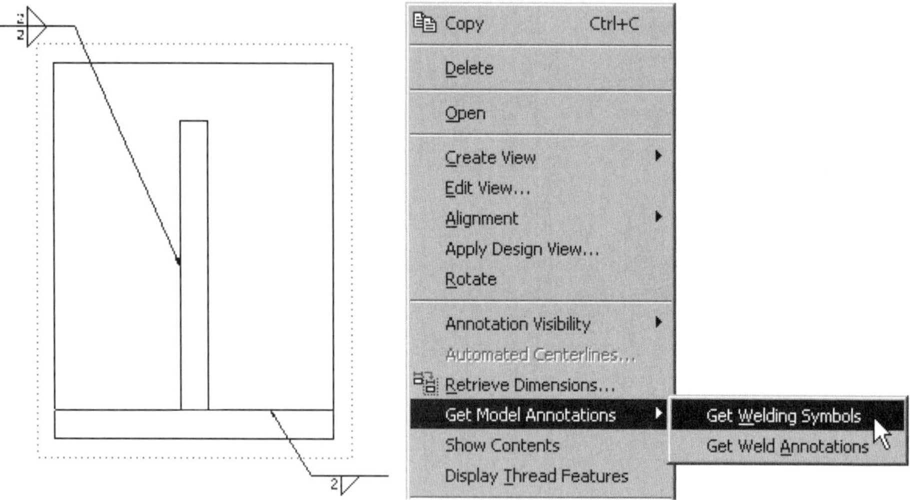


Figure 12-80

## Drawing Weld Documentation

You can add welding symbols, caterpillars, and end treatments to any drawing view. When weld annotations are manually added to the drawing, the display of caterpillars and end treatments is updated when the model changes. Values in 2D welding symbols are not automatically updated.

In the following image, two buttons are available to add caterpillars and end treatments to 2D drawing views.

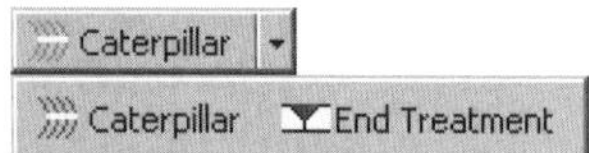


Figure 12-81

## The Weld Caterpillars Dialog Box

Caterpillars are 2D symbols that represent the length, size, and direction of a weld. Caterpillars can be thought of as a type of weld annotation. The caterpillar annotation is added to existing geometry in a drawing view. A dedicated dialog box controls the placement, appearance, and size of the caterpillar annotations. Clicking on the Caterpillar button located on the Drawing Annotation Panel Bar displays the Weld Caterpillars dialog box shown in the following image.

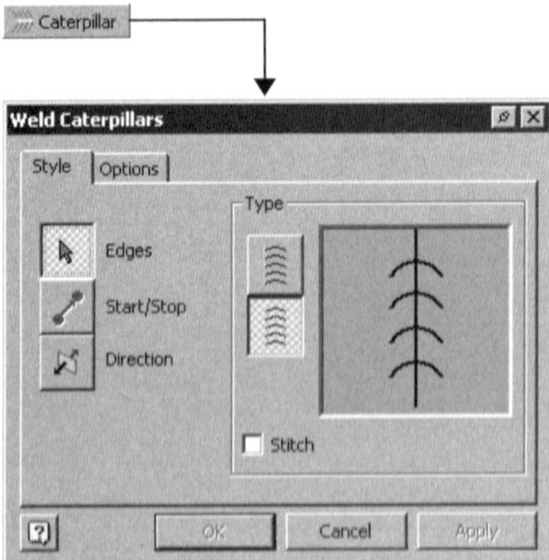


**Figure 12-82**

By default, the dialog box opens in the Style tab, which is used to set the display style for the weld caterpillars. Other settings and options of this tab are described as follows:

**Edges** Use this button to select the edges to which the caterpillar will be applied. Click the button and then click to select the edges in the drawing view.

**Start/Stop** Use this button to set the beginning and end of the caterpillar if it does not extend the full length of the selected geometry. Click the button and then click points on the selected edges to indicate the start and stop points.

**Direction** This button reverses the direction of the caterpillar legs. Click the button to change the direction.

**Type** Use this button to select the caterpillar type. Click the button to select a partial or full caterpillar. When placing a partial caterpillar, click in the graphics area to set the side on which the legs will be displayed.

**Stitch** Check this box to create an intermittent caterpillar. Select Stitch to add spaces to the caterpillar pattern. Clear the checkbox to create a caterpillar with all legs evenly spaced.

Clicking on the Options tab of the Weld Caterpillars dialog box will display information as shown in the following image.

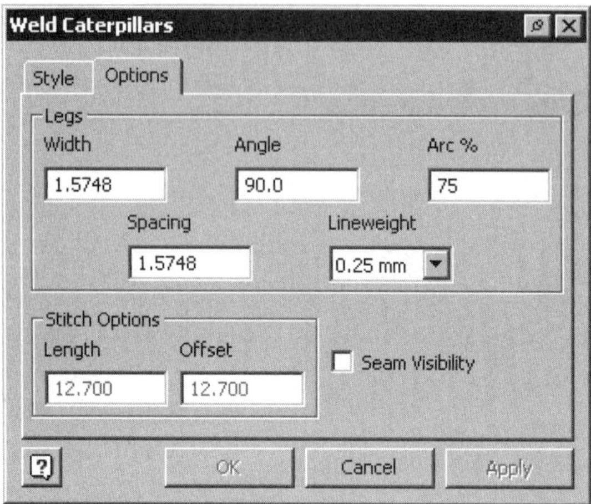


**Figure 12-83**

The settings under this tab control the size of the caterpillar and are described as follows:

**Legs** This section determines the size, angle, and format for the caterpillar legs.

**Width** For a partial caterpillar, this feature determines the distance between the endpoint of the leg and the selected edge. For a full caterpillar, it determines the distance between the endpoints of the legs.

**Angle** This feature sets the angle for an imaginary line between the selected edge and the endpoints of the legs.

**Arc %** This option sets the radius of the arc relative to the length of the arc chord.

**Spacing** This determines the distance between legs in the caterpillar.

**Lineweight** This determines the thickness of the lines in the caterpillar.

**Stitch Options** This option designates the size for intermittent caterpillars.

**Length** This option sets the length of one segment in an intermittent caterpillar.

**Offset** For an intermittent caterpillar, this feature determines the distance between the centers of two adjacent segments.

**Seam Visibility** This feature sets the visibility of the edge on which the weld caterpillar is placed. Select Seam Visibility to show the edge. Clear the checkbox to hide the edge after placing the caterpillar.

**NOTE** The default settings in the dialog box are determined by the active drafting standard.

## Placing a Caterpillar Weld Annotation

Use the Caterpillar button on the Drawing Annotations Panel Bar to add 2D weld caterpillars to geometry in a drawing view. The caterpillar type, size, angle, and other formatting information is determined by the active drafting standard.

To begin placing a caterpillar annotation, click the Caterpillar button on the Drawing Annotation Panel Bar. When the Weld Caterpillars dialog box appears, select an edge in a drawing view, as shown in the following image. Set the type of caterpillar to use in the dialog box, and then click the Options tab to set the size information for the caterpillar if needed.

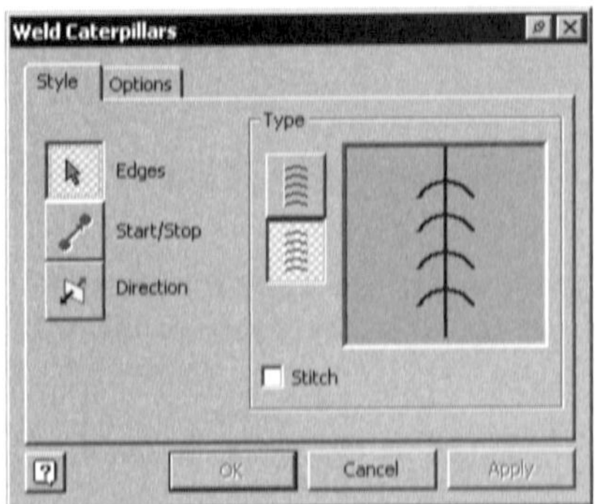


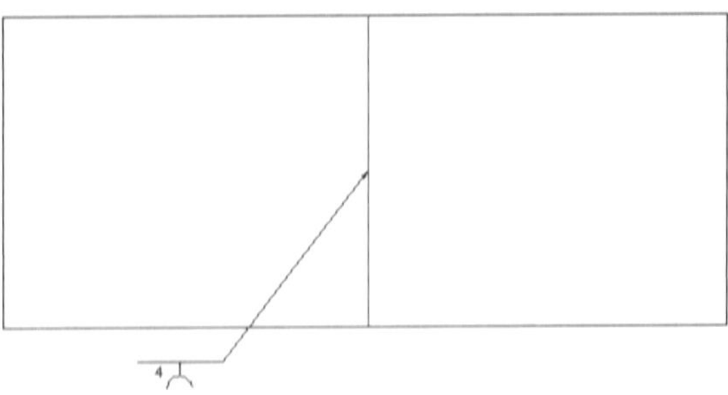


**Figure 12-84**

Click the Apply button to place the caterpillar. The results are displayed in the following image.

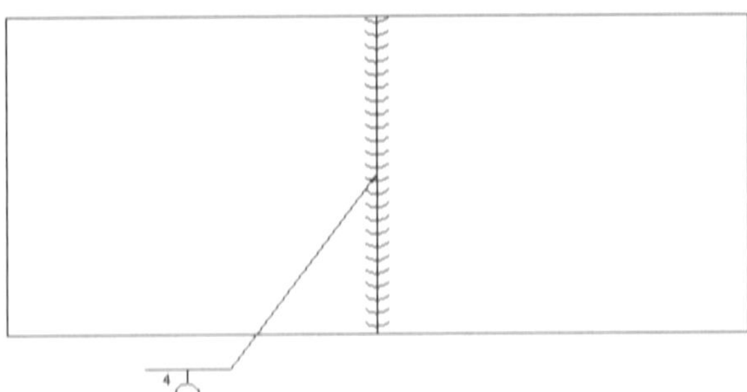


**Figure 12-85**

In the following image, a partial caterpillar annotation has been applied to various edges of the weldment document. After an edge is selected, the

Start/Stop button is used to limit the length of the caterpillar. Once the correct caterpillar annotation length is achieved, click the Apply button to place the caterpillar and continue creating other caterpillars.

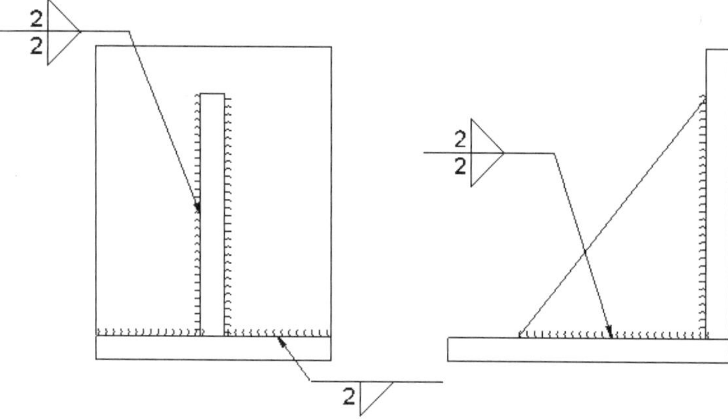


Figure 12-86

## End Treatments

End treatments are end views of a weld that are displayed in a specific drawing view. Like caterpillars, end treatments are considered a type of weld annotation.

Clicking on the End Treatment button, which is located on the Drawing Annotation Panel Bar, will display the End Treatments dialog box, as shown in the following image. The End Treatments dialog box sets the shape for weld end treatments. Click a button to select one of the preset shapes, or click the Custom button to define a custom shape.

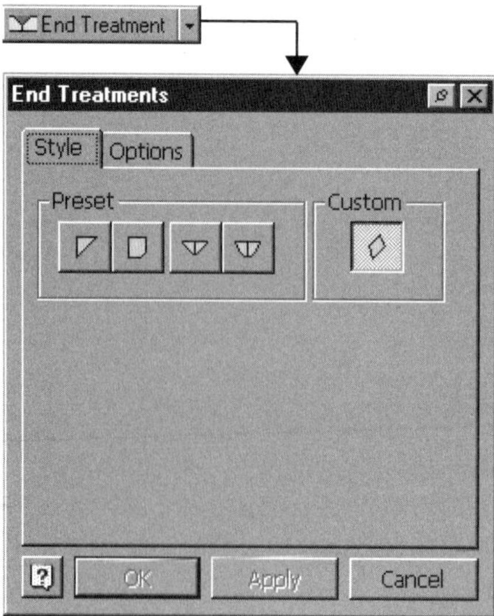


Figure 12-87

Clicking on the Options tab of the End Treatments dialog box will display the information shown in the following image. Use the information under this tab to set the size and fill attributes of the end treatment annotation.

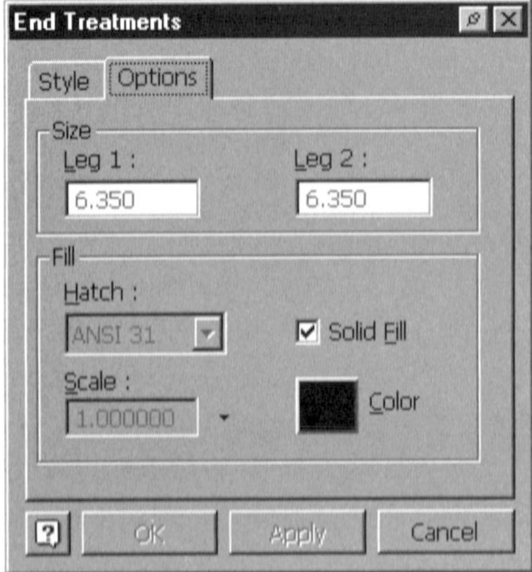


**Figure 12-88**

The settings under this tab control the size and fill of the end treatment annotation and are described as follows:

**Size** This setting determines the length for the legs of the end treatment annotation. Enter the length for each leg into the corresponding box. The Size setting is not available when creating custom end treatments.

**Fill** This feature selects the format for the fill of the end treatment annotation.

**Hatch** This option selects a hatch pattern for the fill. Click the arrow and then select from the available hatches. This feature is not available if the Solid Fill box is checked.

**Solid Fill** This feature adds a solid fill to the end treatment. Select Solid Fill to fill the end treatment. Remove the check mark to use the selected hatch pattern to fill the end treatment.

**Scale** This option determines the scale for the selected hatch pattern. Click the arrow and select the scale. This feature is not available if Solid Fill is checked.

**Color** This setting controls the display color for the hatch or solid fill. Click the color button and then select the color.

## Placing an End Treatment Annotation

Use the End Treatment button on the Drawing Annotations Panel Bar to add 2D end treatment annotations to geometry in a drawing view. The size and other formatting are determined by the active drafting standard.

Begin placing an end treatment by clicking the End Treatment button located in the Drawing Annotation Panel Bar. If the End Treatment button is not visible, click the arrow next to the Caterpillar button, and then click End Treatments. When the End Treatment dialog box is displayed, select the end treatment type as shown in the following image. If needed, click the Options tab and then set the size and fill. In the graphic window, select an edge for the end treatment. You may have to move your cursor to position the end treatment symbol. When the symbol is in the desired position, click to place it on the view.

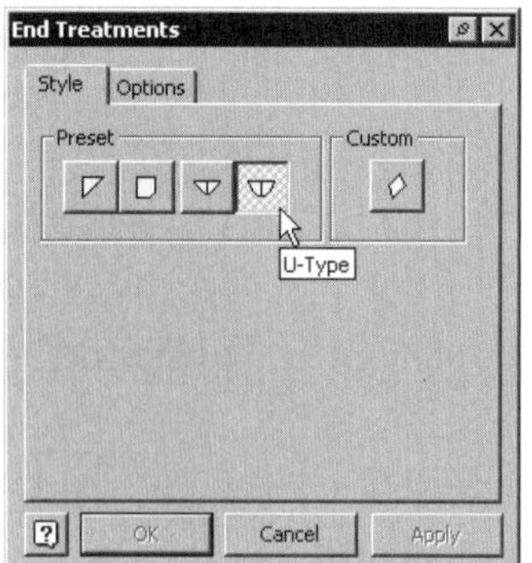


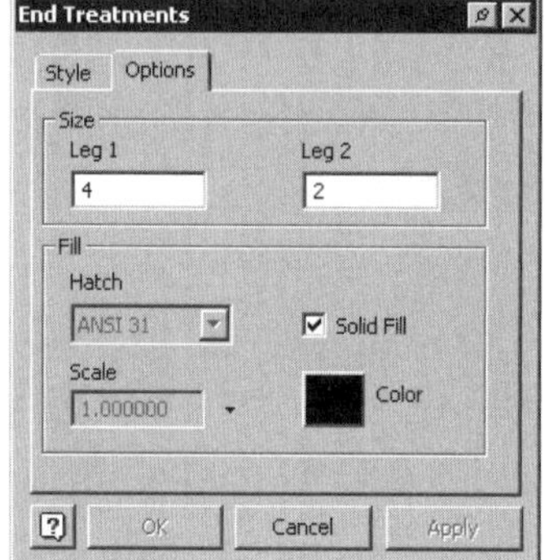


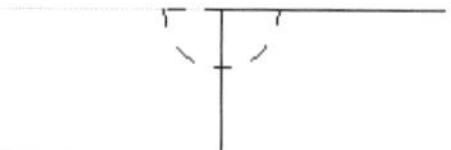

Figure 12-89

Click the Apply button to place the end treatment. The following image shows the results of the manual placement of a caterpillar weld and end treatment annotation.

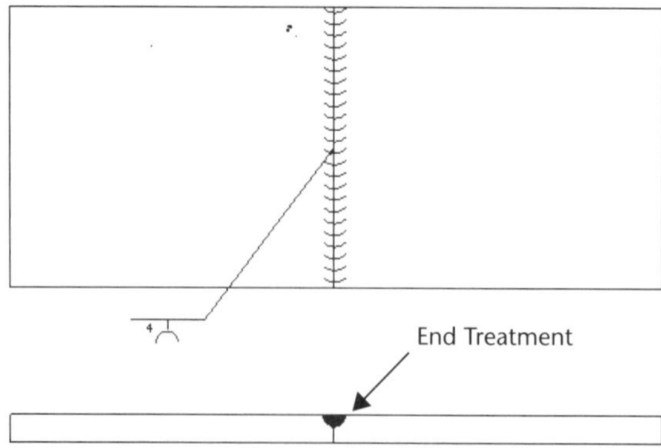


Figure 12-90

**NOTE** If fillet weld beads are generated in the weldment model, this model information can be used to automatically generate caterpillars and end treatments.

## Adding 2D Weld Symbols to a Drawing

You can add weld annotations to a drawing view even if there are no weldments defined in the model. 2D weld annotations are not associated with weldments in the model file and are not updated if the model changes.

To add a 2D weld symbol to a drawing view, use the following steps:

1 Click the Welding Symbol button in the Drawing Annotation Panel Bar, as shown in the following image, to create a welding symbol with a leader line.

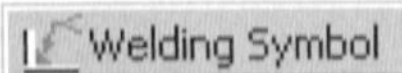


Figure 12-91

2 In the graphics window, select an edge in a drawing view to set the start point for the leader line.

**NOTE** To associate the welding symbol with geometry, move the cursor over the desired edge and then click when the edge is highlighted. If you move the drawing view, symbols that are associated with geometry also move.

3 Move the cursor and click to add a vertex to the leader line.

4 When the symbol indicator is in the desired position, right-click and then select Continue to place the symbol and open the Weld Symbol dialog box, as shown in the following image. The options in the dialog box are determined by the active drafting standard.

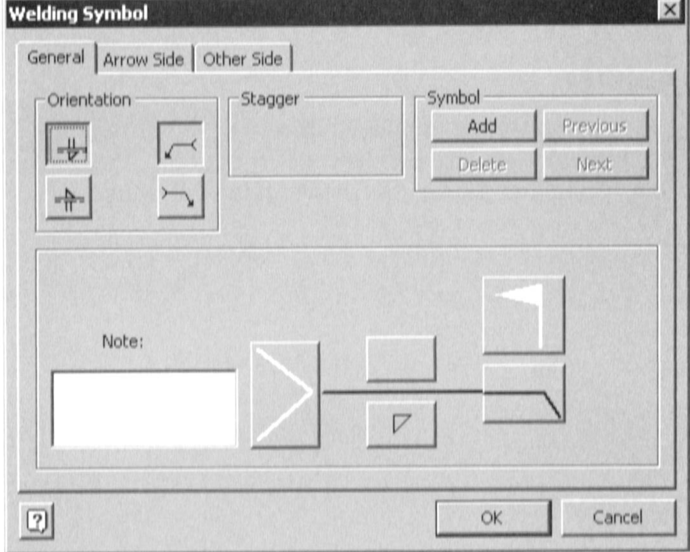


Figure 12-92

5 Set the attributes and values for the symbol.

6 Continue placing welding symbols. When you finish placing symbols, right-click and select Done from the menu to end the operation.

## Drafting Standards for Welds

Use the Welding Symbol tab in the Drafting Standards dialog box, as shown in the following image, to set the defaults for adding 2D welding symbols to a drawing. You activate this dialog box by selecting Format>Standards from the graphic screen menu. On the Welding Symbols tab, select Welding Symbols in the "Show Symbols for:" edit box, and then click the checkbox to display or hide a welding symbol in a drawing.

**NOTE** The drafting standard weld settings also affect the formatting for weld annotations recovered from the model.

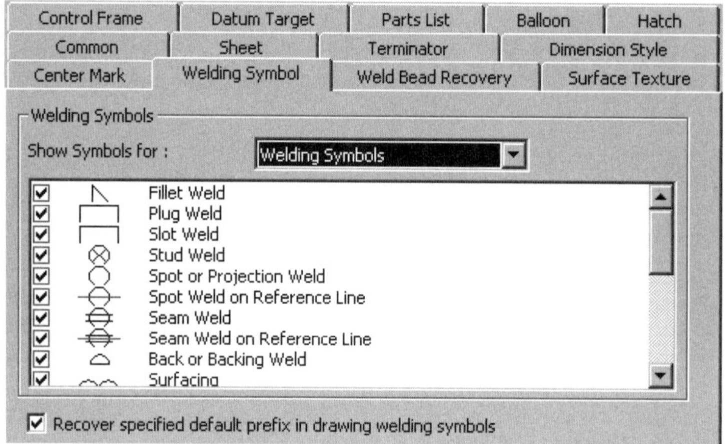


**Figure 12-93**

While in the Welding Symbols tab, select Contour Symbols in the "Show Symbols for:" edit box and then click the checkbox to display or hide a contour symbol in a drawing, as shown in the following image.

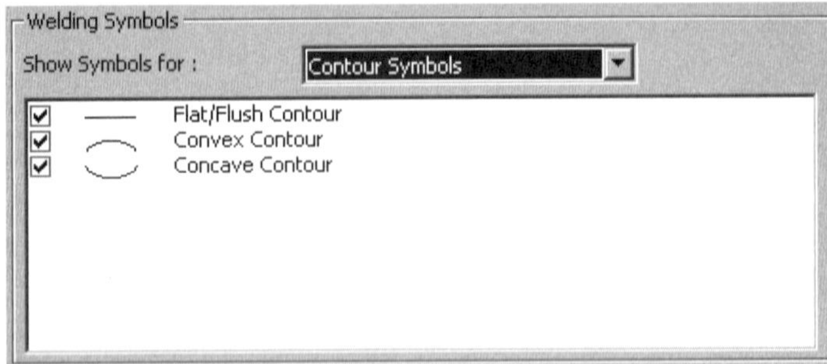


**Figure 12-94**

While in the Welding Symbols tab, select Backing Symbols in the "Show Symbols for:" edit box and then click the checkbox to display or hide a backing symbol in a drawing, as shown in the following image.

Welding Symbols
Show Symbols for :
Backing Symbols
Removable Backing Strip
Melt Through
Consumable Insert

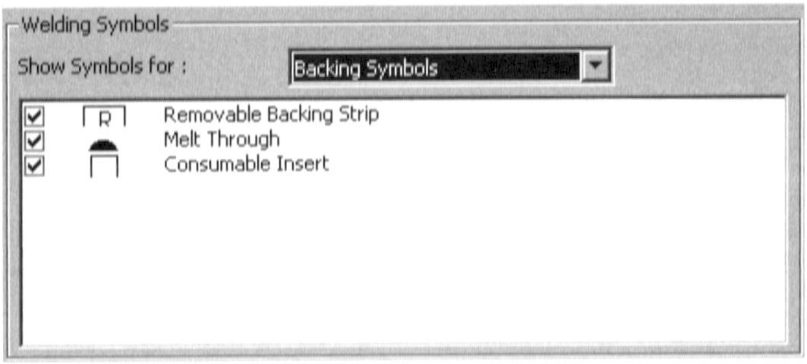

**Figure 12-95**

Use the Weld Bead Recovery tab in the Drafting Standards dialog box, as shown in the following image, to set the defaults for adding caterpillars and end treatments to a drawing. You activate this dialog box by selecting Format>Standards. On the Weld Bead Recovery tab, set the options for caterpillars and end treatments.

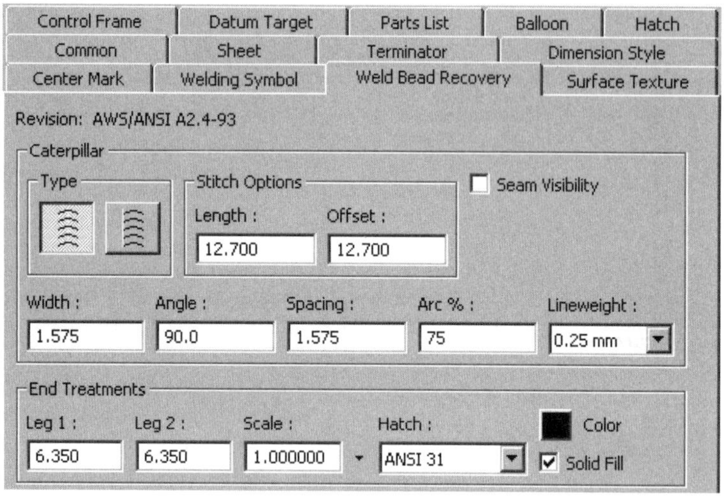

Figure 12-96

## Adding a New Leader to a Welding Symbol

At times, the same welding symbol may be used to identify welds that take place along numerous edges. When a single welding symbol exists in a drawing, you can construct additional leaders to this welding symbol. To perform this operation, right-click on the existing welding symbol to display the menu, as shown in the following image. Click the Add Vertex/Leader option and select an edge from which to begin the new leader. Drag the leader until its endpoint matches the green dot on the existing welding symbol. The results are illustrated in the following image:

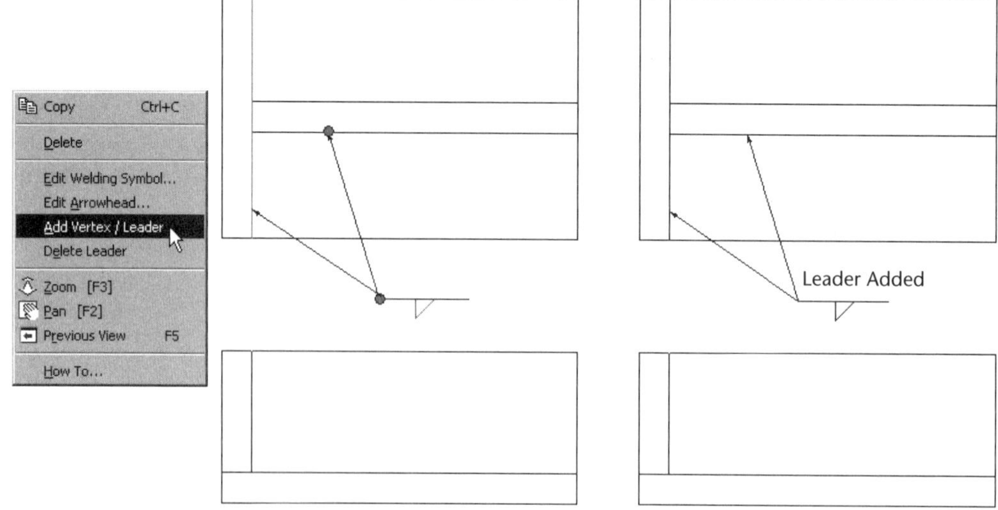


Figure 12-97

# EXERCISE 12-2 Documenting Weldments

In this exercise, you will learn to create drawing views of weldments and add cosmetic weld beads to your drawings. To navigate to the exercise in the *Electronic Student Workbook*, do the following:

1 From the Main TOC page, click Chapter 12.

2 From the TOC for Chapter 12, click Documenting Weldments.

The following image illustrates the completed exercise.

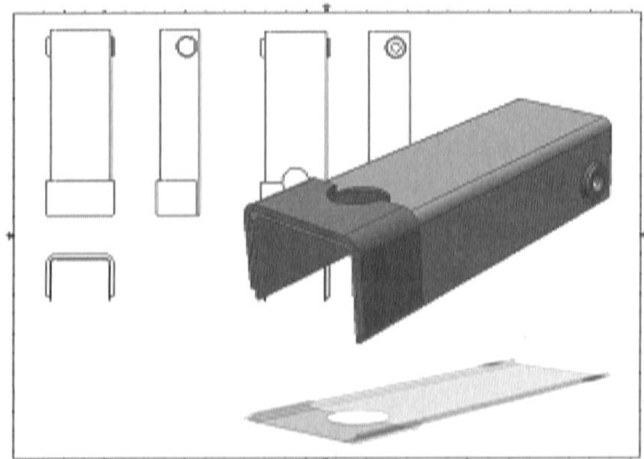
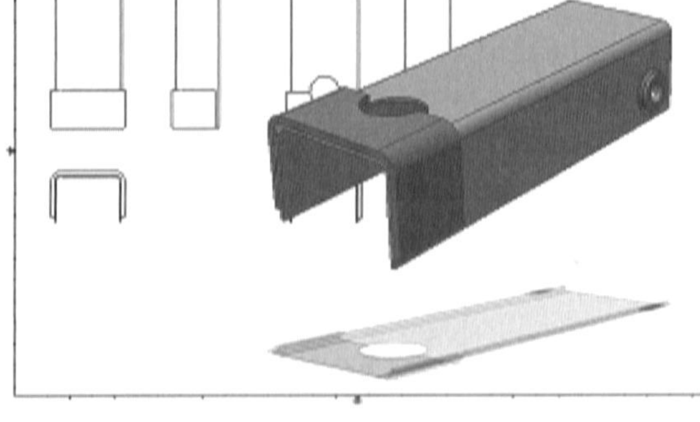

**Figure 12-98 Completed exercise**

# IPARTS

iParts enable design intent and design knowledge to be shared and reused. iParts enable you to store multiple part parameters and properties and then calculate unique part files based on certain configurations of the stored parameters and properties. Using the Parameters dialog box, you can store one value for each parameter. In industry, configured parts have been referred to as tabulated parts, charted parts, or a family of parts. iParts can also be used to create part libraries that allow design data to be reused. An iPart is generated from a standard Autodesk Inventor part file (*.ipt*). When the Create iPart tool is activated, the standard part file is converted into an iPart. You can add individual members (or configurations) to the iPart; it is then referred to as an iPart Factory.

There are two stages to the use of iParts: authoring the iPart and placing the iPart. In the authoring or creation stage, you design the part and establish all possible versions of the design in a table. The rows of the table make up the members of the iPart and an iPart that has multiple members is called an iPart Factory.

In the placement stage, you select a member from the iPart Factory, and an iPart is published for insertion into your assembly.

iParts provide the ability for you to suppress specific features, control iMates and thread properties, add file properties to the different members that are contained within the iPart Factory, and allow user input for specific parameters within a predetermined range or from a list. After creating an iPart, it can be placed into an assembly where a specific member (or configuration) can be selected. There is no limit to the number of members of the same part that can be placed into the assembly. The following sections will explain how to create and then place an iPart into an assembly.

An example of an iPart is a simple bolt. The bolt has a number of different sizes that are associated with it. An iPart allows you to define these different sizes and configurations (material, part properties, and so on) and have them reside within a single file. When placing the bolt in an assembly, you can select which member (or size) of the bolt (or iPart) you want to use. The placed member can then be constrained, similarly to any other part file.

## CREATING IPARTS

There are two types of iParts that can be created: standard and custom.

**Standard iPart (Factories)** Standard iPart values cannot be modified; when placing a standard iPart into an assembly, only the predefined members can be selected. A standard iPart that is placed in an assembly CANNOT have features added to it.

**Custom iPart (Factories)** Custom iParts allow you to place a unique value for at least one variable. A custom iPart that is placed in an assembly CAN have features added to it.

To create an iPart, follow these steps:

1 Create an Autodesk Inventor part or a sheet metal part.
2 Place the dimensions of the geometry of the design to be changed in the member.
3 For easier creation of the members, use descriptive names for the values of the parameters.
4 If parameter names are not used, you will need to determine what each parameter (or d#) represents within the parts geometry.

**TIP** Any parameters that have a name other than d# are automatically added to the list of parameters to include in the creation of the iPart.

5 Issue the Create iPart tool from the Tools menu (as shown in the following image) to add the members or configurations to the iPart.

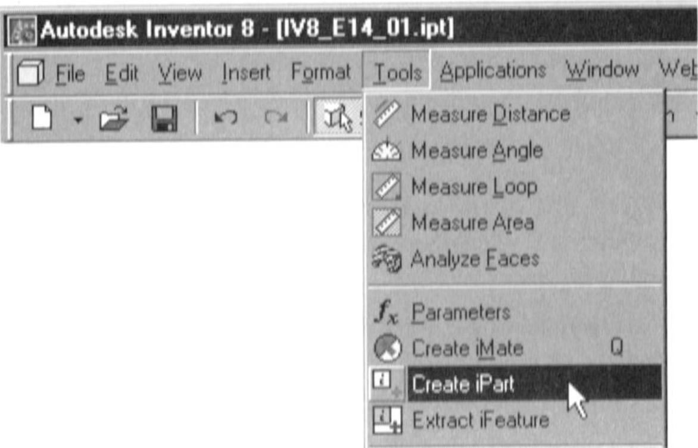


**Figure 12-99**

The iPart Author dialog box is displayed, as shown in the following image.

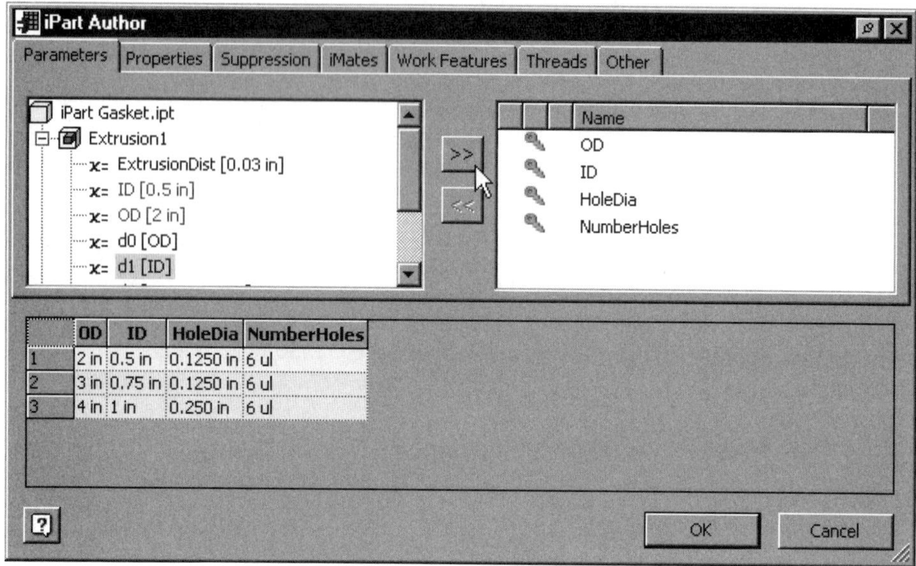


**Figure 12-100**

To create the iPart members, use the steps in the following list:

1. Add to the right side of the dialog box (parameter list) all parameters and dimensions that will be configured. All named parameters are automatically added to the right side.
2. To add a d# dimension to the list on the right side, select it from the left column and then click the >> button, as shown in the previous image. You can also double-click on the parameter to add it.
3. To remove a parameter from the right side of the dialog, select its name and click the << button.
4. After the parameters have been added, the keys need to be defined. Keys identify the column whose values are used to define the iPart member when the part is published (or placed) in an assembly. For example, setting a parameter as a primary key allows the designer placing the part to choose from all available values for that parameter in the selected items list.
5. To specify the key order, click an item in the key column of the selected parameters list to define it as a key, or right-click on the item and select the key sequence number as shown in the following image. Selected keys are blue; items that are not selected as keys have dimmed key symbols. You can add one primary key and up to eight secondary keys. Custom columns cannot be defined as a key.

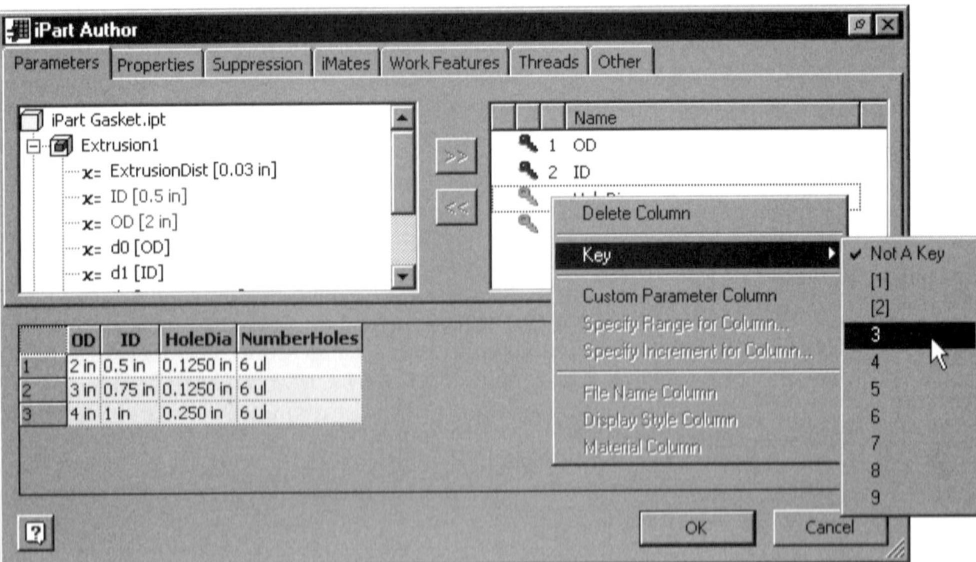


**Figure 12-101**

6 Choose from the other tabs in the dialog box to perform other specific operations. These items are not required to define the iPart factory. They are described in additional detail below:

**Properties Tab** The Properties tab lists all of the summary, project, and physical properties for the file. If you use the material property from this tab to control the material of your iParts, you have to use the Material Column option in the table. This option is described in "Special Table Elements" later in this chapter. The current color of the iPart must also be set to As Material prior to saving the iPart. For part properties to be used in drawings, Bill of Materials (BOMs), and other down-stream purposes, you have to include the properties in the table even if their values do not vary between members.

**Suppression Tab** The Suppression tab gives you the ability to suppress features of specific members of the iPart Factory. If you enter Suppress in a cell for the feature, the feature will be suppressed when it is placed in an assembly. When the cell contains Compute, the feature is not suppressed.

**iMates Tab** From the iMates tab, select one or more iMates to include in the iPart. iMates are included in the iPart if its status is Compute. If the status is suppressed, they are not included in the iPart. You must create the iMates using the Create iMate tool prior to accessing the iPart Author dialog box. Control the iMate properties for each member using the iMates, Parameters, and Suppression tabs. From the iMates tab, you can:

- Define different offset values for different iPart members.
- Suppress iMates or Composite iMates with different iPart members.
- Change the matching name for different assembly configurations.
- Change the sequence of the iMates.

Since iMate names are not unique, each iMate is assigned an indexed tag in the iPart Author. Each iMate property (except the offset value) uses this tag to identify a specific iMate property in the table. Tags are not assigned to offset values because iMates use parameters, which are unique by definition.

**Work Features Tab** From the Work Features tab, select user-defined work features to include in the iPart. Origin work features are automatically included in an iPart Factory and iPart members. The work features are included in the iPart if its status is Include. If the status is Exclude, the work features are not included in the iPart.

**Threads Tab** You can control thread features in the iPart Factory to create table-driven items for regular or tapered thread features. Use the Designation element to control the size and pitch. The Direction can be right or left using R or L in the table. Options to modify the pipe diameter, class, and family are also available.

**Other Tab** You can use the Other tab to create custom table items. You can, for example, add a column that represents the name of the iPart member. For example, you can create a column named Version to identify the appropriate member when it is placed in an assembly. You can also create custom values such as Color if you want to control the display style of the iPart.

**Special Table Elements** You can right-click on a member of the table or a column label to access some additional custom table capabilities. You can define the name of the iPart file when it is published using the File Name Column option. The color of the iPart can be controlled upon publishing using the Display Style Column option. The material of the iPart can also be specified using the Material Column option. These options are only available for part properties and the table elements that are created using the Other tab. They are not available on parameter items.

7 Next, add a member in the iPart table by right-clicking a row at the bottom of the dialog box and selecting Insert Row from the menu, as shown in the following image.

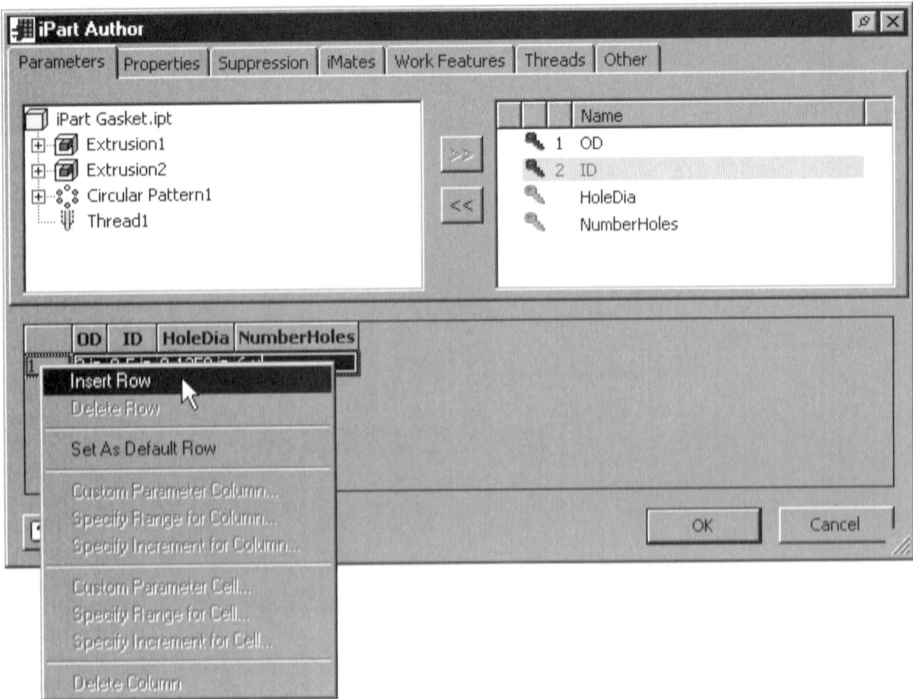


**Figure 12-102**

8 Edit the cell contents by clicking in the cell and typing new values or information as needed. To delete a row, right-click the row number and choose Delete Row from the menu. Each row that is added in the bottom pane of the dialog box represents an additional member within the iPart factory. The table functions similarly to a spreadsheet. In addition to adding or deleting members of the iPart factory, the table can also be used to change members by modifying cell values.

To allow the designer placing the iPart to specify the value of a given column, right-click on the column name and select Custom Parameter Column from the menu, as shown in the following image. To make a specific cell custom, right-click in the individual cell and click Custom Parameter Cell from the menu.

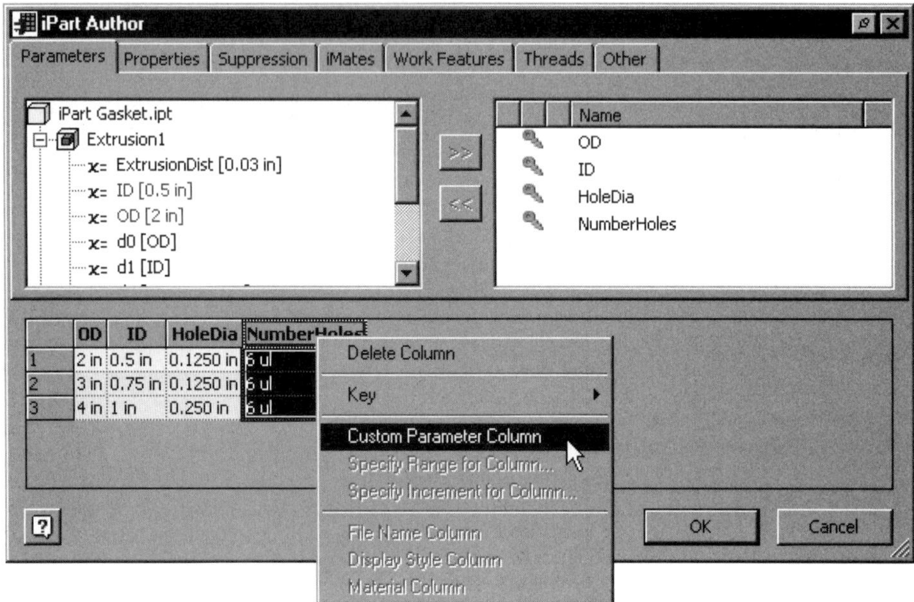


Figure 12-103

After designating a Custom Parameter Column or Cell, you can set a minimum and maximum range of values by right-clicking in the column or cell and selecting Specify Range for Column or Range for Cell from the menu. The Specify Range dialog box is displayed, as in the following image. Select the options as needed. Custom columns and cells are highlighted with a blue background in the table.

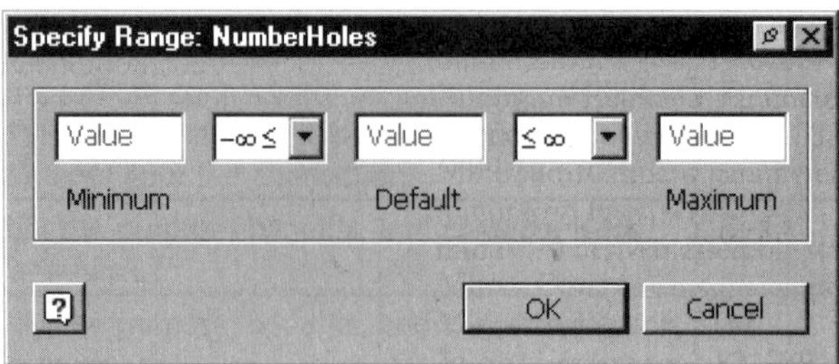


Figure 12-104

9 Set the member that will be the default by right-clicking on the row number and choosing Set As Default Row from the menu. The other options may be selected as needed.

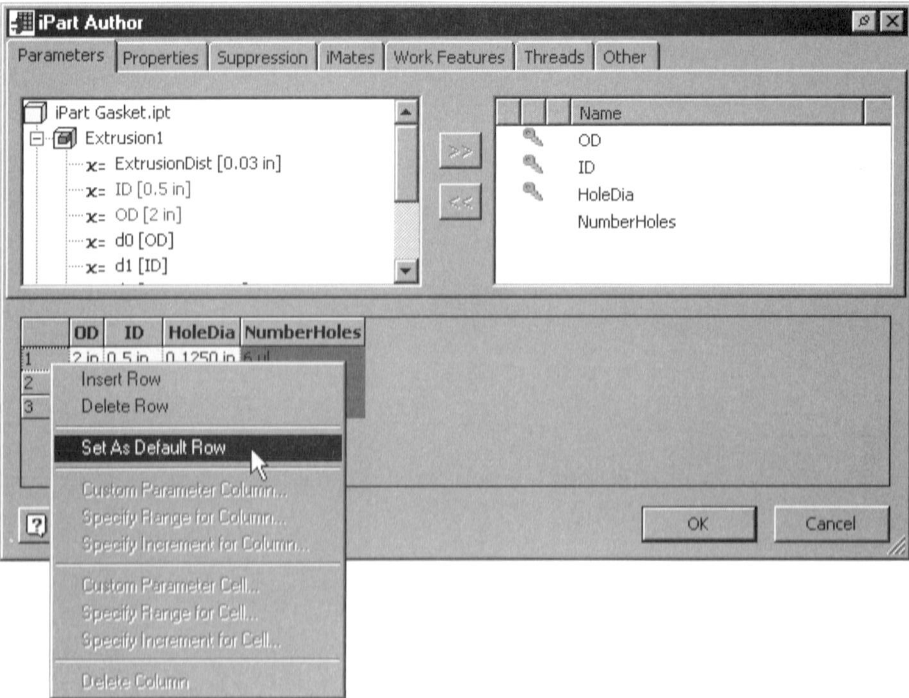


**Figure 12-105**

10 Click OK when you have finished defining the contents of the table. The iPart Author dialog box closes and the part is converted to an iPart Factory. The table is saved in an embedded Microsoft Excel spreadsheet and a table icon is displayed in the Browser. The part icon in the Browser is also changed to display a factory, as shown in the following image.

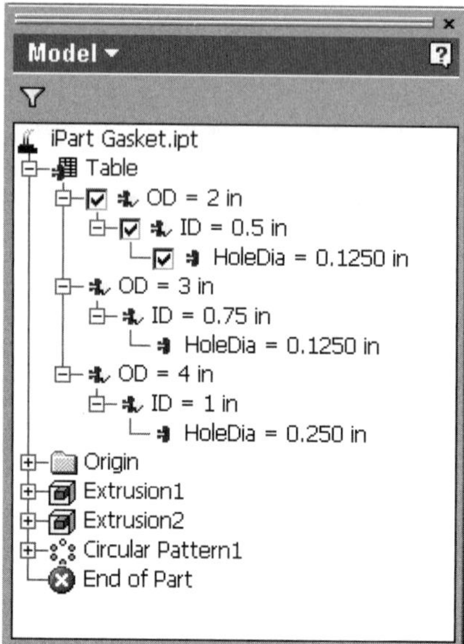


**Figure 12-106**

You can expand the Table icon in the Browser to view the iPart members based on the keys and values that you define. The active version is displayed with a check mark.

## EDITING IPARTS

There are a number of operations that can be performed on an iPart Factory after it has been created. You can delete the table, modify the parameters or properties for individual members, add or delete additional members, and so on. Right-click the Table icon in the Browser to delete the table and convert the iPart Factory back to a part, edit the table with the iPart Author dialog box using the Edit Table option, or edit the table with Microsoft Excel using the Edit via Spread Sheet option, as shown in the following image. Make changes as needed and then save the file. Changes to iParts that have been placed in assemblies are not automatically updated. To update the iParts in an assembly, open the assembly and click the Full Update tool from the Standard toolbar.

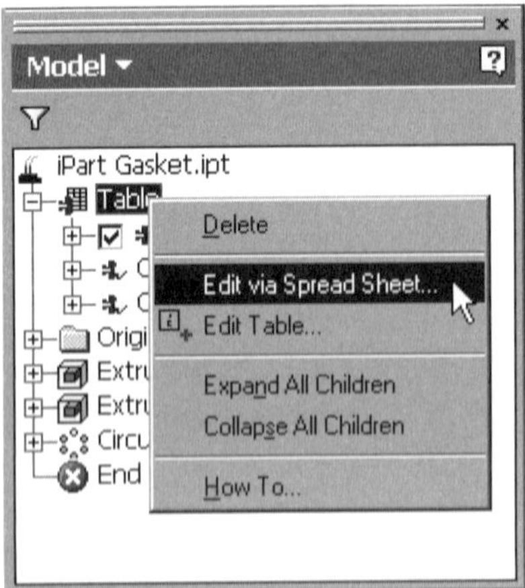


**Figure 12-107**

When editing the spreadsheet using Microsoft Excel, you can incorporate spreadsheet formulas, conditional statements, and multiple sheet data extraction, but you cannot modify spreadsheet formulas and conditional statements from the iPart Author dialog box. These types of cells are inactive and are highlighted in red, as shown in the following image.

| | OD | ID | HoleDia | NumberHoles |
|---|---|---|---|---|
| 1 | 2 in | 0.5 in | 0.1250 in | 6 ul |
| 2 | 3 in | 0.75 in | 0.1250 in | 6 ul |
| 3 | 4 in | 1 | 0.250 in | 6 ul |

**Figure 12-108**

## IPART PLACEMENT

You place Standard iParts in assemblies using the Place Component tool. When you select a Standard iPart Factory, an additional Standard iPart Placement dialog is displayed. It allows you to use the Keys tab to identify an iPart member by selecting the key values, using the Tree to locate a member by expanding the key values, using the Table to identify an iPart member by selecting a row in the table, and placing multiple instances of different iPart members.

To place a Standard iPart into an assembly, follow these steps:

1 Start a new assembly or open an existing assembly in which the iPart will be placed.

2 Use the Place Component tool, navigate to and select the iPart that will be placed in the assembly from the Open dialog box.

3 Select the Open button, and the Place Standard iPart dialog box will be displayed as shown in the following image. If there is a custom cell(s) or column(s), the Place Custom iPart dialog box is displayed.

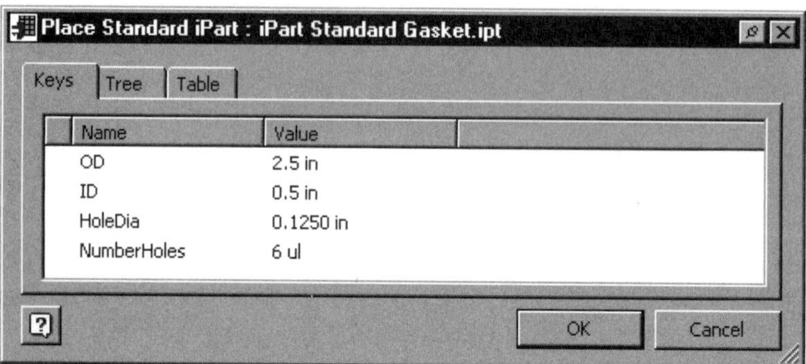


Figure 12-109

4 To select from the member list, select either the Tree or Table tab and then select the member that defines the part you want to place.

5 If a custom part is being placed, select the Keys, Tree, or Table tab and enter a value in the right side of the dialog box, as shown in the following image. The value must fall within the limits set in the iPart factory; otherwise an alert is displayed.

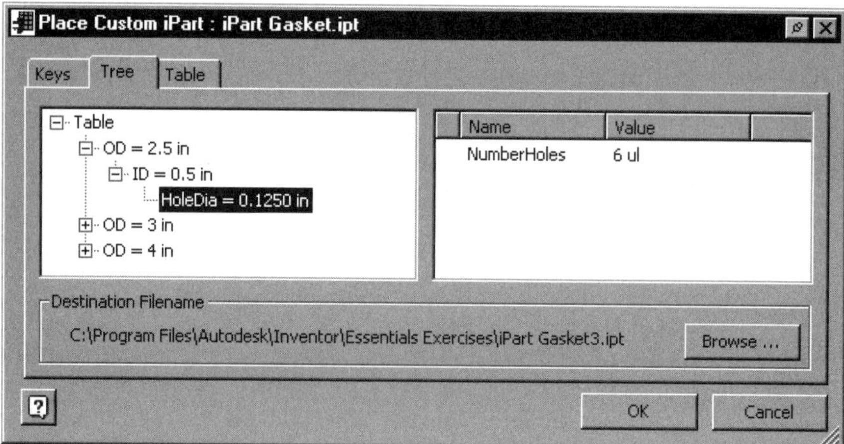


Figure 12-110

6 Place the part in the graphics area.

7 Continue placing instances of the iPart as needed; when you are finished, click the OK button.

8 To change an iPart in an assembly to a different configuration, expand the part in the Browser, right-click on the table name, and choose Change Component. Then select a new member from the Keys, Tree, or Table tab.

The Keys tab displays the primary and any secondary keys defined in the iPart Factory. You can select the values of the keys to identify unique members. The Tree tab displays a hierarchical structure of the keys. If secondary keys exist, the values of the keys are progressively filtered as you choose to expand the key values. The Table tab displays the entire table rather than just the keys. To identify a unique iPart member, select a row then click the OK button to place the iPart.

When you place the first version of a Standard iPart into an assembly, a folder is created with the same name as the iPart Factory. This folder is created in the same folder as the iPart Factory file by default. As you place additional Standard iParts into your assemblies, this folder is checked for existing iPart files prior to creating new iPart files.

## STANDARD IPART LIBRARIES

In a collaborative design environment, the best way to manage iPart Factories and the iParts published from those factories is through the use of libraries. Define a library within your project and place your iPart Factories in this library's path. You can also specify that a Standard iPart Factory publish Standard iParts to a different folder. To do this, create another library using the same name as the iPart Factory library name prefixed with an underscore, as shown in the following image.

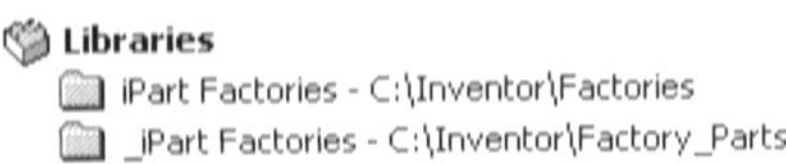


**Figure 12-111**

If you have an iPart Factory for bolts, for example, you could store the iPart Factory in a library named Bolts, and then you could define an additional library where the individual members of the Bolt iPart Factory will be stored. The library should be named _Bolts. Autodesk Inventor will automatically store all iParts published by the factory in the library directory specified by _Bolts. If you store an iPart factory in a workspace, a local search path, or a workgroup search path, the published iParts are stored in a subdirectory with the same name as the factory.

**NOTE** It is recommended that you do not store iParts in the same folders as iPart factories.

## CUSTOM IPARTS

Custom iParts are placed into assemblies in the same way as Standard iParts. If you select an existing Custom iPart file, that member of the part is placed into your assembly with no option to define the value of the custom parameters. When a Custom iPart Factory is selected, the Place Custom iPart dialog box is displayed. Similarly to the Standard iPart Placement dialog, you can use the Keys tab to identify an iPart member by selecting the key value, the Tree tab to expand key values and select a member, or the Table tab to identify an iPart member by selecting a row in the table. The Place Custom iPart dialog box provides additional options so you can enter values for custom parameters, define a destination filename for the Custom iPart, and place multiple instances of different iPart members.

**NOTE** Since each Custom iPart is unique, you have to provide a different filename as you place different members.

After a Custom iPart has been placed in an assembly, you can add more features to the iPart. The following image shows the Browser of a Custom iPart named Gasket-B that has had a fillet and extrusion feature added to it. The capability to add features to an iPart makes the Custom iPart behavior similar to that of a derived part. Derived Parts are discussed later in this chapter.

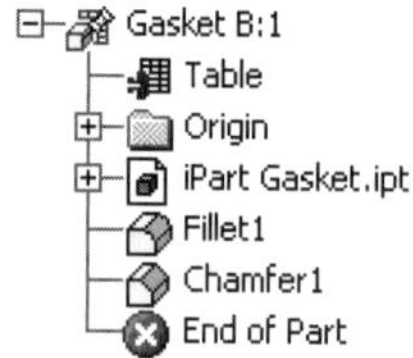


**Figure 12-112**

# EXERCISE 12-3
# Creating and Placing iParts

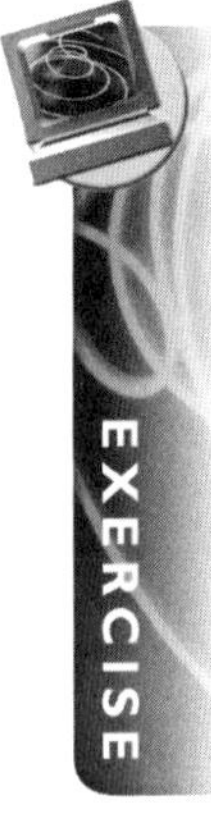


In this exercise, you will create an iPart factory of a spherical bushing. You then place different versions of the iPart into an assembly. To navigate to the exercise in the *Electronic Student Workbook*, do the following:

1. From the Main TOC page, click Chapter 12.
2. From the TOC for Chapter 12, click Creating and Placing iParts.

The following image illustrates the opened exercise.

**Figure 12-113 Opened exercise**

The following image illustrates the completed exercise.

**Figure 12-114 Completed exercise**

## IFEATURES

iFeatures give you the ability to reuse single or multiple features from a part file in other Autodesk Inventor files. iFeatures capture the design intent that is built into a feature(s) that is going to be reused, such as the name or the size and position parameters. You can also embed or attach a file to the iFeature to be used as Placement Help when you place the iFeature into another design.

Reference edges can be included as position geometry in your iFeatures. Reference edges allow you to capture additional design intent, but require that the iFeature be positioned in the same way as it was originally designed in every part in which it is placed. iFeatures are saved in their own type of file that has an *.ide* extension and a unique icon as shown in the following image.

**Figure 12-115**

iFeatures are stored in a catalog. The catalog is a directory on your computer or on a server that is set up to store all of the iFeatures that you create. You can browse the iFeature catalog at any time by selecting the View Catalog tool from the Part Features Panel Bar (or toolbar) as shown in the following image.

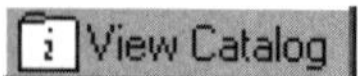


**Figure 12-116**

When the View Catalog tool is selected, Windows Explorer opens to the directory specified in the iFeature Root setting on the iFeature tab of the Options dialog box. From the iFeature Root folder you can view, copy, edit, or delete iFeatures from your catalog.

### CREATE IFEATURES

You create iFeatures using the Extract iFeature tool that is available from the Tools menu, as shown in the following image.

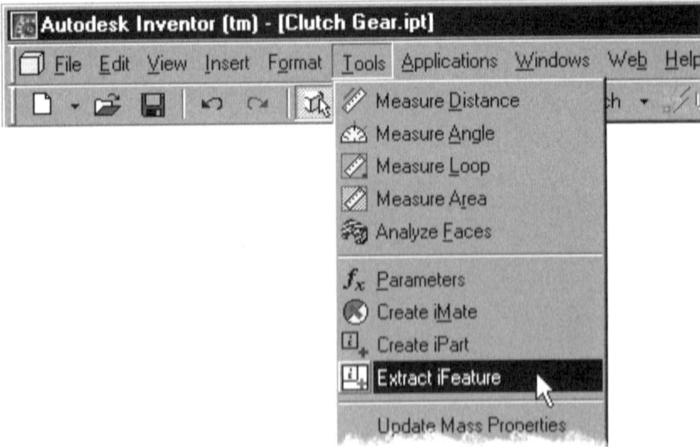


**Figure 12-117**

Once selected, the Create iFeature dialog box is displayed, as shown in the following image.

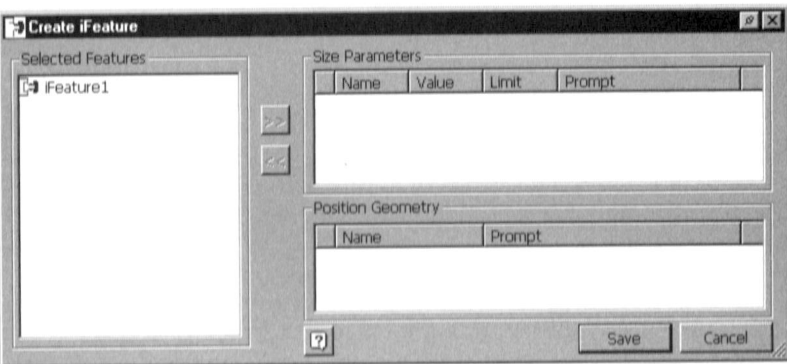


**Figure 12-118**

The Create iFeature dialog box contains the following sections:

## Selected Features

This area of the dialog box lists the features that are selected to be included in the iFeature. Dependent features of a selected feature are included by default, but can be removed from the Selected Features list. Features in the list can be renamed to have more descriptive names to assist in working with the iFeature at a later time.

Use the Add parameters (>>) button and the Remove parameters (<<) button to move parameters from the highlighted features in the Selected Features list to the Size Parameters table.

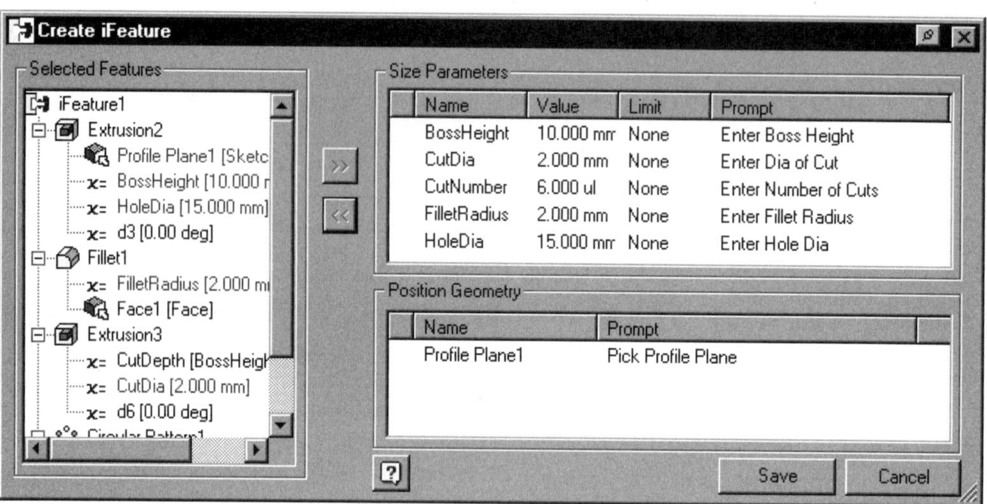


**Figure 12-119**

## Size Parameters

The Size Parameters table lists all of the parameters that will be used for interface when the Feature is placed into another file. You can select the parameters by expanding the features listed in the Selected Features list or directly from the graphics window.

**NOTE** Any parameters that have been given a name in the Parameters dialog box are automatically displayed in the Size Parameters pane of the Create iFeature dialog box. Renaming parameters that you want to include in an iFeature can speed up the process of creating them.

**Name** Specify a descriptive name for the parameter. Use names that describe the purpose of the parameter.

**Value** Place a value to be the default for the parameter when inserting your iFeature into a file. The value is restricted by settings in the Limit column.

**Limit** Place restrictions on the values that are available for the parameter by using one of three options: None, Range, or List, as shown in the following image.

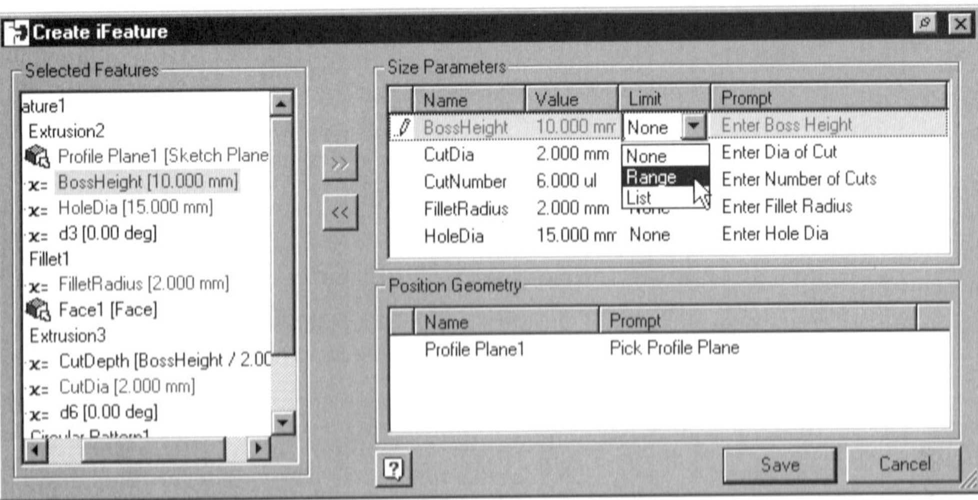


**Figure 12-120**

None specifies that no restrictions be placed on the Value field. Range gives you the ability to specify a minimum and maximum value, including less than, equal to, and infinity (shown on the left in the following image). You can specify the default value to be used. The List option, as shown on the right in the following image, allows you to define a list of values that the user can choose from for the size parameters.

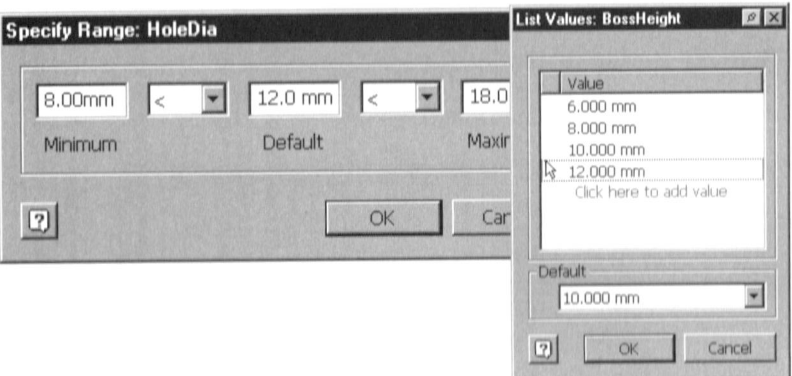


**Figure 12-121**

**Prompt** You can enter descriptive instructions to further explain what the parameter is used for. The text entered in the prompt field is displayed in a dialog box during the iFeature placement.

### Position Geometry

Specify the geometry of the iFeature that is necessary to position it on a part. Geometry can be added or removed to the Position Geometry list from the Selected Features tree by right-clicking the geometry and selecting Expose Geometry. You remove geometry from the Position Geometry list by right-clicking it and selecting Remove Geometry.

**Name** Describes the position geometry.

**Prompt** Instructs the positioning of the iFeature on a part.

You can customize the position geometry by right-clicking it and choosing one of the two additional options available from the menu.

**Make Independent** Separates entries for geometry that is shared by more than one feature.

**Combine Geometry** Combines listings of geometry shared by more than one feature in a single entry.

## INSERT IFEATURES

Inserts iFeatures using the Insert iFeature tool that is available from the Part Features Panel Bar, as shown in the following image.

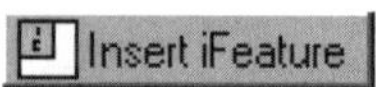


**Figure 12-122**

Once selected, the Insert iFeature dialog box is displayed, as shown in the following image. You can select tasks from the tree structure on the left pane of the dialog box or move forward and backward using the Next and Back buttons of the dialog box.

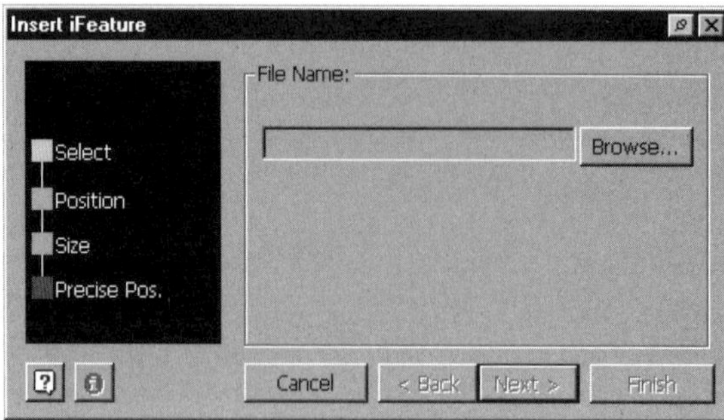


**Figure 12-123**

The Insert iFeature dialog contains the following sections:

**Select** Chooses the iFeature that you want to insert using the Browse button.

**Position** Lists the names of the interface geometries specified during the creation process of the iFeature. After selecting a planar face or work plane you can click the arrowhead on the positioning symbol to either move or rotate the symbol, as shown in the following image.

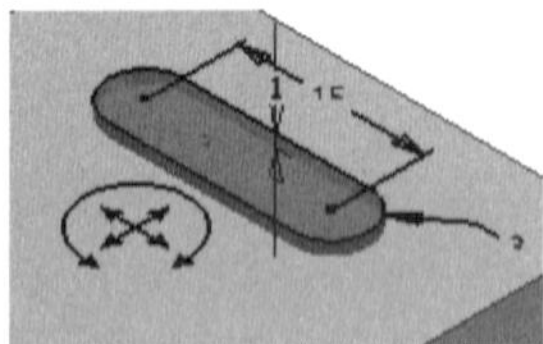

Figure 12-124

You can also specify a precise rotation value directly in the angle field of the dialog box. After the requirement has been approved, a check mark will be placed in the left column as shown in the following image.

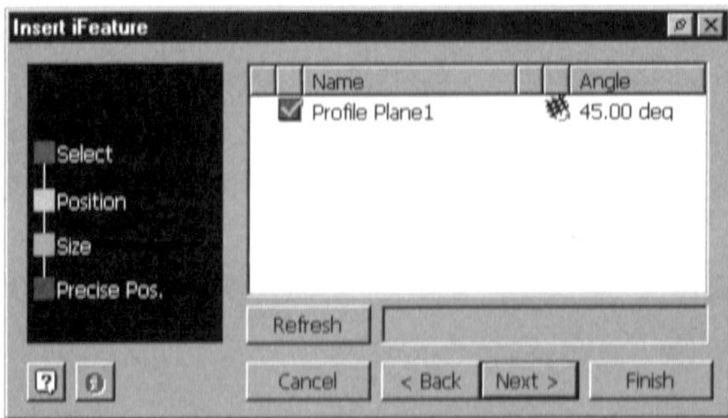


Figure 12-125

**Name** Lists the named interface geometry.

**Angle** Shows the default of the placement geometry on the iFeature.

**Move Coordinate System** Defines horizontal or vertical axes when the iFeature has horizontal or vertical dimensions or constraints included.

## Size

Shows the names and default values specified for the iFeature as shown in the following image. Click in the row to edit the values and then apply them to preview the changes.

**Name** Lists the name of the parameter.

**Value** Lists the value of the parameter.

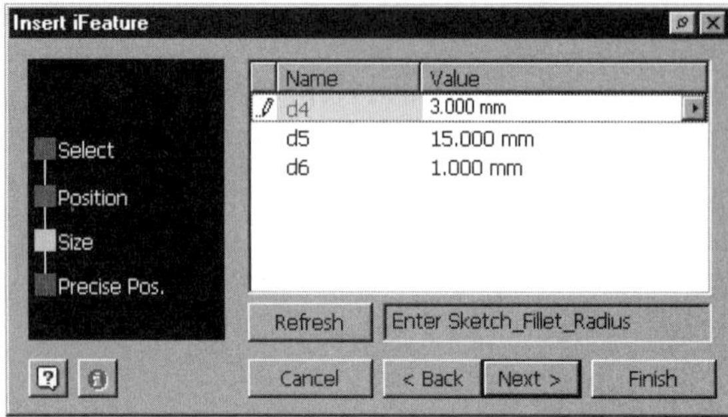


Figure 12-126

## Precise Position

Further refines the position of the iFeature, using either dimensions or constraints after you have placed it.

**Activate Sketch Edit Immediately** Activates the sketch of the iFeature and the 2D Sketch Panel Bar. You can then use dimensions and/or constraints to position the iFeature on the parent feature.

**Do not Activate Sketch Edit** Positions the iFeature without adding further constraints or dimensions.

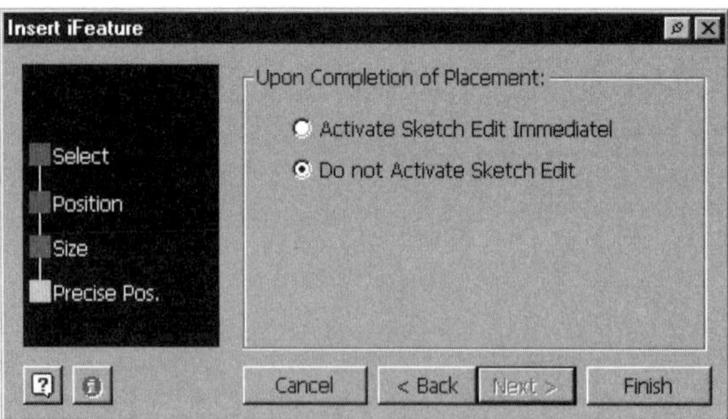


Figure 12-127

### EDITING IFEATURES

You can edit an iFeature by opening the *.ide* file in Autodesk Inventor. There are three tools available in the iFeature environment when the file is opened: Edit iFeature, View Catalog, and iFeature Author Table.

The Edit iFeature tool, shown in the following image, opens the Create iFeature dialog box. You cannot change which parameters are used to define the iFeature after it has been created, but you can modify the size parameters and position geometry by editing the properties for the name, value, limit, and prompt.

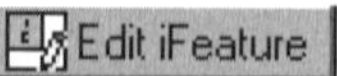


**Figure 12-128**

The iFeature Author Table tool, shown in the following image, makes your iFeatures behave like iParts. You can create iFeatures that are driven by a table.

**Figure 12-129**

Once selected, the iFeature Author dialog box is displayed, as in the following image. The table in the iFeature Author tool behaves the same as in iParts. You can insert, delete, and specify the default row. Custom parameter columns or cells can also be defined and can include a range for the value. The main difference is that there is no option to set a column as a file name, display style, or material. If you are converting an iFeature to a table-driven iFeature that contains a parameter with a list of possible values, the table is automatically populated with the different sizes from the list as rows of the table. The table can also be edited via a spreadsheet.

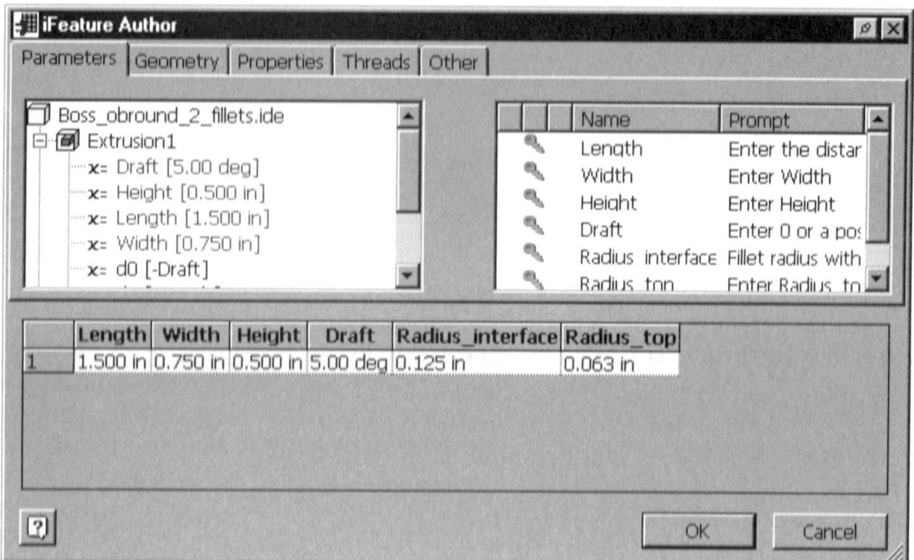


**Figure 12-130**

## Parameters Tab

The Parameters tab (see preceding image) is the default tab that is displayed in the iFeature Author dialog box. This tab lists all of the parameters that were used to create the iFeature. You cannot change which parameters make up the iFeature after it has been created. The Size Parameters pane on the right of the dialog displays all of the parameters used in the definition and their names and prompts. You can define keys for the parameters that function the same as when working with iParts.

## Geometry Tab

The Geometry tab, shown in the following image, allows you to add or remove interface geometry for any iFeatures that are created after the table-driven iFeature is saved. Previously placed iFeatures are not affected. You can also modify the labels during edit.

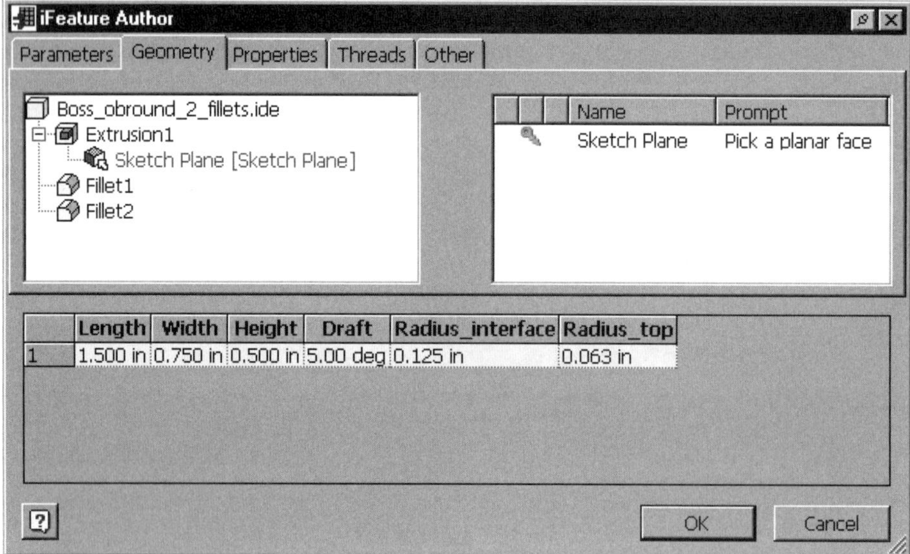


Figure 12-131

## Properties Tab

The Properties tab, shown in the following image, lists all of the Summary, Project, and Physical properties for the file. This tab functions the same as it does when working with iParts, with the exception that the columns cannot be declared to be a File Name, Display Style, or a Material column.

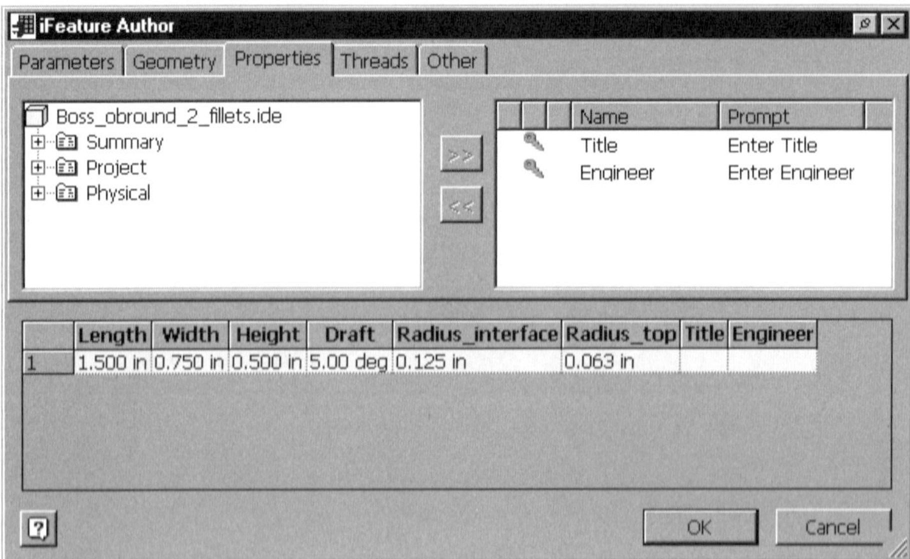


Figure 12-132

## Threads Tab

You can control thread features in the iFeature similarly to how you can when working with iParts. You can create table-driven items for regular or tapered thread features. Use the Designation element to control the size and Pitch. The Direction can be right or left, using the text R or L in the table. Options to modify the pipe diameter, class, and family are also available.

## Other Tab

The Other tab can be used to create custom table items. This tab functions the same as it does when working with iParts, with the exception that the columns cannot be declared to be a File Name, Display Style, or Material column.

### INSERTING TABLE-DRIVEN IFEATURES

You insert table-driven iFeatures the same way that you would insert a typical iFeature. The wizard displays the key parameters as a drop-down list while defining the Size parameters, and any custom parameters use the edit field. In the example shown in the following image, the parameters that are key values display a drop-down list from which you can select the appropriate value. The Draft parameter is a custom parameter, which means you can enter the draft angle directly in the dialog box.

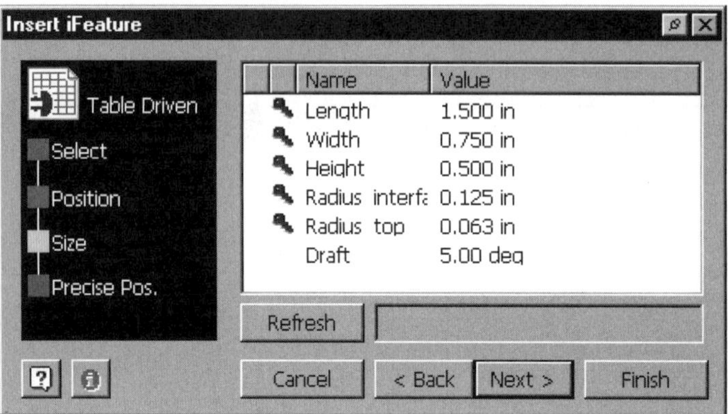


**Figure 12-133**

The Browser displays table-driven iFeatures in a similar manner to that used with iParts.

When a table-driven iFeature is placed in a part file, a new .ide file is not created. The placement is similar to that of an iFeature that is not table driven. The file in which the iFeature is placed contains an independent copy of the iFeature that is not associative to the original *.ide* file. The spreadsheet that contains the table information is not visible in the Browser and cannot be edited once it is placed in the part file.

# EXERCISE 12-4
# Creating and Placing iFeatures

In this exercise, you will create a hex drive iFeature from an existing component and attach a Microsoft Word document to be used as help during the placement of the iFeature. You then insert the iFeature into another part file. To navigate to the exercise in the *Electronic Student Workbook*, do the following:

1 From the Main TOC page, click Chapter 12.

2 From the TOC for Chapter 12, click Creating and Placing iFeatures.

The following image illustrates the opened exercise.

**Figure 12-134 Opened exercise**

The following image illustrates the completed exercise.

**Figure 12-135 Completed exercise**

EXERCISE

## DERIVED PARTS

Derived parts are used to capture the design intent of an entire part, assembly, specific sketch, parameter, or surface by either creating a new part file or by importing any of the aforementioned selections. A derived part is a new part file that uses the selected part, assembly, or sketch as its base feature. Additional features can then be added to the derived part to explore different design scenarios. If the original part changes, the derived part will also update to include any changes that are made to the original file. This capability can also be turned off if you do not need to retain a connection to the original file.

Some possible uses for a derived part or assembly are to:

- Have many different parts that are machined from a single casting blank.
- Scale or mirror a derived part upon creation (this is not for derived assemblies). Derive sketches to be used within an assembly as a layout.
- Derive a part as a work body that can be used to simplify the representation of a part.
- Derive parameters from an existing file to be used within another part file.

You create a derived part using the Derived Component tool located on the Part Features Panel Bar (shown in the following image), or from the Part Features toolbar.

**Figure 12-136**

You then select the Part or Assembly file that you want derived. If a part file is selected, the Derived Part dialog box opens, as shown in the following image. When an assembly file is selected as the derived component, the Derived Assembly dialog box opens to provide you with different options that are discussed later in the chapter.

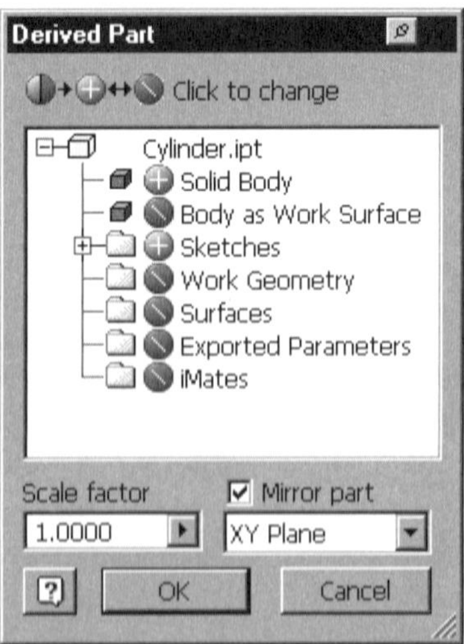


**Figure 12-137**

**NOTE** When the Derived Component tool is not used to create the first feature of the part, you have different options available in the Derived Part dialog box. The ability to Derive a Solid Body or a Body as Work Surface, for example, cannot be selected. You cannot derive an assembly file unless it is derived as the base feature of a part.

## DERIVED PART SYMBOLS

The symbols show you the current status of geometry that will be contained in the derived part and whether it will be included or excluded from the derived part. The available symbols are:

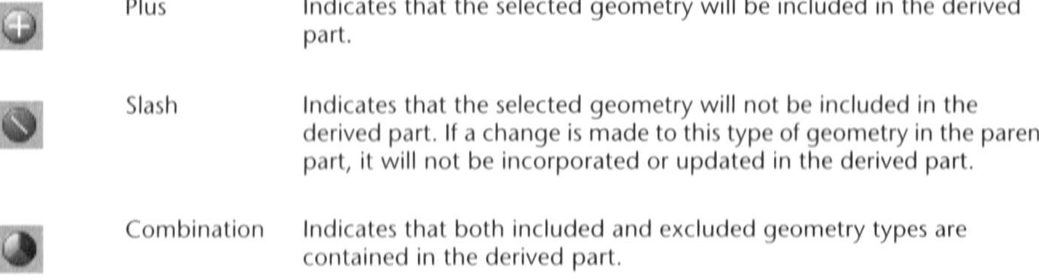

| | |
|---|---|
| Plus | Indicates that the selected geometry will be included in the derived part. |
| Slash | Indicates that the selected geometry will not be included in the derived part. If a change is made to this type of geometry in the parent part, it will not be incorporated or updated in the derived part. |
| Combination | Indicates that both included and excluded geometry types are contained in the derived part. |

## DERIVED PART DIALOG BOX

The Derived Part dialog box contains the following options, as shown in the preceding image.

| | |
|---|---|
| Solid Body | Select to derive the part as a base solid. |
| Body as Work Surface | Select to derive the part body as a work surface. The body is brought in, behaves, and is displayed as an Autodesk Inventor surface. |
| Sketches | Select to include any unconsumed sketches from the original file. |
| Work Geometry | Select to include any work features from the original file. They can then be used to create new geometry. |
| Surfaces | Select to include any surfaces that exist in the original file. |
| Exported Parameters | Select to include any parameters that are designated to be exported parameters in the original file. |

**NOTE** If the original file consists only of surfaces, it will be the only option available.

**NOTE** Designate a parameter to be exported using the check box in the Parameters dialog box, as shown in the following image. The Parameters tool can be accessed from the Part Features toolbar.

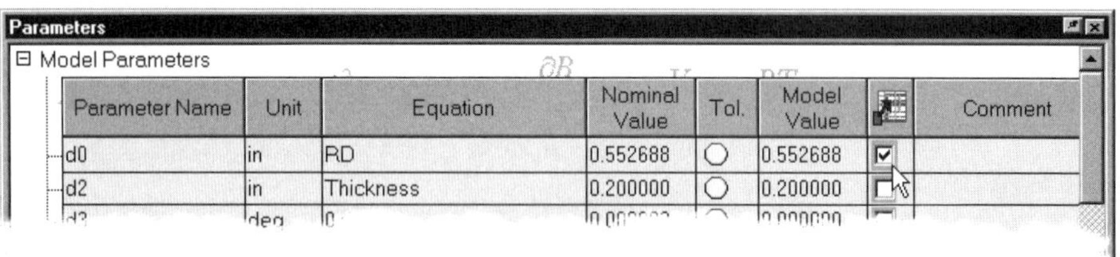

Parameters
Model Parameters
Parameter Name
Unit
Equation
Nominal Value
Tol.
Model Value
Comment
d0
in
RD
0.552688
0.552688
d2
in
Thickness
0.200000
0.200000


**Figure 12-138**

| | |
|---|---|
| iMates | Select to include any iMates that are defined in the original part. |
| Scale Factor | Indicate a scale factor (in percentage) to scale the derived part. The default scale factor is 1.0 (or the same size as the original file). |
| Mirror Part | Select to indicate that the derived part should be mirrored about the XY, XZ, or YZ origin work planes upon creation. The plane to mirror about can be selected from the drop-down list. |

# EXERCISE 12-5 Creating a Derived Part

In this exercise, you derive a part from an existing file to create an opposite hand control arm. You then edit the derived part settings, delete faces, and create a hole to complete the control arm. Complete the exercise by modifying the parent file and updating the derived part. To navigate to the exercise in the *Electronic Student Workbook*, do the following:

1 From the Main TOC page, click Chapter 12.

2 From the TOC for Chapter 12, click Creating a Derived Part.

You will begin by creating a new part file. The completed exercise is shown in the following image.

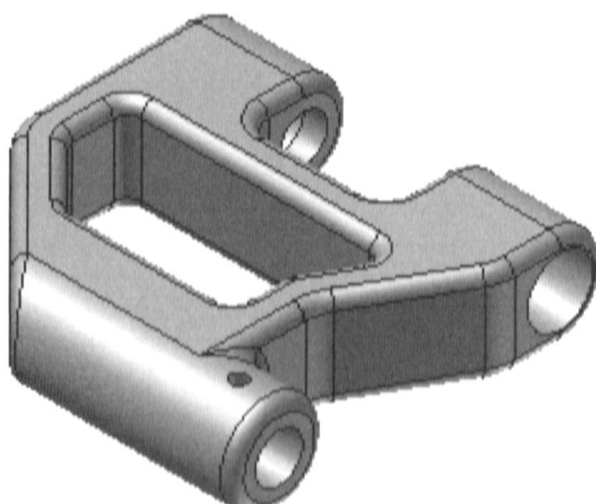

**Figure 12-139 Completed exercise**

## DERIVED ASSEMBLY SYMBOLS

The symbols available when deriving an assembly are different than the symbols available when deriving a part file. The derived assembly symbols are described below:

| | |
|---|---|
| Plus | Indicates that the selected component will be included in the derived part. |
| Minus | Indicates that the selected component will be subtracted from the derived part. If the subtracted component intersects another portion of an included part, the result will be a void or cavity in the derived part. |
| Slash | Indicates that the selected component will not be included in the derived part. Changes to this type of component in the parent assembly will not be incorporated or updated in the derived part. |
| Combination | Indicates that the selected item contains a mixture of the above options. |

## DERIVED ASSEMBLY DIALOG BOX

The derived assembly dialog contains the following options, as shown in the following image.

| | |
|---|---|
| Click to Change | Select the symbol next to each component to designate it to be included or excluded from the derived part. You can also specify whether the component should be subtracted from the resultant derived part. |
| Keep seams between planar faces | Select to create seams between any coincident planar faces of parts. If the box is cleared, the faces are merged. |

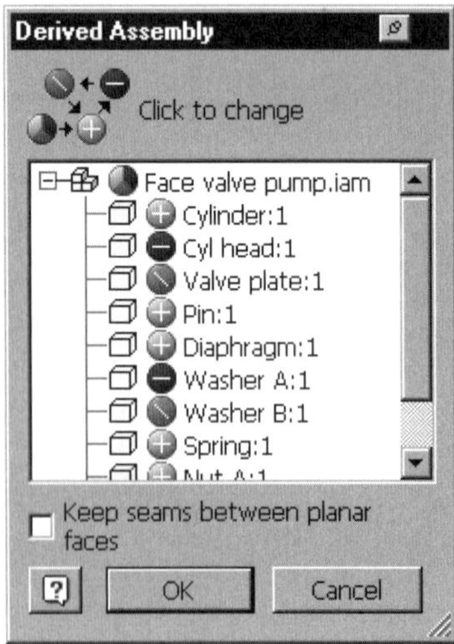


**Figure 12-140**

Because you are using geometry or parts based on another file, any modifications to the parent are incorporated into the derived component. If a parent file has been modified, the icon next to the derived component displays the Autodesk Inventor update symbol, as shown in the following image.

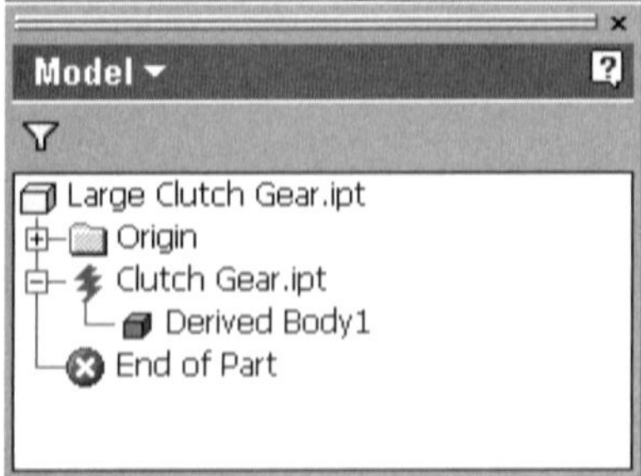


**Figure 12-141**

You can break the link to the parent file by right-clicking on the derived component in the Browser and selecting Break Link With Base Part from the menu. Once the link is broken, the derived component icon will display a broken chain link to signify that the link to the parent file has been broken (see following image). Once the link is broken, it cannot be re-established.

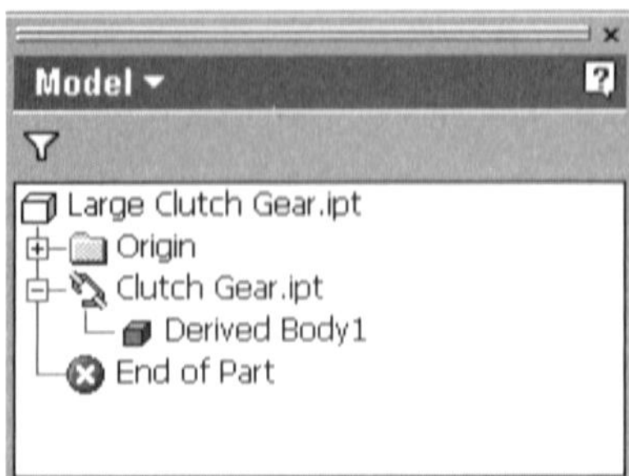


**Figure 12-142**

## CHAPTER SUMMARY

| To | Do This | Tool |
|---|---|---|
| Create a new weldment | Click on the desired weldments icon from the list of templates. | |
| Prepare weldments | Double-click on the Preparations group located in the Browser. | Preparations |
| Activate the Weld Feature dialog box | Click the Weld button located on the Weldment Features Panel Bar. | |
| Create cosmetic welds | Click the Cosmetic Weld button located in the Weld Feature dialog box. | |
| Create weld beads | Click the Weld Bead button located in the Weld Feature dialog box. | |
| Machine weldments | Double-click on the Machining group located in the Browser. | Machining |
| Convert an existing assembly to a weldment | Select the Weldment command found under the Applications menu. | Applications Wind<br>Assembly<br>Weldment |
| Create a caterpillar representation | Select the Caterpillar tool from the Drawing Annotation Panel Bar. | |
| Create an end treatment representation | Select the End Treatment tool from the Drawing Annotation Panel Bar. | |
| Create an iPart | Select Create iPart from the Tools menu. | |
| Create a Custom iPart | Right-click a column in the iPart Author dialog and select Custom Parameter Column. | |
| Create an iFeature | Select Extract iFeature from the Tools menu. | |
| Insert an iFeature | Click Insert iFeature from the Part Features Panel Bar or toolbar. | |
| Create a table-driven iFeature | Click the iFeature Author Table tool from the iFeature Panel Bar or toolbar. | |

# APPLYING YOUR SKILLS

## SKILLS EXERCISE 12-1

In this exercise, you will convert an existing assembly of a spool hand truck into a weldment. You will create fillet and intermittent weld features and then save the model as a new weldment. To navigate to the exercise in the *Electronic Student Workbook*, do the following:

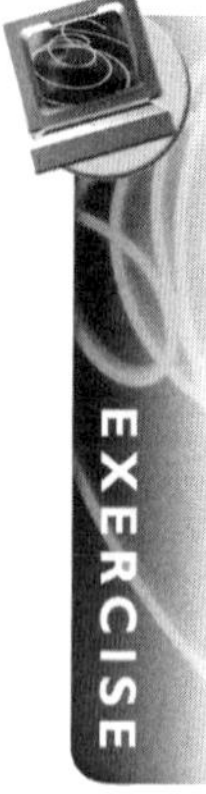


1 From the Main TOC page, click Chapter 12.
2 From the TOC for Chapter 12, click Exercise 1 under Applying Your Skills.

The following image illustrates the opened exercise.

Figure 12-143 Opened exercise

The following image illustrates the completed exercise.

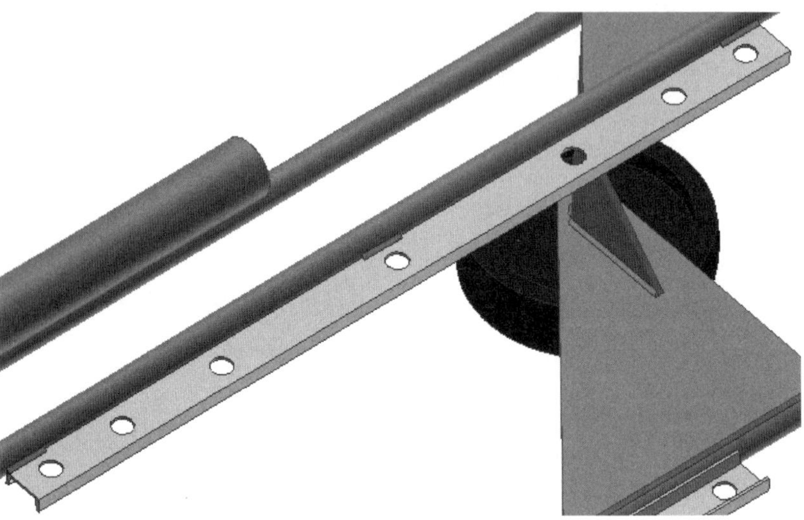

**Figure 12-144 Completed exercise**

### SKILLS EXERCISE 12-2

In this exercise, you open an iPart factory of a rubber membrane for a remote keyless entry device. You add iMates to the iPart factory and then you place two versions of the iPart into an assembly. To navigate to the exercise in the *Electronic Student Workbook*, do the following:

1 From the Main TOC page, click Chapter 12.

2 From the TOC for Chapter 12, click Exercise 2 under Applying Your Skills.

The following image illustrates the opened exercise.

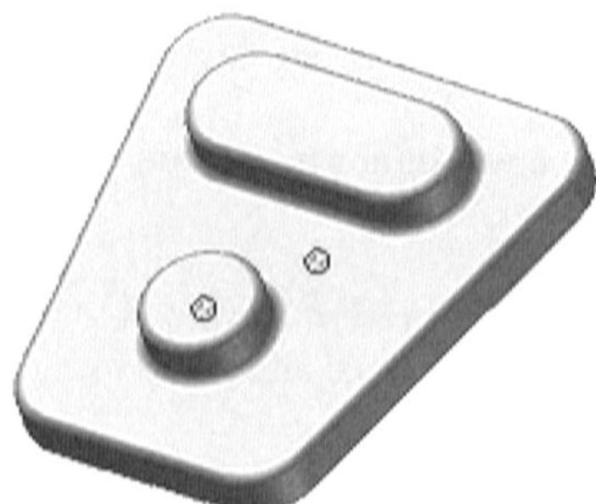

**Figure 12-145 Opened exercise**

The following image illustrates the completed exercise.

**Figure 12-146 Completed exercise**

## CHECKING YOUR SKILLS

Use these questions to test your knowledge of the material covered in this chapter.

1 **True___ False___** The vertical leg of a fillet welding symbol is always drawn to the left of the slanted side of the symbol.

2 **True___ False___** The presence of a black flag on a welding symbol means to terminate all welding operations immediately.

3 **True___ False___** When preparing a weldment, the following operations are supported: extrude-cut, hole, and chamfer.

4 **True___ False___** To identify a weldment in Autodesk Inventor, look for the *.wld* file extension.

5 **True___ False___** The three groups present in the Weldment Browser are Assembly, Preparations, and Machining.

6 **True___ False___** The only weld type available when creating a cosmetic weld is fillet.

7 **True___ False___** Cosmetic welds and weld beads are both available through the Weld Feature dialog box.

8 **True___ False___** Typical operations supported when machining weldments include revolutions, lofts, extrusion joins, and fillets.

9 **True___ False___** Assembly models created in previous versions of Autodesk Inventor cannot be converted to weldments.

10 Describe the difference between a Standard iPart and a Custom iPart.

______________________________________________

______________________________________________

______________________________________________

11 **True___ False___** Multiple versions of an iPart can be placed in an assembly.

12 **True___ False___** When changes are made to an iPart Factory, the changes are automatically updated in iParts that have been placed in assemblies.

**13** **True___** **False___** Standard iParts can have features added to them after they are placed in an assembly.

**14** **True___** **False___** Named parameters are automatically added as Size parameters during iFeature creation and cannot be removed.

**15** What happens when you create a table-driven iFeature and one of the original parameters contains a *list* of values for the parameter?

______________________________________________

______________________________________________

______________________________________________

**16** How do you create a mirrored part file?

______________________________________________

______________________________________________

______________________________________________

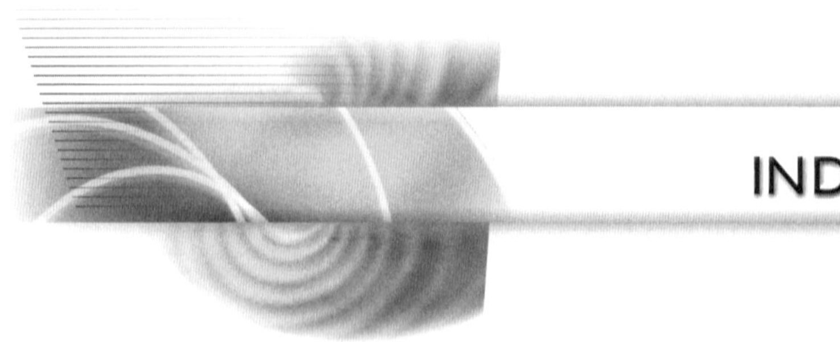


# INDEX

## Symbols



## Numerics



## A

# B

## C

## D

## E

## F

## G



## H

# I



# J



# K

# L



# M

## N



## O

# P

## Q



## R

# S

# T

## U



## V

# LICENSE AGREEMENT FOR AUTODESK PRESS

## A Thomson Learning Company

### Educational Software/Data

You the customer, and Autodesk Press incur certain benefits, rights, and obligations to each other when you open this package and use the software/data it contains. BE SURE YOU READ THE LICENSE AGREEMENT CAREFULLY, SINCE BY USING THE SOFTWARE/DATA YOU INDICATE YOU HAVE READ, UNDERSTOOD, AND ACCEPTED THE TERMS OF THIS AGREEMENT.

### Your rights:

1. You enjoy a non-exclusive license to use the enclosed software/data on a single microcomputer that is not part of a network or multi-machine system in consideration for payment of the required license fee, (which may be included in the purchase price of an accompanying print component), or receipt of this software/data, and your acceptance of the terms and conditions of this agreement.
2. You own the media on which the software/data is recorded, but you acknowledge that you do not own the software/data recorded on them. You also acknowledge that the software/data is furnished "as is," and contains copyrighted and/or proprietary and confidential information of Autodesk Press or its licensors.
3. If you do not accept the terms of this license agreement you may return the media within 30 days. However, you may not use the software during this period.

### There are limitations on your rights:

1. You may not copy or print the software/data for any reason whatsoever, except to install it on a hard drive on a single microcomputer and to make one archival copy, unless copying or printing is expressly permitted in writing or statements recorded on the diskette(s).
2. You may not revise, translate, convert, disassemble or otherwise reverse engineer the software/data except that you may add to or rearrange any data recorded on the media as part of the normal use of the software/data.
3. You may not sell, license, lease, rent, loan, or otherwise distribute or network the software/data except that you may give the software/data to a student or and instructor for use at school or, temporarily at home.

Should you fail to abide by the Copyright Law of the United States as it applies to this software/data your license to use it will become invalid. You agree to erase or otherwise destroy the software/data immediately after receiving note of Autodesk Press' termination of this agreement for violation of its provisions.

Autodesk Press gives you a LIMITED WARRANTY covering the enclosed software/data. The LIMITED WARRANTY can be found in this product and/or the instructor's manual that accompanies it.

This license is the entire agreement between you and Autodesk Press interpreted and enforced under New York law.

### Limited Warranty

Autodesk Press warrants to the original licensee/ purchaser of this copy of microcomputer software/ data and the media on which it is recorded that the media will be free from defects in material and workmanship for ninety (90) days from the date of original purchase. All implied warranties are limited in duration to this ninety (90) day period. THEREAFTER, ANY IMPLIED WARRANTIES, INCLUDING IMPLIED WARRANTIES OF MERCHANTABILITY AND FITNESS FOR A PARTICULAR PURPOSE ARE EXCLUDED. THIS WARRANTY IS IN LIEU OF ALL OTHER WARRANTIES, WHETHER ORAL OR WRITTEN, EXPRESSED OR IMPLIED.

If you believe the media is defective, please return it during the ninety day period to the address shown below. A defective diskette will be replaced without charge provided that it has not been subjected to misuse or damage.

This warranty does not extend to the software or information recorded on the media. The software and information are provided "AS IS." Any statements made about the utility of the software or information are not to be considered as express or implied warranties. Delmar will not be liable for incidental or consequential damages of any kind incurred by you, the consumer, or any other user.

Some states do not allow the exclusion or limitation of incidental or consequential damages, or limitations on the duration of implied warranties, so the above limitation or exclusion may not apply to you. This warranty gives you specific legal rights, and you may also have other rights which vary from state to state. Address all correspondence to:

AutodeskPress
Executive Woods
5 Maxwell Drive
Clifton Park, New York 12065-2919